QUANTUM FIELD THEORY

Quantum field theory is the basic mathematical framework that is used to describe elementary particles. It is a cornerstone of modern physics.

This textbook provides a complete and essential introduction to this subject. Assuming only an undergraduate knowledge of quantum mechanics and special relativity, it is ideal for graduate students beginning the study of elementary particles, and will also be of value to those in related fields such as condensed-matter physics.

The step-by-step presentation begins with basic concepts illustrated by simple examples, and proceeds through historically important results to thorough treatments of modern topics such as the renormalization group, spinor-helicity methods for quark and gluon scattering, magnetic monopoles, instantons, supersymmetry, and the unification of forces.

The book is written in a modular format, with each chapter as self-contained as possible, and with the necessary prerequisite material clearly identified. This structure results in great flexibility, and allows readers to reach topics of specific interest easily. The book is based on a year-long course given by the author and contains extensive problems, with password-protected solutions available to lecturers at www.cambridge.org/9780521864497.

MARK SREDNICKI is Professor of Physics at the University of California, Santa Barbara. He gained his undergraduate degree from Cornell University in 1977, and received a Ph.D. from Stanford University in 1980. Professor Srednicki has held postdoctoral positions at Princeton University and the European Organization for Nuclear Research (CERN).

"This accessible and conceptually structured introduction to quantum field theory will be of value not only to beginning students but also to practicing physicists interested in learning or reviewing specific topics. The book is organized in a modular fashion, which makes it easy to extract the basic information relevant to the reader's area(s) of interest. The material is presented in an intuitively clear and informal style. Foundational topics such as path integrals and Lorentz representations are included early in the exposition, as appropriate for a modern course; later material includes a detailed description of the Standard Model and other advanced topics such as instantons, supersymmetry, and unification, which are essential knowledge for working particle physicists, but which are not treated in most other field theory texts."

Washington Taylor, Massachusetts Institute of Technology

"Over the years I have used parts of Srednicki's book to teach field theory to physics graduate students not specializing in particle physics. This is a vast subject, with many outstanding textbooks. Among these, Srednicki's stands out for its pedagogy. The subject is built logically, rather than historically. The exposition walks the line between getting the idea across and not shying away from a serious calculation. Path integrals enter early, and renormalization theory is pursued from the very start, with the excellent choice of φ^3 in six dimensions as the training workhorse. By the end of the course the student should understand both beta functions and the Standard Model, and be able to carry through a calculation when a perturbative calculation is called for."

Predrag Cvitanović, Georgia Institute of Technology

"This book should become a favorite of quantum field theory students and instructors. The approach is systematic and comprehensive, but the friendly and encouraging voice of the author comes through loud and clear to make the subject feel accessible. Many interesting examples are worked out in pedagogical detail."

Ann Nelson, University of Washington

"I expect that this will be the textbook of choice for many quantum field theory courses. The presentation is straightforward and readable, with the author's easy-going 'voice' coming through in his writing. The organization into a large number of short chapters, with the prerequisites for each chapter clearly marked, makes the book flexible and easy to teach from or to read independently. A large and varied collection of special topics is available, depending on the interests of the instructor and the student."

Joseph Polchinski, University of California, Santa Barbara

QUANTUM FIELD THEORY

MARK SREDNICKI
University of California, Santa Barbara

CAMBRIDGE
UNIVERSITY PRESS

University Printing House, Cambridge CB2 8BS, United Kingdom

One Liberty Plaza, 20th Floor, New York, NY 10006, USA

477 Williamstown Road, Port Melbourne, VIC 3207, Australia

314-321, 3rd Floor, Plot 3, Splendor Forum, Jasola District Centre, New Delhi - 110025, India

79 Anson Road, #06-04/06, Singapore 079906

Cambridge University Press is part of the University of Cambridge.

It furthers the University's mission by disseminating knowledge in the pursuit of education, learning and research at the highest international levels of excellence.

www.cambridge.org
Information on this title: www.cambridge.org/9780521864497

© M. Srednicki 2007

This publication is in copyright. Subject to statutory exception and to the provisions of relevant collective licensing agreements, no reproduction of any part may take place without the written permission of Cambridge University Press.

First published 2007
15th printing 2018

A catalogue record for this publication is available from the British Library

ISBN 978-0-521-86449-7 Hardback

Cambridge University Press has no responsibility for the persistence or accuracy of URLs for external or third-party internet websites referred to in this publication, and does not guarantee that any content on such websites is, or will remain, accurate or appropriate.

To my parents
Casimir and Helen Srednicki
with gratitude

Contents

Preface for students

Quantum field theory is the basic mathematical language that is used to describe and analyze the physics of elementary particles. The goal of this book is to provide a concise, step-by-step introduction to this subject, one that covers all the key concepts that are needed to understand the Standard Model of elementary particles, and some of its proposed extensions.

In order to be prepared to undertake the study of quantum field theory, you should recognize and understand the following equations:

$$\frac{d\sigma}{d\Omega} = |f(\theta,\phi)|^2 ,$$

$$a^\dagger|n\rangle = \sqrt{n{+}1}\,|n{+}1\rangle ,$$

$$J_\pm|j,m\rangle = \sqrt{j(j{+}1){-}m(m{\pm}1)}\,|j,m{\pm}1\rangle ,$$

$$A(t) = e^{+iHt/\hbar}Ae^{-iHt/\hbar} ,$$

$$H = p\dot{q} - L ,$$

$$ct' = \gamma(ct - \beta x) ,$$

$$E = (\mathbf{p}^2c^2 + m^2c^4)^{1/2} ,$$

$$\mathbf{E} = -\dot{\mathbf{A}}/c - \nabla\varphi .$$

This list is not, of course, complete; but if you are familiar with these equations, you probably know enough about quantum mechanics, classical mechanics, special relativity, and electromagnetism to tackle the material in this book.

Quantum field theory has the reputation of being a subject that is hard to learn. The problem, I think, is not so much that its basic ingredients are unusually difficult to master (indeed, the conceptual shift needed to go from quantum mechanics to quantum field theory is not nearly as severe as the one needed to go from classical mechanics to quantum mechanics), but rather that there are a *lot* of these ingredients. Some are fundamental, but many are just technical aspects of an unfamiliar form of perturbation theory.

In this book, I have tried to make the subject as accessible to beginners as possible. There are three main aspects to my approach.

Logical development of the basic concepts. This is, of course, very different from the historical development of quantum field theory, which, like the historical development of most worthwhile subjects, was filled with inspired guesses and brilliant extrapolations of sometimes fuzzy ideas, as well as its fair share of mistakes, misconceptions, and dead ends. None of that is in this book. From this book, you will (I hope) get the impression that the whole subject is effortlessly clear and obvious, with one step following the next like sunshine after refreshing rain.

Illustration of the basic concepts with the simplest examples. In most fields of human endeavor, newcomers are not expected to do the most demanding tasks right away. It takes time, dedication, and lots of practice to work up to what the accomplished masters are doing. There is no reason to expect quantum field theory to be any different in this regard. Therefore, we will start off by analyzing quantum field theories that are not immediately applicable to the real world of electrons, photons, protons, etc., but that will allow us to gain familiarity with the tools we will need, and to practice using them. Then, when we do work up to "real physics," we will be fully ready for the task. To this end, the book is divided into three parts: Spin Zero, Spin One Half, and Spin One. The technical complexities associated with a particular type of particle increase with its spin. We will therefore first learn all we can about spinless particles before moving on to the more difficult (and more interesting) nonzero spins. Once we get to them, we will do a good variety of calculations in (and beyond) the Standard Model of elementary particles.

User friendliness. Each of the three parts is divided into numerous sections. Each section is intended to treat one idea or concept or calculation, and each is written to be as self-contained as possible. For example, when an equation from an earlier section is needed, I usually just repeat it, rather than ask you to leaf back and find it (a reader's task that I've always found annoying). Furthermore, each section is labeled with its immediate prerequisites, so you can tell exactly what you need to have learned in order to proceed. This allows you to construct chains to whatever material may interest you, and to get there as quickly as possible.

That said, I expect that most readers of this book will encounter it as the textbook in a course on quantum field theory. In that case, of course, your reading will be guided by your professor, who I hope will find the above features useful. If, however, you are reading this book on your own, I have two pieces of advice.

The first (and most important) is this: find someone else to read it with you. I promise that it will be far more fun and rewarding that way; talking about a subject to another human being will inevitably improve the depth of your understanding. And you will have someone to work with you on the problems. (As with all physics texts, the problems are a key ingredient. I will not belabor this point, because if you have gotten this far in physics, you already know it well.)

The second piece of advice echoes the novelist and Nobel laureate William Faulkner. An interviewer asked, "Mr. Faulkner, some of your readers claim they still cannot understand your work after reading it two or three times. What approach would you advise them to adopt?" Faulkner replied, "Read it a fourth time."

That's my advice here as well. After the fourth attempt, though, you should consider trying something else. This is, after all, not the only book that has ever been written on the subject. You may find that a different approach (or even the same approach explained in different words) breaks the logjam in your thinking. There are a number of excellent books that you could consult, some of which are listed in the Bibliography. I have also listed particular books that I think could be helpful on specific topics in Reference Notes at the end of some of the sections.

This textbook (like all finite textbooks) has a number of deficiencies. One of these is a rather low level of mathematical rigor. This is partly endemic to the subject; rigorous proofs in quantum field theory are relatively rare, and do not appear in the overwhelming majority of research papers. Even some of the most basic notions lack proof; for example, currently you can get a million dollars from the Clay Mathematics Institute simply for proving that nonabelian gauge theory actually exists and has a unique ground state. Given this general situation, and since this is an introductory book, the proofs that we do have are only outlined.

Another deficiency of this book is that there is no discussion of the application of quantum field theory to condensed matter physics, where it also plays an important role. This connection has been important in the historical development of the subject, and is especially useful if you already know a lot of advanced statistical mechanics. I do not want this to be a prerequisite, however, and so I have chosen to keep the focus on applications within elementary particle physics.

Yet another deficiency is that *there are no references to the original literature*. In this regard, I am following a standard trend: as the foundations of a branch of science retreat into history, textbooks become more and more synthetic and reductionist. For example, it is now rare to see a new textbook on quantum mechanics that refers to the original papers by the famous founders of the subject. For guides to the original literature on quantum field theory, there are a number of other books with extensive references that you can consult; these include *Peskin & Schroeder*, *Weinberg*, and *Siegel*. (Italicized names refer to works listed in the Bibliography.) Unless otherwise noted, experimental numbers are taken from the Review of Particle Properties, available online at http://pdg.lbl.gov. Experimental numbers quoted in this book have an uncertainty of roughly ± 1 in the last significant digit. The Review should be consulted for the most recent experimental results, and for more precise statements of their uncertainty.

To conclude, let me say that you are about to embark on a tour of one of humanity's greatest intellectual endeavors, and certainly the one that has produced the most precise and accurate description of the natural world as we find it. I hope you enjoy the ride.

Preface for instructors

On learning that a new text on quantum field theory has appeared, one is surely tempted to respond with Isidor Rabi's famous comment about the muon: "Who ordered *that*?" After all, many excellent textbooks on quantum field theory are already available. I, for example, would not want to be without my well-worn copies of *Quantum Field Theory* by Lowell S. Brown (Cambridge 1994), *Aspects of Symmetry* by Sidney Coleman (Cambridge 1985), *Introduction to Quantum Field Theory* by Michael E. Peskin and Daniel V. Schroeder (Westview 1995), *Field Theory: A Modern Primer* by Pierre Ramond (Addison-Wesley 1990), *Fields* by Warren Siegel (arXiv.org 2005), *The Quantum Theory of Fields*, Volumes I, II, and III, by Steven Weinberg (Cambridge 1995), and *Quantum Field Theory in a Nutshell* by my colleague Tony Zee (Princeton 2003), to name just a few of the more recent texts. Nevertheless, despite the excellence of these and other books, I have never followed any of them very closely in my twenty years of on-and-off teaching of a year-long course in relativistic quantum field theory.

As discussed in the Preface for Students, this book is based on the notion that quantum field theory is most readily learned by starting with the simplest examples and working through their details in a logical fashion. To this end, I have tried to set things up at the very beginning to anticipate the eventual need for renormalization, and not be cavalier about how the fields are normalized and the parameters defined. I believe that these precautions take a lot of the "hocus pocus" (to quote Feynman) out of the "dippy process" of renormalization. Indeed, with this approach, even the anharmonic oscillator is in need of renormalization; see problem 14.7.

A field theory with many pedagogical virtues is φ^3 theory in six dimensions, where its coupling constant is dimensionless. Perhaps because six dimensions used to seem too outré (though today's prospective string theorists do not even blink), the only introductory textbook I know of that treats this model is *Quantum Field Theory* by George Sterman (Cambridge 1993), though it is also discussed in some more advanced books, such as *Renormalization* by John Collins (Cambridge 1984) and *Foundations of Quantum Chromodynamics* by T. Muta (World Scientific 1998). (There is also a series of lectures by Ed Witten on quantum field theory for mathematicians, available online, that treat φ^3 theory.) The reason φ^3 theory in six dimensions is a nice example is that its Feynman diagrams have a simple structure, but still exhibit the generic phenomena of renormalizable quantum field theory at the one-loop level. (The same cannot be said for φ^4 theory in four dimensions, where momentum-dependent corrections to the propagator do not appear until the two-loop level.) Thus, in Part I of this text, φ^3 theory in six dimensions is the primary example. I use it to give introductory treatments of most aspects of relativistic quantum field theory for spin-zero particles, with a minimum of the technical complications that arise in more realistic theories (like QED) with higher-spin particles.

Although I eventually discuss the Wilson approach to renormalization and effective field theory (in section 29), and use effective field theory extensively for the physics of hadrons in Part III, I do not feel it is pedagogically useful to bring it in at the very beginning, as is sometimes advocated. The problem is that the key notion of the decoupling of physical processes at different length scales is an unfamiliar one for most students; there is nothing in typical courses on quantum mechanics or electromagnetism or classical mechanics to prepare students for this idea (which was deemed worthy of a Nobel Prize for Ken Wilson in 1982). It also does not provide for a simple calculational framework, since one must deal with the infinite number of terms in the effective lagrangian, and then explain why most of them do not matter after all. It is noteworthy that Wilson himself did not spend a lot of time computing properly normalized perturbative S-matrix elements, a skill that we certainly want our students to have; we want them to have it because a great deal of current research still depends on it. Indeed, the vaunted success of quantum field theory as a description of the real world is based almost entirely on our ability to carry out these perturbative calculations. Studying renormalization early on has other pedagogical advantages. With the Nobel Prizes to Gerard 't Hooft and Tini Veltman in 1999 and to David Gross, David Politzer, and Frank Wilczek in 2004, today's students are well aware of beta functions and running couplings, and would like to understand them. I find that they are generally much more excited about this (even in the context of toy models) than they are about learning to reproduce the nearly century-old tree-level calculations of QED. And φ^3 theory in six dimensions is asymptotically free, which ultimately provides for a nice segue to the "real physics" of QCD.

In general, I have tried to present topics so that the more interesting aspects (from a present-day point of view) come first. An example is anomalies; the traditional approach is to start with the $\pi^0 \to \gamma\gamma$ decay rate, but such a low-energy process seems like a dusty relic to most of today's students. I therefore begin by demonstrating that anomalies destroy the self-consistency of the great majority of chiral gauge theories, a fact that strikes me (and, in my experience, most students) as much more interesting and dramatic than an incorrect calculation of the π^0 decay rate. Then, when we do eventually get to this process (in section 90), it appears as a straightforward consequence of what we already learned about anomalies in sections 75–77.

Nevertheless, I want this book to be useful to those who disagree with my pedagogical choices, and so I have tried to structure it to allow for maximum flexibility. Each section treats a particular idea or concept or calculation, and is as self-contained as possible. Each section also lists its immediate prerequisites, so that it is easy to see how to rearrange the material to suit your personal preferences.

In some cases, alternative approaches are developed in the problems. For example, I have chosen to introduce path integrals relatively early (though not before canonical quantization and operator methods are applied to free-field theory), and use them to derive Dyson's expansion. For those who would prefer to delay the introduction of path integrals (but since you will have to cover them eventually, why not get it over with?), problem 9.5 outlines the operator-based derivation in the interaction picture.

Another point worth noting is that a textbook and lectures are ideally complementary. Many sections of this book contain rather tedious mathematical detail that I would not and do not write on the blackboard during a lecture. (Indeed, the earliest origins of this book are supplementary notes that I typed up and handed out.) For example, much of the development of Weyl spinors in sections 34–37 can be left to outside reading. I do encourage you not to eliminate this material entirely, however; pedagogically, the problem with skipping directly to four-component notation is explaining that (in four dimensions) the hermitian conjugate of a left-handed field is right handed, a deeply important fact that is the key to solving problems such as 36.5 and 83.1, which are in turn vital to understanding the structure of the Standard Model and its extensions. A related topic is computing scattering amplitudes for Majorana fields; this is essential for modern research on massive neutrinos and supersymmetric particles, though it could be left out of a time-limited course.

While I have sometimes included more mathematical detail than is ideal for a lecture, I have also tended to omit explanations based on "physical intuition". For example, in section 90, we compute the $\pi^- \to \ell^- \bar{\nu}_\ell$ decay amplitude (where ℓ is a charged lepton) and find that it is proportional to the lepton mass. There is a well-known heuristic explanation of this fact that goes something like this: "The pion has spin zero, and so the lepton and the antineutrino must emerge with opposite spin, and therefore the same helicity. An antineutrino is always right handed, and so the lepton must be as well. But only the left-handed lepton couples to the W^-, so the decay amplitude vanishes if the left- and right-handed leptons are not coupled by a mass term."

This is essentially correct, but the reasoning is a bit more subtle than it first appears. A student may ask, "Why can't there be orbital angular momentum? Then the lepton and the antineutrino could have the same spin." The answer is that orbital angular momentum must be perpendicular to the linear momentum, whereas helicity is (by definition) parallel to the linear momentum; so adding orbital angular momentum cannot change the helicity assignments. (This is explored in a simplified model in problem 48.4.) The larger point is that intuitive explanations can almost always be probed more deeply. This is fine in a classroom, where you are available to answer questions, but a textbook author has a hard time knowing where to stop. Too little detail renders the explanations opaque, and too much can be overwhelming; furthermore, the happy medium tends to differ from student to student. The calculation, on the other hand, is definitive (at least within the framework being explored, and modulo the possibility of mathematical error). As Roger Penrose once said, "The great thing about physical intuition is that it can be adjusted to fit the facts." So, in this book, I have tended to emphasize calculational detail at the expense of heuristic reasoning. Lectures should ideally invert this to some extent.

I should also mention that a section of the book is not intended to coincide exactly with a lecture. The material in some sections could easily be covered in less than an hour, and some would clearly take more. My approach in lecturing is to try to keep to a pace that allows the students to follow the analysis, and then try to come to a more-or-less natural stopping point when class time is up. This sometimes means ending in the middle of a long calculation, but I feel that this is better than trying to artificially speed things along to reach a predetermined destination.

It would take at least three semesters of lectures to cover this entire book, and so a year-long course must omit some. A sequence I might follow (I tend to change things around a bit every year) is 1–23, 26–28, 33–43, 45–48, 51, 52, 54–59, 62–64, 66–68, 24, 69, 70, 44, 53, 71–73, 75–77, 30, 32, 84, 87–89, 29, 82, 83, 90, and, if any time was left, a selection of whatever seemed of most interest to me and the students of the remaining material.

To conclude, I hope you find this book to be a useful tool in working towards our mutual goal of bringing humanity's understanding of the physics of elementary particles to a new audience.

Acknowledgments

Every book is a collaborative effort, even if there is only one author on the title page. Any skills I may have as a teacher were first gleaned as a student in the classes of those who taught me. My first and most important teachers were my parents, Casimir and Helen Srednicki, to whom this book is dedicated. In our small town in Ohio, my excellent public-school teachers included Thelma Kieffaber, Marie Casher, Carol Baird, Jim Chase, Joe Gerin, Hugh Laughlin, and Tom Murphy. In college at Cornell, Don Hartill, Bruce Kusse, Bob Siemann, John Kogut, and Saul Teukolsky taught particularly memorable courses. In graduate school at Stanford, Roberto Peccei gave me my first exposure to quantum field theory, in a superb course that required bicycling in by 8:30 AM (which seemed like a major sacrifice at the time). Everyone in that class very much hoped that Roberto would one day turn his extensive hand-written lecture notes (which he put on reserve in the library) into a book. He never did, but I'd like to think that perhaps a bit of his consummate skill has found its way into this text. I have also used a couple of his jokes.

My thesis advisor at Stanford, Lenny Susskind, taught me how to think about physics without getting bogged down in the details. This book includes a lot of detail that Lenny would no doubt have left out, but while writing it I have tried to keep his exemplary clarity of thought in mind as something to strive for.

During my time in graduate school, and subsequently in postdoctoral positions at Princeton and CERN, and finally as a faculty member at UC Santa Barbara, I was extremely fortunate to be able to interact with many excellent physicists, from whom I learned an enormous amount. These include Stuart Freedman, Eduardo Fradkin, Steve Shenker, Sidney Coleman, Savas Dimopoulos, Stuart Raby, Michael Dine, Willy Fischler, Curt Callan,

David Gross, Malcolm Perry, Sam Treiman, Arthur Wightman, Ed Witten, Hans-Peter Nilles, Daniel Wyler, Dmitri Nanopoulos, John Ellis, Keith Olive, John Cardy, Jose Fulco, Ray Sawyer, Frank Wilczek, Jim Hartle, Gary Horowitz, Andy Strominger, and Tony Zee. I am especially grateful to my Santa Barbara colleagues David Berenstein, Steve Giddings, Don Marolf, Joe Polchinski, and Bob Sugar, who used various drafts of this book while teaching quantum field theory, and made various suggestions for improvement.

I am also grateful to physicists at other institutions who read parts of the manuscript and also made suggestions, including Zvi Bern, Oliver de Wolfe, Lance Dixon, Marcelo Gleiser, Steve Gottlieb, Arkady Tseytlin, and Arkady Vainshtein. I must single out for special thanks Professor Heidi Fearn of Cal State Fullerton, whose careful reading of Parts I and II allowed me to correct many unclear passages and outright errors that would otherwise have slipped by.

Students over the years have suffered through my varied attempts to arrive at a pedagogically acceptable scheme for teaching quantum field theory. I thank all of them for their indulgence. I am especially grateful to Sam Pinansky, Tae Min Hong, and Sho Yaida for their diligence in finding and reporting errors, and to Brian Wignal for help with formatting the manuscript. Also, a number of readers from around the world (as well as Santa Barbara) kindly reported errors in earlier versions; these include Mark Alford, Curtis Asplund, Omri Bahat-Treidel, Hee-Joong Chung, Claudio Coriano, Chris Duston, Daniel J. Feldman, Edson Fernando Ferrari, Gregory Giecold, Julian Heeck, Idse Heemskerk, Tae Min Hong, Ziyang Hu, Nathan Johnson-McDaniel, Nikhil Jayant Joshi, Yevgeny Kats, Sue Ann Koay, Hwasung Lee, Peter Lee, Shu-Ping Lee, Chris Lee Lin, Guilin Liu, Joyce Myers, Matan Mussel, Ahsan Nazer, Hiromichi Nishimura, Chris Pagnutti, Ari Pakman, Jess Riedel, Jorge Rocha, Mauricio Romo, Dusan Simic, Yushu Song, Daniel Vangheluwe, Miles Stoudenmire, Kevin Weil, Masaru Watanabe, Mark Weitzman, Brian Willett, Ting Yu, Ryan Zelen, Jianhui Zhou, and Fabio Zocchi. I thank them for their help, and apologize to anyone who I may have missed.

Throughout this project, the assistance and support of my wife Eloïse and daughter Julia were invaluable. Eloïse read through the manuscript and made suggestions that often clarified the language. Julia offered advice on the cover design (a highly stylized Feynman diagram). And they both kindly indulged the amount of time I spent working on this book that you now hold in your hands.

Part I

Spin Zero

1

Attempts at relativistic quantum mechanics

Prerequisite: none

In order to combine quantum mechanics and relativity, we must first understand what we mean by “quantum mechanics” and “relativity”. Let us begin with quantum mechanics.

Somewhere in most textbooks on the subject, one can find a list of the “axioms of quantum mechanics”. These include statements along the lines of:

> *The state of the system is represented by a vector in Hilbert space.*
>
> *Observables are represented by hermitian operators.*
>
> *The measurement of an observable yields one of its eigenvalues as the result.*

and so on. We do not need to review these closely here. The axiom we need to focus on is the one that says that the time evolution of the state of the system is governed by the Schrödinger equation,

$$i\hbar\frac{\partial}{\partial t}|\psi,t\rangle = H|\psi,t\rangle\ , \tag{1.1}$$

where H is the hamiltonian operator, representing the total energy.

Let us consider a very simple system: a spinless, nonrelativistic particle with no forces acting on it. In this case, the hamiltonian is

$$H = \frac{1}{2m}\mathbf{P}^2\ , \tag{1.2}$$

where m is the particle's mass, and $\mathbf{P}$ is the momentum operator. In the position basis, eq. (1.1) becomes

$$i\hbar\frac{\partial}{\partial t}\psi(\mathbf{x},t) = -\frac{\hbar^2}{2m}\nabla^2\psi(\mathbf{x},t)\ , \tag{1.3}$$

where $\psi(\mathbf{x},t) = \langle \mathbf{x}|\psi,t\rangle$ is the position-space wave function. We would like to generalize this to relativistic motion.

The obvious way to proceed is to take

$$H = +\sqrt{\mathbf{P}^2c^2 + m^2c^4}\ , \tag{1.4}$$

which yields the correct relativistic energy-momentum relation. If we formally expand this hamiltonian in inverse powers of the speed of light c, we get

$$H = mc^2 + \frac{1}{2m}\mathbf{P}^2 + \dots\ . \tag{1.5}$$

This is simply a constant (the rest energy), plus the usual nonrelativistic hamiltonian, eq. (1.2), plus higher-order corrections. With the hamiltonian given by eq. (1.4), the Schrödinger equation becomes

$$i\hbar\frac{\partial}{\partial t}\psi(\mathbf{x},t) = +\sqrt{-\hbar^2c^2\nabla^2 + m^2c^4}\ \psi(\mathbf{x},t)\ . \tag{1.6}$$

Unfortunately, this equation presents us with a number of difficulties. One is that it apparently treats space and time on a different footing: the time derivative appears only on the left, outside the square root, and the space derivatives appear only on the right, under the square root. This asymmetry between space and time is not what we would expect of a relativistic theory. Furthermore, if we expand the square root in powers of ∇^2, we get an infinite number of spatial derivatives acting on $\psi(\mathbf{x},t)$; this implies that eq. (1.6) is not local in space.

We can alleviate these problems by squaring the differential operators on each side of eq. (1.6) before applying them to the wave function. Then we get

$$-\hbar^2\frac{\partial^2}{\partial t^2}\psi(\mathbf{x},t) = \left(-\hbar^2c^2\nabla^2 + m^2c^4\right)\psi(\mathbf{x},t)\ . \tag{1.7}$$

This is the *Klein–Gordon equation*, and it looks a lot nicer than eq. (1.6). It is second-order in both space and time derivatives, and they appear in a symmetric fashion.

To better understand the Klein–Gordon equation, let us consider in more detail what we mean by "relativity". Special relativity tells us that physics looks the same in all inertial frames. To explain what this means, we first suppose that a certain spacetime coordinate system $(ct,\mathbf{x})$ represents (by fiat) an inertial frame. Let us define $x^0 = ct$, and write x^μ, where $\mu = 0,1,2,3$, in place of $(ct,\mathbf{x})$. It is also convenient (for reasons not at all obvious at this point) to define $x_0 = -x^0$ and $x_i = x^i$, where $i = 1,2,3$. This can be expressed more elegantly if we first introduce the *Minkowski metric*,

$$g_{\mu\nu} = \begin{pmatrix} -1 & & & \\ & +1 & & \\ & & +1 & \\ & & & +1 \end{pmatrix} , \tag{1.8}$$

where blank entries are zero. We then have $x_\mu = g_{\mu\nu}x^\nu$, where a repeated index is summed.

To invert this formula, we introduce the inverse of g, which is confusingly also called g, except with both indices up:

$$g^{\mu\nu} = \begin{pmatrix} -1 & & & \\ & +1 & & \\ & & +1 & \\ & & & +1 \end{pmatrix} . \tag{1.9}$$

We then have $g^{\mu\nu}g_{\nu\rho} = \delta^\mu{}_\rho$, where $\delta^\mu{}_\rho$ is the Kronecker delta (equal to one if its two indices take on the same value, zero otherwise). Now we can also write $x^\mu = g^{\mu\nu}x_\nu$.

It is a general rule that any pair of repeated (and therefore summed) indices must consist of one superscript and one subscript; these indices are said to be *contracted*. Also, any unrepeated (and therefore unsummed) indices must match (in both name and height) on the left- and right-hand sides of any valid equation.

Now we are ready to specify what we mean by an inertial frame. If the coordinates x^μ represent an inertial frame (which they do, by assumption), then so do any other coordinates $\bar{x}^\mu$ that are related by

$$\bar{x}^\mu = \Lambda^\mu{}_\nu x^\nu + a^\mu \, , \tag{1.10}$$

where $\Lambda^\mu{}_\nu$ is a *Lorentz transformation matrix* and a^μ is a *translation vector*. Both $\Lambda^\mu{}_\nu$ and a^μ are constant (that is, independent of x^μ). Furthermore, $\Lambda^\mu{}_\nu$ must obey

$$g_{\mu\nu}\Lambda^\mu{}_\rho\Lambda^\nu{}_\sigma = g_{\rho\sigma} \, . \tag{1.11}$$

Eq. (1.11) ensures that the *interval* between two different spacetime points that are labeled by x^μ and x'^μ in one inertial frame, and by $\bar{x}^\mu$ and $\bar{x}'^\mu$ in another, is the same. This interval is defined to be

$$\begin{aligned}(x - x')^2 &\equiv g_{\mu\nu}(x - x')^\mu(x - x')^\nu \\ &= (\mathbf{x} - \mathbf{x}')^2 - c^2(t - t')^2 \, . \end{aligned} \tag{1.12}$$

In the other frame, we have

$$\begin{aligned}(\bar{x}-\bar{x}')^2 &= g_{\mu\nu}(\bar{x}-\bar{x}')^\mu(\bar{x}-\bar{x}')^\nu \\ &= g_{\mu\nu}\Lambda^\mu{}_\rho\Lambda^\nu{}_\sigma(x-x')^\rho(x-x')^\sigma \\ &= g_{\rho\sigma}(x-x')^\rho(x-x')^\sigma \\ &= (x-x')^2\,, \end{aligned} \tag{1.13}$$

as desired.

When we say that *physics looks the same*, we mean that two observers (Alice and Bob, say) using two different sets of coordinates (representing two different inertial frames) should agree on the predicted results of all possible experiments. In the case of quantum mechanics, the simplest possibility is for Alice and Bob to agree on the value of the wave function at a particular spacetime point, a point that is called x by Alice and $\bar{x}$ by Bob. Thus if Alice's predicted wave function is $\psi(x)$, and Bob's is $\bar{\psi}(\bar{x})$, then we should have $\psi(x)=\bar{\psi}(\bar{x})$. Furthermore, in order to maintain $\psi(x)=\bar{\psi}(\bar{x})$ throughout spacetime, $\psi(x)$ and $\bar{\psi}(\bar{x})$ should obey identical equations of motion. Thus a candidate wave equation should take the same form in any inertial frame.

Let us see if this is true of the Klein–Gordon equation. We first introduce some useful notation for spacetime derivatives:

$$\partial_\mu \equiv \frac{\partial}{\partial x^\mu} = \left(+\frac{1}{c}\frac{\partial}{\partial t}, \nabla\right), \tag{1.14}$$

$$\partial^\mu \equiv \frac{\partial}{\partial x_\mu} = \left(-\frac{1}{c}\frac{\partial}{\partial t}, \nabla\right). \tag{1.15}$$

Note that

$$\partial^\mu x^\nu = g^{\mu\nu}\,, \tag{1.16}$$

so that our matching-index-height rule is satisfied.

If $\bar{x}$ and x are related by eq. (1.10), then $\bar{\partial}$ and ∂ are related by

$$\bar{\partial}^\mu = \Lambda^\mu{}_\nu\partial^\nu\,. \tag{1.17}$$

To check this, we note that

$$\bar{\partial}^\rho\bar{x}^\sigma = (\Lambda^\rho{}_\mu\partial^\mu)(\Lambda^\sigma{}_\nu x^\nu + a^\sigma) = \Lambda^\rho{}_\mu\Lambda^\sigma{}_\nu(\partial^\mu x^\nu) = \Lambda^\rho{}_\mu\Lambda^\sigma{}_\nu g^{\mu\nu} = g^{\rho\sigma}\,, \tag{1.18}$$

as expected. The last equality in eq. (1.18) is another form of eq. (1.11); see section 2.

We can now write eq. (1.7) as

$$-\hbar^2 c^2 \partial_0^2 \psi(x) = (-\hbar^2 c^2 \nabla^2 + m^2 c^4)\psi(x) \, . \tag{1.19}$$

After rearranging and identifying $\partial^2 \equiv \partial^\mu \partial_\mu = -\partial_0^2 + \nabla^2$, we have

$$(-\partial^2 + m^2 c^2/\hbar^2)\psi(x) = 0 \, . \tag{1.20}$$

This is Alice's form of the equation. Bob would write

$$(-\bar{\partial}^2 + m^2 c^2/\hbar^2)\bar{\psi}(\bar{x}) = 0 \, . \tag{1.21}$$

Is Bob's equation equivalent to Alice's equation? To see that it is, we set $\bar{\psi}(\bar{x}) = \psi(x)$, and note that

$$\bar{\partial}^2 = g_{\mu\nu}\bar{\partial}^\mu \bar{\partial}^\nu = g_{\mu\nu}\Lambda^\mu{}_\rho \Lambda^\nu{}_\sigma \partial^\rho \partial^\sigma = \partial^2 \, . \tag{1.22}$$

Thus, eq. (1.21) is indeed equivalent to eq. (1.20). The Klein–Gordon equation is therefore manifestly consistent with relativity: it takes the same form in every inertial frame.

This is the good news. The bad news is that the Klein–Gordon equation violates one of the axioms of quantum mechanics: eq. (1.1), the Schrödinger equation in its abstract form. The abstract Schrödinger equation has the fundamental property of being first order in the time derivative, whereas the Klein–Gordon equation is second order. This may not seem too important, but in fact it has drastic consequences. One of these is that the norm of a state,

$$\langle\psi,t|\psi,t\rangle = \int d^3x \, \langle\psi,t|\mathbf{x}\rangle\langle\mathbf{x}|\psi,t\rangle = \int d^3x \, \psi^*(x)\psi(x), \tag{1.23}$$

is not in general time independent. Thus probability is not conserved. The Klein–Gordon equation obeys relativity, but not quantum mechanics.

Dirac attempted to solve this problem (for spin-one-half particles) by introducing an extra discrete label on the wave function, to account for spin: $\psi_a(x)$, $a = 1, 2$. He then tried a Schrödinger equation of the form

$$i\hbar \frac{\partial}{\partial t}\psi_a(x) = \Big(-i\hbar c(\alpha^j)_{ab}\partial_j + mc^2(\beta)_{ab}\Big)\psi_b(x) \, , \tag{1.24}$$

where all repeated indices are summed, and α^j and β are matrices in spin-space. This equation, the *Dirac equation*, is consistent with the abstract Schrödinger equation. The state $|\psi, a, t\rangle$ carries a spin label a, and the hamiltonian is

$$H_{ab} = cP_j(\alpha^j)_{ab} + mc^2(\beta)_{ab} \, , \tag{1.25}$$

where P_j is a component of the momentum operator.

Since the Dirac equation is linear in both time and space derivatives, it has a chance to be consistent with relativity. Note that squaring the hamiltonian yields

$$(H^2)_{ab} = c^2 P_j P_k (\alpha^j \alpha^k)_{ab} + mc^3 P_j (\alpha^j \beta + \beta \alpha^j)_{ab} + (mc^2)^2 (\beta^2)_{ab} \, . \quad (1.26)$$

Since $P_j P_k$ is symmetric on exchange of j and k, we can replace $\alpha^j \alpha^k$ by its symmetric part, $\frac{1}{2}\{\alpha^j, \alpha^k\}$, where $\{A, B\} = AB + BA$ is the anticommutator. Then, if we choose matrices such that

$$\{\alpha^j, \alpha^k\}_{ab} = 2\delta^{jk}\delta_{ab} \, , \quad \{\alpha^j, \beta\}_{ab} = 0 \, , \quad (\beta^2)_{ab} = \delta_{ab} \, , \quad (1.27)$$

we will get

$$(H^2)_{ab} = (\mathbf{P}^2 c^2 + m^2 c^4)\delta_{ab} \, . \quad (1.28)$$

Thus, the eigenstates of H^2 are momentum eigenstates, with H^2 eigenvalue $\mathbf{p}^2 c^2 + m^2 c^4$. This is, of course, the correct relativistic energy-momentum relation. While it is outside the scope of this section to demonstrate it, it turns out that the Dirac equation is fully consistent with relativity, provided the Dirac matrices obey eq. (1.27). So we have apparently succeeded in constructing a quantum mechanical, relativistic theory!

There are, however, some problems. We would like the Dirac matrices to be 2×2, in order to account for electron spin. However, they must in fact be larger. To see this, note that the 2×2 Pauli matrices obey $\{\sigma^i, \sigma^j\} = 2\delta^{ij}$, and are thus candidates for the Dirac α^i matrices. However, there is no fourth matrix that anticommutes with these three (easily proven by writing down the most general 2×2 matrix and working out the three anticommutators explicitly). Also, we can show that the Dirac matrices must be even dimensional; see problem 1.1. Thus their minimum size is 4×4, and it remains for us to interpret the two extra possible "spin" states.

However, these extra states cause a more severe problem than a mere overcounting. Acting on a momentum eigenstate, H becomes the matrix $c\alpha \cdot \mathbf{p} + mc^2\beta$. In problem 1.1, we find that the trace of this matrix is zero. Thus the four eigenvalues must be $+E(\mathbf{p})$, $+E(\mathbf{p})$, $-E(\mathbf{p})$, $-E(\mathbf{p})$, where $E(\mathbf{p}) = +(\mathbf{p}^2 c^2 + m^2 c^4)^{1/2}$. The negative eigenvalues are the problem: they indicate that there is no ground state. In a more elaborate theory that included interactions with photons, there seems to be no reason why a positive energy electron could not emit a photon and drop down into a negative energy state. This downward cascade could continue forever. (The same problem also arises in attempts to interpret the Klein–Gordon equation as a modified form of quantum mechanics.)

Dirac made a wildly brilliant attempt to fix this problem of negative energy states. His solution is based on an empirical fact about electrons: they obey the Pauli exclusion principle. It is impossible to put more than one of them in the same quantum state. What if, Dirac speculated, all the negative energy states were *already occupied*? In this case, a positive energy electron could not drop into one of these states, by Pauli exclusion.

Many questions immediately arise. Why do we not see the negative electric charge of this *Dirac sea* of electrons? Dirac's answer: because we are used to it. (More precisely, the physical effects of a uniform charge density depend on the boundary conditions at infinity that we impose on Maxwell's equations, and there is a choice that renders such a uniform charge density invisible.) However, Dirac noted, if one of these negative energy electrons were excited into a positive energy state (by, say, a sufficiently energetic photon), it would leave behind a *hole* in the sea of negative energy electrons. This hole would appear to have positive charge, and positive energy. Dirac therefore predicted (in 1927) the existence of the *positron*, a particle with the same mass as the electron, but opposite charge. The positron was found experimentally five years later.

However, we have now jumped from an attempt at a quantum description of a *single* relativistic particle to a theory that apparently requires an *infinite* number of particles. Even if we accept this, we still have not solved the problem of how to describe particles like photons or pions or alpha nuclei that do *not* obey Pauli exclusion.

At this point, it is worthwhile to stop and reflect on why it has proven to be so hard to find an acceptable relativistic wave equation for a single quantum particle. Perhaps there is something wrong with our basic approach.

And there is. Recall the axiom of quantum mechanics that says that "Observables are represented by hermitian operators." This is not entirely true. There is one observable in quantum mechanics that is *not* represented by a hermitian operator: time. Time enters into quantum mechanics only when we announce that the "state of the system" depends on an extra parameter t. This parameter is not the eigenvalue of any operator. This is in sharp contrast to the particle's position $\mathbf{x}$, which *is* the eigenvalue of an operator. Thus, space and time are treated very differently, a fact that is obscured by writing the Schrödinger equation in terms of the position-space wave function $\psi(\mathbf{x}, t)$. Since space and time are treated asymmetrically, it is not surprising that we are having trouble incorporating a symmetry that mixes them up.

So, what are we to do?

In principle, the problem could be an intractable one: it might be *impossible* to combine quantum mechanics and relativity. In this case, there would have to be some meta-theory, one that reduces in the nonrelativistic limit to quantum mechanics, and in the classical limit to relativistic particle dynamics, but is actually neither.

This, however, turns out not to be the case. We can solve our problem, but we must put space and time on an equal footing at the outset. There are two ways to do this. One is to demote position from its status as an operator, and render it as an extra label, like time. The other is to promote time to an operator.

Let us discuss the second option first. If time becomes an operator, what do we use as the time parameter in the Schrödinger equation? Happily, in relativistic theories, there is more than one notion of time. We can use the *proper time* τ of the particle (the time measured by a clock that moves with it) as the time parameter. The coordinate time T (the time measured by a stationary clock in an inertial frame) is then promoted to an operator. In the Heisenberg picture (where the state of the system is fixed, but the operators are functions of time that obey the classical equations of motion), we would have operators $X^\mu(\tau)$, where $X^0 = T$. Relativistic quantum mechanics can indeed be developed along these lines, but it is surprisingly complicated to do so. (The many times are the problem; any monotonic function of τ is just as good a candidate as τ itself for the proper time, and this infinite redundancy of descriptions must be understood and accounted for.)

One of the advantages of considering different formalisms is that they may suggest different directions for generalizations. For example, once we have $X^\mu(\tau)$, why not consider adding some more parameters? Then we would have, for example, $X^\mu(\sigma, \tau)$. Classically, this would give us a continuous family of worldlines, what we might call a *worldsheet*, and so $X^\mu(\sigma, \tau)$ would describe a propagating *string*. This is indeed the starting point for string theory.

Thus, promoting time to an operator is a viable option, but is complicated in practice. Let us then turn to the other option, demoting position to a label. The first question is, label on what? The answer is, *on operators*. Thus, consider assigning an operator to each point $\mathbf{x}$ in space; call these operators $\varphi(\mathbf{x})$. A set of operators like this is called a *quantum field*. In the Heisenberg picture, the operators are also time dependent:

$$\varphi(\mathbf{x}, t) = e^{iHt/\hbar} \varphi(\mathbf{x}, 0) e^{-iHt/\hbar} . \tag{1.29}$$

Thus, both position and (in the Heisenberg picture) time are now labels on operators; neither is itself the eigenvalue of an operator.

So, now we have two different approaches to relativistic quantum theory, approaches that might, in principle, yield different results. This, however, is not the case: it turns out that any relativistic quantum physics that can be treated in one formalism can also be treated in the other. Which we use is a matter of convenience and taste. And, *quantum field theory*, the formalism in which position and time are both labels on operators, is much more convenient and efficient for most problems.

There is another useful equivalence: ordinary nonrelativistic quantum mechanics, for a fixed number of particles, can be rewritten as a quantum field theory. This is an informative exercise, since the corresponding physics is already familiar. Let us carry it out.

Begin with the position-basis Schrödinger equation for n particles, all with the same mass m, moving in an external potential $U(\mathbf{x})$, and interacting with each other via an interparticle potential $V(\mathbf{x}_1 - \mathbf{x}_2)$:

$$i\hbar\frac{\partial}{\partial t}\psi = \left[\sum_{j=1}^{n}\left(-\frac{\hbar^2}{2m}\nabla_j^2 + U(\mathbf{x}_j)\right) + \sum_{j=1}^{n}\sum_{k=1}^{j-1} V(\mathbf{x}_j - \mathbf{x}_k)\right]\psi\,, \tag{1.30}$$

where $\psi = \psi(\mathbf{x}_1, \ldots, \mathbf{x}_n; t)$ is the position-space wave function. The quantum mechanics of this system can be rewritten in the abstract form of eq. (1.1) by first introducing (in, for now, the Schrödinger picture) a quantum field $a(\mathbf{x})$ and its hermitian conjugate $a^\dagger(\mathbf{x})$. We take these operators to have the commutation relations

$$\begin{aligned}
[a(\mathbf{x}), a(\mathbf{x}')] &= 0\,, \\
[a^\dagger(\mathbf{x}), a^\dagger(\mathbf{x}')] &= 0\,, \\
[a(\mathbf{x}), a^\dagger(\mathbf{x}')] &= \delta^3(\mathbf{x} - \mathbf{x}')\,,
\end{aligned} \tag{1.31}$$

where $\delta^3(\mathbf{x})$ is the three-dimensional Dirac delta function. Thus, $a^\dagger(\mathbf{x})$ and $a(\mathbf{x})$ behave like harmonic-oscillator creation and annihilation operators that are labeled by a continuous index. In terms of them, we introduce the hamiltonian operator of our quantum field theory,

$$\begin{aligned}
H = &\int d^3x\, a^\dagger(\mathbf{x})\Big(-\tfrac{\hbar^2}{2m}\nabla^2 + U(\mathbf{x})\Big)a(\mathbf{x}) \\
&+ \tfrac{1}{2}\int d^3x\, d^3y\, V(\mathbf{x} - \mathbf{y}) a^\dagger(\mathbf{x}) a^\dagger(\mathbf{y}) a(\mathbf{y}) a(\mathbf{x})\,.
\end{aligned} \tag{1.32}$$

Now consider a time-dependent state of the form

$$|\psi, t\rangle = \int d^3x_1 \ldots d^3x_n\, \psi(\mathbf{x}_1, \ldots, \mathbf{x}_n; t) a^\dagger(\mathbf{x}_1) \ldots a^\dagger(\mathbf{x}_n)|0\rangle\,, \tag{1.33}$$

where $\psi(\mathbf{x}_1, \ldots, \mathbf{x}_n; t)$ is some function of the n particle positions and time, and $|0\rangle$ is the *vacuum state*, the state that is annihilated by all the as,

$$a(\mathbf{x})|0\rangle = 0 \; . \tag{1.34}$$

It is now straightforward (though tedious) to verify that eq. (1.1), the abstract Schrödinger equation, is obeyed if and only if the function ψ satisfies eq. (1.30).

Thus we can interpret the state $|0\rangle$ as a state of "no particles", the state $a^\dagger(\mathbf{x}_1)|0\rangle$ as a state with one particle at position $\mathbf{x}_1$, the state $a^\dagger(\mathbf{x}_1)a^\dagger(\mathbf{x}_2)|0\rangle$ as a state with one particle at position $\mathbf{x}_1$ and another at position $\mathbf{x}_2$, and so on. The operator

$$N = \int d^3x \; a^\dagger(\mathbf{x})a(\mathbf{x}) \tag{1.35}$$

counts the total number of particles. It commutes with the hamiltonian, as is easily checked; thus, if we start with a state of n particles, we remain with a state of n particles at all times.

However, we can imagine generalizations of this version of the theory (generalizations that would not be possible without the field formalism) in which the number of particles is *not* conserved. For example, we could try adding to H a term like

$$\Delta H \propto \int d^3x \left[a^\dagger(\mathbf{x})a^2(\mathbf{x}) + \text{h.c.} \right]. \tag{1.36}$$

This term does *not* commute with N, and so the number of particles would not be conserved with this addition to H.

Theories in which the number of particles can change as time evolves are a good thing: they are needed for correct phenomenology. We are already familiar with the notion that atoms can emit and absorb photons, and so we had better have a formalism that can incorporate this phenomenon. We are less familiar with emission and absorption (that is to say, creation and annihilation) of electrons, but this process also occurs in nature; it is less common because it must be accompanied by the emission or absorption of a positron, antiparticle to the electron. There are not a lot of positrons around to facilitate electron annihilation, while e^+e^- pair creation requires us to have on hand at least $2mc^2$ of energy available for the rest-mass energy of these two particles. The photon, on the other hand, is its own antiparticle, and has zero rest mass; thus photons are easily and copiously produced and destroyed.

There is another important aspect of the quantum theory specified by eqs. (1.32) and (1.33). Because the creation operators commute with each other, only the completely symmetric part of ψ survives the integration in eq. (1.33). Therefore, without loss of generality, we can restrict our attention to ψs of this type:

$$\psi(\ldots \mathbf{x}_i \ldots \mathbf{x}_j \ldots ; t) = +\psi(\ldots \mathbf{x}_j \ldots \mathbf{x}_i \ldots ; t)\ . \tag{1.37}$$

This means that we have a theory of *bosons*, particles that (like photons or pions or alpha nuclei) obey Bose–Einstein statistics. If we want Fermi–Dirac statistics instead, we must replace eq. (1.31) with

$$\begin{aligned} \{a(\mathbf{x}), a(\mathbf{x}')\} &= 0\ , \\ \{a^\dagger(\mathbf{x}), a^\dagger(\mathbf{x}')\} &= 0\ , \\ \{a(\mathbf{x}), a^\dagger(\mathbf{x}')\} &= \delta^3(\mathbf{x} - \mathbf{x}')\ , \end{aligned} \tag{1.38}$$

where again $\{A, B\} = AB + BA$ is the anticommutator. Now only the fully antisymmetric part of ψ survives the integration in eq. (1.33), and so we can restrict our attention to

$$\psi(\ldots \mathbf{x}_i \ldots \mathbf{x}_j \ldots ; t) = -\psi(\ldots \mathbf{x}_j \ldots \mathbf{x}_i \ldots ; t)\ . \tag{1.39}$$

Thus we have a theory of *fermions*. It is straightforward to check that the abstract Schrödinger equation, eq. (1.1), still implies that ψ obeys the differential equation (1.30).[1] Interestingly, there is no simple way to write down a quantum field theory with particles that obey Boltzmann statistics, corresponding to a wave function with no particular symmetry. This is a hint of the *spin-statistics* theorem, which applies to *relativistic* quantum field theory. It says that *interacting* particles with integer spin must be bosons, and *interacting* particles with half-integer spin must be fermions. In our *nonrelativistic* example, the interacting particles clearly have spin zero (because their creation operators carry no labels that could be interpreted as corresponding to different spin states), but can be either bosons or fermions, as we have seen.

Now that we have seen how to rewrite the nonrelativistic quantum mechanics of multiple bosons or fermions as a quantum field theory, it is time to try to construct a relativistic version.

[1] Now, however, the ordering of the a and $a^\dagger$ operators in the last term of eq. (1.32) becomes significant, and must be as written.

Reference notes

The history of the physics of elementary particles is recounted in *Pais.* A brief overview can be found in *Weinberg I.* More details on quantum field theory for nonrelativistic particles can be found in *Brown.*

Problems

1.1) Show that the Dirac matrices must be even dimensional. Hint: show that the eigenvalues of β are all ± 1, and that $\mathrm{Tr}\,\beta = 0$. To show that $\mathrm{Tr}\,\beta = 0$, consider, e.g., $\mathrm{Tr}\,\alpha_1^2\beta$. Similarly, show that $\mathrm{Tr}\,\alpha_i = 0$.

1.2) With the hamiltonian of eq. (1.32), show that the state defined in eq. (1.33) obeys the abstract Schrödinger equation, eq. (1.1), if and only if the wave function obeys eq. (1.30). Your demonstration should apply both to the case of bosons, where the particle creation and annihilation operators obey the commutation relations of eq. (1.31), and to fermions, where the particle creation and annihilation operators obey the anticommutation relations of eq. (1.38).

1.3) Show explicitly that $[N, H] = 0$, where H is given by eq. (1.32) and N by eq. (1.35).

2

Lorentz invariance

Prerequisite: 1

A *Lorentz transformation* is a linear, homogeneous change of coordinates from x^μ to $\bar{x}^\mu$,

$$\bar{x}^\mu = \Lambda^\mu{}_\nu x^\nu , \tag{2.1}$$

that preserves the *interval* x^2 between x^μ and the origin, where

$$x^2 \equiv x^\mu x_\mu = g_{\mu\nu} x^\mu x^\nu = \mathbf{x}^2 - c^2 t^2 . \tag{2.2}$$

This means that the matrix $\Lambda^\mu{}_\nu$ must obey

$$g_{\mu\nu} \Lambda^\mu{}_\rho \Lambda^\nu{}_\sigma = g_{\rho\sigma} , \tag{2.3}$$

where

$$g_{\mu\nu} = \begin{pmatrix} -1 & & & \\ & +1 & & \\ & & +1 & \\ & & & +1 \end{pmatrix} \tag{2.4}$$

is the Minkowski metric.

Note that this set of transformations includes ordinary spatial rotations: take $\Lambda^0{}_0 = 1$, $\Lambda^0{}_i = \Lambda^i{}_0 = 0$, and $\Lambda^i{}_j = R_{ij}$, where R is an orthogonal rotation matrix.

The set of all Lorentz transformations forms a *group*: the product of any two Lorentz transformations is another Lorentz transformation; the product is associative; there is an identity transformation, $\Lambda^\mu{}_\nu = \delta^\mu{}_\nu$; and every Lorentz transformation has an inverse. It is easy to demonstrate these statements explicitly. For example, to find the inverse transformation $(\Lambda^{-1})^\mu{}_\nu$, note that the left-hand side of eq. (2.3) can be written as $\Lambda_{\nu\rho}\Lambda^\nu{}_\sigma$, and that we can raise the ρ index on both sides to get $\Lambda_\nu{}^\rho \Lambda^\nu{}_\sigma = \delta^\rho{}_\sigma$. On the other hand, by definition, $(\Lambda^{-1})^{\rho}{}_{\nu}\Lambda^{\nu}{}_{\sigma} = \delta^{\rho}{}_{\sigma}$. Therefore

$$(\Lambda^{-1})^{\rho}{}_{\nu} = \Lambda_{\nu}{}^{\rho} \,. \tag{2.5}$$

Another useful version of eq. (2.3) is

$$g^{\mu\nu}\Lambda^{\rho}{}_{\mu}\Lambda^{\sigma}{}_{\nu} = g^{\rho\sigma} \,. \tag{2.6}$$

To get eq. (2.6), start with eq. (2.3), but with the inverse transformations $(\Lambda^{-1})^{\mu}{}_{\rho}$ and $(\Lambda^{-1})^{\nu}{}_{\sigma}$. Then use eq. (2.5), raise all down indices, and lower all up indices. The result is eq. (2.6).

For an infinitesimal Lorentz transformation, we can write

$$\Lambda^{\mu}{}_{\nu} = \delta^{\mu}{}_{\nu} + \delta\omega^{\mu}{}_{\nu} \,. \tag{2.7}$$

Eq. (2.3) can be used to show that $\delta\omega$ with both indices down (or up) is antisymmetric:

$$\delta\omega_{\rho\sigma} = -\delta\omega_{\sigma\rho} \,. \tag{2.8}$$

Thus there are six independent infinitesimal Lorentz transformations (in four spacetime dimensions). These can be divided into three rotations ($\delta\omega_{ij} = -\varepsilon_{ijk}\hat{n}_k\delta\theta$ for a rotation by angle $\delta\theta$ about the unit vector $\hat{\mathbf{n}}$) and three boosts ($\delta\omega_{i0} = \hat{n}_i\delta\eta$ for a boost in the direction $\hat{\mathbf{n}}$ by *rapidity* $\delta\eta$).

Not all Lorentz transformations can be reached by compounding infinitesimal ones. If we take the determinant of eq. (2.5), we get $(\det\Lambda)^{-1} = \det\Lambda$, which implies $\det\Lambda = \pm 1$. Transformations with $\det\Lambda = +1$ are *proper*, and transformations with $\det\Lambda = -1$ are *improper*. Note that the product of any two proper Lorentz transformations is proper, and that infinitesimal transformations of the form $\Lambda = 1 + \delta\omega$ are proper. Therefore, any transformation that can be reached by compounding infinitesimal ones is proper. The proper transformations form a *subgroup* of the Lorentz group.

Another subgroup is that of the *orthochronous* Lorentz transformations: those for which $\Lambda^{0}{}_{0} \geq +1$. Note that eq. (2.3) implies $(\Lambda^{0}{}_{0})^2 - \Lambda^{i}{}_{0}\Lambda^{i}{}_{0} = 1$; thus, either $\Lambda^{0}{}_{0} \geq +1$ or $\Lambda^{0}{}_{0} \leq -1$. An infinitesimal transformation is clearly orthochronous, and it is straightforward to show that the product of two orthochronous transformations is also orthochronous.

Thus, the Lorentz transformations that can be reached by compounding infinitesimal ones are both proper and orthochronous, and they form a subgroup. We can introduce two discrete transformations that take us out of this subgroup: *parity* and *time reversal.* The parity transformation is

$$\mathcal{P}^{\mu}{}_{\nu} = (\mathcal{P}^{-1})^{\mu}{}_{\nu} = \begin{pmatrix} +1 & & & \\ & -1 & & \\ & & -1 & \\ & & & -1 \end{pmatrix} . \tag{2.9}$$

It is orthochronous, but improper. The time-reversal transformation is

$$\mathcal{T}^{\mu}{}_{\nu} = (\mathcal{T}^{-1})^{\mu}{}_{\nu} = \begin{pmatrix} -1 & & & \\ & +1 & & \\ & & +1 & \\ & & & +1 \end{pmatrix} . \tag{2.10}$$

It is nonorthochronous and improper.

Generally, when a theory is said to be *Lorentz invariant*, this means under the proper orthochronous subgroup only. Parity and time reversal are treated separately. It is possible for a quantum field theory to be invariant under the proper orthochronous subgroup, but not under parity and/or time-reversal.

From here on, in this section, we will treat the proper orthochronous subgroup only. Parity and time reversal will be treated in section 23.

In quantum theory, symmetries are represented by unitary (or antiunitary) operators. This means that we associate a unitary operator $U(\Lambda)$ to each proper, orthochronous Lorentz transformation Λ. These operators must obey the composition rule

$$U(\Lambda'\Lambda) = U(\Lambda')U(\Lambda) . \tag{2.11}$$

For an infinitesimal transformation, we can write

$$U(1+\delta\omega) = I + \tfrac{i}{2\hbar}\delta\omega_{\mu\nu}M^{\mu\nu} , \tag{2.12}$$

where $M^{\mu\nu} = -M^{\nu\mu}$ is a set of hermitian operators called the *generators of the Lorentz group.* If we start with $U(\Lambda)^{-1}U(\Lambda')U(\Lambda) = U(\Lambda^{-1}\Lambda'\Lambda)$, let $\Lambda' = 1 + \delta\omega'$, and expand both sides to linear order in $\delta\omega$, we get

$$\delta\omega_{\mu\nu}U(\Lambda)^{-1}M^{\mu\nu}U(\Lambda) = \delta\omega_{\mu\nu}\Lambda^{\mu}{}_{\rho}\Lambda^{\nu}{}_{\sigma}M^{\rho\sigma} . \tag{2.13}$$

Then, since $\delta\omega_{\mu\nu}$ is arbitrary (except for being antisymmetric), the antisymmetric part of its coefficient on each side must be the same. In this case, because $M^{\mu\nu}$ is already antisymmetric (by definition), we have

$$U(\Lambda)^{-1}M^{\mu\nu}U(\Lambda) = \Lambda^{\mu}{}_{\rho}\Lambda^{\nu}{}_{\sigma}M^{\rho\sigma} . \tag{2.14}$$

We see that each vector index on $M^{\mu\nu}$ undergoes its own Lorentz transformation. This is a general result: any operator carrying one or more vector indices should behave similarly. For example, consider the energy-momentum four-vector P^μ, where cP^0 is the hamiltonian H and P^i are the components of the total three-momentum operator. We expect

$$U(\Lambda)^{-1}P^\mu U(\Lambda) = \Lambda^\mu{}_\nu P^\nu \; . \tag{2.15}$$

If we now let $\Lambda = 1 + \delta\omega$ in eq. (2.14), expand to linear order in $\delta\omega$, and equate the antisymmetric part of the coefficients of $\delta\omega_{\mu\nu}$, we get the commutation relations

$$[M^{\mu\nu}, M^{\rho\sigma}] = i\hbar\Big(g^{\mu\rho}M^{\nu\sigma} - (\mu{\leftrightarrow}\nu)\Big) - (\rho{\leftrightarrow}\sigma) \; . \tag{2.16}$$

These commutation relations specify the *Lie algebra* of the Lorentz group. We can identify the components of the angular momentum operator $\mathbf{J}$ as $J_i \equiv \frac{1}{2}\varepsilon_{ijk}M^{jk}$, and the components of the boost operator $\mathbf{K}$ as $K_i \equiv M^{i0}$. We then find from eq. (2.16) that

$$\begin{aligned} [J_i, J_j] &= i\hbar\varepsilon_{ijk}J_k \; , \\ [J_i, K_j] &= i\hbar\varepsilon_{ijk}K_k \; , \\ [K_i, K_j] &= -i\hbar\varepsilon_{ijk}J_k \; . \end{aligned} \tag{2.17}$$

The first of these is the usual set of commutators for angular momentum, and the second says that $\mathbf{K}$ transforms as a three-vector under rotations. The third implies that a series of boosts can be equivalent to a rotation.

Similarly, we can let $\Lambda = 1 + \delta\omega$ in eq. (2.15) to get

$$[P^\mu, M^{\rho\sigma}] = i\hbar\Big(g^{\mu\sigma}P^\rho - (\rho{\leftrightarrow}\sigma)\Big) \; , \tag{2.18}$$

which becomes

$$\begin{aligned} [J_i, H] &= 0 \; , \\ [J_i, P_j] &= i\hbar\varepsilon_{ijk}P_k \; , \\ [K_i, H] &= i\hbar c P_i \; , \\ [K_i, P_j] &= i\hbar\delta_{ij}H/c \; . \end{aligned} \tag{2.19}$$

Also, the components of P^μ should commute with each other:

$$\begin{aligned} [P_i, P_j] &= 0 \; , \\ [P_i, H] &= 0 \; . \end{aligned} \tag{2.20}$$

Together, eqs. (2.17), (2.19), and (2.20) form the Lie algebra of the *Poincaré group*.

Let us now consider what should happen to a quantum scalar field $\varphi(x)$ under a Lorentz transformation. We begin by recalling how time evolution works in the Heisenberg picture:

$$e^{+iHt/\hbar}\varphi(\mathbf{x},0)e^{-iHt/\hbar} = \varphi(\mathbf{x},t) \,. \tag{2.21}$$

Obviously, this should have a relativistic generalization,

$$e^{-iPx/\hbar}\varphi(0)e^{+iPx/\hbar} = \varphi(x) \,, \tag{2.22}$$

where $Px = P^\mu x_\mu = \mathbf{P}\cdot\mathbf{x} - Ht$. We can make this a little fancier by defining the unitary *spacetime translation operator*

$$T(a) \equiv \exp(-iP^\mu a_\mu/\hbar) \,. \tag{2.23}$$

Then we have

$$T(a)^{-1}\varphi(x)T(a) = \varphi(x-a) \,. \tag{2.24}$$

For an infinitesimal translation,

$$T(\delta a) = I - \tfrac{i}{\hbar}\delta a_\mu P^\mu \,. \tag{2.25}$$

Comparing eqs. (2.12) and (2.25), we see that eq. (2.24) leads us to expect

$$U(\Lambda)^{-1}\varphi(x)U(\Lambda) = \varphi(\Lambda^{-1}x) \,. \tag{2.26}$$

Derivatives of φ then carry vector indices that transform in the appropriate way, e.g.,

$$U(\Lambda)^{-1}\partial^\mu\varphi(x)U(\Lambda) = \Lambda^\mu{}_\rho\bar{\partial}^\rho\varphi(\Lambda^{-1}x) \,, \tag{2.27}$$

where the bar on a derivative means that it is with respect to the argument $\bar{x} = \Lambda^{-1}x$. Eq. (2.27) also implies

$$U(\Lambda)^{-1}\partial^2\varphi(x)U(\Lambda) = \bar{\partial}^2\varphi(\Lambda^{-1}x) \,, \tag{2.28}$$

so that the Klein–Gordon equation, $(-\partial^2 + m^2/\hbar^2c^2)\varphi = 0$, is Lorentz invariant, as we saw in section 1.

Reference notes

A detailed discussion of quantum Lorentz transformations can be found in *Weinberg I*.

Problems

2.1) Verify that eq. (2.8) follows from eq. (2.3).
2.2) Verify that eq. (2.14) follows from $U(\Lambda)^{-1}U(\Lambda')U(\Lambda) = U(\Lambda^{-1}\Lambda'\Lambda)$.
2.3) Verify that eq. (2.16) follows from eq. (2.14).

2.4) Verify that eq. (2.17) follows from eq. (2.16).
2.5) Verify that eq. (2.18) follows from eq. (2.15).
2.6) Verify that eq. (2.19) follows from eq. (2.18).
2.7) What property should be attributed to the translation operator $T(a)$ that could be used to prove eq. (2.20)?
2.8) a) Let $\Lambda = 1 + \delta\omega$ in eq. (2.26), and show that

$$[\varphi(x), M^{\mu\nu}] = \mathcal{L}^{\mu\nu}\varphi(x) \,, \tag{2.29}$$

where

$$\mathcal{L}^{\mu\nu} \equiv \tfrac{\hbar}{i}(x^\mu\partial^\nu - x^\nu\partial^\mu) \,. \tag{2.30}$$

b) Show that $[[\varphi(x), M^{\mu\nu}], M^{\rho\sigma}] = \mathcal{L}^{\mu\nu}\mathcal{L}^{\rho\sigma}\varphi(x)$.

c) Prove the *Jacobi identity*, $[[A, B], C] + [[B, C], A] + [[C, A], B] = 0$. Hint: write out all the commutators.

d) Use your results from parts (b) and (c) to show that

$$[\varphi(x), [M^{\mu\nu}, M^{\rho\sigma}]] = (\mathcal{L}^{\mu\nu}\mathcal{L}^{\rho\sigma} - \mathcal{L}^{\rho\sigma}\mathcal{L}^{\mu\nu})\varphi(x) \,. \tag{2.31}$$

e) Simplify the right-hand side of eq. (2.31) as much as possible.

f) Use your results from part (e) to verify eq. (2.16), up to the possibility of a term on the right-hand side that commutes with $\varphi(x)$ and its derivatives. (Such a term, called a *central charge*, in fact does not arise for the Lorentz algebra.)

2.9) Let us write

$$\Lambda^\rho{}_\tau = \delta^\rho{}_\tau + \tfrac{i}{2\hbar}\delta\omega_{\mu\nu}(S_{\rm V}^{\mu\nu})^\rho{}_\tau \,, \tag{2.32}$$

where

$$(S_{\rm V}^{\mu\nu})^\rho{}_\tau \equiv \tfrac{\hbar}{i}(g^{\mu\rho}\delta^\nu{}_\tau - g^{\nu\rho}\delta^\mu{}_\tau) \tag{2.33}$$

are matrices which constitute the *vector representation* of the Lorentz generators.

a) Let $\Lambda = 1 + \delta\omega$ in eq. (2.27), and show that

$$[\partial^\rho\varphi(x), M^{\mu\nu}] = \mathcal{L}^{\mu\nu}\partial^\rho\varphi(x) + (S_{\rm V}^{\mu\nu})^\rho{}_\tau\partial^\tau\varphi(x) \,. \tag{2.34}$$

b) Show that the matrices $S_{\rm V}^{\mu\nu}$ must have the same commutation relations as the operators $M^{\mu\nu}$. Hint: see the previous problem.

c) For a rotation by an angle θ about the z axis, we have

$$\Lambda^\mu{}_\nu = \begin{pmatrix} 1 & 0 & 0 & 0 \\ 0 & \cos\theta & -\sin\theta & 0 \\ 0 & \sin\theta & \cos\theta & 0 \\ 0 & 0 & 0 & 1 \end{pmatrix} . \tag{2.35}$$

Show that

$$\Lambda = \exp(-i\theta S_{\rm V}^{12}/\hbar)\ . \tag{2.36}$$

d) For a boost by *rapidity* η in the z direction, we have

$$\Lambda^{\mu}{}_{\nu} = \begin{pmatrix} \cosh\eta & 0 & 0 & \sinh\eta \\ 0 & 1 & 0 & 0 \\ 0 & 0 & 1 & 0 \\ \sinh\eta & 0 & 0 & \cosh\eta \end{pmatrix}. \tag{2.37}$$

Show that

$$\Lambda = \exp(+i\eta S_{\rm V}^{30}/\hbar)\ . \tag{2.38}$$

3

Canonical quantization of scalar fields

Prerequisite: 2

Let us go back and drastically simplify the hamiltonian we constructed in section 1, reducing it to the hamiltonian for free particles:

$$H = \int d^3x \, a^\dagger(\mathbf{x})\left(-\tfrac{1}{2m}\nabla^2\right)a(\mathbf{x})$$
$$= \int d^3p \, \tfrac{1}{2m}\mathbf{p}^2 \, \widetilde{a}^\dagger(\mathbf{p})\widetilde{a}(\mathbf{p}) \, , \tag{3.1}$$

where

$$\widetilde{a}(\mathbf{p}) = \int \frac{d^3x}{(2\pi)^{3/2}} \, e^{-i\mathbf{p}\cdot\mathbf{x}} \, a(\mathbf{x}) \, . \tag{3.2}$$

Here we have simplified our notation by setting

$$\hbar = 1 \, . \tag{3.3}$$

The appropriate factors of $\hbar$ can always be restored in any of our formulae via dimensional analysis. The commutation (or anticommutation) relations of the $\widetilde{a}(\mathbf{p})$ and $\widetilde{a}^\dagger(\mathbf{p})$ operators are

$$[\widetilde{a}(\mathbf{p}), \widetilde{a}(\mathbf{p}')]_\mp = 0 \, ,$$
$$[\widetilde{a}^\dagger(\mathbf{p}), \widetilde{a}^\dagger(\mathbf{p}')]_\mp = 0 \, ,$$
$$[\widetilde{a}(\mathbf{p}), \widetilde{a}^\dagger(\mathbf{p}')]_\mp = \delta^3(\mathbf{p} - \mathbf{p}') \, , \tag{3.4}$$

where $[A, B]_\mp$ is either the commutator (if we want a theory of bosons) or the anticommutator (if we want a theory of fermions). Thus $\widetilde{a}^\dagger(\mathbf{p})$ can be interpreted as creating a state of definite momentum $\mathbf{p}$, and eq. (3.1) describes a theory of free particles. The ground state is the *vacuum* $|0\rangle$; it is annihilated by $\widetilde{a}(\mathbf{p})$,

$$\widetilde{a}(\mathbf{p})|0\rangle = 0 , \tag{3.5}$$

and so its energy eigenvalue is zero. The other eigenstates of H are all of the form $\widetilde{a}^\dagger(\mathbf{p}_1)\ldots\widetilde{a}^\dagger(\mathbf{p}_n)|0\rangle$, and the corresponding energy eigenvalue is $E(\mathbf{p}_1)+\ldots+E(\mathbf{p}_n)$, where $E(\mathbf{p}) = \frac{1}{2m}\mathbf{p}^2$.

It is easy to see how to generalize this theory to a relativistic one; all we need to do is use the relativistic energy formula $E(\mathbf{p}) = +(\mathbf{p}^2c^2 + m^2c^4)^{1/2}$:

$$H = \int d^3p \, (\mathbf{p}^2c^2 + m^2c^4)^{1/2} \, \widetilde{a}^\dagger(\mathbf{p})\widetilde{a}(\mathbf{p}) . \tag{3.6}$$

Now we have a theory of free *relativistic* spin-zero particles, and they can be either bosons or fermions.

Is this theory really Lorentz invariant? We will answer this question (in the affirmative) in a very roundabout way: by constructing it again, from a rather different point of view, a point of view that emphasizes Lorentz invariance from the beginning.

We will start with the *classical* physics of a *real scalar field* $\varphi(x)$. *Real* means that $\varphi(x)$ assigns a real number to every point in spacetime. *Scalar* means that Alice [who uses coordinates x^μ and calls the field $\varphi(x)$] and Bob [who uses coordinates $\bar{x}^\mu$, related to Alice's coordinates by $\bar{x}^\mu = \Lambda^\mu{}_\nu x^\nu + a^\nu$, and calls the field $\bar{\varphi}(\bar{x})$], agree on the numerical value of the field: $\varphi(x) = \bar{\varphi}(\bar{x})$. This then implies that the equation of motion for $\varphi(x)$ must be the same as that for $\bar{\varphi}(\bar{x})$. We have already met an equation of this type: the Klein–Gordon equation,

$$(-\partial^2 + m^2)\varphi(x) = 0 . \tag{3.7}$$

Here we have simplified our notation by setting

$$c = 1 \tag{3.8}$$

in addition to $\hbar = 1$. As with $\hbar$, factors of c can restored, if desired, by dimensional analysis.

We will adopt eq. (3.7) as the equation of motion that we would like $\varphi(x)$ to obey. It should be emphasized at this point that we are doing *classical physics* of a *real scalar field*. We are *not* to think of $\varphi(x)$ as a quantum wave function. Thus, there should not be any factors of $\hbar$ in this version of the Klein–Gordon equation. This means that the parameter m must have dimensions of inverse length; m is *not* (yet) to be thought of as a mass.

The equation of motion can be derived from variation of an action $S = \int dt \, L$, where L is the lagrangian. Since the Klein–Gordon equation is local, we expect that the lagrangian can be written as the space integral of a *lagrangian density* $\mathcal{L}$: $L = \int d^3x\, \mathcal{L}$. Thus, $S = \int d^4x\, \mathcal{L}$. The integration measure d^4x is Lorentz invariant: if we change to coordinates $\bar{x}^\mu = \Lambda^\mu{}_\nu x^\nu$, we have $d^4\bar{x} = |\det \Lambda|\, d^4x = d^4x$. Thus, for the action to be Lorentz invariant, the lagrangian density must be a Lorentz scalar: $\mathcal{L}(x) = \bar{\mathcal{L}}(\bar{x})$. Then we have $\bar{S} = \int d^4\bar{x}\, \bar{\mathcal{L}}(\bar{x}) = \int d^4x\, \mathcal{L}(x) = S$. Any simple function of φ is a Lorentz scalar, and so are products of derivatives with all indices contracted, such as $\partial^\mu\varphi\partial_\mu\varphi$. We will take for $\mathcal{L}$

$$\mathcal{L} = -\tfrac{1}{2}\partial^\mu\varphi\partial_\mu\varphi - \tfrac{1}{2}m^2\varphi^2 + \Omega_0\,, \tag{3.9}$$

where Ω_0 is an arbitrary constant. We find the equation of motion (also known as the *Euler–Lagrange equation*) by making an infinitesimal variation $\delta\varphi(x)$ in $\varphi(x)$, and requiring the corresponding variation of the action to vanish:

$$\begin{aligned} 0 &= \delta S \\ &= \int d^4x \left[-\tfrac{1}{2}\partial^\mu\delta\varphi\partial_\mu\varphi - \tfrac{1}{2}\partial^\mu\varphi\partial_\mu\delta\varphi - m^2\varphi\,\delta\varphi\right] \\ &= \int d^4x \left[+\partial^\mu\partial_\mu\varphi - m^2\varphi\right]\delta\varphi\,. \end{aligned} \tag{3.10}$$

In the last line, we have integrated by parts in each of the first two terms, putting both derivatives on φ. We assume $\delta\varphi(x)$ vanishes at infinity in any direction (spatial or temporal), so that there is no surface term. Since $\delta\varphi$ has an arbitrary x dependence, eq. (3.10) can be true if and only if $(-\partial^2 + m^2)\varphi = 0$.

One solution of the Klein–Gordon equation is a plane wave of the form $\exp(i\mathbf{k}\cdot\mathbf{x} \pm i\omega t)$, where $\mathbf{k}$ is an arbitrary real wave vector, and

$$\omega = +(\mathbf{k}^2 + m^2)^{1/2}\,. \tag{3.11}$$

The general solution (assuming boundary conditions that require φ to remain finite at spatial infinity) is then

$$\varphi(\mathbf{x}, t) = \int \frac{d^3k}{f(k)} \left[a(\mathbf{k})e^{i\mathbf{k}\cdot\mathbf{x} - i\omega t} + b(\mathbf{k})e^{i\mathbf{k}\cdot\mathbf{x} + i\omega t}\right], \tag{3.12}$$

where $a(\mathbf{k})$ and $b(\mathbf{k})$ are arbitrary functions of the wave vector $\mathbf{k}$, and $f(k)$ is a redundant function of the magnitude of $\mathbf{k}$ which we have inserted for later convenience. Note that, *if* we were attempting to interpret $\varphi(x)$ as a quantum wave function (which we most definitely are *not*), then the second term would constitute the "negative energy" contributions to the wave function. This is because a plane-wave solution of the nonrelativistic Schrödinger equation for a single particle looks like $\exp(i\mathbf{p}\cdot\mathbf{x} - iE(\mathbf{p})t)$, with $E(\mathbf{p}) = \frac{1}{2m}\mathbf{p}^2$;

there is a minus sign in front of the positive energy. We *are* trying to interpret eq. (3.12) as a *real* classical field, but this formula does not generically result in φ being real. We must impose $\varphi^*(x) = \varphi(x)$, where

$$\varphi^*(\mathbf{x},t) = \int \frac{d^3k}{f(k)} \left[a^*(\mathbf{k})e^{-i\mathbf{k}\cdot\mathbf{x}+i\omega t} + b^*(\mathbf{k})e^{-i\mathbf{k}\cdot\mathbf{x}-i\omega t}\right]$$
$$= \int \frac{d^3k}{f(k)} \left[a^*(\mathbf{k})e^{-i\mathbf{k}\cdot\mathbf{x}+i\omega t} + b^*(-\mathbf{k})e^{+i\mathbf{k}\cdot\mathbf{x}-i\omega t}\right] . \tag{3.13}$$

In the second term on the second line, we have changed the dummy integration variable from $\mathbf{k}$ to $-\mathbf{k}$. Comparing eqs. (3.12) and (3.13), we see that $\varphi^*(x) = \varphi(x)$ requires $b^*(-\mathbf{k}) = a(\mathbf{k})$. Imposing this condition, we can rewrite φ as

$$\varphi(\mathbf{x},t) = \int \frac{d^3k}{f(k)} \left[a(\mathbf{k})e^{i\mathbf{k}\cdot\mathbf{x}-i\omega t} + a^*(-\mathbf{k})e^{i\mathbf{k}\cdot\mathbf{x}+i\omega t}\right]$$
$$= \int \frac{d^3k}{f(k)} \left[a(\mathbf{k})e^{i\mathbf{k}\cdot\mathbf{x}-i\omega t} + a^*(\mathbf{k})e^{-i\mathbf{k}\cdot\mathbf{x}+i\omega t}\right]$$
$$= \int \frac{d^3k}{f(k)} \left[a(\mathbf{k})e^{ikx} + a^*(\mathbf{k})e^{-ikx}\right] , \tag{3.14}$$

where $kx = \mathbf{k}\cdot\mathbf{x} - \omega t$ is the Lorentz-invariant product of the four-vectors $x^\mu = (t, \mathbf{x})$ and $k^\mu = (\omega, \mathbf{k})$: $kx = k^\mu x_\mu = g_{\mu\nu}k^\mu x^\nu$. Note that

$$k^2 = k^\mu k_\mu = \mathbf{k}^2 - \omega^2 = -m^2 . \tag{3.15}$$

A four-momentum k^μ that obeys $k^2 = -m^2$ is said to be *on the mass shell*, or *on shell* for short.

It is now convenient to choose $f(k)$ so that $d^3k/f(k)$ is Lorentz invariant. An integration measure that is manifestly invariant under orthochronous Lorentz transformations is $d^4k\,\delta(k^2{+}m^2)\,\theta(k^0)$, where $\delta(x)$ is the Dirac delta function, $\theta(x)$ is the unit step function, and k^0 is treated as an independent integration variable. We then have

$$\int_{-\infty}^{+\infty} dk^0\,\delta(k^2{+}m^2)\,\theta(k^0) = \frac{1}{2\omega} . \tag{3.16}$$

Here we have used the rule

$$\int_{-\infty}^{+\infty} dx\,\delta(g(x)) = \sum_i \frac{1}{|g'(x_i)|} , \tag{3.17}$$

where $g(x)$ is any smooth function of x with simple zeros at $x = x_i$; in our case, the only zero is at $k^0 = \omega$.

Thus we see that if we take $f(k) \propto \omega$, then $d^3k/f(k)$ will be Lorentz invariant. We will take $f(k) = (2\pi)^3 2\omega$. It is then convenient to give the corresponding Lorentz-invariant differential its own name:

$$\widetilde{dk} \equiv \frac{d^3k}{(2\pi)^3 2\omega} \; . \tag{3.18}$$

Thus we finally have

$$\varphi(x) = \int \widetilde{dk} \left[a(\mathbf{k}) e^{ikx} + a^*(\mathbf{k}) e^{-ikx} \right] . \tag{3.19}$$

We can also invert this formula to get $a(\mathbf{k})$ in terms of $\varphi(x)$. We have

$$\begin{aligned} \int d^3x \; e^{-ikx} \varphi(x) &= \tfrac{1}{2\omega} a(\mathbf{k}) + \tfrac{1}{2\omega} e^{2i\omega t} a^*(-\mathbf{k}) \; , \\ \int d^3x \; e^{-ikx} \partial_0 \varphi(x) &= -\tfrac{i}{2} a(\mathbf{k}) + \tfrac{i}{2} e^{2i\omega t} a^*(-\mathbf{k}) \; . \end{aligned} \tag{3.20}$$

We can combine these to get

$$\begin{aligned} a(\mathbf{k}) &= \int d^3x \; e^{-ikx} \Big[i\partial_0 \varphi(x) + \omega \varphi(x) \Big] \\ &= i \int d^3x \; e^{-ikx} \overleftrightarrow{\partial_0} \varphi(x) \; , \end{aligned} \tag{3.21}$$

where $f \overleftrightarrow{\partial_\mu} g \equiv f(\partial_\mu g) - (\partial_\mu f) g$, and $\partial_0 \varphi = \partial\varphi/\partial t = \dot{\varphi}$. Note that $a(\mathbf{k})$ is time independent.

Now that we have the lagrangian, we can construct the hamiltonian by the usual rules. Recall that, given a lagrangian $L(q_i, \dot{q}_i)$ as a function of some coordinates q_i and their time derivatives $\dot{q}_i$, the conjugate momenta are given by $p_i = \partial L/\partial \dot{q}_i$, and the hamiltonian by $H = \sum_i p_i \dot{q}_i - L$. In our case, the role of $q_i(t)$ is played by $\varphi(\mathbf{x}, t)$, with $\mathbf{x}$ playing the role of a (continuous) index. The appropriate generalizations are then

$$\Pi(x) = \frac{\partial \mathcal{L}}{\partial \dot{\varphi}(x)} \tag{3.22}$$

and

$$\mathcal{H} = \Pi \dot{\varphi} - \mathcal{L} \; , \tag{3.23}$$

where $\mathcal{H}$ is the *hamiltonian density*, and the hamiltonian itself is $H = \int d^3x \; \mathcal{H}$. In our case, we have

$$\Pi(x) = \dot{\varphi}(x) \tag{3.24}$$

and

$$\mathcal{H} = \tfrac{1}{2}\Pi^2 + \tfrac{1}{2}(\nabla\varphi)^2 + \tfrac{1}{2}m^2\varphi^2 - \Omega_0 \,. \tag{3.25}$$

Using eq. (3.19), we can write H in terms of the $a(\mathbf{k})$ and $a^*(\mathbf{k})$ coefficients:

$$\begin{aligned}
H = -\Omega_0 V + \tfrac{1}{2}\int \widetilde{dk}\,\widetilde{dk}'\,d^3x \\
&\Big[\Big(-i\omega\,a(\mathbf{k})e^{ikx} + i\omega\,a^*(\mathbf{k})e^{-ikx}\Big)\Big(-i\omega'\,a(\mathbf{k}')e^{ik'x} + i\omega'\,a^*(\mathbf{k}')e^{-ik'x}\Big) \\
&+ \Big(+i\mathbf{k}\,a(\mathbf{k})e^{ikx} - i\mathbf{k}\,a^*(\mathbf{k})e^{-ikx}\Big)\cdot\Big(+i\mathbf{k}'\,a(\mathbf{k}')e^{ik'x} - i\mathbf{k}'\,a^*(\mathbf{k}')e^{-ik'x}\Big) \\
&+ m^2\Big(a(\mathbf{k})e^{ikx} + a^*(\mathbf{k})e^{-ikx}\Big)\Big(a(\mathbf{k}')e^{ik'x} + a^*(\mathbf{k}')e^{-ik'x}\Big)\Big] \\
= -\Omega_0 V + \tfrac{1}{2}(2\pi)^3\int \widetilde{dk}\,\widetilde{dk}' \\
&\Big[\delta^3(\mathbf{k}-\mathbf{k}')(+\omega\omega' + \mathbf{k}\cdot\mathbf{k}' + m^2) \\
&\quad\times\Big(a^*(\mathbf{k})a(\mathbf{k}')e^{+i(\omega-\omega')t} + a(\mathbf{k})a^*(\mathbf{k}')e^{-i(\omega-\omega')t}\Big) \\
&+ \delta^3(\mathbf{k}+\mathbf{k}')(-\omega\omega' - \mathbf{k}\cdot\mathbf{k}' + m^2) \\
&\quad\times\Big(a(\mathbf{k})a(\mathbf{k}')e^{-i(\omega+\omega')t} + a^*(\mathbf{k})a^*(\mathbf{k}')e^{+i(\omega+\omega')t}\Big)\Big] \\
= -\Omega_0 V + \tfrac{1}{2}\int \widetilde{dk}\,\tfrac{1}{2\omega} \\
&\Big[(+\omega^2 + \mathbf{k}^2 + m^2)\Big(a^*(\mathbf{k})a(\mathbf{k}) + a(\mathbf{k})a^*(\mathbf{k})\Big) \\
&+ (-\omega^2 + \mathbf{k}^2 + m^2)\Big(a(\mathbf{k})a(-\mathbf{k})e^{-2i\omega t} + a^*(\mathbf{k})a^*(-\mathbf{k})e^{+2i\omega t}\Big)\Big] \\
= -\Omega_0 V + \tfrac{1}{2}\int \widetilde{dk}\,\omega\Big(a^*(\mathbf{k})a(\mathbf{k}) + a(\mathbf{k})a^*(\mathbf{k})\Big)\,,
\end{aligned} \tag{3.26}$$

where V is the volume of space. To get the second equality, we used

$$\int d^3x\; e^{i\mathbf{q}\cdot\mathbf{x}} = (2\pi)^3\delta^3(\mathbf{q})\,. \tag{3.27}$$

To get the third equality, we integrated over $\mathbf{k}'$, using $\widetilde{dk}' = d^3k'/(2\pi)^3 2\omega'$. The last equality then follows from $\omega = (\mathbf{k}^2+m^2)^{1/2}$. Also, we were careful to keep the ordering of $a(\mathbf{k})$ and $a^*(\mathbf{k})$ unchanged throughout, in anticipation of passing to the quantum theory where these classical functions will become operators that may not commute.

Let us take up the quantum theory now. We can go from classical to quantum mechanics via *canonical quantization*. This means that we promote q_i and p_i to operators, with commutation relations $[q_i, q_j] = 0$, $[p_i, p_j] = 0$, and $[q_i, p_j] = i\hbar\delta_{ij}$. In the Heisenberg picture, these operators should be taken at equal times. In our case, where the "index" is continuous (and we have set $\hbar = 1$), we have

$$[\varphi(\mathbf{x},t),\varphi(\mathbf{x}',t)] = 0\,,$$
$$[\Pi(\mathbf{x},t),\Pi(\mathbf{x}',t)] = 0\,,$$
$$[\varphi(\mathbf{x},t),\Pi(\mathbf{x}',t)] = i\delta^3(\mathbf{x}-\mathbf{x}')\,. \tag{3.28}$$

From these *canonical commutation relations*, and from eqs. (3.21) and (3.24), we can deduce

$$[a(\mathbf{k}),a(\mathbf{k}')] = 0\,,$$
$$[a^\dagger(\mathbf{k}),a^\dagger(\mathbf{k}')] = 0\,,$$
$$[a(\mathbf{k}),a^\dagger(\mathbf{k}')] = (2\pi)^3 2\omega\,\delta^3(\mathbf{k}-\mathbf{k}')\,. \tag{3.29}$$

We are now denoting $a^*(\mathbf{k})$ as $a^\dagger(\mathbf{k})$, since $a^\dagger(\mathbf{k})$ is now the hermitian conjugate (rather than the complex conjugate) of the operator $a(\mathbf{k})$. We can now rewrite the hamiltonian as

$$H = \int \widetilde{dk}\;\omega\, a^\dagger(\mathbf{k})a(\mathbf{k}) + (\mathcal{E}_0 - \Omega_0)V\,, \tag{3.30}$$

where

$$\mathcal{E}_0 = \tfrac{1}{2}(2\pi)^{-3}\int d^3k\;\omega \tag{3.31}$$

is the total zero-point energy of all the oscillators per unit volume, and, using eq. (3.27), we have interpreted $(2\pi)^3\delta^3(\mathbf{0})$ as the volume of space V.

If we integrate in eq. (3.31) over the whole range of $\mathbf{k}$, the value of $\mathcal{E}_0$ is infinite. If we integrate only up to a maximum value of Λ, a number known as the *ultraviolet cutoff*, we find

$$\mathcal{E}_0 = \frac{\Lambda^4}{16\pi^2}\,, \tag{3.32}$$

where we have assumed $\Lambda \gg m$. This is physically justified if, in the real world, the formalism of quantum field theory breaks down at some large energy scale. For now, we simply note that the value of Ω_0 is arbitrary, and so we are free to choose $\Omega_0 = \mathcal{E}_0$. With this choice, the ground state has energy eigenvalue zero. Now, if we like, we can take the limit $\Lambda \to \infty$, with no further consequences. (We will meet more of these *ultraviolet divergences* after we introduce interactions.)

The hamiltonian of eq. (3.30) is now the same as that of eq. (3.6), with $a(\mathbf{k}) = [(2\pi)^3 2\omega]^{1/2}\,\widetilde{a}(\mathbf{k})$. The commutation relations (3.4) and (3.29) are also equivalent, if we choose commutators (rather than anticommutators) in eq. (3.4). Thus, *we have rederived the hamiltonian of free relativistic bosons by quantization of a scalar field whose equation of motion is the Klein–Gordon equation.* The parameter m in the lagrangian is now seen to be the mass of the particle in the quantum theory. (More precisely, since m has dimensions of inverse length, the particle mass is $\hbar m/c$.)

What if we want fermions? Then we should use anticommutators in eqs. (3.28) and (3.29). There is a problem, though; eq. (3.26) does not then become eq. (3.30). Instead, we get $H = -\Omega_0 V$, a simple constant. Clearly there is something wrong with using anticommutators. This is another hint of the spin-statistics theorem, which we will take up in section 4.

Next, we would like to add Lorentz-invariant interactions to our theory. With the formalism we have developed, this is easy to do. Any local function of $\varphi(x)$ is a Lorentz scalar, and so if we add a term like φ^3 or φ^4 to the lagrangian density $\mathcal{L}$, the resulting action will still be Lorentz invariant. Now, however, we will have interactions among the particles. Our next task is to deduce the consequences of these interactions.

However, we already have enough tools at our disposal to prove the spin-statistics theorem for spin-zero particles, and that is what we turn to next.

Problems

3.1) Derive eq. (3.29) from eqs. (3.21), (3.24).

3.2) Use the commutation relations, eq. (3.29), to show explicitly that a state of the form

$$|k_1 \dots k_n\rangle \equiv a^\dagger(\mathbf{k}_1) \dots a^\dagger(\mathbf{k}_n)|0\rangle \tag{3.33}$$

is an eigenstate of the hamiltonian, eq. (3.30), with eigenvalue $\omega_1 + \dots + \omega_n$. The vacuum $|0\rangle$ is annihilated by $a(\mathbf{k})$, $a(\mathbf{k})|0\rangle = 0$, and we take $\Omega_0 = \mathcal{E}_0$ in eq. (3.30).

3.3) Use $U(\Lambda)^{-1}\varphi(x)U(\Lambda) = \varphi(\Lambda^{-1}x)$ to show that

$$U(\Lambda)^{-1}a(\mathbf{k})U(\Lambda) = a(\Lambda^{-1}\mathbf{k}) \; ,$$
$$U(\Lambda)^{-1}a^\dagger(\mathbf{k})U(\Lambda) = a^\dagger(\Lambda^{-1}\mathbf{k}) \; , \tag{3.34}$$

and hence that

$$U(\Lambda)|k_1 \dots k_n\rangle = |\Lambda k_1 \dots \Lambda k_n\rangle \; , \tag{3.35}$$

where $|k_1 \dots k_n\rangle = a^\dagger(\mathbf{k}_1)\dots a^\dagger(\mathbf{k}_n)|0\rangle$ is a state of n particles with momenta $k_1, \dots, k_n$.

3.4) Recall that $T(a)^{-1}\varphi(x)T(a) = \varphi(x - a)$, where $T(a) \equiv \exp(-iP^\mu a_\mu)$ is the spacetime translation operator, and P^0 is identified as the hamiltonian H.

a) Let a^μ be infinitesimal, and derive an expression for $[P^\mu, \varphi(x)]$.

b) Show that the time component of your result is equivalent to the Heisenberg equation of motion $\dot{\varphi} = i[H, \varphi]$.

c) For a free field, use the Heisenberg equation to derive the Klein–Gordon equation.

d) Define a spatial momentum operator

$$\mathbf{P} \equiv -\int d^3x\, \Pi(x) \nabla \varphi(x) \,. \tag{3.36}$$

Use the canonical commutation relations to show that $\mathbf{P}$ obeys the relation you derived in part (a).

e) Express $\mathbf{P}$ in terms of $a(\mathbf{k})$ and $a^\dagger(\mathbf{k})$.

3.5) Consider a complex (that is, nonhermitian) scalar field φ with lagrangian density

$$\mathcal{L} = -\partial^\mu \varphi^\dagger \partial_\mu \varphi - m^2 \varphi^\dagger \varphi + \Omega_0 \,. \tag{3.37}$$

a) Show that φ obeys the Klein–Gordon equation.

b) Treat φ and $\varphi^\dagger$ as independent fields, and find the conjugate momentum for each. Compute the hamiltonian density in terms of these conjugate momenta and the fields themselves (but not their time derivatives).

c) Write the mode expansion of φ as

$$\varphi(x) = \int \widetilde{dk} \left[a(\mathbf{k}) e^{ikx} + b^\dagger(\mathbf{k}) e^{-ikx} \right] \,. \tag{3.38}$$

Express $a(\mathbf{k})$ and $b(\mathbf{k})$ in terms of φ and $\varphi^\dagger$ and their time derivatives.

d) Assuming canonical commutation relations for the fields and their conjugate momenta, find the commutation relations obeyed by $a(\mathbf{k})$ and $b(\mathbf{k})$ and their hermitian conjugates.

e) Express the hamiltonian in terms of $a(\mathbf{k})$ and $b(\mathbf{k})$ and their hermitian conjugates. What value must Ω_0 have in order for the ground state to have zero energy?

4

The spin-statistics theorem

Prerequisite: 3

Let us consider a theory of free, spin-zero particles specified by the hamiltonian

$$H_0 = \int \widetilde{dk}\, \omega\, a^\dagger(\mathbf{k})a(\mathbf{k}) , \tag{4.1}$$

where $\omega = (\mathbf{k}^2 + m^2)^{1/2}$, and either the commutation or anticommutation relations

$$\begin{aligned}
[a(\mathbf{k}), a(\mathbf{k}')]_\mp &= 0 , \\
[a^\dagger(\mathbf{k}), a^\dagger(\mathbf{k}')]_\mp &= 0 , \\
[a(\mathbf{k}), a^\dagger(\mathbf{k}')]_\mp &= (2\pi)^3 2\omega\, \delta^3(\mathbf{k} - \mathbf{k}') .
\end{aligned} \tag{4.2}$$

Of course, if we want a theory of bosons, we should use commutators, and if we want fermions, we should use anticommutators.

Now let us consider adding terms to the hamiltonian that will result in *local, Lorentz invariant* interactions. In order to do this, it is convenient to define a nonhermitian field,

$$\varphi^+(\mathbf{x}, 0) \equiv \int \widetilde{dk}\, e^{i\mathbf{k}\cdot\mathbf{x}}\, a(\mathbf{k}) , \tag{4.3}$$

and its hermitian conjugate

$$\varphi^-(\mathbf{x}, 0) \equiv \int \widetilde{dk}\, e^{-i\mathbf{k}\cdot\mathbf{x}}\, a^\dagger(\mathbf{k}) . \tag{4.4}$$

These are then time-evolved with H_0:

$$\begin{aligned}
\varphi^+(\mathbf{x}, t) &= e^{iH_0 t}\varphi^+(\mathbf{x}, 0)e^{-iH_0 t} = \int \widetilde{dk}\, e^{ikx}\, a(\mathbf{k}) , \\
\varphi^-(\mathbf{x}, t) &= e^{iH_0 t}\varphi^-(\mathbf{x}, 0)e^{-iH_0 t} = \int \widetilde{dk}\, e^{-ikx}\, a^\dagger(\mathbf{k}) .
\end{aligned} \tag{4.5}$$

Note that the usual hermitian free field $\varphi(x)$ is just the sum of these: $\varphi(x) = \varphi^+(x) + \varphi^-(x)$.

For a proper orthochronous Lorentz transformation Λ, we have

$$U(\Lambda)^{-1}\varphi(x)U(\Lambda) = \varphi(\Lambda^{-1}x) \,. \tag{4.6}$$

This implies that the particle creation and annihilation operators transform as

$$\begin{aligned} U(\Lambda)^{-1}a(\mathbf{k})U(\Lambda) &= a(\Lambda^{-1}\mathbf{k}) \,, \\ U(\Lambda)^{-1}a^\dagger(\mathbf{k})U(\Lambda) &= a^\dagger(\Lambda^{-1}\mathbf{k}) \,. \end{aligned} \tag{4.7}$$

This, in turn, implies that $\varphi^+(x)$ and $\varphi^-(x)$ are Lorentz scalars:

$$U(\Lambda)^{-1}\varphi^\pm(x)U(\Lambda) = \varphi^\pm(\Lambda^{-1}x) \,. \tag{4.8}$$

We will then have local, Lorentz invariant interactions if we take the interaction lagrangian density $\mathcal{L}_1$ to be a hermitian function of $\varphi^+(x)$ and $\varphi^-(x)$.

To proceed we need to recall some facts about time-dependent perturbation theory in quantum mechanics. The transition amplitude $\mathcal{T}_{f\leftarrow i}$ to start with an initial state $|i\rangle$ at time $t = -\infty$ and end with a final state $|f\rangle$ at time $t = +\infty$ is

$$\mathcal{T}_{f\leftarrow i} = \langle f|\,\mathrm{T}\exp\left[-i\int_{-\infty}^{+\infty} dt\; H_I(t)\right]|i\rangle \,, \tag{4.9}$$

where $H_I(t)$ is the perturbing hamiltonian in the *interaction picture*,

$$H_I(t) = \exp(+iH_0t)\,H_1\exp(-iH_0t) \,, \tag{4.10}$$

H_1 is the perturbing hamiltonian in the Schrödinger picture, and T is the *time ordering symbol*: a product of operators to its right is to be ordered, not as written, but with operators at later times to the left of those at earlier times. We write $H_1 = \int d^3x\,\mathcal{H}_1(\mathbf{x},0)$, and specify $\mathcal{H}_1(\mathbf{x},0)$ as a hermitian function of $\varphi^+(\mathbf{x},0)$ and $\varphi^-(\mathbf{x},0)$. Then, using eqs. (4.5) and (4.10), we can see that, in the interaction picture, the perturbing hamiltonian density $\mathcal{H}_I(\mathbf{x},t)$ is simply given by the same function of $\varphi^+(\mathbf{x},t)$ and $\varphi^-(\mathbf{x},t)$.

Now we come to the key point: for the transition amplitude $\mathcal{T}_{f\leftarrow i}$ to be Lorentz invariant, the time ordering must be *frame independent*. The time ordering of two spacetime points x and x' is frame independent if their separation is *timelike*; this means that the interval between them is negative, $(x-x')^2 < 0$. Two spacetime points whose separation is *spacelike*, $(x-x')^2 > 0$, can have different temporal ordering in different frames. In order to avoid $\mathcal{T}_{f\leftarrow i}$ being different in different frames, we must then require

$$[\mathcal{H}_I(x), \mathcal{H}_I(x')] = 0 \quad \text{whenever} \quad (x - x')^2 > 0 \,. \tag{4.11}$$

Obviously, $[\varphi^+(x), \varphi^+(x')]_\mp = [\varphi^-(x), \varphi^-(x')]_\mp = 0$. However,

$$\begin{aligned}
[\varphi^+(x), \varphi^-(x')]_\mp &= \int \widetilde{dk}\, \widetilde{dk}'\; e^{i(kx-k'x')}[a(\mathbf{k}), a^\dagger(\mathbf{k}')]_\mp \\
&= \int \widetilde{dk}\; e^{ik(x-x')} \\
&= \frac{m}{4\pi^2 r} K_1(mr) \\
&\equiv C(r) \,.
\end{aligned} \tag{4.12}$$

In the next-to-last line, we have taken $(x - x')^2 = r^2 > 0$, and $K_1(z)$ is a modified Bessel function. (This Lorentz-invariant integral is most easily evaluated in the frame where $t' = t$.) The function $C(r)$ is *not* zero for any $r > 0$. (Not even when $m = 0$; in this case, $C(r) = 1/4\pi^2 r^2$.) On the other hand, $\mathcal{H}_I(x)$ must involve both $\varphi^+(x)$ and $\varphi^-(x)$, by hermiticity. Thus, generically, we will not be able to satisfy eq. (4.11).

To resolve this problem, let us try using only particular linear combinations of $\varphi^+(x)$ and $\varphi^-(x)$. Define

$$\begin{aligned}
\varphi_\lambda(x) &\equiv \varphi^+(x) + \lambda\varphi^-(x) \,, \\
\varphi_\lambda^\dagger(x) &\equiv \varphi^-(x) + \lambda^*\varphi^+(x) \,,
\end{aligned} \tag{4.13}$$

where λ is an arbitrary complex number. We then have

$$\begin{aligned}
[\varphi_\lambda(x), \varphi_\lambda^\dagger(x')]_\mp &= [\varphi^+(x), \varphi^-(x')]_\mp + |\lambda|^2[\varphi^-(x), \varphi^+(x')]_\mp \\
&= (1 \mp |\lambda|^2)\, C(r)
\end{aligned} \tag{4.14}$$

and

$$\begin{aligned}
[\varphi_\lambda(x), \varphi_\lambda(x')]_\mp &= \lambda[\varphi^+(x), \varphi^-(x')]_\mp + \lambda[\varphi^-(x), \varphi^+(x')]_\mp \\
&= \lambda(1 \mp 1)\, C(r) \,.
\end{aligned} \tag{4.15}$$

Thus, if we want $\varphi_\lambda(x)$ to either commute or anticommute with both $\varphi_\lambda(x')$ and $\varphi_\lambda^\dagger(x')$ at spacelike separations, we must choose $|\lambda| = 1$, *and* we must choose commutators. Then (and only then), we can build a suitable $\mathcal{H}_I(x)$ by making it a hermitian function of $\varphi_\lambda(x)$.

But this has simply returned us to the theory of a real scalar field, because, for $\lambda = e^{i\alpha}$, $e^{-i\alpha/2}\varphi_\lambda(x)$ is hermitian. In fact, if we make the replacements

$a(\mathbf{k}) \to e^{+i\alpha/2}a(\mathbf{k})$ and $a^\dagger(\mathbf{k}) \to e^{-i\alpha/2}a^\dagger(\mathbf{k})$, then the commutation relations of eq. (4.2) are unchanged, and $e^{-i\alpha/2}\varphi_\lambda(x) = \varphi(x) = \varphi^+(x) + \varphi^-(x)$. Thus, our attempt to start with the creation and annihilation operators $a^\dagger(\mathbf{k})$ and $a(\mathbf{k})$ as the fundamental objects has simply led us back to the real, commuting, scalar field $\varphi(x)$ as the fundamental object.

Let us return to thinking of $\varphi(x)$ as fundamental, with a lagrangian density given by some function of the Lorentz scalars $\varphi(x)$ and $\partial^\mu\varphi(x)\partial_\mu\varphi(x)$. Then, quantization will result in $[\varphi(x), \varphi(x')]_\mp = 0$ for $t = t'$. If we choose anticommutators, then $[\varphi(x)]^2 = 0$ and $[\partial_\mu\varphi(x)]^2 = 0$, resulting in a trivial $\mathcal{L}$ that is at most linear in φ, and independent of $\dot{\varphi}$. This clearly does not lead to the correct physics.

This situation turns out to generalize to fields of higher spin, in any number of spacetime dimensions. One choice of quantization (commutators or anticommutators) always leads to a trivial $\mathcal{L}$, and so this choice is disallowed. Furthermore, the allowed choice is always commutators for fields of integer spin, and anticommutators for fields of half-integer spin. If we try treating the particle creation and annihilation operators as fundamental, rather than the fields, we find a situation similar to that of the spin-zero case, and are led to the reconstruction of a field that must obey the appropriate quantization scheme.

Reference notes

This discussion of the spin-statistics theorem follows that of *Weinberg I*, which has more details.

Problems

4.1) Verify eq. (4.12). Verify its limit as $m \to 0$.

5

The LSZ reduction formula

Prerequisite: 3

Let us now consider how to construct appropriate initial and final states for scattering experiments. In the free theory, we can create a state of one particle by acting on the vacuum state with a creation operator

$$|k\rangle = a^\dagger(\mathbf{k})|0\rangle \ , \tag{5.1}$$

where

$$a^\dagger(\mathbf{k}) = -i \int d^3x \, e^{ikx} \overleftrightarrow{\partial_0} \varphi(x) \ . \tag{5.2}$$

The vacuum state $|0\rangle$ is annihilated by every $a(\mathbf{k})$,

$$a(\mathbf{k})|0\rangle = 0 \ , \tag{5.3}$$

and has unit norm,

$$\langle 0|0\rangle = 1 \ . \tag{5.4}$$

The one-particle state $|k\rangle$ then has the Lorentz-invariant normalization

$$\langle k|k'\rangle = (2\pi)^3 \, 2\omega \, \delta^3(\mathbf{k} - \mathbf{k}') \ , \tag{5.5}$$

where $\omega = (\mathbf{k}^2 + m^2)^{1/2}$.

Next, let us define a time-independent operator that (in the free theory) creates a particle localized in momentum space near $\mathbf{k}_1$, and localized in position space near the origin:

$$a_1^\dagger \equiv \int d^3k \ f_1(\mathbf{k}) a^\dagger(\mathbf{k}) \ , \tag{5.6}$$

where

$$f_1(\mathbf{k}) \propto \exp[-(\mathbf{k} - \mathbf{k}_1)^2/4\sigma^2] \tag{5.7}$$

is an appropriate wave packet, and σ is its width in momentum space. Consider the state $a_1^\dagger|0\rangle$. If we time evolve this state in the Schrödinger picture, the wave packet will propagate (and spread out). The particle is thus localized far from the origin as $t \to \pm\infty$. If we consider instead a state of the form $a_1^\dagger a_2^\dagger|0\rangle$, where $\mathbf{k}_1 \neq \mathbf{k}_2$, then the two particles are widely separated in the far past.

Let us guess that this still works in the interacting theory. One complication is that $a^\dagger(\mathbf{k})$ will no longer be time independent, and so $a_1^\dagger$, eq. (5.6), becomes time dependent as well. Our guess for a suitable initial state of a scattering experiment is then

$$|i\rangle = \lim_{t\to-\infty} a_1^\dagger(t)a_2^\dagger(t)|0\rangle \ . \tag{5.8}$$

By appropriately normalizing the wave packets, we can make $\langle i|i\rangle = 1$, and we will assume that this is the case. Similarly, we can consider a final state

$$|f\rangle = \lim_{t\to+\infty} a_{1'}^\dagger(t)a_{2'}^\dagger(t)|0\rangle \ , \tag{5.9}$$

where $\mathbf{k}_1' \neq \mathbf{k}_2'$, and $\langle f|f\rangle = 1$. This describes two widely separated particles in the far future. (We could also consider acting with more creation operators, if we are interested in the production of some extra particles in the collision of two.) Now the scattering amplitude is simply given by $\langle f|i\rangle$.

We need to find a more useful expression for $\langle f|i\rangle$. To this end, let us note that

$$\begin{aligned}
a_1^\dagger(+\infty) - a_1^\dagger(-\infty) &= \int_{-\infty}^{+\infty} dt\ \partial_0 a_1^\dagger(t) \\
&= -i\int d^3k\ f_1(\mathbf{k})\int d^4x\ \partial_0\Big(e^{ikx}\overleftrightarrow{\partial_0}\varphi(x)\Big) \\
&= -i\int d^3k\ f_1(\mathbf{k})\int d^4x\ e^{ikx}(\partial_0^2+\omega^2)\varphi(x) \\
&= -i\int d^3k\ f_1(\mathbf{k})\int d^4x\ e^{ikx}(\partial_0^2+\mathbf{k}^2+m^2)\varphi(x) \\
&= -i\int d^3k\ f_1(\mathbf{k})\int d^4x\ e^{ikx}(\partial_0^2-\overleftarrow{\nabla}^2+m^2)\varphi(x) \\
&= -i\int d^3k\ f_1(\mathbf{k})\int d^4x\ e^{ikx}(\partial_0^2-\overrightarrow{\nabla}^2+m^2)\varphi(x) \\
&= -i\int d^3k\ f_1(\mathbf{k})\int d^4x\ e^{ikx}(-\partial^2+m^2)\varphi(x) \ .
\end{aligned} \tag{5.10}$$

The first equality is just the fundamental theorem of calculus. To get the second, we substituted the definition of $a_1^\dagger(t)$, and combined the d^3x from this definition with the dt to get d^4x. The third comes from straightforward evaluation of the time derivatives. The fourth uses $\omega^2 = \mathbf{k}^2 + m^2$. The fifth writes $\mathbf{k}^2$ as $-\nabla^2$ acting on $e^{i\mathbf{k}\cdot\mathbf{x}}$. The sixth uses integration by parts to move the ∇^2 onto the field $\varphi(x)$; here the wave packet is needed to avoid a surface term. The seventh simply identifies $\partial_0^2 - \nabla^2$ as $-\partial^2$.

In free-field theory, the right-hand side of eq. (5.10) is zero, since $\varphi(x)$ obeys the Klein–Gordon equation. In an interacting theory, with (say) $\mathcal{L}_1 = \frac{1}{6}g\varphi^3$, we have instead $(-\partial^2 + m^2)\varphi = \frac{1}{2}g\varphi^2$. Thus the right-hand side of eq. (5.10) is not zero in an interacting theory.

Rearranging eq. (5.10), we have

$$a_1^\dagger(-\infty) = a_1^\dagger(+\infty) + i\int d^3k\, f_1(\mathbf{k}) \int d^4x\, e^{ikx}(-\partial^2 + m^2)\varphi(x)\,. \quad (5.11)$$

We will also need the hermitian conjugate of this formula, which (after a little more rearranging) reads

$$a_1(+\infty) = a_1(-\infty) + i\int d^3k\, f_1(\mathbf{k}) \int d^4x\, e^{-ikx}(-\partial^2 + m^2)\varphi(x)\,. \quad (5.12)$$

Let us return to the scattering amplitude,

$$\langle f|i\rangle = \langle 0|a_{1'}(+\infty)a_{2'}(+\infty)a_1^\dagger(-\infty)a_2^\dagger(-\infty)|0\rangle\,. \quad (5.13)$$

Note that the operators are in time order. Thus, if we feel like it, we can put in a *time-ordering symbol* without changing anything:

$$\langle f|i\rangle = \langle 0|\mathrm{T}a_{1'}(+\infty)a_{2'}(+\infty)a_1^\dagger(-\infty)a_2^\dagger(-\infty)|0\rangle\,. \quad (5.14)$$

The symbol T means the product of operators to its right is to be ordered, not as written, but with operators at later times to the left of those at earlier times.

Now let us use eqs. (5.11) and (5.12) in eq. (5.14). The time-ordering symbol automatically moves all $a_{i'}(-\infty)$s to the right, where they annihilate $|0\rangle$. Similarly, all $a_i^\dagger(+\infty)$s move to the left, where they annihilate $\langle 0|$.

The wave packets no longer play a key role, and we can take the $\sigma \to 0$ limit in eq. (5.7), so that $f_1(\mathbf{k}) = \delta^3(\mathbf{k} - \mathbf{k}_1)$. The initial and final states now have a delta-function normalization, the multiparticle generalization of eq. (5.5). We are left with

$$\begin{aligned}\langle f|i\rangle = i^{n+n'}\int & d^4x_1\, e^{ik_1x_1}(-\partial_1^2 + m^2)\ldots \\ & d^4x'_1\, e^{-ik'_1x'_1}(-\partial_{1'}^2 + m^2)\ldots \\ & \times \langle 0|\mathrm{T}\varphi(x_1)\ldots\varphi(x'_1)\ldots|0\rangle\,. \end{aligned} \quad (5.15)$$

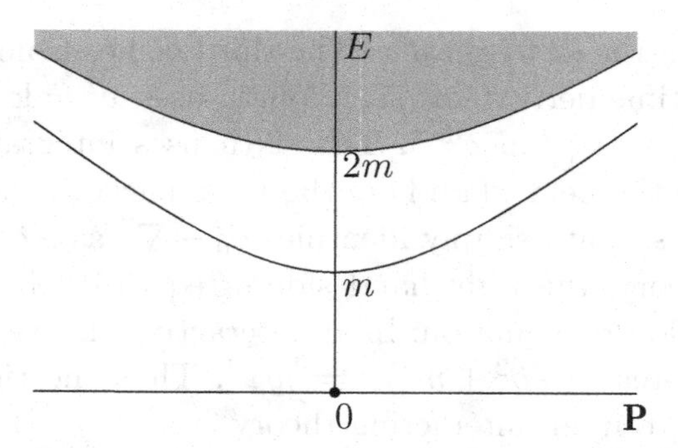

Figure 5.1. The exact energy eigenstates in the $(\mathbf{P}, E)$ plane. The ground state is isolated at $(\mathbf{0}, 0)$, the one-particle states form an isolated hyperbola that passes through $(\mathbf{0}, m)$, and the multiparticle continuum lies at and above the hyperbola that passes through $(\mathbf{0}, 2m)$.

This formula has been written to apply to the more general case of n incoming particles and n' outgoing particles; the ellipses stand for similar factors for each of the other incoming and outgoing particles.

Eq. (5.15) is the *Lehmann–Symanzik–Zimmermann reduction formula*, or LSZ formula for short. It is one of the key equations of quantum field theory.

However, our derivation of the LSZ formula relied on the supposition that the creation operators of *free* field theory would work comparably in the *interacting* theory. This is a rather suspect assumption, and so we must review it.

Let us consider what we can deduce about the energy and momentum eigenstates of the interacting theory on physical grounds. First, we assume that there is a unique ground state $|0\rangle$, with zero energy and momentum. The first excited state is a state of a single particle with mass m. This state can have an arbitrary three-momentum $\mathbf{k}$; its energy is then $E = \omega = (\mathbf{k}^2 + m^2)^{1/2}$. The next excited state is that of two particles. These two particles could form a bound state with energy *less* than $2m$ (like the hydrogen atom in quantum electrodynamics), but, to keep things simple, let us assume that there are no such bound states. Then the lowest possible energy of a two-particle state is $2m$. However, a two-particle state with zero total three-momentum can have *any* energy above $2m$, because the two particles could have some *relative* momentum that contributes to their total energy. Thus we are led to a picture of the states of theory as shown in fig. 5.1.

Now let us consider what happens when we act on the ground state with the field operator $\varphi(x)$. To this end, it is helpful to write

$$\varphi(x) = \exp(-iP^\mu x_\mu)\varphi(0)\exp(+iP^\mu x_\mu) , \tag{5.16}$$

where P^μ is the energy-momentum four-vector. (This equation, introduced in section 2, is just the relativistic generalization of the Heisenberg equation.) Now let us sandwich $\varphi(x)$ between the ground state (on the right) and other possible states (on the left). For example, let us put the ground state on the left as well. Then we have

$$\begin{aligned}\langle 0|\varphi(x)|0\rangle &= \langle 0|e^{-iPx}\varphi(0)e^{+iPx}|0\rangle \\ &= \langle 0|\varphi(0)|0\rangle \,. \end{aligned} \tag{5.17}$$

To get the second line, we used $P^\mu|0\rangle = 0$. The final expression is just a Lorentz-invariant number. Since $|0\rangle$ is the exact ground state of the interacting theory, we have (in general) no idea what this number is.

We would like $\langle 0|\varphi(0)|0\rangle$ to be zero. This is because we would like $a_1^\dagger(\pm\infty)$, when acting on $|0\rangle$, to create a single particle state. We do *not* want $a_1^\dagger(\pm\infty)$ to create a linear combination of a single particle state and the ground state. But this is precisely what will happen if $\langle 0|\varphi(0)|0\rangle$ is not zero.

So, if $v \equiv \langle 0|\varphi(0)|0\rangle$ is not zero, we will shift the field $\varphi(x)$ by the constant v. This means that we go back to the lagrangian, and replace $\varphi(x)$ everywhere by $\varphi(x) + v$. This is just a change of the name of the operator of interest, and does not affect the physics. However, the shifted $\varphi(x)$ obeys, by construction, $\langle 0|\varphi(x)|0\rangle = 0$.

Let us now consider $\langle p|\varphi(x)|0\rangle$, where $|p\rangle$ is a one-particle state with four-momentum p, normalized according to eq. (5.5). Again using eq. (5.16), we have

$$\begin{aligned}\langle p|\varphi(x)|0\rangle &= \langle p|e^{-iPx}\varphi(0)e^{+iPx}|0\rangle \\ &= e^{-ipx}\langle p|\varphi(0)|0\rangle \,, \end{aligned} \tag{5.18}$$

where $\langle p|\varphi(0)|0\rangle$ is a Lorentz-invariant number. It is a function of p, but the only Lorentz-invariant functions of p are functions of p^2, and p^2 is just the constant $-m^2$. So $\langle p|\varphi(0)|0\rangle$ is just some number that depends on m and (presumably) the other parameters in the lagrangian.

We would like $\langle p|\varphi(0)|0\rangle$ to be one. That is what it is in free-field theory, and we know that, in free-field theory, $a_1^\dagger(\pm\infty)$ creates a correctly normalized one-particle state. Thus, for $a_1^\dagger(\pm\infty)$ to create a correctly normalized one-particle state in the interacting theory, we must have $\langle p|\varphi(0)|0\rangle = 1$.

So, if $\langle p|\varphi(0)|0\rangle$ is not equal to one, we will rescale (or, one might say, *renormalize*) $\varphi(x)$ by a multiplicative constant. This is just a change of the name of the operator of interest, and does not affect the physics. However, the rescaled $\varphi(x)$ obeys, by construction, $\langle p|\varphi(0)|0\rangle = 1$.

Finally, consider $\langle p,n|\varphi(x)|0\rangle$, where $|p,n\rangle$ is a multiparticle state with total three-momentum $\mathbf{p}$, and n is short for all other labels (such as mass and relative momenta) needed to specify this state. We have

$$\begin{aligned}\langle p,n|\varphi(x)|0\rangle &= \langle p,n|e^{-iPx}\varphi(0)e^{+iPx}|0\rangle \\ &= e^{-ipx}\langle p,n|\varphi(0)|0\rangle \\ &= e^{-ipx}A_n(\mathbf{p})\ , \end{aligned} \tag{5.19}$$

where $A_n(\mathbf{p})$ is a function of Lorentz-invariant products of the various (relative and total) four-momenta needed to specify the state. Note that, from fig. (5.1), $p^0 = (\mathbf{p}^2 + M^2)^{1/2}$ with $M \geq 2m$. The invariant mass M is one of the parameters included in the set n.

We would like $\langle p,n|\varphi(x)|0\rangle$ to be zero, because we would like $a_1^\dagger(\pm\infty)$, when acting on $|0\rangle$, to create a single particle state. We do *not* want $a_1^\dagger(\pm\infty)$ to create any multiparticle states. But this is precisely what may happen if $\langle p,n|\varphi(x)|0\rangle$ is not zero.

Actually, we are being a little too strict. We really need $\langle p,n|a_1^\dagger(\pm\infty)|0\rangle$ to be zero, and perhaps it will be zero even if $\langle p,n|\varphi(x)|0\rangle$ is not. Also, we really should test $a_1^\dagger(\pm\infty)|0\rangle$ only against *normalizable* states. Mathematically, non-normalizable states cause all sorts of trouble; mathematicians don't consider them to be states at all. In physics, this usually doesn't bother us, but here we must be especially careful. So let us write

$$|\psi\rangle = \sum_n \int d^3p\ \psi_n(\mathbf{p})|p,n\rangle\ , \tag{5.20}$$

where the $\psi_n(\mathbf{p})$s are wave packets for the total three-momentum $\mathbf{p}$. Note that eq. (5.20) is highly schematic; the sum over n includes integrals over continuous parameters like relative momenta.

Now we want to examine

$$\langle\psi|a_1^\dagger(t)|0\rangle = -i\sum_n \int d^3p\ \psi_n^*(\mathbf{p})\int d^3k\ f_1(\mathbf{k})\int d^3x\ e^{ikx}\overleftrightarrow{\partial_0}\langle p,n|\varphi(x)|0\rangle\ . \tag{5.21}$$

We will take the limit $t\to\pm\infty$ in a moment. Using eq. (5.19), eq. (5.21) becomes

$$\begin{aligned}\langle\psi|a_1^\dagger(t)|0\rangle &= -i\sum_n \int d^3p\ \psi_n^*(\mathbf{p})\int d^3k\ f_1(\mathbf{k})\int d^3x\Big(e^{ikx}\overleftrightarrow{\partial_0}e^{-ipx}\Big)A_n(\mathbf{p}) \\ &= \sum_n \int d^3p\ \psi_n^*(\mathbf{p})\int d^3k\ f_1(\mathbf{k})\int d^3x\ (p^0{+}k^0)e^{i(k-p)x}A_n(\mathbf{p})\ . \end{aligned} \tag{5.22}$$

Next we use $\int d^3x\, e^{i(\mathbf{k}-\mathbf{p})\cdot\mathbf{x}} = (2\pi)^3\delta^3(\mathbf{k}-\mathbf{p})$ to get

$$\langle\psi|a_1^\dagger(t)|0\rangle = \sum_n \int d^3p\,(2\pi)^3(p^0+k^0)\psi_n^*(\mathbf{p})f_1(\mathbf{p})A_n(\mathbf{p})e^{i(p^0-k^0)t}\,, \tag{5.23}$$

where $p^0 = (\mathbf{p}^2+M^2)^{1/2}$ and $k^0 = (\mathbf{p}^2+m^2)^{1/2}$.

Now comes the key point. Note that p^0 is strictly greater than k^0, because $M \ge 2m > m$. Thus the integrand of eq. (5.23) contains a phase factor that oscillates more and more rapidly as $t \to \pm\infty$. Therefore, by the *Riemann–Lebesgue lemma*, the right-hand side of eq. (5.23) vanishes as $t \to \pm\infty$.

Physically, this means that a one-particle wave packet spreads out differently than a multiparticle wave packet, and the overlap between them goes to zero as the elapsed time goes to infinity. Thus, even though our operator $a_1^\dagger(t)$ creates some multiparticle states that we don't want, we can "follow" the one-particle state that we do want by using an appropriate wave packet. By waiting long enough, we can make the multiparticle contribution to the scattering amplitude as small as we like.

Let us recap. The basic formula for a scattering amplitude in terms of the fields of an interacting quantum field theory is the LSZ formula, which is worth writing down again:

$$\begin{aligned}\langle f|i\rangle = i^{n+n'}\int & d^4x_1\, e^{ik_1x_1}(-\partial_1^2+m^2)\ldots \\ & d^4x_{1'}\, e^{-ik'_1x'_1}(-\partial_{1'}^2+m^2)\ldots \\ & \times \langle 0|\mathrm{T}\varphi(x_1)\ldots\varphi(x'_1)\ldots|0\rangle\,.\end{aligned} \tag{5.24}$$

The LSZ formula is valid *provided* that the field obeys

$$\langle 0|\varphi(x)|0\rangle = 0 \qquad \text{and} \qquad \langle k|\varphi(x)|0\rangle = e^{-ikx}\,. \tag{5.25}$$

These normalization conditions may conflict with our original choice of field and parameter normalization in the lagrangian. Consider, for example, a lagrangian originally specified as

$$\mathcal{L} = -\tfrac{1}{2}\partial^\mu\varphi\partial_\mu\varphi - \tfrac{1}{2}m^2\varphi^2 + \tfrac{1}{6}g\varphi^3\,. \tag{5.26}$$

After shifting and rescaling (and renaming some parameters), we will have instead

$$\mathcal{L} = -\tfrac{1}{2}Z_\varphi\partial^\mu\varphi\partial_\mu\varphi - \tfrac{1}{2}Z_m m^2\varphi^2 + \tfrac{1}{6}Z_g g\varphi^3 + Y\varphi\,. \tag{5.27}$$

Here the three Zs and Y are as yet unknown constants. They must be chosen to ensure the validity of eq. (5.25); this gives us two conditions in four unknowns. We fix the parameter m by requiring it to be equal to the actual mass of the particle (equivalently, the energy of the first excited state relative to the ground state), and we fix the parameter g by requiring some particular scattering cross section to depend on g in some particular way. (For example, in quantum electrodynamics, the parameter analogous to g is the electron charge e. The low-energy Coulomb scattering cross section is proportional to e^4, with a definite constant of proportionality and no higher-order corrections; this relationship defines e.) Thus we have four conditions in four unknowns, and it is possible to calculate Y and the three Zs order by order in powers of g.

Next, we must develop the tools needed to compute the correlation functions $\langle 0|\mathrm{T}\varphi(x_1)\ldots|0\rangle$ in an interacting quantum field theory.

Reference notes

Useful discussions of the LSZ reduction formula can be found in *Brown*, *Itzykson & Zuber*, *Peskin & Schroeder*, and *Weinberg I*.

Problems

5.1) Work out the LSZ reduction formula for the complex scalar field that was introduced in problem 3.5. Note that we must specify the type (a or b) of each incoming and outgoing particle.

6

Path integrals in quantum mechanics

Prerequisite: none

Consider the nonrelativistic quantum mechanics of one particle in one dimension; the hamiltonian is

$$H(P,Q) = \tfrac{1}{2m}P^2 + V(Q) \,, \tag{6.1}$$

where P and Q are operators obeying $[Q,P] = i$. (We set $\hbar = 1$ for notational convenience.) We wish to evaluate the probability amplitude for the particle to start at position q' at time t', and end at position q'' at time t''. This amplitude is $\langle q''|e^{-iH(t''-t')}|q'\rangle$, where $|q'\rangle$ and $|q''\rangle$ are eigenstates of the position operator Q.

We can also formulate this question in the Heisenberg picture, where operators are time dependent and the state of the system is time independent, as opposed to the more familiar Schrödinger picture. In the Heisenberg picture, we write $Q(t) = e^{iHt}Qe^{-iHt}$. We can then define an instantaneous eigenstate of $Q(t)$ via $Q(t)|q,t\rangle = q|q,t\rangle$. These instantaneous eigenstates can be expressed explicitly as $|q,t\rangle = e^{+iHt}|q\rangle$, where $Q|q\rangle = q|q\rangle$. Then our transition amplitude can be written as $\langle q'',t''|q',t'\rangle$ in the Heisenberg picture.

To evaluate $\langle q'',t''|q',t'\rangle$, we begin by dividing the time interval $T \equiv t'' - t'$ into $N+1$ equal pieces of duration $\delta t = T/(N+1)$. Then introduce N complete sets of position eigenstates to get

$$\langle q'',t''|q',t'\rangle = \int \prod_{j=1}^{N} dq_j \; \langle q''|e^{-iH\delta t}|q_N\rangle\langle q_N|e^{-iH\delta t}|q_{N-1}\rangle \ldots \langle q_1|e^{-iH\delta t}|q'\rangle \,. \tag{6.2}$$

The integrals over the qs all run from $-\infty$ to $+\infty$.

Now consider $\langle q_2|e^{-iH\delta t}|q_1\rangle$. We can use the Campbell–Baker–Hausdorf formula

$$\exp(A+B) = \exp(A)\exp(B)\exp(-\tfrac{1}{2}[A,B]+\ldots) \tag{6.3}$$

to write

$$\exp(-iH\delta t) = \exp[-i(\delta t/2m)P^2]\exp[-i\delta t V(Q)]\exp[O(\delta t^2)]\,. \tag{6.4}$$

Then, in the limit of small δt, we should be able to ignore the final exponential. Inserting a complete set of momentum states then gives

$$\begin{aligned}\langle q_2|e^{-iH\delta t}|q_1\rangle &= \int dp_1\, \langle q_2|e^{-i(\delta t/2m)P^2}|p_1\rangle\langle p_1|e^{-i\delta t V(Q)}|q_1\rangle \\ &= \int dp_1\, e^{-i(\delta t/2m)p_1^2}\, e^{-i\delta t V(q_1)}\, \langle q_2|p_1\rangle\langle p_1|q_1\rangle \\ &= \int \frac{dp_1}{2\pi}\, e^{-i(\delta t/2m)p_1^2}\, e^{-i\delta t V(q_1)}\, e^{ip_1(q_2-q_1)} \\ &= \int \frac{dp_1}{2\pi}\, e^{-iH(p_1,q_1)\delta t}\, e^{ip_1(q_2-q_1)}\,. \end{aligned} \tag{6.5}$$

To get the third line, we used $\langle q|p\rangle = (2\pi)^{-1/2}\exp(ipq)$.

If we happen to be interested in more general hamiltonians than eq. (6.1), then we must worry about the ordering of the P and Q operators in any term that contains both. If we adopt *Weyl ordering*, where the quantum hamiltonian $H(P,Q)$ is given in terms of the classical hamiltonian $H(p,q)$ by

$$H(P,Q) \equiv \int \frac{dx}{2\pi}\frac{dk}{2\pi}\, e^{ixP+ikQ} \int dp\,dq\, e^{-ixp-ikq}\, H(p,q)\,, \tag{6.6}$$

then eq. (6.5) is not quite correct; in the last line, $H(p_1,q_1)$ should be replaced with $H(p_1,\bar{q}_1)$, where $\bar{q}_1 = \frac{1}{2}(q_1+q_2)$. For the hamiltonian of eq. (6.1), which is Weyl ordered, this replacement makes no difference in the limit $\delta t \to 0$.

Adopting Weyl ordering for the general case, we now have

$$\langle q'',t''|q',t'\rangle = \int \prod_{k=1}^{N} dq_k \prod_{j=0}^{N} \frac{dp_j}{2\pi}\, e^{ip_j(q_{j+1}-q_j)}\, e^{-iH(p_j,\bar{q}_j)\delta t}\,, \tag{6.7}$$

where $\bar{q}_j = \frac{1}{2}(q_j+q_{j+1})$, $q_0 = q'$, and $q_{N+1} = q''$. If we now define $\dot{q}_j \equiv (q_{j+1}-q_j)/\delta t$, and take the formal limit of $\delta t \to 0$, we get

$$\langle q'',t''|q',t'\rangle = \int \mathcal{D}q\,\mathcal{D}p\, \exp\left[i\int_{t'}^{t''} dt\Big(p(t)\dot{q}(t) - H(p(t),q(t))\Big)\right]. \tag{6.8}$$

The integration is to be understood as over all paths in phase space that start at $q(t') = q'$ (with an arbitrary value of the initial momentum) and end at $q(t'') = q''$ (with an arbitrary value of the final momentum).

If $H(p, q)$ is no more than quadratic in the momenta [as is the case for eq. (6.1)], then the integral over p is gaussian, and can be done in closed form. If the term that is quadratic in p is independent of q [as is the case for eq. (6.1)], then the prefactors generated by the gaussian integrals are all constants, and can be absorbed into the definition of $\mathcal{D}q$. The result of integrating out p is then

$$\langle q'', t'' | q', t' \rangle = \int \mathcal{D}q \, \exp\left[i \int_{t'}^{t''} dt \, L(\dot{q}(t), q(t)) \right] , \tag{6.9}$$

where $L(\dot{q}, q)$ is computed by first finding the stationary point of the p integral by solving

$$0 = \frac{\partial}{\partial p} \Big(p\dot{q} - H(p, q) \Big) = \dot{q} - \frac{\partial H(p, q)}{\partial p} \tag{6.10}$$

for p in terms of $\dot{q}$ and q, and then plugging this solution back into $p\dot{q} - H$ to get L. We recognize this procedure from classical mechanics: we are passing from the hamiltonian formulation to the lagrangian formulation.

Now that we have eqs. (6.8) and (6.9), what are we going to do with them? Let us begin by considering some generalizations; let us examine, for example, $\langle q'', t'' | Q(t_1) | q', t' \rangle$, where $t' < t_1 < t''$. This is given by

$$\langle q'', t'' | Q(t_1) | q', t' \rangle = \langle q'' | e^{-iH(t'' - t_1)} Q e^{-iH(t_1 - t')} | q' \rangle . \tag{6.11}$$

In the path integral formula, the extra operator Q inserted at time t_1 will simply result in an extra factor of $q(t_1)$. Thus

$$\langle q'', t'' | Q(t_1) | q', t' \rangle = \int \mathcal{D}p \, \mathcal{D}q \; q(t_1) \, e^{iS} , \tag{6.12}$$

where $S = \int_{t'}^{t''} dt \, (p\dot{q} - H)$. Now let us go in the other direction; consider $\int \mathcal{D}p \, \mathcal{D}q \, q(t_1) q(t_2) e^{iS}$. This clearly requires the operators $Q(t_1)$ and $Q(t_2)$, but their order depends on whether $t_1 < t_2$ or $t_2 < t_1$. Thus we have

$$\int \mathcal{D}p \, \mathcal{D}q \; q(t_1) q(t_2) \, e^{iS} = \langle q'', t'' | \mathrm{T} Q(t_1) Q(t_2) | q', t' \rangle . \tag{6.13}$$

where T is the *time ordering symbol*: a product of operators to its right is to be ordered, not as written, but with operators at later times to the left of those at earlier times. This is significant, because time-ordered products enter into the LSZ formula for scattering amplitudes.

To further develop these methods, we need another trick: *functional derivatives.* We define the functional derivative $\delta/\delta f(t)$ via

$$\frac{\delta}{\delta f(t_1)}\, f(t_2) = \delta(t_1 - t_2)\ , \tag{6.14}$$

where $\delta(t)$ is the Dirac delta function. Also, functional derivatives are defined to satisfy all the usual rules of derivatives (product rule, chain rule, etc.). Eq. (6.14) can be thought of as the continuous generalization of $(\partial/\partial x_i)x_j = \delta_{ij}$.

Now, consider modifying the lagrangian of our theory by including external forces acting on the particle:

$$H(p,q) \to H(p,q) - f(t)q(t) - h(t)p(t)\ , \tag{6.15}$$

where $f(t)$ and $h(t)$ are specified functions. In this case we will write

$$\langle q'',t''|q',t'\rangle_{f,h} = \int \mathcal{D}p\,\mathcal{D}q\ \exp\left[i\int_{t'}^{t''} dt\,\Big(p\dot{q} - H + fq + hp\Big)\right] \tag{6.16}$$

where H is the original hamiltonian. Then we have

$$\begin{aligned}
\frac{1}{i}\frac{\delta}{\delta f(t_1)}\,\langle q'',t''|q',t'\rangle_{f,h} &= \int \mathcal{D}p\,\mathcal{D}q\ q(t_1)\, e^{i\int dt\,[p\dot{q}-H+fq+hp]}\ ,\\
\frac{1}{i}\frac{\delta}{\delta f(t_1)}\frac{1}{i}\frac{\delta}{\delta f(t_2)}\,\langle q'',t''|q',t'\rangle_{f,h} &= \int \mathcal{D}p\,\mathcal{D}q\ q(t_1)q(t_2)\, e^{i\int dt\,[p\dot{q}-H+fq+hp]}\ ,\\
\frac{1}{i}\frac{\delta}{\delta h(t_1)}\,\langle q'',t''|q',t'\rangle_{f,h} &= \int \mathcal{D}p\,\mathcal{D}q\ p(t_1)\, e^{i\int dt\,[p\dot{q}-H+fq+hp]}\ ,
\end{aligned} \tag{6.17}$$

and so on. After we have finished bringing down as many factors of $q(t_i)$ or $p(t_i)$ as we like, we can set $f(t) = h(t) = 0$, and return to the original hamiltonian. Thus,

$$\begin{aligned}
&\langle q'',t''|\mathrm{T}Q(t_1)\ldots P(t_n)\ldots|q',t'\rangle\\
&\qquad = \frac{1}{i}\frac{\delta}{\delta f(t_1)}\cdots\frac{1}{i}\frac{\delta}{\delta h(t_n)}\cdots\langle q'',t''|q',t'\rangle_{f,h}\bigg|_{f=h=0}\ .
\end{aligned} \tag{6.18}$$

Suppose we are also interested in initial and final states other than position eigenstates. Then we must multiply by the wave functions for these states, and integrate. We will be interested, in particular, in the ground state as both the initial and final state. Also, we will take the limits $t' \to -\infty$ and $t'' \to +\infty$. The object of our attention is then

$$\langle 0|0\rangle_{f,h} = \lim_{\substack{t'\to-\infty \\ t''\to+\infty}} \int dq''\,dq'\;\psi_0^*(q'')\;\langle q'',t''|q',t'\rangle_{f,h}\;\psi_0(q')\;, \tag{6.19}$$

where $\psi_0(q) = \langle q|0\rangle$ is the ground-state wave function. Eq. (6.19) is a rather cumbersome formula, however. We will, therefore, employ a trick to simplify it.

Let $|n\rangle$ denote an eigenstate of H with eigenvalue E_n. We will suppose that $E_0 = 0$; if this is not the case, we will shift H by an appropriate constant. Next we write

$$\begin{aligned} |q',t'\rangle &= e^{iHt'}|q'\rangle \\ &= \sum_{n=0}^{\infty} e^{iHt'}|n\rangle\langle n|q'\rangle \\ &= \sum_{n=0}^{\infty} \psi_n^*(q')e^{iE_nt'}|n\rangle\;, \end{aligned} \tag{6.20}$$

where $\psi_n(q) = \langle q|n\rangle$ is the wave function of the nth eigenstate. Now, replace H with $(1-i\epsilon)H$ in eq. (6.20), where ϵ is a small positive infinitesimal. Then, take the limit $t' \to -\infty$ of eq. (6.20) with ϵ held fixed. Every state except the ground state is then multiplied by a vanishing exponential factor, and so the limit is simply $\psi_0^*(q')|0\rangle$. Next, multiply by an arbitrary function $\chi(q')$, and integrate over q'. The only requirement is that $\langle 0|\chi\rangle \neq 0$. We then have a constant times $|0\rangle$, and this constant can be absorbed into the normalization of the path integral. A similar analysis of $\langle q'',t''| = \langle q''|e^{-iHt''}$ shows that the replacement $H \to (1-i\epsilon)H$ also picks out the ground state as the final state in the $t'' \to +\infty$ limit.

What all this means is that if we use $(1-i\epsilon)H$ instead of H, we can be cavalier about the boundary conditions on the endpoints of the path. Any reasonable boundary conditions will result in the ground state as both the initial and final state. Thus we have

$$\langle 0|0\rangle_{f,h} = \int \mathcal{D}p\,\mathcal{D}q\,\exp\left[i\int_{-\infty}^{+\infty} dt\left(p\dot{q} - (1-i\epsilon)H + fq + hp\right)\right]. \tag{6.21}$$

Now let us suppose that $H = H_0 + H_1$, where we can solve for the eigenstates and eigenvalues of H_0, and H_1 can be treated as a perturbation.

Suppressing the $i\epsilon$, eq. (6.21) can be written as

$$\langle 0|0\rangle_{f,h} = \int \mathcal{D}p\,\mathcal{D}q\, \exp\left[i\int_{-\infty}^{+\infty} dt\,\Big(p\dot{q} - H_0(p,q) - H_1(p,q) + fq + hp\Big)\right]$$

$$= \exp\left[-i\int_{-\infty}^{+\infty} dt\, H_1\!\left(\frac{1}{i}\frac{\delta}{\delta h(t)}, \frac{1}{i}\frac{\delta}{\delta f(t)}\right)\right]$$

$$\times \int \mathcal{D}p\,\mathcal{D}q\, \exp\left[i\int_{-\infty}^{+\infty} dt\,\Big(p\dot{q} - H_0(p,q) + fq + hp\Big)\right]. \qquad (6.22)$$

To understand the second line of this equation, take the exponential prefactor inside the path integral. Then the functional derivatives (that appear as the arguments of H_1) just pull out appropriate factors of $p(t)$ and $q(t)$, generating the right-hand side of the first line. We assume that we can compute the functional integral in the second line, since it involves only the solvable hamiltonian H_0. The exponential prefactor can then be expanded in powers of H_1 to generate a perturbation series.

If H_1 depends only on q (and not on p), and if we are only interested in time-ordered products of Qs (and not Ps), and if H is no more than quadratic in P, and if the term quadratic in P does not involve Q, *then* eq. (6.22) can be simplified to

$$\langle 0|0\rangle_{f} = \exp\left[i\int_{-\infty}^{+\infty} dt\, L_1\!\left(\frac{1}{i}\frac{\delta}{\delta f(t)}\right)\right]$$

$$\times \int \mathcal{D}q\, \exp\left[i\int_{-\infty}^{+\infty} dt\,\Big(L_0(\dot{q},q) + fq\Big)\right], \qquad (6.23)$$

where $L_1(q) = -H_1(q)$.

Reference notes

Brown and *Ramond I* have especially clear treatments of various aspects of path integrals. For a careful derivation of the *midpoint rule* of eq. (6.7), see *Berry & Mount.*

Problems

6.1) a) Find an explicit formula for $\mathcal{D}q$ in eq. (6.9). Your formula should be of the form $\mathcal{D}q = C\prod_{j=1}^{N} dq_j$, where C is a constant that you should compute.

b) For the case of a free particle, $V(Q) = 0$, evaluate the path integral of eq. (6.9) explicitly. Hint: integrate over q_1, then q_2, etc., and look for a pattern. Express you final answer in terms of q', t', q'', t'', and m. Restore $\hbar$ by dimensional analysis.

c) Compute $\langle q'', t''|q', t'\rangle = \langle q''|e^{-iH(t''-t')}|q'\rangle$ by inserting a complete set of momentum eigenstates, and performing the integral over the momentum. Compare with your result in part (b).

7

The path integral for the harmonic oscillator

Prerequisite: 6

Consider a harmonic oscillator with hamiltonian

$$H(P,Q) = \tfrac{1}{2m}P^2 + \tfrac{1}{2}m\omega^2 Q^2 \,. \tag{7.1}$$

We begin with the formula from section 6 for the ground state to ground state transition amplitude in the presence of an external force, specialized to the case of a harmonic oscillator:

$$\langle 0|0\rangle_f = \int \mathcal{D}p\,\mathcal{D}q\, \exp i\int_{-\infty}^{+\infty} dt \Big[p\dot{q} - (1{-}i\epsilon)H + fq\Big] \,. \tag{7.2}$$

Looking at eq. (7.1), we see that multiplying H by $1{-}i\epsilon$ is equivalent to the replacements $m^{-1} \to (1{-}i\epsilon)m^{-1}$ [or, equivalently, $m \to (1{+}i\epsilon)m$] and $m\omega^2 \to (1{-}i\epsilon)m\omega^2$. Passing to the lagrangian formulation then gives

$$\langle 0|0\rangle_f = \int \mathcal{D}q\, \exp i\int_{-\infty}^{+\infty} dt \Big[\tfrac{1}{2}(1{+}i\epsilon)m\dot{q}^2 - \tfrac{1}{2}(1{-}i\epsilon)m\omega^2 q^2 + fq\Big] \,. \tag{7.3}$$

From now on, we will simplify the notation by setting $m = 1$.

Next, let us use Fourier-transformed variables,

$$\widetilde{q}(E) = \int_{-\infty}^{+\infty} dt\, e^{iEt}\, q(t) \,, \qquad q(t) = \int_{-\infty}^{+\infty} \frac{dE}{2\pi}\, e^{-iEt}\, \widetilde{q}(E) \,. \tag{7.4}$$

The expression in square brackets in eq. (7.3) becomes

$$\Big[\cdots\Big] = \frac{1}{2}\int_{-\infty}^{+\infty} \frac{dE}{2\pi}\,\frac{dE'}{2\pi}\, e^{-i(E+E')t}\Big[\Big(-(1{+}i\epsilon)EE' - (1{-}i\epsilon)\omega^2\Big)\widetilde{q}(E)\widetilde{q}(E') + \widetilde{f}(E)\widetilde{q}(E') + \widetilde{f}(E')\widetilde{q}(E)\Big] \,. \tag{7.5}$$

Note that the only t dependence is now in the prefactor. Integrating over t then generates a factor of $2\pi\delta(E+E')$. Then we can easily integrate over E' to get

$$
\begin{aligned}
S &= \int_{-\infty}^{+\infty} dt \left[\cdots\right] \\
&= \frac{1}{2}\int_{-\infty}^{+\infty} \frac{dE}{2\pi} \Big[\Big((1+i\epsilon)E^2 - (1-i\epsilon)\omega^2\Big)\widetilde{q}(E)\widetilde{q}(-E) \\
&\qquad\qquad + \widetilde{f}(E)\widetilde{q}(-E) + \widetilde{f}(-E)\widetilde{q}(E)\Big] \,. \qquad (7.6)
\end{aligned}
$$

The factor in large parentheses is equal to $E^2 - \omega^2 + i(E^2+\omega^2)\epsilon$, and we can absorb the positive coefficient into ϵ to get $E^2 - \omega^2 + i\epsilon$.

Now it is convenient to change integration variables to

$$
\widetilde{x}(E) = \widetilde{q}(E) + \frac{\widetilde{f}(E)}{E^2-\omega^2+i\epsilon} \,. \qquad (7.7)
$$

Then we get

$$
S = \frac{1}{2}\int_{-\infty}^{+\infty} \frac{dE}{2\pi} \left[\widetilde{x}(E)(E^2-\omega^2+i\epsilon)\widetilde{x}(-E) - \frac{\widetilde{f}(E)\widetilde{f}(-E)}{E^2-\omega^2+i\epsilon}\right] . \qquad (7.8)
$$

Furthermore, because eq. (7.7) is just a shift by a constant, $\mathcal{D}q = \mathcal{D}x$. Now we have

$$
\begin{aligned}
\langle 0|0\rangle_f &= \exp\left[\frac{i}{2}\int_{-\infty}^{+\infty} \frac{dE}{2\pi}\, \frac{\widetilde{f}(E)\widetilde{f}(-E)}{-E^2+\omega^2-i\epsilon}\right] \\
&\quad \times \int \mathcal{D}x \, \exp\left[\frac{i}{2}\int_{-\infty}^{+\infty} \frac{dE}{2\pi}\, \widetilde{x}(E)(E^2-\omega^2+i\epsilon)\widetilde{x}(-E)\right] . \qquad (7.9)
\end{aligned}
$$

Now comes the key point. The path integral on the second line of eq. (7.9) is what we get for $\langle 0|0\rangle_f$ in the case $f=0$. On the other hand, if there is no external force, a system in its ground state will remain in its ground state, and so $\langle 0|0\rangle_{f=0} = 1$. Thus $\langle 0|0\rangle_f$ is given by the first line of eq. (7.9),

$$
\langle 0|0\rangle_f = \exp\left[\frac{i}{2}\int_{-\infty}^{+\infty} \frac{dE}{2\pi}\, \frac{\widetilde{f}(E)\widetilde{f}(-E)}{-E^2+\omega^2-i\epsilon}\right] . \qquad (7.10)
$$

We can also rewrite $\langle 0|0\rangle_f$ in terms of time-domain variables as

$$
\langle 0|0\rangle_f = \exp\left[\frac{i}{2}\int_{-\infty}^{+\infty} dt\, dt'\, f(t)G(t-t')f(t')\right] , \qquad (7.11)
$$

where

$$G(t-t') = \int_{-\infty}^{+\infty} \frac{dE}{2\pi} \, \frac{e^{-iE(t-t')}}{-E^2+\omega^2-i\epsilon} \; . \tag{7.12}$$

Note that $G(t-t')$ is a Green's function for the oscillator equation of motion:

$$\left(\frac{\partial^2}{\partial t^2}+\omega^2\right) G(t-t') = \delta(t-t') \; . \tag{7.13}$$

This can be seen directly by plugging eq. (7.12) into eq. (7.13) and then taking the $\epsilon \to 0$ limit. We can also evaluate $G(t-t')$ explicitly by treating the integral over E on the right-hand side of eq. (7.12) as a contour integral in the complex E plane, and then evaluating it via the residue theorem. The result is

$$G(t-t') = \frac{i}{2\omega} \exp\Bigl(-i\omega|t-t'|\Bigr) \; . \tag{7.14}$$

Consider now the formula from section 6 for the time-ordered product of operators. In the case of initial and final ground states, it becomes

$$\langle 0|\mathrm{T}Q(t_1)\ldots|0\rangle = \frac{1}{i}\frac{\delta}{\delta f(t_1)} \ldots \langle 0|0\rangle_f \Big|_{f=0} \; . \tag{7.15}$$

Using our explicit formula, eq. (7.11), we have

$$\begin{aligned}
\langle 0|\mathrm{T}Q(t_1)Q(t_2)|0\rangle &= \frac{1}{i}\frac{\delta}{\delta f(t_1)}\frac{1}{i}\frac{\delta}{\delta f(t_2)} \langle 0|0\rangle_f \Big|_{f=0} \\
&= \frac{1}{i}\frac{\delta}{\delta f(t_1)} \left[\int_{-\infty}^{+\infty} dt' \, G(t_2-t')f(t')\right] \langle 0|0\rangle_f \Big|_{f=0} \\
&= \left[\tfrac{1}{i}G(t_2-t_1) + (\text{term with } f\text{s})\right] \langle 0|0\rangle_f \Big|_{f=0} \\
&= \tfrac{1}{i}G(t_2-t_1) \; .
\end{aligned} \tag{7.16}$$

We can continue in this way to compute the ground-state expectation value of the time-ordered product of more $Q(t)$s. If the number of $Q(t)$s is odd, then there is always a left-over $f(t)$ in the prefactor, and so the result is zero. If the number of $Q(t)$s is even, then we must pair up the functional derivatives in an appropriate way to get a nonzero result. Thus, for example,

$$\begin{aligned}
\langle 0|\mathrm{T}Q(t_1)Q(t_2)Q(t_3)Q(t_4)|0\rangle = \frac{1}{i^2}\Bigl[&G(t_1-t_2)G(t_3-t_4) \\
&+ G(t_1-t_3)G(t_2-t_4) \\
&+ G(t_1-t_4)G(t_2-t_3)\Bigr] \; .
\end{aligned} \tag{7.17}$$

More generally,

$$\langle 0|\mathrm{T}Q(t_1)\ldots Q(t_{2n})|0\rangle = \frac{1}{i^n}\sum_{\text{pairings}} G(t_{i_1}-t_{i_2})\ldots G(t_{i_{2n-1}}-t_{i_{2n}})\,. \qquad (7.18)$$

Problems

7.1) Starting with eq. (7.12), do the contour integral to verify eq. (7.14).

7.2) Starting with eq. (7.14), verify eq. (7.13).

7.3) a) Use the Heisenberg equation of motion, $\dot{A} = i[H, A]$, to find explicit expressions for $\dot{Q}$ and $\dot{P}$. Solve these to get the Heisenberg-picture operators $Q(t)$ and $P(t)$ in terms of the Schrödinger-picture operators Q and P.

b) Write the Schrödinger-picture operators Q and P in terms of the creation and annihilation operators a and $a^\dagger$, where $H = \hbar\omega(a^\dagger a + \frac{1}{2})$. Then, using your result from part (a), write the Heisenberg-picture operators $Q(t)$ and $P(t)$ in terms of a and $a^\dagger$.

c) Using your result from part (b), and $a|0\rangle = \langle 0|a^\dagger = 0$, verify eqs. (7.16) and (7.17).

7.4) Consider a harmonic oscillator in its ground state at $t = -\infty$. It is then subjected to an external force $f(t)$. Compute the probability $|\langle 0|0\rangle_f|^2$ that the oscillator is still in its ground state at $t = +\infty$. Write your answer as a manifestly real expression, and in terms of the Fourier transform $\widetilde{f}(E) = \int_{-\infty}^{+\infty} dt\, e^{iEt} f(t)$. Your answer should not involve any other unevaluated integrals.

8

The path integral for free-field theory

Prerequisites: 3, 7

Our results for the harmonic oscillator can be straightforwardly generalized to a free-field theory with hamiltonian density

$$\mathcal{H}_0 = \tfrac{1}{2}\Pi^2 + \tfrac{1}{2}(\nabla\varphi)^2 + \tfrac{1}{2}m^2\varphi^2 \; . \tag{8.1}$$

The dictionary we need is

$$\begin{aligned} q(t) &\longrightarrow \varphi(\mathbf{x},t) \quad \text{(classical field)} \\ Q(t) &\longrightarrow \varphi(\mathbf{x},t) \quad \text{(operator field)} \\ f(t) &\longrightarrow J(\mathbf{x},t) \quad \text{(classical } \textit{source}\text{)} \; . \end{aligned} \tag{8.2}$$

The distinction between the classical field $\varphi(x)$ and the corresponding operator field should be clear from the context.

To employ the ϵ trick, we multiply $\mathcal{H}_0$ by $1 - i\epsilon$. The results are equivalent to replacing m^2 in $\mathcal{H}_0$ with $m^2 - i\epsilon$. From now on, for notational simplicity, we will write m^2 when we really mean $m^2 - i\epsilon$.

Let us write down the path integral (also called the *functional integral*) for our free-field theory:

$$Z_0(J) \equiv \langle 0|0\rangle_J = \int \mathcal{D}\varphi \; e^{i\int d^4x[\mathcal{L}_0 + J\varphi]} \; , \tag{8.3}$$

where

$$\mathcal{L}_0 = -\tfrac{1}{2}\partial^\mu\varphi\partial_\mu\varphi - \tfrac{1}{2}m^2\varphi^2 \tag{8.4}$$

is the lagrangian density, and

$$\mathcal{D}\varphi \propto \prod_x d\varphi(x) \tag{8.5}$$

is the *functional measure*. Note that when we say *path integral*, we now mean a path in the space of field configurations.

We can evaluate $Z_0(J)$ by mimicking what we did for the harmonic oscillator in section 7. We introduce four-dimensional Fourier transforms,

$$\widetilde{\varphi}(k) = \int d^4x\, e^{-ikx}\, \varphi(x)\,, \qquad \varphi(x) = \int \frac{d^4k}{(2\pi)^4}\, e^{ikx}\, \widetilde{\varphi}(k)\,, \tag{8.6}$$

where $kx = -k^0t + \mathbf{k}\cdot\mathbf{x}$, and k^0 is an integration variable. Then, starting with $S_0 = \int d^4x\, [\mathcal{L}_0 + J\varphi]$, we get

$$S_0 = \frac{1}{2}\int \frac{d^4k}{(2\pi)^4}\Big[-\widetilde{\varphi}(k)(k^2+m^2)\widetilde{\varphi}(-k) + \widetilde{J}(k)\widetilde{\varphi}(-k) + \widetilde{J}(-k)\widetilde{\varphi}(k)\Big]\,, \tag{8.7}$$

where $k^2 = \mathbf{k}^2 - (k^0)^2$. We now change path integration variables to

$$\widetilde{\chi}(k) = \widetilde{\varphi}(k) - \frac{\widetilde{J}(k)}{k^2+m^2}\,. \tag{8.8}$$

Since this is merely a shift by a constant, we have $\mathcal{D}\varphi = \mathcal{D}\chi$. The action becomes

$$S_0 = \frac{1}{2}\int \frac{d^4k}{(2\pi)^4}\left[\frac{\widetilde{J}(k)\widetilde{J}(-k)}{k^2+m^2} - \widetilde{\chi}(k)(k^2+m^2)\widetilde{\chi}(-k)\right]. \tag{8.9}$$

Just as for the harmonic oscillator, the integral over χ simply yields a factor of $Z_0(0) = \langle 0|0\rangle_{J=0} = 1$. Therefore

$$\begin{aligned} Z_0(J) &= \exp\left[\frac{i}{2}\int \frac{d^4k}{(2\pi)^4}\,\frac{\widetilde{J}(k)\widetilde{J}(-k)}{k^2+m^2-i\epsilon}\right] \\ &= \exp\left[\frac{i}{2}\int d^4x\, d^4x'\; J(x)\Delta(x-x')J(x')\right]. \end{aligned} \tag{8.10}$$

Here we have defined the *Feynman propagator*,

$$\Delta(x-x') = \int \frac{d^4k}{(2\pi)^4}\,\frac{e^{ik(x-x')}}{k^2+m^2-i\epsilon}\,. \tag{8.11}$$

The Feynman propagator is a Green's function for the Klein–Gordon equation,

$$(-\partial_x^2 + m^2)\Delta(x-x') = \delta^4(x-x')\,. \tag{8.12}$$

This can be seen directly by plugging eq. (8.11) into eq. (8.12) and then taking the $\epsilon \to 0$ limit. We can also evaluate $\Delta(x-x')$ explicitly by treating the k^0 integral on the right-hand side of eq. (8.11) as a contour integral in the complex k^0 plane, and then evaluating it via the residue theorem. The result is

$$\begin{aligned}\Delta(x-x') &= i\int \widetilde{dk}\, e^{i\mathbf{k}\cdot(\mathbf{x}-\mathbf{x}')-i\omega|t-t'|} \\ &= i\theta(t-t')\int \widetilde{dk}\, e^{ik(x-x')} + i\theta(t'-t)\int \widetilde{dk}\, e^{-ik(x-x')}\,, \qquad (8.13)\end{aligned}$$

where $\theta(t)$ is the unit step function. The integral over $\widetilde{dk}$ can also be performed in terms of Bessel functions; see section 4.

Now, by analogy with the formula for the ground-state expectation value of a time-ordered product of operators for the harmonic oscillator, we have

$$\langle 0|\mathrm{T}\varphi(x_1)\ldots|0\rangle = \frac{1}{i}\frac{\delta}{\delta J(x_1)}\ldots Z_0(J)\Big|_{J=0}\,. \qquad (8.14)$$

Using our explicit formula, eq. (8.10), we have

$$\begin{aligned}\langle 0|\mathrm{T}\varphi(x_1)\varphi(x_2)|0\rangle &= \frac{1}{i}\frac{\delta}{\delta J(x_1)}\frac{1}{i}\frac{\delta}{\delta J(x_2)}\,Z_0(J)\Big|_{J=0} \\ &= \frac{1}{i}\frac{\delta}{\delta J(x_1)}\left[\int d^4x'\,\Delta(x_2-x')J(x')\right] Z_0(J)\Big|_{J=0} \\ &= \left[\tfrac{1}{i}\Delta(x_2-x_1) + (\text{term with } J\text{s})\right] Z_0(J)\Big|_{J=0} \\ &= \tfrac{1}{i}\Delta(x_2-x_1)\,. \qquad (8.15)\end{aligned}$$

We can continue in this way to compute the ground-state expectation value of the time-ordered product of more φs. If the number of φs is odd, then there is always a left-over J in the prefactor, and so the result is zero. If the number of φs is even, then we must pair up the functional derivatives in an appropriate way to get a nonzero result. Thus, for example,

$$\begin{aligned}\langle 0|\mathrm{T}\varphi(x_1)\varphi(x_2)\varphi(x_3)\varphi(x_4)|0\rangle = \frac{1}{i^2}\Big[&\Delta(x_1-x_2)\Delta(x_3-x_4) \\ &+\Delta(x_1-x_3)\Delta(x_2-x_4) \\ &+\Delta(x_1-x_4)\Delta(x_2-x_3)\Big]. \qquad (8.16)\end{aligned}$$

More generally,

$$\langle 0|\mathrm{T}\varphi(x_1)\ldots\varphi(x_{2n})|0\rangle = \frac{1}{i^n}\sum_{\text{pairings}}\Delta(x_{i_1}-x_{i_2})\ldots\Delta(x_{i_{2n-1}}-x_{i_{2n}})\,. \qquad (8.17)$$

This result is known as *Wick's theorem.*

Problems

8.1) Starting with eq. (8.11), verify eq. (8.12).

8.2) Starting with eq. (8.11), verify eq. (8.13).

8.3) Starting with eq. (8.13), verify eq. (8.12). Note that the time derivatives in the Klein–Gordon wave operator can act on either the field (which obeys the Klein–Gordon equation) or the time-ordering step functions.

8.4) Use eqs. (3.19), (3.29), and (5.3) (and its hermitian conjugate) to verify the last line of eq. (8.15).

8.5) The retarded and advanced Green's functions for the Klein–Gordon wave operator satisfy $\Delta_{\rm ret}(x-y)=0$ for $x^0 \ge y^0$ and $\Delta_{\rm adv}(x-y)=0$ for $x^0 \le y^0$. Find the pole prescriptions on the right-hand side of eq. (8.11) that yield these Green's functions.

8.6) Let $Z_0(J)=\exp iW_0(J)$, and evaluate the real and imaginary parts of $W_0(J)$.

8.7) Repeat the analysis of this section for the complex scalar field that was introduced in problem 3.5, and further studied in problem 5.1. Write your source term in the form $J^\dagger\varphi + J\varphi^\dagger$, and find an explicit formula, analogous to eq. (8.10), for $Z_0(J^\dagger,J)$. Write down the appropriate generalization of eq. (8.14), and use it to compute $\langle 0|{\rm T}\varphi(x_1)\varphi(x_2)|0\rangle$, $\langle 0|{\rm T}\varphi^\dagger(x_1)\varphi(x_2)|0\rangle$, and $\langle 0|{\rm T}\varphi^\dagger(x_1)\varphi^\dagger(x_2)|0\rangle$. Then verify your results by using the method of problem 8.4. Finally, give the appropriate generalization of eq. (8.17).

8.8) A harmonic oscillator (in units with $m=\hbar=1$) has a ground-state wave function $\langle q|0\rangle \propto e^{-\omega q^2/2}$. Now consider a real scalar field $\varphi(x)$, and define a *field eigenstate* $|A\rangle$ that obeys

$$\varphi(\mathbf{x},0)|A\rangle = A(\mathbf{x})|A\rangle\ , \tag{8.18}$$

where the function $A(\mathbf{x})$ is everywhere real. For a free-field theory specified by the hamiltonian of eq. (8.1), show that the *ground-state wave functional* is

$$\langle A|0\rangle \propto \exp\left[-\frac{1}{2}\int\frac{d^3k}{(2\pi)^3}\,\omega(\mathbf{k})\tilde{A}(\mathbf{k})\tilde{A}(-\mathbf{k})\right]\ , \tag{8.19}$$

where $\tilde{A}(\mathbf{k}) \equiv \int d^3x\, e^{-i\mathbf{k}\cdot\mathbf{x}}A(\mathbf{x})$ and $\omega(\mathbf{k}) \equiv (\mathbf{k}^2+m^2)^{1/2}$.

9

The path integral for interacting field theory

Prerequisite: 8

Let us consider an interacting quantum field theory specified by a lagrangian of the form

$$\mathcal{L} = -\tfrac{1}{2}Z_\varphi \partial^\mu\varphi\partial_\mu\varphi - \tfrac{1}{2}Z_m m^2\varphi^2 + \tfrac{1}{6}Z_g g\varphi^3 + Y\varphi\,. \tag{9.1}$$

As we discussed at the end of section 5, we fix the parameter m by requiring it to be equal to the actual mass of the particle (equivalently, the energy of the first excited state relative to the ground state), and we fix the parameter g by requiring some particular scattering cross section to depend on g in some particular way. (We will have more to say about this after we have learned to calculate cross sections.) We also assume that the field is normalized by

$$\langle 0|\varphi(x)|0\rangle = 0 \qquad \text{and} \qquad \langle k|\varphi(x)|0\rangle = e^{-ikx}\,. \tag{9.2}$$

Here $|0\rangle$ is the ground state, normalized via $\langle 0|0\rangle = 1$, and $|k\rangle$ is a state of one particle with four-momentum k^μ, where $k^2 = k^\mu k_\mu = -m^2$, normalized via

$$\langle k'|k\rangle = (2\pi)^3 2k^0 \delta^3(\mathbf{k}' - \mathbf{k})\,. \tag{9.3}$$

Thus we have four conditions (the specified values of m, g, $\langle 0|\varphi|0\rangle$, and $\langle k|\varphi|0\rangle$), and we will use these four conditions to determine the values of the four remaining parameters (Y and the three Zs) that appear in $\mathcal{L}$.

Before going further, we should note that this theory (known as φ^3 theory, pronounced "phi-cubed") actually has a fatal flaw. The hamiltonian density is

$$\mathcal{H} = \tfrac{1}{2}Z_\varphi^{-1}\Pi^2 + \tfrac{1}{2}Z_\varphi(\nabla\varphi)^2 - Y\varphi + \tfrac{1}{2}Z_m m^2\varphi^2 - \tfrac{1}{6}Z_g g\varphi^3\,. \tag{9.4}$$

Classically, we can make this arbitrarily negative by choosing an arbitrarily large value for φ. Quantum mechanically, this means that this hamiltonian has no ground state. If we start off near $\varphi = 0$, we can tunnel through the potential barrier to large φ, and then "roll down the hill". However, this process is invisible in perturbation theory in g. The situation is exactly analogous to the problem of a harmonic oscillator perturbed by a q^3 term. This system also has no ground state, but perturbation theory (both time dependent and time independent) does not "know" this. We will be interested in eq. (9.1) only as an example of how to do perturbation expansions in a simple context, and so we will overlook this problem.

We would like to evaluate the path integral for this theory,

$$Z(J) \equiv \langle 0|0\rangle_J = \int \mathcal{D}\varphi \; e^{i\int d^4x[\mathcal{L}_0+\mathcal{L}_1+J\varphi]} \; . \tag{9.5}$$

We can evaluate $Z(J)$ by mimicking what we did for quantum mechanics at the end of section 6. Specifically, we can rewrite eq. (9.5) as

$$\begin{aligned} Z(J) &= e^{i\int d^4x \; \mathcal{L}_1\left(\frac{1}{i}\frac{\delta}{\delta J(x)}\right)} \int \mathcal{D}\varphi \; e^{i\int d^4x[\mathcal{L}_0+J\varphi]} \\ &\propto e^{i\int d^4x \; \mathcal{L}_1\left(\frac{1}{i}\frac{\delta}{\delta J(x)}\right)} \; Z_0(J) \; , \end{aligned} \tag{9.6}$$

where $Z_0(J)$ is the result in free-field theory,

$$Z_0(J) = \exp\left[\frac{i}{2}\int d^4x \, d^4x' \; J(x)\Delta(x-x')J(x')\right] . \tag{9.7}$$

We have written $Z(J)$ as proportional to (rather than equal to) the right-hand side of eq. (9.6) because the ϵ trick does not give us the correct overall normalization; instead, we must require $Z(0) = 1$, and enforce this by hand.

Note that, in eq. (9.7), we have implicitly assumed that

$$\mathcal{L}_0 = -\tfrac{1}{2}\partial^\mu\varphi\partial_\mu\varphi - \tfrac{1}{2}m^2\varphi^2 \; , \tag{9.8}$$

since this is the $\mathcal{L}_0$ that gives us eq. (9.7). Therefore, the rest of $\mathcal{L}$ must be included in $\mathcal{L}_1$. We write

$$\begin{aligned} \mathcal{L}_1 &= \tfrac{1}{6}Z_g g\varphi^3 + \mathcal{L}_{\text{ct}} \; , \\ \mathcal{L}_{\text{ct}} &= -\tfrac{1}{2}(Z_\varphi - 1)\partial^\mu\varphi\partial_\mu\varphi - \tfrac{1}{2}(Z_m - 1)m^2\varphi^2 + Y\varphi \; , \end{aligned} \tag{9.9}$$

where $\mathcal{L}_{\text{ct}}$ is called the *counterterm* lagrangian. We expect that, as $g \to 0$, $Y \to 0$ and $Z_i \to 1$. In fact, as we will see, $Y = O(g)$ and $Z_i = 1 + O(g^2)$.

In order to make use of eq. (9.6), we will have to compute lots and lots of functional derivatives of $Z_0(J)$. Let us begin by ignoring the counterterms.

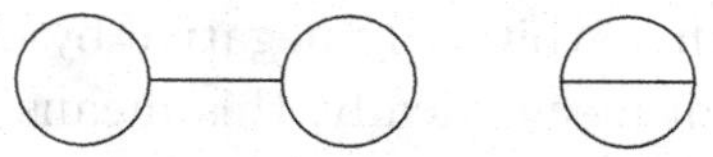

$S = 2^3$ $S = 2 \times 3!$

Figure 9.1. All connected diagrams with $E = 0$ and $V = 2$.

We define

$$Z_1(J) \propto \exp\left[\frac{i}{6} Z_g g \int d^4x \left(\frac{1}{i}\frac{\delta}{\delta J(x)}\right)^3\right] Z_0(J) , \tag{9.10}$$

where the constant of proportionality is fixed by $Z_1(0) = 1$. We now make a dual Taylor expansion in powers of g and J to get

$$Z_1(J) \propto \sum_{V=0}^{\infty} \frac{1}{V!}\left[\frac{iZ_g g}{6}\int d^4x \left(\frac{1}{i}\frac{\delta}{\delta J(x)}\right)^3\right]^V$$
$$\times \sum_{P=0}^{\infty} \frac{1}{P!}\left[\frac{i}{2}\int d^4y\, d^4z\, J(y)\Delta(y-z)J(z)\right]^P . \tag{9.11}$$

If we focus on a term in eq. (9.11) with particular values of V and P, then the number of surviving sources (after we take all the functional derivatives) is $E = 2P - 3V$. (Here E stands for *external*, a terminology that should become clear by the end of the next section; V stands for *vertex* and P for *propagator*.) The overall phase factor of such a term is then $i^V (1/i)^{3V} i^P = i^{V+E-P}$, and the $3V$ functional derivatives can act on the $2P$ sources in $(2P)!/(2P-3V)!$ different combinations. However, many of the resulting expressions are algebraically identical.

To organize them, we introduce *Feynman diagrams*. In these diagrams, a line segment (straight or curved) stands for a propagator $\frac{1}{i}\Delta(x-y)$, a filled circle at one end of a line segment for a source $i\int d^4x\, J(x)$, and a vertex joining three line segments for $iZ_g g \int d^4x$. Sets of diagrams with different values of E and V are shown in figs. 9.1–9.11.

To count the number of terms on the right-hand side of eq. (9.11) that result in a particular diagram, we first note that, in each diagram, the number of lines is P and the number of vertices is V. We can rearrange the three functional derivatives from a particular vertex without changing the resulting diagram; this yields a counting factor of $3!$ for each vertex. Also, we can rearrange the vertices themselves; this yields a counting factor of $V!$. Similarly, we can rearrange the two sources at the ends of a particular propagator without changing the resulting diagram; this yields a counting factor of $2!$ for each propagator. Also, we can rearrange the propagators

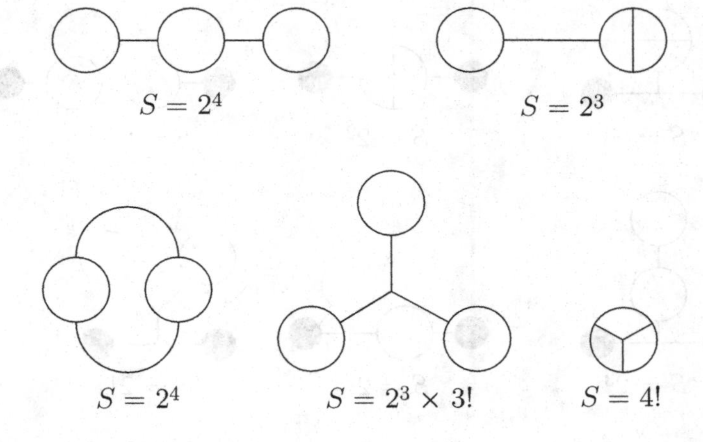

Figure 9.2. All connected diagrams with $E = 0$ and $V = 4$.

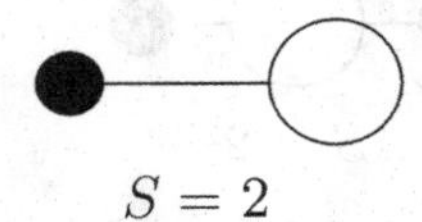

Figure 9.3. All connected diagrams with $E = 1$ and $V = 1$.

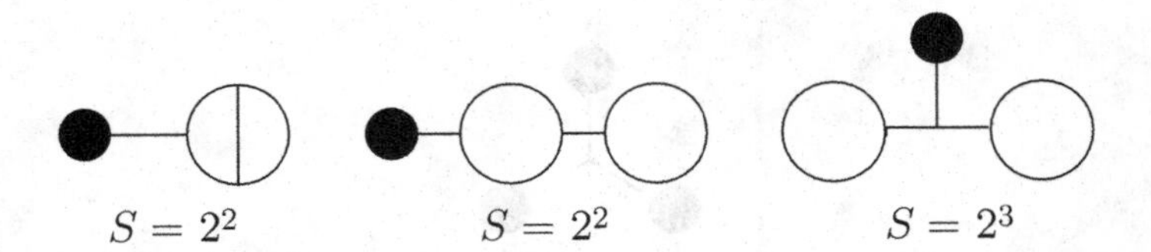

Figure 9.4. All connected diagrams with $E = 1$ and $V = 3$.

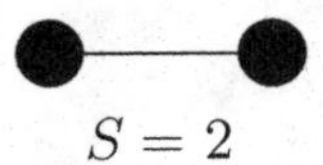

Figure 9.5. All connected diagrams with $E = 2$ and $V = 0$.

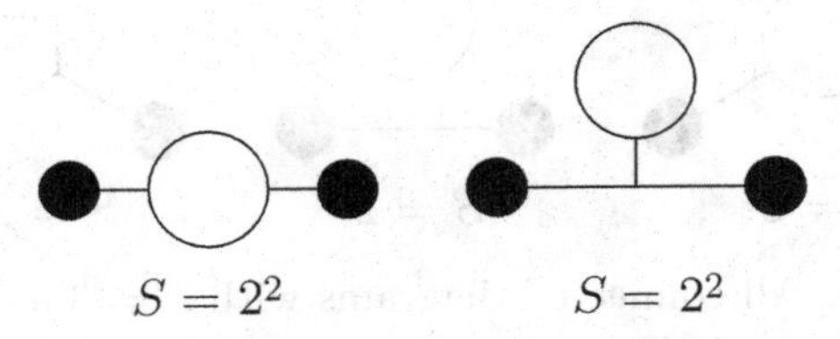

Figure 9.6. All connected diagrams with $E = 2$ and $V = 2$.

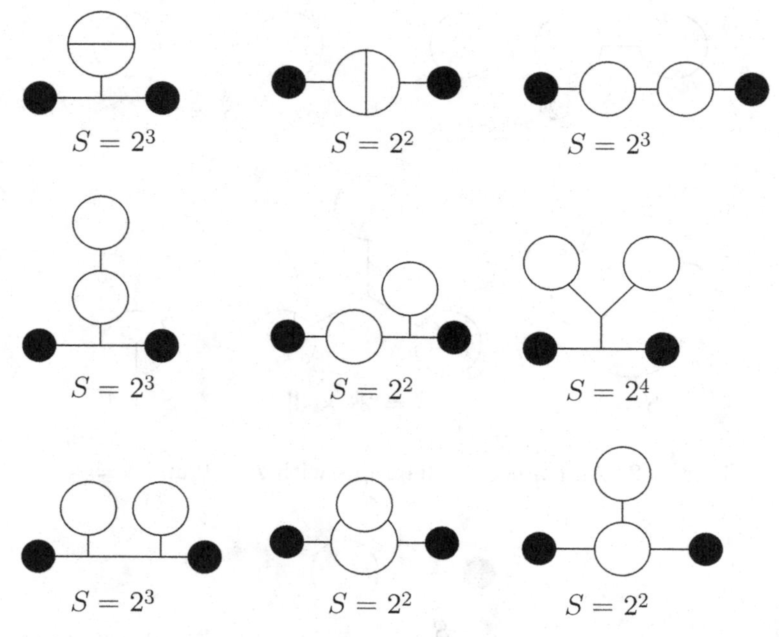

Figure 9.7. All connected diagrams with $E = 2$ and $V = 4$.

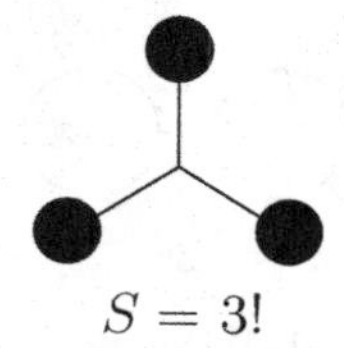

Figure 9.8. All connected diagrams with $E = 3$ and $V = 1$.

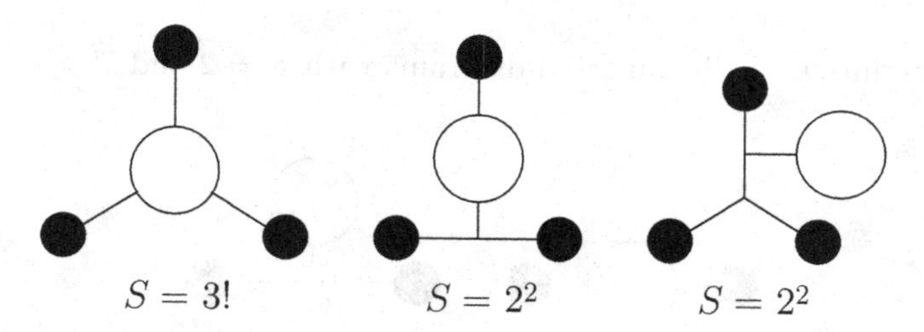

Figure 9.9. All connected diagrams with $E = 3$ and $V = 3$.

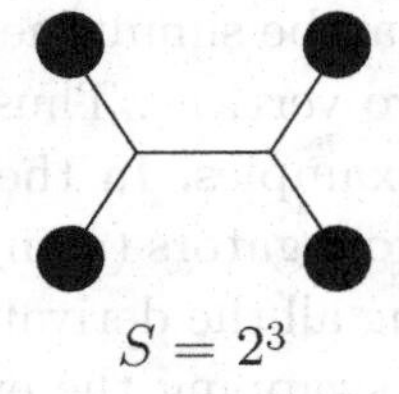

Figure 9.10. All connected diagrams with $E = 4$ and $V = 2$.

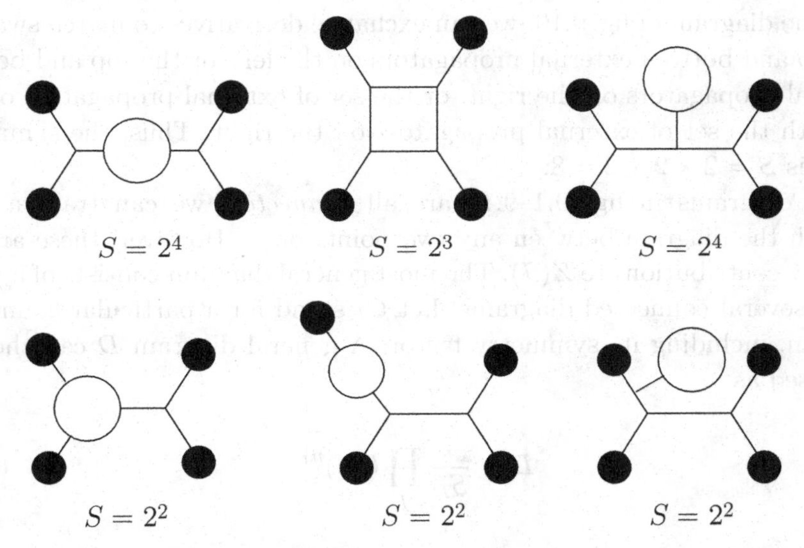

Figure 9.11. All connected diagrams with $E = 4$ and $V = 4$.

themselves; this yields a counting factor of $P!$. All together, these counting factors neatly cancel the numbers from the dual Taylor expansions in eq. (9.11).

However, this procedure generally results in an overcounting of the number of terms that give identical results. This happens when some rearrangement of derivatives gives the *same* match-up to sources as some rearrangement of sources. This possibility is always connected to some symmetry property of the diagram, and so the factor by which we have overcounted is called the *symmetry factor*. The figures show the symmetry factor S of each diagram.

Consider, for example, the second diagram of fig. 9.1. The three propagators can be rearranged in 3! ways, and all these rearrangements can be duplicated by exchanging the derivatives at the vertices. Furthermore, the endpoints of each propagator can be simultaneously swapped, and the effect duplicated by swapping the two vertices. Thus, $S = 2 \times 3! = 12$.

Let us consider two more examples. In the first diagram of fig. 9.6, the exchange of the two external propagators (along with their attached sources) can be duplicated by exchanging all the derivatives at one vertex for those at the other, and simultaneously swapping the endpoints of each semicircular propagator. Also, the effect of swapping the top and bottom semicircular propagators can be duplicated by swapping the corresponding derivatives at each vertex. Thus, the symmetry factor is $S = 2 \times 2 = 4$.

In the diagram of fig. 9.10, we can exchange derivatives to match swaps of the top and bottom external propagators on the left, or the top and bottom external propagators on the right, or the set of external propagators on the left with the set of external propagators on the right. Thus, the symmetry factor is $S = 2 \times 2 \times 2 = 8$.

The diagrams in figs. 9.1–9.11 are all *connected:* we can trace a path through the diagram between any two points on it. However, these are not the only contributions to $Z(J)$. The most general diagram consists of a product of several connected diagrams. Let C_I stand for a particular connected diagram, including its symmetry factor. A general diagram D can then be expressed as

$$D = \frac{1}{S_D} \prod_I (C_I)^{n_I} , \qquad (9.12)$$

where n_I is an integer that counts the number of C_Is in D, and S_D is the *additional* symmetry factor for D (that is, the part of the symmetry factor that is not already accounted for by the symmetry factors already included in each of the connected diagrams). We now need to determine S_D.

Since we have already accounted for propagator and vertex rearrangements *within* each C_I, we only need to consider exchanges of propagators and vertices among *different* connected diagrams. These can leave the total diagram D unchanged only if (1) the exchanges are made among different but *identical* connected diagrams, and only if (2) the exchanges involve *all* of the propagators and vertices in a given connected diagram. If there are n_I factors of C_I in D, there are $n_I!$ ways to make these rearrangements. Overall, then, we have

$$S_D = \prod_I n_I! \, . \qquad (9.13)$$

Now $Z_1(J)$ is given (up to an overall normalization) by summing all diagrams D, and each D is labeled by the integers n_I. Therefore

$$\begin{aligned} Z_1(J) &\propto \sum_{\{n_I\}} D \\ &\propto \sum_{\{n_I\}} \prod_I \frac{1}{n_I!}(C_I)^{n_I} \\ &\propto \prod_I \sum_{n_I=0}^{\infty} \frac{1}{n_I!}(C_I)^{n_I} \\ &\propto \prod_I \exp(C_I) \\ &\propto \exp\left(\textstyle\sum_I C_I\right) . \end{aligned} \tag{9.14}$$

Thus we have a remarkable result: $Z_1(J)$ is given by the exponential of the sum of *connected* diagrams. This makes it easy to impose the normalization $Z_1(0) = 1$: we simply omit the *vacuum diagrams* (those with no sources), like those of figs. 9.1 and 9.2. We then have

$$Z_1(J) = \exp[iW_1(J)] , \tag{9.15}$$

where we have defined

$$iW_1(J) \equiv \sum_{I \neq \{0\}} C_I , \tag{9.16}$$

and the notation $I \neq \{0\}$ means that the vacuum diagrams are omitted from the sum, so that $W_1(0) = 0$.

Were it not for the counterterms in $\mathcal{L}_1$, we would have $Z(J) = Z_1(J)$. Let us see what we would get if this was, in fact, the case. In particular, let us compute the vacuum expectation value of the field $\varphi(x)$, which is given by

$$\begin{aligned} \langle 0|\varphi(x)|0\rangle &= \frac{1}{i}\frac{\delta}{\delta J(x)} Z_1(J)\bigg|_{J=0} \\ &= \frac{\delta}{\delta J(x)} W_1(J)\bigg|_{J=0} . \end{aligned} \tag{9.17}$$

This expression is then the sum of all diagrams (such as those in figs. 9.3 and 9.4) that have a single source, with the source removed:

$$\langle 0|\varphi(x)|0\rangle = \tfrac{1}{2}ig \int d^4y\, \tfrac{1}{i}\Delta(x-y)\tfrac{1}{i}\Delta(y-y) + O(g^3) . \tag{9.18}$$

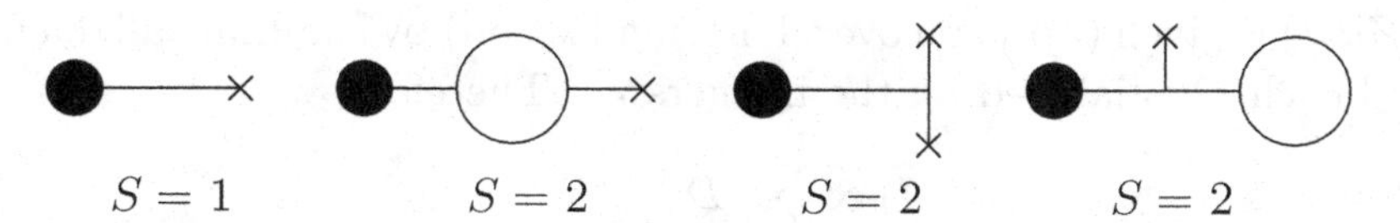

Figure 9.12. All connected diagrams with $E=1$, $X \geq 1$ (where X is the number of one-point vertices from the linear counterterm), and $V+X \leq 3$.

Here we have set $Z_g = 1$ in the first term, since $Z_g = 1 + O(g^2)$. We see the vacuum-expectation value of $\varphi(x)$ is not zero, as is required for the validity of the LSZ formula. To fix this, we must introduce the counterterm $Y\varphi$. Including this term in the interaction lagrangian $\mathcal{L}_1$ introduces a new kind of vertex, one where a single line segment ends; the corresponding vertex factor is $iY \int d^4y$. The simplest diagrams including this new vertex are shown in fig. 9.12, with a cross symbolizing the vertex.

Assuming $Y = O(g)$, only the first diagram in fig. 9.12 contributes at $O(g)$, and we have

$$\langle 0|\varphi(x)|0\rangle = \big(iY + \tfrac{1}{2}(ig)\tfrac{1}{i}\Delta(0)\big) \int d^4y\, \tfrac{1}{i}\Delta(x-y) + O(g^3)\,. \tag{9.19}$$

Thus, in order to have $\langle 0|\varphi(x)|0\rangle = 0$, we should choose

$$Y = \tfrac{1}{2} ig\Delta(0) + O(g^3)\,. \tag{9.20}$$

The factor of i is disturbing, because Y must be a real number: it is the coefficient of a hermitian operator in the hamiltonian, as seen in eq. (9.4). Therefore, $\Delta(0)$ must be purely imaginary, or we are in trouble. We have

$$\Delta(0) = \int \frac{d^4k}{(2\pi)^4} \frac{1}{k^2 + m^2 - i\epsilon}\,. \tag{9.21}$$

From eq. (9.21), it is not immediately obvious whether or not $\Delta(0)$ is purely imaginary, but eq. (9.21) does reveal another problem: the integral diverges at large k. This is another example of an *ultraviolet divergence*, similar to the one we encountered in section 3 when we computed the zero-point energy of the field.

To make some progress, we introduce an *ultraviolet cutoff* Λ, which we assume is much larger than m and any other energy of physical interest. Modifications to the propagator above some cutoff may be well justified physically; for example, quantum fluctuations in spacetime itself should become important above the *Planck scale*, which is given by the inverse square root of Newton's constant, and has the numerical value of 10^{19} GeV (compared to, say, the proton mass, which is 1 GeV).

In order to retain the Lorentz-transformation properties of the propagator, we implement the ultraviolet cutoff in a more subtle way than we did in section 3; specifically, we make the replacement

$$\Delta(x-y) \to \int \frac{d^4k}{(2\pi)^4} \frac{e^{ik(x-y)}}{k^2+m^2-i\epsilon} \left(\frac{\Lambda^2}{k^2+\Lambda^2-i\epsilon}\right)^2 . \tag{9.22}$$

The integral is now convergent, and we can evaluate the modified $\Delta(0)$ with the methods of section 14; for $\Lambda \gg m$, the result is

$$\Delta(0) = \frac{i}{16\pi^2} \Lambda^2 . \tag{9.23}$$

Thus Y is real, as required. If we like, we can now formally take the limit $\Lambda \to \infty$. The parameter Y becomes infinite, but $\langle 0|\varphi(x)|0\rangle$ remains zero, at least to this order in g.

It may be disturbing to have a parameter in the lagrangian that is formally infinite. However, such parameters are not directly measurable, and so need not obey our preconceptions about their magnitudes. Also, it is important to remember that Y includes a factor of g; this means that we can expand in powers of Y as part of our general expansion in powers of g. When we compute something measurable (like a scattering cross section), all the formally infinite numbers will cancel in a well-defined way, leaving behind finite coefficients for the various powers of g. We will see how this works in detail in sections 14–20.

As we go to higher orders in g, things become more complicated, but in principle the procedure is the same. Thus, at $O(g^3)$, we sum up the diagrams of figs. 9.4 and 9.12, and then add to Y whatever $O(g^3)$ term is needed to maintain $\langle 0|\varphi(x)|0\rangle = 0$. In this way we can determine the value of Y order by order in powers of g.

Once this is done, there is a remarkable simplification. Our adjustment of Y to keep $\langle 0|\varphi(x)|0\rangle = 0$ means that the sum of all connected diagrams with a single source is zero. Consider now that same infinite set of diagrams, but replace the single source in each of them with some other subdiagram. Here is the point: *no matter what this replacement subdiagram is, the sum of all these diagrams is still zero.* Therefore, we need not bother to compute any of them! The rule is this: ignore any diagram that, when a single line is cut, falls into two parts, *one of which has no sources.* All of these diagrams (known as *tadpoles*) are canceled by the Y counterterm, no matter what subdiagram they are attached to. The diagrams that remain (and need to be computed!) are shown in fig. 9.13.

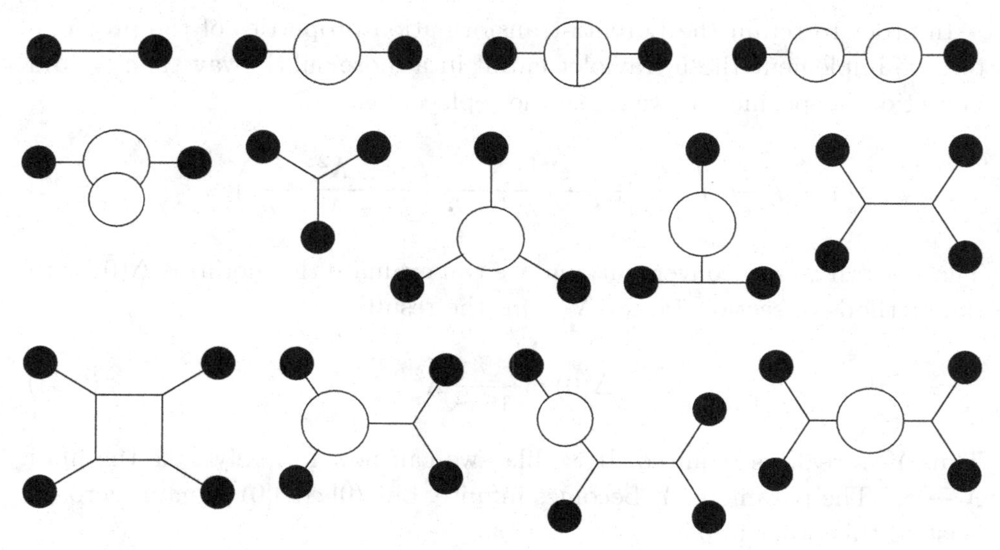

Figure 9.13. All connected diagrams without tadpoles with $E \le 4$ and $V \le 4$.

We turn next to the remaining two counterterms. For notational simplicity we define

$$A = Z_\varphi - 1\,, \qquad B = Z_m - 1\,, \tag{9.24}$$

and recall that we expect each of these to be $O(g^2)$. We now have

$$Z(J) = \exp\left[-\frac{i}{2}\int d^4x \left(\frac{1}{i}\frac{\delta}{\delta J(x)}\right)\left(-A\partial_x^2 + Bm^2\right)\left(\frac{1}{i}\frac{\delta}{\delta J(x)}\right)\right] Z_Y(J)\,, \tag{9.25}$$

where $Z_Y(J)$ includes the Y counterterm. We have integrated by parts to put both ∂_xs onto one $\delta/\delta J(x)$. (Note that the time derivatives in this interaction should really be treated by including an extra source term for the conjugate momentum $\Pi = \dot{\varphi}$. However, the space derivatives are correctly treated, and then the time derivatives must work out comparably by Lorentz invariance.)

Eq. (9.25) results in a new vertex at which two lines meet. The corresponding vertex factor is $(-i)\int d^4x\,(-A\partial_x^2 + Bm^2)$; the ∂_x^2 acts on the x in one or the other (but not both) propagators. (Which one does not matter, and can be changed via integration by parts.) Diagramatically, all we need do is sprinkle these new vertices onto the propagators in our existing diagrams. How many of these vertices we need to add depends on the order in g we are working to achieve.

This completes our calculation of $Z(J)$ in φ^3 theory. We express it as

$$Z(J) = \exp[iW(J)] \, , \tag{9.26}$$

where $W(J)$ is given by the sum of all connected diagrams with no tadpoles and at least two sources, and including the counterterm vertices just discussed.

Now that we have $Z(J)$, we must find out what we can do with it.

Problems

9.1) Compute the symmetry factor for each diagram in fig. 9.13. (You can then check your answers by consulting the earlier figures.)

9.2) Consider a real scalar field with $\mathcal{L} = \mathcal{L}_0 + \mathcal{L}_1$, where

$$\begin{aligned}
\mathcal{L}_0 &= -\tfrac{1}{2}\partial^\mu\varphi\partial_\mu\varphi - \tfrac{1}{2}m^2\varphi^2 \, , \\
\mathcal{L}_1 &= -\tfrac{1}{24}Z_\lambda\lambda\varphi^4 + \mathcal{L}_{\text{ct}} \, , \\
\mathcal{L}_{\text{ct}} &= -\tfrac{1}{2}(Z_\varphi - 1)\partial^\mu\varphi\partial_\mu\varphi - \tfrac{1}{2}(Z_m - 1)m^2\varphi^2 \, .
\end{aligned}$$

a) What kind of vertex appears in the diagrams for this theory (that is, how many line segments does it join?), and what is the associated vertex factor?

b) Ignoring the counterterms, draw all the connected diagrams with $1 \le E \le 4$ and $0 \le V \le 2$, and find their symmetry factors.

c) Explain why we did not have to include a counterterm linear in φ to cancel tadpoles.

9.3) Consider a complex scalar field (see problems 3.5, 5.1, and 8.7) with $\mathcal{L} = \mathcal{L}_0 + \mathcal{L}_1$, where

$$\begin{aligned}
\mathcal{L}_0 &= -\partial^\mu\varphi^\dagger\partial_\mu\varphi - m^2\varphi^\dagger\varphi \, , \\
\mathcal{L}_1 &= -\tfrac{1}{4}Z_\lambda\lambda(\varphi^\dagger\varphi)^2 + \mathcal{L}_{\text{ct}} \, , \\
\mathcal{L}_{\text{ct}} &= -(Z_\varphi - 1)\partial^\mu\varphi^\dagger\partial_\mu\varphi - (Z_m - 1)m^2\varphi^\dagger\varphi \, .
\end{aligned}$$

This theory has two kinds of sources, J and $J^\dagger$, and so we need a way to tell which is which when we draw the diagrams. Rather than labeling the source blobs with a J or $J^\dagger$, we will indicate which is which by putting an arrow on the attached propagator that points *towards* the source if it is a $J^\dagger$, and *away* from the source if it is a J.

a) What kind of vertex appears in the diagrams for this theory, and what is the associated vertex factor? Hint: your answer should involve those arrows!

b) Ignoring the counterterms, draw all the connected diagrams with $1 \le E \le 4$ and $0 \le V \le 2$, and find their symmetry factors. Hint: the arrows are important!

9.4) Consider the integral

$$\exp W(g,J) \equiv \frac{1}{\sqrt{2\pi}} \int_{-\infty}^{+\infty} dx \, \exp\Big[-\tfrac{1}{2}x^2 + \tfrac{1}{6}gx^3 + Jx\Big] . \tag{9.27}$$

This integral does not converge, but it can be used to generate a joint power series in g and J,

$$W(g,J) = \sum_{V=0}^{\infty} \sum_{E=0}^{\infty} C_{V,E} \, g^V J^E . \tag{9.28}$$

a) Show that

$$C_{V,E} = \sum_I \frac{1}{S_I} , \tag{9.29}$$

where the sum is over all connected Feynman diagrams with E sources and V three-point vertices, and S_I is the symmetry factor for each diagram.

b) Use eqs. (9.27) and (9.28) to compute $C_{V,E}$ for $V \le 4$ and $E \le 5$. (This is most easily done with a symbolic manipulation program like Mathematica.) Verify that the symmetry factors given in figs. 9.1–9.11 satisfy the sum rule of eq. (9.29).

c) Now consider $W(g, J{+}Y)$, with Y fixed by the "no tadpole" condition

$$\frac{\partial}{\partial J} W(g, J{+}Y)\bigg|_{J=0} = 0 . \tag{9.30}$$

Then write

$$W(g, J{+}Y) = \sum_{V=0}^{\infty} \sum_{E=0}^{\infty} \widetilde{C}_{V,E} \, g^V J^E . \tag{9.31}$$

Show that

$$\widetilde{C}_{V,E} = \sum_I \frac{1}{S_I} , \tag{9.32}$$

where the sum is over all connected Feynman diagrams with E sources and V three-point vertices and *no tadpoles*, and S_I is the symmetry factor for each diagram.

d) Let $Y = a_1 g + a_3 g^3 + \ldots$, and use eq. (9.30) to determine a_1 and a_3. Compute $\widetilde{C}_{V,E}$ for $V \le 4$ and $E \le 4$. Verify that the symmetry factors for the diagrams in fig. (9.13) satisfy the sum rule of eq. (9.32).

9.5) *The interaction picture.* In this problem, we will derive a formula for $\langle 0|\mathrm{T}\varphi(x_n)\ldots\varphi(x_1)|0\rangle$ without using path integrals. Suppose we have a hamiltonian density $\mathcal{H} = \mathcal{H}_0 + \mathcal{H}_1$, where $\mathcal{H}_0 = \frac{1}{2}\Pi^2 + \frac{1}{2}(\nabla\varphi)^2 + \frac{1}{2}m^2\varphi^2$, and $\mathcal{H}_1$ is a function of $\Pi(\mathbf{x},0)$ and $\varphi(\mathbf{x},0)$ and their spatial derivatives. (It should be chosen to preserve Lorentz invariance, but we will not be concerned with this issue.) We add a constant to H so that $H|0\rangle = 0$. Let $|\emptyset\rangle$ be the ground state of H_0, with a constant added to H_0 so that $H_0|\emptyset\rangle = 0$. (H_1 is then defined as $H - H_0$.) The Heisenberg-picture field is

$$\varphi(\mathbf{x},t) \equiv e^{iHt}\varphi(\mathbf{x},0)e^{-iHt} \,. \tag{9.33}$$

We now define the *interaction-picture* field

$$\varphi_I(\mathbf{x},t) \equiv e^{iH_0t}\varphi(\mathbf{x},0)e^{-iH_0t} \,. \tag{9.34}$$

a) Show that $\varphi_I(x)$ obeys the Klein–Gordon equation, and hence is a free field.
b) Show that $\varphi(x) = U^\dagger(t)\varphi_I(x)U(t)$, where $U(t) \equiv e^{iH_0t}e^{-iHt}$ is unitary.
c) Show that $U(t)$ obeys the differential equation $i\frac{d}{dt}U(t) = H_I(t)U(t)$, where $H_I(t) = e^{iH_0t}H_1e^{-iH_0t}$ is the interaction hamiltonian in the interaction picture, and the boundary condition $U(0) = 1$.
d) If $\mathcal{H}_1$ is specified by a particular function of the Schrödinger-picture fields $\Pi(\mathbf{x},0)$ and $\varphi(\mathbf{x},0)$, show that $\mathcal{H}_I(t)$ is given by the same function of the interaction-picture fields $\Pi_I(\mathbf{x},t)$ and $\varphi_I(\mathbf{x},t)$.
e) Show that, for $t > 0$,

$$U(t) = \mathrm{T}\exp\left[-i\int_0^t dt'\; H_I(t')\right] \tag{9.35}$$

obeys the differential equation and boundary condition of part (c). What is the comparable expression for $t < 0$? Hint: you may need to define a new ordering symbol.

f) Define $U(t_2,t_1) \equiv U(t_2)U^\dagger(t_1)$. Show that, for $t_2 > t_1$,

$$U(t_2,t_1) = \mathrm{T}\exp\left[-i\int_{t_1}^{t_2} dt'\; H_I(t')\right] . \tag{9.36}$$

What is the comparable expression for $t_1 > t_2$?

g) For any time ordering, show that $U(t_3,t_1) = U(t_3,t_2)U(t_2,t_1)$ and that $U^\dagger(t_1,t_2) = U(t_2,t_1)$.
h) Show that

$$\begin{aligned}\varphi(x_n)\ldots\varphi(x_1) = U^\dagger(t_n,0)\varphi_I(x_n)U(t_n,t_{n-1})\varphi_I(x_{n-1})\\ \ldots U(t_2,t_1)\varphi_I(x_1)U(t_1,0) \,.\end{aligned} \tag{9.37}$$

i) Show that $U^\dagger(t_n,0) = U^\dagger(\infty,0)U(\infty,t_n)$ and also that $U(t_1,0) = U(t_1,-\infty)U(-\infty,0)$.

(j) Replace H_0 with $(1-i\epsilon)H_0$, and show that $\langle 0|U^\dagger(\infty,0) = \langle 0|\emptyset\rangle\langle\emptyset|$ and that $U(-\infty,0)|0\rangle = |\emptyset\rangle\langle\emptyset|0\rangle$.

(k) Show that

$$\begin{aligned}\langle 0|\varphi(x_n)\ldots\varphi(x_1)|0\rangle = \langle\emptyset|U(\infty,t_n)\varphi_I(x_n)U(t_n,t_{n-1})\varphi_I(x_{n-1})\ldots\\ U(t_2,t_1)\varphi_I(x_1)U(t_1,-\infty)|\emptyset\rangle\\ \times|\langle\emptyset|0\rangle|^2 \,.\end{aligned} \tag{9.38}$$

(l) Show that

$$\langle 0|\mathrm{T}\varphi(x_n)\ldots\varphi(x_1)|0\rangle = \langle\emptyset|\mathrm{T}\varphi_I(x_n)\ldots\varphi_I(x_1)e^{-i\int d^4x\,\mathcal{H}_I(x)}|\emptyset\rangle \times |\langle\emptyset|0\rangle|^2 \,. \tag{9.39}$$

(m) Show that

$$|\langle\emptyset|0\rangle|^2 = 1/\langle\emptyset|\mathrm{T}e^{-i\int d^4x\,\mathcal{H}_I(x)}|\emptyset\rangle \,. \tag{9.40}$$

Thus we have

$$\langle 0|\mathrm{T}\varphi(x_n)\ldots\varphi(x_1)|0\rangle = \frac{\langle\emptyset|\mathrm{T}\varphi_I(x_n)\ldots\varphi_I(x_1)e^{-i\int d^4x\,\mathcal{H}_I(x)}|\emptyset\rangle}{\langle\emptyset|\mathrm{T}e^{-i\int d^4x\,\mathcal{H}_I(x)}|\emptyset\rangle} \,. \tag{9.41}$$

We can now Taylor expand the exponentials on the right-hand side of eq. (9.41), and use free-field theory to compute the resulting correlation functions.

10

Scattering amplitudes and the Feynman rules

Prerequisites: 5, 9

Now that we have an expression for $Z(J) = \exp iW(J)$, we can take functional derivatives to compute vacuum expectation values of time-ordered products of fields. Consider the case of two fields; we define the exact propagator via

$$\tfrac{1}{i}\mathbf{\Delta}(x_1 - x_2) \equiv \langle 0|\mathrm{T}\varphi(x_1)\varphi(x_2)|0\rangle \ . \tag{10.1}$$

For notational simplicity let us define

$$\delta_j \equiv \frac{1}{i}\,\frac{\delta}{\delta J(x_j)} \ . \tag{10.2}$$

Then we have

$$\begin{aligned}\langle 0|\mathrm{T}\varphi(x_1)\varphi(x_2)|0\rangle &= \delta_1\delta_2 Z(J)\Big|_{J=0} \\ &= \delta_1\delta_2 iW(J)\Big|_{J=0} + \delta_1 iW(J)\Big|_{J=0}\,\delta_2 iW(J)\Big|_{J=0} \\ &= \delta_1\delta_2 iW(J)\Big|_{J=0} \ .\end{aligned} \tag{10.3}$$

To get the last line we used $\delta_j iW(J)|_{J=0} = \langle 0|\varphi(x_j)|0\rangle = 0$. Diagramatically, δ_1 removes a source, and labels the propagator endpoint x_1. Thus $\frac{1}{i}\mathbf{\Delta}(x_1{-}x_2)$ is given by the sum of diagrams with two sources, with those sources removed and the endpoints labeled x_1 and x_2. (The labels must be applied in both ways. If the diagram was originally symmetric on exchange of the two sources, the associated symmetry factor of 2 is then canceled by the double labeling.) At lowest order, the only contribution is the "barbell" diagram of fig. 9.5 with the sources removed. Thus we recover the obvious fact that $\frac{1}{i}\mathbf{\Delta}(x_1{-}x_2) = \frac{1}{i}\Delta(x_1{-}x_2) + O(g^2)$. We will take up the subject of the $O(g^2)$ corrections in section 14.

For now, let us go on to compute

$$\begin{aligned}\langle 0|\mathrm{T}\varphi(x_1)\varphi(x_2)\varphi(x_3)\varphi(x_4)|0\rangle &= \delta_1\delta_2\delta_3\delta_4 Z(J)\Big|_{J=0} \\ &= \Big[\,\delta_1\delta_2\delta_3\delta_4 iW \\ &\qquad + (\delta_1\delta_2 iW)(\delta_3\delta_4 iW) \\ &\qquad + (\delta_1\delta_3 iW)(\delta_2\delta_4 iW) \\ &\qquad + (\delta_1\delta_4 iW)(\delta_2\delta_3 iW)\,\Big]_{J=0}\,. \end{aligned} \tag{10.4}$$

We have dropped terms that contain a factor of $\langle 0|\varphi(x)|0\rangle = 0$. According to eq. (10.3), the last three terms in eq. (10.4) simply give products of the exact propagators.

Let us see what happens when these terms are inserted into the LSZ formula for two incoming and two outgoing particles,

$$\begin{aligned}\langle f|i\rangle = i^4 \int d^4x_1\, d^4x_2\, d^4x_1'\, d^4x_2'\, e^{i(k_1x_1+k_2x_2-k_1'x_1'-k_2'x_2')} \\ \times(-\partial_1^2+m^2)(-\partial_2^2+m^2)(-\partial_{1'}^2+m^2)(-\partial_{2'}^2+m^2) \\ \times\langle 0|\mathrm{T}\varphi(x_1)\varphi(x_2)\varphi(x_1')\varphi(x_2')|0\rangle\,. \end{aligned} \tag{10.5}$$

If we consider, for example, $\frac{1}{i}\mathbf{\Delta}(x_1-x_1')\frac{1}{i}\mathbf{\Delta}(x_2-x_2')$ as one term in the correlation function in eq. (10.5), we get from this term

$$\begin{aligned}\int d^4x_1\, d^4x_2\, d^4x_1'\, d^4x_2'\, e^{i(k_1x_1+k_2x_2-k_1'x_1'-k_2'x_2')} F(x_{11'})F(x_{22'}) \\ = (2\pi)^4\delta^4(k_1-k_1')\,(2\pi)^4\delta^4(k_2-k_2')\,\widetilde{F}(\bar{k}_{11'})\,\widetilde{F}(\bar{k}_{22'})\,, \end{aligned} \tag{10.6}$$

where $F(x_{ij}) \equiv (-\partial_i^2+m^2)(-\partial_j^2+m^2)\mathbf{\Delta}(x_{ij})$, $\widetilde{F}(k)$ is its Fourier transform, $x_{ij'} \equiv x_i - x_j'$, and $\bar{k}_{ij'} \equiv (k_i+k_j')/2$. The important point is the two delta functions: these tell us that the four-momenta of the two outgoing particles ($1'$ and $2'$) are equal to the four-momenta of the two incoming particles (1 and 2). In other words, no scattering has occurred. This is not the event whose probability we wish to compute! The other two similar terms in eq. (10.4) either contribute to "no scattering" events, or vanish due to factors like $\delta^4(k_1+k_2)$ (which is zero because $k_1^0+k_2^0 \ge 2m > 0$). In general, the diagrams that contribute to the scattering process of interest are only those that are *fully connected*: every endpoint can be reached from every other endpoint by tracing through the diagram. These are the diagrams that arise from all the δs acting on a single factor of W. Therefore, from here on, we restrict our attention to those diagrams alone. We define the *connected*

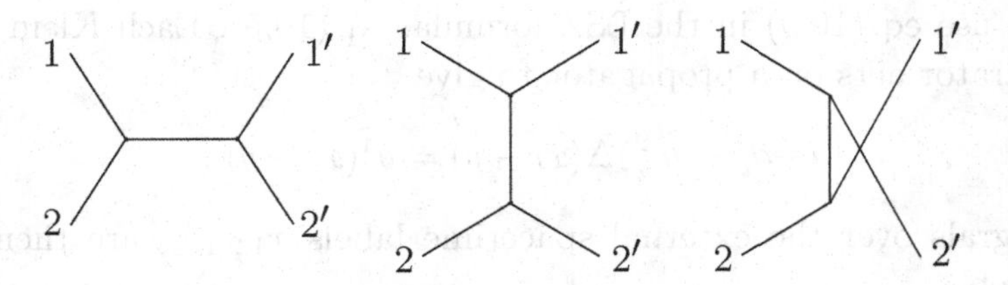

Figure 10.1. The three tree-level Feynman diagrams that contribute to the connected correlation function $\langle 0|\mathrm{T}\varphi(x_1)\varphi(x_2)\varphi(x_1')\varphi(x_2')|0\rangle_{\mathrm{C}}$.

correlation functions via

$$\langle 0|\mathrm{T}\varphi(x_1)\ldots\varphi(x_E)|0\rangle_{\mathrm{C}} \equiv \delta_1\ldots\delta_E iW(J)\Big|_{J=0}\,, \tag{10.7}$$

and use these instead of $\langle 0|\mathrm{T}\varphi(x_1)\ldots\varphi(x_E)|0\rangle$ in the LSZ formula.

Returning to eq. (10.4), we have

$$\langle 0|\mathrm{T}\varphi(x_1)\varphi(x_2)\varphi(x_1')\varphi(x_2')|0\rangle_{\mathrm{C}} = \delta_1\delta_2\delta_{1'}\delta_{2'} iW\Big|_{J=0}\,. \tag{10.8}$$

The lowest-order (in g) nonzero contribution to this comes from the diagram of fig. 9.10, which has four sources and two vertices. The four δs remove the four sources; there are 4! ways of matching up the δs to the sources. These 24 diagrams can then be collected into three groups of eight diagrams each; the eight diagrams in each group are identical. The three distinct diagrams are shown in fig. 10.1. Note that the factor of eight neatly cancels the symmetry factor $S = 8$ of the diagram with sources.

This is a general result for *tree diagrams* (those with no closed loops): once the sources have been stripped off and the endpoints labeled, each diagram with a distinct endpoint labeling has an overall symmetry factor of one. The tree diagrams for a given process represent the lowest-order (in g) nonzero contribution to that process.

We now have

$$\begin{aligned}
&\langle 0|\mathrm{T}\varphi(x_1)\varphi(x_2)\varphi(x_1')\varphi(x_2')|0\rangle_{\mathrm{C}} \\
&\qquad = (ig)^2\left(\tfrac{1}{i}\right)^5 \int d^4y\,d^4z\,\Delta(y-z) \\
&\qquad\quad \times\Big[\,\Delta(x_1-y)\Delta(x_2-y)\Delta(x_1'-z)\Delta(x_2'-z) \\
&\qquad\quad + \Delta(x_1-y)\Delta(x_1'-y)\Delta(x_2-z)\Delta(x_2'-z) \\
&\qquad\quad + \Delta(x_1-y)\Delta(x_2'-y)\Delta(x_2-z)\Delta(x_1'-z)\Big] \\
&\qquad\quad + O(g^4)\,.
\end{aligned} \tag{10.9}$$

Next, we use eq. (10.9) in the LSZ formula, eq. (10.5). Each Klein–Gordon wave operator acts on a propagator to give

$$(-\partial_i^2 + m^2)\Delta(x_i - y) = \delta^4(x_i - y)\ . \qquad (10.10)$$

The integrals over the external spacetime labels $x_{1,2,1',2'}$ are then trivial, and we get

$$\begin{aligned}\langle f|i\rangle = (ig)^2\left(\tfrac{1}{i}\right)\int d^4y\, d^4z\,\Delta(y-z)\Big[\, & e^{i(k_1y+k_2y-k_1'z-k_2'z)} \\ & + e^{i(k_1y+k_2z-k_1'y-k_2'z)} \\ & + e^{i(k_1y+k_2z-k_1'z-k_2'y)}\Big] + O(g^4)\ . \qquad (10.11)\end{aligned}$$

This can be simplified by substituting

$$\Delta(y-z) = \int \frac{d^4k}{(2\pi)^4}\,\frac{e^{ik(y-z)}}{k^2+m^2-i\epsilon} \qquad (10.12)$$

into eq. (10.9). Then the spacetime arguments appear only in phase factors, and we can integrate them to get delta functions:

$$\begin{aligned}\langle f|i\rangle &= ig^2\int\frac{d^4k}{(2\pi)^4}\,\frac{1}{k^2+m^2-i\epsilon} \\ &\qquad\times\Big[(2\pi)^4\delta^4(k_1{+}k_2{+}k)\,(2\pi)^4\delta^4(k_1'{+}k_2'{+}k) \\ &\qquad + (2\pi)^4\delta^4(k_1{-}k_1'{+}k)\,(2\pi)^4\delta^4(k_2'{-}k_2{+}k) \\ &\qquad + (2\pi)^4\delta^4(k_1{-}k_2'{+}k)\,(2\pi)^4\delta^4(k_1'{-}k_2{+}k)\Big] + O(g^4) \\ &= ig^2\,(2\pi)^4\delta^4(k_1{+}k_2{-}k_1'{-}k_2') \\ &\qquad\times\left[\frac{1}{(k_1{+}k_2)^2+m^2}+\frac{1}{(k_1{-}k_1')^2+m^2}+\frac{1}{(k_1{-}k_2')^2+m^2}\right] \\ &\qquad + O(g^4)\ . \qquad (10.13)\end{aligned}$$

In eq. (10.13), we have left out the $i\epsilon$s for notational convenience only; m^2 is really $m^2 - i\epsilon$. The overall delta function in eq. (10.13) tells that that four-momentum is conserved in the scattering process, which we should, of course, expect. For a general scattering process, it is then convenient to define a scattering matrix element $\mathcal{T}$ via

$$\langle f|i\rangle = (2\pi)^4\delta^4(k_{\rm in}{-}k_{\rm out})\,i\mathcal{T}\ , \qquad (10.14)$$

where $k_{\rm in}$ and $k_{\rm out}$ are the total four-momenta of the incoming and outgoing particles, respectively.

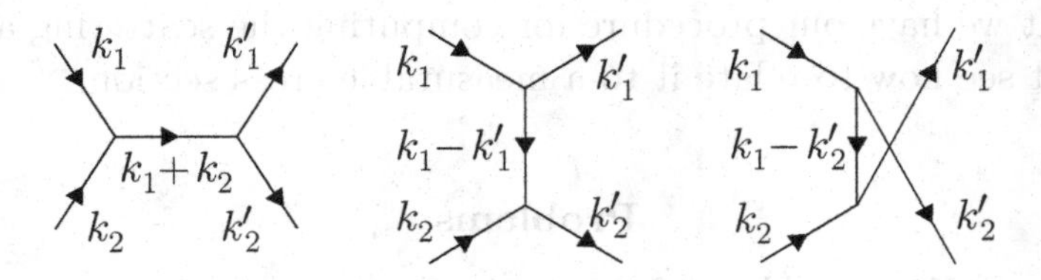

Figure 10.2. The tree-level s-, t-, and u-channel diagrams contributing to $i\mathcal{T}$ for two-particle scattering.

Examining the calculation which led to eq. (10.13), we can take away some universal features that lead to a simple set of *Feynman rules* for computing contributions to $i\mathcal{T}$ for a given scattering process. The Feynman rules are given below.

1) Draw lines (called *external lines*) for each incoming and each outgoing particle.
2) Leave one end of each external line free, and attach the other to a vertex at which exactly three lines meet. Include extra *internal lines* in order to do this. In this way, draw all possible diagrams that are *topologically inequivalent.*
3) On each incoming line, draw an arrow pointing towards the vertex. On each outgoing line, draw an arrow pointing away from the vertex. On each internal line, draw an arrow with an arbitrary direction.
4) Assign each line its own four-momentum. The four-momentum of an external line should be the four-momentum of the corresponding particle.
5) Think of the four-momenta as flowing along the arrows, and conserve four-momentum at each vertex. For a tree diagram, this fixes the momenta on all the internal lines.
6) The value of a diagram consists of the following factors:
   for each external line, 1;
   for each internal line with momentum k, $-i/(k^2+m^2-i\epsilon)$;
   for each vertex, $iZ_g g$.
7) A diagram with L closed loops will have L internal momenta that are not fixed by rule no. 5. Integrate over each of these momenta ℓ_i with measure $d^4\ell_i/(2\pi)^4$.
8) A loop diagram may have some left-over symmetry factors if there are exchanges of *internal* propagators and vertices that leave the diagram unchanged; in this case, divide the value of the diagram by the symmetry factor associated with exchanges of internal propagators and vertices.
9) Include diagrams with the *counterterm vertex* that connects two propagators, each with the same four-momentum k. The value of this vertex is $-i(Ak^2+Bm^2)$, where $A=Z_\varphi-1$ and $B=Z_m-1$, and each is $O(g^2)$.
10) The value of $i\mathcal{T}$ is given by a sum over the values of all these diagrams.

For the two-particle scattering process, the tree diagrams resulting from these rules are shown in fig. 10.2.

Now that we have our procedure for computing the scattering amplitude $\mathcal{T}$, we must see how to relate it to a measurable cross section.

Problems

10.1) Use eq. (9.41) of problem 9.5 to rederive eq. (10.9).

10.2) Write down the Feynman rules for the complex scalar field of problem 9.3. Remember that there are two kinds of particles now (which we can think of as positively and negatively charged), and that your rules must have a way of distinguishing them. Hint: the most direct approach requires two kinds of arrows: momentum arrows (as discussed in this section) and what we might call "charge" arrows (as discussed in problem 9.3). Try to find a more elegant approach that requires only one kind of arrow.

10.3) Consider a complex scalar field φ that interacts with a real scalar field χ via $\mathcal{L}_1 = g\chi\varphi^\dagger\varphi$. Use a solid line for the φ propagator and a dashed line for the χ propagator. Draw the vertex (remember the arrows!), and find the associated vertex factor.

10.4) Consider a real scalar field with $\mathcal{L}_1 = \frac{1}{2}g\varphi\partial^\mu\varphi\partial_\mu\varphi$. Find the associated vertex factor.

10.5) The scattering amplitudes should be unchanged if we make a *field redefinition*. Suppose, for example, we have

$$\mathcal{L} = -\tfrac{1}{2}\partial^\mu\varphi\partial_\mu\varphi - \tfrac{1}{2}m^2\varphi^2 \,, \tag{10.15}$$

and we make the field redefinition

$$\varphi \to \varphi + \lambda\varphi^2 \,. \tag{10.16}$$

Work out the lagrangian in terms of the redefined field, and the corresponding Feynman rules. Compute (at tree level) the $\varphi\varphi \to \varphi\varphi$ scattering amplitude. You should get zero, because this is a free-field theory in disguise. (At the loop level, we also have to take into account the transformation of the functional measure $\mathcal{D}\varphi$; see section 85.)

11

Cross sections and decay rates

Prerequisite: 10

Now that we have a method for computing the scattering amplitude $\mathcal{T}$, we must convert it into something that could be measured in an experiment.

In practice, we are almost always concerned with one of two generic cases: one incoming particle, for which we compute a *decay rate*, or two incoming particles, for which we compute a *cross section*. We begin with the latter.

Let us also specialize, for now, to the case of two outgoing particles as well as two incoming particles. In φ^3 theory, we found in section 10 that in this case we have

$$\mathcal{T} = g^2\left[\frac{1}{(k_1{+}k_2)^2+m^2} + \frac{1}{(k_1{-}k_1')^2+m^2} + \frac{1}{(k_1{-}k_2')^2+m^2}\right] + O(g^4)\ , \tag{11.1}$$

where k_1 and k_2 are the four-momenta of the two incoming particles, k_1' and k_2' are the four-momenta of the two outgoing particles, and $k_1 + k_2 = k_1' + k_2'$. Also, these particles are all *on shell*: $k_i^2 = -m_i^2$. (Here, for later use, we allow for the possibility that the particles all have different masses.)

Let us think about the kinematics of this process. In the *center-of-mass frame*, or *CM frame* for short, we take $\mathbf{k}_1 + \mathbf{k}_2 = \mathbf{0}$, and choose $\mathbf{k}_1$ to be in the $+z$ direction. Now the only variable left to specify about the initial state is the magnitude of $\mathbf{k}_1$. Equivalently, we could specify the total energy in the CM frame, $E_1 + E_2$. However, it is even more convenient to define a Lorentz scalar $s \equiv -(k_1 + k_2)^2$. In the CM frame, s reduces to $(E_1 + E_2)^2$; s is therefore called the *center-of-mass energy squared*. Then, since $E_1 = (\mathbf{k}_1^2 + m_1^2)^{1/2}$ and $E_2 = (\mathbf{k}_1^2 + m_2^2)^{1/2}$, we can solve for $|\mathbf{k}_1|$ in terms of s, with the result

$$|\mathbf{k}_1| = \frac{1}{2\sqrt{s}}\sqrt{s^2 - 2(m_1^2+m_2^2)s + (m_1^2-m_2^2)^2} \quad \text{(CM frame)}\ . \tag{11.2}$$

Now consider the two outgoing particles. Since momentum is conserved, we must have $\mathbf{k}_1' + \mathbf{k}_2' = \mathbf{0}$, and since energy is conserved, we must also have $(E_1' + E_2')^2 = s$. Then we find

$$|\mathbf{k}_1'| = \frac{1}{2\sqrt{s}}\sqrt{s^2 - 2(m_{1'}^2 + m_{2'}^2)s + (m_{1'}^2 - m_{2'}^2)^2} \quad \text{(CM frame)} . \qquad (11.3)$$

Now the only variable left to specify about the final state is the angle θ between $\mathbf{k}_1$ and $\mathbf{k}_1'$. However, it is often more convenient to work with the Lorentz scalar $t \equiv -(k_1 - k_1')^2$, which is related to θ by

$$t = m_1^2 + m_{1'}^2 - 2E_1E_1' + 2|\mathbf{k}_1||\mathbf{k}_1'|\cos\theta . \qquad (11.4)$$

This formula is valid in any frame.

The Lorentz scalars s and t are two of the three *Mandelstam variables*, defined as

$$\begin{aligned} s &\equiv -(k_1{+}k_2)^2 = -(k_1'{+}k_2')^2 , \\ t &\equiv -(k_1{-}k_1')^2 = -(k_2{-}k_2')^2 , \\ u &\equiv -(k_1{-}k_2')^2 = -(k_2{-}k_1')^2 . \end{aligned} \qquad (11.5)$$

The three Mandelstam variables are not independent; they satisfy the linear relation

$$s + t + u = m_1^2 + m_2^2 + m_{1'}^2 + m_{2'}^2 . \qquad (11.6)$$

In terms of s, t, and u, we can rewrite eq. (11.1) as

$$\mathcal{T} = g^2\left[\frac{1}{m^2 - s} + \frac{1}{m^2 - t} + \frac{1}{m^2 - u}\right] + O(g^4) , \qquad (11.7)$$

which demonstrates the notational utility of the Mandelstam variables.

Now let us consider a different frame, the *fixed target* or *FT frame* (also sometimes called the *lab frame*), in which particle no. 2 is initially at rest: $\mathbf{k}_2 = \mathbf{0}$. In this case we have

$$|\mathbf{k}_1| = \frac{1}{2m_2}\sqrt{s^2 - 2(m_1^2 + m_2^2)s + (m_1^2 - m_2^2)^2} \quad \text{(FT frame)} . \qquad (11.8)$$

Note that, from eqs. (11.8) and (11.2),

$$m_2|\mathbf{k}_1|_{\text{FT}} = \sqrt{s}\,|\mathbf{k}_1|_{\text{CM}} . \qquad (11.9)$$

This will be useful later.

We would now like to derive a formula for the differential scattering cross section. In order to do so, we assume that the whole experiment is taking place in a big box of volume V, and lasts for a large time T. We should really think about wave packets coming together, but we will use some simple shortcuts instead. Also, to get a more general answer, we will let the number of outgoing particles be arbitrary.

Recall from section 10 that the overlap between the initial and final states is given by

$$\langle f|i\rangle = (2\pi)^4\delta^4(k_{\rm in}-k_{\rm out})i\mathcal{T} \; . \tag{11.10}$$

To get a probability, we must square $\langle f|i\rangle$, and divide by the norms of the initial and final states:

$$P = \frac{|\langle f|i\rangle|^2}{\langle f|f\rangle\langle i|i\rangle} \; . \tag{11.11}$$

The numerator of this expression is

$$|\langle f|i\rangle|^2 = [(2\pi)^4\delta^4(k_{\rm in}-k_{\rm out})]^2\,|\mathcal{T}|^2 \, . \tag{11.12}$$

We write the square of the delta function as

$$[(2\pi)^4\delta^4(k_{\rm in}-k_{\rm out})]^2 = (2\pi)^4\delta^4(k_{\rm in}-k_{\rm out}) \times (2\pi)^4\delta^4(0) \; , \tag{11.13}$$

and note that

$$(2\pi)^4\delta^4(0) = \int d^4x\, e^{i0\cdot x} = VT \; . \tag{11.14}$$

Also, the norm of a single particle state is given by

$$\langle k|k\rangle = (2\pi)^3 2k^0\delta^3(\mathbf{0}) = 2k^0 V \; . \tag{11.15}$$

Thus we have

$$\langle i|i\rangle = 4E_1E_2V^2 \; , \tag{11.16}$$

$$\langle f|f\rangle = \prod_{j=1}^{n'} 2k_j'^0 V \; , \tag{11.17}$$

where n' is the number of outgoing particles.

If we now divide eq. (11.11) by the elapsed time T, we get a probability per unit time

$$\dot{P} = \frac{(2\pi)^4\delta^4(k_{\rm in}-k_{\rm out})\, V\, |\mathcal{T}|^2}{4E_1E_2V^2\prod_{j=1}^{n'} 2k_j'^0 V} \; . \tag{11.18}$$

This is the probability per unit time to scatter into a set of outgoing particles with precise momenta. To get something measurable, we should sum each outgoing three-momentum $\mathbf{k}_j'$ over some small range. Due to the box, all three-momenta are quantized: $\mathbf{k}_j' = (2\pi/L)\mathbf{n}_j'$, where $V = L^3$, and $\mathbf{n}_j'$ is a three-vector with integer entries. (Here we have assumed periodic boundary conditions, but this choice does not affect the final result.) In the limit of large L, we have

$$\sum_{\mathbf{n}'_j} \to \frac{V}{(2\pi)^3} \int d^3k'_j \,. \tag{11.19}$$

Thus we should multiply $\dot{P}$ by a factor of $Vd^3k'_j/(2\pi)^3$ for each outgoing particle. Then we get

$$\dot{P} = \frac{(2\pi)^4\delta^4(k_{\text{in}}-k_{\text{out}})}{4E_1E_2V} |\mathcal{T}|^2 \prod_{j=1}^{n'} \widetilde{dk}'_j \,, \tag{11.20}$$

where we have identified the Lorentz-invariant phase-space differential

$$\widetilde{dk} \equiv \frac{d^3k}{(2\pi)^3 2k^0} \tag{11.21}$$

that we first introduced in section 3.

To convert $\dot{P}$ to a differential cross section $d\sigma$, we must divide by the incident flux. Let us see how this works in the FT frame, where particle no. 2 is at rest. The incident flux is the number of particles per unit volume that are striking the target particle (no. 2) times their speed. We have one incident particle (no. 1) in a volume V with speed $v = |\mathbf{k}_1|/E_1$, and so the incident flux is $|\mathbf{k}_1|/E_1V$. Dividing eq. (11.20) by this flux cancels the last factor of V, and replaces E_1 in the denominator with $|\mathbf{k}_1|$. We also set $E_2 = m_2$ and note that eq. (11.8) gives $|\mathbf{k}_1|m_2$ as a function of s; $d\sigma$ will be Lorentz invariant if, in other frames, we simply use this function as the value of $|\mathbf{k}_1|m_2$. Adopting this convention, and using eq. (11.9), we have

$$d\sigma = \frac{1}{4|\mathbf{k}_1|_{\text{CM}}\sqrt{s}} |\mathcal{T}|^2 \, d\text{LIPS}_{n'}(k_1+k_2) \,, \tag{11.22}$$

where $|\mathbf{k}_1|_{\text{CM}}$ is given as a function of s by eq. (11.2), and we have defined the n'-body Lorentz-invariant phase-space measure

$$d\text{LIPS}_{n'}(k) \equiv (2\pi)^4\delta^4(k-\textstyle\sum_{i=1}^{n'} k'_i) \displaystyle\prod_{j=1}^{n'} \widetilde{dk}'_j \,. \tag{11.23}$$

Eq. (11.22) is our final result for the differential cross section for the scattering of two incoming particles into n' outgoing particles.

Let us now specialize to the case of two outgoing particles. We need to evaluate

$$d\text{LIPS}_2(k) = (2\pi)^4 \delta^4(k - k_1' - k_2')\, \widetilde{dk}_1' \widetilde{dk}_2' \,, \tag{11.24}$$

where $k = k_1 + k_2$. Since $d\text{LIPS}_2(k)$ is Lorentz invariant, we can compute it in any convenient frame. Let us work in the CM frame, where $\mathbf{k} = \mathbf{k}_1 + \mathbf{k}_2 = \mathbf{0}$ and $k^0 = E_1 + E_2 = \sqrt{s}$; then we have

$$d\text{LIPS}_2(k) = \frac{1}{4(2\pi)^2 E_1' E_2'}\, \delta(E_1' + E_2' - \sqrt{s}\,)\, \delta^3(\mathbf{k}_1' + \mathbf{k}_2')\, d^3k_1' d^3k_2' \,. \tag{11.25}$$

We can use the spatial part of the delta function to integrate over d^3k_2', with the result

$$d\text{LIPS}_2(k) = \frac{1}{4(2\pi)^2 E_1' E_2'}\, \delta(E_1' + E_2' - \sqrt{s}\,)\, d^3k_1' \,, \tag{11.26}$$

where now

$$E_1' = \sqrt{\mathbf{k}_1'^{\,2} + m_{1'}^2} \quad \text{and} \quad E_2' = \sqrt{\mathbf{k}_1'^{\,2} + m_{2'}^2} \,. \tag{11.27}$$

Next, let us write

$$d^3k_1' = |\mathbf{k}_1'|^2\, d|\mathbf{k}_1'|\, d\Omega_{\text{CM}} \,, \tag{11.28}$$

where $d\Omega_{\text{CM}} = \sin\theta\, d\theta\, d\phi$ is the differential solid angle, and θ is the angle between $\mathbf{k}_1$ and $\mathbf{k}_1'$ in the CM frame. We can carry out the integral over the magnitude of $\mathbf{k}_1'$ in eq. (11.26) using $\int dx\, \delta(f(x)) = \sum_i |f'(x_i)|^{-1}$, where x_i satisfies $f(x_i) = 0$. In our case, the argument of the delta function vanishes at just one value of $|\mathbf{k}_1'|$, the value given by eq. (11.3). Also, the derivative of that argument with respect to $|\mathbf{k}_1'|$ is

$$\begin{aligned} \frac{\partial}{\partial |\mathbf{k}_1'|}\left(E_1' + E_2' - \sqrt{s}\right) &= \frac{|\mathbf{k}_1'|}{E_1'} + \frac{|\mathbf{k}_1'|}{E_2'} \\ &= |\mathbf{k}_1'| \left(\frac{E_1' + E_2'}{E_1'\, E_2'}\right) \\ &= \frac{|\mathbf{k}_1'|\sqrt{s}}{E_1' E_2'} \,. \end{aligned} \tag{11.29}$$

Putting all of this together, we get

$$d\text{LIPS}_2(k) = \frac{|\mathbf{k}_1'|}{16\pi^2 \sqrt{s}}\, d\Omega_{\text{CM}} \,. \tag{11.30}$$

Combining this with eq. (11.22), we have

$$\frac{d\sigma}{d\Omega_{\rm CM}} = \frac{1}{64\pi^2 s} \frac{|\mathbf{k}_1'|}{|\mathbf{k}_1|} |\mathcal{T}|^2 \,, \tag{11.31}$$

where $|\mathbf{k}_1|$ and $|\mathbf{k}_1'|$ are the functions of s given by eqs. (11.2) and (11.3), and $d\Omega_{\rm CM}$ is the differential solid angle in the CM frame.

The differential cross section can also be expressed in a frame-independent manner by noting that, in the CM frame, we can take the differential of eq. (11.4) at fixed s to get

$$dt = 2\,|\mathbf{k}_1|\,|\mathbf{k}_1'|\,d\cos\theta \tag{11.32}$$

$$= 2\,|\mathbf{k}_1|\,|\mathbf{k}_1'|\,\frac{d\Omega_{\rm CM}}{2\pi} \,. \tag{11.33}$$

Now we can rewrite eq. (11.31) as

$$\frac{d\sigma}{dt} = \frac{1}{64\pi s|\mathbf{k}_1|^2} |\mathcal{T}|^2 \,, \tag{11.34}$$

where $|\mathbf{k}_1|$ is given as a function of s by eq. (11.2).

We can now transform $d\sigma/dt$ into $d\sigma/d\Omega$ in any frame we might like (such as the FT frame) by taking the differential of eq. (11.4) in that frame. In general, though, $|\mathbf{k}_1'|$ depends on θ as well as s, so the result is more complicated than it is in eq. (11.32) for the CM frame.

Returning to the general case of n' outgoing particles, we can define a Lorentz invariant *total cross section* by integrating completely over all the outgoing momenta, and dividing by an appropriate *symmetry factor* S. If there are n_i' identical outgoing particles of type i, then

$$S = \prod_i n_i'! \,, \tag{11.35}$$

and

$$\sigma = \frac{1}{S} \int d\sigma \,, \tag{11.36}$$

where $d\sigma$ is given by eq. (11.22). We need the symmetry factor because merely integrating over all the outgoing momenta in $d\text{LIPS}_{n'}$ treats the final state as being labeled by an *ordered* list of these momenta. But if some outgoing particles are identical, this is not correct; the momenta of the identical particles should be specified by an *unordered* list (because, for example, the state $a_1^\dagger a_2^\dagger|0\rangle$ is identical to the state $a_2^\dagger a_1^\dagger|0\rangle$). The symmetry factor provides the appropriate correction.

In the case of two outgoing particles, eq. (11.36) becomes

$$\sigma = \frac{1}{S}\int d\Omega_{\rm CM}\,\frac{d\sigma}{d\Omega_{\rm CM}} \tag{11.37}$$

$$= \frac{2\pi}{S}\int_{-1}^{+1} d\cos\theta\,\frac{d\sigma}{d\Omega_{\rm CM}}\;, \tag{11.38}$$

where $S = 2$ if the two outgoing particles are identical, and $S = 1$ if they are distinguishable. Equivalently, we can compute σ from eq. (11.34) via

$$\sigma = \frac{1}{S}\int_{t_{\rm min}}^{t_{\rm max}} dt\,\frac{d\sigma}{dt}\;, \tag{11.39}$$

where $t_{\rm min}$ and $t_{\rm max}$ are given by eq. (11.4) in the CM frame with $\cos\theta = -1$ and $+1$, respectively. To compute σ with eq. (11.38), we should first express t and u in terms of s and θ via eqs. (11.4) and (11.6), and then integrate over θ at fixed s. To compute σ with eq. (11.39), we should first express u in terms of s and t via eq. (11.6), and then integrate over t at fixed s.

Let us see how all this works for the scattering amplitude of φ^3 theory, eq. (11.7). In this case, all the masses are equal, and so, in the CM frame, $E = \frac{1}{2}\sqrt{s}$ for all four particles, and $|\mathbf{k}_1'| = |\mathbf{k}_1| = \frac{1}{2}(s-4m^2)^{1/2}$. Then eq. (11.4) becomes

$$t = -\tfrac{1}{2}(s-4m^2)(1-\cos\theta)\;. \tag{11.40}$$

From eq. (11.6), we also have

$$u = -\tfrac{1}{2}(s-4m^2)(1+\cos\theta)\;. \tag{11.41}$$

Thus $|\mathcal{T}|^2$ is quite a complicated function of s and θ. In the nonrelativistic limit, $|\mathbf{k}_1| \ll m$ or equivalently $s - 4m^2 \ll m^2$, we have

$$\mathcal{T} = \frac{5g^2}{3m^2}\left[1 - \frac{8}{15}\left(\frac{s-4m^2}{m^2}\right) + \frac{5}{18}\left(1+\frac{27}{25}\cos^2\theta\right)\left(\frac{s-4m^2}{m^2}\right)^2 + \ldots\right] + O(g^4)\;. \tag{11.42}$$

Thus the differential cross section is nearly isotropic. In the extreme relativistic limit, $|\mathbf{k}_1| \gg m$ or equivalently $s \gg m^2$, we have

$$\mathcal{T} = \frac{g^2}{s\sin^2\theta}\left[3+\cos^2\theta - \left(\frac{(3+\cos^2\theta)^2}{\sin^2\theta} - 16\right)\frac{m^2}{s} + \ldots\right] + O(g^4)\;. \tag{11.43}$$

Now the differential cross section is sharply peaked in the forward ($\theta = 0$) and backward ($\theta = \pi$) directions.

We can compute the total cross section σ from eq. (11.39). We have in this case $t_{\rm min} = -(s-4m^2)$ and $t_{\rm max} = 0$. Since the two outgoing particles are identical, the symmetry factor is $S = 2$. Then setting $u = 4m^2 - s - t$, and performing the integral in eq. (11.39) over t at fixed s, we get

$$\sigma = \frac{g^4}{32\pi s(s-4m^2)}\Biggl[\frac{2}{m^2} + \frac{s-4m^2}{(s-m^2)^2} - \frac{2}{s-3m^2} + \frac{4m^2}{(s-m^2)(s-2m^2)}\ln\left(\frac{s-3m^2}{m^2}\right)\Biggr] + O(g^6)\,. \tag{11.44}$$

In the nonrelativistic limit, this becomes

$$\sigma = \frac{25g^4}{1152\pi m^6}\left[1 - \frac{79}{60}\left(\frac{s-4m^2}{m^2}\right) + \ldots\right] + O(g^6)\,. \tag{11.45}$$

In the extreme relativistic limit, we get

$$\sigma = \frac{g^4}{16\pi m^2 s^2}\left[1 + \frac{7}{2}\frac{m^2}{s} + \ldots\right] + O(g^6)\,. \tag{11.46}$$

These results illustrate how even a very simple quantum field theory can yield specific predictions for cross sections that could be tested experimentally.

Let us now turn to the other basic problem mentioned at the beginning of this section: the case of a single incoming particle that *decays* to n' other particles.

We have an immediate conceptual problem. According to our development of the LSZ formula in section 5, each incoming and outgoing particle should correspond to a single-particle state that is an exact eigenstate of the exact hamiltonian. This is clearly not the case for a particle that can decay. Referring to fig. 5.1, the hyperbola of such a particle must lie above the continuum threshold. Strictly speaking, then, the LSZ formula is not applicable.

A proper understanding of this issue requires a study of loop corrections that we will undertake in section 25. For now, we will simply assume that the LSZ formula continues to hold for a single incoming particle. Then we can retrace the steps from eq. (11.11) to eq. (11.20); the only change is that the norm of the initial state is now

$$\langle i|i\rangle = 2E_1 V \tag{11.47}$$

instead of eq. (11.16). Identifying the differential decay rate $d\Gamma$ with $\dot{P}$ then gives

$$d\Gamma = \frac{1}{2E_1}\,|\mathcal{T}|^2\,d\mathrm{LIPS}_{n'}(k_1)\,, \tag{11.48}$$

where now $s = -k_1^2 = m_1^2$. In the CM frame (which is now the rest frame of the initial particle), we have $E_1 = m_1$; in other frames, the relative factor of E_1/m_1 in $d\Gamma$ accounts for relativistic time dilation of the decay rate.

We can also define a total decay rate by integrating over all the outgoing momenta, and dividing by the symmetry factor of eq. (11.35):

$$\Gamma = \frac{1}{S}\int d\Gamma\,. \tag{11.49}$$

We will compute a decay rate in problem 11.1.

Reference notes

For a derivation with wave packets, see *Brown*, *Itzykson & Zuber*, or *Peskin & Schroeder*.

Problems

11.1) a) Consider a theory of two real scalar fields A and B with an interaction $\mathcal{L}_1 = gAB^2$. Assuming that $m_A > 2m_B$, compute the total decay rate of the A particle at tree level.

b) Consider a theory of a real scalar field φ and a complex scalar field χ with $\mathcal{L}_1 = g\varphi\chi^\dagger\chi$. Assuming that $m_\varphi > 2m_\chi$, compute the total decay rate of the φ particle at tree level.

11.2) Consider *Compton scattering*, in which a massless photon is scattered by an electron, initially at rest. (This is the FT frame.) In problem 59.1, we will compute $|\mathcal{T}|^2$ for this process (summed over the possible spin states of the scattered photon and electron, and averaged over the possible spin states of the initial photon and electron), with the result

$$|\mathcal{T}|^2 = 32\pi^2\alpha^2\left[\frac{m^4+m^2(3s+u)-su}{(m^2-s)^2}+\frac{m^4+m^2(3u+s)-su}{(m^2-u)^2}\right.$$
$$\left.+\frac{2m^2(s+u+2m^2)}{(m^2-s)(m^2-u)}\right]+O(\alpha^4)\,, \tag{11.50}$$

where $\alpha = 1/137.036$ is the fine-structure constant.

a) Express the Mandelstam variables s and u in terms of the initial and final photon energies ω and ω'.

b) Express the scattering angle $\theta_{\rm FT}$ between the initial and final photon three-momenta in terms of ω and ω'.

c) Express the differential scattering cross section $d\sigma/d\Omega_{\rm FT}$ in terms of ω and ω'. Show that your result is equivalent to the *Klein–Nishina formula*

$$\frac{d\sigma}{d\Omega_{\rm FT}} = \frac{\alpha^2}{2m^2}\frac{\omega'^2}{\omega^2}\left[\frac{\omega}{\omega'} + \frac{\omega'}{\omega} - \sin^2\theta_{\rm FT}\right] . \tag{11.51}$$

11.3) Consider the process of *muon decay*, $\mu^- \to e^-\overline{\nu}_e\nu_\mu$. In section 88, we will compute $|\mathcal{T}|^2$ for this process (summed over the possible spin states of the decay products, and averaged over the possible spin states of the initial muon), with the result

$$|\mathcal{T}|^2 = 64G_{\rm F}^2(k_1\cdot k_2')(k_1'\cdot k_3') , \tag{11.52}$$

where $G_{\rm F}$ is the *Fermi constant*, k_1 is the four-momentum of the muon, and $k'_{1,2,3}$ are the four-momenta of the $\overline{\nu}_e$, ν_μ, and e^-, respectively. In the rest frame of the muon, its decay rate is therefore

$$\Gamma = \frac{32G_{\rm F}^2}{m}\int (k_1\cdot k_2')(k_1'\cdot k_3')\; d\text{LIPS}_3(k_1) , \tag{11.53}$$

where $k_1 = (m, \mathbf{0})$, and m is the muon mass. The neutrinos are massless, and the electron mass is 200 times less than the muon mass, so we can take the electron to be massless as well. To evaluate Γ, we perform the following analysis.

a) Show that

$$\Gamma = \frac{32G_{\rm F}^2}{m}\int \widetilde{dk_3'}\; k_{1\mu}k'_{3\nu}\int k_2'^{\mu}k_1'^{\nu}\; d\text{LIPS}_2(k_1{-}k_3') . \tag{11.54}$$

b) Use Lorentz invariance to argue that, for $m_{1'} = m_{2'} = 0$,

$$\int k_1'^{\mu}k_2'^{\nu}\; d\text{LIPS}_2(k) = Ak^2g^{\mu\nu} + Bk^\mu k^\nu , \tag{11.55}$$

where A and B are numerical constants.

c) Show that, for $m_{1'} = m_{2'} = 0$,

$$\int d\text{LIPS}_2(k) = \frac{1}{8\pi} . \tag{11.56}$$

d) By contracting both sides of eq. (11.55) with $g_{\mu\nu}$ and with $k_\mu k_\nu$, and using eq. (11.56), evaluate A and B.

e) Use the results of parts (b) and (d) in eq. (11.54). Set $k_1 = (m, \mathbf{0})$, and compute $d\Gamma/dE_e$; here $E_e \equiv E_3'$ is the energy of the electron. Note that the maximum value of E_e is reached when the electron is emitted in one direction, and the two neutrinos in the opposite direction; what is this maximum value?

f) Perform the integral over E_e to obtain the muon decay rate Γ.

g) The measured lifetime of the muon is 2.197×10^{-6} s. The muon mass is 105.66 MeV. Determine the value of G_F in GeV^{-2}. (Your answer is too low by about 0.2%, due to loop corrections to the decay rate.)

h) Define the *energy spectrum of the electron* $P(E_e) \equiv \Gamma^{-1}\, d\Gamma/dE_e$. Note that $P(E_e)\, dE_e$ is the probability for the electron to be emitted with energy between E_e and $E_e + dE_e$. Draw a graph of $P(E_e)$ versus E_e/m_μ.

11.4) Consider a theory of three real scalar fields (A, B, and C) with

$$\begin{aligned}\mathcal{L} = &-\tfrac{1}{2}\partial^\mu A \partial_\mu A - \tfrac{1}{2}m_A^2 A^2 \\ &-\tfrac{1}{2}\partial^\mu B \partial_\mu B - \tfrac{1}{2}m_B^2 B^2 \\ &-\tfrac{1}{2}\partial^\mu C \partial_\mu C - \tfrac{1}{2}m_C^2 C^2 \\ &+ gABC\,. \end{aligned} \tag{11.57}$$

Write down the tree-level scattering amplitude (given by the sum of the contributing tree diagrams) for each of the following processes:

$$\begin{aligned} AA &\to AA\,, \\ AA &\to AB\,, \\ AA &\to BB\,, \\ AA &\to BC\,, \\ AB &\to AB\,, \\ AB &\to AC\,. \end{aligned} \tag{11.58}$$

Your answers should take the form

$$\mathcal{T} = g^2 \left[\frac{c_s}{m_s^2 - s} + \frac{c_t}{m_t^2 - t} + \frac{c_u}{m_u^2 - u} \right], \tag{11.59}$$

where, in each case, each c_i is a positive integer, and each m_i^2 is m_A^2 or m_B^2 or m_C^2. Hint: $\mathcal{T}$ may be zero for some processes.

12

Dimensional analysis with $\hbar = c = 1$

Prerequisite: 3

We have set $\hbar = c = 1$. This allows us to convert a time T to a length L via $T = c^{-1}L$, and a length L to an inverse mass M^{-1} via $L = \hbar c^{-1} M^{-1}$. Thus any quantity A can be thought of as having units of mass to some power (positive, negative, or zero) that we will call $[A]$. For example,

$$[m] = +1\,, \tag{12.1}$$

$$[x^\mu] = -1\,, \tag{12.2}$$

$$[\partial^\mu] = +1\,, \tag{12.3}$$

$$[d^d x] = -d\,. \tag{12.4}$$

In the last line, we have generalized our considerations to theories in d spacetime dimensions.

Let us now consider a scalar field in d spacetime dimensions with lagrangian density

$$\mathcal{L} = -\tfrac{1}{2}\partial^\mu\varphi\partial_\mu\varphi - \tfrac{1}{2}m^2\varphi^2 - \sum_{n=3}^{N} \tfrac{1}{n!} g_n \varphi^n\,. \tag{12.5}$$

The action is

$$S = \int d^d x\, \mathcal{L}\,, \tag{12.6}$$

and the path integral is

$$Z(J) = \int \mathcal{D}\varphi\, \exp\left[i \int d^d x\, (\mathcal{L} + J\varphi) \right]. \tag{12.7}$$

From eq. (12.7), we see that the action S must be dimensionless, because it appears as the argument of the exponential function. Therefore

$$[S] = 0 \,. \tag{12.8}$$

Combining eqs. (12.4) and (12.8) yields

$$[\mathcal{L}] = d \,. \tag{12.9}$$

Then, from eqs. (12.9) and (12.3), and the fact that $\partial^\mu\varphi\partial_\mu\varphi$ is a term in $\mathcal{L}$, we see that we must have

$$[\varphi] = \tfrac{1}{2}(d-2) \,. \tag{12.10}$$

Then, since $g_n\varphi^n$ is also a term in $\mathcal{L}$, we must have

$$[g_n] = d - \tfrac{1}{2}n(d-2) \,. \tag{12.11}$$

In particular, for the φ^3 theory we have been working with, we have

$$[g_3] = \tfrac{1}{2}(6-d) \,. \tag{12.12}$$

Thus we see that the coupling constant of φ^3 theory is dimensionless in $d = 6$ spacetime dimensions.

Theories with dimensionless couplings tend to be more interesting than theories with dimensionful couplings. This is because any nontrivial dependence of a scattering amplitude on a coupling must be expressed as a function of a dimensionless parameter. If the coupling is itself dimensionful, this parameter must be the ratio of the coupling to the appropriate power of either the particle mass m (if it is not zero) or, in the high-energy regime $s \gg m^2$, the Mandelstam variable s. Thus the relevant parameter is $g\, s^{-[g]/2}$. If $[g]$ is negative [and it usually is: see eq. (12.11)], then $g\, s^{-[g]/2}$ blows up at high energies, and the perturbative expansion breaks down. This behavior is connected to the *nonrenormalizability* of theories with couplings with negative mass dimension, a subject we will take up in section 18. It turns out that such theories require an infinite number of input parameters to make sense; see section 29. In the opposite case, $[g]$ positive, the theory becomes trivial at high energy, because $g\, s^{-[g]/2}$ goes rapidly to zero.

Thus the case of $[g] = 0$ is just right: scattering amplitudes can have a nontrivial dependence on g at all energies.

Therefore, from here on, we will be primarily interested in φ^3 theory in $d = 6$ spacetime dimensions, where $[g_3] = 0$.

Problems

12.1) Express $\hbar c$ in GeV fm, where 1 fm = 1 Fermi = 10^{-13} cm.

12.2) Express the masses of the proton, neutron, pion, electron, muon, and tau in GeV.

12.3) The proton is a strongly interacting blob of quarks and gluons. It has a nonzero *charge radius* r_p, given by $r_p^2 = \int d^3x\,\rho(r)r^2$, where $\rho(r)$ is the quantum expectation value of the electric charge distribution inside the proton. Estimate the value of r_p, and then look up its measured value. How accurate was your estimate?

13

The Lehmann–Källén form of the exact propagator

Prerequisite: 9

Before turning to the subject of loop corrections to scattering amplitudes, it will be helpful to consider what we can learn about the exact propagator $\mathbf{\Delta}(x-y)$ from general principles. We define the exact propagator via

$$\mathbf{\Delta}(x-y) \equiv i\langle 0|\mathrm{T}\varphi(x)\varphi(y)|0\rangle \; . \tag{13.1}$$

We take the field $\varphi(x)$ to be normalized so that

$$\langle 0|\varphi(x)|0\rangle = 0 \qquad \text{and} \qquad \langle k|\varphi(x)|0\rangle = e^{-ikx} \; . \tag{13.2}$$

In d spacetime dimensions, the one-particle state $|k\rangle$ has the normalization

$$\langle k|k'\rangle = (2\pi)^{d-1}\, 2\omega\, \delta^{d-1}(\mathbf{k}-\mathbf{k}') \; , \tag{13.3}$$

with $\omega = (\mathbf{k}^2+m^2)^{1/2}$. The corresponding completeness statement is

$$\int \widetilde{dk}\; |k\rangle\langle k| = I_1 \; , \tag{13.4}$$

where I_1 is the identity operator in the one-particle subspace, and

$$\widetilde{dk} \equiv \frac{d^{d-1}k}{(2\pi)^{d-1}2\omega} \tag{13.5}$$

is the Lorentz invariant phase-space differential. We also define the exact momentum-space propagator $\tilde{\mathbf{\Delta}}(k^2)$ via

$$\mathbf{\Delta}(x-y) \equiv \int \frac{d^dk}{(2\pi)^d}\, e^{ik(x-y)}\, \tilde{\mathbf{\Delta}}(k^2) \; . \tag{13.6}$$

In free-field theory, the momentum-space propagator is

$$\tilde{\Delta}(k^2) = \frac{1}{k^2+m^2-i\epsilon} \; . \tag{13.7}$$

It has an isolated pole at $k^2 = -m^2$ with residue one; m is the actual, physical mass of the particle, the mass that enters into the energy-momentum relation.

We begin our analysis with eq. (13.1). We take $x^0 > y^0$, and insert a complete set of energy eigenstates between the two fields. Recall from section 5 that there are three general classes of energy eigenstates.

1) The ground state or vacuum $|0\rangle$, which is a single state with zero energy and momentum.
2) The one particle states $|k\rangle$, specified by a three-momentum $\mathbf{k}$ and with energy $\omega = (\mathbf{k}^2 + m^2)^{1/2}$.
3) States in the multiparticle continuum $|k, n\rangle$, specified by a three-momentum $\mathbf{k}$ and other parameters (such as relative momenta among the different particles) that we will collectively denote as n. The energy of each of these states is $\omega = (\mathbf{k}^2 + M^2)^{1/2}$, where $M \geq 2m$; M is one of the parameters in the set n.

Thus we get

$$\begin{aligned}\langle 0|\varphi(x)\varphi(y)|0\rangle = \langle 0|\varphi(x)|0\rangle\langle 0|\varphi(y)|0\rangle \\ + \int \widetilde{dk}\, \langle 0|\varphi(x)|k\rangle\langle k|\varphi(y)|0\rangle \\ + \sum_n \int \widetilde{dk}\, \langle 0|\varphi(x)|k,n\rangle\langle k,n|\varphi(y)|0\rangle \,. \qquad (13.8)\end{aligned}$$

The sum over n is schematic, and includes integrals over continuous parameters like relative momenta.

The first two terms in eq. (13.8) can be simplified via eq. (13.2). Also, writing the field as $\varphi(x) = \exp(-iP^\mu x_\mu)\varphi(0)\exp(+iP^\mu x_\mu)$, where P^μ is the energy-momentum operator, gives us

$$\langle k,n|\varphi(x)|0\rangle = e^{-ikx}\langle k,n|\varphi(0)|0\rangle \,, \qquad (13.9)$$

where $k^0 = (\mathbf{k}^2 + M^2)^{1/2}$. We now have

$$\langle 0|\varphi(x)\varphi(y)|0\rangle = \int \widetilde{dk}\, e^{ik(x-y)} + \sum_n \int \widetilde{dk}\, e^{ik(x-y)} |\langle k,n|\varphi(0)|0\rangle|^2 \,. \qquad (13.10)$$

Next, we define the *spectral density*

$$\rho(s) \equiv \sum_n |\langle k,n|\varphi(0)|0\rangle|^2\, \delta(s - M^2) \,. \qquad (13.11)$$

Obviously, $\rho(s) \geq 0$ for $s \geq 4m^2$, and $\rho(s) = 0$ for $s < 4m^2$. Now we have

$$\langle 0|\varphi(x)\varphi(y)|0\rangle = \int \widetilde{dk}\, e^{ik(x-y)} + \int_{4m^2}^{\infty} ds\, \rho(s) \int \widetilde{dk}\, e^{ik(x-y)} \,. \qquad (13.12)$$

In the first term, $k^0 = (\mathbf{k}^2 + m^2)^{1/2}$, and in the second term, $k^0 = (\mathbf{k}^2 + s)^{1/2}$. Clearly we can also swap x and y to get

$$\langle 0|\varphi(y)\varphi(x)|0\rangle = \int \widetilde{dk}\, e^{-ik(x-y)} + \int_{4m^2}^{\infty} ds\, \rho(s) \int \widetilde{dk}\, e^{-ik(x-y)} \tag{13.13}$$

as well. We can then combine eqs. (13.12) and (13.13) into a formula for the time-ordered product

$$\langle 0|\mathrm{T}\varphi(x)\varphi(y)|0\rangle = \theta(x^0{-}y^0)\langle 0|\varphi(x)\varphi(y)|0\rangle + \theta(y^0{-}x^0)\langle 0|\varphi(y)\varphi(x)|0\rangle, \tag{13.14}$$

where $\theta(t)$ is the unit step function, by means of the identity

$$\int \frac{d^dk}{(2\pi)^d} \frac{e^{ik(x-y)}}{k^2 + m^2 - i\epsilon} = i\theta(x^0{-}y^0) \int \widetilde{dk}\, e^{ik(x-y)} + i\theta(y^0{-}x^0) \int \widetilde{dk}\, e^{-ik(x-y)}\ ; \tag{13.15}$$

the derivation of eq. (13.15) was sketched in section 8. Combining eqs. (13.12)–(13.15), we get

$$i\langle 0|\mathrm{T}\varphi(x)\varphi(y)|0\rangle = \int \frac{d^dk}{(2\pi)^d}\, e^{ik(x-y)} \left[\frac{1}{k^2 + m^2 - i\epsilon} + \int_{4m^2}^{\infty} ds\, \rho(s) \frac{1}{k^2 + s - i\epsilon} \right]. \tag{13.16}$$

Comparing eqs. (13.1), (13.6), and (13.16), we see that

$$\tilde{\mathbf{\Delta}}(k^2) = \frac{1}{k^2 + m^2 - i\epsilon} + \int_{4m^2}^{\infty} ds\, \rho(s) \frac{1}{k^2 + s - i\epsilon}\ . \tag{13.17}$$

This is the *Lehmann–Källén form* of the exact momentum-space propagator $\tilde{\mathbf{\Delta}}(k^2)$. We note in particular that *$\tilde{\mathbf{\Delta}}(k^2)$ has an isolated pole at $k^2 = -m^2$ with residue one*, just like the propagator in free-field theory.

Problems

13.1) Consider an interacting scalar field theory in d spacetime dimensions,

$$\mathcal{L} = -\tfrac{1}{2} Z_\varphi \partial^\mu\varphi\partial_\mu\varphi - \tfrac{1}{2} Z_m m^2\varphi^2 - \mathcal{L}_1(\varphi)\ , \tag{13.18}$$

where $\mathcal{L}_1(\varphi)$ is a function of φ (and not its derivatives). The exact momentum-space propagator for φ can be expressed in Lehmann–Källén form by eq. (13.17). Find a formula for the renormalizing factor Z_φ in terms of $\rho(s)$. Hint: consider the commutator $[\varphi(x), \dot{\varphi}(y)]$.

14

Loop corrections to the propagator

Prerequisites: 10, 12, 13

In section 10, we wrote the exact propagator as

$$\tfrac{1}{i}\mathbf{\Delta}(x_1{-}x_2) \equiv \langle 0|\mathrm{T}\varphi(x_1)\varphi(x_2)|0\rangle = \delta_1\delta_2 iW(J)\Big|_{J=0} , \tag{14.1}$$

where $iW(J)$ is the sum of connected diagrams, and δ_i acts to remove a source from a diagram and label the corresponding propagator endpoint x_i. In φ^3 theory, the $O(g^2)$ corrections to $\frac{1}{i}\mathbf{\Delta}(x_1{-}x_2)$ come from the diagrams of fig. 14.1. To compute them, it is simplest to work directly in momentum space, following the Feynman rules of section 10. An appropriate assignment of momenta to the lines is shown in fig. 14.1; we then have

$$\tfrac{1}{i}\tilde{\mathbf{\Delta}}(k^2) = \tfrac{1}{i}\tilde{\Delta}(k^2) + \tfrac{1}{i}\tilde{\Delta}(k^2)\Big[i\Pi(k^2)\Big]\tfrac{1}{i}\tilde{\Delta}(k^2) + O(g^4) , \tag{14.2}$$

where

$$\tilde{\Delta}(k^2) = \frac{1}{k^2 + m^2 - i\epsilon} \tag{14.3}$$

is the free-field propagator, and

$$\begin{aligned} i\Pi(k^2) = {} & \tfrac{1}{2}(ig)^2 \left(\tfrac{1}{i}\right)^2 \int \frac{d^d\ell}{(2\pi)^d}\, \tilde{\Delta}((\ell{+}k)^2)\tilde{\Delta}(\ell^2) \\ & - i(Ak^2 + Bm^2) + O(g^4) \end{aligned} \tag{14.4}$$

is the *self-energy*. Here we have written the integral appropriate for d spacetime dimensions; for now we will leave d arbitrary, but later we will want to focus on $d = 6$, where the coupling g is dimensionless.

In the first term in eq. (14.4), the factor of one-half is the symmetry factor associated with exchanging the top and bottom semicircular propagators. Also, we have written the vertex factor as ig rather than $iZ_g g$ because we expect $Z_g = 1 + O(g^2)$, and so the $Z_g - 1$ contribution can be lumped into

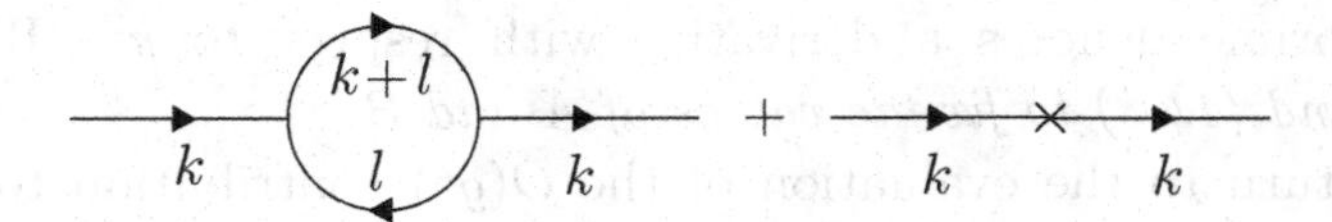

Figure 14.1. The $O(g^2)$ corrections to the propagator.

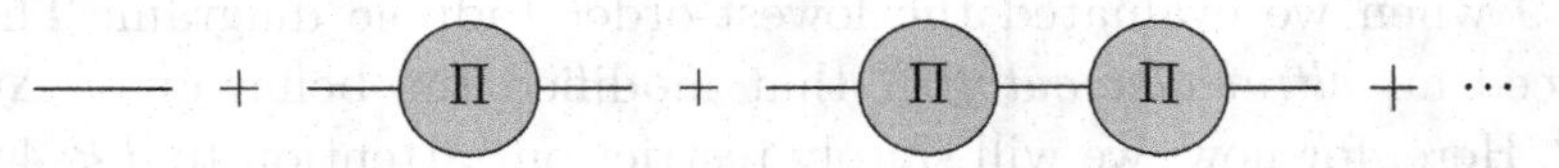

Figure 14.2. The geometric series for the exact propagator.

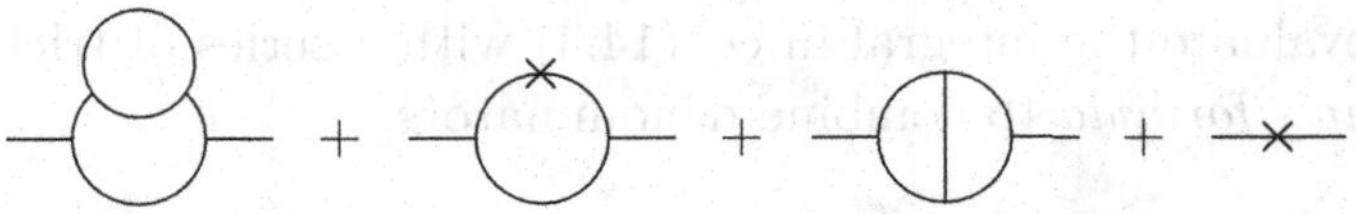

Figure 14.3. The $O(g^4)$ contributions to $i\Pi(k^2)$.

the $O(g^4)$ term. In the second term, $A = Z_\varphi - 1$ and $B = Z_m - 1$ are both expected to be $O(g^2)$.

It will prove convenient to define $\Pi(k^2)$ to all orders via the geometric series

$$
\begin{aligned}
\tfrac{1}{i}\tilde{\mathbf{\Delta}}(k^2) = \tfrac{1}{i}\tilde{\Delta}(k^2) &+ \tfrac{1}{i}\tilde{\Delta}(k^2)\Big[i\Pi(k^2)\Big]\tfrac{1}{i}\tilde{\Delta}(k^2) \\
&+ \tfrac{1}{i}\tilde{\Delta}(k^2)\Big[i\Pi(k^2)\Big]\tfrac{1}{i}\tilde{\Delta}(k^2)\Big[i\Pi(k^2)\Big]\tfrac{1}{i}\tilde{\Delta}(k^2) \\
&+ \ldots \,.
\end{aligned} \tag{14.5}
$$

This is illustrated in fig. 14.2. The sum in eq. (14.5) will include *all* the diagrams that contribute to $\tilde{\mathbf{\Delta}}(k^2)$ if we take $i\Pi(k^2)$ to be given by the sum of all diagrams that are *one-particle irreducible*, or 1PI for short. A diagram is 1PI if it is still connected after any one line is cut. The 1PI diagrams that make an $O(g^4)$ contribution to $i\Pi(k^2)$ are shown in fig. 14.3. When writing down the value of one of these diagrams, we omit the two external propagators.

If we sum up the series in eq. (14.5), we get

$$
\tilde{\mathbf{\Delta}}(k^2) = \frac{1}{k^2 + m^2 - i\epsilon - \Pi(k^2)} \,. \tag{14.6}
$$

In section 13, we learned that the exact propagator has a pole at $k^2 = -m^2$ with residue one. This is consistent with eq. (14.6) if and only if

$$
\Pi(-m^2) = 0 \,, \tag{14.7}
$$

$$
\Pi'(-m^2) = 0 \,, \tag{14.8}
$$

where the prime denotes a derivative with respect to k^2. *We will use eqs. (14.7) and (14.8) to fix the values of A and B.*

Next we turn to the evaluation of the $O(g^2)$ contribution to $i\Pi(k^2)$ in eq. (14.4). We have the immediate problem that the integral on the right-hand side diverges at large ℓ for $d \geq 4$. We faced a similar situation in section 9 when we evaluated the lowest-order tadpole diagram. There we introduced an *ultraviolet cutoff* Λ that modified the behavior of $\tilde{\Delta}(\ell^2)$ at large ℓ^2. Here, for now, we will simply restrict our attention to $d < 4$, where the integral in eq. (14.4) is finite. Later we will see what we can say about larger values of d.

We will evaluate the integral in eq. (14.4) with a series of tricks. We first use *Feynman's formula* to combine denominators,

$$\frac{1}{A_1 \ldots A_n} = \int dF_n \, (x_1 A_1 + \ldots + x_n A_n)^{-n} \,, \tag{14.9}$$

where the integration measure over the *Feynman parameters* x_i is

$$\int dF_n = (n{-}1)! \int_0^1 dx_1 \ldots dx_n \; \delta(x_1 + \ldots + x_n - 1) \,. \tag{14.10}$$

This measure is normalized so that

$$\int dF_n \, 1 = 1 \,. \tag{14.11}$$

We will prove eq. (14.9) in problem 14.1.

In the case at hand, we have

$$\begin{aligned}
\tilde{\Delta}((k{+}\ell)^2)\tilde{\Delta}(\ell^2) &= \frac{1}{(\ell^2 + m^2)((\ell + k)^2 + m^2)} \\
&= \int_0^1 dx \left[x((\ell + k)^2 + m^2) + (1{-}x)(\ell^2 + m^2) \right]^{-2} \\
&= \int_0^1 dx \left[\ell^2 + 2x\ell{\cdot}k + xk^2 + m^2 \right]^{-2} \\
&= \int_0^1 dx \left[(\ell + xk)^2 + x(1{-}x)k^2 + m^2 \right]^{-2} \\
&= \int_0^1 dx \left[q^2 + D \right]^{-2} ,
\end{aligned} \tag{14.12}$$

where we have suppressed the $i\epsilon$s for notational convenience; they can be restored via the replacement $m^2 \to m^2{-}i\epsilon$. In the last line we have defined

$$q \equiv \ell + xk \tag{14.13}$$

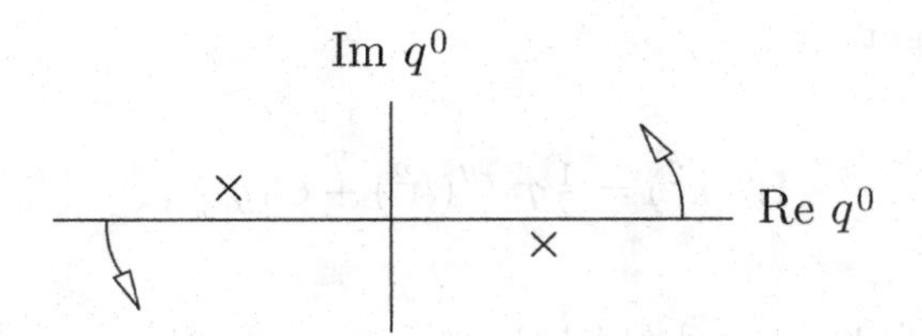

Figure 14.4. The q^0 integration contour along the real axis can be rotated to the imaginary axis without passing through the poles at $q^0 = -\omega + i\epsilon$ and $q^0 = +\omega - i\epsilon$.

and

$$D \equiv x(1-x)k^2 + m^2 \,. \tag{14.14}$$

We then change the integration variable in eq. (14.4) from ℓ to q; the jacobian is trivial, and we have $d^d\ell = d^dq$.

Next, think of the integral over q^0 from $-\infty$ to $+\infty$ as a contour integral in the complex q^0 plane. If the integrand vanishes fast enough as $|q^0| \to \infty$, we can rotate this contour counterclockwise by 90°, as shown in fig. 14.4, so that it runs from $-i\infty$ to $+i\infty$. In making this *Wick rotation*, the contour does not pass over any poles. (The $i\epsilon$s are needed to make this statement unambiguous.) Thus the value of the integral is unchanged. It is now convenient to define a euclidean d-dimensional vector $\bar{q}$ via $q^0 = i\bar{q}_d$ and $q_j = \bar{q}_j$; then $q^2 = \bar{q}^2$, where

$$\bar{q}^2 = \bar{q}_1^2 + \ldots + \bar{q}_d^2 \,. \tag{14.15}$$

Also, $d^dq = i\, d^d\bar{q}$. Therefore, in general,

$$\int d^dq \; f(q^2 - i\epsilon) = i \int d^d\bar{q} \; f(\bar{q}^2) \tag{14.16}$$

as long as $f(\bar{q}^2) \to 0$ faster than $1/\bar{q}^d$ as $\bar{q} \to \infty$.

Now we can write

$$\Pi(k^2) = \tfrac{1}{2} g^2 I(k^2) - Ak^2 - Bm^2 + O(g^4) \,, \tag{14.17}$$

where

$$I(k^2) \equiv \int_0^1 dx \int \frac{d^d\bar{q}}{(2\pi)^d} \frac{1}{(\bar{q}^2 + D)^2} \,. \tag{14.18}$$

It is now straightforward to evaluate the d-dimensional integral over $\bar{q}$ in spherical coordinates.

Before we perform this calculation, however, let us introduce another trick, one that can simplify the task of fixing A and B through the imposition of eqs. (14.7) and (14.8). Here is the trick: differentiate $\Pi(k^2)$ twice with respect to k^2 to get

$$\Pi''(k^2) = \tfrac{1}{2}g^2 I''(k^2) + O(g^4) \,, \tag{14.19}$$

where, from eqs. (14.18) and (14.14),

$$I''(k^2) = \int_0^1 dx\, 6x^2(1-x)^2 \int \frac{d^d\bar{q}}{(2\pi)^d} \frac{1}{(\bar{q}^2 + D)^4} \,. \tag{14.20}$$

Then, after we evaluate these integrals, we can get $\Pi(k^2)$ by integrating with respect to k^2, subject to the boundary conditions of eqs. (14.7) and (14.8). In this way we can construct $\Pi(k^2)$ without ever explicitly computing A and B.

Notice that this trick does something else for us as well. The integral over $\bar{q}$ in eq. (14.20) is finite for any $d < 8$, whereas the original integral in eq. (14.18) is finite only for $d < 4$. This expanded range of d now includes the value of greatest interest, $d = 6$.

How did this happen? We can gain some insight by making a Taylor expansion of $\Pi(k^2)$ about $k^2 = -m^2$:

$$\begin{aligned}\Pi(k^2) = {} & \left[\tfrac{1}{2}g^2 I(-m^2) + (A-B)m^2\right] \\ & + \left[\tfrac{1}{2}g^2 I'(-m^2) - A\right](k^2+m^2) \\ & + \tfrac{1}{2!}\left[\tfrac{1}{2}g^2 I''(-m^2)\right](k^2+m^2)^2 + \ldots \\ & + O(g^4) \,.\end{aligned} \tag{14.21}$$

From eqs. (14.18) and (14.14), it is straightforward to see that $I(-m^2)$ is divergent for $d \geq 4$, $I'(-m^2)$ is divergent for $d \geq 6$, and, in general, $I^{(n)}(-m^2)$ is divergent for $d \geq 4 + 2n$. We can use the $O(g^2)$ terms in A and B to cancel off the $\frac{1}{2}g^2 I(-m^2)$ and $\frac{1}{2}g^2 I'(-m^2)$ terms in $\Pi(k^2)$, whether or not they are divergent. But if we are to end up with a finite $\Pi(k^2)$, all of the remaining terms must be finite, since we have no more free parameters left to adjust. This is the case for $d < 8$.

Of course, for $4 \leq d < 8$, the values of A and B (and hence the lagrangian coefficients $Z_\varphi = 1 + A$ and $Z_m = 1 + B$) are formally infinite, and this may be disturbing. However, these coefficients are not directly measurable, and so need not obey our preconceptions about their magnitudes. Also, it is important to remember that A and B each includes a factor of g^2; this means that we can expand in powers of A and B as part of our general expansion in powers of g. When we compute $\Pi(k^2)$ (which enters into observable cross sections), all the formally infinite numbers cancel in a well-defined way, provided $d < 8$.

For $d \geq 8$, this procedure breaks down, and we do not obtain a finite expression for $\Pi(k^2)$. In this case, we say that the theory is *nonrenormalizable*. We will discuss the criteria for renormalizability of a theory in detail in section 18. It turns out that φ^3 theory is renormalizable for $d \leq 6$. (The problem with $6 < d < 8$ arises from higher-order corrections, as we will see in section 18.)

Now let us return to the calculation of $\Pi(k^2)$. Rather than using the trick of first computing $\Pi''(k^2)$, we will instead evaluate $\Pi(k^2)$ directly from eq. (14.18) as a function of d for $d < 4$. Then we will analytically continue the result to arbitrary d. This procedure is known as *dimensional regularization.* Then we will fix A and B by imposing eqs. (14.7) and (14.8), and finally take the limit $d \to 6$.

We could just as well use the method of section 9. Making the replacement

$$\tilde{\Delta}(p^2) \to \frac{1}{p^2 + m^2 - i\epsilon} \frac{\Lambda^2}{p^2 + \Lambda^2 - i\epsilon}, \tag{14.22}$$

where Λ is the ultraviolet cutoff, renders the $O(g^2)$ term in $\Pi(k^2)$ finite for $d < 8$; This procedure is known as *Pauli–Villars regularization.* We then evaluate $\Pi(k^2)$ as a function of Λ, fix A and B by imposing eqs. (14.7) and (14.8), and take the $\Lambda \to \infty$ limit. Calculations with Pauli–Villars regularization are generally much more cumbersome than they are with dimensional regularization. However, the final result for $\Pi(k^2)$ is the same. Eq. (14.21) demonstrates that *any* regularization scheme will give the same result for $d < 8$, at least as long as it preserves the Lorentz invariance of the integrals.

We therefore turn to the evaluation of $I(k^2)$, eq. (14.18). The angular part of the integral over $\bar{q}$ yields the area Ω_d of the unit sphere in d dimensions, which is

$$\Omega_d = \frac{2\pi^{d/2}}{\Gamma(\frac{1}{2}d)}; \tag{14.23}$$

this is most easily verified by computing the gaussian integral $\int d^d\bar{q}\, e^{-\bar{q}^2}$ in both cartesian and spherical coordinates. Here $\Gamma(x)$ is the Euler gamma function; for a nonnegative integer n and small x,

$$\Gamma(n+1) = n!\,, \tag{14.24}$$

$$\Gamma(n+\tfrac{1}{2}) = \frac{(2n)!}{n!2^{(2n)}}\sqrt{\pi}\,, \tag{14.25}$$

$$\Gamma(-n+x) = \frac{(-1)^n}{n!}\left[\frac{1}{x} - \gamma + \sum\nolimits_{k=1}^{n} k^{-1} + O(x)\right], \tag{14.26}$$

where $\gamma = 0.5772\ldots$ is the Euler–Mascheroni constant.

The radial part of the $\bar{q}$ integral can also be evaluated in terms of gamma functions. The overall result (generalized slightly) is

$$\int \frac{d^d\bar{q}}{(2\pi)^d}\,\frac{(\bar{q}^2)^a}{(\bar{q}^2 + D)^b} = \frac{\Gamma(b-a-\frac{1}{2}d)\Gamma(a+\frac{1}{2}d)}{(4\pi)^{d/2}\Gamma(b)\Gamma(\frac{1}{2}d)}\,D^{-(b-a-d/2)}\,. \tag{14.27}$$

We will make frequent use of this formula throughout this book. In the case of interest, eq. (14.18), we have $a = 0$ and $b = 2$.

There is one more complication to deal with. Recall that we want to focus on $d = 6$ because in that case g is dimensionless. However, for general d, g has mass dimension $\varepsilon/2$, where

$$\varepsilon \equiv 6 - d\,. \tag{14.28}$$

To account for this, we introduce a new parameter $\tilde{\mu}$ with dimensions of mass, and make the replacement

$$g \to g\tilde{\mu}^{\varepsilon/2}\,. \tag{14.29}$$

In this way g remains dimensionless for all ε. Of course, $\tilde{\mu}$ is not an actual parameter of the $d = 6$ theory. Therefore, nothing measurable (like a cross section) can depend on it.

This seemingly innocuous statement is actually quite powerful, and will eventually serve as the foundation of the renormalization group.

We now return to eq. (14.18), use eq. (14.27), and set $d = 6 - \varepsilon$; we get

$$I(k^2) = \frac{\Gamma(-1+\frac{\varepsilon}{2})}{(4\pi)^3}\int_0^1 dx\,D\left(\frac{4\pi}{D}\right)^{\varepsilon/2}\,. \tag{14.30}$$

Hence, with the substitution of eq. (14.29), and defining

$$\alpha \equiv \frac{g^2}{(4\pi)^3} \tag{14.31}$$

for notational convenience, we have

$$\Pi(k^2) = \tfrac{1}{2}\alpha\,\Gamma(-1+\tfrac{\varepsilon}{2})\int_0^1 dx\, D\left(\frac{4\pi\tilde{\mu}^2}{D}\right)^{\varepsilon/2} - Ak^2 - Bm^2 + O(\alpha^2)\,. \tag{14.32}$$

Now we can take the $\varepsilon \to 0$ limit, using eq. (14.26) and

$$A^{\varepsilon/2} = 1 + \tfrac{\varepsilon}{2}\ln A + O(\varepsilon^2)\,. \tag{14.33}$$

The result is

$$\Pi(k^2) = -\tfrac{1}{2}\alpha\left[(\tfrac{2}{\varepsilon}+1)(\tfrac{1}{6}k^2+m^2) + \int_0^1 dx\, D\ln\left(\frac{4\pi\tilde{\mu}^2}{e^{\gamma}D}\right)\right] - Ak^2 - Bm^2 + O(\alpha^2)\,. \tag{14.34}$$

Here we have used $\int_0^1 dx\, D = \frac{1}{6}k^2 + m^2$. It is now convenient to define

$$\mu \equiv \sqrt{4\pi}\, e^{-\gamma/2}\,\tilde{\mu}\,, \tag{14.35}$$

and rearrange things to get

$$\begin{aligned}\Pi(k^2) = {} & \tfrac{1}{2}\alpha\int_0^1 dx\, D\ln(D/m^2) \\ & - \{\tfrac{1}{6}\alpha[\tfrac{1}{\varepsilon} + \ln(\mu/m) + \tfrac{1}{2}] + A\}\,k^2 \\ & - \{\ \alpha[\tfrac{1}{\varepsilon} + \ln(\mu/m) + \tfrac{1}{2}] + B\}\,m^2 + O(\alpha^2)\,.\end{aligned} \tag{14.36}$$

If we take A and B to have the form

$$A = -\tfrac{1}{6}\alpha\big[\tfrac{1}{\varepsilon} + \ln(\mu/m) + \tfrac{1}{2} + \kappa_A\big] + O(\alpha^2)\,, \tag{14.37}$$

$$B = -\ \alpha\big[\tfrac{1}{\varepsilon} + \ln(\mu/m) + \tfrac{1}{2} + \kappa_B\big] + O(\alpha^2)\,, \tag{14.38}$$

where κ_A and κ_B are purely numerical constants, then we get

$$\Pi(k^2) = \tfrac{1}{2}\alpha\int_0^1 dx\, D\ln(D/m^2) + \alpha\Big(\tfrac{1}{6}\kappa_A k^2 + \kappa_B m^2\Big) + O(\alpha^2)\,. \tag{14.39}$$

Thus this choice of A and B renders $\Pi(k^2)$ finite and independent of μ, as required.

To fix κ_A and κ_B, we must still impose the conditions $\Pi(-m^2) = 0$ and $\Pi'(-m^2) = 0$. The easiest way to do this is to first note that, schematically,

$$\Pi(k^2) = \tfrac{1}{2}\alpha\int_0^1 dx\, D\ln D + \text{linear in } k^2 \text{ and } m^2 + O(\alpha^2)\,, \tag{14.40}$$

We can then impose $\Pi(-m^2) = 0$ via

$$\Pi(k^2) = \tfrac{1}{2}\alpha \int_0^1 dx\, D \ln(D/D_0) + \text{linear in } (k^2 + m^2) + O(\alpha^2) , \qquad (14.41)$$

where

$$D_0 \equiv D\Big|_{k^2=-m^2} = [1-x(1-x)]m^2 . \qquad (14.42)$$

Now it is straightforward to differentiate eq. (14.41) with respect to k^2, and find that $\Pi'(-m^2)$ vanishes for

$$\Pi(k^2) = \tfrac{1}{2}\alpha \int_0^1 dx\, D \ln(D/D_0) - \tfrac{1}{12}\alpha(k^2 + m^2) + O(\alpha^2) . \qquad (14.43)$$

The integral over x can be done in closed form; the result is

$$\Pi(k^2) = \tfrac{1}{12}\alpha\Big[c_1 k^2 + c_2 m^2 + 2k^2 f(r)\Big] + O(\alpha^2) , \qquad (14.44)$$

where $c_1 = 3-\pi\sqrt{3}$, $c_2 = 3-2\pi\sqrt{3}$, and

$$f(r) = r^3 \tanh^{-1}(1/r) , \qquad (14.45)$$

$$r = (1 + 4m^2/k^2)^{1/2} . \qquad (14.46)$$

There is a square-root branch point at $k^2 = -4m^2$, and $\Pi(k^2)$ acquires an imaginary part for $k^2 < -4m^2$; we will discuss this further in the next section.

We can write the exact propagator as

$$\tilde{\mathbf{\Delta}}(k^2) = \left(\frac{1}{1 - \Pi(k^2)/(k^2 + m^2)}\right) \frac{1}{k^2 + m^2 - i\epsilon} . \qquad (14.47)$$

In fig. 14.5, we plot the real and imaginary parts of $\Pi(k^2)/(k^2 + m^2)$ in units of α. We see that its values are quite modest for the plotted range. For much larger values of $|k^2|$, we have

$$\frac{\Pi(k^2)}{k^2 + m^2} \simeq \tfrac{1}{12}\alpha\Big[\ln(k^2/m^2) + c_1\Big] + O(\alpha^2) . \qquad (14.48)$$

If we had kept track of the $i\epsilon$s, k^2 would be $k^2 - i\epsilon$; when k^2 is negative, we have $\ln(k^2 - i\epsilon) = \ln|k^2| - i\pi$. The imaginary part of $\Pi(k^2)/(k^2 + m^2)$ therefore approaches the asymptotic value of $-\frac{1}{12}\pi\alpha + O(\alpha^2)$ when k^2 is large and negative. The real part of $\Pi(k^2)/(k^2 + m^2)$, however, continues to increase logarithmically with $|k^2|$ when $|k^2|$ is large. We will begin to address the meaning of this in section 26.

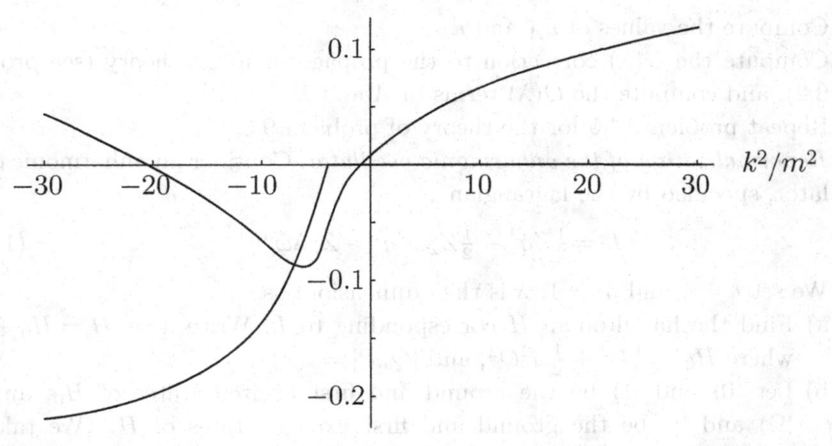

Figure 14.5. The real and imaginary parts of $\Pi(k^2)/(k^2+m^2)$ in units of α.

Problems

14.1) Derive a generalization of Feynman's formula,

$$\frac{1}{A_1^{\alpha_1}\ldots A_n^{\alpha_n}} = \frac{\Gamma\left(\sum_i \alpha_i\right)}{\prod_i \Gamma(\alpha_i)}\,\frac{1}{(n-1)!}\int dF_n\,\frac{\prod_i x_i^{\alpha_i-1}}{\left(\sum_i x_i A_i\right)^{\sum_i \alpha_i}}\,. \tag{14.49}$$

Hint: start with

$$\frac{\Gamma(\alpha)}{A^\alpha} = \int_0^\infty dt\; t^{\alpha-1}\,e^{-At}\,, \tag{14.50}$$

which defines the gamma function. Put an index on A, α, and t, and take the product. Then multiply on the right-hand side by

$$1 = \int_0^\infty ds\,\delta\left(s-\textstyle\sum_i t_i\right)\,. \tag{14.51}$$

Make the change of variable $t_i = sx_i$, and carry out the integral over s.

14.2) Verify eq. (14.23).

14.3) a) Show that

$$\int d^dq\; q^\mu f(q^2) = 0\,, \tag{14.52}$$

$$\int d^dq\; q^\mu q^\nu f(q^2) = C_2\, g^{\mu\nu}\int d^dq\; q^2 f(q^2)\,, \tag{14.53}$$

and evaluate the constant C_2 in terms of d. Hint: use Lorentz symmetry to argue for the general structure, and evaluate C_2 by contracting with $g_{\mu\nu}$.

b) Similarly evaluate $\int d^dq\; q^\mu q^\nu q^\rho q^\sigma f(q^2)$.

14.4) Compute the values of κ_A and κ_B.

14.5) Compute the $O(\lambda)$ correction to the propagator in φ^4 theory (see problem 9.2), and compute the $O(\lambda)$ terms in A and B.

14.6) Repeat problem 14.5 for the theory of problem 9.3.

14.7) *Renormalization of the anharmonic oscillator.* Consider an anharmonic oscillator, specified by the lagrangian

$$L = \tfrac{1}{2}Z\dot{q}^2 - \tfrac{1}{2}Z_\omega \omega^2 q^2 - Z_\lambda \lambda \omega^3 q^4 \,. \tag{14.54}$$

We set $\hbar = 1$ and $m = 1$; λ is then dimensionless.

a) Find the hamiltonian H corresponding to L. Write it as $H = H_0 + H_1$, where $H_0 = \frac{1}{2}P^2 + \frac{1}{2}\omega^2 Q^2$, and $[Q, P] = i$.

b) Let $|0\rangle$ and $|1\rangle$ be the ground and first excited states of H_0, and let $|\Omega\rangle$ and $|\mathrm{I}\rangle$ be the ground and first excited states of H. (We take all these eigenstates to have unit norm.) We define ω to be the excitation energy of H, $\omega \equiv E_\mathrm{I} - E_\Omega$. We normalize the position operator Q by setting $\langle \mathrm{I}|Q|\Omega\rangle = \langle 1|Q|0\rangle = (2\omega)^{-1/2}$. Finally, to make things mathematically simpler, we set Z_λ equal to one, rather than using a more physically motivated definition. Write $Z = 1 + A$ and $Z_\omega = 1 + B$, where $A = \kappa_A \lambda + O(\lambda^2)$ and $B = \kappa_B \lambda + O(\lambda^2)$. Use Rayleigh–Schrödinger perturbation theory to compute the $O(\lambda)$ corrections to the unperturbed energy eigenvalues and eigenstates.

c) Find the numerical values of κ_A and κ_B that yield $\omega = E_\mathrm{I} - E_\Omega$ and $\langle \mathrm{I}|Q|\Omega\rangle = (2\omega)^{-1/2}$.

d) Now think of the lagrangian of eq. (14.54) as specifying a quantum field theory in $d = 1$ dimensions. Compute the $O(\lambda)$ correction to the propagator. Fix κ_A and κ_B by requiring the propagator to have a pole at $k^2 = -\omega^2$ with residue one. Do your results agree with those of part (c)? Should they?

15

The one-loop correction in Lehmann–Källén form

Prerequisite: 14

In section 13, we found that the exact propagator could be written in Lehmann–Källén form as

$$\tilde{\mathbf{\Delta}}(k^2) = \frac{1}{k^2 + m^2 - i\epsilon} + \int_{4m^2}^{\infty} ds\, \rho(s) \frac{1}{k^2 + s - i\epsilon}\,, \tag{15.1}$$

where the *spectral density* $\rho(s)$ is real and nonnegative. In section 14, on the other hand, we found that the exact propagator could be written as

$$\tilde{\mathbf{\Delta}}(k^2) = \frac{1}{k^2 + m^2 - i\epsilon - \Pi(k^2)} \tag{15.2}$$

and that, to $O(g^2)$ in φ^3 theory in six dimensions,

$$\Pi(k^2) = \tfrac{1}{2}\alpha \int_0^1 dx\, D \ln(D/D_0) - \tfrac{1}{12}\alpha(k^2 + m^2) + O(\alpha^2)\,, \tag{15.3}$$

where

$$\alpha \equiv g^2/(4\pi)^3\,, \tag{15.4}$$

$$D = x(1-x)k^2 + m^2 - i\epsilon\,, \tag{15.5}$$

$$D_0 = [1-x(1-x)]m^2\,. \tag{15.6}$$

In this section, we will attempt to reconcile eqs. (15.2) and (15.3) with eq. (15.1).

Let us begin by considering the imaginary part of the propagator. We will always take k^2 and m^2 to be real, and explicitly include the appropriate factors of $i\epsilon$ whenever they are needed.

We can use eq. (15.1) and the identity

$$\frac{1}{x-i\epsilon} = \frac{x}{x^2+\epsilon^2} + \frac{i\epsilon}{x^2+\epsilon^2}$$

$$= P\frac{1}{x} + i\pi\delta(x)\ , \tag{15.7}$$

where P means the principal part, to write

$$\text{Im}\,\tilde{\mathbf{\Delta}}(k^2) = \pi\delta(k^2+m^2) + \int_{4m^2}^{\infty} ds\,\rho(s)\,\pi\delta(k^2+s)$$

$$= \pi\delta(k^2+m^2) + \pi\rho(-k^2)\ , \tag{15.8}$$

where $\rho(s) \equiv 0$ for $s < 4m^2$. Thus we have

$$\pi\rho(s) = \text{Im}\,\tilde{\mathbf{\Delta}}(-s) \quad \text{for} \quad s \geq 4m^2\ . \tag{15.9}$$

Let us now suppose that $\text{Im}\,\Pi(k^2) = 0$ for some range of k^2. (In section 14, we saw that the $O(\alpha)$ contribution to $\Pi(k^2)$ is purely real for $k^2 > -4m^2$.) Then, from eqs. (15.2) and (15.7), we get

$$\text{Im}\,\tilde{\mathbf{\Delta}}(k^2) = \pi\delta(k^2+m^2-\Pi(k^2)) \quad \text{for} \quad \text{Im}\,\Pi(k^2) = 0\ . \tag{15.10}$$

From $\Pi(-m^2) = 0$, we know that the argument of the delta function vanishes at $k^2 = -m^2$, and from $\Pi'(-m^2) = 0$, we know that the derivative of this argument with respect to k^2 equals one at $k^2 = -m^2$. Therefore

$$\text{Im}\,\tilde{\mathbf{\Delta}}(k^2) = \pi\delta(k^2+m^2) \quad \text{for} \quad \text{Im}\,\Pi(k^2) = 0\ . \tag{15.11}$$

Comparing this with eq. (15.8), we see that $\rho(-k^2) = 0$ if $\text{Im}\,\Pi(k^2) = 0$.

Now suppose $\text{Im}\,\Pi(k^2)$ is *not* zero for some range of k^2. (In section 14, we saw that the $O(\alpha)$ contribution to $\Pi(k^2)$ has a nonzero imaginary part for $k^2 < -4m^2$.) Then we can ignore the $i\epsilon$ in eq. (15.2), and

$$\text{Im}\,\tilde{\mathbf{\Delta}}(k^2) = \frac{\text{Im}\,\Pi(k^2)}{(k^2+m^2-\text{Re}\,\Pi(k^2))^2 + (\text{Im}\,\Pi(k^2))^2} \quad \text{for} \quad \text{Im}\,\Pi(k^2) \neq 0\ . \tag{15.12}$$

Comparing this with eq. (15.8) we see that

$$\pi\rho(s) = \frac{\text{Im}\,\Pi(-s)}{(-s+m^2-\text{Re}\,\Pi(-s))^2 + (\text{Im}\,\Pi(-s))^2}\ . \tag{15.13}$$

Since we know $\rho(s) = 0$ for $s < 4m^2$, this tells us that we must also have $\text{Im}\,\Pi(-s) = 0$ for $s < 4m^2$, or equivalently $\text{Im}\,\Pi(k^2) = 0$ for $k^2 > -4m^2$.

This is just what we found for the $O(\alpha)$ contribution to $\Pi(k^2)$ in section 14.

We can also see this directly from eq. (15.3), without doing the integral over x. The integrand in this formula is real as long as the argument of the logarithm is real and positive. From eq. (15.5), we see that D is real and positive if and only if $x(1-x)k^2 > -m^2$. The maximum value of $x(1-x)$ is $1/4$, and so the argument of the logarithm is real and positive for the whole integration range $0 \le x \le 1$ if and only if $k^2 > -4m^2$. In this regime, $\operatorname{Im}\Pi(k^2) = 0$. On the other hand, for $k^2 < -4m^2$, the argument of the logarithm becomes negative for some of the integration range, and so $\operatorname{Im}\Pi(k^2) \ne 0$ for $k^2 < -4m^2$. This is exactly what we need to reconcile eqs. (15.2) and (15.3) with eq. (15.1).

Problems

15.1) In this problem we will verify the result of problem 13.1 to $O(\alpha)$.

a) Let $\Pi_{\rm loop}(k^2)$ be given by the first line of eq. (14.32), with $\varepsilon > 0$. Show that, up to $O(\alpha^2)$ corrections,

$$A = \Pi'_{\rm loop}(-m^2) \; . \tag{15.14}$$

Then use Cauchy's integral formula to write this as

$$A = \oint \frac{dw}{2\pi i} \frac{\Pi_{\rm loop}(w)}{(w+m^2)^2} \; , \tag{15.15}$$

where the contour of integration is a small counterclockwise circle around $-m^2$ in the complex w plane.

b) By examining eq. (14.32), show that the only singularity of $\Pi_{\rm loop}(k^2)$ is a branch point at $k^2 = -4m^2$. Take the cut to run along the negative real axis.

c) Distort the contour in eq. (15.15) to a circle at infinity with a detour around the branch cut. Examine eq. (14.32) to show that, for $\varepsilon > 0$, the circle at infinity does not contribute. The contour around the branch cut then yields

$$A = \int_{-\infty}^{-4m^2} \frac{dw}{2\pi i} \frac{1}{(w+m^2)^2} \Big[\Pi_{\rm loop}(w{+}i\epsilon) - \Pi_{\rm loop}(w{-}i\epsilon)\Big] \; , \tag{15.16}$$

where ϵ is infinitesimal (and is not to be confused with $\varepsilon = 6{-}d$).

d) Examine eq. (14.32) to show that the real part of $\Pi_{\rm loop}(w)$ is continuous across the branch cut, and that the imaginary part changes sign, so that

$$\Pi_{\rm loop}(w{+}i\epsilon) - \Pi_{\rm loop}(w{-}i\epsilon) = -2i \operatorname{Im}\Pi_{\rm loop}(w{-}i\epsilon) \; . \tag{15.17}$$

e) Let $w = -s$ in eq. (15.16) and use eq. (15.17) to get

$$A = -\frac{1}{\pi}\int_{4m^2}^{\infty} ds\,\frac{\text{Im}\,\Pi_{\text{loop}}(-s-i\epsilon)}{(s-m^2)^2}\,. \tag{15.18}$$

Use this to verify the result of problem 13.1 to $O(\alpha)$.

15.2) *Dispersion relations.* Consider the exact $\Pi(k^2)$, with $\varepsilon = 0$. Assume that its only singularity is a branch point at $k^2 = -4m^2$, that it obeys eq. (15.17), and that $\Pi(k^2)$ grows more slowly than $|k^2|^2$ at large $|k^2|$. By recapitulating the analysis in the previous problem, show that

$$\Pi''(k^2) = \frac{2}{\pi}\int_{4m^2}^{\infty} ds\,\frac{\text{Im}\,\Pi(-s-i\varepsilon)}{(k^2+s)^3}\,. \tag{15.19}$$

This is a *twice subtracted dispersion relation.* It gives $\Pi''(k^2)$ throughout the complex k^2 plane in terms of the values of the imaginary part of $\Pi(k^2)$ along the branch cut.

16

Loop corrections to the vertex

Prerequisite: 14

Consider the $O(g^3)$ diagram of fig. 16.1, which corrects the φ^3 vertex. In this section we will evaluate this diagram.

We can define an exact three-point vertex function $i\mathbf{V}_3(k_1, k_2, k_3)$ as the sum of one-particle irreducible diagrams with three external lines carrying momenta k_1, k_2, and k_3, all incoming, with $k_1 + k_2 + k_3 = 0$ by momentum conservation. (In adopting this convention, we allow k_i^0 to have either sign; if k_i is the momentum of an external particle, then the sign of k_i^0 is positive if the particle is incoming, and negative if it is outgoing.) The original vertex $iZ_g g$ is the first term in this sum, and the diagram of fig. 16.1 is the second. Thus we have

$$i\mathbf{V}_3(k_1, k_2, k_3) = iZ_g g + (ig)^3 \left(\tfrac{1}{i}\right)^3 \int \frac{d^d\ell}{(2\pi)^d} \tilde{\Delta}((\ell - k_1)^2)\tilde{\Delta}((\ell + k_2)^2)\tilde{\Delta}(\ell^2) + O(g^5) \,. \tag{16.1}$$

In the second term, we have set $Z_g = 1 + O(g^2)$. We proceed immediately to the evaluation of this integral, using the series of tricks from section 14.

First we use Feynman's formula to write

$$\tilde{\Delta}((\ell - k_1)^2)\tilde{\Delta}((\ell + k_2)^2)\tilde{\Delta}(\ell^2) = \int dF_3 \Big[x_1(\ell - k_1)^2 + x_2(\ell + k_2)^2 + x_3\ell^2 + m^2 \Big]^{-3} , \tag{16.2}$$

where

$$\int dF_3 = 2\int_0^1 dx_1\, dx_2\, dx_3\, \delta(x_1 + x_2 + x_3 - 1) \,. \tag{16.3}$$

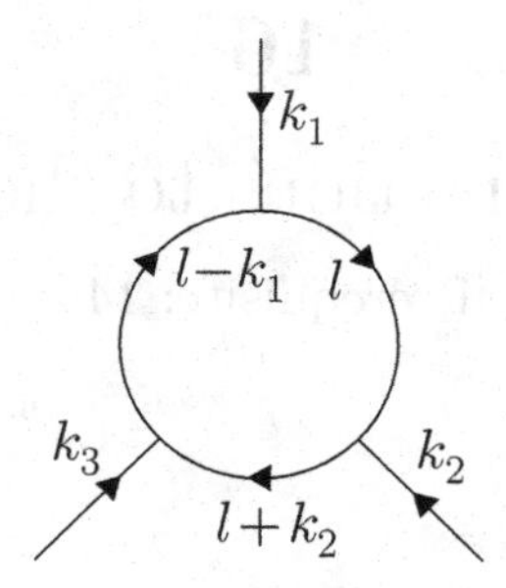

Figure 16.1. The $O(g^3)$ correction to the vertex $i\mathbf{V}_3(k_1,k_2,k_3)$.

We manipulate the right-hand side of eq. (16.2) to get

$$\begin{aligned}
&\tilde{\Delta}((\ell-k_1)^2)\tilde{\Delta}((\ell+k_2)^2)\tilde{\Delta}(\ell^2) \\
&\quad = \int dF_3 \Big[\ell^2 - 2\ell\cdot(x_1k_1 - x_2k_2) + x_1k_1^2 + x_2k_2^2 + m^2\Big]^{-3} \\
&\quad = \int dF_3 \Big[(\ell - x_1k_1 + x_2k_2)^2 + x_1(1-x_1)k_1^2 + x_2(1-x_2)k_2^2 \\
&\qquad\qquad\qquad + 2x_1x_2k_1\cdot k_2 + m^2\Big]^{-3} \\
&\quad = \int dF_3 \Big[q^2 + D\Big]^{-3} .
\end{aligned} \tag{16.4}$$

In the last line, we have defined $q \equiv \ell - x_1k_1 + x_2k_2$, and

$$\begin{aligned}
D &\equiv x_1(1-x_1)k_1^2 + x_2(1-x_2)k_2^2 + 2x_1x_2k_1\cdot k_2 + m^2 \\
&= x_3x_1k_1^2 + x_3x_2k_2^2 + x_1x_2k_3^2 + m^2 ,
\end{aligned} \tag{16.5}$$

where we used $k_3^2 = (k_1 + k_2)^2$ and $x_1 + x_2 + x_3 = 1$ to simplify the second line.

After making a Wick rotation of the q^0 contour, we have

$$\mathbf{V}_3(k_1,k_2,k_3)/g = Z_g + g^2 \int dF_3 \int \frac{d^d\bar{q}}{(2\pi)^d} \frac{1}{(\bar{q}^2 + D)^3} + O(g^4) , \tag{16.6}$$

where $\bar{q}$ is a euclidean vector. This integral diverges for $d \geq 6$. We therefore evaluate it for $d < 6$, using the general formula from section 14; the result is

$$\int \frac{d^d\bar{q}}{(2\pi)^d} \frac{1}{(\bar{q}^2 + D)^3} = \frac{\Gamma(3-\frac{1}{2}d)}{2(4\pi)^{d/2}} D^{-(3-d/2)} . \tag{16.7}$$

Now we set $d = 6 - \varepsilon$. To keep g dimensionless, we make the replacement $g \to g\tilde{\mu}^{\varepsilon/2}$. Then we have

$$\mathbf{V}_3(k_1, k_2, k_3)/g = Z_g + \tfrac{1}{2}\alpha\,\Gamma(\tfrac{\varepsilon}{2}) \int dF_3 \left(\frac{4\pi\tilde{\mu}^2}{D}\right)^{\varepsilon/2} + O(\alpha^2)\,, \tag{16.8}$$

where $\alpha = g^2/(4\pi)^3$. Now we can take the $\varepsilon \to 0$ limit. The result is

$$\mathbf{V}_3(k_1, k_2, k_3)/g = Z_g + \tfrac{1}{2}\alpha\left[\frac{2}{\varepsilon} + \int dF_3 \ln\left(\frac{4\pi\tilde{\mu}^2}{e^\gamma D}\right)\right] + O(\alpha^2)\,, \tag{16.9}$$

where we have used $\int dF_3 = 1$. We now let $\mu^2 = 4\pi e^{-\gamma}\tilde{\mu}^2$, set

$$Z_g = 1 + C\,, \tag{16.10}$$

and rearrange to get

$$\begin{aligned}\mathbf{V}_3(k_1, k_2, k_3)/g = 1 &+ \{\alpha[\tfrac{1}{\varepsilon} + \ln(\mu/m)] + C\} \\ &- \tfrac{1}{2}\alpha \int dF_3 \ln(D/m^2) \\ &+ O(\alpha^2)\,.\end{aligned} \tag{16.11}$$

If we take C to have the form

$$C = -\alpha\big[\tfrac{1}{\varepsilon} + \ln(\mu/m) + \kappa_C\big] + O(\alpha^2)\,, \tag{16.12}$$

where κ_C is a purely numerical constant, we get

$$\mathbf{V}_3(k_1, k_2, k_3)/g = 1 - \tfrac{1}{2}\alpha \int dF_3 \ln(D/m^2) - \kappa_C\alpha + O(\alpha^2)\,. \tag{16.13}$$

Thus this choice of C renders $\mathbf{V}_3(k_1, k_2, k_3)$ finite and independent of μ, as required.

We now need a condition, analogous to $\Pi(-m^2) = 0$ and $\Pi'(-m^2) = 0$, to fix the value of κ_C. These conditions on $\Pi(k^2)$ were mandated by known properties of the exact propagator, but there is nothing directly comparable for the vertex. Different choices of κ_C correspond to different definitions of the coupling g. This is because, in order to measure g, we would measure a cross section that depends on g; these cross sections also depend on κ_C. Thus we can use any value for κ_C that we might fancy, as long as we all agree on that value when we compare our calculations with experimental measurements. It is then most convenient to simply set $\kappa_C = 0$. This corresponds to the condition

$$\mathbf{V}_3(0, 0, 0) = g\,. \tag{16.14}$$

This condition can then also be used to fix the higher-order (in g) terms in Z_g.

The integrals over the Feynman parameters in eq. (16.13) cannot be done in closed form, but it is easy to see that if (for example) $|k_1^2| \gg m^2$, then

$$\mathbf{V}_3(k_1,k_2,k_3)/g \simeq 1 - \tfrac{1}{2}\alpha\Big[\ln(k_1^2/m^2) + O(1)\Big] + O(\alpha^2) \,. \qquad (16.15)$$

Thus the magnitude of the one-loop correction to the vertex function increases logarithmically with $|k_i^2|$ when $|k_i^2| \gg m^2$. This is the same behavior that we found for $\Pi(k^2)/(k^2+m^2)$ in section 14.

Problems

16.1) Compute the $O(\lambda^2)$ correction to $\mathbf{V}_4$ in φ^4 theory (see problem 9.2). Take $\mathbf{V}_4 = -\lambda$ when all four external momenta are on shell, and $s = 4m^2$. What is the $O(\lambda)$ contribution to C?

16.2) Repeat problem 16.1 for the theory of problem 9.3.

17

Other 1PI vertices

Prerequisite: 16

In section 16, we defined the three-point vertex function $i\mathbf{V}_3(k_1, k_2, k_3)$ as the sum of all one-particle irreducible diagrams with three external lines, with the external propagators removed. We can extend this definition to the n-point vertex $i\mathbf{V}_n(k_1, \ldots, k_n)$.

There are two key differences between $\mathbf{V}_{n>3}$ and $\mathbf{V}_3$ in φ^3 theory. The first is that there is no tree-level contribution to $\mathbf{V}_{n>3}$. The second is that the one-loop contribution to $\mathbf{V}_{n>3}$ is finite for $d < 2n$. In particular, the one-loop contribution to $\mathbf{V}_{n>3}$ is finite for $d = 6$.

Let us see how this works for the case $n = 4$. We treat all the external momenta as incoming, so that $k_1 + k_2 + k_3 + k_4 = 0$. One of the three contributing one-loop diagrams is shown in fig. 17.1; in this diagram, the k_3 vertex is opposite to the k_1 vertex. Two other inequivalent diagrams are then obtained by swapping $k_3 \leftrightarrow k_2$ and $k_3 \leftrightarrow k_4$. We then have

$$
\begin{aligned}
i\mathbf{V}_4 = g^4 \int \frac{d^6\ell}{(2\pi)^6}\, &\tilde{\Delta}((\ell{-}k_1)^2)\tilde{\Delta}((\ell{+}k_2)^2)\tilde{\Delta}((\ell{+}k_2{+}k_3)^2)\tilde{\Delta}(\ell^2) \\
&+ (k_3 \leftrightarrow k_2) + (k_3 \leftrightarrow k_4) \\
&+ O(g^6)\,. \qquad (17.1)
\end{aligned}
$$

Feynman's formula gives

$$
\begin{aligned}
&\tilde{\Delta}((\ell{-}k_1)^2)\tilde{\Delta}((\ell{+}k_2)^2)\tilde{\Delta}((\ell{+}k_2{+}k_3)^2)\tilde{\Delta}(\ell^2) \\
&\quad = \int dF_4 \Big[x_1(\ell{-}k_1)^2 + x_2(\ell{+}k_2)^2 + x_3(\ell{+}k_2{+}k_3)^2 + x_4\ell^2 + m^2 \Big]^{-4} \\
&\quad = \int dF_4 \Big[q^2 + D_{1234} \Big]^{-4}, \qquad (17.2)
\end{aligned}
$$

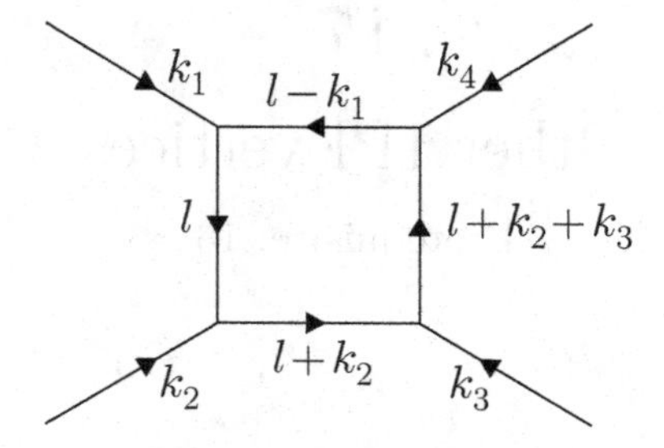

Figure 17.1. One of the three one-loop Feynman diagrams contributing to the four-point vertex $i\mathbf{V}_4(k_1, k_2, k_3, k_4)$; the other two are obtained by swapping $k_3 \leftrightarrow k_2$ and $k_3 \leftrightarrow k_4$.

where $q = \ell - x_1k_1 + x_2k_2 + x_3(k_2+k_3)$ and, after making repeated use of $x_1+x_2+x_3+x_4 = 1$ and $k_1+k_2+k_3+k_4 = 0$,

$$
\begin{aligned}
D_{1234} = {} & x_1x_4k_1^2 + x_2x_4k_2^2 + x_2x_3k_3^2 + x_1x_3k_4^2 \\
& + x_1x_2(k_1+k_2)^2 + x_3x_4(k_2+k_3)^2 + m^2 \,. \qquad (17.3)
\end{aligned}
$$

We see that the integral over q is finite for $d < 8$, and in particular for $d = 6$. After a Wick rotation of the q^0 contour and applying the general formula of section 14, we find

$$
\int \frac{d^6q}{(2\pi)^6} \frac{1}{(q^2 + D)^4} = \frac{i}{6(4\pi)^3 D} \,. \qquad (17.4)
$$

Thus we get

$$
\mathbf{V}_4 = \frac{g^4}{6(4\pi)^3} \int dF_4 \left(\frac{1}{D_{1234}} + \frac{1}{D_{1324}} + \frac{1}{D_{1243}} \right) + O(g^6) \,. \qquad (17.5)
$$

This expression is finite and well-defined; the same is true for the one-loop contribution to $\mathbf{V}_n$ for all $n > 3$.

Problems

17.1) Verify eq. (17.3).

18

Higher-order corrections and renormalizability

Prerequisite: 17

In sections 14–17, we computed the one-loop diagrams with two, three, and four external lines for φ^3 theory in six dimensions. We found that the first two involved divergent momentum integrals, but that these divergences could be absorbed into the coefficients of terms in the lagrangian. If this is true for all higher-order (in g) contributions to the propagator and to the one-particle irreducible vertex functions (with $n \geq 3$ external lines), then we say that the theory is *renormalizable.* If this is not the case, and further divergences arise, it may be possible to absorb them by adding some new terms to the lagrangian. If a finite number of such new terms is required, the theory is still said to be renormalizable. However, if an infinite number of new terms is required, then the theory is said to be *nonrenormalizable.* Despite the infinite number of parameters needed to specify it, a nonrenormalizable theory is generally able to make useful predictions at energies below some *ultraviolet cutoff* Λ; we will discuss this in section 29.

In this section, we will deduce the necessary conditions for renormalizability. As an example, we will analyze a scalar field theory in d spacetime dimensions of the form

$$\mathcal{L} = -\tfrac{1}{2}Z_\varphi \partial^\mu\varphi\partial_\mu\varphi - \tfrac{1}{2}Z_m m^2\varphi^2 - \sum_{n=3}^{\infty} \tfrac{1}{n!}Z_n g_n \varphi^n \ . \tag{18.1}$$

Consider a Feynman diagram with E external lines, I internal lines, L closed loops, and V_n vertices that connect n lines. (Here V_n is just a number, not to be confused with the vertex function $\mathbf{V}_n$.) Do the momentum integrals associated with this diagram diverge?

We begin by noting that each closed loop gives a factor of $d^d\ell_i$, and each internal propagator gives a factor of $1/(p^2 + m^2)$, where p is some linear combination of external momenta k_i and loop momenta ℓ_i. The diagram would then appear to have an ultraviolet divergence at large ℓ_i if there are more ℓs in the numerator than there are in the denominator. The number of ℓs in the numerator minus the number of ℓs in the denominator is the diagram's *superficial degree of divergence*

$$D \equiv dL - 2I \,, \tag{18.2}$$

and the diagram appears to be divergent if

$$D \geq 0 \,. \tag{18.3}$$

Next we derive a more useful formula for D. The diagram has E external lines, so another contributing diagram is the tree diagram where all the lines are joined by a single vertex, with vertex factor $-iZ_E g_E$; this is, in fact, the value of this entire diagram, which then has mass dimension $[g_E]$. (The Zs are all dimensionless, by definition.) Therefore, the original diagram also has mass dimension $[g_E]$, since both are contributions to the same scattering amplitude:

$$[\text{diagram}] = [g_E] \,. \tag{18.4}$$

On the other hand, the mass dimension of any diagram is given by the sum of the mass dimensions of its components, namely

$$[\text{diagram}] = dL - 2I + \sum_{n=3}^{\infty} V_n [g_n] \,. \tag{18.5}$$

From eqs. (18.2), (18.4), and (18.5), we get

$$D = [g_E] - \sum_{n=3}^{\infty} V_n [g_n] \,. \tag{18.6}$$

This is the formula we need.

From eq. (18.6), it is immediately clear that if any $[g_n] < 0$, we expect uncontrollable divergences, since D increases with every added vertex of this type. Therefore, *a theory with any $[g_n] < 0$ is nonrenormalizable.*

According to our results in section 12, the coupling constants have mass dimension

$$[g_n] = d - \tfrac{1}{2}n(d-2) \,, \tag{18.7}$$

and so we have

$$[g_n] < 0 \quad \text{if} \quad n > \frac{2d}{d-2} \,. \tag{18.8}$$

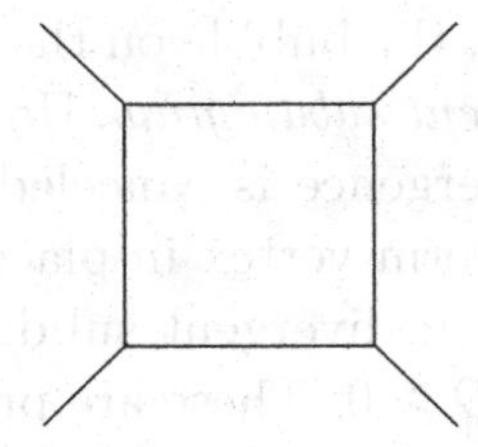

Figure 18.1. The one-loop contribution to $\mathbf{V}_4$.

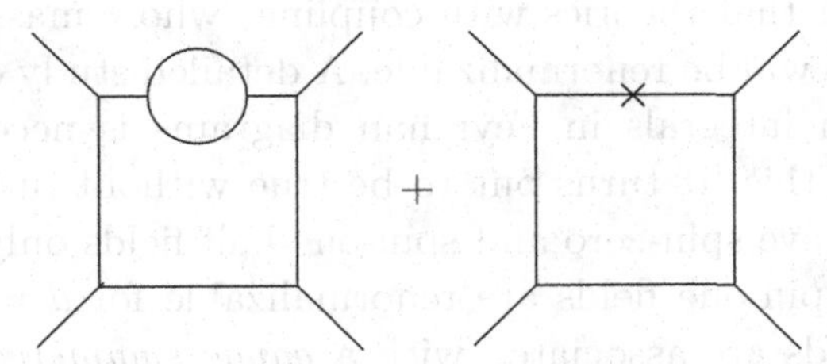

Figure 18.2. A two-loop contribution to $\mathbf{V}_4$, and the corresponding counterterm insertion.

Thus we are limited to powers no higher than φ^4 in four dimensions, and no higher than φ^3 in six dimensions.

The same criterion applies to more complicated theories as well: *a theory is nonrenormalizable if any coefficient of any term in the lagrangian has negative mass dimension.*

What about theories with couplings with only positive or zero mass dimension? We see from eq. (18.6) that the only dangerous diagrams (those with $D \geq 0$) are those for which $[g_E] \geq 0$. But in this case, we can absorb the divergence simply by adjusting the value of Z_E. This discussion also applies to the propagator; we can think of $\Pi(k^2)$ as representing the loop-corrected counterterm vertex $Ak^2 + Bm^2$, with A and Bm^2 playing the roles of two couplings. We have $[A] = 0$ and $[Bm^2] = 2$, so the contributing diagrams are expected to be divergent (as we have already seen in detail), and the divergences must be absorbed into A and Bm^2.

D is called the *superficial* degree of divergence because a diagram might diverge even if $D < 0$, or might be finite even if $D \geq 0$. The latter can happen if there are cancellations among ℓs in the numerator. Quantum electrodynamics provides an example of this phenomenon that we will encounter in Part III; see problem 62.3. For now we turn our attention to the case of diagrams with $D < 0$ that nevertheless diverge.

Consider, for example, the diagrams of figs. 18.1 and 18.2. The one-loop diagram of fig. 18.1 with $E = 4$ is finite, but the two-loop correction from the first diagram of fig. 18.2 is not: the bubble on the upper propagator diverges. This is an example of a *divergent subdiagram.* However, this is not a problem in this case, because this divergence is canceled by the second diagram of fig. 18.2, which has a counterterm vertex in place of the bubble.

This is the generic situation: divergent subdiagrams are diagrams that, considered in isolation, have $D \geq 0$. These are precisely the diagrams whose divergences can be canceled by adjusting the Z factor of the corresponding tree diagram (in theories where $[g_n] \geq 0$ for all nonzero g_n).

Thus, we expect that theories with couplings whose mass dimensions are all positive or zero will be renormalizable. A detailed study of the properties of the momentum integrals in Feynman diagrams is necessary to give a complete proof of this. It turns out to be true without further restrictions for theories that have spin-zero and spin-one-half fields only.

Theories with spin-one fields are renormalizable for $d = 4$ if and only if these spin-one fields are associated with a *gauge symmetry.* We will study this in Part III.

Theories of fields with spin greater than one are never renormalizable for $d \geq 4$.

Reference notes

Explicit two-loop calculations in φ^3 theory can be found in *Collins*, *Muta*, and *Sterman.*

Problems

18.1) In any number d of spacetime dimensions, a *Dirac field* $\Psi_\alpha(x)$ carries a spin index α, and has a kinetic term of the form $i\overline{\Psi}\gamma^\mu\partial_\mu\Psi$, where we have suppressed the spin indices; the *gamma matrices* γ^μ are dimensionless, and $\overline{\Psi} = \Psi^\dagger\gamma^0$.

a) What is the mass dimension $[\Psi]$ of the field Ψ?

b) Consider interactions of the form $g_n(\overline{\Psi}\Psi)^n$, where $n \geq 2$ is an integer. What is the mass dimension $[g_n]$ of g_n?

c) Consider interactions of the form $g_{m,n}\varphi^m(\overline{\Psi}\Psi)^n$, where φ is a scalar field, and $m \geq 1$ and $n \geq 1$ are integers. What is the mass dimension $[g_{m,n}]$ of $g_{m,n}$?

d) In $d = 4$ spacetime dimensions, which of these interactions are allowed in a renormalizable theory?

19

Perturbation theory to all orders

Prerequisite: 18

In section 18, we found that, generally, a theory is renormalizable if all of its lagrangian coefficients have positive or zero mass dimension. In this section, using φ^3 theory in six dimensions as our example, we will see how to construct a finite expression for a scattering amplitude to arbitrarily high order in the φ^3 coupling g.

We begin by summing all one-particle irreducible diagrams with two external lines; this gives us the self-energy $\Pi(k^2)$. We next sum all 1PI diagrams with three external lines; this gives us the three-point vertex function $\mathbf{V}_3(k_1, k_2, k_3)$. Order by order in g, we must adjust the value of the lagrangian coefficients Z_φ, Z_m, and Z_g to maintain the conditions $\Pi(-m^2) = 0$, $\Pi'(-m^2) = 0$, and $\mathbf{V}_3(0,0,0) = g$.

Next we will construct the n-point vertex functions $\mathbf{V}_n(k_1, \ldots, k_n)$ with $4 \le n \le E$, where E is the number of external lines in the process of interest. We compute these using a *skeleton expansion.* This means that we draw all the contributing 1PI diagrams, but omit diagrams that include either propagator or three-point vertex corrections. That is, we omit any 1PI diagram that contains a subdiagram with two or three external lines that is more complicated than a single tree-level propagator (for a subdiagram with two external lines) or tree-level vertex (for a subdiagram with three external lines). Then we take the propagators and vertices in these diagrams to be given by the *exact* propagator $\tilde{\mathbf{\Delta}}(k^2) = (k^2 + m^2 - \Pi(k^2))^{-1}$ and vertex $\mathbf{V}_3(k_1, k_2, k_3)$, rather than by the tree-level propagator $\tilde{\Delta}(k^2) = (k^2 + m^2)^{-1}$ and vertex g. We then sum these *skeleton diagrams* to get $\mathbf{V}_n$ for $4 \le n \le E$. Order by order in g, this procedure is equivalent to computing $\mathbf{V}_n$ by summing the usual set of contributing 1PI diagrams.

Next we draw all *tree-level* diagrams that contribute to the process of interest (which has E external lines), including not only three-point vertices, but also n-point vertices for $n = 3, 4, \ldots, E$. Then we evaluate these diagrams using the exact propagator $\tilde{\mathbf{\Delta}}(k^2)$ for internal lines, and the exact 1PI vertices $\mathbf{V}_n$; external lines are assigned a factor of one.[1] We sum these tree diagrams to get the scattering amplitude; loop corrections have all been accounted for already in $\tilde{\mathbf{\Delta}}(k^2)$ and $\mathbf{V}_n$. Order by order in g, this procedure is equivalent to computing the scattering amplitude by summing the usual set of contributing diagrams.

Thus we now know how to compute an arbitrary scattering amplitude to arbitrarily high order. The procedure is the same in any quantum field theory; only the form of the propagators and vertices change, depending on the spins of the fields.

The tree-level diagrams of the final step can be thought of as the Feynman diagrams of a *quantum action* (or *effective action*, or *quantum effective action*) $\Gamma(\varphi)$. There is a simple and interesting relationship between the effective action $\Gamma(\varphi)$ and the sum of connected diagrams with sources $iW(J)$. We derive it in section 21.

Reference notes

The detailed procedure for renormalization at higher orders is discussed in *Coleman*, *Collins*, *Muta*, and *Sterman.*

[1] This is because, in the LSZ formula, each Klein–Gordon wave operator becomes (in momentum space) a factor of $k_i^2 + m^2$ that multiplies each external propagator, leaving behind only the residue of the pole in that propagator at $k_i^2 = -m^2$; by construction, this residue is one.

20

Two-particle elastic scattering at one loop

Prerequisite: 19

We now illustrate the general rules of section 19 by computing the two-particle elastic scattering amplitude, including all one-loop corrections, in φ^3 theory in six dimensions. *Elastic* means that the number of outgoing particles (of each species, in more general contexts) is the same as the number of incoming particles (of each species).

We computed the amplitude for this process at tree level in section 10, with the result

$$i\mathcal{T}_{\text{tree}} = \tfrac{1}{i}(ig)^2\Big[\tilde{\Delta}(-s) + \tilde{\Delta}(-t) + \tilde{\Delta}(-u)\Big] \,, \tag{20.1}$$

where $\tilde{\Delta}(-s) = 1/(-s + m^2 - i\epsilon)$ is the free-field propagator, and s, t, and u are the Mandelstam variables. Later we will need to remember that s is positive, that t and u are negative, and that $s + t + u = 4m^2$.

The exact scattering amplitude is given by the diagrams of fig. 20.1, with all propagators and vertices interpreted as *exact* propagators and vertices. (Recall, however, that each external propagator contributes only the residue of the pole at $k^2 = -m^2$, and that this residue is one; thus the factor associated with each external line is simply one.) We get the one-loop approximation to the exact amplitude by using the one-loop expressions for the internal propagators and vertices. We thus have

$$\begin{aligned} i\mathcal{T}_{1\text{-loop}} = \tfrac{1}{i}\Big([i\mathbf{V}_3(s)]^2\tilde{\mathbf{\Delta}}(-s) + [i\mathbf{V}_3(t)]^2\tilde{\mathbf{\Delta}}(-t) + [i\mathbf{V}_3(u)]^2\tilde{\mathbf{\Delta}}(-u)\Big) \\ + i\mathbf{V}_4(s,t,u) \,, \end{aligned} \tag{20.2}$$

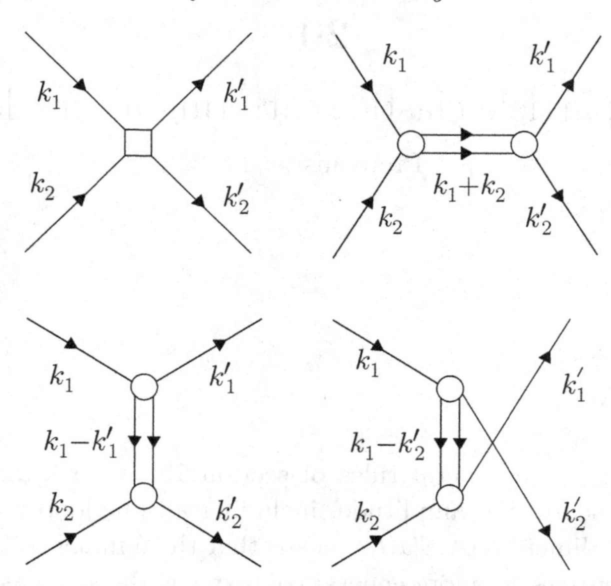

Figure 20.1. The Feynman diagrams contributing to the two-particle elastic scattering amplitude; a double line stands for the exact propagator $\frac{1}{i}\tilde{\mathbf{\Delta}}(k)$, a circle for the exact three-point vertex $\mathbf{V}_3(k_1, k_2, k_3)$, and a square for the exact four-point vertex $\mathbf{V}_4(k_1, k_2, k_3, k_4)$. An external line stands for the unit residue of the pole at $k^2 = -m^2$.

where, suppressing the $i\epsilon$s,

$$\tilde{\mathbf{\Delta}}(-s) = \frac{1}{-s + m^2 - \Pi(-s)}\,, \tag{20.3}$$

$$\Pi(-s) = \tfrac{1}{2}\alpha \int_0^1 dx\, D_2(s) \ln\Big(D_2(s)/D_0\Big) - \tfrac{1}{12}\alpha(-s + m^2)\,, \tag{20.4}$$

$$\mathbf{V}_3(s)/g = 1 - \tfrac{1}{2}\alpha \int dF_3 \, \ln\Big(D_3(s)/m^2\Big)\,, \tag{20.5}$$

$$\mathbf{V}_4(s,t,u) = \tfrac{1}{6} g^2 \alpha \int dF_4 \left[\frac{1}{D_4(s,t)} + \frac{1}{D_4(t,u)} + \frac{1}{D_4(u,s)}\right]. \tag{20.6}$$

Here $\alpha = g^2/(4\pi)^3$, the Feynman integration measure is

$$\begin{aligned}\int dF_n\, f(x) &= (n{-}1)! \int_0^1 dx_1 \ldots dx_n\, \delta(x_1{+}\ldots{+}x_n{-}1) f(x) \\ &= (n{-}1)! \int_0^1 dx_1 \int_0^{1-x_1} dx_2 \ldots \int_0^{1-x_1-\ldots-x_{n-2}} dx_{n-1} \\ &\qquad \times f(x)\Big|_{x_n = 1-x_1-\ldots-x_{n-1}}\,,\end{aligned} \tag{20.7}$$

and we have defined

$$D_2(s) = -x(1-x)s + m^2 \,, \tag{20.8}$$

$$D_0 = +[1-x(1-x)]m^2 \,, \tag{20.9}$$

$$D_3(s) = -x_1x_2s + [1-(x_1+x_2)x_3]m^2 \,, \tag{20.10}$$

$$D_4(s,t) = -x_1x_2s - x_3x_4t + [1-(x_1+x_2)(x_3+x_4)]m^2 \,. \tag{20.11}$$

We obtain $\mathbf{V}_3(s)$ from the general three-point function $\mathbf{V}_3(k_1,k_2,k_3)$ by setting two of the three k_i^2 to $-m^2$, and the third to $-s$. We obtain $\mathbf{V}_4(s,t,u)$ from the general four-point function $\mathbf{V}_4(k_1,\ldots,k_4)$ by setting all four k_i^2 to $-m^2$, $(k_1+k_2)^2$ to $-s$, $(k_1+k_3)^2$ to $-t$, and $(k_1+k_4)^2$ to $-u$. (Recall that the vertex functions are defined with all momenta treated as incoming; here we have identified $-k_3$ and $-k_4$ as the outgoing momenta.)

Eqs. (20.2)–(20.11) are formidable expressions. To gain some intuition about them, let us consider the limit of high-energy, fixed angle scattering, where we take s, $|t|$, and $|u|$ all much larger than m^2. Equivalently, we are considering the amplitude in the limit of zero particle mass.

We can then set $m^2 = 0$ in $D_2(s)$, $D_3(s)$, and $D_4(s,t)$. For the self-energy, we get

$$\begin{aligned}\Pi(-s) &= -\tfrac{1}{2}\alpha\, s \int_0^1 dx\, x(1-x)\left[\ln\!\left(\frac{-s}{m^2}\right) + \ln\!\left(\frac{x(1-x)}{1-x(1-x)}\right)\right] + \tfrac{1}{12}\alpha\, s \\ &= -\tfrac{1}{12}\alpha\, s\Big[\ln(-s/m^2) + 3 - \pi\sqrt{3}\,\Big] \,. \end{aligned}\tag{20.12}$$

Thus,

$$\begin{aligned}\tilde{\mathbf{\Delta}}(-s) &= \frac{1}{-s - \Pi(-s)} \\ &= -\frac{1}{s}\Big(1 + \tfrac{1}{12}\alpha\Big[\ln(-s/m^2) + 3 - \pi\sqrt{3}\,\Big]\Big) + O(\alpha^2) \,. \end{aligned}\tag{20.13}$$

The appropriate branch of the logarithm is found by replacing s by $s + i\epsilon$. For s real and positive, $-s$ lies just below the negative real axis, and so

$$\ln(-s) = \ln s - i\pi \,. \tag{20.14}$$

For t (or u), which is negative, we have instead

$$\begin{aligned}\ln(-t) &= \ln|t| \,, \\ \ln t &= \ln|t| + i\pi \,. \end{aligned}\tag{20.15}$$

For the three-point vertex, we get

$$\mathbf{V}_3(s)/g = 1 - \tfrac{1}{2}\alpha \int dF_3 \Big[\ln(-s/m^2) + \ln(x_1 x_2) \Big]$$
$$= 1 - \tfrac{1}{2}\alpha \Big[\ln(-s/m^2) - 3 \Big] , \tag{20.16}$$

where the same comments about the appropriate branch apply.

For the four-point vertex, the integral over the Feynman parameters can be done in closed form, with the result

$$\int \frac{dF_4}{D_4(s,t)} = -\frac{3}{s+t}\left(\pi^2 + \Big[\ln(s/t)\Big]^2\right)$$
$$= +\frac{3}{u}\left(\pi^2 + \Big[\ln(s/t)\Big]^2\right) , \tag{20.17}$$

where the second line follows from $s + t + u = 0$.

Putting all of this together, we have

$$\mathcal{T}_{\text{1-loop}} = g^2 \Big[F(s,t,u) + F(t,u,s) + F(u,s,t) \Big] , \tag{20.18}$$

where

$$F(s,t,u) \equiv -\frac{1}{s}\left(1 - \tfrac{11}{12}\alpha\Big[\ln(-s/m^2) + c\Big] - \tfrac{1}{2}\alpha\Big[\ln(t/u)\Big]^2\right) , \tag{20.19}$$

and $c = (6\pi^2 + \pi\sqrt{3} - 39)/11 = 2.33$. This is a typical result of a loop calculation: the original tree-level amplitude is corrected by powers of logarithms of kinematic variables.

Problems

20.1) Verify eq. (20.17).

20.2) Compute the $O(\alpha)$ correction to the two-particle scattering amplitude *at threshold*, that is, for $s = 4m^2$ and $t = u = 0$, corresponding to zero three-momentum for both the incoming and outgoing particles.

21

The quantum action

Prerequisite: 19

In section 19, we saw how to compute (in φ^3 theory in $d=6$ dimensions) the 1PI vertex functions $\mathbf{V}_n(k_1,\ldots,k_n)$ for $n \geq 4$ via the *skeleton expansion*: draw all Feynman diagrams with n external lines that are one-, two-, and three-particle irreducible, and compute them using the exact propagator $\tilde{\mathbf{\Delta}}(k^2)$ and three-point vertex function $\mathbf{V}_3(k_1,k_2,k_3)$.

We now define the *quantum action* (or *effective action*, or *quantum effective action*)

$$\Gamma(\varphi) \equiv -\frac{1}{2}\int \frac{d^dk}{(2\pi)^d}\,\widetilde{\varphi}(-k)\Big(k^2+m^2-\Pi(k^2)\Big)\widetilde{\varphi}(k)$$
$$+\sum_{n=3}^{\infty}\frac{1}{n!}\int \frac{d^dk_1}{(2\pi)^d}\cdots\frac{d^dk_n}{(2\pi)^d}\,(2\pi)^d\delta^d(k_1+\ldots+k_n)$$
$$\times \mathbf{V}_n(k_1,\ldots,k_n)\,\widetilde{\varphi}(k_1)\ldots\widetilde{\varphi}(k_n)\,, \qquad (21.1)$$

where $\widetilde{\varphi}(k) = \int d^dx\, e^{-ikx}\varphi(x)$. The quantum action has the property that the *tree-level* Feynman diagrams it generates give the *complete* scattering amplitude of the original theory.

In this section, we will determine the relationship between $\Gamma(\varphi)$ and the sum of connected diagrams with sources, $iW(J)$, introduced in section 9. Recall that $W(J)$ is related to the path integral

$$Z(J) = \int \mathcal{D}\varphi\, \exp\left[\, iS(\varphi) + i\int d^dx\, J\varphi \right], \qquad (21.2)$$

where $S = \int d^dx\, \mathcal{L}$ is the action, via

$$Z(J) = \exp[iW(J)]\,. \qquad (21.3)$$

Consider now the path integral

$$Z_\Gamma(J) \equiv \int \mathcal{D}\varphi \, \exp\left[i\Gamma(\varphi) + i\int d^d x \, J\varphi \right] \tag{21.4}$$

$$= \exp[iW_\Gamma(J)] \, . \tag{21.5}$$

$W_\Gamma(J)$ is given by the sum of connected diagrams (with sources) in which each line represents the exact propagator, and each n-point vertex represents the exact 1PI vertex $\mathbf{V}_n$. $W_\Gamma(J)$ would be equal to $W(J)$ if we included only tree diagrams in $W_\Gamma(J)$.

We can isolate the tree-level contribution to a path integral by means of the following trick. Introduce a dimensionless parameter that we will call $\hbar$, and the path integral

$$Z_{\Gamma,\hbar}(J) \equiv \int \mathcal{D}\varphi \, \exp\left[\frac{i}{\hbar}\left(\Gamma(\varphi) + \int d^d x \, J\varphi \right) \right] \tag{21.6}$$

$$= \exp[iW_{\Gamma,\hbar}(J)] \, . \tag{21.7}$$

In a given connected diagram with sources, every propagator (including those that connect to sources) is multiplied by $\hbar$, every source by $1/\hbar$, and every vertex by $1/\hbar$. The overall factor of $\hbar$ is then $\hbar^{P-E-V}$, where V is the number of vertices, E is the number of sources (equivalently, the number of external lines after we remove the sources), and P is the number of propagators (external and internal). We next note that $P{-}E{-}V$ is equal to $L{-}1$, where L is the number of closed loops. This can be seen by counting the number of internal momenta and the constraints among them. Specifically, assign an unfixed momentum to each internal line; there are $P{-}E$ of these momenta. Then the V vertices provide V constraints. One linear combination of these constraints gives overall momentum conservation, and so does not constrain the internal momenta. Therefore, the number of internal momenta left unfixed by the vertex constraints is $(P{-}E){-}(V{-}1)$, and the number of unfixed momenta is the same as the number of loops L.

So, $W_{\Gamma,\hbar}(J)$ can be expressed as a power series in $\hbar$ of the form

$$W_{\Gamma,\hbar}(J) = \sum_{L=0}^{\infty} \hbar^{L-1} \, W_{\Gamma,L}(J) \, . \tag{21.8}$$

If we take the formal limit of $\hbar \to 0$, the dominant term is the one with $L = 0$, which is given by the sum of tree diagrams only. This is just what we want. We conclude that

$$W(J) = W_{\Gamma,L=0}(J) \, . \tag{21.9}$$

Next we perform the path integral in eq. (21.6) by the method of stationary phase. We find the point (actually, the field configuration) at which the exponent is stationary; this is given by the solution of the *quantum equation of motion*

$$\frac{\delta}{\delta\varphi(x)}\Gamma(\varphi) = -J(x) \; . \tag{21.10}$$

Let $\varphi_J(x)$ denote the solution of eq. (21.10) with a specified source function $J(x)$. Then the stationary-phase approximation to $Z_{\Gamma,\hbar}(J)$ is

$$Z_{\Gamma,\hbar}(J) = \exp\left[\frac{i}{\hbar}\left(\Gamma(\varphi_J) + \int d^dx\, J\varphi_J\right) + O(\hbar^0)\right] . \tag{21.11}$$

Combining the results of eqs. (21.7)–(21.9) and (21.11), we find

$$W(J) = \Gamma(\varphi_J) + \int d^dx\, J\varphi_J \, . \tag{21.12}$$

This is the main result of this section.

Let us explore it further. Recall from section 9 that the vacuum expectation value of the field operator $\varphi(x)$ is given by

$$\langle 0|\varphi(x)|0\rangle = \frac{\delta}{\delta J(x)} W(J)\bigg|_{J=0} \; . \tag{21.13}$$

Now consider what we get if we do not set $J = 0$ after taking the derivative:

$$\langle 0|\varphi(x)|0\rangle_J \equiv \frac{\delta}{\delta J(x)} W(J) \; . \tag{21.14}$$

This is the vacuum expectation value of $\varphi(x)$ in the presence of a nonzero source function $J(x)$. We can get some more information about it by using eq. (21.12) for $W(J)$. Making use of the product rule for derivatives, we have

$$\langle 0|\varphi(x)|0\rangle_J = \frac{\delta}{\delta J(x)}\Gamma(\varphi_J) + \varphi_J(x) + \int d^dy\, J(y)\frac{\delta\varphi_J(y)}{\delta J(x)} \; . \tag{21.15}$$

We can evaluate the first term on the right-hand side by using the chain rule,

$$\frac{\delta}{\delta J(x)}\Gamma(\varphi_J) = \int d^dy\, \frac{\delta\Gamma(\varphi_J)}{\delta\varphi_J(y)}\frac{\delta\varphi_J(y)}{\delta J(x)} \; . \tag{21.16}$$

Then we can combine the first and third terms on the right-hand side of eq. (21.15) to get

$$\langle 0|\varphi(x)|0\rangle_J = \int d^dy \left[\frac{\delta\Gamma(\varphi_J)}{\delta\varphi_J(y)} + J(y)\right]\frac{\delta\varphi_J(y)}{\delta J(x)} + \varphi_J(x) \; . \tag{21.17}$$

Now we note from eq. (21.10) that the factor in large brackets on the right-hand side of eq. (21.17) vanishes, and so

$$\langle 0|\varphi(x)|0\rangle_J = \varphi_J(x) \; . \tag{21.18}$$

That is, the vacuum expectation value of the field operator $\varphi(x)$ in the presence of a nonzero source function is also the solution to the quantum equation of motion, eq. (21.10).

We can also write the quantum action in terms of a *derivative expansion*,

$$\Gamma(\varphi) = \int d^d x \left[-\mathcal{U}(\varphi) - \tfrac{1}{2}\mathcal{Z}(\varphi)\partial^\mu\varphi\partial_\mu\varphi + \ldots \right], \tag{21.19}$$

where the ellipses stand for an infinite number of terms with more and more derivatives, and $\mathcal{U}(\varphi)$ and $\mathcal{Z}(\varphi)$ are ordinary functions (not functionals) of $\varphi(x)$. $\mathcal{U}(\varphi)$ is called the *quantum potential* (or *effective potential*, or *quantum effective potential*), and it plays an important conceptual role in theories with spontaneous symmetry breaking; see section 31. However, it is rarely necessary to compute it explicitly, except in those cases where we are unable to do so.

Reference notes

Construction of the quantum action is discussed in *Coleman*, *Itzykson & Zuber*, *Peskin & Schroeder*, and *Weinberg II*.

Problems

21.1) Show that

$$\Gamma(\varphi) = W(J_\varphi) - \int d^d x \, J_\varphi \varphi \, , \tag{21.20}$$

where $J_\varphi(x)$ is the solution of

$$\frac{\delta}{\delta J(x)} W(J) = \varphi(x) \tag{21.21}$$

for a specified $\varphi(x)$.

21.2) *Symmetries of the quantum action.* Suppose that we have a set of fields $\varphi_a(x)$, and that both the classical action $S(\varphi)$ and the integration measure $\mathcal{D}\varphi$ are invariant under

$$\varphi_a(x) \to \int d^d y \; R_{ab}(x,y)\varphi_b(y) \tag{21.22}$$

for some particular function $R_{ab}(x,y)$. Typically $R_{ab}(x,y)$ is a constant matrix times $\delta^d(x-y)$, or a finite number of derivatives of $\delta^d(x-y)$; see sections 22–24 for some examples.

a) Show that $W(J)$ is invariant under

$$J_a(x) \to \int d^dy \; J_b(y) R_{ba}(y,x) \; . \tag{21.23}$$

b) Use eqs. (21.20) and (21.23) to show that the quantum action $\Gamma(\varphi)$ is invariant under eq. (21.22). This is an important result that we will use frequently.

21.3) Consider performing the path integral in the presence of a *background field* $\bar{\varphi}(x)$; we define

$$\exp[iW(J;\bar{\varphi})] \equiv \int \mathcal{D}\varphi \; \exp\left[iS(\varphi{+}\bar{\varphi}) + i\int d^dx \; J\varphi \right] . \tag{21.24}$$

Then $W(J;0)$ is the original $W(J)$ of eq. (21.3). We also define the quantum action in the presence of the background field,

$$\Gamma(\varphi;\bar{\varphi}) \equiv W(J_\varphi;\bar{\varphi}) - \int d^dx \; J_\varphi \varphi \; , \tag{21.25}$$

where $J_\varphi(x)$ is the solution of

$$\frac{\delta}{\delta J(x)} W(J;\bar{\varphi}) = \varphi(x) \tag{21.26}$$

for a specified $\varphi(x)$. Show that

$$\Gamma(\varphi;\bar{\varphi}) = \Gamma(\varphi{+}\bar{\varphi};0) \; , \tag{21.27}$$

where $\Gamma(\varphi;0)$ is the original quantum action of eq. (21.1).

22

Continuous symmetries and conserved currents

Prerequisite: 8

Suppose we have a set of scalar fields $\varphi_a(x)$, and a lagrangian density $\mathcal{L}(x) = \mathcal{L}(\varphi_a(x), \partial_\mu\varphi_a(x))$. Consider what happens to $\mathcal{L}(x)$ if we make an infinitesimal change $\varphi_a(x) \to \varphi_a(x) + \delta\varphi_a(x)$ in each field. We have $\mathcal{L}(x) \to \mathcal{L}(x) + \delta\mathcal{L}(x)$, where $\delta\mathcal{L}(x)$ is given by the chain rule,

$$\delta\mathcal{L}(x) = \frac{\partial\mathcal{L}}{\partial\varphi_a(x)}\,\delta\varphi_a(x) + \frac{\partial\mathcal{L}}{\partial(\partial_\mu\varphi_a(x))}\,\partial_\mu\delta\varphi_a(x)\,. \tag{22.1}$$

Next consider the classical equations of motion (also known as the Euler–Lagrange equations, or the *field equations*), given by the action principle

$$\frac{\delta S}{\delta\varphi_a(x)} = 0\,, \tag{22.2}$$

where $S = \int d^4y\,\mathcal{L}(y)$ is the action, and $\delta/\delta\varphi_a(x)$ is a functional derivative. (For definiteness, we work in four spacetime dimensions, though our results will apply in any number.) We have (with repeated indices implicitly summed)

$$\begin{aligned}
\frac{\delta S}{\delta\varphi_a(x)} &= \int d^4y\,\frac{\delta\mathcal{L}(y)}{\delta\varphi_a(x)} \\
&= \int d^4y\left[\frac{\partial\mathcal{L}(y)}{\partial\varphi_b(y)}\,\frac{\delta\varphi_b(y)}{\delta\varphi_a(x)} + \frac{\partial\mathcal{L}(y)}{\partial(\partial_\mu\varphi_b(y))}\,\frac{\delta(\partial_\mu\varphi_b(y))}{\delta\varphi_a(x)}\right] \\
&= \int d^4y\left[\frac{\partial\mathcal{L}(y)}{\partial\varphi_b(y)}\,\delta_{ba}\delta^4(y-x) + \frac{\partial\mathcal{L}(y)}{\partial(\partial_\mu\varphi_b(y))}\,\delta_{ba}\partial_\mu\delta^4(y-x)\right] \\
&= \frac{\partial\mathcal{L}(x)}{\partial\varphi_a(x)} - \partial_\mu\,\frac{\partial\mathcal{L}(x)}{\partial(\partial_\mu\varphi_a(x))}\,.
\end{aligned} \tag{22.3}$$

We can use this result to make the replacement

$$\frac{\partial\mathcal{L}(x)}{\partial\varphi_a(x)} \to \partial_\mu \frac{\partial\mathcal{L}(x)}{\partial(\partial_\mu\varphi_a(x))} + \frac{\delta S}{\delta\varphi_a(x)} \tag{22.4}$$

in eq. (22.1). Then, combining two of the terms, we get

$$\delta\mathcal{L}(x) = \partial_\mu\left(\frac{\partial\mathcal{L}(x)}{\partial(\partial_\mu\varphi_a(x))}\,\delta\varphi_a(x)\right) + \frac{\delta S}{\delta\varphi_a(x)}\delta\varphi_a(x)\,. \tag{22.5}$$

Next we identify the object in large parentheses in eq. (22.5) as the *Noether current*

$$j^\mu(x) \equiv \frac{\partial\mathcal{L}(x)}{\partial(\partial_\mu\varphi_a(x))}\,\delta\varphi_a(x)\,. \tag{22.6}$$

Eq. (22.5) then implies

$$\partial_\mu j^\mu(x) = \delta\mathcal{L}(x) - \frac{\delta S}{\delta\varphi_a(x)}\delta\varphi_a(x)\,. \tag{22.7}$$

If the classical field equations are satisfied, then the second term on the right-hand side of eq. (22.7) vanishes.

The Noether current plays a special role if we can find a set of infinitesimal field transformations that leaves the lagrangian unchanged, or *invariant.* In this case, we have $\delta\mathcal{L} = 0$, and we say that the lagrangian has a *continuous symmetry.* From eq. (22.7), we then have $\partial_\mu j^\mu = 0$ whenever the field equations are satisfied, and we say that the Noether current is *conserved.* In terms of its space and time components, this means that

$$\frac{\partial}{\partial t}j^0(x) + \nabla\cdot\mathbf{j}(x) = 0\,. \tag{22.8}$$

If we interpret $j^0(x)$ as a *charge density*, and $\mathbf{j}(x)$ as the corresponding *current density*, then eq. (22.8) expresses the local conservation of this charge.

Let us see an example of this. Consider a theory of a *complex* scalar field with lagrangian

$$\mathcal{L} = -\partial^\mu\varphi^\dagger\partial_\mu\varphi - m^2\varphi^\dagger\varphi - \tfrac{1}{4}\lambda(\varphi^\dagger\varphi)^2\,. \tag{22.9}$$

We can also rewrite $\mathcal{L}$ in terms of two real scalar fields by setting $\varphi = (\varphi_1 + i\varphi_2)/\sqrt{2}$ to get

$$\mathcal{L} = -\tfrac{1}{2}\partial^\mu\varphi_1\partial_\mu\varphi_1 - \tfrac{1}{2}\partial^\mu\varphi_2\partial_\mu\varphi_2 - \tfrac{1}{2}m^2(\varphi_1^2+\varphi_2^2) - \tfrac{1}{16}\lambda(\varphi_1^2+\varphi_2^2)^2\,. \tag{22.10}$$

In the form of eq. (22.9), it is obvious that $\mathcal{L}$ is left invariant by the transformation

$$\varphi(x) \to e^{-i\alpha}\varphi(x) \,, \tag{22.11}$$

where α is a real number. This is called a *U(1) transformation*, a transformation by a unitary 1×1 matrix. In terms of φ_1 and φ_2, this transformation reads

$$\begin{pmatrix} \varphi_1(x) \\ \varphi_2(x) \end{pmatrix} \to \begin{pmatrix} \cos\alpha & \sin\alpha \\ -\sin\alpha & \cos\alpha \end{pmatrix} \begin{pmatrix} \varphi_1(x) \\ \varphi_2(x) \end{pmatrix} . \tag{22.12}$$

If we think of (φ_1, φ_2) as a two-component vector, then eq. (22.12) is just a rotation of this vector in the plane by angle α. Eq. (22.12) is called an *SO(2) transformation*, a transformation by an orthogonal 2×2 matrix with a special value of the determinant (namely $+1$, as opposed to -1, the only other possibility for an orthogonal matrix). We have learned that a U(1) transformation can be mapped into an SO(2) transformation.

The infinitesimal form of eq. (22.11) is

$$\begin{aligned} \varphi(x) &\to \varphi(x) - i\alpha\varphi(x) \,, \\ \varphi^\dagger(x) &\to \varphi^\dagger(x) + i\alpha\varphi^\dagger(x) \,, \end{aligned} \tag{22.13}$$

where α is now infinitesimal. In eq. (22.6), we should treat φ and $\varphi^\dagger$ as independent fields. It is also conventional to scale the infinitesimal parameter out of the current, so that we have

$$\begin{aligned} \alpha\, j^\mu &= \frac{\partial\mathcal{L}}{\partial(\partial_\mu\varphi)}\,\delta\varphi + \frac{\partial\mathcal{L}}{\partial(\partial_\mu\varphi^\dagger)}\,\delta\varphi^\dagger \\ &= (-\partial^\mu\varphi^\dagger)(-i\alpha\varphi) + (-\partial^\mu\varphi)(+i\alpha\varphi^\dagger) \\ &= \alpha\,\mathrm{Im}(\varphi^\dagger \overleftrightarrow{\partial^\mu} \varphi) \,, \end{aligned} \tag{22.14}$$

where $A\overleftrightarrow{\partial^\mu}B \equiv A\partial^\mu B - (\partial^\mu A)B$. Canceling out α, we find that the Noether current is

$$j^\mu = \mathrm{Im}(\varphi^\dagger \overleftrightarrow{\partial^\mu} \varphi) \,. \tag{22.15}$$

We can also repeat this exercise using the SO(2) form of the transformation. For infinitesimal α, eq. (22.12) becomes $\delta\varphi_1 = +\alpha\varphi_2$ and $\delta\varphi_2 = -\alpha\varphi_1$.

Then the Noether current is given by

$$\begin{aligned}\alpha\, j^\mu &= \frac{\partial\mathcal{L}}{\partial(\partial_\mu\varphi_1)}\delta\varphi_1 + \frac{\partial\mathcal{L}}{\partial(\partial_\mu\varphi_2)}\delta\varphi_2 \\ &= (-\partial^\mu\varphi_1)(+\alpha\varphi_2) + (-\partial^\mu\varphi_2)(-\alpha\varphi_1) \\ &= \alpha\,(\varphi_1\overleftrightarrow{\partial^\mu}\varphi_2)\,, \qquad (22.16)\end{aligned}$$

which is (hearteningly) equivalent to eq. (22.14).

Let us define the Noether charge

$$Q \equiv \int d^3x\, j^0(x) = \int d^3x\ \mathrm{Im}(\varphi^\dagger\overleftrightarrow{\partial^0}\varphi)\,, \qquad (22.17)$$

and investigate its properties. If we integrate eq. (22.8) over d^3x, use Gauss's law to write the volume integral of $\nabla\cdot\mathbf{j}$ as a surface integral, and assume that the boundary conditions at infinity fix $\mathbf{j}(x) = 0$ on that surface, then we find that Q is constant in time. To get a better idea of the physical implications of this, let us rewrite Q using the free-field expansions

$$\begin{aligned}\varphi(x) &= \int \widetilde{dk}\,\Big[a(\mathbf{k})e^{ikx} + b^*(\mathbf{k})e^{-ikx}\Big]\,, \\ \varphi^\dagger(x) &= \int \widetilde{dk}\,\Big[b(\mathbf{k})e^{ikx} + a^*(\mathbf{k})e^{-ikx}\Big]\,. \qquad (22.18)\end{aligned}$$

We have written $a^*(\mathbf{k})$ and $b^*(\mathbf{k})$ rather than $a^\dagger(\mathbf{k})$ and $b^\dagger(\mathbf{k})$ because so far our discussion has been about the classical field theory. In a theory with interactions, these formulae (and their first time derivatives) are valid at any one particular time (say, $t = -\infty$). Then, we can plug them into eq. (22.17), and find (after some manipulation similar to what we did for the hamiltonian in section 3)

$$Q = \int \widetilde{dk}\,\Big[a^*(\mathbf{k})a(\mathbf{k}) - b(\mathbf{k})b^*(\mathbf{k})\Big]\,. \qquad (22.19)$$

In the quantum theory, this becomes an operator that counts the number of a particles minus the number of b particles. This number is then time-independent, and so the scattering amplitude vanishes identically for any process that changes the value of Q. This can be seen directly from the Feynman rules, which conserve Q at every vertex.

To better understand the implications of the Noether current in the quantum theory, we begin by considering the path integral,

$$Z(J) = \int \mathcal{D}\varphi\, e^{i[S+\int d^4y\, J_a\varphi_a]}\,. \qquad (22.20)$$

The value of $Z(J)$ is unchanged if we make the change of variable $\varphi_a(x) \to \varphi_a(x) + \delta\varphi_a(x)$, with $\delta\varphi_a(x)$ an arbitrary infinitesimal shift that (we assume) leaves the measure $\mathcal{D}\varphi$ invariant. Thus we have

$$\begin{aligned} 0 &= \delta Z(J) \\ &= i \int \mathcal{D}\varphi\, e^{i[S+\int d^4y\, J_b\varphi_b]} \int d^4x \left(\frac{\delta S}{\delta\varphi_a(x)} + J_a(x) \right) \delta\varphi_a(x) \,. \end{aligned} \tag{22.21}$$

We can now take n functional derivatives with respect to $J_{a_j}(x_j)$, and then set $J = 0$, to get

$$\begin{aligned} 0 = \int \mathcal{D}\varphi\, e^{iS} \int d^4x \biggl[i\, &\frac{\delta S}{\delta\varphi_a(x)} \varphi_{a_1}(x_1) \ldots \varphi_{a_n}(x_n) \\ &+ \sum_{j=1}^{n} \varphi_{a_1}(x_1) \ldots \delta_{aa_j} \delta^4(x{-}x_j) \ldots \varphi_{a_n}(x_n) \biggr] \delta\varphi_a(x) \,. \end{aligned} \tag{22.22}$$

Since $\delta\varphi_a(x)$ is arbitrary, we can drop it (and the integral over d^4x). Then, since the path integral computes the vacuum expectation value of the time-ordered product, we have

$$\begin{aligned} 0 = i\langle 0|\mathrm{T}\, &\frac{\delta S}{\delta\varphi_a(x)} \varphi_{a_1}(x_1) \ldots \varphi_{a_n}(x_n)|0\rangle \\ &+ \sum_{j=1}^{n} \langle 0|\mathrm{T}\varphi_{a_1}(x_1) \ldots \delta_{aa_j} \delta^4(x{-}x_j) \ldots \varphi_{a_n}(x_n)|0\rangle \,. \end{aligned} \tag{22.23}$$

These are the *Schwinger–Dyson equations* for the theory.

To get a feel for them, let us look at free-field theory for a single real scalar field, for which $\delta S/\delta\varphi(x) = (\partial_x^2 - m^2)\varphi(x)$. For $n = 1$ we get

$$(-\partial_x^2 + m^2)i\langle 0|\mathrm{T}\varphi(x)\varphi(x_1)|0\rangle = \delta^4(x{-}x_1) \,. \tag{22.24}$$

That the Klein–Gordon wave operator should sit outside the time-ordered product (and hence act on the time-ordering step functions) is clear from the path integral form of eq. (22.22). We see from eq. (22.24) that the free-field propagator, $\Delta(x{-}x_1) = i\langle 0|\mathrm{T}\varphi(x)\varphi(x_1)|0\rangle$, is a Green's function for the Klein–Gordon wave operator, a fact we first learned in section 8.

More generally, we can write

$$\langle 0|\mathrm{T} \frac{\delta S}{\delta\varphi_a(x)} \varphi_{a_1}(x_1) \ldots \varphi_{a_n}(x_n)|0\rangle = 0 \quad \text{for} \quad x \neq x_{1,\ldots,n} \,. \tag{22.25}$$

We see that the classical equation of motion is satisfied by a quantum field inside a correlation function, as long as its spacetime argument differs from those of all the other fields. When this is not the case, we get extra *contact terms.*

Let us now consider a theory that has a continuous symmetry and a corresponding Noether current. Take eq. (22.22), and set $\delta\varphi_a(x)$ to be the infinitesimal change in $\varphi_a(x)$ that results in $\delta\mathcal{L}(x) = 0$. Now sum over the index a, and use eq. (22.7). The result is the *Ward* (or *Ward–Takahashi*) *identity*

$$0 = \partial_\mu \langle 0|\mathrm{T}j^\mu(x)\varphi_{a_1}(x_1)\ldots\varphi_{a_n}(x_n)|0\rangle \\ + i\sum_{j=1}^{n}\langle 0|\mathrm{T}\varphi_{a_1}(x_1)\ldots\delta\varphi_{a_j}(x)\delta^4(x-x_j)\ldots\varphi_{a_n}(x_n)|0\rangle \,. \tag{22.26}$$

Thus, conservation of the Noether current holds in the quantum theory, with the current inside a correlation function, up to contact terms with a specific form that depends on the details of the infinitesimal transformation that leaves $\mathcal{L}$ invariant.

The Noether current is also useful in a slightly more general context. Suppose we have a transformation of the fields such that $\delta\mathcal{L}(x)$ is not zero, but instead is a total divergence: $\delta\mathcal{L}(x) = \partial_\mu K^\mu(x)$ for some $K^\mu(x)$. Then there is still a conserved current, now given by

$$j^\mu(x) = \frac{\partial\mathcal{L}(x)}{\partial(\partial_\mu\varphi_a(x))}\,\delta\varphi_a(x) - K^\mu(x) \,. \tag{22.27}$$

An example of this is provided by the symmetry of *spacetime translations.* We transform the fields via $\varphi_a(x) \to \varphi_a(x-a)$, where a^μ is a constant four-vector. The infinitesimal version of this is $\varphi_a(x) \to \varphi_a(x) - a^\nu\partial_\nu\varphi_a(x)$, and so we have $\delta\varphi_a(x) = -a^\nu\partial_\nu\varphi_a(x)$. Under this transformation, we obviously have $\mathcal{L}(x) \to \mathcal{L}(x-a)$, and so $\delta\mathcal{L}(x) = -a^\nu\partial_\nu\mathcal{L}(x) = -\partial_\nu(a^\nu\mathcal{L}(x))$. Thus in this case $K^\nu(x) = -a^\nu\mathcal{L}(x)$, and the conserved current is

$$j^\mu(x) = \frac{\partial\mathcal{L}(x)}{\partial(\partial_\mu\varphi_a(x))}(-a^\nu\partial_\nu\varphi_a(x)) + a^\mu\mathcal{L}(x) \\ = a_\nu T^{\mu\nu}(x) \,, \tag{22.28}$$

where we have defined the *stress-energy* or *energy-momentum tensor*

$$T^{\mu\nu}(x) \equiv -\frac{\partial\mathcal{L}(x)}{\partial(\partial_\mu\varphi_a(x))}\,\partial^\nu\varphi_a(x) + g^{\mu\nu}\mathcal{L}(x) \,. \tag{22.29}$$

For a renormalizable theory of a set of real scalar fields $\varphi_a(x)$, the lagrangian takes the form

$$\mathcal{L} = -\tfrac{1}{2}\partial^\mu\varphi_a\partial_\mu\varphi_a - V(\varphi) \,, \tag{22.30}$$

where $V(\varphi)$ is a polynomial in the φ_as. In this case

$$T^{\mu\nu} = \partial^\mu\varphi_a\partial^\nu\varphi_a + g^{\mu\nu}\mathcal{L} \,. \tag{22.31}$$

In particular,

$$T^{00} = \tfrac{1}{2}\Pi_a^2 + \tfrac{1}{2}(\nabla\varphi_a)^2 + V(\varphi) \,, \tag{22.32}$$

where $\Pi_a = \partial_0\varphi_a$ is the canonical momentum conjugate to the field φ_a. We recognize T^{00} as the *hamiltonian density* $\mathcal{H}$ that corresponds to the lagrangian density of eq. (22.30). Then, by Lorentz symmetry, T^{0j} must be the corresponding momentum density. We have

$$T^{0j} = \partial^0\varphi_a\partial^j\varphi_a = -\Pi_a\nabla^j\varphi_a \,. \tag{22.33}$$

To check that this is a sensible result, we use the free-field expansion for a set of real scalar fields [the same as eq. (22.18), but with $b(\mathbf{k}) = a(\mathbf{k})$ for each field]; then we find that the momentum operator is given by

$$P^j = \int d^3x\, T^{0j}(x) = \int \widetilde{dk}\; k^j\, a_a^\dagger(\mathbf{k})a_a(\mathbf{k}) \,, \tag{22.34}$$

which is just what we would expect. We therefore identify the *energy-momentum four-vector* as

$$P^\mu = \int d^3x\, T^{0\mu}(x) \,. \tag{22.35}$$

Recall that in section 2 we defined the *spacetime translation operator* as

$$T(a) \equiv \exp(-iP^\mu a_\mu) \,, \tag{22.36}$$

and announced that it had the property that

$$T(a)^{-1}\varphi_a(x)T(a) = \varphi_a(x-a) \,. \tag{22.37}$$

Now that we have an explicit formula for P^μ, we can check this. This is easiest to do for infinitesimal a^μ; then eq. (22.37) becomes

$$[\varphi_a(x), P^\mu] = \tfrac{1}{i}\partial^\mu\varphi_a(x) \,. \tag{22.38}$$

This can indeed be verified by using the canonical commutation relations for $\varphi_a(x)$ and $\Pi_a(x)$.

One more symmetry we can investigate is Lorentz symmetry. If we make an infinitesimal Lorentz transformation, we have $\varphi_a(x) \to \varphi_a(x + \delta\omega{\cdot}x)$, where $\delta\omega{\cdot}x$ is shorthand for $\delta\omega^{\nu}{}_{\rho}x^{\rho}$. This case is very similar to that of spacetime translations; the only difference is that the translation parameter a^{ν} is now x dependent, $a^{\nu} \to \delta\omega^{\nu}{}_{\rho}x^{\rho}$. The resulting conserved current is

$$\mathcal{M}^{\mu\nu\rho}(x) = x^{\nu}T^{\mu\rho}(x) - x^{\rho}T^{\mu\nu}(x) , \tag{22.39}$$

and it obeys $\partial_{\mu}\mathcal{M}^{\mu\nu\rho} = 0$, with the derivative contracted with the first index. $\mathcal{M}^{\mu\nu\rho}$ is antisymmetric on its second two indices; this comes about because $\delta\omega^{\nu\rho}$ is antisymmetric. The conserved charges associated with this current are

$$M^{\nu\rho} = \int d^3x\, \mathcal{M}^{0\nu\rho}(x) , \tag{22.40}$$

and these are the *generators of the Lorentz group* that were introduced in section 2. Again, we can use the canonical commutation relations for the fields to check that the Lorentz generators have the right commutation relations, both with the fields and with each other.

Reference notes

The path-integral approach to Ward identities is treated in more detail in *Peskin & Schroeder*. An operator-based derivation can be found in *Weinberg I*.

Problems

22.1) For the Noether current of eq. (22.6), and assuming that $\delta\varphi_a$ does not involve time derivatives, use the canonical commutation relations to show that

$$[\varphi_a, Q] = i\delta\varphi_a , \tag{22.41}$$

where Q is the Noether charge.

22.2) Use the canonical commutation relations to verify eq. (22.38).

22.3) a) With $T^{\mu\nu}$ given by eq. (22.31), compute the equal-time ($x^0 = y^0$) commutators $[T^{00}(x), T^{00}(y)]$, $[T^{0i}(x), T^{00}(y)]$, and $[T^{0i}(x), T^{0j}(y)]$.

b) Use your results to verify eqs. (2.17), (2.19), and (2.20).

23

Discrete symmetries: P, T, C and Z

Prerequisite: 22

In section 2, we studied the *proper orthochronous* Lorentz transformations, which are continuously connected to the identity. In this section, we will consider the effects of *parity*,

$$\mathcal{P}^{\mu}{}_{\nu} = (\mathcal{P}^{-1})^{\mu}{}_{\nu} = \begin{pmatrix} +1 & & & \\ & -1 & & \\ & & -1 & \\ & & & -1 \end{pmatrix} . \tag{23.1}$$

and *time reversal*,

$$\mathcal{T}^{\mu}{}_{\nu} = (\mathcal{T}^{-1})^{\mu}{}_{\nu} = \begin{pmatrix} -1 & & & \\ & +1 & & \\ & & +1 & \\ & & & +1 \end{pmatrix} . \tag{23.2}$$

We will also consider certain other discrete transformations such as *charge conjugation*.

Recall from section 2 that for every proper orthochronous Lorentz transformation $\Lambda^{\mu}{}_{\nu}$ there is an associated unitary operator $U(\Lambda)$ with the property that

$$U(\Lambda)^{-1}\varphi(x)U(\Lambda) = \varphi(\Lambda^{-1}x) . \tag{23.3}$$

Thus for parity and time-reversal, we expect that there are corresponding unitary operators

$$P \equiv U(\mathcal{P}) , \tag{23.4}$$

$$T \equiv U(\mathcal{T}) , \tag{23.5}$$

such that

$$P^{-1}\varphi(x)P = \varphi(\mathcal{P}x) \,, \tag{23.6}$$
$$T^{-1}\varphi(x)T = \varphi(\mathcal{T}x) \,. \tag{23.7}$$

There is, however, an extra possible complication. Since the $\mathcal{P}$ and $\mathcal{T}$ matrices are their own inverses, a second parity or time-reversal transformation should transform all observables back into themselves. Using eqs. (23.6) and (23.7), along with $\mathcal{P}^2 = 1$ and $\mathcal{T}^2 = 1$, we see that

$$P^{-2}\varphi(x)P^2 = \varphi(x) \,, \tag{23.8}$$
$$T^{-2}\varphi(x)T^2 = \varphi(x) \,. \tag{23.9}$$

Since $\varphi(x)$ is a hermitian operator, it is in principle an observable, and so eqs. (23.8) and (23.9) are just what we expect. However, another possibility for the parity transformation of the field, different from eqs. (23.6) and (23.7), but nevertheless consistent with eqs. (23.8) and (23.9), is

$$P^{-1}\varphi(x)P = -\varphi(\mathcal{P}x) \,, \tag{23.10}$$
$$T^{-1}\varphi(x)T = -\varphi(\mathcal{T}x) \,. \tag{23.11}$$

This possible extra minus sign cannot arise for proper orthochronous Lorentz transformations, because they are continuously connected to the identity, and for the identity transformation (that is, no transformation at all), we must obviously have the plus sign.

If the minus sign appears on the right-hand side, we say that the field is *odd under parity* (or time reversal). If a scalar field is odd under parity, we sometimes say that it is a *pseudoscalar*.[1]

So, how do we know which is right, eqs. (23.6) and (23.7), or eqs. (23.10) and (23.11)? The general answer is that we get to choose, but there is a key principle to guide our choice: if at all possible, we want to define P and T so that the lagrangian density is even,

$$P^{-1}\mathcal{L}(x)P = +\mathcal{L}(\mathcal{P}x) \,, \tag{23.12}$$
$$T^{-1}\mathcal{L}(x)T = +\mathcal{L}(\mathcal{T}x) \,. \tag{23.13}$$

Then, after we integrate over d^4x to get the action S, the action will be invariant. This means that parity and time-reversal are *conserved*.

For theories with spin-zero fields only, it is clear that the choice of eqs. (23.6) and (23.7) always leads to eqs. (23.12) and (23.13), and so there is

[1] It is still a scalar under proper orthochronous Lorentz transformations; that is, eq. (23.3) still holds. Thus the appellation *scalar* often means eq. (23.3), and *either* eq. (23.6) *or* eq. (23.10), and that is how we will use the term.

no reason to flirt with eqs. (23.10) and (23.11). For theories that also include spin-one-half fields, certain scalar bilinears in these fields are necessarily odd under parity and time reversal, as we will see in section 40. If a scalar field couples to such a bilinear, then eqs. (23.12) and (23.13) will hold if and only if we choose eqs. (23.10) and (23.11) for that scalar, and so that is what we must do.

There is one more interesting fact about the time-reversal operator T: it is *antiunitary*, rather than unitary. *Antiunitary* means that $T^{-1}iT = -i$.

To see why this must be the case, consider a Lorentz transformation of the energy-momentum four-vector,

$$U(\Lambda)^{-1}P^\mu U(\Lambda) = \Lambda^\mu{}_\nu P^\nu \ . \tag{23.14}$$

For parity and time-reversal, we therefore expect

$$P^{-1}P^\mu P = \mathcal{P}^\mu{}_\nu P^\nu \ , \tag{23.15}$$

$$T^{-1}P^\mu T = \mathcal{T}^\mu{}_\nu P^\nu \ . \tag{23.16}$$

In particular, for $\mu = 0$, we expect $P^{-1}HP = +H$ and $T^{-1}HT = -H$. The first of these is fine; it says the hamiltonian is invariant under parity, which is what we want.[2] However, eq. (23.16) is a disaster: it says that the hamiltonian is invariant under time-reversal if and only if $H = -H$, which is possible only if $H = 0$.

Can we just put an extra minus sign on the right-hand side of eq. (23.16), as we did for eq. (23.11)? The answer is no. We constructed P^μ explicitly in terms of the fields in section 22, and it is easy to check that choosing eq. (23.11) for the fields does not yield an extra minus sign in eq. (23.16) for the energy-momentum four-vector.

Let us reconsider the origin of eq. (23.14). We first recall that the spacetime translation operator

$$T(a) = \exp(-iP\cdot a) \ , \tag{23.17}$$

(which should not be confused with the time-reversal operator T) transforms a scalar field according to

$$T(a)^{-1}\varphi(x)T(a) = \varphi(x-a) \ . \tag{23.18}$$

The spacetime translation operator is a scalar with a spacetime coordinate as a label; by analogy with eq. (23.3), we should have

$$U(\Lambda)^{-1}T(a)U(\Lambda) = T(\Lambda^{-1}a) \ . \tag{23.19}$$

[2] When spin-one-half fields are present, it may be that no operator exists that satisfies either eq. (23.6) or eq. (23.10) and also eq. (23.15); in this case we say that parity is *explicitly broken*.

Now, treat a^μ as infinitesimal in eq. (23.19) to get

$$U(\Lambda)^{-1}(I - ia_\mu P^\mu)U(\Lambda) = I - i(\Lambda^{-1})_\nu{}^\mu a_\mu P^\nu$$
$$= I - i\Lambda^\mu{}_\nu a_\mu P^\nu \; . \qquad (23.20)$$

For time-reversal, this becomes

$$T^{-1}(I - ia_\mu P^\mu)T = I - iT^\mu{}_\nu a_\mu P^\nu \; . \qquad (23.21)$$

If we now identify the coefficients of $-ia_\mu$ on each side, we get eq. (23.16), which is bad. In order to get the extra minus sign that we need, we must impose the antiunitary condition

$$T^{-1} i T = -i \; . \qquad (23.22)$$

We then find

$$T^{-1}P^\mu T = -T^\mu{}_\nu P^\nu \qquad (23.23)$$

instead of eq. (23.16). This yields $T^{-1}HT = +H$, which is the correct expression of time-reversal invariance.

We turn now to other unitary operators that change the signs of scalar fields, but do nothing to their spacetime arguments. Suppose we have a theory with real scalar fields $\varphi_a(x)$, and a unitary operator Z that obeys

$$Z^{-1}\varphi_a(x)Z = \eta_a\varphi_a(x) \; , \qquad (23.24)$$

where η_a is either $+1$ or -1 for each field. We will call Z a Z_2 *operator*, because Z_2 is the additive group of the integers modulo 2, which is equivalent to the multiplicative group of $+1$ and -1. This also implies that $Z^2 = 1$, and so $Z^{-1} = Z$. (For theories with spin-zero fields only, the same is also true of P and T, but things are more subtle for higher spin, as we will see in Part II.)

Consider the theory of a complex scalar field $\varphi = (\varphi_1 + i\varphi_2)/\sqrt{2}$ that was introduced in section 22, with lagrangian

$$\mathcal{L} = -\partial^\mu\varphi^\dagger\partial_\mu\varphi - m^2\varphi^\dagger\varphi - \tfrac{1}{4}\lambda(\varphi^\dagger\varphi)^2 \qquad (23.25)$$

$$= -\tfrac{1}{2}\partial^\mu\varphi_1\partial_\mu\varphi_1 - \tfrac{1}{2}\partial^\mu\varphi_2\partial_\mu\varphi_2 - \tfrac{1}{2}m^2(\varphi_1^2 + \varphi_2^2) - \tfrac{1}{16}\lambda(\varphi_1^2 + \varphi_2^2)^2 . \qquad (23.26)$$

In the form of eq. (23.25), $\mathcal{L}$ is obviously invariant under the U(1) transformation

$$\varphi(x) \to e^{-i\alpha}\varphi(x) \; . \qquad (23.27)$$

In the form of eq. (23.26), $\mathcal{L}$ is obviously invariant under the equivalent SO(2) transformation,

$$\begin{pmatrix} \varphi_1(x) \\ \varphi_2(x) \end{pmatrix} \to \begin{pmatrix} \cos\alpha & \sin\alpha \\ -\sin\alpha & \cos\alpha \end{pmatrix} \begin{pmatrix} \varphi_1(x) \\ \varphi_2(x) \end{pmatrix} . \tag{23.28}$$

However, it is also obvious that $\mathcal{L}$ has an additional discrete symmetry,

$$\varphi(x) \leftrightarrow \varphi^\dagger(x) \tag{23.29}$$

in the form of eq. (23.25), or equivalently

$$\begin{pmatrix} \varphi_1(x) \\ \varphi_2(x) \end{pmatrix} \to \begin{pmatrix} +1 & 0 \\ 0 & -1 \end{pmatrix} \begin{pmatrix} \varphi_1(x) \\ \varphi_2(x) \end{pmatrix} \tag{23.30}$$

in the form of eq. (23.26). This discrete symmetry is called *charge conjugation.* It always occurs as a companion to a continuous U(1) symmetry. In terms of the two real fields, it enlarges the group from SO(2) (the group of 2×2 orthogonal matrices with determinant $+1$) to O(2) (the group of 2×2 orthogonal matrices).

We can implement charge conjugation by means of a particular Z_2 operator C that obeys

$$C^{-1}\varphi(x)C = \varphi^\dagger(x) , \tag{23.31}$$

or equivalently

$$C^{-1}\varphi_1(x)C = +\varphi_1(x) , \tag{23.32}$$
$$C^{-1}\varphi_2(x)C = -\varphi_2(x) . \tag{23.33}$$

We then have

$$C^{-1}\mathcal{L}(x)C = \mathcal{L}(x) , \tag{23.34}$$

and so charge conjugation is a symmetry of the theory. Physically, it implies that the scattering amplitudes are unchanged if we exchange all the a-type particles (which have charge $+1$) with all the b-type particles (which have charge -1). This means, in particular, that the a and b particles must have exactly the same mass. We say that b is a's *antiparticle.*

More generally, we can also have Z_2 symmetries that are not related to antiparticles. Consider, for example, φ^4 theory, where φ is a real scalar field with lagrangian

$$\mathcal{L} = -\tfrac{1}{2}\partial^\mu\varphi\partial_\mu\varphi - \tfrac{1}{2}m^2\varphi^2 - \tfrac{1}{24}\lambda\varphi^4 . \tag{23.35}$$

If we define the Z_2 operator Z via

$$Z^{-1}\varphi(x)Z = -\varphi(x) , \tag{23.36}$$

then $\mathcal{L}$ is obviously invariant. We therefore have $Z^{-1}HZ = H$, or equivalently $[Z, H] = 0$, where H is the hamiltonian. If we assume that (as usual) the ground state is unique, then, since Z commutes with H, the ground state must also be an eigenstate of Z. We can fix the phase of Z [which is undetermined by eq. (23.36)] via

$$Z|0\rangle = Z^{-1}|0\rangle = +|0\rangle . \tag{23.37}$$

Then, using eqs. (23.36) and (23.37), we have

$$\begin{aligned}\langle 0|\varphi(x)|0\rangle &= \langle 0|ZZ^{-1}\varphi(x)ZZ^{-1}|0\rangle \\ &= -\langle 0|\varphi(x)|0\rangle .\end{aligned} \tag{23.38}$$

Since $\langle 0|\varphi(x)|0\rangle$ is equal to minus itself, it must be zero. Thus, as long as the ground state is unique, the Z_2 symmetry of φ^4 theory guarantees that the field has zero vacuum expectation value. We therefore do not need to enforce this condition with a counterterm $Y\varphi$, as we did in φ^3 theory. (The assumption of a unique ground state does not necessarily hold, however, as we will see in section 30.)

24

Nonabelian symmetries

Prerequisite: 22

Consider the theory (introduced in section 22) of two real scalar fields φ_1 and φ_2 with

$$\mathcal{L} = -\tfrac{1}{2}\partial^\mu\varphi_1\partial_\mu\varphi_1 - \tfrac{1}{2}\partial^\mu\varphi_2\partial_\mu\varphi_2 - \tfrac{1}{2}m^2(\varphi_1^2+\varphi_2^2) - \tfrac{1}{16}\lambda(\varphi_1^2+\varphi_2^2)^2 \, . \quad (24.1)$$

We can generalize this to the case of N real scalar fields φ_i with

$$\mathcal{L} = -\tfrac{1}{2}\partial^\mu\varphi_i\partial_\mu\varphi_i - \tfrac{1}{2}m^2\varphi_i\varphi_i - \tfrac{1}{16}\lambda(\varphi_i\varphi_i)^2 \, , \quad (24.2)$$

where a repeated index is summed. This lagrangian is clearly invariant under the SO(N) transformation

$$\varphi_i(x) \to R_{ij}\varphi_j(x) \, , \quad (24.3)$$

where R is an orthogonal matrix with a positive determinant: $R^T = R^{-1}$, $\det R = +1$. This lagrangian is also clearly invariant under the Z_2 transformation $\varphi_i(x) \to -\varphi_i(x)$, which enlarges SO($N$) to O($N$); see section 23. However, in this section we will be concerned only with the continuous SO(N) part of the symmetry.

Next we will need some results from group theory. Consider an infinitesimal SO(N) transformation,

$$R_{ij} = \delta_{ij} + \theta_{ij} + O(\theta^2) \, . \quad (24.4)$$

Orthogonality of R_{ij} implies that θ_{ij} is real and antisymmetric. It is convenient to express θ_{ij} in terms of a basis set of hermitian matrices $(T^a)_{ij}$. The index a runs from 1 to $\frac{1}{2}N(N-1)$, the number of linearly independent, hermitian, antisymmetric, $N \times N$ matrices. We can, for example, choose each T^a to have a single nonzero entry $-i$ above the main diagonal, and a corresponding $+i$ below the main diagonal. These matrices obey the normalization condition

$$\mathrm{Tr}(T^a T^b) = 2\delta^{ab} \ . \tag{24.5}$$

In terms of them, we can write

$$\theta_{jk} = -i\theta^a (T^a)_{jk} \ , \tag{24.6}$$

where θ^a is a set of $\frac{1}{2}N(N-1)$ real, infinitesimal parameters.

The T^as are the *generator matrices* of SO(N). The product of any two SO(N) transformations is another SO(N) transformation; this implies (see problem 24.2) that the commutator of any two generator matrices must be a linear combination of generator matrices,

$$[T^a, T^b] = i f^{abc} T^c \ . \tag{24.7}$$

The numerical factors f^{abc} are the *structure coefficients* of the group, and eq. (24.7) specifies its *Lie algebra*. If $f^{abc} = 0$, the group is *abelian*. Otherwise, it is *nonabelian*. Thus, U(1) and SO(2) are abelian groups (since they each have only one generator that obviously must commute with itself), and SO(N) for $N \geq 3$ is nonabelian.

If we multiply eq. (24.7) on the right by T^d, take the trace, and use eq. (24.5), we find

$$f^{abd} = -\tfrac{1}{2} i \,\mathrm{Tr}\Big([T^a, T^b] T^d\Big) \ . \tag{24.8}$$

Using the cyclic property of the trace, we find that f^{abd} must be completely antisymmetric. Taking the complex conjugate of eq. (24.8) (and remembering that the T^as are hermitian matrices), we find that f^{abd} must be real.

The simplest nonabelian group is SO(3). In this case, we can choose $(T^a)_{ij} = -i\varepsilon^{aij}$, where ε^{ijk} is the completely antisymmetric Levi-Civita symbol, with $\varepsilon^{123} = +1$. The commutation relations become

$$[T^a, T^b] = i\varepsilon^{abc} T^c \ . \tag{24.9}$$

That is, the structure coefficients of SO(3) are given by $f^{abc} = \varepsilon^{abc}$.

Consider now a theory with N *complex* scalar fields φ_i, and a lagrangian

$$\mathcal{L} = -\partial^\mu \varphi_i^\dagger \partial_\mu \varphi_i - m^2 \varphi_i^\dagger \varphi_i - \tfrac{1}{4}\lambda(\varphi_i^\dagger \varphi_i)^2 \ , \tag{24.10}$$

where a repeated index is summed. This lagrangian is clearly invariant under the U(N) transformation

$$\varphi_i(x) \to U_{ij}\varphi_j(x) \ , \tag{24.11}$$

where U is a unitary matrix: $U^\dagger = U^{-1}$. We can write $U_{ij} = e^{-i\theta}\widetilde{U}_{ij}$, where θ is a real parameter and $\det \widetilde{U}_{ij} = +1$; $\widetilde{U}_{ij}$ is called a *special* unitary matrix. Clearly the product of two special unitary matrices is another special unitary matrix; the $N \times N$ special unitary matrices form the group SU(N). The group U(N) is the *direct product* of the group U(1) and the group SU(N); we write U(N) = U(1) × SU(N).

Consider an infinitesimal SU(N) transformation,

$$\widetilde{U}_{ij} = \delta_{ij} - i\theta^a (T^a)_{ij} + O(\theta^2)\,, \tag{24.12}$$

where θ^a is a set of real, infinitesimal parameters. Unitarity of $\widetilde{U}$ implies that the generator matrices T^a are hermitian, and $\det \widetilde{U} = +1$ implies that each T^a is traceless. (This follows from the general matrix formula $\ln\det A = \mathrm{Tr}\ln A$.) The index a runs from 1 to $N^2{-}1$, the number of linearly independent, hermitian, traceless, $N \times N$ matrices. We can choose these matrices to obey the normalization condition of eq. (24.5). For SU(2), the generators can be chosen to be the Pauli matrices; the structure coefficients of SU(2) then turn out to be $f^{abc} = 2\varepsilon^{abc}$, the same as those of SO(3), up to an irrelevant overall factor [which could be removed by changing the numerical factor on the right-hand side of eq. (24.5) from 2 to $\frac{1}{2}$].

For SU(N), we can choose the T^as in the following way. First, there are the SO(N) generators, with one $-i$ above the main diagonal, a corresponding $+i$ below; there are $\frac{1}{2}N(N{-}1)$ of these. Next, we get another set by putting one $+1$ above the main diagonal and a corresponding $+1$ below; there are $\frac{1}{2}N(N{-}1)$ of these. Finally, there are diagonal matrices with n 1s along the main diagonal, followed a single entry $-n$, followed by zeros [times an overall normalization constant to enforce eq. (24.5)]; there are $N{-}1$ of these. The total is $N^2{-}1$, as required.

However, if we examine the lagrangian of eq. (24.10) more closely, we find that it is actually invariant under a larger symmetry group, namely SO($2N$). To see this, write each complex scalar field in terms of two real scalar fields, $\varphi_j = (\varphi_{j1} + i\varphi_{j2})/\sqrt{2}$. Then

$$\varphi_j^\dagger \varphi_j = \tfrac{1}{2}(\varphi_{11}^2 + \varphi_{12}^2 + \ldots + \varphi_{N1}^2 + \varphi_{N2}^2)\,. \tag{24.13}$$

Thus, we have $2N$ real scalar fields that enter $\mathcal{L}$ symmetrically, and so the actual symmetry group of eq. (24.10) is SO($2N$), rather than just the obvious subgroup U(N).

We will, however, meet the SU(N) groups again in Parts II and III, where they will play a more important role.

Problems

24.1) Show that θ_{ij} in eq. (24.4) must be antisymmetric if R is orthogonal.

24.2) By considering the SO(N) transformation $R'^{-1}R^{-1}R'R$, where R and R' are independent infinitesimal SO(N) transformations, prove eq. (24.7).

24.3) a) Find the Noether current $j^{a\mu}$ for the transformation of eq. (24.6).

b) Show that $[\varphi_i, Q^a] = (T^a)_{ij}\varphi_j$, where Q^a is the Noether charge.

c) Use this result, eq. (24.7), and the Jacobi identity (see problem 2.8) to show that $[Q^a, Q^b] = if^{abc}Q^c$.

24.4) The elements of the group SO(N) can be defined as $N \times N$ matrices R that satisfy

$$R_{ii'}R_{jj'}\delta_{i'j'} = \delta_{ij} \; . \tag{24.14}$$

The elements of the *symplectic group* Sp($2N$) can be defined as $2N \times 2N$ matrices S that satisfy

$$S_{ii'}S_{jj'}\eta_{i'j'} = \eta_{ij} \; , \tag{24.15}$$

where the *symplectic metric* η_{ij} is antisymmetric, $\eta_{ij} = -\eta_{ji}$, and squares to minus the identity: $\eta^2 = -I$. One way to write η is

$$\eta = \begin{pmatrix} 0 & I \\ -I & 0 \end{pmatrix} , \tag{24.16}$$

where I is the $N \times N$ identity matrix. Find the number of generators of Sp($2N$).

25

Unstable particles and resonances

Prerequisite: 14

Consider a theory of two real scalar fields, φ and χ, with lagrangian

$$\mathcal{L} = -\tfrac{1}{2}\partial^\mu\varphi\partial_\mu\varphi - \tfrac{1}{2}m_\varphi^2\varphi^2 - \tfrac{1}{2}\partial^\mu\chi\partial_\mu\chi - \tfrac{1}{2}m_\chi^2\chi^2 + \tfrac{1}{2}g\varphi\chi^2 + \tfrac{1}{6}h\varphi^3 \; . \tag{25.1}$$

This theory is renormalizable in six dimensions, where g and h are dimensionless coupling constants.

Let us assume that $m_\varphi > 2m_\chi$. Then it is kinematically possible for the φ particle to decay into two χ particles. The amplitude for this process is given at tree level by the Feynman diagram of fig. 25.1, and is simply $\mathcal{T} = g$. We can also choose to define g as the value of the exact $\varphi\chi^2$ vertex function $\mathbf{V}_3(k, k_1', k_2')$ when all three particles are on shell: $k^2 = -m_\varphi^2$, $k_1'^2 = k_2'^2 = -m_\chi^2$. This implies that

$$\mathcal{T} = g \tag{25.2}$$

exactly.

According to the formulae of section 11, the differential decay rate (in the rest frame of the initial φ particle) is

$$d\Gamma = \frac{1}{2m_\varphi}\, d\text{LIPS}_2 \, |\mathcal{T}|^2 \; , \tag{25.3}$$

where $d\text{LIPS}_2$ is the Lorentz-invariant phase-space differential for two outgoing particles, introduced in section 11. We must make a slight adaptation for six dimensions:

$$d\text{LIPS}_2 \equiv (2\pi)^6\delta^6(k_1'{+}k_2'{-}k)\, \widetilde{dk_1'}\, \widetilde{dk_2'} \; . \tag{25.4}$$

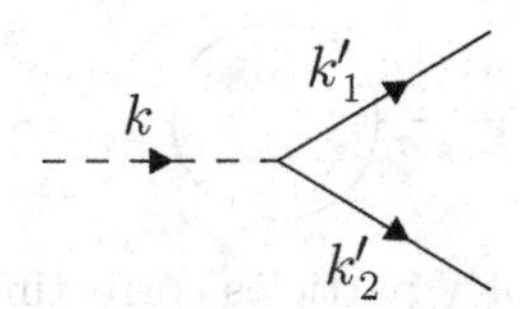

Figure 25.1. The tree-level Feynman diagram for the decay of a φ particle (dashed line) into two χ particles (solid lines).

Here $k = (m_\varphi, \mathbf{0})$ is the energy-momentum of the decaying particle, and

$$\widetilde{dk} = \frac{d^5k}{(2\pi)^5 2\omega} \tag{25.5}$$

is the Lorentz-invariant phase-space differential for one particle. Recall that we can also write it as

$$\widetilde{dk} = \frac{d^6k}{(2\pi)^6}\, 2\pi\delta(k^2 + m_\chi^2)\, \theta(k^0)\ , \tag{25.6}$$

where $\theta(x)$ is the unit step function. Performing the integral over k^0 turns eq. (25.6) into eq. (25.5).

Repeating for six dimensions what we did in section 11 for four dimensions, we find

$$d\text{LIPS}_2 = \frac{|\mathbf{k}_1'|^3}{4(2\pi)^4 m_\varphi}\, d\Omega\ , \tag{25.7}$$

where $|\mathbf{k}_1'| = \frac{1}{2}(m_\varphi^2 - 4m_\chi^2)^{1/2}$ is the magnitude of the spatial momentum of one of the outgoing particles. We can now plug this into eq. (25.3), and use $\int d\Omega = \Omega_5 = 2\pi^{5/2}/\Gamma(\frac{5}{2}) = \frac{8}{3}\pi^2$. We also need to divide by a symmetry factor of two, due to the presence of two identical particles in the final state. The result is

$$\Gamma = \frac{1}{2}\cdot\frac{1}{2m_\varphi}\int d\text{LIPS}_2\, |\mathcal{T}|^2 \tag{25.8}$$

$$= \tfrac{1}{12}\pi\alpha(1 - 4m_\chi^2/m_\varphi^2)^{3/2}\, m_\varphi\ , \tag{25.9}$$

where $\alpha = g^2/(4\pi)^3$.

However, as we discussed in section 11, we have a conceptual problem. According to our development of the LSZ formula in section 5, each incoming and outgoing particle should correspond to a single-particle state that is an

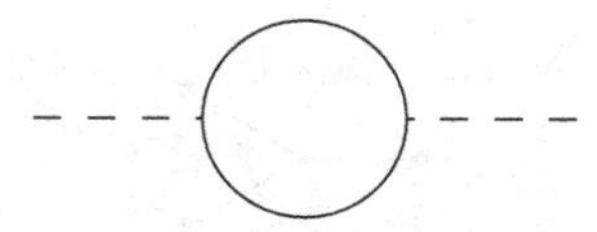

Figure 25.2. A loop of χ particles correcting the φ propagator.

exact eigenstate of the exact hamiltonian. This is clearly not the case for a particle that can decay.

Let us, then, compute something else instead: the correction to the φ propagator from a loop of χ particles, as shown in fig. 25.2. The diagram is the same as the one we already analyzed in section 14, except that the internal propagators contain m_χ instead of m_φ. (There is also a contribution from a loop of φ particles, but we can ignore it if we assume that $h \ll g$.) We have

$$\Pi(k^2) = \tfrac{1}{2}\alpha \int_0^1 dx\, D \ln D - A'k^2 - B'm_\varphi^2 \,, \tag{25.10}$$

where

$$D = x(1{-}x)k^2 + m_\chi^2 - i\epsilon \,, \tag{25.11}$$

and A' and B' are the finite counterterm coefficients that remain after the infinities have been absorbed. We now try to fix A' and B' by imposing the usual on-shell conditions $\Pi(-m_\varphi^2) = 0$ and $\Pi'(-m_\varphi^2) = 0$.

But, we have a problem. For $k^2 = -m_\varphi^2$ and $m_\varphi > 2m_\chi$, D is negative for part of the range of x. Therefore $\ln D$ has an imaginary part. This imaginary part cannot be canceled by A' and B', since A' and B' must be real: they are coefficients of hermitian operators in the lagrangian. The best we can do is $\mathrm{Re}\,\Pi(-m_\varphi^2) = 0$ and $\mathrm{Re}\,\Pi'(-m_\varphi^2) = 0$. Imposing these gives

$$\Pi(k^2) = \tfrac{1}{2}\alpha \int_0^1 dx\, D \ln(D/|D_0|) - \tfrac{1}{12}\alpha(k^2 + m_\varphi^2) \,, \tag{25.12}$$

where

$$D_0 = -x(1{-}x)m_\varphi^2 + m_\chi^2 \,. \tag{25.13}$$

Now let us compute the imaginary part of $\Pi(k^2)$. This arises from the integration range $x_- < x < x_+$, where $x_\pm = \frac{1}{2} \pm \frac{1}{2}(1 + 4m_\chi^2/k^2)^{1/2}$ are the roots of $D = 0$ when $k^2 < -4m_\chi^2$. In this range, $\mathrm{Im}\ln D = -\pi$; the minus sign arises because, according to eq. (25.11), D has a small negative imaginary part. Now we have

$$\begin{aligned} \operatorname{Im}\Pi(k^2) &= -\tfrac{1}{2}\pi\alpha \int_{x_-}^{x_+} dx\, D \\ &= -\tfrac{1}{12}\pi\alpha(1 + 4m_\chi^2/k^2)^{3/2}\, k^2 \end{aligned} \tag{25.14}$$

when $k^2 < -4m_\chi^2$. Evaluating eq. (25.14) at $k^2 = -m_\varphi^2$, we get

$$\operatorname{Im}\Pi(-m_\varphi^2) = \tfrac{1}{12}\pi\alpha(1 - 4m_\chi^2/m_\varphi^2)^{3/2}\, m_\varphi^2\ . \tag{25.15}$$

From this and eq. (25.9), we see that

$$\operatorname{Im}\Pi(-m_\varphi^2) = m_\varphi\Gamma\ . \tag{25.16}$$

This is not an accident. Instead, it is a general rule. We will argue this in two ways: first, from the mathematics of Feynman diagrams, and second, from the physics of resonant scattering in quantum mechanics.

We begin with the mathematics of Feynman diagrams. Return to the diagrammatic expression for $\Pi(k^2)$, before we evaluated any of the integrals:

$$\begin{aligned} \Pi(k^2) = -\tfrac{1}{2}ig^2 \int \frac{d^6\ell_1}{(2\pi)^6}\, \frac{d^6\ell_2}{(2\pi)^6}\, &(2\pi)^6\delta^6(\ell_1{+}\ell_2{-}k) \\ &\times \frac{1}{\ell_1^2 + m_\chi^2 - i\epsilon}\, \frac{1}{\ell_2^2 + m_\chi^2 - i\epsilon} \\ &- (Ak^2 + Bm_\varphi^2)\ . \end{aligned} \tag{25.17}$$

Here, for later convenience, we have assigned the internal lines momenta ℓ_1 and ℓ_2, and explicitly included the momentum-conserving delta function that fixes one of them. We can take the imaginary part of $\Pi(k^2)$ by using the identity

$$\frac{1}{x - i\epsilon} = P\frac{1}{x} + i\pi\delta(x)\ , \tag{25.18}$$

where P means the principal part. We then get, in a shorthand notation,

$$\operatorname{Im}\Pi(k^2) = -\tfrac{1}{2}g^2 \int \Bigl(P_1P_2 - \pi^2\delta_1\delta_2\Bigr)\ . \tag{25.19}$$

Next, we note that the integral in eq. (25.17) is the Fourier transform of $[\Delta(x-y)]^2$, where

$$\Delta(x-y) = \int \frac{d^6k}{(2\pi)^6} \frac{e^{ik(x-y)}}{k^2 + m_\chi^2 - i\epsilon} \tag{25.20}$$

is the Feynman propagator. Recall (from problem 8.5) that we can get the retarded or advanced propagator (rather than the Feynman propagator) by replacing the ϵ in eq. (25.20) with, respectively, $-s\epsilon$ or $+s\epsilon$, where $s \equiv \text{sign}(k^0)$. Therefore, in eq. (25.19), replacing δ_1 with $-s_1\delta_1$ and δ_2 with $+s_2\delta_2$ yields an integral that is the real part of the Fourier transform of $\Delta_{\text{ret}}(x-y)\Delta_{\text{adv}}(x-y)$. But this product is zero, because the first factor vanishes when $x^0 \geq y^0$, and the second when $x^0 \leq y^0$. So we can subtract the modified integrand from the original without changing the value of the integral. Thus we have

$$\text{Im}\,\Pi(k^2) = \tfrac{1}{2}g^2\pi^2 \int (1 + s_1 s_2)\delta_1\delta_2 \,. \tag{25.21}$$

The factor of $1 + s_1 s_2$ vanishes if ℓ_1^0 and ℓ_2^0 have opposite signs, and equals 2 if they have the same sign. Because the delta function in eq. (25.17) enforces $\ell_1^0 + \ell_2^0 = k^0$, and $k^0 = m_\varphi$ is positive, both ℓ_1^0 and ℓ_2^0 must be positive. So we can replace the factor of $1 + s_1 s_2$ in eq. (25.21) with $2\theta(\ell_1^0)\theta(\ell_2^0)$. Rearranging the numerical factors, we have

$$\begin{aligned} \text{Im}\,\Pi(k^2) = \tfrac{1}{4}g^2 \int & \frac{d^6\ell_1}{(2\pi)^6} \frac{d^6\ell_2}{(2\pi)^6} (2\pi)^6\delta^6(\ell_1+\ell_2-k) \\ & \times 2\pi\delta(\ell_1^2 + m_\chi^2)\theta(\ell_1^0)\, 2\pi\delta(\ell_2^2 + m_\chi^2)\theta(\ell_2^0) \,. \end{aligned} \tag{25.22}$$

If we now set $k^2 = -m_\varphi^2$, use eqs. (25.4) and (25.6), and recall that $\mathcal{T} = g$ is the decay amplitude, we can rewrite eq. (25.22) as

$$\text{Im}\,\Pi(-m_\varphi^2) = \tfrac{1}{4} \int d\text{LIPS}_2\, |\mathcal{T}|^2 \,. \tag{25.23}$$

Comparing eqs. (25.8) and (25.23), we see that we indeed have

$$\text{Im}\,\Pi(-m_\varphi^2) = m_\varphi \Gamma \,. \tag{25.24}$$

This relation persists at higher orders in perturbation theory. Our analysis can be generalized to give the *Cutkosky rules* for computing the imaginary part of any Feynman diagram, but this is beyond the scope of our current interest.

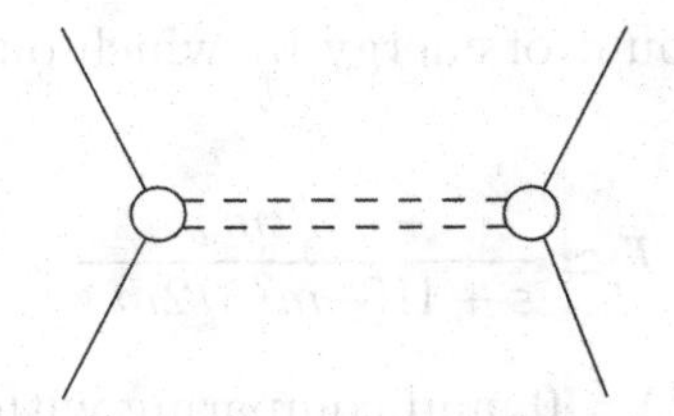

Figure 25.3. For s near m_φ^2, the dominant contribution to $\chi\chi$ scattering is s-channel φ exchange.

To get a more physical understanding of this result, recall that in nonrelativistic quantum mechanics, a metastable state with energy E_0 and angular momentum quantum number ℓ shows up as a *resonance* in the partial-wave scattering amplitude,

$$f_\ell(E) \sim \frac{1}{E - E_0 + i\Gamma/2} \,. \tag{25.25}$$

If we imagine convolving this amplitude with a wave packet $\widetilde{\psi}(E)e^{-iEt}$, we will find a time dependence

$$\begin{aligned}\psi(t) &\sim \int dE\, \frac{1}{E - E_0 + i\Gamma/2}\, \widetilde{\psi}(E)e^{-iEt} \\ &\sim e^{-iE_0 t - \Gamma t/2} \,.\end{aligned} \tag{25.26}$$

Therefore $|\psi(t)|^2 \sim e^{-\Gamma t}$, and we identify Γ as the inverse lifetime of the metastable state.

In the relativistic case, consider the scattering process $\chi\chi \to \chi\chi$. The contributing diagrams from the effective action are those of fig. 20.1, where the exact internal propagator is that of the φ field. Suppose that the center-of-mass energy squared s is close to m_φ^2. Since the φ propagator has a pole near $s = m_\varphi^2$, s-channel φ exchange, shown in fig. 25.3, makes the dominant contribution to the $\chi\chi$ scattering amplitude. We then have

$$\mathcal{T} \simeq \frac{g^2}{-s + m_\varphi^2 - \Pi(-s)} \,. \tag{25.27}$$

Here we have used the fact that the exact $\varphi\chi\chi$ vertex has the value g when all three particles are on-shell. Now let us write

$$s = (m_\varphi + \varepsilon)^2 \simeq m_\varphi^2 + 2m_\varphi\varepsilon \,, \tag{25.28}$$

where $\varepsilon \ll m_\varphi$ is the amount of energy by which our incoming particles are *off resonance*. We find

$$\mathcal{T} \simeq \frac{-g^2/2m_\varphi}{\varepsilon + \Pi(-m_\varphi^2)/2m_\varphi} \,. \tag{25.29}$$

Recalling that $\operatorname{Re}\Pi(-m_\varphi^2) = 0$, and comparing with eq. (25.25), we see that we should make the identification of eq. (25.24).

Reference notes

The Cutkosky rules are discussed in more detail in *Peskin & Schroeder*. More details on resonances can be found in *Weinberg I*.

26

Infrared divergences

Prerequisite: 20

In section 20, we computed the $\varphi\varphi \to \varphi\varphi$ scattering amplitude in φ^3 theory in six dimensions in the high-energy limit (s, $|t|$, and $|u|$ all much larger than m^2). We found that

$$\mathcal{T} = \mathcal{T}_0 \left[1 - \tfrac{11}{12}\alpha\Big(\ln(s/m^2) + O(m^0)\Big) + O(\alpha^2) \right], \qquad (26.1)$$

where $\mathcal{T}_0 = -g^2(s^{-1} + t^{-1} + u^{-1})$ is the tree-level result, and the $O(m^0)$ term includes everything without a *large logarithm* that blows up in the limit $m \to 0$.[1]

Suppose we are interested in the limit of massless particles. The large log is then problematic, since it blows up in this limit. What does this mean?

It means we have made a mistake. Actually, two mistakes. In this section, we will remedy one of them.

Throughout the physical sciences, it is necessary to make various idealizations in order to make progress. (Recall the "massless springs" and "frictionless planes" of freshman mechanics.) Sometimes these idealizations can lead us into trouble, and that is one of the things that has gone wrong here.

We have assumed that we can isolate individual particles. The reasoning behind this was explained in section 5, and it depends on the existence of an energy gap between the one-particle states and the multiparticle continuum. However, this gap vanishes if the theory includes massless particles. In this case, it is possible that the scattering process involved the creation of some extra very low energy (or *soft*) particles that escaped detection. Or, there may have been some extra soft particles hiding in the initial state that discreetly participated in the scattering process. Or, what was seen as a

[1] In writing $\mathcal{T}$ in this form, we have traded factors of $\ln t$ and $\ln u$ for $\ln s$ by first using $\ln t = \ln s + \ln(t/s)$, and then hiding the $\ln(t/s)$ terms in the $O(m^0)$ catch-all.

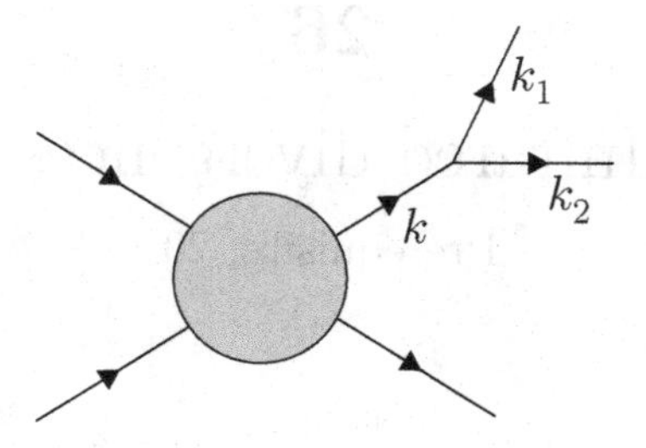

Figure 26.1. An outgoing particle splits into two. The gray circle stands for the sum of all diagrams contributing to the original amplitude $i\mathcal{T}$.

single high-energy particle may actually have been two or more particles that were moving colinearly and sharing the energy.

Let us, then, correct our idealization of a perfect detector and account for these possibilities. We will work with φ^3 theory, initially in d spacetime dimensions.

Let $\mathcal{T}$ be the amplitude for some scattering process in φ^3 theory. Now consider the possibility that one of the outgoing particles in this process splits into two, as shown in fig. 26.1. The amplitude for this new process is given in terms of $\mathcal{T}$ by

$$\mathcal{T}_{\rm split} = ig\,\frac{-i}{k^2+m^2}\,\mathcal{T}\;, \tag{26.2}$$

where $k = k_1 + k_2$, and k_1 and k_2 are the on-shell four-momenta of the two particles produced by the split. (For notational convenience, we drop our usual primes on the outgoing momenta.) The key point is this: in the massless limit, it is possible for $1/(k^2+m^2)$ to diverge.

To understand the physical consequences of this possibility, we should compute an appropriate cross section. To get the cross section for the original process (without the split), we multiply $|\mathcal{T}|^2$ by $\widetilde{dk}$ (as well as by similar differentials for other outgoing particles, and by an overall energy-momentum delta function). For the process with the split, we multiply $|\mathcal{T}_{\rm split}|^2$ by $\frac{1}{2}\widetilde{dk}_1\widetilde{dk}_2$ instead of $\widetilde{dk}$. (The factor of one-half is for counting of identical particles.) If we assume that (due to some imperfection) our detector cannot tell whether or not the one particle actually split into two, then we should (according to the usual rules of quantum mechanics) add the probabilities for the two events, which are distinguishable in principle. We can therefore define an effectively observable squared-amplitude via

$$|\mathcal{T}|^2_{\rm obs}\,\widetilde{dk} = |\mathcal{T}|^2\,\widetilde{dk} + |\mathcal{T}_{\rm split}|^2\,\tfrac{1}{2}\widetilde{dk}_1\widetilde{dk}_2 + \ldots\;. \tag{26.3}$$

Here the ellipses stand for all other similar processes involving emission of one or more extra particles in the final state, or absorption of one or more extra particles in the initial state.

We can simplify eq. (26.3) by including a factor of

$$1 = (2\pi)^{d-1}\, 2\omega\, \delta^{d-1}(\mathbf{k}_1{+}\mathbf{k}_2{-}\mathbf{k})\, \widetilde{dk} \tag{26.4}$$

in the second term. Now all terms in eq. (26.3) include a factor of $\widetilde{dk}$, so we can drop it. Then, using eq. (26.2), we get

$$|\mathcal{T}|^2_{\text{obs}} \equiv |\mathcal{T}|^2 \left[1 + \frac{g^2}{(k^2+m^2)^2} (2\pi)^{d-1}\, 2\omega\, \delta^{d-1}(\mathbf{k}_1{+}\mathbf{k}_2{-}\mathbf{k})\, \tfrac{1}{2}\widetilde{dk}_1 \widetilde{dk}_2 + \ldots \right] . \tag{26.5}$$

Now we come to the point: in the massless limit, the phase space integral in the second term in eq. (26.5) can diverge. This is because, for $m = 0$,

$$k^2 = (k_1 + k_2)^2 = -4\omega_1\omega_2 \sin^2(\theta/2) \,, \tag{26.6}$$

where θ is the angle between the spatial momenta $\mathbf{k}_1$ and $\mathbf{k}_2$, and $\omega_{1,2} = |\mathbf{k}_{1,2}|$. Also, for $m = 0$,

$$\widetilde{dk}_1 \widetilde{dk}_2 \sim (\omega_1^{d-3}\, d\omega_1)\, (\omega_2^{d-3}\, d\omega_2)\, (\sin^{d-3}\theta\, d\theta) \,. \tag{26.7}$$

Therefore, for small θ,

$$\frac{\widetilde{dk}_1 \widetilde{dk}_2}{(k^2)^2} \sim \frac{d\omega_1}{\omega_1^{5-d}}\, \frac{d\omega_2}{\omega_2^{5-d}}\, \frac{d\theta}{\theta^{7-d}} \,. \tag{26.8}$$

Thus the integral over each ω diverges at the low end for $d \le 4$, and the integral over θ diverges at the low end for $d \le 6$. These divergent integrals would be cut off (and rendered finite) if we kept the mass m nonzero, as we will see below.

Our discussion leads us to expect that the $m \to 0$ divergence in the second term of eq. (26.5) should *cancel* the $m \to 0$ divergence in the loop correction to $|\mathcal{T}|^2$. We will now see how this works (or fails to work) in detail for the familiar case of two-particle scattering in six spacetime dimensions, where $\mathcal{T}$ is given by eq. (26.1). For $d = 6$, there is no problem with soft particles (corresponding to the small-ω divergence), but there is a problem with collinear particles (corresponding to the small-θ divergence).

Let us assume that our imperfect detector cannot tell one particle from two nearly collinear particles if the angle θ between their spatial momenta is less than some small angle δ. Since we ultimately want to take the $m \to 0$ limit, we will evaluate eq. (26.5) with $m^2/\mathbf{k}^2 \ll \delta^2 \ll 1$.

We can immediately integrate over d^5k_2 using the delta function, which results in setting $\mathbf{k}_2 = \mathbf{k} - \mathbf{k}_1$ everywhere. Let β then be the angle between $\mathbf{k}_1$ (which is still to be integrated over) and $\mathbf{k}$ (which is fixed). For two-particle scattering, $|\mathbf{k}| = \frac{1}{2}\sqrt{s}$ in the limit $m \to 0$. We then have

$$(2\pi)^5\, 2\omega\, \delta^5(\mathbf{k}_1{+}\mathbf{k}_2{-}\mathbf{k})\, \tfrac{1}{2}\widetilde{dk}_1\widetilde{dk}_2 \to \frac{\Omega_4}{4(2\pi)^5}\, \frac{\omega}{\omega_1\omega_2}\, |\mathbf{k}_1|^4\, d|\mathbf{k}_1|\, \sin^3\!\beta\, d\beta\;, \quad (26.9)$$

where $\Omega_4 = 2\pi^2$ is the area of the unit four-sphere. Now let γ be the angle between $\mathbf{k}_2$ and $\mathbf{k}$. The geometry of this trio of vectors implies $\theta = \beta + \gamma$, $|\mathbf{k}_1| = (\sin\gamma/\sin\theta)|\mathbf{k}|$, and $|\mathbf{k}_2| = (\sin\beta/\sin\theta)|\mathbf{k}|$. All three of the angles are small and positive, and it then is useful to write $\beta = x\theta$ and $\gamma = (1{-}x)\theta$, with $0 \le x \le 1$ and $\theta \le \delta \ll 1$.

In the low mass limit, we can safely set $m = 0$ everywhere in eq. (26.5) except in the propagator, $1/(k^2 + m^2)$. Then, expanding to leading order in both θ and m, we find (after some algebra)

$$k^2 + m^2 \simeq -x(1{-}x)\mathbf{k}^2 \left[\theta^2 + (m^2/\mathbf{k}^2)f(x)\right], \quad (26.10)$$

where $f(x) = (1{-}x{+}x^2)/(x{-}x^2)^2$. Everywhere else in eq. (26.5), we can safely set $\omega_1 = |\mathbf{k}_1| = (1{-}x)|\mathbf{k}|$ and $\omega_2 = |\mathbf{k}_2| = x|\mathbf{k}|$. Then, changing the integration variables in eq. (26.9) from $|\mathbf{k}_1|$ and β to x and θ, we get

$$|\mathcal{T}|^2_{\rm obs} = |\mathcal{T}|^2 \left[1 + \frac{g^2\Omega_4}{4(2\pi)^5} \int_0^1 x(1{-}x)dx \int_0^\delta \frac{\theta^3\, d\theta}{[\theta^2 + (m^2/\mathbf{k}^2)f(x)]^2} + \ldots\right]. \quad (26.11)$$

Performing the integral over θ yields $\frac{1}{2}[\ln(\delta^2\mathbf{k}^2/m^2) - \ln f(x) - 1]$. Then, performing the integral over x and using $\Omega_4 = 2\pi^2$ and $\alpha = g^2/(4\pi)^3$, we get

$$|\mathcal{T}|^2_{\rm obs} = |\mathcal{T}|^2 \left[1 + \tfrac{1}{12}\alpha\Big(\ln(\delta^2\mathbf{k}^2/m^2) + c\Big) + \ldots\right], \quad (26.12)$$

where $c = (4 - 3\sqrt{3}\pi)/3 = -4.11$.

The displayed correction term accounts for the possible splitting of *one* of the two outgoing particles. Obviously, there is an identical correction for the other outgoing particle. Less obviously (but still true), there is an identical correction for each of the two incoming particles. (A glib explanation is that we are computing an effective amplitude-squared, and this is the same for the reverse process, with in and outgoing particles switched. So in and out particles should be treated symmetrically.) Then, since we have a total of four in and out particles (before accounting for any splitting),

$$|\mathcal{T}|^2_{\rm obs} = |\mathcal{T}|^2 \left[1 + \tfrac{4}{12}\alpha\Big(\ln(\delta^2\mathbf{k}^2/m^2) + c\Big) + O(\alpha^2)\right] . \qquad (26.13)$$

We have now accounted for the $O(\alpha)$ corrections due to the failure of our detector to separate two particles whose spatial momenta are nearly parallel. Combining this with eq. (26.1), and recalling that $\mathbf{k}^2 = \frac{1}{4}s$, we get

$$\begin{aligned} |\mathcal{T}|^2_{\rm obs} &= |\mathcal{T}_0|^2 \left[1 - \tfrac{11}{6}\alpha\Big(\ln(s/m^2) + O(m^0)\Big) + O(\alpha^2)\right] \\ &\quad\times \left[1 + \tfrac{1}{3}\alpha\Big(\ln(\delta^2 s/m^2) + O(m^0)\Big) + O(\alpha^2)\right] \\ &= |\mathcal{T}_0|^2 \Big[1 - \alpha\Big(\tfrac{3}{2}\ln(s/m^2) + \tfrac{1}{3}\ln(1/\delta^2) + O(m^0)\Big) \\ &\qquad + O(\alpha^2)\Big] . \end{aligned} \qquad (26.14)$$

We now have two kinds of large logs. One is $\ln(1/\delta^2)$; this factor depends on the properties of our detector. If we build a very good detector, one for which $\alpha\ln(1/\delta^2)$ is not small, then we will have to do more work, and calculate higher-order corrections to eq. (26.14).

The other large log is our original nemesis $\ln(s/m^2)$. This factor blows up in the massless limit. This means that there is still a mistake hidden somewhere in our analysis.

Reference notes

Infrared divergences in quantum electrodynamics are discussed in *Brown* and *Peskin & Schroeder*. More general treatments can be found in *Sterman* and *Weinberg I*.

27

Other renormalization schemes

Prerequisite: 26

To find the remaining mistake in eq. (26.15), we must review our renormalization procedure. Recall our result from section 14 for the one-loop correction to the propagator,

$$\begin{aligned}\Pi(k^2) = &-\Big[A + \tfrac{1}{6}\alpha\Big(\tfrac{1}{\varepsilon}+\tfrac{1}{2}\Big)\Big]k^2 - \Big[B + \alpha\Big(\tfrac{1}{\varepsilon}+\tfrac{1}{2}\Big)\Big]m^2 \\ &+ \tfrac{1}{2}\alpha\int_0^1 dx\, D\ln(D/\mu^2) + O(\alpha^2)\,, \end{aligned} \tag{27.1}$$

where $\alpha = g^2/(4\pi)^3$ and $D = x(1-x)k^2 + m^2$. The derivative of $\Pi(k^2)$ with respect to k^2 is

$$\begin{aligned}\Pi'(k^2) = &-\Big[A + \tfrac{1}{6}\alpha\Big(\tfrac{1}{\varepsilon}+\tfrac{1}{2}\Big)\Big] \\ &+ \tfrac{1}{2}\alpha\int_0^1 dx\, x(1-x)\Big[\ln(D/\mu^2)+1\Big] + O(\alpha^2)\,. \end{aligned} \tag{27.2}$$

We previously determined A and B via the requirements $\Pi(-m^2) = 0$ and $\Pi'(-m^2) = 0$. The first condition ensures that the exact propagator $\tilde{\mathbf{\Delta}}(k^2)$ has a pole at $k^2 = -m^2$, and the second ensures that the residue of this pole is one. Recall that the field must be normalized in this way for the validity of the LSZ formula.

We now consider the massless limit. We have $D = x(1-x)k^2$, and we should apparently try to impose $\Pi(0) = \Pi'(0) = 0$. However, $\Pi(0)$ is now automatically zero for any values of A and B, while $\Pi'(0)$ is ill defined.

Physically, the problem is that the one-particle states are no longer separated from the multiparticle continuum by a finite gap in energy. Mathematically, the pole in $\tilde{\mathbf{\Delta}}(k^2)$ at $k^2 = -m^2$ merges with the branch point at $k^2 = -4m^2$, and is no longer a simple pole.

The only way out of this difficulty is to change the *renormalization scheme*. Let us first see what this means in the case $m \neq 0$, where we know what we are doing.

Let us try making a different choice of A and B. Specifically, let

$$A = -\tfrac{1}{6}\alpha\tfrac{1}{\varepsilon} + O(\alpha^2)\,,$$

$$B = -\alpha\tfrac{1}{\varepsilon} + O(\alpha^2)\,. \tag{27.3}$$

Here we have chosen A and B to cancel the infinities, and nothing more; we say that A and B have no *finite parts.* This choice represents a different *renormalization scheme.* Our original choice (which, up until now, we have pretended was inescapable!) is called the *on-shell* or OS scheme. The choice of eq. (27.3) is called the *modified minimal-subtraction* or $\overline{\text{MS}}$ (pronounced "emm-ess-bar") scheme. ("Modified" because we introduced μ via $g \to g\tilde{\mu}^{\varepsilon/2}$, with $\mu^2 = 4\pi e^{-\gamma}\tilde{\mu}^2$; had we set $\mu = \tilde{\mu}$ instead, the scheme would be just plain *minimal subtraction* or MS.) Now we have

$$\Pi_{\overline{\text{MS}}}(k^2) = -\tfrac{1}{12}\alpha(k^2 + 6m^2) + \tfrac{1}{2}\alpha \int_0^1 dx\, D \ln(D/\mu^2) + O(\alpha^2)\,, \tag{27.4}$$

as compared to our old result in the on-shell scheme,

$$\Pi_{\text{OS}}(k^2) = -\tfrac{1}{12}\alpha(k^2 + m^2) + \tfrac{1}{2}\alpha \int_0^1 dx\, D \ln(D/D_0) + O(\alpha^2)\,, \tag{27.5}$$

where again $D = x(1-x)k^2 + m^2$, and $D_0 = [-x(1-x)+1]m^2$. Note that $\Pi_{\overline{\text{MS}}}(k^2)$ has a well-defined $m \to 0$ limit, whereas $\Pi_{\text{OS}}(k^2)$ does not. On the other hand, $\Pi_{\overline{\text{MS}}}(k^2)$ depends explicitly on the fake parameter μ, whereas $\Pi_{\text{OS}}(k^2)$ does not.

What does this all mean?

First, in the $\overline{\text{MS}}$ scheme, the propagator $\boldsymbol{\Delta}_{\overline{\text{MS}}}(k^2)$ will no longer have a pole at $k^2 = -m^2$. The pole will be somewhere else. However, *by definition*, the actual physical mass m_{ph} of the particle is determined by the location of this pole: $k^2 = -m^2_{\text{ph}}$. Thus, the lagrangian parameter m is no longer the same as m_{ph}.

Furthermore, the residue of this pole is no longer one. Let us call the residue R. The LSZ formula must now be corrected by multiplying its right-hand side by a factor of $R^{-1/2}$ for each external particle (incoming or outgoing). This is because it is the field $R^{-1/2}\varphi(x)$ that now has unit amplitude to create a one-particle state.

Note also that, in the LSZ formula, each Klein–Gordon wave operator should be $-\partial^2 + m^2_{\text{ph}}$, and not $-\partial^2 + m^2$; also, each external four-momentum should square to $-m_{\rm ph}^2$, and not $-m^2$. A review of the derivation of the LSZ formula clearly shows that each of these mass parameters must be the actual particle mass, and not the parameter in the lagrangian.

Finally, in the LSZ formula, each external line will contribute a factor of R when the associated Klein–Gordon wave operator hits the external propagator and cancels its momentum-space pole, leaving behind the residue R. Combined with the correction factor of $R^{-1/2}$ for each field, we get a net factor of $R^{1/2}$ for each external line when using the $\overline{\rm MS}$ scheme. Internal lines each contribute a factor of $(-i)/(k^2+m^2)$, where m is the lagrangian-parameter mass, and each vertex contributes a factor of $iZ_g g$, where g is the lagrangian-parameter coupling.

Let us now compute the relation between m and $m_{\rm ph}$, and then compute R. We have

$$\mathbf{\Delta}_{\overline{\rm MS}}(k^2)^{-1} = k^2 + m^2 - \Pi_{\overline{\rm MS}}(k^2)\ , \tag{27.6}$$

and, by definition,

$$\mathbf{\Delta}_{\overline{\rm MS}}(-m_{\rm ph}^2)^{-1} = 0\ . \tag{27.7}$$

Setting $k^2 = -m_{\rm ph}^2$ in eq. (27.6), using eq. (27.7), and rearranging, we find

$$m_{\rm ph}^2 = m^2 - \Pi_{\overline{\rm MS}}(-m_{\rm ph}^2)\ . \tag{27.8}$$

Since $\Pi_{\overline{\rm MS}}(k^2)$ is $O(\alpha)$, we see that the difference between $m_{\rm ph}^2$ and m^2 is $O(\alpha)$. Therefore, on the right-hand side, we can replace $m_{\rm ph}^2$ with m^2, and only make an error of $O(\alpha^2)$. Thus

$$m_{\rm ph}^2 = m^2 - \Pi_{\overline{\rm MS}}(-m^2) + O(\alpha^2)\ . \tag{27.9}$$

Working this out, we get

$$m_{\rm ph}^2 = m^2 - \tfrac{1}{2}\alpha\left[\tfrac{1}{6}m^2 - m^2 + \int_0^1 dx\, D_0 \ln(D_0/\mu^2)\right] + O(\alpha^2)\ , \tag{27.10}$$

where $D_0 = [1{-}x(1{-}x)]m^2$. Doing the integrals yields

$$m_{\rm ph}^2 = m^2\left[1 + \tfrac{5}{12}\alpha\Big(\ln(\mu^2/m^2) + c'\Big) + O(\alpha^2)\right], \tag{27.11}$$

where $c' = (34 - 3\pi\sqrt{3})/15 = 1.18$.

Now, physics should be independent of the fake parameter μ. However, the right-hand side of eq. (27.11) depends explicitly on μ. It must be, then, that m and α take on different numerical values as μ is varied, in just the right way to leave physical quantities (like $m_{\rm ph}$) unchanged.

We can use this information to find differential equations that tell us how m and α change with μ. For example, take the logarithm of eq. (27.11) and divide by two to get

$$\ln m_{\rm ph} = \ln m + \tfrac{5}{12}\alpha\Big(\ln(\mu/m) + \tfrac{1}{2}c'\Big) + O(\alpha^2) \,. \tag{27.12}$$

Now differentiate with respect to $\ln\mu$ and require $m_{\rm ph}$ to remain fixed:

$$\begin{aligned} 0 &= \frac{d}{d\ln\mu}\ln m_{\rm ph} \\ &= \frac{1}{m}\frac{dm}{d\ln\mu} + \tfrac{5}{12}\alpha + O(\alpha^2) \,. \end{aligned} \tag{27.13}$$

To get the second line, we had to assume that $d\alpha/d\ln\mu = O(\alpha^2)$, which we will verify shortly. Then, rearranging eq. (27.13) gives

$$\frac{dm}{d\ln\mu} = \big(-\tfrac{5}{12}\alpha + O(\alpha^2)\big)\, m \,. \tag{27.14}$$

The factor in large parentheses on the right is called the *anomalous dimension* of the mass parameter, and it is often given the name $\gamma_m(\alpha)$.

Turning now to the residue R, we have

$$R^{-1} = \frac{d}{dk^2}\Big[\mathbf{\Delta}_{\overline{\rm MS}}(k^2)^{-1}\Big]\bigg|_{k^2=-m_{\rm ph}^2} \,. \tag{27.15}$$

Using eq. (27.6), we get

$$\begin{aligned} R^{-1} &= 1 - \Pi'_{\overline{\rm MS}}(-m_{\rm ph}^2) \\ &= 1 - \Pi'_{\overline{\rm MS}}(-m^2) + O(\alpha^2) \\ &= 1 + \tfrac{1}{12}\alpha\Big(\ln(\mu^2/m^2) + c''\Big) + O(\alpha^2) \,, \end{aligned} \tag{27.16}$$

where $c'' = (17 - 3\pi\sqrt{3})/3 = 0.23$.

We can also use $\overline{\rm MS}$ to define the vertex function. We take

$$C = -\alpha\tfrac{1}{\varepsilon} + O(\alpha^2) \,, \tag{27.17}$$

and so

$$\mathbf{V}_{3,\overline{\rm MS}}(k_1,k_2,k_3) = g\left[1 - \tfrac{1}{2}\alpha\int dF_3\,\ln(D/\mu^2) + O(\alpha^2)\right] , \tag{27.18}$$

where $D = xyk_1^2 + yzk_2^2 + zxk_3^2 + m^2$.

Let us now compute the $\varphi\varphi \to \varphi\varphi$ scattering amplitude in our fancy new renormalization scheme. In the low-mass limit, repeating the steps that led to eq. (26.1), and including the LSZ correction factor $(R^{1/2})^4$, we get

$$\mathcal{T} = R^2\,\mathcal{T}_0\left[1 - \tfrac{11}{12}\alpha\Big(\ln(s/\mu^2) + O(m^0)\Big) + O(\alpha^2)\right], \qquad (27.19)$$

where $\mathcal{T}_0 = -g^2(s^{-1} + t^{-1} + u^{-1})$ is the tree-level result. Now using R from eq. (27.16), we find

$$\mathcal{T} = \mathcal{T}_0\left[1 - \alpha\Big(\tfrac{11}{12}\ln(s/\mu^2) + \tfrac{1}{6}\ln(\mu^2/m^2) + O(m^0)\Big) + O(\alpha^2)\right]. \qquad (27.20)$$

To get an observable amplitude-squared with an imperfect detector, we must square eq. (27.20) and multiply it by the correction factor we derived in section 26,

$$|\mathcal{T}|^2_{\rm obs} = |\mathcal{T}|^2\left[1 + \tfrac{1}{3}\alpha\Big(\ln(\delta^2 s/m^2) + O(m^0)\Big) + O(\alpha^2)\right], \qquad (27.21)$$

where δ is the angular resolution of the detector. Combining this with eq. (27.20), we get

$$|\mathcal{T}|^2_{\rm obs} = |\mathcal{T}_0|^2\left[1 - \alpha\Big(\tfrac{3}{2}\ln(s/\mu^2) + \tfrac{1}{3}\ln(1/\delta^2) + O(m^0)\Big) + O(\alpha^2)\right]. \qquad (27.22)$$

All factors of $\ln m^2$ have disappeared! Finally, we have obtained an expression that has a well-defined $m \to 0$ limit.

Of course, μ is still a fake parameter, and so $|\mathcal{T}|^2_{\rm obs}$ cannot depend on it. It must be, then, that the explicit dependence on μ in eq. (27.22) is canceled by the implicit μ dependence of α. We can use this information to figure out how α must vary with μ. Noting that $|\mathcal{T}_0|^2 = O(g^4) = O(\alpha^2)$, we have

$$\ln|\mathcal{T}|^2_{\rm obs} = C_1 + 2\ln\alpha + 3\alpha(\ln\mu + C_2) + O(\alpha^2)\,, \qquad (27.23)$$

where C_1 and C_2 are independent of μ and α (but depend on the Mandelstam variables). Differentiating with respect to $\ln\mu$ then gives

$$\begin{aligned} 0 &= \frac{d}{d\ln\mu}\ln|\mathcal{T}|^2_{\rm obs} \\ &= \frac{2}{\alpha}\frac{d\alpha}{d\ln\mu} + 3\alpha + O(\alpha^2)\,, \end{aligned} \qquad (27.24)$$

or, after rearranging,

$$\frac{d\alpha}{d\ln\mu} = -\tfrac{3}{2}\alpha^2 + O(\alpha^3)\,. \qquad (27.25)$$

The right-hand side of this equation is called the *beta function.*

Returning to eq. (27.22), we are free to choose any convenient value of μ that we might like. To avoid introducing unnecessary large logs, we should choose $\mu^2 \sim s$.

To compare the results at different values of s, we need to solve eq. (27.25). Keeping only the leading term in the beta function, the solution is

$$\alpha(\mu_2) = \frac{\alpha(\mu_1)}{1 + \frac{3}{2}\alpha(\mu_1)\ln(\mu_2/\mu_1)} \,. \tag{27.26}$$

Thus, as μ increases, $\alpha(\mu)$ decreases. A theory with this property is said to be *asymptotically free*. In this case, the tree-level approximation (in the $\overline{\text{MS}}$ scheme with $\mu^2 \sim s$) becomes better and better at higher and higher energies.

Of course, the opposite is true as well: as μ decreases, $\alpha(\mu)$ increases. As we go to lower and lower energies, the theory becomes more and more strongly coupled.

If the particle mass is nonzero, this process stops at $\mu \sim m$. This is because the minimum value of s is $4m^2$, and so the factor of $\ln(s/\mu^2)$ becomes an unwanted large log for $\mu \ll m$. We should therefore not use values of μ below m. Perturbation theory is still good at these low energies if $\alpha(m) \ll 1$.

If the particle mass is zero, $\alpha(\mu)$ continues to increase at lower and lower energies, and eventually perturbation theory breaks down. This is a signal that the low-energy physics may be quite different from what we expect on the basis of a perturbative analysis.

In the case of φ^3 theory, we know what the correct low-energy physics is: the perturbative ground state is unstable against tunneling through the potential barrier, and there is no true ground state. Asymptotic freedom is, in this case, a signal of this impending disaster.

Much more interesting is asymptotic freedom in a theory that *does* have a true ground state, such as quantum chromodynamics. In this example, the particle excitations are colorless hadrons, rather than the quarks and gluons we would expect from examining the lagrangian.

If the sign of the beta function is positive, then the theory is *infrared free*. The coupling increases as μ increases, and, at sufficiently high energy, perturbation theory breaks down. On the other hand, the coupling decreases as we go to lower energies. Once again, though, we should stop this process at $\mu \sim m$ if the particles have nonzero mass. Quantum electrodynamics with massive electrons (but, of course, massless photons) is in this category.

Still more complicated behaviors are possible if the beta function has a zero at a nonzero value of α. We briefly consider this case in the next section.

Reference notes

Minimal subtraction is treated in more detail in *Brown*, *Collins*, and *Ramond I.*

Problems

27.1) Suppose that we have a theory with

$$\beta(\alpha) = b_1\alpha^2 + O(\alpha^3) , \tag{27.27}$$

$$\gamma_m(\alpha) = c_1\alpha + O(\alpha^2) . \tag{27.28}$$

Neglecting the higher-order terms, show that

$$m(\mu_2) = \left[\frac{\alpha(\mu_2)}{\alpha(\mu_1)}\right]^{c_1/b_1} m(\mu_1) . \tag{27.29}$$

28

The renormalization group

Prerequisite: 27

In section 27 we introduced the $\overline{\text{MS}}$ renormalization scheme, and used the fact that physical observables must be independent of the fake parameter μ to figure out how the lagrangian parameters m and g must change with μ. In this section we rederive these results from a much more formal (but calculationally simpler) point of view, and see how they extend to all orders of perturbation theory. Equations that tell us how the lagrangian parameters (and other objects that are not directly measurable, like correlation functions) vary with μ are collectively called the equations of the *renormalization group*.

Let us recall the lagrangian of our theory, and write it in two different ways. In $d = 6 - \varepsilon$ dimensions, we have

$$\mathcal{L} = -\tfrac{1}{2} Z_\varphi \partial^\mu \varphi \partial_\mu \varphi - \tfrac{1}{2} Z_m m^2 \varphi^2 + \tfrac{1}{6} Z_g g \tilde{\mu}^{\varepsilon/2} \varphi^3 + Y\varphi \tag{28.1}$$

and

$$\mathcal{L} = -\tfrac{1}{2} \partial^\mu \varphi_0 \partial_\mu \varphi_0 - \tfrac{1}{2} m_0^2 \varphi_0^2 + \tfrac{1}{6} g_0 \varphi_0^3 + Y_0 \varphi_0 \,. \tag{28.2}$$

The fields and parameters in eq. (28.1) are the *renormalized* fields and parameters. (And in particular, they are renormalized using the $\overline{\text{MS}}$ scheme, with $\mu^2 = 4\pi e^{-\gamma} \tilde{\mu}^2$.) The fields and parameters in eq. (28.2) are the *bare* fields and parameters. Comparing eqs. (28.1) and (28.2) gives us the relationships between them:

$$\varphi_0(x) = Z_\varphi^{1/2} \varphi(x) \,, \tag{28.3}$$

$$m_0 = Z_\varphi^{-1/2} Z_m^{1/2} m \,, \tag{28.4}$$

$$g_0 = Z_\varphi^{-3/2} Z_g g \tilde{\mu}^{\varepsilon/2} \,, \tag{28.5}$$

$$Y_0 = Z_\varphi^{-1/2} Y \,. \tag{28.6}$$

Recall that, after using dimensional regularization, the infinities coming from loop integrals take the form of inverse powers of $\varepsilon = 6 - d$. In the $\overline{\text{MS}}$ renormalization scheme, we choose the Zs to cancel off these powers of $1/\varepsilon$, and nothing more. Therefore the Zs can be written as

$$Z_\varphi = 1 + \sum_{n=1}^{\infty} \frac{a_n(\alpha)}{\varepsilon^n} , \tag{28.7}$$

$$Z_m = 1 + \sum_{n=1}^{\infty} \frac{b_n(\alpha)}{\varepsilon^n} , \tag{28.8}$$

$$Z_g = 1 + \sum_{n=1}^{\infty} \frac{c_n(\alpha)}{\varepsilon^n} , \tag{28.9}$$

where $\alpha = g^2/(4\pi)^3$. Computing $\Pi(k^2)$ and $\mathbf{V}_3(k_1, k_2, k_3)$ in perturbation theory in the $\overline{\text{MS}}$ scheme gives us Taylor series in α for $a_n(\alpha)$, $b_n(\alpha)$, and $c_n(\alpha)$. So far we have found

$$a_1(\alpha) = -\tfrac{1}{6}\alpha + O(\alpha^2) , \tag{28.10}$$

$$b_1(\alpha) = -\alpha + O(\alpha^2) , \tag{28.11}$$

$$c_1(\alpha) = -\alpha + O(\alpha^2) , \tag{28.12}$$

and that $a_n(\alpha)$, $b_n(\alpha)$, and $c_n(\alpha)$ are all at least $O(\alpha^2)$ for $n \geq 2$.

Next we turn to the trick that we will employ to compute the beta function for α, the anomalous dimension of m, and other useful things. This is the trick: *bare fields and parameters must be independent of* μ.

Why is this so? Recall that we introduced μ when we found that we had to regularize the theory to avoid infinities in the loop integrals of Feynman diagrams. We argued at the time (and ever since) that physical quantities had to be independent of μ. Thus μ is not really a parameter of the theory, but just a crutch that we had to introduce at an intermediate stage of the calculation. In principle, the theory is completely specified by the values of the bare parameters, and, if we were smart enough, we would be able to compute the exact scattering amplitudes in terms of them, without ever introducing μ. The point is this: since the exact scattering amplitudes are independent of μ, the bare parameters must be as well.

Let us start with g_0. It is convenient to define

$$\alpha_0 \equiv g_0^2/(4\pi)^3 = Z_g^2 Z_\varphi^{-3} \tilde{\mu}^\varepsilon \alpha , \tag{28.13}$$

and also

$$G(\alpha, \varepsilon) \equiv \ln(Z_g^2 Z_\varphi^{-3}) . \tag{28.14}$$

From the general structure of eqs. (28.7) and (28.9), we have

$$G(\alpha,\varepsilon) = \sum_{n=1}^{\infty} \frac{G_n(\alpha)}{\varepsilon^n} , \tag{28.15}$$

where, in particular,

$$\begin{aligned} G_1(\alpha) &= 2c_1(\alpha) - 3a_1(\alpha) \\ &= -\tfrac{3}{2}\alpha + O(\alpha^2) . \end{aligned} \tag{28.16}$$

The logarithm of eq. (28.13) can now be written as

$$\ln \alpha_0 = G(\alpha,\varepsilon) + \ln \alpha + \varepsilon \ln \tilde{\mu} . \tag{28.17}$$

Next, differentiate eq. (28.17) with respect to $\ln \mu$, and require α_0 to be independent of it:

$$\begin{aligned} 0 &= \frac{d}{d\ln\mu} \ln \alpha_0 \\ &= \frac{\partial G(\alpha,\varepsilon)}{\partial \alpha} \frac{d\alpha}{d\ln\mu} + \frac{1}{\alpha}\frac{d\alpha}{d\ln\mu} + \varepsilon . \end{aligned} \tag{28.18}$$

Now regroup the terms, multiply by α, and use eq. (28.15) to get

$$0 = \left(1 + \frac{\alpha G_1'(\alpha)}{\varepsilon} + \frac{\alpha G_2'(\alpha)}{\varepsilon^2} + \ldots\right) \frac{d\alpha}{d\ln\mu} + \varepsilon\alpha . \tag{28.19}$$

Next we use some physical reasoning: $d\alpha/d\ln\mu$ is the rate at which α must change to compensate for a small change in $\ln\mu$. If compensation is possible at all, this rate should be finite in the $\varepsilon \to 0$ limit. Therefore, in a renormalizable theory, we should have

$$\frac{d\alpha}{d\ln\mu} = -\varepsilon\alpha + \beta(\alpha) . \tag{28.20}$$

The first term, $-\varepsilon\alpha$, is fixed by matching the $O(\varepsilon)$ terms in eq. (28.19). The second term, the beta function $\beta(\alpha)$, is similarly determined by matching the $O(\varepsilon^0)$ terms; the result is

$$\beta(\alpha) = \alpha^2 G_1'(\alpha) . \tag{28.21}$$

Terms that are higher-order in $1/\varepsilon$ must also cancel, and this determines all the other $G_n'(\alpha)$s in terms of $G_1'(\alpha)$. Thus, for example, cancellation of the $O(\varepsilon^{-1})$ terms fixes $G_2'(\alpha) = \alpha G_1'(\alpha)^2$. These relations among the $G_n'(\alpha)$s can of course be checked order by order in perturbation theory.

From eq. (28.21) and eq. (28.16), we find that the beta function is

$$\beta(\alpha) = -\tfrac{3}{2}\alpha^2 + O(\alpha^3)\,. \tag{28.22}$$

Hearteningly, this is the same result we found in section 27 by requiring the observed scattering cross section $|\mathcal{T}|^2_{\rm obs}$ to be independent of μ. However, simply as a matter of practical calculation, it is much easier to compute $G_1(\alpha)$ than it is to compute $|\mathcal{T}|^2_{\rm obs}$.

Next consider the invariance of m_0. We begin by defining

$$\begin{aligned} M(\alpha,\varepsilon) &\equiv \ln(Z_m^{1/2} Z_\varphi^{-1/2}) \\ &= \sum_{n=1}^{\infty} \frac{M_n(\alpha)}{\varepsilon^n}\,. \end{aligned} \tag{28.23}$$

From eqs. (28.10) and (28.11) we have

$$\begin{aligned} M_1(\alpha) &= \tfrac{1}{2}b_1(\alpha) - \tfrac{1}{2}a_1(\alpha) \\ &= -\tfrac{5}{12}\alpha + O(\alpha^2)\,. \end{aligned} \tag{28.24}$$

Then, from eq. (28.4), we have

$$\ln m_0 = M(\alpha,\varepsilon) + \ln m\,. \tag{28.25}$$

Take the derivative with respect to $\ln\mu$ and require m_0 to be unchanged:

$$\begin{aligned} 0 &= \frac{d}{d\ln\mu}\ln m_0 \\ &= \frac{\partial M(\alpha,\varepsilon)}{\partial\alpha}\,\frac{d\alpha}{d\ln\mu} + \frac{1}{m}\frac{dm}{d\ln\mu} \\ &= \frac{\partial M(\alpha,\varepsilon)}{\partial\alpha}\Big(-\varepsilon\alpha + \beta(\alpha)\Big) + \frac{1}{m}\,\frac{dm}{d\ln\mu}\,. \end{aligned} \tag{28.26}$$

Rearranging, we find

$$\begin{aligned} \frac{1}{m}\,\frac{dm}{d\ln\mu} &= \Big(\varepsilon\alpha - \beta(\alpha)\Big)\sum_{n=1}^{\infty}\frac{M_n'(\alpha)}{\varepsilon^n} \\ &= \alpha M_1'(\alpha) + \ldots\,, \end{aligned} \tag{28.27}$$

where the ellipses stand for terms with powers of $1/\varepsilon$. In a renormalizable theory, $dm/d\ln\mu$ should be finite in the $\varepsilon \to 0$ limit, and so these terms must actually all be zero. Therefore, the anomalous dimension of the mass, defined via

$$\gamma_m(\alpha) \equiv \frac{1}{m}\frac{dm}{d\ln\mu}\,, \tag{28.28}$$

is given by

$$\begin{aligned}\gamma_m(\alpha) &= \alpha M_1'(\alpha) \\ &= -\tfrac{5}{12}\alpha + O(\alpha^2)\,. \end{aligned}\tag{28.29}$$

Comfortingly, this is just what we found in section 27.

Let us now consider the propagator in the $\overline{\text{MS}}$ renormalization scheme,

$$\tilde{\mathbf{\Delta}}(k^2) = i\int d^6x\, e^{-ikx}\langle 0|\mathrm{T}\varphi(x)\varphi(0)|0\rangle\,. \tag{28.30}$$

The bare propagator,

$$\tilde{\mathbf{\Delta}}_0(k^2) = i\int d^6x\, e^{-ikx}\langle 0|\mathrm{T}\varphi_0(x)\varphi_0(0)|0\rangle\,, \tag{28.31}$$

should be (by the now-familiar argument) independent of μ. The bare and renormalized propagators are related by

$$\tilde{\mathbf{\Delta}}_0(k^2) = Z_\varphi\tilde{\mathbf{\Delta}}(k^2)\,. \tag{28.32}$$

Taking the logarithm and differentiating with respect to $\ln\mu$, we get

$$\begin{aligned}0 &= \frac{d}{d\ln\mu}\ln\tilde{\mathbf{\Delta}}_0(k^2) \\ &= \frac{d\ln Z_\varphi}{d\ln\mu} + \frac{d}{d\ln\mu}\ln\tilde{\mathbf{\Delta}}(k^2) \\ &= \frac{d\ln Z_\varphi}{d\ln\mu} + \frac{1}{\tilde{\mathbf{\Delta}}(k^2)}\left(\frac{\partial}{\partial\ln\mu} + \frac{d\alpha}{d\ln\mu}\frac{\partial}{\partial\alpha} + \frac{dm}{d\ln\mu}\frac{\partial}{\partial m}\right)\tilde{\mathbf{\Delta}}(k^2). \end{aligned}\tag{28.33}$$

We can write

$$\ln Z_\varphi = \frac{a_1(\alpha)}{\varepsilon} + \frac{a_2(\alpha) - \frac{1}{2}a_1^2(\alpha)}{\varepsilon^2} + \ldots\,. \tag{28.34}$$

Then we have

$$\begin{aligned}\frac{d\ln Z_\varphi}{d\ln\mu} &= \frac{\partial\ln Z_\varphi}{\partial\alpha}\frac{d\alpha}{d\ln\mu} \\ &= \left(\frac{a_1'(\alpha)}{\varepsilon} + \ldots\right)\Big(-\varepsilon\alpha + \beta(\alpha)\Big) \\ &= -\alpha a_1'(\alpha) + \ldots\,, \end{aligned}\tag{28.35}$$

where the ellipses in the last line stand for terms with powers of $1/\varepsilon$. Since $\tilde{\mathbf{\Delta}}(k^2)$ should vary smoothly with μ in the $\varepsilon \to 0$ limit, these must all be zero. We then define the *anomalous dimension of the field*

$$\gamma_\varphi(\alpha) \equiv \frac{1}{2}\frac{d\ln Z_\varphi}{d\ln\mu}\,. \tag{28.36}$$

From eq. (28.35) we find

$$\begin{aligned}\gamma_\varphi(\alpha) &= -\tfrac{1}{2}\alpha a_1'(\alpha) \\ &= +\tfrac{1}{12}\alpha + O(\alpha^2)\,.\end{aligned} \tag{28.37}$$

Eq. (28.33) can now be written as

$$\left(\frac{\partial}{\partial\ln\mu} + \beta(\alpha)\frac{\partial}{\partial\alpha} + \gamma_m(\alpha)m\frac{\partial}{\partial m} + 2\gamma_\varphi(\alpha)\right)\tilde{\mathbf{\Delta}}(k^2) = 0 \tag{28.38}$$

in the $\varepsilon \to 0$ limit. This is the *Callan–Symanzik equation* for the propagator.

The Callan–Symanzik equation is most interesting in the massless limit, and for a theory with a zero of the beta function at a nonzero value of α. So, let us suppose that $\beta(\alpha_*) = 0$ for some $\alpha_* \neq 0$. Then, for $\alpha = \alpha_*$ and $m = 0$, the Callan–Symanzik equation becomes

$$\left(\frac{\partial}{\partial\ln\mu} + 2\gamma_\varphi(\alpha_*)\right)\tilde{\mathbf{\Delta}}(k^2) = 0\,. \tag{28.39}$$

The solution is

$$\tilde{\mathbf{\Delta}}(k^2) = \frac{C(\alpha_*)}{k^2}\left(\frac{\mu^2}{k^2}\right)^{-\gamma_\varphi(\alpha_*)}, \tag{28.40}$$

where $C(\alpha_*)$ is an integration constant. (We used the fact that $\tilde{\mathbf{\Delta}}(k^2)$ has mass dimension -2 to get the k^2 dependence in addition to the μ dependence.) Thus the naive scaling law $\tilde{\mathbf{\Delta}}(k^2) \sim k^{-2}$ is changed to $\tilde{\mathbf{\Delta}}(k^2) \sim k^{-2[1-\gamma_\varphi(\alpha_*)]}$. This has applications in the theory of critical phenomena, which is beyond the scope of this book.

Reference notes

The formal development of the renormalization group is explored in more detail in *Brown*, *Collins*, and *Ramond I*.

Problems

28.1) Consider φ^4 theory,

$$\mathcal{L} = -\tfrac{1}{2}Z_\varphi\partial^\mu\varphi\partial_\mu\varphi - \tfrac{1}{2}Z_m m^2\varphi^2 - \tfrac{1}{24}Z_\lambda\lambda\tilde{\mu}^\varepsilon\varphi^4\,, \tag{28.41}$$

in $d = 4 - \varepsilon$ dimensions. Compute the beta function to $O(\lambda^2)$, the anomalous dimension of m to $O(\lambda)$, and the anomalous dimension of φ to $O(\lambda)$.

28.2) Repeat problem 28.1 for the theory of problem 9.3.

28.3) Consider the lagrangian density,

$$\begin{aligned}\mathcal{L} = &-\tfrac{1}{2}Z_\varphi \partial^\mu\varphi\partial_\mu\varphi - \tfrac{1}{2}Z_m m^2\varphi^2 + Y\varphi \\ &-\tfrac{1}{2}Z_\chi \partial^\mu\chi\partial_\mu\chi - \tfrac{1}{2}Z_M M^2\chi^2 \\ &+\tfrac{1}{6}Z_g g\tilde{\mu}^{\varepsilon/2}\varphi^3 + \tfrac{1}{2}Z_h h\tilde{\mu}^{\varepsilon/2}\varphi\chi^2 \,,\end{aligned} \tag{28.42}$$

in $d = 6 - \varepsilon$ dimensions, where φ and χ are real scalar fields, and Y is adjusted to make $\langle 0|\varphi(x)|0\rangle = 0$. (Why is no such term needed for χ?)

a) Compute the one-loop contributions to each of the Zs in the $\overline{\text{MS}}$ renormalization scheme.

b) The bare couplings are related to the renormalized ones via

$$g_0 = Z_\varphi^{-3/2} Z_g g \tilde{\mu}^{\varepsilon/2} \,, \tag{28.43}$$

$$h_0 = Z_\varphi^{-1/2} Z_\chi^{-1} Z_h h \tilde{\mu}^{\varepsilon/2} \,. \tag{28.44}$$

Define

$$G(g,h,\varepsilon) = \textstyle\sum_{n=1}^{\infty} G_n(g,h)\varepsilon^{-n} \equiv \ln\left(Z_\varphi^{-3/2} Z_g\right) , \tag{28.45}$$

$$H(g,h,\varepsilon) = \textstyle\sum_{n=1}^{\infty} H_n(g,h)\varepsilon^{-n} \equiv \ln\left(Z_\varphi^{-1/2} Z_\chi^{-1} Z_h\right) . \tag{28.46}$$

By requiring g_0 and h_0 to be independent of μ, and by assuming that $dg/d\mu$ and $dh/d\mu$ are finite as $\varepsilon \to 0$, show that

$$\mu\frac{dg}{d\mu} = -\tfrac{1}{2}\varepsilon g + \tfrac{1}{2}g\left(g\frac{\partial G_1}{\partial g} + h\frac{\partial G_1}{\partial h}\right) , \tag{28.47}$$

$$\mu\frac{dh}{d\mu} = -\tfrac{1}{2}\varepsilon h + \tfrac{1}{2}h\left(g\frac{\partial H_1}{\partial g} + h\frac{\partial H_1}{\partial h}\right) . \tag{28.48}$$

c) Use your results from part (a) to compute the beta functions $\beta_g(g,h) \equiv \lim_{\varepsilon\to 0}\mu\, dg/d\mu$ and $\beta_h(g,h) \equiv \lim_{\varepsilon\to 0}\mu\, dh/d\mu$. You should find terms of order g^3, gh^2, and h^3 in β_g, and terms of order g^2h, gh^2, and h^3 in β_h.

d) Without loss of generality, we can choose g to be positive; h can then be positive or negative, and the difference is physically significant. (You should understand why this is true.) For what numerical range(s) of h/g are β_g and β_h/h both negative? Why is this an interesting question?

29

Effective field theory

Prerequisite: 28

So far we have been discussing only renormalizable theories. In this section, we investigate what meaning can be assigned to nonrenormalizable theories, following an approach pioneered by Ken Wilson.

We will begin by analyzing a renormalizable theory from a new point of view. Consider, as an example, φ^4 theory in four spacetime dimensions:

$$\mathcal{L} = -\tfrac{1}{2}Z_\varphi \partial^\mu\varphi\partial_\mu\varphi - \tfrac{1}{2}Z_m m_{\text{ph}}^2\varphi^2 - \tfrac{1}{24}Z_\lambda\lambda_{\text{ph}}\varphi^4 \,. \tag{29.1}$$

(This example is actually problematic, because this theory is *trivial*, a technical term that we will explain later. For now we proceed with a perturbative analysis.) We take the renormalizing Z factors to be defined in an on-shell scheme, and have emphasized this by writing the particle mass as m_{ph} and the coupling constant as λ_{ph}. We define λ_{ph} as the value of the exact 1PI four-point vertex with zero external four-momenta:

$$\lambda_{\text{ph}} \equiv \mathbf{V}_4(0,0,0,0) \,. \tag{29.2}$$

The path integral is given by

$$Z(J) = \int \mathcal{D}\varphi \; e^{iS+i\int J\varphi} \,, \tag{29.3}$$

where $S = \int d^4x\, \mathcal{L}$ and $\int J\varphi$ is short for $\int d^4x\, J\varphi$.

Our first step in analyzing this theory will be to perform the Wick rotation (applied to loop integrals in section 14) directly on the action. We define a euclidean time $\tau \equiv it$. Then we have

$$Z(J) = \int \mathcal{D}\varphi \; e^{-S_{\text{E}}+\int J\varphi} \,, \tag{29.4}$$

where $S_\mathrm{E} = \int d^4x\, \mathcal{L}_\mathrm{E}$, $d^4x = d^3x\, d\tau$,

$$\mathcal{L}_\mathrm{E} = \tfrac{1}{2} Z_\varphi \partial_\mu \varphi \partial_\mu \varphi + \tfrac{1}{2} Z_m m_\mathrm{ph}^2 \varphi^2 + \tfrac{1}{24} Z_\lambda \lambda_\mathrm{ph} \varphi^4 \,, \tag{29.5}$$

and

$$\partial_\mu \varphi \partial_\mu \varphi = (\partial\varphi/\partial\tau)^2 + (\nabla\varphi)^2 \,. \tag{29.6}$$

Note that each term in S_E is always positive (or zero) for any field configuration $\varphi(x)$. This is the advantage of working in euclidean space: eq. (29.4), the *euclidean path integral*, is strongly damped (rather than rapidly oscillating) at large values of the field and/or its derivatives, and this makes its convergence properties more obvious.

Next, we Fourier transform to (euclidean) momentum space via

$$\varphi(x) = \int \frac{d^4k}{(2\pi)^4}\, e^{ikx}\, \widetilde{\varphi}(k) \,. \tag{29.7}$$

The euclidean action becomes

$$\begin{aligned} S_\mathrm{E} = \frac{1}{2} \int \frac{d^4k}{(2\pi)^4}\, \widetilde{\varphi}(-k) \Big(Z_\varphi k^2 + Z_m m_\mathrm{ph}^2 \Big) \widetilde{\varphi}(k) \\ + \frac{1}{24} Z_\lambda \lambda_\mathrm{ph} \int \frac{d^4k_1}{(2\pi)^4} \cdots \frac{d^4k_4}{(2\pi)^4}\, (2\pi)^4 \delta^4(k_1{+}k_2{+}k_3{+}k_4) \\ \cdot \times \widetilde{\varphi}(k_1)\widetilde{\varphi}(k_2)\widetilde{\varphi}(k_3)\widetilde{\varphi}(k_4) \,. \end{aligned} \tag{29.8}$$

Note that $k^2 = \mathbf{k}^2 + k_\tau^2 \geq 0$.

We now introduce an *ultraviolet cutoff* Λ. It should be much larger than the particle mass m_ph, or any other energy scale of practical interest. Then we perform the path integral over all $\widetilde{\varphi}(k)$ with $|k| > \Lambda$. We also take $\widetilde{J}(k) = 0$ for $|k| > \Lambda$. Then we find

$$Z(J) = \int \mathcal{D}\varphi_{|k|<\Lambda}\, e^{-S_\mathrm{eff}(\varphi;\Lambda) + \int J\varphi} \,, \tag{29.9}$$

where

$$e^{-S_\mathrm{eff}(\varphi;\Lambda)} = \int \mathcal{D}\varphi_{|k|>\Lambda}\, e^{-S_\mathrm{E}(\varphi)} \,. \tag{29.10}$$

$S_\mathrm{eff}(\varphi;\Lambda)$ is called the *Wilsonian effective action.* We can write the corresponding lagrangian density as

$$\begin{aligned} \mathcal{L}_\mathrm{eff}(\varphi;\Lambda) = \tfrac{1}{2} Z(\Lambda) \partial_\mu \varphi \partial_\mu \varphi + \tfrac{1}{2} m^2(\Lambda) \varphi^2 + \tfrac{1}{24} \lambda(\Lambda) \varphi^4 \\ + \sum_{d\geq 6} \sum_i c_{d,i}(\Lambda) \mathcal{O}_{d,i} \,, \end{aligned} \tag{29.11}$$

where the Fourier components of $\varphi(x)$ are now cut off at $|k| > \Lambda$:

$$\varphi(x) = \int_0^\Lambda \frac{d^4k}{(2\pi)^4}\, e^{ikx}\, \widetilde{\varphi}(k)\ . \tag{29.12}$$

The operators $\mathcal{O}_{d,i}$ in eq. (29.11) consist of all terms that have mass dimension $d \geq 6$ and that are even under $\varphi \leftrightarrow -\varphi$; i is an index that distinguishes operators of the same dimension that are inequivalent after integrations by parts of any derivatives that act on the fields. (The operators must be even under $\varphi \leftrightarrow -\varphi$ in order to respect the $\varphi \leftrightarrow -\varphi$ symmetry of the original lagrangian.)

The coefficients $Z(\Lambda)$, $m^2(\Lambda)$, $\lambda(\Lambda)$, and $c_{d,i}(\Lambda)$ in eq. (29.11) are all *finite* functions of Λ. This is established by the following argument. We can differentiate eq. (29.9) with respect to $J(x)$ to compute correlation functions of the renormalized field $\varphi(x)$, and correlation functions of renormalized fields are finite. Using eq. (29.9), we can compute these correlation functions as a series of Feynman diagrams, with Feynman rules based on $\mathcal{L}_{\rm eff}$. These rules include an ultraviolet cutoff Λ on the loop momenta, since the fields with higher momenta have already been *integrated out.* Thus all of the loop integrals in these diagrams are finite. Therefore the other parameters that enter the diagrams—$Z(\Lambda)$, $m^2(\Lambda)$, $\lambda(\Lambda)$, and $c_{d,i}(\Lambda)$—must be finite as well, in order to end up with finite correlation functions.

To compute these parameters, we can think of eq. (29.8) as the action for two kinds of fields, those with $|k| < \Lambda$ and those with $|k| > \Lambda$. Then we draw all 1PI diagrams with external lines for $|k| < \Lambda$ fields only. For $\lambda_{\rm ph} \ll 1$, the dominant contribution to $c_{d,i}(\Lambda)$ for an operator $\mathcal{O}_{d,i}$ with $2n$ fields and $d-2n$ derivatives is then given by a one-loop diagram with $2n$ external lines (representing $|k| < \Lambda$ fields), n vertices, and a $|k| > \Lambda$ field circulating in the loop; see fig. 29.1.

The simplest case to consider is $\mathcal{O}_{2n,1} \equiv \varphi^{2n}$. With $2n$ external lines, there are $(2n)!$ ways of assigning the external momenta to the lines, but $2^n \times n \times 2$ of these give the same diagram: 2^n for exchanging the two external lines that meet at any one vertex; n for rotations of the diagram; and 2 for reflection of the diagram. Since there are no derivatives on the external fields, we can set all of the external momenta to zero; then all $(2n)!/(2^n 2n)$ diagrams have the same value. With a euclidean action, each internal line contributes a factor of $1/(k^2+m^2_{\rm ph})$, and each vertex contributes a factor of $-Z_\lambda\lambda_{\rm ph} = -\lambda_{\rm ph} + O(\lambda^2_{\rm ph})$. The vertex factor associated with the term

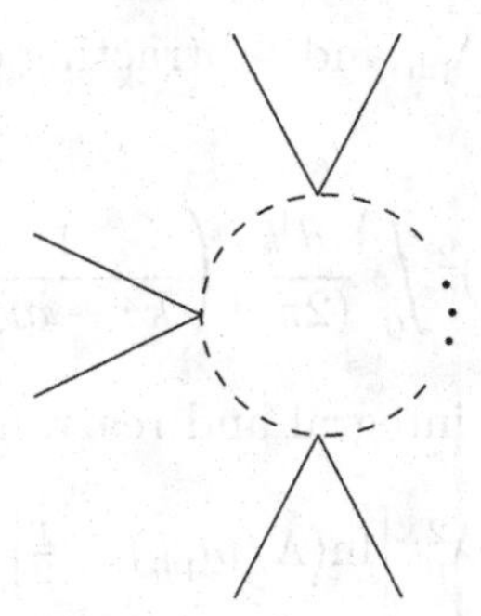

Figure 29.1. A one-loop 1PI diagram with $2n$ external lines. Each external line represents a field with $|k| < \Lambda$. The internal (dashed) line represents a field with $|k| > \Lambda$.

$c_{2n,1}(\Lambda)\varphi^{2n}$ in $\mathcal{L}_{\text{eff}}$ is $-(2n)!\,c_{2n,1}(\Lambda)$. Thus we have

$$-(2n)!\,c_{2n,1}(\Lambda) = \frac{(-\lambda_{\text{ph}})^n(2n)!}{2^n 2n}\int_\Lambda^\infty \frac{d^4k}{(2\pi)^4}\left(\frac{1}{k^2+m_{\text{ph}}^2}\right)^n + O(\lambda_{\text{ph}}^{n+1})\,. \tag{29.13}$$

For $2n \ge 6$, the integral converges, and we find

$$c_{2n,1}(\Lambda) = -\frac{(-\lambda_{\text{ph}}/2)^n}{32\pi^2 n(n-2)}\,\frac{1}{\Lambda^{2n-4}} + O(\lambda_{\text{ph}}^{n+1})\,. \tag{29.14}$$

We have taken $\Lambda \gg m_{\text{ph}}$, and dropped terms down by powers of m_{ph}/Λ.

For $2n = 4$, we have to include the tree-level vertex; in this case, we have

$$-\lambda(\Lambda) = -Z_\lambda\lambda_{\text{ph}} + \frac{3}{2}(-\lambda_{\text{ph}})^2\int_\Lambda^\infty \frac{d^4k}{(2\pi)^4}\left(\frac{1}{k^2+m_{\text{ph}}^2}\right)^2 + O(\lambda_{\text{ph}}^3)\,. \tag{29.15}$$

This integral diverges. To evaluate it, we note that the one-loop contribution to the exact four-point vertex is given by the *same* diagram, but with fields of *all* momenta circulating in the loop. Thus we have

$$-\mathbf{V}_4(0,0,0,0) = -Z_\lambda\lambda_{\text{ph}} + \frac{3}{2}(-\lambda_{\text{ph}})^2\int_0^\infty \frac{d^4k}{(2\pi)^4}\left(\frac{1}{k^2+m_{\text{ph}}^2}\right)^2 + O(\lambda_{\text{ph}}^3)\,. \tag{29.16}$$

Then, using $\mathbf{V}_4(0,0,0,0) = \lambda_{\rm ph}$ and subtracting eq. (29.15) from eq. (29.16), we get

$$-\lambda_{\rm ph} + \lambda(\Lambda) = \frac{3}{2}(-\lambda_{\rm ph})^2 \int_0^\Lambda \frac{d^4k}{(2\pi)^4} \left(\frac{1}{k^2 + m_{\rm ph}^2}\right)^2 + O(\lambda_{\rm ph}^3) \,. \qquad (29.17)$$

Evaluating the (now finite!) integral and rearranging, we have

$$\lambda(\Lambda) = \lambda_{\rm ph} + \frac{3}{16\pi^2}\lambda_{\rm ph}^2 \Big[\ln(\Lambda/m_{\rm ph}) - \tfrac{1}{2}\Big] + O(\lambda_{\rm ph}^3) \,. \qquad (29.18)$$

Note that this result has the problem of a *large log*; the second term is smaller than the first only if $\lambda_{\rm ph} \ln(\Lambda/m_{\rm ph}) \ll 1$. To cure this problem, we must change the renormalization scheme. We will take up this issue shortly, but first let us examine the case of two external lines while continuing to use the on-shell scheme.

For the case of two external lines, the one-loop diagram has just one vertex, and by momentum conservation, the loop integral is completely indepedent of the external momentum. This implies that the one-loop contribution to $Z(\Lambda)$ vanishes, and so we have

$$Z(\Lambda) = 1 + O(\lambda_{\rm ph}^2) \,. \qquad (29.19)$$

The one-loop diagram does, however, give a nonzero contribution to $m^2(\Lambda)$; after including the tree-level term, we find

$$-m^2(\Lambda) = -Z_m m_{\rm ph}^2 + \frac{1}{2}(-\lambda_{\rm ph}) \int_\Lambda^\infty \frac{d^4k}{(2\pi)^4} \frac{1}{k^2 + m_{\rm ph}^2} + O(\lambda_{\rm ph}^2) \,. \qquad (29.20)$$

This integral diverges. To evaluate it, recall that the one-loop contribution to the exact particle mass-squared is given by the *same* diagram, but with fields of *all* momenta circulating in the loop. Thus we have

$$-m_{\rm ph}^2 = -Z_m m_{\rm ph}^2 + \frac{1}{2}(-\lambda_{\rm ph}) \int_0^\infty \frac{d^4k}{(2\pi)^4} \frac{1}{k^2 + m_{\rm ph}^2} + O(\lambda_{\rm ph}^2) \,. \qquad (29.21)$$

Then, subtracting eq. (29.20) from eq. (29.21), we get

$$-m_{\rm ph}^2 + m^2(\Lambda) = \frac{1}{2}(-\lambda_{\rm ph}) \int_0^\Lambda \frac{d^4k}{(2\pi)^4} \frac{1}{k^2 + m_{\rm ph}^2} + O(\lambda_{\rm ph}^2) \,. \qquad (29.22)$$

Evaluating the (now finite!) integral and rearranging, we have

$$m^2(\Lambda) = m_{\rm ph}^2 - \frac{\lambda_{\rm ph}}{32\pi^2}\Big[\Lambda^2 - m_{\rm ph}^2 \ln(\Lambda^2/m_{\rm ph}^2)\Big] + O(\lambda_{\rm ph}^2) \,. \qquad (29.23)$$

We see that we now have an even worse situation than we did with the large log in $\lambda(\Lambda)$: the correction term is *quadratically divergent*.

As already noted, to fix these problems we must change the renormalization scheme. In the context of an effective action with a specific value of the cutoff Λ_0, there is a simple way to do so: we simply treat this effective action as the fundamental starting point, with $Z(\Lambda_0)$, $m^2(\Lambda_0)$, $\lambda(\Lambda_0)$, and $c_{d,i}(\Lambda_0)$ as input parameters. We then see what physics emerges at energy scales well below Λ_0. We can set $Z(\Lambda_0) = 1$, with the understanding that the field no longer has the LSZ normalization (and that we will have to correct the LSZ formula to account for this). We will also assume that the parameters $\lambda(\Lambda_0)$, $m^2(\Lambda_0)$, and $c_{d,i}(\Lambda_0)$ are all small when measured in units of the cutoff:

$$\lambda(\Lambda_0) \ll 1\ , \tag{29.24}$$

$$|m^2(\Lambda_0)| \ll \Lambda_0^2\ , \tag{29.25}$$

$$c_{d,i}(\Lambda_0) \ll \Lambda_0^{-(d-4)}\ . \tag{29.26}$$

The proposal to treat the effective action as the fundamental starting point may not seem very appealing. For one thing, we now have an infinite number of parameters to specify, rather than two! Also, we now have an explicit cutoff in place, rather than trying to have a theory that works at all energy scales.

On the other hand, it may well be that quantum field theory does not work at arbitrarily high energies. For example, quantum fluctuations in spacetime itself should become important above the *Planck scale*, which is given by the inverse square root of Newton's constant, and has a numerical value of $\sim 10^{19}$ GeV (compared to, say, the proton mass, which is ~ 1 GeV).

So, let us leave the cutoff Λ_0 in place for now. We will then make a two-pronged analysis. First, we will see what happens at much lower energies. Then, we will see what happens if we try to take the limit $\Lambda_0 \to \infty$.

We begin by examining lower energies. To make things more tractable, we will set

$$c_{d,i}(\Lambda_0) = 0\ ; \tag{29.27}$$

later we will examine the effects of a more general choice.

A nice way to see what happens at lower energies is to integrate out some more high-energy degrees of freedom. Let us, then, perform the functional integral over Fourier modes $\widetilde{\varphi}(k)$ with $\Lambda < |k| < \Lambda_0$; we have

$$e^{-S_{\rm eff}(\varphi;\Lambda)} = \int \mathcal{D}\varphi_{\Lambda<|k|<\Lambda_0}\, e^{-S_{\rm eff}(\varphi;\Lambda_0)} \, . \tag{29.28}$$

We can do this calculation in perturbation theory, mimicking the procedure that we used earlier. We find

$$m^2(\Lambda) = m^2(\Lambda_0) + \frac{1}{2}\lambda(\Lambda_0)\int_\Lambda^{\Lambda_0} \frac{d^4k}{(2\pi)^4}\frac{1}{k^2+m^2(\Lambda_0)} + \ldots \, , \tag{29.29}$$

$$\lambda(\Lambda) = \lambda(\Lambda_0) - \frac{3}{2}\lambda^2(\Lambda_0)\int_\Lambda^{\Lambda_0} \frac{d^4k}{(2\pi)^4}\left(\frac{1}{k^2+m^2(\Lambda_0)}\right)^2 + \ldots \, , \tag{29.30}$$

$$c_{2n,1}(\Lambda) = -\frac{(-1)^n}{2^n 2n}\lambda^n(\Lambda_0)\int_\Lambda^{\Lambda_0} \frac{d^4k}{(2\pi)^4}\left(\frac{1}{k^2+m^2(\Lambda_0)}\right)^n + \ldots \, , \tag{29.31}$$

where the ellipses stand for higher-order corrections. For Λ not too much less than Λ_0 (and, in particular, for $|m^2(\Lambda_0)| \ll \Lambda^2$), we find

$$m^2(\Lambda) = m^2(\Lambda_0) + \frac{1}{32\pi^2}\lambda(\Lambda_0)\left(\Lambda_0^2 - \Lambda^2\right) + \ldots \, , \tag{29.32}$$

$$\lambda(\Lambda) = \lambda(\Lambda_0) - \frac{3}{16\pi^2}\lambda^2(\Lambda_0)\ln\frac{\Lambda_0}{\Lambda} + \ldots \, , \tag{29.33}$$

$$c_{2n,1}(\Lambda) = -\frac{(-1)^n}{32\pi^2 2^n n(n-2)}\lambda^n(\Lambda_0)\left(\frac{1}{\Lambda^{2n-4}} - \frac{1}{\Lambda_0^{2n-4}}\right) + \ldots \, . \tag{29.34}$$

We see from this that the corrections to $m^2(\Lambda)$, which is the only coefficient with positive mass dimension, are dominated by contributions from the *high* end of the integral. On the other hand, the corrections to $c_{d,i}(\Lambda)$, coefficients with negative mass dimension, are dominated by contributions from the *low* end of the integral. And the corrections to $\lambda(\Lambda)$, which is dimensionless, come equally from all portions of the range of integration.

For the $c_{d,i}(\Lambda)$, this means that their starting values at Λ_0 were not very important, as long as eq. (29.26) is obeyed. Nonzero starting values would contribute another term of order $1/\Lambda_0^{2n-4}$ to $c_{2n,1}(\Lambda)$, but all such terms are less important than the one of order $1/\Lambda^{2n-4}$ that comes from doing the integral down to $|k| = \Lambda$.

Similarly, nonzero values of $c_{d,i}(\Lambda_0)$ would make subdominant contributions to $\lambda(\Lambda)$. As an example, consider the contribution of the diagram in fig. 29.2. Ignoring numerical factors, the vertex factor is $c_{6,1}(\Lambda_0)$, and the loop integral is the same as the one that enters into $m^2(\Lambda)$; it yields a factor of $\Lambda_0^2 - \Lambda^2 \sim \Lambda_0^2$. Thus the contribution of this diagram to $\lambda(\Lambda)$ is of order $c_{6,1}(\Lambda_0)\Lambda_0^2$. This is a pure number that, according to eq. (29.26), is small.

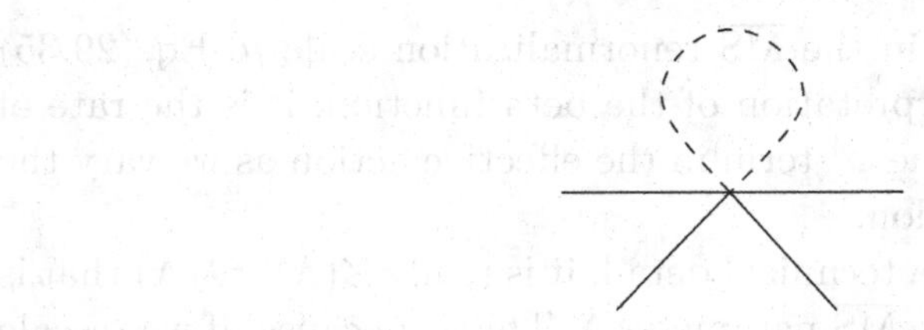

Figure 29.2. A one-loop contribution to the φ^4 vertex for fields with $|k| < \Lambda$. The internal (dashed) line represents a field with $|k| > \Lambda$.

This contribution is missing the logarithmic enhancement factor $\ln(\Lambda_0/\Lambda)$ that we see in eq. (29.33).

On the other hand, for $m^2(\Lambda)$, there are infinitely many contributions of order Λ_0^2 when $c_{d,i}(\Lambda_0) \neq 0$. These must add up to give $m^2(\Lambda)$ a value that is much smaller. Indeed, we want to continue the process down to lower and lower values of Λ, with $m^2(\Lambda)$ dropping until it becomes of order $m_{\rm ph}^2$ at $\Lambda \sim m_{\rm ph}$. For this to happen, there must be very precise cancellations among all the terms of order Λ_0^2 that contribute to $m^2(\Lambda)$. In some sense, it is more "natural" to have $m_{\rm ph}^2 \sim \lambda(\Lambda_0)\Lambda_0^2$, rather than to arrange for these very precise cancellations. This philosophical issue is called the *fine-tuning problem*, and it generically arises in theories with spin-zero fields.

In theories with higher-spin fields only, the action typically has more symmetry when these fields are massless, and this typically prevents divergences that are worse than logarithmic. These theories are said to be *technically natural*, while theories with spin-zero fields (with physical masses well below the cutoff) generally are not. (The only exceptions are theories where *supersymmetry* relates spin-zero and spin-one-half fields; the spin-zero fields then inherit the technical naturalness of their spin-one-half partners.) For now, in φ^4 theory, we will simply accept the necessity of fine-tuning in order to have $m_{\rm ph} \ll \Lambda$.

Returning to eqs. (29.32)–(29.34), we can recast them as differential equations that tell us how these parameters change with the value of the cutoff Λ. In particular, let us do this for $\lambda(\Lambda)$. We take the derivative of eq. (29.33) with respect to Λ, multiply by Λ, and then set $\Lambda_0 = \Lambda$ to get

$$\frac{d}{d\ln\Lambda}\lambda(\Lambda) = \frac{3}{16\pi^2}\lambda^2(\Lambda) + \ldots \,. \tag{29.35}$$

Note that the right-hand side of eq. (29.35) is apparently the same as the *beta function* $\beta(\lambda) \equiv d\lambda/d\ln\mu$ that we calculated in problem 28.1, where it represented the rate of change in the $\overline{\rm MS}$ parameter λ that was needed to compensate for a change in the $\overline{\text{MS}}$ renormalization scale μ. Eq. (29.35) gives us a new physical interpretation of the beta function: it is the rate of change in the coefficient of the φ^4 term in the effective action as we vary the ultraviolet cutoff in that action.

Actually, though, there is a technical detail: it is really $Z(\Lambda)^{-2}\lambda(\Lambda)$ that is most closely analogous to the $\overline{\text{MS}}$ parameter λ. This is because, if we rescale φ so that it has a canonical kinetic term of $\frac{1}{2}\partial_\mu\varphi\partial_\mu\varphi$, then the coefficient of the φ^4 term is $Z(\Lambda)^{-2}\lambda(\Lambda)$. Since $Z(\Lambda) = 1 + O(\lambda^2(\Lambda))$, this has no effect at the one-loop level, but it does matter at two loops. We can account for the effect of this *wave function renormalization* (in all the couplings) by writing, instead of eq. (29.11),

$$\mathcal{L}_{\text{eff}}(\varphi;\Lambda) = \tfrac{1}{2}Z(\Lambda)\partial_\mu\varphi\partial_\mu\varphi + \tfrac{1}{2}Z(\Lambda)m^2(\Lambda)\varphi^2 + \tfrac{1}{24}Z^2(\Lambda)\lambda(\Lambda)\varphi^4 + \sum_{d\geq 6}\sum_i Z^{n_{d,i}/2}(\Lambda)c_{d,i}(\Lambda)\mathcal{O}_{d,i}\,, \tag{29.36}$$

where $n_{d,i}$ is the number of fields in the operator $\mathcal{O}_{d,i}$. Now the beta function for λ is universal up through two loops; see problem 29.1. At three and higher loops, differences with the $\overline{\text{MS}}$ beta function can arise, due to the different underlying definitions of the coupling λ in the cutoff scheme and the $\overline{\text{MS}}$ scheme.

We now have the overall picture of Wilson's approach to quantum field theory. First, define a quantum field theory via an action with an explicit momentum cutoff in place.[1] Then, lower the cutoff by integrating out higher-momentum degrees of freedom. As a result, the coefficients in the effective action will change. If the field theory is weakly coupled—which in practice means eqs. (29.24)–(29.26) are obeyed—then the coefficients with negative mass dimension will start to take on the values we would have computed for them in perturbation theory, regardless of their precise initial values. If we continuously rescale the fields to have canonical kinetic terms, then the dimensionless coupling constant(s) will change according to their beta functions. The final results, at an energy scale E well below the initial cutoff Λ_0, are the same as we would predict via renormalized perturbation theory, up to small corrections by powers of E/Λ_0.

The advantage of the Wilson scheme is that it gives a nonperturbative definition of the theory which is applicable even if the theory is *not* weakly

[1] This can be done in various ways: for example, we could replace continuous spacetime with a discrete *lattice* of points with *lattice spacing* a; then there is an effective largest momentum of order $1/a$.

coupled. With a spacetime lattice providing the cutoff, other techniques (typically requiring large-scale computer calculations) can be brought to bear on strongly-coupled theories.

The Wilson scheme also allows us to give physical meaning to nonrenormalizable theories. Given an action for a nonrenormalizable theory, we can regard it as an effective action. We should then impose a momentum cutoff Λ_0, where Λ_0 can be defined by saying that the coefficient of every operator $\mathcal{O}_i$ with mass dimension $D_i > 4$ is given by $c_i/\Lambda_0^{D_i-4}$ with $c_i \lesssim 1$. Then we can use this theory for physics at energies below Λ_0. At energies E far below Λ_0, the effective theory will look like a renormalizable one, up to corrections by powers of E/Λ_0. (This renormalizable theory might simply be a free-field theory with *no* interactions, or no theory at all if there are no particles with physical masses well below Λ_0.)

We now turn to the final issue: can we remove the cutoff completely?

Returning to the example of φ^4 theory, let us suppose that we are somehow able to compute the exact beta function. Then we can integrate the renormalization-group equation $d\lambda/d\ln\Lambda = \beta(\lambda)$ from $\Lambda = m_{\rm ph}$ to $\Lambda = \Lambda_0$ to get

$$\int_{\lambda(m_{\rm ph})}^{\lambda(\Lambda_0)} \frac{d\lambda}{\beta(\lambda)} = \ln \frac{\Lambda_0}{m_{\rm ph}} \,. \tag{29.37}$$

We would like to take the limit $\Lambda_0 \to \infty$. Obviously, the right-hand side of eq. (29.37) becomes infinite in this limit, and so the left-hand side must as well.

However, it may not. Recall that, for small λ, $\beta(\lambda)$ is positive, and it increases faster than λ. If this is true for all λ, then the left-hand side of eq. (29.37) will approach a fixed, finite value as we take the upper limit of integration to infinity. This yields a maximum possible value for the initial cutoff, given by

$$\ln \frac{\Lambda_{\rm max}}{m_{\rm ph}} \equiv \int_{\lambda(m_{\rm ph})}^{\infty} \frac{d\lambda}{\beta(\lambda)} \,. \tag{29.38}$$

If we approximate the exact beta function with its leading term, $3\lambda^2/16\pi^2$, and use the leading term in eq. (29.18) to get $\lambda(m_{\rm ph}) = \lambda_{\rm ph}$, then we find

$$\Lambda_{\rm max} \simeq m_{\rm ph}\, e^{16\pi^2/3\lambda_{\rm ph}} \,. \tag{29.39}$$

The existence of a maximum possible value for the cutoff means that we cannot take the limit as the cutoff goes to infinity; we *must* use an effective action with a cutoff as our starting point. If we insist on taking the cutoff to infinity, then the only possible value of $\lambda_{\rm ph}$ is $\lambda_{\rm ph} = 0$. Thus, φ^4 theory is *trivial* in the limit of infinite cutoff: there are no interactions. (There is much evidence for this, but as yet no rigorous proof. The same is true of quantum electrodynamics, as was first conjectured by Landau; in this case, $\Lambda_{\rm max}$ is known as the location of the *Landau pole.*)

However, the cutoff *can* be removed if the beta function grows no faster than λ at large λ; then the left-hand side of eq. (29.37) would diverge as we take the upper limit of integration to infinity. Or, $\beta(\lambda)$ could drop to zero (and then become negative) at some finite value λ_*. Then, if $\lambda_{\rm ph} < \lambda_*$, the left-hand side of eq. (29.37) would diverge as the upper limit of integration approaches λ_*. In this case, the effective coupling at higher and higher energies would remain fixed at λ_*, and $\lambda = \lambda_*$ is called an *ultraviolet fixed point* of the renormalization group.

If the beta function is negative for $\lambda = \lambda(m_{\rm ph})$, the theory is said to be *asymptotically free*, and $\lambda(\Lambda)$ *decreases* as the cutoff is increased. In this case, there is no barrier to taking the limit $\Lambda \to \infty$. In four spacetime dimensions, the only asymptotically free theories are nonabelian gauge theories; see section 69.

Reference notes

Effective field theory is discussed in *Georgi*, *Peskin & Schroeder*, and *Weinberg I.* An introduction to lattice theory can be found in *Smit.*

Problems

29.1) Consider a theory with a single dimensionless coupling g whose beta function takes the form $\beta(g) = b_1 g^2 + b_2 g^3 + \ldots$. Now consider a new definition of the coupling $\tilde{g}$ that agrees with the original definition at lowest order, so that we have $\tilde{g} = g + c_2 g^2 + \ldots$.

a) Show that $\beta(\tilde{g}) = b_1 \tilde{g}^2 + b_2 \tilde{g}^3 + \ldots$.

b) Generalize this result to the case of multiple dimensionless couplings.

29.2) Consider φ^3 theory in six euclidean spacetime dimensions, with lagrangian

$$\mathcal{L} = \tfrac{1}{2} Z(\Lambda_0) \partial_\mu \varphi \partial_\mu \varphi + \tfrac{1}{24} Z^{3/2}(\Lambda_0) g(\Lambda_0) \varphi^3 \,. \tag{29.40}$$

We assume that we have fine-tuned to keep $m^2(\Lambda) \ll \Lambda^2$, and so we neglect the mass term.

a) Show that

$$Z(\Lambda) = Z(\Lambda_0)\left(1 - \frac{1}{2}\,g^2(\Lambda_0)\,\frac{d}{dk^2}\left[\int_\Lambda^{\Lambda_0} \frac{d^6\ell}{(2\pi)^6}\,\frac{1}{(k+\ell)^2\ell^2}\right]_{k^2=0} + \ldots\right),$$

$$g(\Lambda) = \frac{Z^{3/2}(\Lambda_0)}{Z^{3/2}(\Lambda)}\,g(\Lambda_0)\left(1 + g^2(\Lambda_0)\int_\Lambda^{\Lambda_0} \frac{d^6\ell}{(2\pi)^6}\,\frac{1}{(\ell^2)^3} + \ldots\right).$$

Hint: note that the tree-level propagator is $\tilde{\Delta}(k) = [Z(\Lambda_0)k^2]^{-1}$.

b) Use your results to compute the beta function

$$\beta(g(\Lambda)) \equiv \frac{d}{d\ln\Lambda}\,g(\Lambda)\,, \tag{29.41}$$

and compare with the beta function in section 27.

30

Spontaneous symmetry breaking

Prerequisite: 21

Consider φ^4 theory, where φ is a real scalar field with lagrangian

$$\mathcal{L} = -\tfrac{1}{2}\partial^\mu\varphi\partial_\mu\varphi - \tfrac{1}{2}m^2\varphi^2 - \tfrac{1}{24}\lambda\varphi^4 \,. \tag{30.1}$$

As we discussed in section 23, this theory has a Z_2 symmetry: $\mathcal{L}$ is invariant under $\varphi(x) \to -\varphi(x)$, and we can define a unitary operator Z that implements this:

$$Z^{-1}\varphi(x)Z = -\varphi(x) \,. \tag{30.2}$$

We also have $Z^2 = 1$, and so $Z^{-1} = Z$. Since unitarity implies $Z^{-1} = Z^\dagger$, this makes Z hermitian as well as unitary.

Now suppose that the parameter m^2 is, in spite of its name, negative rather than positive. We can write $\mathcal{L}$ in the form

$$\mathcal{L} = -\tfrac{1}{2}\partial^\mu\varphi\partial_\mu\varphi - V(\varphi) \,, \tag{30.3}$$

where the potential is

$$\begin{aligned} V(\varphi) &= \tfrac{1}{2}m^2\varphi^2 + \tfrac{1}{24}\lambda\varphi^4 \\ &= \tfrac{1}{24}\lambda(\varphi^2 - v^2)^2 - \tfrac{1}{24}\lambda v^4 \,. \end{aligned} \tag{30.4}$$

In the second line, we have defined

$$v \equiv +(6|m^2|/\lambda)^{1/2} \,. \tag{30.5}$$

We can (and will) drop the last, constant, term in eq. (30.4).

From eq. (30.4) it is clear that there are two classical field configurations that minimize the energy: $\varphi(x) = +v$ and $\varphi(x) = -v$. This is in contrast to the usual case of positive m^2, for which the minimum-energy classical field configuration is $\varphi(x) = 0$.

We can expect that the quantum theory will follow suit. For $m^2 < 0$, there will be two ground states, $|0+\rangle$ and $|0-\rangle$, with the property that

$$\langle 0+|\varphi(x)|0+\rangle = +v \, ,$$
$$\langle 0-|\varphi(x)|0-\rangle = -v \, , \tag{30.6}$$

up to quantum corrections from loop diagrams that we will treat in detail below. These two ground states are exchanged by the operator Z,

$$Z|0+\rangle = |0-\rangle \, , \tag{30.7}$$

and they are orthogonal: $\langle 0+|0-\rangle = 0$.

This last claim requires some comment. Consider a similar problem in quantum mechanics,

$$H = \tfrac{1}{2}p^2 + \tfrac{1}{24}\lambda(x^2 - v^2)^2 \, . \tag{30.8}$$

There are two approximate ground states in this case, specified by the approximate wave functions

$$\psi_\pm(x) = \langle x|0\pm\rangle \sim \exp[-\omega(x \mp v)^2/2] \, , \tag{30.9}$$

where $\omega = (\lambda v^2/3)^{1/2}$ is the frequency of small oscillations about the minimum. However, the true ground state is a symmetric linear combination of these. The antisymmetric linear combination has a slightly higher energy, due to the effects of quantum tunneling.

We can regard a field theory as an infinite set of oscillators, one for each point in space, each with a hamiltonian like eq. (30.8), and coupled together by the $(\nabla\varphi)^2$ term in the field-theory hamiltonian. There is a tunneling amplitude for each oscillator, but to turn the field-theoretic state $|0+\rangle$ into $|0-\rangle$, *all* the oscillators have to tunnel, and so the tunneling amplitude gets raised to the power of the number of oscillators, that is, to the power of infinity (more precisely, to a power that scales like the volume of space). Therefore, in the limit of infinite volume, $\langle 0+|0-\rangle$ vanishes.

Thus we can pick either $|0+\rangle$ or $|0-\rangle$ to use as the ground state. Let us choose $|0+\rangle$. Then we can define a shifted field,

$$\rho(x) = \varphi(x) - v \, , \tag{30.10}$$

which obeys $\langle 0+|\rho(x)|0+\rangle = 0$. (We must still worry about loop corrections, which we will do at the end of this section.) The potential becomes

$$\begin{aligned} V(\varphi) &= \tfrac{1}{24}\lambda[(\rho + v)^2 - v^2]^2 \\ &= \tfrac{1}{6}\lambda v^2\rho^2 + \tfrac{1}{6}\lambda v\rho^3 + \tfrac{1}{24}\lambda\rho^4 \, , \end{aligned} \tag{30.11}$$

and so the lagrangian is now

$$\mathcal{L} = -\tfrac{1}{2}\partial^\mu\rho\partial_\mu\rho - \tfrac{1}{6}\lambda v^2\rho^2 - \tfrac{1}{6}\lambda v\rho^3 - \tfrac{1}{24}\lambda\rho^4 \,. \tag{30.12}$$

We see that the coefficient of the ρ^2 term is $\frac{1}{6}\lambda v^2 = |m^2|$. This coefficient should be identified as $\frac{1}{2}m_\rho^2$, where m_ρ is the mass of the corresponding ρ particle. Also, we see that the shifted field now has a cubic as well as a quartic interaction.

Eq. (30.12) specifies a perfectly sensible, renormalizable quantum field theory, but it no longer has an obvious Z_2 symmetry. We say that the Z_2 symmetry is *spontaneously broken.*

This leads to a question about renormalization. If we include renormalizing Z factors in the original lagrangian, we get

$$\mathcal{L} = -\tfrac{1}{2}Z_\varphi\partial^\mu\varphi\partial_\mu\varphi - \tfrac{1}{2}Z_m m^2\varphi^2 - \tfrac{1}{24}Z_\lambda\lambda\varphi^4 \,. \tag{30.13}$$

For positive m^2, these three Z factors are sufficient to absorb infinities for $d \le 4$, where the mass dimension of λ is positive or zero. On the other hand, looking at the lagrangian for negative m^2 after the shift, eq. (30.12), we would seem to need an extra Z factor for the ρ^3 term. Also, once we have a ρ^3 term, we would expect to need to add a ρ term to cancel tadpoles. So, the question is, are the original three Z factors sufficient to absorb all the divergences in the Feynman diagrams derived from eq. (30.13)?

The answer is yes. To see why, consider the quantum action (introduced in section 21)

$$\begin{aligned}\Gamma(\varphi) = \frac{1}{2}\int\frac{d^4k}{(2\pi)^4}\,\widetilde{\varphi}(-k)\Big(k^2+m^2-\Pi(k^2)\Big)\widetilde{\varphi}(k) \\ +\sum_{n=3}^{\infty}\frac{1}{n!}\int\frac{d^4k_1}{(2\pi)^4}\cdots\frac{d^4k_n}{(2\pi)^4}\,(2\pi)^4\delta^4(k_1{+}\ldots{+}k_n) \\ \times\mathbf{V}_n(k_1,\ldots,k_n)\,\widetilde{\varphi}(k_1)\ldots\widetilde{\varphi}(k_n)\,,\end{aligned} \tag{30.14}$$

computed with $m^2 > 0$. The ingredients of $\Gamma(\varphi)$—the self-energy $\Pi(k^2)$ and the exact 1PI vertices $\mathbf{V}_n$—are all made finite and well-defined (in, say, the $\overline{\text{MS}}$ renormalization scheme) by adjusting the three Z factors in eq. (30.13). Furthermore, for $m^2 > 0$, the quantum action inherits the Z_2 symmetry of the classical action. To see this directly, we note that $\mathbf{V}_n$ must be zero for odd n, simply because there is no way to draw a 1PI diagram with an odd number of external lines using only a four-point vertex. Thus $\Gamma(\varphi)$ also has the Z_2 symmetry. This is a simple example of a more general result that we proved in problem 21.2: the quantum action inherits any linear symmetry of the classical action, provided that it is also a symmetry of the integration measure $\mathcal{D}\varphi$. (*Linear* means that the transformed fields are linear functions of the original ones.) The integration measure is almost always invariant; when it is not, the symmetry is said to be *anomalous.* We will meet an anomalous symmetry in section 75.

Once we have computed the quantum action for $m^2 > 0$, we can go ahead and consider the case of $m^2 < 0$. Recall from section 21 that the quantum equation of motion in the presence of a source is $\delta\Gamma/\delta\varphi(x) = -J(x)$, and that the solution of this equation is also the vacuum expectation value of $\varphi(x)$. Now set $J(x) = 0$, and look for a translationally invariant (that is, constant) solution $\varphi(x) = v$. If there is more than one such solution, we want the one(s) with the lowest energy. This is equivalent to minimizing the quantum potential $\mathcal{U}(\varphi)$, where

$$\Gamma(\varphi) = \int d^4x \left[-\mathcal{U}(\varphi) - \tfrac{1}{2}\mathcal{Z}(\varphi)\partial^\mu\varphi\partial_\mu\varphi + \ldots \right], \tag{30.15}$$

and the ellipses denote terms with more derivatives. In a weakly coupled theory, we can expect the loop-corrected potential $\mathcal{U}(\varphi)$ to be qualitatively similar to the classical potential $V(\varphi)$. Therefore, for $m^2 < 0$, we expect that there are two minima of $\mathcal{U}(\varphi)$ with equal energy, located at $\varphi(x) = \pm v$, where $v = \langle 0|\varphi(x)|0\rangle$ is the exact vacuum expectation value of the field.

Thus we have a description of spontaneous symmetry breaking in the quantum theory based on the quantum action, and the quantum action is made finite by adjusting only the three Z factors that appear in the original, symmetric form of the lagrangian.

In the next section, we will see how this works in explicit calculations.

31

Broken symmetry and loop corrections

Prerequisite: 30

Consider φ^4 theory, where φ is a real scalar field with lagrangian

$$\mathcal{L} = -\tfrac{1}{2}Z_\varphi \partial^\mu \varphi \partial_\mu \varphi - \tfrac{1}{2}Z_m m^2 \varphi^2 - \tfrac{1}{24}Z_\lambda \lambda \varphi^4 \,. \tag{31.1}$$

In $d = 4$ spacetime dimensions, the coupling λ is dimensionless.

We begin by considering the case $m^2 > 0$, where the Z_2 symmetry of $\mathcal{L}$ under $\varphi \to -\varphi$ is manifest. We wish to compute the three renormalizing Z factors. We work in $d = 4 - \varepsilon$ dimensions, and take $\lambda \to \lambda\tilde{\mu}^\varepsilon$ (where $\tilde{\mu}$ has dimensions of mass) so that λ remains dimensionless.

The propagator correction $\Pi(k^2)$ is given by the diagrams of fig. 31.1, which yield

$$i\Pi(k^2) = \tfrac{1}{2}(-i\lambda\tilde{\mu}^\varepsilon)\tfrac{1}{i}\Delta(0) - i(Ak^2 + Bm^2) \,, \tag{31.2}$$

where $A = Z_\varphi - 1$ and $B = Z_m - 1$, and

$$\Delta(0) = \int \frac{d^d\ell}{(2\pi)^d}\,\frac{1}{\ell^2 + m^2} \,. \tag{31.3}$$

Using the usual bag of tricks from section 14, we find

$$\tilde{\mu}^\varepsilon \Delta(0) = \frac{-i}{(4\pi)^2}\left[\frac{2}{\varepsilon} + 1 + \ln(\mu^2/m^2)\right] m^2 \,, \tag{31.4}$$

where $\mu^2 = 4\pi e^{-\gamma}\tilde{\mu}^2$. Thus

$$\Pi(k^2) = \frac{\lambda}{2(4\pi)^2}\left[\frac{2}{\varepsilon} + 1 + \ln(\mu^2/m^2)\right] m^2 - Ak^2 - Bm^2 \,. \tag{31.5}$$

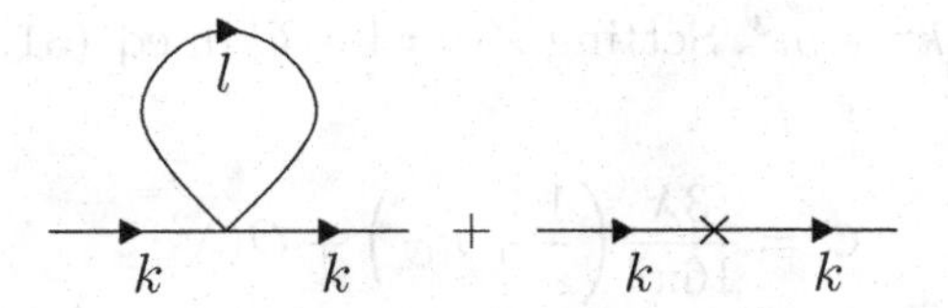

Figure 31.1. $O(\lambda)$ corrections to $\Pi(k^2)$.

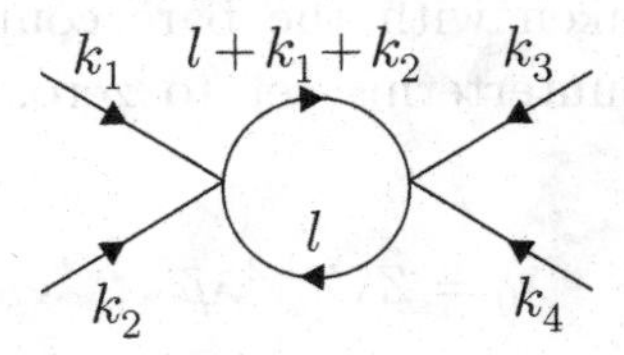

Figure 31.2. The $O(\lambda^2)$ correction to $\mathbf{V}_4(k_1, k_2, k_3, k_4)$. Two other diagrams, obtained from this one via $k_2 \leftrightarrow k_3$ and $k_2 \leftrightarrow k_4$, also contribute.

From eq. (31.5) we see that we must have

$$A = O(\lambda^2) , \tag{31.6}$$

$$B = \frac{\lambda}{16\pi^2}\left(\frac{1}{\varepsilon} + \kappa_B\right) + O(\lambda^2) , \tag{31.7}$$

where κ_B is a finite constant (that may depend on μ). In the $\overline{\text{MS}}$ renormalization scheme, we take $\kappa_B = 0$, but we will leave κ_B arbitrary for now.

Next we turn to the vertex correction, given by the diagram of fig. 31.2, plus two others with $k_2 \leftrightarrow k_3$ and $k_2 \leftrightarrow k_4$; all momenta are treated as incoming. We have

$$\begin{aligned} i\mathbf{V}_4(k_1, k_2, k_3, k_4) = -iZ_\lambda\lambda + \tfrac{1}{2}(-i\lambda)^2 \left(\tfrac{1}{i}\right)^2 \Big[iF(-s) + iF(-t) + iF(-u)\Big] \\ + O(\lambda^3) . \end{aligned} \tag{31.8}$$

Here we have defined $s = -(k_1 + k_2)^2$, $t = -(k_1 + k_3)^2$, $u = -(k_1 + k_4)^2$, and

$$\begin{aligned} iF(k^2) &\equiv \tilde{\mu}^\varepsilon \int \frac{d^d\ell}{(2\pi)^d} \frac{1}{((\ell{+}k)^2 + m^2)(\ell^2 + m^2)} \\ &= \frac{i}{16\pi^2}\left[\frac{2}{\varepsilon} + \int_0^1 dx\, \ln(\mu^2/D)\right] , \end{aligned} \tag{31.9}$$

where $D = x(1-x)k^2 + m^2$. Setting $Z_\lambda = 1 + C$ in eq. (31.8), we see that we need

$$C = \frac{3\lambda}{16\pi^2}\left(\frac{1}{\varepsilon} + \kappa_C\right) + O(\lambda^2)\,, \tag{31.10}$$

where κ_C is a finite constant.

We may as well pause to compute the beta function, $\beta(\lambda) = d\lambda/d\ln\mu$, where the derivative is taken with the bare coupling λ_0 held fixed, and the finite parts of the counterterms set to zero, in accord with the $\overline{\text{MS}}$ prescription. We have

$$\lambda_0 = Z_\lambda Z_\varphi^{-2}\lambda\tilde{\mu}^\varepsilon\,, \tag{31.11}$$

with

$$\ln\Bigl(Z_\lambda Z_\varphi^{-2}\Bigr) = \frac{3\lambda}{16\pi^2}\,\frac{1}{\varepsilon} + O(\lambda^2)\,. \tag{31.12}$$

Let $L_1(\lambda)$ be the coefficient of $1/\varepsilon$ in eq. (31.12). Our analysis in section 28 shows that the beta function is then given by $\beta(\lambda) = \lambda^2 L_1'(\lambda)$. Thus we find

$$\beta(\lambda) = \frac{3\lambda^2}{16\pi^2} + O(\lambda^3)\,. \tag{31.13}$$

The beta function is positive, which means that the theory becomes more and more strongly coupled at higher and higher energies.

Now we consider the more interesting case of $m^2 < 0$, which results in the spontaneous breakdown of the Z_2 symmetry.

Following the procedure of section 30, we set $\varphi(x) = \rho(x) + v$, where $v = (6|m^2|/\lambda)^{1/2}$ minimizes the potential (without Z factors). Then the lagrangian becomes (with Z factors)

$$\begin{aligned}\mathcal{L} = &-\tfrac{1}{2}Z_\varphi\partial^\mu\rho\partial_\mu\rho - \tfrac{1}{2}(\tfrac{3}{2}Z_\lambda - \tfrac{1}{2}Z_m)m_\rho^2\rho^2 \\ &+ \tfrac{1}{2}(Z_m - Z_\lambda)(3/\lambda\tilde{\mu}^\varepsilon)^{1/2}m_\rho^3\rho \\ &- \tfrac{1}{6}Z_\lambda(3\lambda\tilde{\mu}^\varepsilon)^{1/2}m_\rho\rho^3 - \tfrac{1}{24}Z_\lambda\lambda\tilde{\mu}^\varepsilon\rho^4\,,\end{aligned} \tag{31.14}$$

where $m_\rho^2 = 2|m^2|$. Now we can compute various one-loop corrections.

We begin with the vacuum expectation value of ρ. The $O(\lambda)$ correction is given by the diagrams of fig. 31.3. The three-point vertex factor is $-iZ_\lambda g_3$, where g_3 can be read off eq. (31.14):

$$g_3 = (3\lambda\tilde{\mu}^\varepsilon)^{1/2}m_\rho\,. \tag{31.15}$$

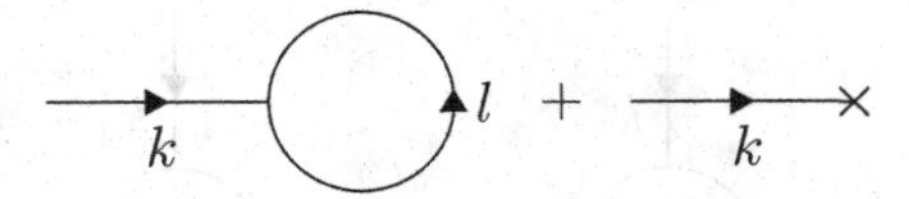

Figure 31.3. The $O(\lambda)$ correction to the vacuum expectation value of the ρ field.

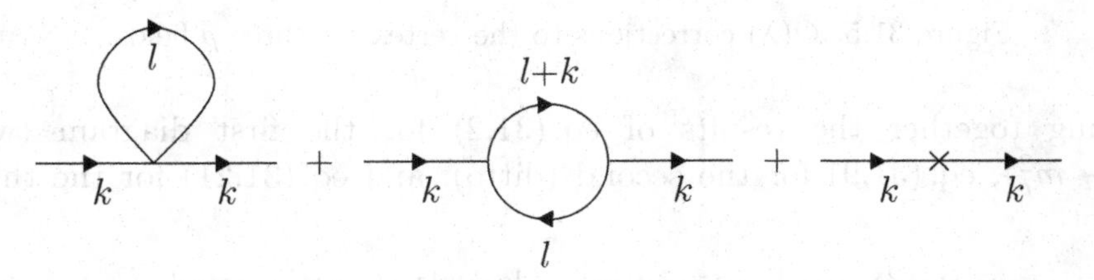

Figure 31.4. $O(\lambda)$ corrections to the ρ propagator.

The one-point vertex factor is iY, where Y can also be read off eq. (31.14):

$$Y = \tfrac{1}{2}(Z_m - Z_\lambda)(3/\lambda\tilde{\mu}^\varepsilon)^{1/2} m_\rho^3 \,. \tag{31.16}$$

Following the discussion of section 9, we then find that

$$\langle 0|\rho(x)|0\rangle = \left(iY + \tfrac{1}{2}(-iZ_\lambda g_3)\tfrac{1}{i}\Delta(0)\right)\int d^4y\, \tfrac{1}{i}\Delta(x-y) \,, \tag{31.17}$$

plus higher-order corrections. Using eqs. (31.15) and (31.16), and eq. (31.4) with $m^2 \to m_\rho^2$, the factor in large parentheses in eq. (31.17) becomes

$$\frac{i}{2}(3/\lambda)^{1/2} m_\rho^3 \left(Z_m - Z_\lambda + \frac{\lambda}{16\pi^2}\left[\frac{2}{\varepsilon} + 1 + \ln(\mu^2/m_\rho^2)\right] + O(\lambda^2)\right) \,. \tag{31.18}$$

Using $Z_m = 1 + B$ and $Z_\lambda = 1 + C$, with B and C from eqs. (31.7) and (31.10), the factor in large parentheses in eqs. (31.18) becomes

$$\frac{\lambda}{16\pi^2}\left[\kappa_B - 3\kappa_C + 1 + \ln(\mu^2/m_\rho^2)\right] . \tag{31.19}$$

All the $1/\varepsilon$s have canceled. The remaining finite vacuum expectation value for $\rho(x)$ can now be removed by choosing

$$\kappa_B - 3\kappa_C = -1 - \ln(\mu^2/m_\rho^2) \,. \tag{31.20}$$

This will also cancel all diagrams with one-loop tadpoles.

Next we consider the ρ propagator. The diagrams contributing to the $O(\lambda)$ correction are shown in fig. 31.4. The counterterm insertion is $-iX$, where, again reading off of eq. (31.14),

$$X = Ak^2 + (\tfrac{3}{2}C - \tfrac{1}{2}B)m_\rho^2 \,. \tag{31.21}$$

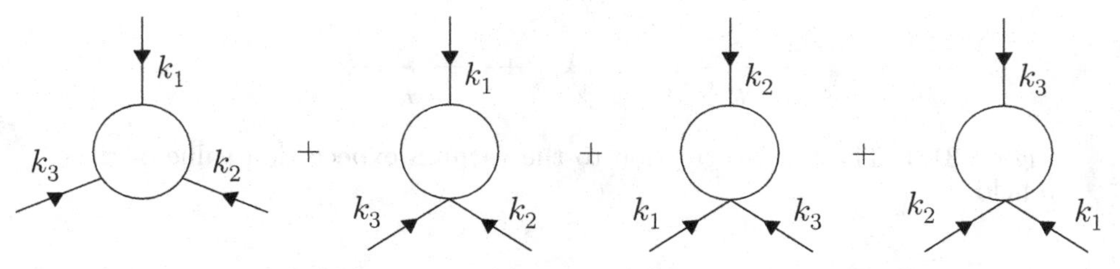

Figure 31.5. $O(\lambda)$ corrections to the vertex for three ρ fields.

Putting together the results of eq. (31.2) for the first diagram (with $m^2 \to m_\rho^2$), eq. (31.9) for the second (ditto), and eq. (31.21) for the third, we get

$$\begin{aligned}\Pi(k^2) &= -\tfrac{1}{2}(\lambda\tilde{\mu}^\varepsilon)\tfrac{1}{i}\Delta(0) + \tfrac{1}{2}g_3^2 F(k^2) - X + O(\lambda^2) \\ &= \frac{\lambda}{32\pi^2}m_\rho^2\left[\frac{2}{\varepsilon} + 1 + \ln(\mu^2/m_\rho^2)\right] \\ &\quad + \frac{3\lambda}{32\pi^2}m_\rho^2\left[\frac{2}{\varepsilon} + \int_0^1 dx\ \ln(\mu^2/D)\right] \\ &\quad - Ak^2 - (\tfrac{3}{2}C - \tfrac{1}{2}B)m_\rho^2 + O(\lambda^2)\ . \end{aligned} \tag{31.22}$$

Again using eqs. (31.7) and (31.10) for B and C, we see that all the $1/\varepsilon$s cancel, and we're left with

$$\begin{aligned}\Pi(k^2) = \frac{\lambda}{32\pi^2}m_\rho^2&\left[1 + \ln(\mu^2/m_\rho^2) + 3\int_0^1 dx\ \ln(\mu^2/D) - (9\kappa_C - \kappa_B)\right] \\ &- Ak^2 + O(\lambda^2)\ . \end{aligned} \tag{31.23}$$

We can now choose to work in an OS scheme, where we require $\Pi(-m_\rho^2) = 0$ and $\Pi'(-m_\rho^2) = 0$. Since $\Pi(k^2)$ depends on k^2 through D, we must choose a nonzero (but finite) value of A to make $\Pi'(-m_\rho^2)$ vanish. Then we can choose $9\kappa_C - \kappa_B$ to fix $\Pi(-m_\rho^2) = 0$. Together with eq. (31.20), this determines the values of κ_B and κ_C to this order in λ.

Next we consider the one-loop correction to the three-point vertex, given by the diagrams of fig. 31.5. We wish to show that the infinities are canceled by the value of $Z_\lambda = 1 + C$ that we have already determined. The first diagram in fig. 31.5 is finite, and so for our purposes we can ignore it. The remaining three, plus the original vertex, sum to give

$$\begin{aligned}i\mathbf{V}_3(k_1,k_2,k_3)_{\text{div}} = -iZ_\lambda g_3 &+ \tfrac{1}{2}(-i\lambda)(-ig_3)\left(\tfrac{1}{i}\right)^2 \\ &\times \left[iF(k_1^2) + iF(k_2^2) + iF(k_3^2)\right] \\ &+ O(\lambda^{5/2})\ , \end{aligned} \tag{31.24}$$

where the subscript div means that we are keeping only the divergent part. Using eq. (31.9), we have

$$\mathbf{V}_3(k_1, k_2, k_3)_{\text{div}} = -g_3\left(1 + C - \frac{3\lambda}{16\pi^2}\frac{1}{\varepsilon} + O(\lambda^2)\right) . \tag{31.25}$$

From eq. (31.10), we see that the divergent terms do indeed cancel to this order in λ.

Finally, we have the correction to the four-point vertex. In this case, the divergent diagrams are just those of fig. 31.2, and so the calculation of the divergent part of $\mathbf{V}_4$ is exactly the same as it is when $m^2 > 0$ (but with m_ρ in place of m). Since we have already done that calculation (it was how we determined C in the first place), we need not repeat it.

We have thus seen how we can compute the divergent parts of the counterterms in the simpler case of $m^2 > 0$, where the Z_2 symmetry is unbroken, and that these counterterms will also serve to cancel the divergences in the more complicated case of $m^2 < 0$, where the Z_2 symmetry is spontaneously broken. This is a general rule for renormalizable theories with spontaneous symmetry breaking, regardless of the nature of the symmetry group.

Reference notes

Another example of renormalization of a spontaneously broken theory is worked out in *Peskin & Schroeder*.

32

Spontaneous breaking of continuous symmetries

Prerequisites: 22, 30

Consider the theory (introduced in section 22) of a complex scalar field φ with

$$\mathcal{L} = -\partial^\mu \varphi^\dagger \partial_\mu \varphi - m^2 \varphi^\dagger \varphi - \tfrac{1}{4}\lambda(\varphi^\dagger \varphi)^2 \,. \tag{32.1}$$

This lagrangian is obviously invariant under the U(1) transformation

$$\varphi(x) \to e^{-i\alpha}\varphi(x) \,, \tag{32.2}$$

where α is a real number.

Now suppose that m^2 is negative. The minimum of the potential of eq. (32.1) is achieved for

$$\varphi(x) = \tfrac{1}{\sqrt{2}} v e^{-i\theta} \,, \tag{32.3}$$

where

$$v = (4|m^2|/\lambda)^{1/2} \,, \tag{32.4}$$

and the phase θ is arbitrary; the factor of root-two in eq. (32.3) is conventional. Thus we have a continuous family of minima of the potential, parameterized by θ. Under the U(1) transformation of eq. (32.2), θ changes to $\theta + \alpha$; thus the different minimum-energy field configurations are all related to each other by the symmetry.

In the quantum theory, we therefore expect to find a continuous family of ground states, labeled by θ, with the property that

$$\langle\theta|\varphi(x)|\theta\rangle = \tfrac{1}{\sqrt{2}} v e^{-i\theta} \,. \tag{32.5}$$

Also, according to the discussion in section 30, we expect $\langle\theta'|\theta\rangle = 0$ for $\theta' \neq \theta$.

Returning to classical language, there is a *flat direction* in field space that we can move along without changing the energy. The physical consequence of this is the existence of a massless particle called a *Goldstone boson.*

Let us see how this works in more detail. We first choose the phase $\theta = 0$, and then write

$$\varphi(x) = \tfrac{1}{\sqrt{2}}[v + a(x) + ib(x)] \; , \qquad (32.6)$$

where a and b are real scalar fields. Substituting eq. (32.6) into eq. (32.1), we find

$$\begin{aligned} \mathcal{L} = &-\tfrac{1}{2}\partial^\mu a \partial_\mu a - \tfrac{1}{2}\partial^\mu b \partial_\mu b \\ &- |m^2| a^2 - \tfrac{1}{2}\lambda^{1/2}|m| a(a^2+b^2) - \tfrac{1}{16}\lambda(a^2+b^2)^2 \; . \end{aligned} \qquad (32.7)$$

We see from this that the a field has a mass given by $\frac{1}{2}m_a^2 = |m^2|$. The b field, on the other hand, is massless, and we identify it as the Goldstone boson.

A different parameterization brings out the role of the massless field more clearly. We write

$$\varphi(x) = \tfrac{1}{\sqrt{2}}(v + \rho(x))e^{-i\chi(x)/v} \; , \qquad (32.8)$$

where ρ and χ are real scalar fields. Substituting eq. (32.8) into eq. (32.1), we get

$$\begin{aligned} \mathcal{L} = &-\tfrac{1}{2}\partial^\mu \rho \partial_\mu \rho - \tfrac{1}{2}\Big(1 + \frac{\rho}{v}\Big)^2 \partial^\mu \chi \partial_\mu \chi \\ &- |m^2|\rho^2 - \tfrac{1}{2}\lambda^{1/2}|m|\rho^3 - \tfrac{1}{16}\lambda\rho^4 \; . \end{aligned} \qquad (32.9)$$

We see from this that the ρ field has a mass given by $\frac{1}{2}m_\rho^2 = |m^2|$, and that the χ field is massless. These are the same particle masses we found using the parameterization of eq. (32.6). This is not an accident: the particle masses and scattering amplitudes are independent of field redefinitions.

Note that the χ field does not appear in the potential at all. Thus it parameterizes the flat direction. In terms of the ρ and χ fields, the U(1) transformation takes the simple form $\chi(x) \to \chi(x) + v\alpha$.

Does the masslessness of the χ field survive loop corrections? It does. To see this, we note that if the χ field remains massless, its exact propagator $\tilde{\mathbf{\Delta}}_\chi(k^2)$ should have a pole at $k^2 = 0$; equivalently, the self-energy $\Pi_\chi(k^2)$, related to the propagator by $\tilde{\mathbf{\Delta}}_\chi(k^2) = 1/[k^2 - \Pi_\chi(k^2)]$, should satisfy $\Pi_\chi(0) = 0$.

We can evaluate $\Pi_\chi(0)$ by summing all 1PI diagrams with two external χ lines, each with four-momentum $k = 0$. We note from eq. (32.9) that the derivatives acting on the χ fields imply that the vertex factors for the $\rho\chi\chi$ and $\rho\rho\chi\chi$ vertices are each proportional to $k_1 \cdot k_2$, where k_1 and k_2 are the momenta of the two χ lines that meet at that vertex. Since the external lines have zero momentum, the attached vertices vanish; hence, $\Pi_\chi(0) = 0$, and the χ particle remains massless.

The same conclusion can be reached by considering the quantum action $\Gamma(\varphi)$, which includes all loop corrections. According to our discussion in section 29, the quantum action has the same symmetries as the classical action. Therefore, in the case at hand,

$$\Gamma(\varphi) = \Gamma(e^{-i\alpha}\varphi) \ . \tag{32.10}$$

Spontaneous symmetry breaking occurs if the minimum of $\Gamma(\varphi)$ is at a constant, nonzero value of φ. Because of eq. (32.10), the phase of this constant is arbitrary. Therefore, there must be a flat direction in field space, corresponding to the phase of $\varphi(x)$. The physical consequence of this flat direction is a massless particle, the Goldstone boson.

All of this has a straightforward extension to the nonabelian case. Consider

$$\mathcal{L} = -\tfrac{1}{2}\partial^\mu\varphi_i\partial_\mu\varphi_i - \tfrac{1}{2}m^2\varphi_i\varphi_i - \tfrac{1}{16}\lambda(\varphi_i\varphi_i)^2 \ , \tag{32.11}$$

where a repeated index is summed. This lagrangian is invariant under the infinitesimal SO(N) transformation

$$\delta\varphi_i = -i\theta^a(T^a)_{ij}\varphi_j \ , \tag{32.12}$$

where a runs from 1 to $\frac{1}{2}N(N-1)$, θ^a is a set of $\frac{1}{2}N(N-1)$ real, infinitesimal parameters, and each antisymmetric generator matrix T^a has a single nonzero entry $-i$ above the main diagonal, and a corresponding $+i$ below the main diagonal.

Now let us take $m^2 < 0$ in eq. (32.11). The minimum of the potential is achieved for $\varphi_i(x) = v_i$, where $v^2 = v_i v_i = 4|m^2|/\lambda$, and the direction in which the N-component vector $\vec{v}$ points is arbitrary. In the quantum theory, we interpret v_i as the vacuum expectation value (VEV for short) of the quantum field $\varphi_i(x)$. We can choose our coordinate system so that $v_i = v\delta_{iN}$; that is, the VEV lies entirely in the last component.

Now consider making an infinitesimal SO(N) transformation. This changes the VEV; we have

$$\begin{aligned} v_i &\to v_i - i\theta^a(T^a)_{ij}v_j \\ &= v\delta_{iN} - i\theta^a(T^a)_{iN}v \ . \end{aligned} \tag{32.13}$$

For some choices of θ^a, the second term on the right-hand side of eq. (32.13) vanishes. This happens if the corresponding T^a has no nonzero entry in the last column. There are $N-1$ T^as with a nonzero entry in the last column: those with the $-i$ in the first row and last column, in the second row and last column, etc., down to the $N-1$th row and last column. These T^as are said to be *broken* generators. A generator is broken if $(T^a)_{ij}v_j \neq 0$, and unbroken if $(T^a)_{ij}v_j = 0$.

An infinitesimal SO(N) transformation that involves a broken generator changes the VEV of the field, but not the energy. Thus, each broken generator corresponds to a *flat direction* in field space. Each flat direction implies the existence of a corresponding massless particle. This is *Goldstone's theorem*: there is one massless Goldstone boson for each broken generator.

The unbroken generators, on the other hand, do not change the VEV of the field. Therefore, after rewriting the lagrangian in terms of shifted fields (each with zero VEV), there should still be a manifest symmetry corresponding to the set of unbroken generators. In the present case, the number of unbroken generators is $\frac{1}{2}N(N-1) - (N-1) = \frac{1}{2}(N-1)(N-2)$. This is the number of generators of SO($N-1$). Therefore, we expect SO($N-1$) to be an obvious symmetry of the lagrangian after it is written in terms of shifted fields.

Let us see how this works in the present case. We can rewrite eq. (32.11) as

$$\mathcal{L} = -\tfrac{1}{2}\partial^\mu\varphi_i\partial_\mu\varphi_i - V(\varphi), \tag{32.14}$$

with

$$V(\varphi) = \tfrac{1}{16}\lambda(\varphi_i\varphi_i - v^2)^2 \,, \tag{32.15}$$

where $v = (4|m^2|/\lambda)^{1/2}$, and the repeated index i is implicitly summed from 1 to N. Now let $\varphi_N(x) = v + \rho(x)$, and plug this into eq. (32.14). With the repeated index i now implicitly summed from 1 to $N-1$, we have

$$\mathcal{L} = -\tfrac{1}{2}\partial^\mu\varphi_i\partial_\mu\varphi_i - \tfrac{1}{2}\partial^\mu\rho\partial_\mu\rho - V(\rho,\varphi) \,, \tag{32.16}$$

where

$$\begin{aligned} V(\rho,\varphi) &= \tfrac{1}{16}\lambda[(v+\rho)^2 + \varphi_i\varphi_i - v^2]^2 \\ &= \tfrac{1}{16}\lambda(2v\rho + \rho^2 + \varphi_i\varphi_i)^2 \\ &= \tfrac{1}{4}\lambda v^2\rho^2 + \tfrac{1}{4}\lambda v\rho(\rho^2 + \varphi_i\varphi_i) + \tfrac{1}{16}\lambda(\rho^2 + \varphi_i\varphi_i)^2 \,. \end{aligned} \tag{32.17}$$

There is indeed a manifest SO($N-1$) symmetry in eqs. (32.16) and (32.17). Also, the $N-1$ φ_i fields are massless: they are the expected $N-1$ Goldstone bosons.

Reference notes

Further discussion of Goldstone's theorem can be found in *Georgi*, *Peskin & Schroeder*, and *Weinberg II*.

Problems

32.1) Consider the Noether current j^μ for the U(1) symmetry of eq. (32.1), and the corresponding charge Q.

a) Show that $e^{-i\alpha Q}\,\varphi\, e^{+i\alpha Q} = e^{+i\alpha}\varphi$.

b) Use eq. (32.5) to show that $e^{-i\alpha Q}|\theta\rangle = |\theta+\alpha\rangle$.

c) Show that $Q|0\rangle \neq 0$; that is, the charge does not annihilate the $\theta = 0$ vacuum. Contrast this with the case of an unbroken symmetry.

32.2) In problem 24.3, we showed that $[\varphi_i, Q^a] = (T^a)_{ij}\varphi_j$, where Q^a is the Noether charge in the SO(N) symmetric theory. Use this result to show that $Q^a|0\rangle \neq 0$ if and only if Q^a is broken.

32.3) We define the *decay constant* f of the Goldstone boson via

$$\langle k|j^\mu(x)|0\rangle = ifk^\mu e^{-ikx}\ , \tag{32.18}$$

where $|k\rangle$ is the state of a single Goldstone boson with four-momentum k, normalized in the usual way, $|0\rangle$ is the $\theta = 0$ vacuum, and $j^\mu(x)$ is the Noether current.

a) Compute f at tree level. (That is, express j^μ in terms of the ρ and χ fields, and then use free field theory to compute the matrix element.) A nonvanishing value of f indicates that the corresponding current is spontaneously broken.

b) Discuss how your result would be modified by higher-order corrections.

Part II

Spin One Half

33

Representations of the Lorentz group

Prerequisite: 2

In section 2, we saw that we could define a unitary operator $U(\Lambda)$ that implemented a Lorentz transformation on a scalar field $\varphi(x)$ via

$$U(\Lambda)^{-1}\varphi(x)U(\Lambda) = \varphi(\Lambda^{-1}x) . \tag{33.1}$$

As shown in section 2, this implies that the derivative of the field transforms as

$$U(\Lambda)^{-1}\partial^\mu\varphi(x)U(\Lambda) = \Lambda^\mu{}_\rho\bar{\partial}^\rho\varphi(\Lambda^{-1}x) , \tag{33.2}$$

where the bar on the derivative means that it is with respect to the argument $\bar{x} = \Lambda^{-1}x$.

Eq. (33.2) suggests that we could define a *vector field* $A^\mu(x)$ that would transform as

$$U(\Lambda)^{-1}A^\mu(x)U(\Lambda) = \Lambda^\mu{}_\rho A^\rho(\Lambda^{-1}x) , \tag{33.3}$$

or a *tensor field* $B^{\mu\nu}(x)$ that would transform as

$$U(\Lambda)^{-1}B^{\mu\nu}(x)U(\Lambda) = \Lambda^\mu{}_\rho\Lambda^\nu{}_\sigma B^{\rho\sigma}(\Lambda^{-1}x) . \tag{33.4}$$

Note that if $B^{\mu\nu}$ is either symmetric, $B^{\mu\nu}(x) = B^{\nu\mu}(x)$, or antisymmetric, $B^{\mu\nu}(x) = -B^{\nu\mu}(x)$, then the symmetry is preserved by the Lorentz transformation. Also, if we take the trace to get $T(x) \equiv g_{\mu\nu}B^{\mu\nu}(x)$, then, using $g_{\mu\nu}\Lambda^\mu{}_\rho\Lambda^\nu{}_\sigma = g_{\rho\sigma}$, we find that $T(x)$ transforms like a scalar field,

$$U(\Lambda)^{-1}T(x)U(\Lambda) = T(\Lambda^{-1}x) . \tag{33.5}$$

Thus, given a tensor field $B^{\mu\nu}(x)$ with no particular symmetry, we can write

$$B^{\mu\nu}(x) = A^{\mu\nu}(x) + S^{\mu\nu}(x) + \tfrac{1}{4}g^{\mu\nu}T(x) , \tag{33.6}$$

where $A^{\mu\nu}$ is antisymmetric ($A^{\mu\nu} = -A^{\nu\mu}$) and $S^{\mu\nu}$ is symmetric ($S^{\mu\nu} = S^{\nu\mu}$) and traceless ($g_{\mu\nu}S^{\mu\nu} = 0$). The key point is that the fields $A^{\mu\nu}$, $S^{\mu\nu}$, and T do not mix with each other under Lorentz transformations.

Is it possible to further break apart these fields into still smaller sets that do not mix under Lorentz transformations? How do we make this decomposition into *irreducible representations* of the Lorentz group for a field carrying n vector indices? Are there any other kinds of indices we could consistently assign to a field? If so, how do these behave under a Lorentz transformation?

The answers to these questions are to be found in the theory of *group representations.* Let us see how this works for the Lorentz group (in four spacetime dimensions).

Consider a field (not necessarily hermitian) that carries a generic Lorentz index, $\varphi_A(x)$. Under a Lorentz transformation, we have

$$U(\Lambda)^{-1}\varphi_A(x)U(\Lambda) = L_A{}^B(\Lambda)\varphi_B(\Lambda^{-1}x) \,, \tag{33.7}$$

where $L_A{}^B(\Lambda)$ is a matrix that depends on Λ. These finite-dimensional matrices must obey the group composition rule

$$L_A{}^B(\Lambda')L_B{}^C(\Lambda) = L_A{}^C(\Lambda'\Lambda) \,. \tag{33.8}$$

We say that the matrices $L_A{}^B(\Lambda)$ form a *representation* of the Lorentz group.

For an infinitesimal transformation $\Lambda^\mu{}_\nu = \delta^\mu{}_\nu + \delta\omega^\mu{}_\nu$, we can write

$$U(1{+}\delta\omega) = I + \tfrac{i}{2}\delta\omega_{\mu\nu}M^{\mu\nu} \,, \tag{33.9}$$

where the operators $M^{\mu\nu}$ are the *generators* of the Lorentz group. As shown in section 2, the generators obey the commutation relations

$$[M^{\mu\nu}, M^{\rho\sigma}] = i\Big(g^{\mu\rho}M^{\nu\sigma} - (\mu{\leftrightarrow}\nu)\Big) - (\rho{\leftrightarrow}\sigma) \,, \tag{33.10}$$

which specify the *Lie algebra* of the Lorentz group.

We can identify the components of the angular momentum operator $\vec{J}$ as $J_i \equiv \frac{1}{2}\varepsilon_{ijk}M^{jk}$ and the components of the boost operator $\vec{K}$ as $K_i \equiv M^{i0}$. We then find from eq. (33.10) that

$$[J_i, J_j] = +i\varepsilon_{ijk}J_k \,, \tag{33.11}$$

$$[J_i, K_j] = +i\varepsilon_{ijk}K_k \,, \tag{33.12}$$

$$[K_i, K_j] = -i\varepsilon_{ijk}J_k \,. \tag{33.13}$$

For an infinitesimal transformation, we also have

$$L_A{}^B(1{+}\delta\omega) = \delta_A{}^B + \tfrac{i}{2}\delta\omega_{\mu\nu}(S^{\mu\nu})_A{}^B \,, \tag{33.14}$$

Eq. (33.7) then becomes

$$[\varphi_A(x), M^{\mu\nu}] = \mathcal{L}^{\mu\nu}\varphi_A(x) + (S^{\mu\nu})_A{}^B\varphi_B(x) \, , \tag{33.15}$$

where $\mathcal{L}^{\mu\nu} \equiv \frac{1}{i}(x^\mu\partial^\nu - x^\nu\partial^\mu)$. Both the differential operators $\mathcal{L}^{\mu\nu}$ and the representation matrices $(S^{\mu\nu})_A{}^B$ must separately obey the same commutation relations as the generators themselves; see problems 2.8 and 2.9.

Our problem now is to find all possible sets of finite-dimensional matrices that obey eq. (33.10), or equivalently eqs. (33.11)–(33.13). Although the operators $M^{\mu\nu}$ must be hermitian, the matrices $(S^{\mu\nu})_A{}^B$ need not be.

If we restrict our attention to eq. (33.11) alone, we know (from standard results in the quantum mechanics of angular momentum) that we can find three $(2j{+}1) \times (2j{+}1)$ hermitian matrices $\mathcal{J}_1$, $\mathcal{J}_2$, and $\mathcal{J}_3$ that obey eq. (33.11), and that the eigenvalues of (say) $\mathcal{J}_3$ are $-j, -j{+}1, \ldots, +j$, where j has the possible values $0, \frac{1}{2}, 1, \ldots$ We further know that these matrices constitute all of the inequivalent, irreducible representations of the Lie algebra of SO(3), the rotation group in three dimensions. *Inequivalent* means not related by a unitary transformation; *irreducible* means cannot be made block-diagonal by a unitary transformation. (The standard derivation assumes that the matrices are hermitian, but allowing nonhermitian matrices does not enlarge the set of solutions.) Also, when j is a half integer, a rotation by 2π results in an overall minus sign; these representations of the *Lie algebra* of SO(3) are therefore actually not representations of the *group* SO(3), since a 2π rotation should be equivalent to no rotation. As we saw in section 24, the Lie algebra of SO(3) is the same as the Lie algebra of SU(2); the half-integer representations of this Lie algebra *do* qualify as representations of the group SU(2).

We would like to extend these conclusions to encompass the full set of eqs. (33.11)–(33.13). In order to do so, it is helpful to define some nonhermitian operators whose physical significance is obscure, but which simplify the commutation relations. These are

$$N_i \equiv \tfrac{1}{2}(J_i - iK_i) \, , \tag{33.16}$$

$$N_i^\dagger \equiv \tfrac{1}{2}(J_i + iK_i) \, . \tag{33.17}$$

In terms of N_i and $N_i^\dagger$, eqs. (33.11)–(33.13) become

$$[N_i, N_j] = i\varepsilon_{ijk}N_k \, , \tag{33.18}$$

$$[N_i^\dagger, N_j^\dagger] = i\varepsilon_{ijk}N_k^\dagger \, , \tag{33.19}$$

$$[N_i, N_j^\dagger] = 0 \, . \tag{33.20}$$

We see that we have two different SU(2) Lie algebras that are exchanged by hermitian conjugation. As we just discussed, a representation of the SU(2) Lie algebra is specified by an integer or half integer; we therefore conclude that a representation of the Lie algebra of the Lorentz group in four spacetime dimensions is specified by *two* integers or half-integers n and n'.

We will label these representations as $(2n{+}1, 2n'{+}1)$; the number of components of a representation is then $(2n{+}1)(2n'{+}1)$. Different components within a representation can also be labeled by their angular momentum representations. To do this, we first note that, from eqs. (33.16) and (33.17), we have $J_i = N_i + N_i^\dagger$. Thus, deducing the allowed values of j given n and n' becomes a standard problem in the addition of angular momenta. The general result is that the allowed values of j are $|n{-}n'|, |n{-}n'|{+}1, \ldots, n{+}n'$, and each of these values appears exactly once.

The four simplest and most often encountered representations are $(1,1)$, $(2,1)$, $(1,2)$, and $(2,2)$. These are given special names:

$$\begin{aligned}(1,1) &= \textit{scalar} \text{ or } \textit{singlet} \\ (2,1) &= \textit{left-handed spinor} \\ (1,2) &= \textit{right-handed spinor} \\ (2,2) &= \textit{vector}. \end{aligned} \tag{33.21}$$

It may seem a little surprising that $(2,2)$ is to be identified as the vector representation. To see that this must be the case, we first note that the vector representation is irreducible: all the components of a four-vector mix with each other under a general Lorentz transformation. Secondly, the vector representation has four components. The only candidate irreducible representations are $(4,1)$, $(1,4)$, and $(2,2)$. The first two of these contain angular momenta $j = \frac{3}{2}$ only, whereas $(2,2)$ contains $j = 0$ and 1. This is just right for a four-vector, whose time component is a scalar under spatial rotations, and whose space components are a three-vector.

In order to gain a better understanding of what it means for $(2,2)$ to be the vector representation, we must first investigate the spinor representations $(1,2)$ and $(2,1)$, which contain angular momenta $j = \frac{1}{2}$ only.

Reference notes

An extended treatment of representations of the Lorentz group in four dimensions can be found in *Weinberg I*.

Problems

33.1) Express $A^{\mu\nu}(x)$, $S^{\mu\nu}(x)$, and $T(x)$ in terms of $B^{\mu\nu}(x)$.

33.2) Verify that eqs. (33.18)–(33.20) follow from eqs. (33.11)–(33.13).

34

Left- and right-handed spinor fields

Prerequisites: 3, 33

Consider a *left-handed spinor field* $\psi_a(x)$, also known as a *left-handed Weyl field*, which is in the $(2,1)$ representation of the Lie algebra of the Lorentz group. Here the index a is a *left-handed spinor index* that takes on two possible values. Under a Lorentz transformation, we have

$$U(\Lambda)^{-1}\psi_a(x)U(\Lambda) = L_a{}^b(\Lambda)\psi_b(\Lambda^{-1}x)\,, \tag{34.1}$$

where $L_a{}^b(\Lambda)$ is a matrix in the $(2,1)$ representation. These matrices satisfy the group composition rule

$$L_a{}^b(\Lambda')L_b{}^c(\Lambda) = L_a{}^c(\Lambda'\Lambda)\,. \tag{34.2}$$

For an infinitesimal transformation $\Lambda^\mu{}_\nu = \delta^\mu{}_\nu + \delta\omega^\mu{}_\nu$, we can write

$$L_a{}^b(1{+}\delta\omega) = \delta_a{}^b + \tfrac{i}{2}\delta\omega_{\mu\nu}(S_{\rm L}^{\mu\nu})_a{}^b\,, \tag{34.3}$$

where $(S_{\rm L}^{\mu\nu})_a{}^b = -(S_{\rm L}^{\nu\mu})_a{}^b$ is a set of 2×2 matrices that obey the same commutation relations as the generators $M^{\mu\nu}$, namely

$$[S_{\rm L}^{\mu\nu}, S_{\rm L}^{\rho\sigma}] = i\Big(g^{\mu\rho}S_{\rm L}^{\nu\sigma} - (\mu{\leftrightarrow}\nu)\Big) - (\rho{\leftrightarrow}\sigma)\,. \tag{34.4}$$

Using

$$U(1{+}\delta\omega) = I + \tfrac{i}{2}\delta\omega_{\mu\nu}M^{\mu\nu}\,, \tag{34.5}$$

eq. (34.1) becomes

$$[\psi_a(x), M^{\mu\nu}] = \mathcal{L}^{\mu\nu}\psi_a(x) + (S_{\rm L}^{\mu\nu})_a{}^b\psi_b(x)\,, \tag{34.6}$$

where $\mathcal{L}^{\mu\nu} = \frac{1}{i}(x^\mu\partial^\nu - x^\nu\partial^\mu)$. The $\mathcal{L}^{\mu\nu}$ term in eq. (34.6) would also be present for a scalar field, and is not the focus of our current interest; we will suppress it by evaluating the fields at the spacetime origin, $x^\mu = 0$.

Recalling that $M^{ij} = \varepsilon^{ijk} J_k$, where J_k is the angular momentum operator, we have

$$\varepsilon^{ijk}[\psi_a(0), J_k] = (S_{\rm L}^{ij})_a{}^b \psi_b(0) \ . \tag{34.7}$$

Recall that the $(2,1)$ representation of the Lorentz group includes angular momentum $j = \frac{1}{2}$ only. For a spin-one-half operator, the standard convention is that the matrix on the right-hand side of eq. (34.7) is $\frac{1}{2}\varepsilon^{ijk}\sigma_k$, where we have suppressed the row index a and the column index b, and where σ_k is a Pauli matrix:

$$\sigma_1 = \begin{pmatrix} 0 & 1 \\ 1 & 0 \end{pmatrix}, \quad \sigma_2 = \begin{pmatrix} 0 & -i \\ i & 0 \end{pmatrix}, \quad \sigma_3 = \begin{pmatrix} 1 & 0 \\ 0 & -1 \end{pmatrix}. \tag{34.8}$$

We therefore conclude that

$$(S_{\rm L}^{ij})_a{}^b = \tfrac{1}{2}\varepsilon^{ijk}\sigma_k \ . \tag{34.9}$$

Thus, for example, setting $i{=}1$ and $j{=}2$ yields $(S_{\rm L}^{12})_a{}^b = \frac{1}{2}\varepsilon^{12k}\sigma_k = \frac{1}{2}\sigma_3$, and so $(S_{\rm L}^{12})_1{}^1 = +\frac{1}{2}$, $(S_{\rm L}^{12})_2{}^2 = -\frac{1}{2}$, and $(S_{\rm L}^{12})_1{}^2 = (S_{\rm L}^{12})_2{}^1 = 0$.

Once we have the $(2,1)$ representation matrices for the angular momentum operator J_i, we can easily get them for the boost operator $K_k = M^{k0}$. This is because $J_k = N_k + N_k^\dagger$ and $K_k = i(N_k - N_k^\dagger)$, and, acting on a field in the $(2,1)$ representation, $N_k^\dagger$ is zero. Therefore, the representation matrices for K_k are simply i times those for J_k, and so

$$(S_{\rm L}^{k0})_a{}^b = \tfrac{1}{2} i\sigma_k \ . \tag{34.10}$$

Now consider taking the hermitian conjugate of the left-handed spinor field $\psi_a(x)$. Recall that hermitian conjugation swaps the two SU(2) Lie algebras that comprise the Lie algebra of the Lorentz group. Therefore, the hermitian conjugate of a field in the $(2,1)$ representation should be a field in the $(1,2)$ representation; such a field is called a *right-handed spinor field* or a *right-handed Weyl field.* We will distinguish the indices of the $(1,2)$ representation from those of the $(2,1)$ representation by putting dots over them. Thus, we write

$$[\psi_a(x)]^\dagger = \psi^\dagger_{\dot a}(x) \ . \tag{34.11}$$

Under a Lorentz transformation, we have

$$U(\Lambda)^{-1}\psi^\dagger_{\dot a}(x)U(\Lambda) = R_{\dot a}{}^{\dot b}(\Lambda)\psi^\dagger_{\dot b}(\Lambda^{-1}x) \ , \tag{34.12}$$

where $R_{\dot{a}}{}^{\dot{b}}(\Lambda)$ is a matrix in the $(1,2)$ representation. These matrices satisfy the group composition rule

$$R_{\dot{a}}{}^{\dot{b}}(\Lambda')R_{\dot{b}}{}^{\dot{c}}(\Lambda) = R_{\dot{a}}{}^{\dot{c}}(\Lambda'\Lambda) \,. \tag{34.13}$$

For an infinitesimal transformation $\Lambda^{\mu}{}_{\nu} = \delta^{\mu}{}_{\nu} + \delta\omega^{\mu}{}_{\nu}$, we can write

$$R_{\dot{a}}{}^{\dot{b}}(1{+}\delta\omega) = \delta_{\dot{a}}{}^{\dot{b}} + \tfrac{i}{2}\delta\omega_{\mu\nu}(S_{\rm R}^{\mu\nu})_{\dot{a}}{}^{\dot{b}} \,, \tag{34.14}$$

where $(S_{\rm R}^{\mu\nu})_{\dot{a}}{}^{\dot{b}} = -(S_{\rm R}^{\nu\mu})_{\dot{a}}{}^{\dot{b}}$ is a set of 2×2 matrices that obey the same commutation relations as the generators $M^{\mu\nu}$. We then have

$$[\psi^{\dagger}_{\dot{a}}(0), M^{\mu\nu}] = (S_{\rm R}^{\mu\nu})_{\dot{a}}{}^{\dot{b}}\psi^{\dagger}_{\dot{b}}(0) \,. \tag{34.15}$$

Taking the hermitian conjugate of this equation, we get

$$[M^{\mu\nu}, \psi_a(0)] = [(S_{\rm R}^{\mu\nu})_{\dot{a}}{}^{\dot{b}}]^{*}\psi_b(0) \,. \tag{34.16}$$

Comparing this with eq. (34.6), we see that

$$(S_{\rm R}^{\mu\nu})_{\dot{a}}{}^{\dot{b}} = -[(S_{\rm L}^{\mu\nu})_{a}{}^{b}]^{*} \,. \tag{34.17}$$

In the previous section, we examined the Lorentz-transformation properties of a field carrying two vector indices. To help us get better acquainted with the properties of spinor indices, let us now do the same for a field that carries two $(2,1)$ indices. Call this field $C_{ab}(x)$. Under a Lorentz transformation, we have

$$U(\Lambda)^{-1}C_{ab}(x)U(\Lambda) = L_a{}^{c}(\Lambda)L_b{}^{d}(\Lambda)C_{cd}(\Lambda^{-1}x) \,. \tag{34.18}$$

The question we wish to address is whether or not the four components of C_{ab} can be grouped into smaller sets that do not mix with each other under Lorentz transformations.

To answer this question, recall from quantum mechanics that two spin-one-half particles can be in a state of total spin zero, or total spin one. Furthermore, the single spin-zero state is the unique *antisymmetric* combination of the two spin-one-half states, and the three spin-one states are the three *symmetric* combinations of the two spin-one-half states. We can write this schematically as $2\otimes 2 = 1_{\rm A}\oplus 3_{\rm S}$, where we label the representation of SU(2) by the number of its components, and the subscripts S and A indicate whether that representation appears in the symmetric or antisymmetric combination of the two 2s. For the Lorentz group, the relevant relation is $(2,1)\otimes(2,1) = (1,1)_{\rm A}\oplus(3,1)_{\rm S}$. This implies that we should be able to write

$$C_{ab}(x) = \varepsilon_{ab} D(x) + G_{ab}(x)\ , \tag{34.19}$$

where $D(x)$ is a scalar field, $\varepsilon_{ab} = -\varepsilon_{ba}$ is an antisymmetric set of constants, and $G_{ab}(x) = G_{ba}(x)$. The symbol ε_{ab} is uniquely determined by its symmetry properties up to an overall constant; we will choose $\varepsilon_{21} = -\varepsilon_{12} = +1$.

Since $D(x)$ is a Lorentz scalar, eq. (34.19) is consistent with eq. (34.18) only if

$$L_a{}^c(\Lambda) L_b{}^d(\Lambda) \varepsilon_{cd} = \varepsilon_{ab}\ . \tag{34.20}$$

This means that ε_{ab} is an *invariant symbol* of the Lorentz group: it does not change under a Lorentz transformation that acts on all of its indices. In this way, ε_{ab} is analogous to the metric $g_{\mu\nu}$, which is also an invariant symbol, since

$$\Lambda_\mu{}^\rho \Lambda_\nu{}^\sigma g_{\rho\sigma} = g_{\mu\nu}\ . \tag{34.21}$$

We use $g_{\mu\nu}$ and its inverse $g^{\mu\nu}$ to raise and lower vector indices, and we can use ε_{ab} and its inverse ε^{ab} to raise and lower left-handed spinor indices. Here we define ε^{ab} via

$$\varepsilon^{12} = \varepsilon_{21} = +1\ , \qquad \varepsilon^{21} = \varepsilon_{12} = -1\ . \tag{34.22}$$

With this definition, we have

$$\varepsilon_{ab}\varepsilon^{bc} = \delta_a{}^c\ , \qquad \varepsilon^{ab}\varepsilon_{bc} = \delta^a{}_c\ . \tag{34.23}$$

We can then define

$$\psi^a(x) \equiv \varepsilon^{ab}\psi_b(x)\ . \tag{34.24}$$

We also have (suppressing the spacetime argument of the field)

$$\psi_a = \varepsilon_{ab}\psi^b = \varepsilon_{ab}\varepsilon^{bc}\psi_c = \delta_a{}^c\psi_c\ , \tag{34.25}$$

as we would expect. However, the antisymmetry of ε^{ab} means that we must be careful with minus signs; for example, eq. (34.24) can be written in various ways, such as

$$\psi^a = \varepsilon^{ab}\psi_b = -\varepsilon^{ba}\psi_b = -\psi_b\varepsilon^{ba} = \psi_b\varepsilon^{ab}\ . \tag{34.26}$$

We must also be careful about signs when we contract indices, since

$$\psi^a\chi_a = \varepsilon^{ab}\psi_b\chi_a = -\varepsilon^{ba}\psi_b\chi_a = -\psi_b\chi^b\ . \tag{34.27}$$

In section 35, we will (mercifully) develop an index-free notation that automatically keeps track of these essential (but annoying) minus signs.

An exactly analogous discussion applies to the second SU(2) factor; from the group-theoretic relation $(1,2) \otimes (1,2) = (1,1)_{\rm A} \oplus (1,3)_{\rm S}$, we can deduce the existence of an invariant symbol $\varepsilon_{\dot a\dot b} = -\varepsilon_{\dot b\dot a}$. We will normalize $\varepsilon^{\dot a\dot b}$ according to eq. (34.22). Then eqs. (34.23)–(34.27) hold if *all* the undotted indices are replaced by dotted indices.

Now consider a field carrying one undotted and one dotted index, $A_{a\dot a}(x)$. Such a field is in the $(2,2)$ representation, and in section 33 we concluded that the $(2,2)$ representation was the vector representation. We would more naturally write a field in the vector representation as $A^\mu(x)$. There must, then, be a dictionary that gives us the components of $A_{a\dot a}(x)$ in terms of the components of $A^\mu(x)$; we can write this as

$$A_{a\dot a}(x) = \sigma^\mu_{a\dot a} A_\mu(x) \,, \tag{34.28}$$

where $\sigma^\mu_{a\dot a}$ is another invariant symbol. That such a symbol must exist can be deduced from the group-theoretic relation

$$(2,1) \otimes (1,2) \otimes (2,2) = (1,1) \oplus \ldots \,. \tag{34.29}$$

As we will see in section 35, it turns out to be consistent with our already established conventions for $S_{\rm L}^{\mu\nu}$ and $S_{\rm R}^{\mu\nu}$ to choose

$$\sigma^\mu_{a\dot a} = (I, \vec\sigma) \,. \tag{34.30}$$

Thus, for example, $\sigma^3_{1\dot 1} = +1$, $\sigma^3_{2\dot 2} = -1$, $\sigma^3_{1\dot 2} = \sigma^3_{2\dot 1} = 0$.

In general, whenever the product of a set of representations includes the singlet, there is a corresponding invariant symbol. For example, we can deduce the existence of $g_{\mu\nu} = g_{\nu\mu}$ from

$$(2,2) \otimes (2,2) = (1,1)_{\rm S} \oplus (1,3)_{\rm A} \oplus (3,1)_{\rm A} \oplus (3,3)_{\rm S} \,. \tag{34.31}$$

Another invariant symbol, the *Levi-Civita symbol*, follows from

$$(2,2) \otimes (2,2) \otimes (2,2) \otimes (2,2) = (1,1)_{\rm A} \oplus \ldots \,, \tag{34.32}$$

where the subscript A denotes the completely antisymmetric part. The Levi-Civita symbol is $\varepsilon^{\mu\nu\rho\sigma}$, which is antisymmetric on exchange of any pair of its indices, and is normalized via $\varepsilon^{0123} = +1$. To see that $\varepsilon^{\mu\nu\rho\sigma}$ is invariant, we note that $\Lambda^\mu{}_\alpha \Lambda^\nu{}_\beta \Lambda^\rho{}_\gamma \Lambda^\sigma{}_\delta \varepsilon^{\alpha\beta\gamma\delta}$ is antisymmetric on exchange of any two of its uncontracted indices, and therefore must be proportional to $\varepsilon^{\mu\nu\rho\sigma}$. The constant of proportionality works out to be $\det\Lambda$, which is $+1$ for a proper Lorentz transformation.

We are finally ready to answer a question we posed at the beginning of section 33. There we considered a field $B^{\mu\nu}(x)$ carrying two vector indices, and we decomposed it as

$$B^{\mu\nu}(x) = A^{\mu\nu}(x) + S^{\mu\nu}(x) + \tfrac{1}{4}g^{\mu\nu}T(x)\,, \tag{34.33}$$

where $A^{\mu\nu}$ is antisymmetric ($A^{\mu\nu} = -A^{\nu\mu}$) and $S^{\mu\nu}$ is symmetric ($S^{\mu\nu} = S^{\nu\mu}$) and traceless ($g_{\mu\nu}S^{\mu\nu} = 0$). We asked whether further decomposition into still smaller irreducible representations was possible. The answer to this question can be found in eq. (34.31). Obviously, $T(x)$ corresponds to $(1,1)$, and $S^{\mu\nu}(x)$ to $(3,3)$.[1] But, according to eq. (34.31), the antisymmetric field $A^{\mu\nu}(x)$ should correspond to $(3,1)\oplus(1,3)$. A field in the $(3,1)$ representation carries a symmetric pair of left-handed (undotted) spinor indices; its hermitian conjugate is a field in the $(1,3)$ representation that carries a symmetric pair of right-handed (dotted) spinor indices. We should, then, be able to find a mapping, analogous to eq. (34.28), that gives $A^{\mu\nu}(x)$ in terms of a field $G_{ab}(x)$ and its hermitian conjugate $G^\dagger_{\dot a\dot b}(x)$.

This mapping is provided by the generator matrices $S_{\rm L}^{\mu\nu}$ and $S_{\rm R}^{\mu\nu}$. We first note that the Pauli matrices are traceless, and so eqs. (34.9) and (34.10) imply that $(S_{\rm L}^{\mu\nu})_a{}^a = 0$. Using eq. (34.24), we can rewrite this as $\varepsilon^{ab}(S_{\rm L}^{\mu\nu})_{ab} = 0$. Since ε^{ab} is antisymmetric, $(S_{\rm L}^{\mu\nu})_{ab}$ must be symmetric on exchange of its two spinor indices. An identical argument shows that $(S_{\rm R}^{\mu\nu})_{\dot a\dot b}$ must be symmetric on exchange of its two spinor indices. Furthermore, according to eqs. (34.9) and (34.10), we have

$$(S_{\rm L}^{10})_a{}^b = +i(S_{\rm L}^{23})_a{}^b\,. \tag{34.34}$$

This can be written covariantly with the Levi-Cevita symbol as

$$(S_{\rm L}^{\mu\nu})_a{}^b = -\tfrac{i}{2}\varepsilon^{\mu\nu\rho\sigma}(S_{{\rm L}\,\rho\sigma})_a{}^b\,. \tag{34.35}$$

Similarly,

$$(S_{\rm R}^{\mu\nu})_{\dot a}{}^{\dot b} = +\tfrac{i}{2}\varepsilon^{\mu\nu\rho\sigma}(S_{{\rm R}\,\rho\sigma})_{\dot a}{}^{\dot b}\,. \tag{34.36}$$

Eq. (34.36) follows from taking the complex conjugate of eq. (34.35) and using eq. (34.17).

Now, given a field $G_{ab}(x)$ in the $(3,1)$ representation, we can map it into a *self-dual antisymmetric tensor* $G^{\mu\nu}(x)$ via

$$G^{\mu\nu}(x) \equiv (S_{\rm L}^{\mu\nu})^{ab}G_{ab}(x)\,. \tag{34.37}$$

[1] Note that a symmetric traceless tensor has three independent diagonal components, and six independent off-diagonal components, for a total of nine, the number of components of the $(3,3)$ representation.

By *self-dual*, we mean that $G^{\mu\nu}(x)$ obeys

$$G^{\mu\nu}(x) = -\tfrac{i}{2}\varepsilon^{\mu\nu\rho\sigma}G_{\rho\sigma}(x)\,. \tag{34.38}$$

Taking the hermitian conjugate of eq. (34.37), and using eq. (34.17), we get

$$G^{\dagger\mu\nu}(x) = -(S_{\rm R}^{\mu\nu})^{\dot a\dot b}G^{\dagger}_{\dot a\dot b}(x)\,, \tag{34.39}$$

which is *anti-self-dual*,

$$G^{\dagger\mu\nu}(x) = +\tfrac{i}{2}\varepsilon^{\mu\nu\rho\sigma}G^{\dagger}_{\rho\sigma}(x)\,. \tag{34.40}$$

Given a hermitian antisymmetric tensor field $A^{\mu\nu}(x)$, we can extract its self-dual and anti-self-dual parts via

$$G^{\mu\nu}(x) = \tfrac{1}{2}A^{\mu\nu}(x) - \tfrac{i}{4}\varepsilon^{\mu\nu\rho\sigma}A_{\rho\sigma}(x)\,, \tag{34.41}$$

$$G^{\dagger\mu\nu}(x) = \tfrac{1}{2}A^{\mu\nu}(x) + \tfrac{i}{4}\varepsilon^{\mu\nu\rho\sigma}A_{\rho\sigma}(x)\,. \tag{34.42}$$

Then we have

$$A^{\mu\nu}(x) = G^{\mu\nu}(x) + G^{\dagger\mu\nu}(x)\,. \tag{34.43}$$

The field $G^{\mu\nu}(x)$ is in the $(3,1)$ representation, and the field $G^{\dagger\mu\nu}(x)$ is in the $(1,3)$ representation; these do not mix under Lorentz transformations.

Problems

34.1) Verify that eq. (34.6) follows from eq. (34.1).

34.2) Verify that eqs. (34.9) and (34.10) obey eq. (34.4).

34.3) Show that the Levi-Civita symbol obeys

$$\begin{aligned}\varepsilon^{\mu\nu\rho\sigma}\varepsilon_{\alpha\beta\gamma\sigma} = &-\delta^{\mu}{}_{\alpha}\delta^{\nu}{}_{\beta}\delta^{\rho}{}_{\gamma} - \delta^{\mu}{}_{\beta}\delta^{\nu}{}_{\gamma}\delta^{\rho}{}_{\alpha} - \delta^{\mu}{}_{\gamma}\delta^{\nu}{}_{\alpha}\delta^{\rho}{}_{\beta} \\ &+ \delta^{\mu}{}_{\beta}\delta^{\nu}{}_{\alpha}\delta^{\rho}{}_{\gamma} + \delta^{\mu}{}_{\alpha}\delta^{\nu}{}_{\gamma}\delta^{\rho}{}_{\beta} + \delta^{\mu}{}_{\gamma}\delta^{\nu}{}_{\beta}\delta^{\rho}{}_{\alpha}\,,\end{aligned} \tag{34.44}$$

$$\varepsilon^{\mu\nu\rho\sigma}\varepsilon_{\alpha\beta\rho\sigma} = -2(\delta^{\mu}{}_{\alpha}\delta^{\nu}{}_{\beta} - \delta^{\mu}{}_{\beta}\delta^{\nu}{}_{\alpha})\,, \tag{34.45}$$

$$\varepsilon^{\mu\nu\rho\sigma}\varepsilon_{\alpha\nu\rho\sigma} = -6\,\delta^{\mu}{}_{\alpha}\,. \tag{34.46}$$

34.4) Consider a field $C^{a\ldots c\dot a\ldots\dot c}(x)$, with N undotted indices and M dotted indices, that is furthermore symmetric on exchange of any pair of undotted indices, and also symmetric on exchange of any pair of dotted indices. Show that this field corresponds to a single irreducible representation $(2n{+}1, 2n'{+}1)$ of the Lorentz group, and identify n and n'.

35

Manipulating spinor indices

Prerequisite: 34

In section 34 we introduced the invariant symbols ε_{ab}, ε^{ab}, $\varepsilon_{\dot{a}\dot{b}}$, and $\varepsilon^{\dot{a}\dot{b}}$, where

$$\varepsilon^{12} = \varepsilon^{\dot{1}\dot{2}} = \varepsilon_{21} = \varepsilon_{\dot{2}\dot{1}} = +1\,, \qquad \varepsilon^{21} = \varepsilon^{\dot{2}\dot{1}} = \varepsilon_{12} = \varepsilon_{\dot{1}\dot{2}} = -1\,. \tag{35.1}$$

We use the ε symbols to raise and lower spinor indices, contracting the second index on the ε. (If we contract the first index instead, then there is an extra minus sign.)

Another invariant symbol is

$$\sigma^{\mu}_{a\dot{a}} = (I, \vec{\sigma})\,, \tag{35.2}$$

where I is the 2×2 identity matrix, and

$$\sigma_1 = \begin{pmatrix} 0 & 1 \\ 1 & 0 \end{pmatrix}, \quad \sigma_2 = \begin{pmatrix} 0 & -i \\ i & 0 \end{pmatrix}, \quad \sigma_3 = \begin{pmatrix} 1 & 0 \\ 0 & -1 \end{pmatrix} \tag{35.3}$$

are the Pauli matrices.

Now let us consider some combinations of invariant symbols with some indices contracted, such as $g_{\mu\nu}\sigma^{\mu}_{a\dot{a}}\sigma^{\nu}_{b\dot{b}}$. This object must also be invariant. Then, since it carries two undotted and two dotted spinor indices, it must be proportional to $\varepsilon_{ab}\varepsilon_{\dot{a}\dot{b}}$. Using eqs. (35.1) and (35.2), we can laboriously check this; it turns out to be correct.[1] The proportionality constant works out to be minus two:

$$\sigma^{\mu}_{a\dot{a}}\sigma_{\mu b\dot{b}} = -2\varepsilon_{ab}\varepsilon_{\dot{a}\dot{b}}\,. \tag{35.4}$$

[1] If it did not turn out to be correct, then eq. (35.2) would not be a viable choice of numerical values for this symbol.

Similarly, $\varepsilon^{ab}\varepsilon^{\dot{a}\dot{b}}\sigma^{\mu}_{a\dot{a}}\sigma^{\nu}_{b\dot{b}}$ must be proportional to $g^{\mu\nu}$, and the proportionality constant is again minus two:

$$\varepsilon^{ab}\varepsilon^{\dot{a}\dot{b}}\sigma^{\mu}_{a\dot{a}}\sigma^{\nu}_{b\dot{b}} = -2g^{\mu\nu} . \tag{35.5}$$

Next, let us see what we can learn about the generator matrices $(S^{\mu\nu}_{\rm L})_a{}^b$ and $(S^{\mu\nu}_{\rm R})_{\dot{a}}{}^{\dot{b}}$ from the fact that ε_{ab}, $\varepsilon_{\dot{a}\dot{b}}$, and $\sigma^{\mu}_{a\dot{a}}$ are all invariant symbols. Begin with

$$\varepsilon_{ab} = L(\Lambda)_a{}^c L(\Lambda)_b{}^d \varepsilon_{cd} , \tag{35.6}$$

which expresses the Lorentz invariance of ε_{ab}. For an infinitesimal transformation $\Lambda^{\mu}{}_{\nu} = \delta^{\mu}{}_{\nu} + \delta\omega^{\mu}{}_{\nu}$, we have

$$L_a{}^b(1{+}\delta\omega) = \delta_a{}^b + \tfrac{i}{2}\delta\omega_{\mu\nu}(S^{\mu\nu}_{\rm L})_a{}^b , \tag{35.7}$$

and eq. (35.6) becomes

$$\begin{aligned} \varepsilon_{ab} &= \varepsilon_{ab} + \tfrac{i}{2}\delta\omega_{\mu\nu}\Big[(S^{\mu\nu}_{\rm L})_a{}^c\varepsilon_{cb} + (S^{\mu\nu}_{\rm L})_b{}^d\varepsilon_{ad}\Big] + O(\delta\omega^2) \\ &= \varepsilon_{ab} + \tfrac{i}{2}\delta\omega_{\mu\nu}\Big[-(S^{\mu\nu}_{\rm L})_{ab} + (S^{\mu\nu}_{\rm L})_{ba}\Big] + O(\delta\omega^2) . \end{aligned} \tag{35.8}$$

Since eq. (35.8) holds for any choice of $\delta\omega_{\mu\nu}$, it must be that the factor in square brackets vanishes. Thus we conclude that $(S^{\mu\nu}_{\rm L})_{ab} = (S^{\mu\nu}_{\rm L})_{ba}$, which we had already deduced in section 34 by a different method. Similarly, starting from the Lorentz invariance of $\varepsilon_{\dot{a}\dot{b}}$, we can show that $(S^{\mu\nu}_{\rm R})_{\dot{a}\dot{b}} = (S^{\mu\nu}_{\rm R})_{\dot{b}\dot{a}}$.

Next, start from

$$\sigma^{\rho}_{a\dot{a}} = \Lambda^{\rho}{}_{\tau} L(\Lambda)_a{}^b R(\Lambda)_{\dot{a}}{}^{\dot{b}} \sigma^{\tau}_{b\dot{b}} , \tag{35.9}$$

which expresses the Lorentz invariance of $\sigma^{\rho}_{a\dot{a}}$. For an infinitesimal transformation, we have

$$\Lambda^{\rho}{}_{\tau} = \delta^{\rho}{}_{\tau} + \tfrac{i}{2}\delta\omega_{\mu\nu}(S^{\mu\nu}_{\rm V})^{\rho}{}_{\tau} , \tag{35.10}$$

$$L_a{}^b(1{+}\delta\omega) = \delta_a{}^b + \tfrac{i}{2}\delta\omega_{\mu\nu}(S^{\mu\nu}_{\rm L})_a{}^b , \tag{35.11}$$

$$R_{\dot{a}}{}^{\dot{b}}(1{+}\delta\omega) = \delta_{\dot{a}}{}^{\dot{b}} + \tfrac{i}{2}\delta\omega_{\mu\nu}(S^{\mu\nu}_{\rm R})_{\dot{a}}{}^{\dot{b}} , \tag{35.12}$$

where

$$(S^{\mu\nu}_{\rm V})^{\rho}{}_{\tau} \equiv \tfrac{1}{i}(g^{\mu\rho}\delta^{\nu}{}_{\tau} - g^{\nu\rho}\delta^{\mu}{}_{\tau}) . \tag{35.13}$$

Substituting eqs. (35.10)–(35.13) into eq. (35.9) and isolating the coefficient of $\delta\omega_{\mu\nu}$ yields

$$(g^{\mu\rho}\delta^{\nu}{}_{\tau} - g^{\nu\rho}\delta^{\mu}{}_{\tau})\sigma^{\tau}_{a\dot{a}} + i(S^{\mu\nu}_{\rm L})_a{}^b\sigma^{\rho}_{b\dot{a}} + i(S^{\mu\nu}_{\rm R})_{\dot{a}}{}^{\dot{b}}\sigma^{\rho}_{a\dot{b}} = 0 . \tag{35.14}$$

Now multiply by $\sigma_{\rho c\dot{c}}$ to get

$$\sigma^{\mu}_{c\dot{c}}\sigma^{\nu}_{a\dot{a}} - \sigma^{\nu}_{c\dot{c}}\sigma^{\mu}_{a\dot{a}} + i(S^{\mu\nu}_{\rm L})_a{}^b\sigma^{\rho}_{b\dot{a}}\sigma_{\rho c\dot{c}} + i(S^{\mu\nu}_{\rm R})_{\dot{a}}{}^{\dot{b}}\sigma^{\rho}_{a\dot{b}}\sigma_{\rho c\dot{c}} = 0\ . \tag{35.15}$$

Next use eq. (35.4) in each of the last two terms to get

$$\sigma^{\mu}_{c\dot{c}}\sigma^{\nu}_{a\dot{a}} - \sigma^{\nu}_{c\dot{c}}\sigma^{\mu}_{a\dot{a}} + 2i(S^{\mu\nu}_{\rm L})_{ac}\varepsilon_{\dot{a}\dot{c}} + 2i(S^{\mu\nu}_{\rm R})_{\dot{a}\dot{c}}\varepsilon_{ac} = 0\ . \tag{35.16}$$

If we multiply eq. (35.16) by $\varepsilon^{\dot{a}\dot{c}}$, and remember that $\varepsilon^{\dot{a}\dot{c}}(S^{\mu\nu}_{\rm R})_{\dot{a}\dot{c}} = 0$ and that $\varepsilon^{\dot{a}\dot{c}}\varepsilon_{\dot{a}\dot{c}} = -2$, we get a formula for $(S^{\mu\nu}_{\rm L})_{ac}$, namely

$$(S^{\mu\nu}_{\rm L})_{ac} = \tfrac{i}{4}\varepsilon^{\dot{a}\dot{c}}(\sigma^{\mu}_{a\dot{a}}\sigma^{\nu}_{c\dot{c}} - \sigma^{\nu}_{a\dot{a}}\sigma^{\mu}_{c\dot{c}})\ . \tag{35.17}$$

Similarly, if we multiply eq. (35.16) by ε^{ac}, we get

$$(S^{\mu\nu}_{\rm R})_{\dot{a}\dot{c}} = \tfrac{i}{4}\varepsilon^{ac}(\sigma^{\mu}_{a\dot{a}}\sigma^{\nu}_{c\dot{c}} - \sigma^{\nu}_{a\dot{a}}\sigma^{\mu}_{c\dot{c}})\ . \tag{35.18}$$

These formulae can be made to look a little nicer if we define

$$\bar{\sigma}^{\mu\dot{a}a} \equiv \varepsilon^{ab}\varepsilon^{\dot{a}\dot{b}}\sigma^{\mu}_{b\dot{b}}\ . \tag{35.19}$$

Numerically, it turns out that

$$\bar{\sigma}^{\mu\dot{a}a} = (I, -\vec{\sigma})\ . \tag{35.20}$$

Using $\bar{\sigma}^{\mu}$, we can write eqs. (35.17) and (35.18) as

$$(S^{\mu\nu}_{\rm L})_a{}^b = +\tfrac{i}{4}(\sigma^{\mu}\bar{\sigma}^{\nu} - \sigma^{\nu}\bar{\sigma}^{\mu})_a{}^b\ , \tag{35.21}$$

$$(S^{\mu\nu}_{\rm R})^{\dot{a}}{}_{\dot{b}} = -\tfrac{i}{4}(\bar{\sigma}^{\mu}\sigma^{\nu} - \bar{\sigma}^{\nu}\sigma^{\mu})^{\dot{a}}{}_{\dot{b}}\ . \tag{35.22}$$

In eq. (35.22), we have suppressed a contracted pair of undotted indices arranged as ${}^c{}_c$, and in eq. (35.21), we have suppressed a contracted pair of dotted indices arranged as ${}_{\dot{c}}{}^{\dot{c}}$.

We will adopt this as a general convention: a missing pair of contracted, undotted indices is understood to be written as ${}^c{}_c$, and a missing pair of contracted, dotted indices is understood to be written as ${}_{\dot{c}}{}^{\dot{c}}$. Thus, if χ and ψ are two left-handed Weyl fields, we have

$$\chi\psi = \chi^a\psi_a \quad \text{and} \quad \chi^\dagger\psi^\dagger = \chi^\dagger_{\dot{a}}\psi^{\dagger\dot{a}}\ . \tag{35.23}$$

We expect Weyl fields to describe spin-one-half particles, and (by the spin-statistics theorem) these particles must be *fermions*. Therefore the corresponding fields must *anticommute*, rather than commute. That is, we should have

$$\chi_a(x)\psi_b(y) = -\psi_b(y)\chi_a(x) . \tag{35.24}$$

Thus we can rewrite eq. (35.23) as

$$\chi\psi = \chi^a\psi_a = -\psi_a\chi^a = \psi^a\chi_a = \psi\chi . \tag{35.25}$$

The second equality follows from anticommutation of the fields, and the third from switching ${}_a{}^a$ to ${}^a{}_a$ (which introduces an extra minus sign). Eq. (35.25) tells us that $\chi\psi = \psi\chi$, which is a nice feature of this notation. Furthermore, if we take the hermitian conjugate of $\chi\psi$, we get

$$(\chi\psi)^\dagger = (\chi^a\psi_a)^\dagger = (\psi_a)^\dagger(\chi^a)^\dagger = \psi^\dagger_{\dot{a}}\chi^{\dagger\dot{a}} = \psi^\dagger\chi^\dagger . \tag{35.26}$$

That $(\chi\psi)^\dagger = \psi^\dagger\chi^\dagger$ is just what we would expect if we ignored the indices completely. Of course, by analogy with eq. (35.25), we also have $\psi^\dagger\chi^\dagger = \chi^\dagger\psi^\dagger$.

In order to tell whether a spinor field is left- or right-handed when its spinor index is suppressed, we will adopt the convention that a right-handed field is always written as the hermitian conjugate of a left-handed field. Thus, a right-handed field is always written with a dagger, and a left-handed field is always written without a dagger.

Let us try computing the hermitian conjugate of something a little more complicated:

$$\psi^\dagger\bar{\sigma}^\mu\chi = \psi^\dagger_{\dot{a}}\bar{\sigma}^{\mu\dot{a}c}\chi_c . \tag{35.27}$$

This behaves like a vector field under Lorentz transformations,

$$U(\Lambda)^{-1}[\psi^\dagger\bar{\sigma}^\mu\chi]U(\Lambda) = \Lambda^\mu{}_\nu[\psi^\dagger\bar{\sigma}^\nu\chi] . \tag{35.28}$$

(To avoid clutter, we suppressed the spacetime argument of the fields; as usual, it is x on the left-hand side and $\Lambda^{-1}x$ on the right.) The hermitian conjugate of eq. (35.27) is

$$\begin{aligned}
[\psi^\dagger\bar{\sigma}^\mu\chi]^\dagger &= [\psi^\dagger_{\dot{a}}\bar{\sigma}^{\mu\dot{a}c}\chi_c]^\dagger \\
&= \chi^\dagger_{\dot{c}}(\bar{\sigma}^{\mu a\dot{c}})^*\psi_a \\
&= \chi^\dagger_{\dot{c}}\bar{\sigma}^{\mu\dot{c}a}\psi_a \\
&= \chi^\dagger\bar{\sigma}^\mu\psi .
\end{aligned} \tag{35.29}$$

In the third line, we used the hermiticity of the matrices $\bar{\sigma}^\mu = (I, -\vec{\sigma})$.

We will get considerably more practice with this notation in the following sections.

Problems

35.1) Verify that eq. (35.20) follows from eqs. (35.2) and (35.19). Hint: write everything in "matrix multiplication" order, and note that, numerically, $\varepsilon^{ab} = -\varepsilon_{ab} = i\sigma_2$. Then make use of the properties of the Pauli matrices.

35.2) Verify that eq. (35.21) is consistent with eqs. (34.9) and (34.10).

35.3) Verify that eq. (35.22) is consistent with eq. (34.17).

35.4) Verify eq. (35.5).

36

Lagrangians for spinor fields

Prerequisites: 22, 35

Suppose we have a left-handed spinor field ψ_a. We would like to find a suitable lagrangian for it. This lagrangian must be Lorentz invariant, and it must be hermitian. We would also like it to be quadratic in ψ_a and its hermitian conjugate $\psi^\dagger_{\dot a}$, because this will lead to a linear equation of motion, with plane-wave solutions. We want plane-wave solutions because these describe free particles, the starting point for a theory of interacting particles.

Let us begin with terms with no derivatives. The only possibility is $\psi\psi = \psi^a\psi_a = \varepsilon^{ab}\psi_b\psi_a$, plus its hermitian conjugate. Because of anticommutation of the fields ($\psi_b\psi_a = -\psi_a\psi_b$), this expression does not vanish (as it would if the fields commuted), and so we can use it as a term in $\mathcal{L}$.

Next we need a term with derivatives. The obvious choice is $\partial^\mu\psi\partial_\mu\psi$, plus its hermitian conjugate. This, however, yields a hamiltonian that is unbounded below, which is unacceptable. To get a bounded hamiltonian, the kinetic term must involve both ψ and $\psi^\dagger$. A candidate is $i\psi^\dagger\bar\sigma^\mu\partial_\mu\psi$. This is not hermitian, but

$$
\begin{aligned}
(i\psi^\dagger\bar\sigma^\mu\partial_\mu\psi)^\dagger &= (i\psi^\dagger_{\dot a}\,\bar\sigma^{\mu\dot a c}\partial_\mu\psi_c)^\dagger \\
&= -i\partial_\mu\psi^\dagger_{\dot c}\,(\bar\sigma^{\mu a\dot c})^*\psi_a \\
&= -i\partial_\mu\psi^\dagger_{\dot c}\,\bar\sigma^{\mu\dot c a}\psi_a \\
&= i\psi^\dagger_{\dot c}\,\bar\sigma^{\mu\dot c a}\partial_\mu\psi_a - i\partial_\mu(\psi^\dagger_{\dot c}\,\bar\sigma^{\mu\dot c a}\psi_a) \\
&= i\psi^\dagger\bar\sigma^\mu\partial_\mu\psi - i\partial_\mu(\psi^\dagger\bar\sigma^\mu\psi)\;.
\end{aligned}
\tag{36.1}
$$

In the third line, we used the hermiticity of the matrices $\bar\sigma^\mu = (I, -\vec\sigma)$. In the fourth line, we used $-(\partial A)B = A\partial B - \partial(AB)$. In the last line, the second term is a total divergence, and vanishes (with suitable boundary conditions on the fields at infinity) when we integrate it over d^4x to get the action S. Thus $i\psi^\dagger\bar{\sigma}^\mu\partial_\mu\psi$ has the hermiticity properties necessary for a term in $\mathcal{L}$.

Our complete lagrangian for ψ is then

$$\mathcal{L} = i\psi^\dagger\bar{\sigma}^\mu\partial_\mu\psi - \tfrac{1}{2}m\psi\psi - \tfrac{1}{2}m^*\psi^\dagger\psi^\dagger \,, \tag{36.2}$$

where m is a complex parameter with dimensions of mass. The phase of m is actually irrelevant: if $m = |m|e^{i\alpha}$, we can set $\psi = e^{-i\alpha/2}\,\tilde{\psi}$ in eq. (36.2); then we get a lagrangian for $\tilde{\psi}$ that is identical to eq. (36.2), but with m replaced by $|m|$. So we can, without loss of generality, take m to be real and positive in the first place, and that is what we will do, setting $m^* = m$ in eq. (36.2).

The equation of motion for ψ is then

$$0 = -\frac{\delta S}{\delta\psi^\dagger} = -i\bar{\sigma}^\mu\partial_\mu\psi + m\psi^\dagger \,. \tag{36.3}$$

Restoring the spinor indices, this reads

$$0 = -i\bar{\sigma}^{\mu\dot{a}c}\partial_\mu\psi_c + m\psi^{\dagger\dot{a}} \,. \tag{36.4}$$

Taking the hermitian conjugate (or, equivalently, computing $-\delta S/\delta\psi$), we get

$$\begin{aligned} 0 &= +i(\bar{\sigma}^{\mu a\dot{c}})^*\,\partial_\mu\psi^\dagger_{\dot{c}} + m\psi^a \\ &= +i\bar{\sigma}^{\mu\dot{c}a}\partial_\mu\psi^\dagger_{\dot{c}} + m\psi^a \\ &= -i\sigma^\mu_{a\dot{c}}\,\partial_\mu\psi^{\dagger\dot{c}} + m\psi_a \,. \end{aligned} \tag{36.5}$$

In the second line, we used the hermiticity of the matrices $\bar{\sigma}^\mu = (I, -\vec{\sigma})$. In the third, we lowered the undotted index, and switched ${}^{\dot{c}}{}_{\dot{c}}$ to ${}_{\dot{c}}{}^{\dot{c}}$, which gives an extra minus sign.

Eqs. (36.5) and (36.4) can be combined to read

$$\begin{pmatrix} m\delta_a{}^c & -i\sigma^\mu_{a\dot{c}}\,\partial_\mu \\ -i\bar{\sigma}^{\mu\dot{a}c}\,\partial_\mu & m\delta^{\dot{a}}{}_{\dot{c}} \end{pmatrix} \begin{pmatrix} \psi_c \\ \psi^{\dagger\dot{c}} \end{pmatrix} = 0 \,. \tag{36.6}$$

We can write this more compactly by introducing the 4×4 *gamma matrices*

$$\gamma^\mu \equiv \begin{pmatrix} 0 & \sigma^\mu_{a\dot{c}} \\ \bar{\sigma}^{\mu\dot{a}c} & 0 \end{pmatrix} . \tag{36.7}$$

Using the sigma-matrix relations,

$$\begin{aligned} (\sigma^\mu\bar{\sigma}^\nu + \sigma^\nu\bar{\sigma}^\mu)_a{}^c &= -2g^{\mu\nu}\delta_a{}^c \,, \\ (\bar{\sigma}^\mu\sigma^\nu + \bar{\sigma}^\nu\sigma^\mu)^{\dot{a}}{}_{\dot{c}} &= -2g^{\mu\nu}\delta^{\dot{a}}{}_{\dot{c}} \,, \end{aligned} \tag{36.8}$$

which are most easily derived from the numerical formulae $\sigma^{\mu}_{a\dot{a}} = (I, \vec{\sigma})$ and $\bar{\sigma}^{\mu\dot{a}a} = (I, -\vec{\sigma})$, we see that the gamma matrices obey

$$\{\gamma^\mu, \gamma^\nu\} = -2g^{\mu\nu} , \tag{36.9}$$

where $\{A, B\} \equiv AB + BA$ denotes the anticommutator, and there is an understood 4×4 identity matrix on the right-hand side. We also introduce a four-component *Majorana field*

$$\Psi \equiv \begin{pmatrix} \psi_c \\ \psi^{\dagger\dot{c}} \end{pmatrix} . \tag{36.10}$$

Then eq. (36.6) becomes

$$(-i\gamma^\mu \partial_\mu + m)\Psi = 0 . \tag{36.11}$$

This is the *Dirac equation.* We first encountered it in section 1, where the gamma matrices were given different names ($\beta = \gamma^0$ and $\alpha^k = \gamma^0\gamma^k$). Also, in section 1 we were trying (and failing) to interpret Ψ as a wave function, rather than as a quantum field.

Now consider a theory of two left-handed spinor fields with an SO(2) symmetry,

$$\mathcal{L} = i\psi_i^\dagger \bar{\sigma}^\mu \partial_\mu \psi_i - \tfrac{1}{2} m\psi_i\psi_i - \tfrac{1}{2} m\psi_i^\dagger \psi_i^\dagger , \tag{36.12}$$

where the spinor indices are suppressed and $i = 1, 2$ is implicitly summed. As in the analogous case of two scalar fields discussed in sections 22 and 23, this lagrangian is invariant under the SO(2) transformation

$$\begin{pmatrix} \psi_1 \\ \psi_2 \end{pmatrix} \to \begin{pmatrix} \cos\alpha & \sin\alpha \\ -\sin\alpha & \cos\alpha \end{pmatrix} \begin{pmatrix} \psi_1 \\ \psi_2 \end{pmatrix} . \tag{36.13}$$

We can write the lagrangian so that the SO(2) symmetry appears as a U(1) symmetry instead; let

$$\chi = \tfrac{1}{\sqrt{2}}(\psi_1 + i\psi_2) , \tag{36.14}$$

$$\xi = \tfrac{1}{\sqrt{2}}(\psi_1 - i\psi_2) . \tag{36.15}$$

In terms of these fields, we have

$$\mathcal{L} = i\chi^\dagger \bar{\sigma}^\mu \partial_\mu \chi + i\xi^\dagger \bar{\sigma}^\mu \partial_\mu \xi - m\chi\xi - m\xi^\dagger\chi^\dagger . \tag{36.16}$$

Eq. (36.16) is invariant under the U(1) version of eq. (36.13),

$$\begin{aligned} \chi &\to e^{-i\alpha}\chi , \\ \xi &\to e^{+i\alpha}\xi . \end{aligned} \tag{36.17}$$

Next, let us derive the equations of motion that we get from eq. (36.16), following the same procedure that ultimately led to eq. (36.6). The result is

$$\begin{pmatrix} m\delta_a{}^c & -i\sigma^\mu_{a\dot c}\,\partial_\mu \\ -i\bar\sigma^{\mu\dot a c}\,\partial_\mu & m\delta^{\dot a}{}_{\dot c} \end{pmatrix} \begin{pmatrix} \chi_c \\ \xi^{\dagger\dot c} \end{pmatrix} = 0\,. \tag{36.18}$$

We can now define a four-component *Dirac field*

$$\Psi \equiv \begin{pmatrix} \chi_c \\ \xi^{\dagger\dot c} \end{pmatrix}, \tag{36.19}$$

which obeys the *Dirac equation*, eq. (36.11). (We have annoyingly used the same symbol Ψ to denote both a Majorana field and a Dirac field; these are different objects, and so we must always announce which is meant when we write Ψ.)

We can also write the lagrangian, eq. (36.16), in terms of the Dirac field Ψ, eq. (36.19). First we take the hermitian conjugate of Ψ to get

$$\Psi^\dagger = (\chi^\dagger_{\dot a}\,,\ \xi^a)\,. \tag{36.20}$$

Introduce the matrix

$$\beta \equiv \begin{pmatrix} 0 & \delta^{\dot a}{}_{\dot c} \\ \delta_a{}^c & 0 \end{pmatrix}. \tag{36.21}$$

Numerically, $\beta = \gamma^0$. However, the spinor index structure of β and γ^0 is different, and so we will distinguish them. Given β, we define

$$\overline\Psi \equiv \Psi^\dagger\beta = (\xi^a,\ \chi^\dagger_{\dot a})\,. \tag{36.22}$$

Then we have

$$\overline\Psi\Psi = \xi^a\chi_a + \chi^\dagger_{\dot a}\xi^{\dagger\dot a}\,. \tag{36.23}$$

Also,

$$\overline\Psi\gamma^\mu\partial_\mu\Psi = \xi^a\sigma^\mu_{a\dot c}\,\partial_\mu\xi^{\dagger\dot c} + \chi^\dagger_{\dot a}\,\bar\sigma^{\mu\dot a c}\,\partial_\mu\chi_c\,. \tag{36.24}$$

Using $A\partial B = -(\partial A)B + \partial(AB)$, the first term on the right-hand side of eq. (36.24) can be rewritten as

$$\xi^a\sigma^\mu_{a\dot c}\,\partial_\mu\xi^{\dagger\dot c} = -(\partial_\mu\xi^a)\sigma^\mu_{a\dot c}\,\xi^{\dagger\dot c} + \partial_\mu(\xi^a\sigma^\mu_{a\dot c}\,\xi^{\dagger\dot c})\,. \tag{36.25}$$

Then the first term on the right-hand side of eq. (36.25) can be rewritten as

$$-(\partial_\mu\xi^a)\sigma^\mu_{a\dot c}\,\xi^{\dagger\dot c} = +\xi^{\dagger\dot c}\sigma^\mu_{a\dot c}\,\partial_\mu\xi^a = +\xi^\dagger_{\dot c}\,\bar\sigma^{\mu\dot c a}\partial_\mu\xi_a\,. \tag{36.26}$$

Here we used anticommutation of the fields to get the first equality, and switched ${}^{\dot c}{}_{\dot c}$ to ${}_{\dot c}{}^{\dot c}$ and ${}_{a}{}^{a}$ to ${}^{a}{}_{a}$ (thus generating two minus signs) to get the second. Combining eqs. (36.24)–(36.26), we get

$$\overline{\Psi}\gamma^\mu\partial_\mu\Psi = \chi^\dagger\bar\sigma^\mu\partial_\mu\chi + \xi^\dagger\bar\sigma^\mu\partial_\mu\xi + \partial_\mu(\xi\sigma^\mu\xi^\dagger) . \tag{36.27}$$

Therefore, up to an irrelevant total divergence, we have

$$\mathcal{L} = i\overline{\Psi}\gamma^\mu\partial_\mu\Psi - m\overline{\Psi}\Psi . \tag{36.28}$$

This form of the lagrangian is invariant under the U(1) transformation

$$\begin{aligned}\Psi &\to e^{-i\alpha}\,\Psi ,\\ \overline{\Psi} &\to e^{+i\alpha}\,\overline{\Psi} ,\end{aligned} \tag{36.29}$$

which, given eq. (36.19), is the same as eq. (36.17). The Noether current associated with this symmetry is

$$j^\mu = \overline{\Psi}\gamma^\mu\Psi = \chi^\dagger\bar\sigma^\mu\chi - \xi^\dagger\bar\sigma^\mu\xi . \tag{36.30}$$

In quantum electrodynamics, the electromagnetic current is $e\overline{\Psi}\gamma^\mu\Psi$, where e is the charge of the electron.

As in the case of a complex scalar field with a U(1) symmetry, there is an additional discrete symmetry, called *charge conjugation*, that enlarges SO(2) to O(2). Charge conjugation simply exchanges χ and ξ. We can define a unitary *charge conjugation operator* C that implements this,

$$\begin{aligned}C^{-1}\chi_a(x)C &= \xi_a(x) ,\\ C^{-1}\xi_a(x)C &= \chi_a(x) ,\end{aligned} \tag{36.31}$$

where, for the sake of precision, we have restored the spinor index and spacetime argument. We then have $C^{-1}\mathcal{L}(x)C = \mathcal{L}(x)$.

To express eq. (36.31) in terms of the Dirac field, eq. (36.19), we first introduce the *charge conjugation matrix*

$$\mathcal{C} \equiv \begin{pmatrix} \varepsilon_{ac} & 0 \\ 0 & \varepsilon^{\dot a\dot c} \end{pmatrix} . \tag{36.32}$$

Next we notice that, if we take the transpose of $\overline{\Psi}$, eq. (36.22), we get

$$\overline{\Psi}^{\mathrm{T}} = \begin{pmatrix} \xi^a \\ \chi^\dagger_{\dot a} \end{pmatrix} . \tag{36.33}$$

Then, if we multiply by $\mathcal{C}$, we get a field that we will call $\Psi^{\rm C}$, the *charge conjugate* of Ψ,

$$\Psi^{\rm C} \equiv \mathcal{C}\overline{\Psi}^{\rm T} = \begin{pmatrix} \xi_a \\ \chi^{\dagger\dot{a}} \end{pmatrix} . \tag{36.34}$$

We see that $\Psi^{\rm C}$ is the same as the original field Ψ, eq. (36.19), except that the roles of χ and ξ have been switched. We therefore have

$$C^{-1}\Psi(x)C = \Psi^{\rm C}(x) \tag{36.35}$$

for a Dirac field.

The charge conjugation matrix has a number of useful properties. As a numerical matrix, it obeys

$$\mathcal{C}^{\rm T} = \mathcal{C}^{\dagger} = \mathcal{C}^{-1} = -\mathcal{C} \, , \tag{36.36}$$

and we can also write it as

$$\mathcal{C} = \begin{pmatrix} -\varepsilon^{ac} & 0 \\ 0 & -\varepsilon_{\dot{a}\dot{c}} \end{pmatrix} . \tag{36.37}$$

A result that we will need later is

$$\begin{aligned} \mathcal{C}^{-1}\gamma^{\mu}\mathcal{C} &= \begin{pmatrix} \varepsilon^{ab} & 0 \\ 0 & \varepsilon_{\dot{a}\dot{b}} \end{pmatrix} \begin{pmatrix} 0 & \sigma^{\mu}_{b\dot{c}} \\ \bar{\sigma}^{\mu\dot{b}c} & 0 \end{pmatrix} \begin{pmatrix} \varepsilon_{ce} & 0 \\ 0 & \varepsilon^{\dot{c}\dot{e}} \end{pmatrix} \\ &= \begin{pmatrix} 0 & \varepsilon^{ab}\sigma^{\mu}_{b\dot{c}}\varepsilon^{\dot{c}\dot{e}} \\ \varepsilon_{\dot{a}\dot{b}}\bar{\sigma}^{\mu\dot{b}c}\varepsilon_{ce} & 0 \end{pmatrix} \\ &= \begin{pmatrix} 0 & -\bar{\sigma}^{\mu a\dot{e}} \\ -\sigma^{\mu}_{\dot{a}e} & 0 \end{pmatrix} . \end{aligned} \tag{36.38}$$

The minus signs in the last line come from raising or lowering an index by contracting with the first (rather than the second) index of an ε symbol. Comparing with

$$\gamma^{\mu} = \begin{pmatrix} 0 & \sigma^{\mu}_{e\dot{a}} \\ \bar{\sigma}^{\mu\dot{e}a} & 0 \end{pmatrix} , \tag{36.39}$$

we see that

$$\mathcal{C}^{-1}\gamma^{\mu}\mathcal{C} = -(\gamma^{\mu})^{\rm T} \, . \tag{36.40}$$

Now let us return to the Majorana field, eq. (36.10). It is obvious that a Majorana field is its own charge conjugate, that is, $\Psi^{\rm C} = \Psi$. This condition is analogous to the condition $\varphi^\dagger = \varphi$ that is satisfied by a real scalar field. A Dirac field, with its U(1) symmetry, is analogous to a complex scalar field, while a Majorana field, which has no U(1) symmetry, is analogous to a real scalar field.

We can write our original lagrangian for a single left-handed spinor field, eq. (36.2), in terms of a Majorana field, eq. (36.10), by retracing eqs. (36.20)–(36.28) with $\chi \to \psi$ and $\xi \to \psi$. The result is

$$\mathcal{L} = \tfrac{i}{2}\overline{\Psi}\gamma^\mu\partial_\mu\Psi - \tfrac{1}{2}m\overline{\Psi}\Psi \ . \tag{36.41}$$

However, we cannot yet derive the equation of motion from eq. (36.41) because it does not yet incorporate the Majorana condition $\Psi^{\mathrm{C}} = \Psi$. To remedy this, we use eq. (36.36) to write the Majorana condition $\Psi = \mathcal{C}\overline{\Psi}^{\mathrm{T}}$ as $\overline{\Psi} = \Psi^{\mathrm{T}}\mathcal{C}$. Then we can replace $\overline{\Psi}$ in eq. (36.41) by $\Psi^{\mathrm{T}}\mathcal{C}$ to get

$$\mathcal{L} = \tfrac{i}{2}\Psi^{\mathrm{T}}\mathcal{C}\gamma^\mu\partial_\mu\Psi - \tfrac{1}{2}m\Psi^{\mathrm{T}}\mathcal{C}\Psi \ . \tag{36.42}$$

The equation of motion that follows from this lagrangian is once again the Dirac equation.

We can also recover the Weyl components of a Dirac or Majorana field by means of a suitable projection matrix. Define

$$\gamma_5 \equiv \begin{pmatrix} -\delta_a{}^c & 0 \\ 0 & +\delta^{\dot a}{}_{\dot c} \end{pmatrix} \tag{36.43}$$

where the subscript 5 is simply part of the traditional name of this matrix, rather than the value of some index. Then we can define left and right projection matrices

$$P_{\mathrm{L}} \equiv \tfrac{1}{2}(1-\gamma_5) = \begin{pmatrix} \delta_a{}^c & 0 \\ 0 & 0 \end{pmatrix},$$
$$P_{\mathrm{R}} \equiv \tfrac{1}{2}(1+\gamma_5) = \begin{pmatrix} 0 & 0 \\ 0 & \delta^{\dot a}{}_{\dot c} \end{pmatrix}. \tag{36.44}$$

Thus we have, for a Dirac field,

$$P_{\mathrm{L}}\Psi = \begin{pmatrix} \chi_c \\ 0 \end{pmatrix},$$
$$P_{\mathrm{R}}\Psi = \begin{pmatrix} 0 \\ \xi^{\dagger\dot c} \end{pmatrix}. \tag{36.45}$$

The matrix γ_5 can also be expressed as

$$\begin{aligned}\gamma_5 &= i\gamma^0\gamma^1\gamma^2\gamma^3 \\ &= -\tfrac{i}{24}\varepsilon_{\mu\nu\rho\sigma}\gamma^\mu\gamma^\nu\gamma^\rho\gamma^\sigma \,, \end{aligned} \tag{36.46}$$

where $\varepsilon_{0123} = -1$.

Finally, let us consider the behavior of a Dirac or Majorana field under a Lorentz transformation. Recall that left- and right-handed spinor fields transform according to

$$U(\Lambda)^{-1}\psi_a(x)U(\Lambda) = L(\Lambda)_a{}^c\,\psi_c(\Lambda^{-1}x) \,, \tag{36.47}$$

$$U(\Lambda)^{-1}\psi^\dagger_{\dot a}(x)U(\Lambda) = R(\Lambda)_{\dot a}{}^{\dot c}\,\psi^\dagger_{\dot c}(\Lambda^{-1}x) \,, \tag{36.48}$$

where, for an infinitesimal transformation $\Lambda^\mu{}_\nu = \delta^\mu{}_\nu + \delta\omega^\mu{}_\nu$,

$$L(1{+}\delta\omega)_a{}^c = \delta_a{}^c + \tfrac{i}{2}\delta\omega_{\mu\nu}(S^{\mu\nu}_{\rm L})_a{}^c \,, \tag{36.49}$$

$$R(1{+}\delta\omega)_{\dot a}{}^{\dot c} = \delta_{\dot a}{}^{\dot c} + \tfrac{i}{2}\delta\omega_{\mu\nu}(S^{\mu\nu}_{\rm R})_{\dot a}{}^{\dot c} \,, \tag{36.50}$$

and where

$$(S^{\mu\nu}_{\rm L})_a{}^c = +\tfrac{i}{4}(\sigma^\mu\bar\sigma^\nu - \sigma^\nu\bar\sigma^\mu)_a{}^c \,, \tag{36.51}$$

$$(S^{\mu\nu}_{\rm R})^{\dot a}{}_{\dot c} = -\tfrac{i}{4}(\bar\sigma^\mu\sigma^\nu - \bar\sigma^\nu\sigma^\mu)^{\dot a}{}_{\dot c} \,. \tag{36.52}$$

From these formulae, and the definition of γ^μ, eq. (36.7), we can see that

$$\tfrac{i}{4}[\gamma^\mu,\gamma^\nu] = \begin{pmatrix} +(S^{\mu\nu}_{\rm L})_a{}^c & 0 \\ 0 & -(S^{\mu\nu}_{\rm R})^{\dot a}{}_{\dot c} \end{pmatrix} \equiv S^{\mu\nu} \,. \tag{36.53}$$

Then, for either a Dirac or Majorana field Ψ, we can write

$$U(\Lambda)^{-1}\Psi(x)U(\Lambda) = D(\Lambda)\Psi(\Lambda^{-1}x) \,, \tag{36.54}$$

where, for an infinitesimal transformation, the 4×4 matrix $D(\Lambda)$ is

$$D(1{+}\delta\omega) = 1 + \tfrac{i}{2}\delta\omega_{\mu\nu}S^{\mu\nu} \,, \tag{36.55}$$

with $S^{\mu\nu}$ given by eq. (36.53). The minus sign in front of $S^{\mu\nu}_{\rm R}$ in eq. (36.53) is compensated by the switch from a ${}^{\dot c}{}_{\dot c}$ contraction in eq. (36.50) to a ${}_{\dot c}{}^{\dot c}$ contraction in eq. (36.54).

Problems

36.1) Using the results of problem 2.9, show that, for a rotation by an angle θ about the z axis, we have

$$D(\Lambda) = \exp(-i\theta S^{12}) , \tag{36.56}$$

and that, for a boost by rapidity η in the z direction, we have

$$D(\Lambda) = \exp(+i\eta S^{30}) . \tag{36.57}$$

36.2) Verify that eq. (36.46) is consistent with eq. (36.43).

36.3) a) Prove the *Fierz identities*

$$(\chi_1^\dagger \bar{\sigma}^\mu \chi_2)(\chi_3^\dagger \bar{\sigma}_\mu \chi_4) = -2(\chi_1^\dagger \chi_3^\dagger)(\chi_2 \chi_4) , \tag{36.58}$$

$$(\chi_1^\dagger \bar{\sigma}^\mu \chi_2)(\chi_3^\dagger \bar{\sigma}_\mu \chi_4) = (\chi_1^\dagger \bar{\sigma}^\mu \chi_4)(\chi_3^\dagger \bar{\sigma}_\mu \chi_2) . \tag{36.59}$$

b) Define the Dirac fields

$$\Psi_i \equiv \begin{pmatrix} \chi_i \\ \xi_i^\dagger \end{pmatrix}, \qquad \Psi_i^{\rm C} \equiv \begin{pmatrix} \xi_i \\ \chi_i^\dagger \end{pmatrix}. \tag{36.60}$$

Use eqs. (36.58) and (36.59) to prove the Dirac form of the Fierz identities,

$$(\overline{\Psi}_1 \gamma^\mu P_{\rm L} \Psi_2)(\overline{\Psi}_3 \gamma_\mu P_{\rm L} \Psi_4) = -2(\overline{\Psi}_1 P_{\rm R} \Psi_3^{\rm C})(\overline{\Psi}_4^{\rm C} P_{\rm L} \Psi_2) , \tag{36.61}$$

$$(\overline{\Psi}_1 \gamma^\mu P_{\rm L} \Psi_2)(\overline{\Psi}_3 \gamma_\mu P_{\rm L} \Psi_4) = (\overline{\Psi}_1 \gamma^\mu P_{\rm L} \Psi_4)(\overline{\Psi}_3 \gamma_\mu P_{\rm L} \Psi_2) . \tag{36.62}$$

c) By writing both sides out in terms of Weyl fields, show that

$$\overline{\Psi}_1 \gamma^\mu P_{\rm R} \Psi_2 = -\overline{\Psi}_2^{\rm C} \gamma^\mu P_{\rm L} \Psi_1^{\rm C} , \tag{36.63}$$

$$\overline{\Psi}_1 P_{\rm L} \Psi_2 = +\overline{\Psi}_2^{\rm C} P_{\rm L} \Psi_1^{\rm C} , \tag{36.64}$$

$$\overline{\Psi}_1 P_{\rm R} \Psi_2 = +\overline{\Psi}_2^{\rm C} P_{\rm R} \Psi_1^{\rm C} . \tag{36.65}$$

Combining eqs. (36.63)–(36.65) with eqs. (36.61) and (36.62) yields more useful forms of the Fierz identities.

36.4) Consider a field $\varphi_A(x)$ in an unspecified representation of the Lorentz group, indexed by A, that obeys

$$U(\Lambda)^{-1} \varphi_A(x) U(\Lambda) = L_A{}^B(\Lambda) \varphi_B(\Lambda^{-1}x) . \tag{36.66}$$

For an infinitesimal transformation,

$$L_A{}^B(1+\delta\omega) = \delta_A{}^B + \tfrac{i}{2}\delta\omega_{\mu\nu}(S^{\mu\nu})_A{}^B . \tag{36.67}$$

a) Following the procedure of section 22, show that the energy-momentum tensor is

$$T^{\mu\nu} = g^{\mu\nu}\mathcal{L} - \frac{\partial \mathcal{L}}{\partial(\partial_\mu \varphi_A)}\, \partial^\nu \varphi_A \,. \tag{36.68}$$

b) Show that the Noether current corresponding to a Lorentz transformation is

$$\mathcal{M}^{\mu\nu\rho} = x^\nu T^{\mu\rho} - x^\rho T^{\mu\nu} + B^{\mu\nu\rho} \,, \tag{36.69}$$

where

$$B^{\mu\nu\rho} \equiv -i\frac{\partial \mathcal{L}}{\partial(\partial_\mu \varphi_A)}(S^{\nu\rho})_A{}^B \varphi_B \,. \tag{36.70}$$

c) Use the conservation laws $\partial_\mu T^{\mu\nu} = 0$ and $\partial_\mu \mathcal{M}^{\mu\nu\rho} = 0$ to show that

$$T^{\nu\rho} - T^{\rho\nu} + \partial_\mu B^{\mu\nu\rho} = 0 \,. \tag{36.71}$$

d) Define the *improved energy-momentum tensor* or *Belinfante tensor*

$$\Theta^{\mu\nu} \equiv T^{\mu\nu} + \tfrac{1}{2}\partial_\rho (B^{\rho\mu\nu} - B^{\mu\rho\nu} - B^{\nu\rho\mu}) \,. \tag{36.72}$$

Show that $\Theta^{\mu\nu}$ is symmetric: $\Theta^{\mu\nu} = \Theta^{\nu\mu}$. Also show that $\Theta^{\mu\nu}$ is conserved, $\partial_\mu \Theta^{\mu\nu} = 0$, and that $\int d^3x\, \Theta^{0\nu} = \int d^3x\, T^{0\nu} = P^\nu$, where P^ν is the energy-momentum four-vector. (In general relativity, it is the Belinfante tensor that couples to gravity.)

e) Show that the improved tensor

$$\Xi^{\mu\nu\rho} \equiv x^\nu \Theta^{\mu\rho} - x^\rho \Theta^{\mu\nu} \tag{36.73}$$

obeys $\partial_\mu \Xi^{\mu\nu\rho} = 0$, and that $\int d^3x\, \Xi^{0\nu\rho} = \int d^3x\, \mathcal{M}^{0\nu\rho} = M^{\nu\rho}$, where $M^{\nu\rho}$ are the Lorentz generators.

f) Compute $\Theta^{\mu\nu}$ for a left-handed Weyl field with $\mathcal{L}$ given by eq. (36.2), and for a Dirac field with $\mathcal{L}$ given by eq. (36.28).

36.5) *Symmetries of fermion fields.* (Prerequisite: 24.) Consider a theory with N massless Weyl fields ψ_j,

$$\mathcal{L} = i\psi_j^\dagger \bar{\sigma}^\mu \partial_\mu \psi_j \,, \tag{36.74}$$

where the repeated index j is summed. This lagrangian is clearly invariant under the U(N) transformation,

$$\psi_j \to U_{jk}\psi_k \,, \tag{36.75}$$

where U is a unitary matrix. State the invariance group for the following cases:

a) N Weyl fields with a common mass m,

$$\mathcal{L} = i\psi_j^\dagger \bar{\sigma}^\mu \partial_\mu \psi_j - \tfrac{1}{2}m(\psi_j\psi_j + \psi_j^\dagger \psi_j^\dagger) \,. \tag{36.76}$$

b) N massless Majorana fields,

$$\mathcal{L} = \tfrac{i}{2}\Psi_j^{\mathrm{T}}\mathcal{C}\gamma^\mu\partial_\mu\Psi_j \;. \tag{36.77}$$

c) N Majorana fields with a common mass m,

$$\mathcal{L} = \tfrac{i}{2}\Psi_j^{\mathrm{T}}\mathcal{C}\gamma^\mu\partial_\mu\Psi_j - \tfrac{1}{2}m\Psi_j^{\mathrm{T}}\mathcal{C}\Psi_j \;. \tag{36.78}$$

d) N massless Dirac fields,

$$\mathcal{L} = i\overline{\Psi}_j\gamma^\mu\partial_\mu\Psi_j \;. \tag{36.79}$$

e) N Dirac fields with a common mass m,

$$\mathcal{L} = i\overline{\Psi}_j\gamma^\mu\partial_\mu\Psi_j - m\overline{\Psi}_j\Psi_j \;. \tag{36.80}$$

37

Canonical quantization of spinor fields I

Prerequisite: 36

Consider a left-handed Weyl field ψ with lagrangian

$$\mathcal{L} = i\psi^\dagger\bar{\sigma}^\mu\partial_\mu\psi - \tfrac{1}{2}m(\psi\psi + \psi^\dagger\psi^\dagger) \; . \tag{37.1}$$

The canonically conjugate momentum to the field $\psi_a(x)$ is then[1]

$$\begin{aligned}\pi^a(x) &\equiv \frac{\partial\mathcal{L}}{\partial(\partial_0\psi_a(x))} \\ &= i\psi^\dagger_{\dot{a}}(x)\bar{\sigma}^{0\dot{a}a} \; .\end{aligned} \tag{37.2}$$

The hamiltonian is

$$\begin{aligned}\mathcal{H} &= \pi^a\partial_0\psi_a - \mathcal{L} \\ &= i\psi^\dagger_{\dot{a}}\bar{\sigma}^{0\dot{a}a}\partial_0\psi_a - \mathcal{L} \\ &= -i\psi^\dagger\bar{\sigma}^i\partial_i\psi + \tfrac{1}{2}m(\psi\psi + \psi^\dagger\psi^\dagger) \; .\end{aligned} \tag{37.3}$$

The appropriate canonical *anticommutation* relations are

$$\{\psi_a(\mathbf{x},t), \psi_c(\mathbf{y},t)\} = 0 \; , \tag{37.4}$$

$$\{\psi_a(\mathbf{x},t), \pi^c(\mathbf{y},t)\} = i\delta_a{}^c\,\delta^3(\mathbf{x}-\mathbf{y}) \; . \tag{37.5}$$

Substituting in eq. (37.2) for π^c, we get

$$\{\psi_a(\mathbf{x},t), \psi^\dagger_{\dot{c}}(\mathbf{y},t)\}\bar{\sigma}^{0\dot{c}c} = \delta_a{}^c\,\delta^3(\mathbf{x}-\mathbf{y}) \; . \tag{37.6}$$

[1] Here we gloss over a subtlety about differentiating with respect to an anticommuting object; we will take up this topic in section 44, and for now simply assume that eq. (37.2) is correct.

Then, using $\bar{\sigma}^0 = \sigma^0 = I$, we have

$$\{\psi_a(\mathbf{x},t), \psi^\dagger_{\dot{c}}(\mathbf{y},t)\} = \sigma^0_{a\dot{c}}\, \delta^3(\mathbf{x}-\mathbf{y}) \;, \tag{37.7}$$

or, equivalently,

$$\{\psi^a(\mathbf{x},t), \psi^{\dagger\dot{c}}(\mathbf{y},t)\} = \bar{\sigma}^{0\dot{c}a}\, \delta^3(\mathbf{x}-\mathbf{y}) \;. \tag{37.8}$$

We can also translate this into four-component notation for either a Dirac or a Majorana field. A Dirac field is defined in terms of two left-handed Weyl fields χ and ξ via

$$\Psi \equiv \begin{pmatrix} \chi_c \\ \xi^{\dagger\dot{c}} \end{pmatrix} . \tag{37.9}$$

We also define

$$\overline{\Psi} \equiv \Psi^\dagger \beta = (\xi^a, \chi^\dagger_{\dot{a}}) \;, \tag{37.10}$$

where

$$\beta \equiv \begin{pmatrix} 0 & \delta^{\dot{a}}{}_{\dot{c}} \\ \delta_a{}^c & 0 \end{pmatrix} . \tag{37.11}$$

The lagrangian is

$$\begin{aligned} \mathcal{L} &= i\chi^\dagger\bar{\sigma}^\mu\partial_\mu\chi + i\xi^\dagger\bar{\sigma}^\mu\partial_\mu\xi - m(\chi\xi + \xi^\dagger\chi^\dagger) \\ &= i\overline{\Psi}\gamma^\mu\partial_\mu\Psi - m\overline{\Psi}\Psi \;. \end{aligned} \tag{37.12}$$

The fields χ and ξ each obey the canonical anticommutation relations of eqs. (37.4) and (37.5). This translates into

$$\{\Psi_\alpha(\mathbf{x},t), \Psi_\beta(\mathbf{y},t)\} = 0 \;, \tag{37.13}$$

$$\{\Psi_\alpha(\mathbf{x},t), \overline{\Psi}_\beta(\mathbf{y},t)\} = (\gamma^0)_{\alpha\beta}\, \delta^3(\mathbf{x}-\mathbf{y}) \;, \tag{37.14}$$

where α and β are four-component spinor indices, and

$$\gamma^\mu \equiv \begin{pmatrix} 0 & \sigma^\mu_{a\dot{c}} \\ \bar{\sigma}^{\mu\dot{a}c} & 0 \end{pmatrix} . \tag{37.15}$$

Eqs. (37.13) and (37.14) can also be derived directly from the four-component form of the lagrangian, eq. (37.12), by noting that the canonically conjugate momentum to the field Ψ is $\partial\mathcal{L}/\partial(\partial_0\Psi) = i\overline{\Psi}\gamma^0$, and that $(\gamma^0)^2 = 1$.

A Majorana field is defined in terms of a single left-handed Weyl field ψ via

$$\Psi \equiv \begin{pmatrix} \psi_c \\ \psi^{\dagger\dot{c}} \end{pmatrix} . \tag{37.16}$$

We also define

$$\overline{\Psi} \equiv \Psi^\dagger \beta = (\psi^a, \psi^\dagger_{\dot{a}}) . \tag{37.17}$$

A Majorana field obeys the Majorana condition

$$\overline{\Psi} = \Psi^{\mathrm{T}} \mathcal{C} , \tag{37.18}$$

where

$$\mathcal{C} \equiv \begin{pmatrix} -\varepsilon^{ac} & 0 \\ 0 & -\varepsilon_{\dot{a}\dot{c}} \end{pmatrix} \tag{37.19}$$

is the charge conjugation matrix. The lagrangian is

$$\begin{aligned} \mathcal{L} &= i\psi^\dagger \bar{\sigma}^\mu \partial_\mu \psi - \tfrac{1}{2} m(\psi\psi + \psi^\dagger\psi^\dagger) \\ &= \tfrac{i}{2} \overline{\Psi} \gamma^\mu \partial_\mu \Psi - \tfrac{1}{2} m \overline{\Psi} \Psi \\ &= \tfrac{i}{2} \Psi^{\mathrm{T}} \mathcal{C} \gamma^\mu \partial_\mu \Psi - \tfrac{1}{2} m \Psi^{\mathrm{T}} \mathcal{C} \Psi . \end{aligned} \tag{37.20}$$

The field ψ obeys the canonical anticommutation relations of eqs. (37.4) and (37.5). This translates into

$$\{\Psi_\alpha(\mathbf{x},t), \Psi_\beta(\mathbf{y},t)\} = (\mathcal{C}\gamma^0)_{\alpha\beta}\, \delta^3(\mathbf{x}-\mathbf{y}) , \tag{37.21}$$

$$\{\Psi_\alpha(\mathbf{x},t), \overline{\Psi}_\beta(\mathbf{y},t)\} = (\gamma^0)_{\alpha\beta}\, \delta^3(\mathbf{x}-\mathbf{y}) , \tag{37.22}$$

where α and β are four-component spinor indices. To derive eqs. (37.21) and (37.22) directly from the four-component form of the lagrangian, eq. (37.20), requires new formalism for the quantization of *constrained systems*. This is because the canonically conjugate momentum to the field Ψ is $\partial\mathcal{L}/\partial(\partial_0\Psi) = \frac{i}{2}\Psi^{\mathrm{T}}\mathcal{C}\gamma^0$, and this is linearly related to Ψ itself; this relation constitutes a constraint that must be solved before imposition of the anticommutation relations. In this case, solving the constraint simply returns us to the Weyl formalism with which we began.

The equation of motion that follows from either eq. (37.12) or eq. (37.20) is the Dirac equation,

$$(-i\partial\!\!\!/ + m)\Psi = 0 . \tag{37.23}$$

Here we have introduced the *Feynman slash*: given any four-vector a^μ, we define

$$\not{a} \equiv a_\mu \gamma^\mu \ . \tag{37.24}$$

To solve the Dirac equation, we first note that if we act on it with $i\not{\partial} + m$, we get

$$\begin{aligned} 0 &= (i\not{\partial} + m)(-i\not{\partial} + m)\Psi \\ &= (\not{\partial}\not{\partial} + m^2)\Psi \\ &= (-\partial^2 + m^2)\Psi \ . \end{aligned} \tag{37.25}$$

Here we have used

$$\begin{aligned} \not{a}\not{a} &= a_\mu a_\nu \gamma^\mu \gamma^\nu \\ &= a_\mu a_\nu \Big(\tfrac{1}{2}\{\gamma^\mu, \gamma^\nu\} + \tfrac{1}{2}[\gamma^\mu, \gamma^\nu] \Big) \\ &= a_\mu a_\nu \Big(-g^{\mu\nu} + \tfrac{1}{2}[\gamma^\mu, \gamma^\nu] \Big) \\ &= -a_\mu a_\nu g^{\mu\nu} + 0 \\ &= -a^2 \ . \end{aligned} \tag{37.26}$$

From eq. (37.25), we see that Ψ obeys the Klein–Gordon equation. Therefore, the Dirac equation has plane-wave solutions. Let us consider a specific solution of the form

$$\Psi(x) = u(\mathbf{p})e^{ipx} + v(\mathbf{p})e^{-ipx} \ . \tag{37.27}$$

where $p^0 = \omega \equiv (\mathbf{p}^2 + m^2)^{1/2}$, and $u(\mathbf{p})$ and $v(\mathbf{p})$ are four-component constant spinors. Plugging eq. (37.27) into eq. (37.23), we get

$$(\not{p} + m)u(\mathbf{p})e^{ipx} + (-\not{p} + m)v(\mathbf{p})e^{-ipx} = 0 \ . \tag{37.28}$$

Thus we require

$$\begin{aligned} (\not{p} + m)u(\mathbf{p}) &= 0 \ , \\ (-\not{p} + m)v(\mathbf{p}) &= 0 \ . \end{aligned} \tag{37.29}$$

Each of these equations has two linearly independent solutions that we will call $u_\pm(\mathbf{p})$ and $v_\pm(\mathbf{p})$; their detailed properties will be worked out in the next section. The general solution of the Dirac equation can then be written as

$$\Psi(x) = \sum_{s=\pm} \int \widetilde{dp} \, \Big[b_s(\mathbf{p})u_s(\mathbf{p})e^{ipx} + d_s^\dagger(\mathbf{p})v_s(\mathbf{p})e^{-ipx} \Big] \ , \tag{37.30}$$

where the integration measure is as usual

$$\widetilde{dp} \equiv \frac{d^3p}{(2\pi)^3 2\omega} \,. \tag{37.31}$$

Problems

37.1) Verify that eqs. (37.13) and (37.14) follow from eqs. (37.4) and (37.5).

38

Spinor technology

Prerequisite: 37

The four-component spinors $u_s(\mathbf{p})$ and $v_s(\mathbf{p})$ obey the equations

$$
\begin{aligned}
(\not p + m)u_s(\mathbf{p}) &= 0 , \\
(-\not p + m)v_s(\mathbf{p}) &= 0 .
\end{aligned}
\tag{38.1}
$$

Each of these equations has two solutions, which we label via $s = +$ and $s = -$. For $m \neq 0$, we can go to the rest frame, $\mathbf{p} = \mathbf{0}$. We will then distinguish the two solutions by the eigenvalue of the spin matrix

$$
S_z = \tfrac{i}{4}[\gamma^1, \gamma^2] = \tfrac{i}{2}\gamma^1\gamma^2 = \begin{pmatrix} \frac{1}{2}\sigma_3 & 0 \\ 0 & \frac{1}{2}\sigma_3 \end{pmatrix} . \tag{38.2}
$$

Specifically, we will require

$$
\begin{aligned}
S_z u_\pm(\mathbf{0}) &= \pm\tfrac{1}{2}u_\pm(\mathbf{0}) , \\
S_z v_\pm(\mathbf{0}) &= \mp\tfrac{1}{2}v_\pm(\mathbf{0}) .
\end{aligned}
\tag{38.3}
$$

The reason for the opposite sign for the v spinor is that this choice results in

$$
\begin{aligned}
[J_z, b^\dagger_\pm(\mathbf{0})] &= \pm\tfrac{1}{2}b^\dagger_\pm(\mathbf{0}) , \\
[J_z, d^\dagger_\pm(\mathbf{0})] &= \pm\tfrac{1}{2}d^\dagger_\pm(\mathbf{0}) ,
\end{aligned}
\tag{38.4}
$$

where J_z is the z component of the angular momentum operator. Eq. (38.4) implies that $b^\dagger_+(\mathbf{0})$ and $d^\dagger_+(\mathbf{0})$ each creates a particle with spin up along the z axis. We will verify eq. (38.4) in problem 39.2.

For $\mathbf{p} = \mathbf{0}$, we have $\not{p} = -m\gamma^0$, where

$$\gamma^0 = \begin{pmatrix} 0 & I \\ I & 0 \end{pmatrix} . \tag{38.5}$$

Eqs. (38.1) and (38.3) are then easy to solve. Choosing (for later convenience) a specific normalization and phase for each of $u_\pm(\mathbf{0})$ and $v_\pm(\mathbf{0})$, we get

$$u_+(\mathbf{0}) = \sqrt{m} \begin{pmatrix} 1 \\ 0 \\ 1 \\ 0 \end{pmatrix} , \qquad u_-(\mathbf{0}) = \sqrt{m} \begin{pmatrix} 0 \\ 1 \\ 0 \\ 1 \end{pmatrix} ,$$

$$v_+(\mathbf{0}) = \sqrt{m} \begin{pmatrix} 0 \\ 1 \\ 0 \\ -1 \end{pmatrix} , \qquad v_-(\mathbf{0}) = \sqrt{m} \begin{pmatrix} -1 \\ 0 \\ 1 \\ 0 \end{pmatrix} . \tag{38.6}$$

For later use we also compute the barred spinors

$$\begin{aligned} \overline{u}_s(\mathbf{p}) &\equiv u_s^\dagger(\mathbf{p})\beta , \\ \overline{v}_s(\mathbf{p}) &\equiv v_s^\dagger(\mathbf{p})\beta , \end{aligned} \tag{38.7}$$

where

$$\beta = \begin{pmatrix} 0 & I \\ I & 0 \end{pmatrix} \tag{38.8}$$

satisfies

$$\beta^{\mathrm{T}} = \beta^\dagger = \beta^{-1} = \beta . \tag{38.9}$$

We get

$$\begin{aligned} \overline{u}_+(\mathbf{0}) &= \sqrt{m}\,(1,\,0,\,1,\,0) , \\ \overline{u}_-(\mathbf{0}) &= \sqrt{m}\,(0,\,1,\,0,\,1) , \\ \overline{v}_+(\mathbf{0}) &= \sqrt{m}\,(0,\,-1,\,0,\,1) , \\ \overline{v}_-(\mathbf{0}) &= \sqrt{m}\,(1,\,0,\,-1,\,0) . \end{aligned} \tag{38.10}$$

We can now find the spinors corresponding to an arbitrary three-momentum $\mathbf{p}$ by applying to $u_s(\mathbf{0})$ and $v_s(\mathbf{0})$ the matrix $D(\Lambda)$ that corresponds to an appropriate boost. This is given by

$$D(\Lambda) = \exp(i\eta\,\hat{\mathbf{p}}\cdot\mathbf{K}) , \tag{38.11}$$

where $\hat{\mathbf{p}}$ is a unit vector in the $\mathbf{p}$ direction, $K^j = \frac{i}{4}[\gamma^j, \gamma^0] = \frac{i}{2}\gamma^j\gamma^0$ is the boost matrix, and $\eta \equiv \sinh^{-1}(|\mathbf{p}|/m)$ is the *rapidity* (see problem 2.9). Thus we have

$$\begin{aligned} u_s(\mathbf{p}) &= \exp(i\eta\,\hat{\mathbf{p}}\cdot\mathbf{K})u_s(\mathbf{0})\,, \\ v_s(\mathbf{p}) &= \exp(i\eta\,\hat{\mathbf{p}}\cdot\mathbf{K})v_s(\mathbf{0})\,. \end{aligned} \tag{38.12}$$

We also have

$$\begin{aligned} \overline{u}_s(\mathbf{p}) &= \overline{u}_s(\mathbf{0})\exp(-i\eta\,\hat{\mathbf{p}}\cdot\mathbf{K})\,, \\ \overline{v}_s(\mathbf{p}) &= \overline{v}_s(\mathbf{0})\exp(-i\eta\,\hat{\mathbf{p}}\cdot\mathbf{K})\,. \end{aligned} \tag{38.13}$$

This follows from $\overline{K^j} = K^j$, where, for any general combination of gamma matrices,

$$\overline{A} \equiv \beta A^\dagger \beta\,. \tag{38.14}$$

In particular, it turns out that

$$\begin{aligned} \overline{\gamma^\mu} &= \gamma^\mu\,, \\ \overline{S^{\mu\nu}} &= S^{\mu\nu}\,, \\ \overline{i\gamma_5} &= i\gamma_5\,, \\ \overline{\gamma^\mu\gamma_5} &= \gamma^\mu\gamma_5\,, \\ \overline{i\gamma_5 S^{\mu\nu}} &= i\gamma_5 S^{\mu\nu}\,. \end{aligned} \tag{38.15}$$

The barred spinors satisfy the equations

$$\begin{aligned} \overline{u}_s(\mathbf{p})(\not{p} + m) &= 0\,, \\ \overline{v}_s(\mathbf{p})(-\not{p} + m) &= 0\,. \end{aligned} \tag{38.16}$$

It is not very hard to work out $u_s(\mathbf{p})$ and $v_s(\mathbf{p})$ from eq. (38.12), but it is even easier to use various tricks that will sidestep any need for the explicit formulae. Consider, for example, $\overline{u}_{s'}(\mathbf{p})u_s(\mathbf{p})$; from eqs. (38.12) and (38.13), we see that $\overline{u}_{s'}(\mathbf{p})u_s(\mathbf{p}) = \overline{u}_{s'}(\mathbf{0})u_s(\mathbf{0})$, and this is easy to compute from eqs. (38.6) and (38.10). We find

$$\begin{aligned} \overline{u}_{s'}(\mathbf{p})u_s(\mathbf{p}) &= +2m\,\delta_{s's}\,, \\ \overline{v}_{s'}(\mathbf{p})v_s(\mathbf{p}) &= -2m\,\delta_{s's}\,, \\ \overline{u}_{s'}(\mathbf{p})v_s(\mathbf{p}) &= 0\,, \\ \overline{v}_{s'}(\mathbf{p})u_s(\mathbf{p}) &= 0\,. \end{aligned} \tag{38.17}$$

Also useful are the *Gordon identities*,

$$
\begin{aligned}
2m\,\overline{u}_{s'}(\mathbf{p}')\gamma^\mu u_s(\mathbf{p}) &= \overline{u}_{s'}(\mathbf{p}')\Big[(p'+p)^\mu - 2iS^{\mu\nu}(p'-p)_\nu\Big]u_s(\mathbf{p})\ , \\
-2m\,\overline{v}_{s'}(\mathbf{p}')\gamma^\mu v_s(\mathbf{p}) &= \overline{v}_{s'}(\mathbf{p}')\Big[(p'+p)^\mu - 2iS^{\mu\nu}(p'-p)_\nu\Big]v_s(\mathbf{p})\ . \qquad (38.18)
\end{aligned}
$$

To derive them, start with

$$\gamma^\mu \not{p} = \tfrac{1}{2}\{\gamma^\mu, \not{p}\} + \tfrac{1}{2}[\gamma^\mu, \not{p}\,] = -p^\mu - 2iS^{\mu\nu}p_\nu\ , \qquad (38.19)$$

$$\not{p}'\gamma^\mu = \tfrac{1}{2}\{\gamma^\mu, \not{p}'\} - \tfrac{1}{2}[\gamma^\mu, \not{p}'] = -p'^\mu + 2iS^{\mu\nu}p'_\nu\ . \qquad (38.20)$$

Add eqs. (38.19) and (38.20), sandwich them between $\overline{u}'$ and u spinors (or $\overline{v}'$ and v spinors), and use eqs. (38.1) and (38.16). An important special case is $p' = p$; then, using eq. (38.17), we find

$$
\begin{aligned}
\overline{u}_{s'}(\mathbf{p})\gamma^\mu u_s(\mathbf{p}) &= 2p^\mu \delta_{s's}\ , \\
\overline{v}_{s'}(\mathbf{p})\gamma^\mu v_s(\mathbf{p}) &= 2p^\mu \delta_{s's}\ . \qquad (38.21)
\end{aligned}
$$

With a little more effort, we can also show

$$
\begin{aligned}
\overline{u}_{s'}(\mathbf{p})\gamma^0 v_s(-\mathbf{p}) &= 0\ , \\
\overline{v}_{s'}(\mathbf{p})\gamma^0 u_s(-\mathbf{p}) &= 0\ . \qquad (38.22)
\end{aligned}
$$

We will need eqs. (38.21) and (38.22) in the next section.

Consider now the spin sums $\sum_{s=\pm} u_s(\mathbf{p})\overline{u}_s(\mathbf{p})$ and $\sum_{s=\pm} v_s(\mathbf{p})\overline{v}_s(\mathbf{p})$, each of which is a 4×4 matrix. The sum over eigenstates of S_z should remove any memory of the spin-quantization axis, and so the result should be expressible in terms of the four-vector p^μ and various gamma matrices, with all vector indices contracted. In the rest frame, $\not{p} = -m\gamma^0$, and it is easy to check that $\sum_{s=\pm} u_s(\mathbf{0})\overline{u}_s(\mathbf{0}) = m\gamma^0 + m$ and $\sum_{s=\pm} v_s(\mathbf{0})\overline{v}_s(\mathbf{0}) = m\gamma^0 - m$. We therefore conclude that

$$
\begin{aligned}
\sum_{s=\pm} u_s(\mathbf{p})\overline{u}_s(\mathbf{p}) &= -\not{p} + m\ , \\
\sum_{s=\pm} v_s(\mathbf{p})\overline{v}_s(\mathbf{p}) &= -\not{p} - m\ . \qquad (38.23)
\end{aligned}
$$

We will make extensive use of eq. (38.23) when we calculate scattering cross sections for spin-one-half particles.

From eq. (38.23), we can get $u_+(\mathbf{p})\overline{u}_+(\mathbf{p})$, etc., by applying appropriate spin projection matrices. In the rest frame, we have

$$
\begin{aligned}
\tfrac{1}{2}(1 + 2sS_z)u_{s'}(\mathbf{0}) &= \delta_{ss'}\,u_{s'}(\mathbf{0})\ , \\
\tfrac{1}{2}(1 - 2sS_z)v_{s'}(\mathbf{0}) &= \delta_{ss'}\,v_{s'}(\mathbf{0})\ . \qquad (38.24)
\end{aligned}
$$

In order to boost these projection matrices to a more general frame, we first recall that

$$\gamma_5 \equiv i\gamma^0\gamma^1\gamma^2\gamma^3 = \begin{pmatrix} -I & 0 \\ 0 & I \end{pmatrix} . \tag{38.25}$$

This allows us to write $S_z = \frac{i}{2}\gamma^1\gamma^2$ as $S_z = -\frac{1}{2}\gamma_5\gamma^3\gamma^0$. In the rest frame, we can write γ^0 as $-\not{p}/m$, and γ^3 as $\not{z}$, where $z^\mu = (0, \hat{\mathbf{z}})$; thus we have

$$S_z = \tfrac{1}{2m}\gamma_5 \not{z} \not{p} \, . \tag{38.26}$$

Now we can boost S_z to any other frame simply by replacing $\not{z}$ and $\not{p}$ with their values in that frame. (Note that, in any frame, z^μ satisfies $z^2 = 1$ and $z{\cdot}p = 0$.) Boosting eq. (38.24) then yields

$$\begin{aligned} \tfrac{1}{2}(1 - s\gamma_5\not{z})u_{s'}(\mathbf{p}) &= \delta_{ss'}\, u_{s'}(\mathbf{p}) \, , \\ \tfrac{1}{2}(1 - s\gamma_5\not{z})v_{s'}(\mathbf{p}) &= \delta_{ss'}\, v_{s'}(\mathbf{p}) \, , \end{aligned} \tag{38.27}$$

where we have used eq. (38.1) to eliminate $\not{p}$. Combining eqs. (38.23) and (38.27) we get

$$\begin{aligned} u_s(\mathbf{p})\overline{u}_s(\mathbf{p}) &= \tfrac{1}{2}(1 - s\gamma_5\not{z})(-\not{p} + m) \, , \\ v_s(\mathbf{p})\overline{v}_s(\mathbf{p}) &= \tfrac{1}{2}(1 - s\gamma_5\not{z})(-\not{p} - m) \, . \end{aligned} \tag{38.28}$$

It is interesting to consider the extreme relativistic limit of this formula. Let us take the three-momentum to be in the z direction, so that it is parallel to the spin-quantization axis. The component of the spin in the direction of the three-momentum is called the *helicity*. A fermion with helicity $+1/2$ is said to be *right-handed*, and a fermion with helicity $-1/2$ is said to be *left-handed*. For rapidity η, we have

$$\begin{aligned} \tfrac{1}{m}p^\mu &= (\cosh\eta, 0, 0, \sinh\eta) \, , \\ z^\mu &= (\sinh\eta, 0, 0, \cosh\eta) \, . \end{aligned} \tag{38.29}$$

The first equation is simply the definition of η, and the second follows from $z^2 = 1$ and $p{\cdot}z = 0$ (along with the knowledge that a boost of a four-vector in the z direction does not change its x and y components). In the limit of large η, we see that

$$z^\mu = \tfrac{1}{m}p^\mu + O(e^{-\eta}) \, . \tag{38.30}$$

Hence, in eq. (38.28), we can replace $\not{z}$ with $\not{p}/m$, and then use the matrix relation $(\not{p}/m)(-\not{p} \pm m) = \mp(-\not{p} \pm m)$, which holds for $p^2 = -m^2$. For consistency, we should then also drop the m relative to $\not{p}$, since it is down by a factor of $O(e^{-\eta})$. We get

$$\begin{aligned} u_s(\mathbf{p})\overline{u}_s(\mathbf{p}) &\to \tfrac{1}{2}(1+s\gamma_5)(-\not{p}) \,, \\ v_s(\mathbf{p})\overline{v}_s(\mathbf{p}) &\to \tfrac{1}{2}(1-s\gamma_5)(-\not{p}) \,. \end{aligned} \tag{38.31}$$

The spinor corresponding to a right-handed fermion (helicity $+1/2$) is $u_+(\mathbf{p})$ for a b-type particle and $v_-(\mathbf{p})$ for a d-type particle. According to eq. (38.31), either of these is projected by $\frac{1}{2}(1+\gamma_5) = \text{diag}(0,0,1,1)$ onto the lower two components only. In terms of the Dirac field $\Psi(x)$, this is the part that corresponds to the right-handed Weyl field. Similarly, left-handed fermions are projected (in the extreme relativistic limit) onto the upper two spinor components only, corresponding to the left-handed Weyl field.

The case of a massless particle follows from the extreme relativistic limit of a massive particle. In particular, eqs. (38.1), (38.16), (38.17), and (38.21)–(38.23) are all valid with $m = 0$, and eq. (38.31) becomes exact.

Finally, for our discussion of parity, time reversal, and charge conjugation in section 40, we will need a number of relationships among the u and v spinors. First, note that $\beta u_s(\mathbf{0}) = +u_s(\mathbf{0})$ and $\beta v_s(\mathbf{0}) = -v_s(\mathbf{0})$. Also, $\beta K^j = -K^j\beta$. We then have

$$\begin{aligned} u_s(-\mathbf{p}) &= +\beta u_s(\mathbf{p}) \,, \\ v_s(-\mathbf{p}) &= -\beta v_s(\mathbf{p}) \,. \end{aligned} \tag{38.32}$$

Next, we need the charge conjugation matrix

$$\mathcal{C} = \begin{pmatrix} 0 & -1 & 0 & 0 \\ +1 & 0 & 0 & 0 \\ 0 & 0 & 0 & +1 \\ 0 & 0 & -1 & 0 \end{pmatrix} , \tag{38.33}$$

which obeys

$$\mathcal{C}^{\mathrm{T}} = \mathcal{C}^{\dagger} = \mathcal{C}^{-1} = -\mathcal{C} \,, \tag{38.34}$$

$$\beta\mathcal{C} = -\mathcal{C}\beta \,, \tag{38.35}$$

$$\mathcal{C}^{-1}\gamma^{\mu}\mathcal{C} = -(\gamma^{\mu})^{\mathrm{T}} \,. \tag{38.36}$$

Using eqs. (38.6), (38.10), and (38.33), we can show that $\mathcal{C}\overline{u}_s(\mathbf{0})^{\mathrm{T}} = v_s(\mathbf{0})$ and $\mathcal{C}\overline{v}_s(\mathbf{0})^{\mathrm{T}} = u_s(\mathbf{0})$. Also, eq. (38.36) implies $\mathcal{C}^{-1}K^j\mathcal{C} = -(K^j)^{\mathrm{T}}$. From this we can conclude that

$$\begin{aligned} \mathcal{C}\overline{u}_s(\mathbf{p})^{\mathrm{T}} &= v_s(\mathbf{p}) \,, \\ \mathcal{C}\overline{v}_s(\mathbf{p})^{\mathrm{T}} &= u_s(\mathbf{p}) \,. \end{aligned} \tag{38.37}$$

Taking the complex conjugate of eq. (38.37), and using $\overline{u}^{\mathrm{T}*} = \overline{u}^{\dagger} = \beta u$, we get

$$\begin{aligned} u_s^*(\mathbf{p}) &= \mathcal{C}\beta v_s(\mathbf{p}) \,, \\ v_s^*(\mathbf{p}) &= \mathcal{C}\beta u_s(\mathbf{p}) \,. \end{aligned} \tag{38.38}$$

Next, note that $\gamma_5 u_s(\mathbf{0}) = +s\, v_{-s}(\mathbf{0})$ and $\gamma_5 v_s(\mathbf{0}) = -s\, u_{-s}(\mathbf{0})$, and that $\gamma_5 K^j = K^j \gamma_5$. Therefore

$$\begin{aligned} \gamma_5 u_s(\mathbf{p}) &= +s\, v_{-s}(\mathbf{p}) \,, \\ \gamma_5 v_s(\mathbf{p}) &= -s\, u_{-s}(\mathbf{p}) \,. \end{aligned} \tag{38.39}$$

Combining eqs. (38.32), (38.38), and (38.39) results in

$$\begin{aligned} u_{-s}^*(-\mathbf{p}) &= -s\,\mathcal{C}\gamma_5 u_s(\mathbf{p}) \,, \\ v_{-s}^*(-\mathbf{p}) &= -s\,\mathcal{C}\gamma_5 v_s(\mathbf{p}) \,. \end{aligned} \tag{38.40}$$

We will need eq. (38.32) in our discussion of parity, eq. (38.37) in our discussion of charge conjugation, and eq. (38.40) in our discussion of time reversal.

Problems

38.1) Use eq. (38.12) to compute $u_s(\mathbf{p})$ and $v_s(\mathbf{p})$ explicity. Hint: show that the matrix $2i\hat{\mathbf{p}}\cdot\mathbf{K}$ has eigenvalues ± 1, and that, for any matrix A with eigenvalues ± 1, $e^{cA} = (\cosh c) + (\sinh c)A$, where c is an arbitrary complex number.

38.2) Verify eq. (38.15).

38.3) Verify eq. (38.22).

38.4) Derive the Gordon identities

$$\begin{aligned} \overline{u}_{s'}(\mathbf{p}')\Big[(p'+p)^{\mu} - 2iS^{\mu\nu}(p'-p)_{\nu}\Big]\gamma_5 u_s(\mathbf{p}) &= 0 \,, \\ \overline{v}_{s'}(\mathbf{p}')\Big[(p'+p)^{\mu} - 2iS^{\mu\nu}(p'-p)_{\nu}\Big]\gamma_5 v_s(\mathbf{p}) &= 0 \,. \end{aligned} \tag{38.41}$$

39

Canonical quantization of spinor fields II

Prerequisite: 38

A Dirac field Ψ with lagrangian

$$\mathcal{L} = i\overline{\Psi}\partial\!\!\!/\Psi - m\overline{\Psi}\Psi \tag{39.1}$$

obeys the canonical anticommutation relations

$$\{\Psi_\alpha(\mathbf{x},t), \Psi_\beta(\mathbf{y},t)\} = 0 \ , \tag{39.2}$$

$$\{\Psi_\alpha(\mathbf{x},t), \overline{\Psi}_\beta(\mathbf{y},t)\} = (\gamma^0)_{\alpha\beta}\,\delta^3(\mathbf{x}-\mathbf{y}) \ , \tag{39.3}$$

and has the Dirac equation

$$(-i\partial\!\!\!/ + m)\Psi = 0 \tag{39.4}$$

as its equation of motion. The general solution is

$$\Psi(x) = \sum_{s=\pm} \int \widetilde{dp}\ \Big[b_s(\mathbf{p})u_s(\mathbf{p})e^{ipx} + d_s^\dagger(\mathbf{p})v_s(\mathbf{p})e^{-ipx}\Big] \ , \tag{39.5}$$

where $b_s(\mathbf{p})$ and $d_s^\dagger(\mathbf{p})$ are operators; the properties of the four-component spinors $u_s(\mathbf{p})$ and $v_s(\mathbf{p})$ were belabored in the previous section.

Let us express $b_s(\mathbf{p})$ and $d_s^\dagger(\mathbf{p})$ in terms of $\Psi(x)$ and $\overline{\Psi}(x)$. We begin with

$$\int d^3x\ e^{-ipx}\Psi(x) = \sum_{s'=\pm}\Big[\tfrac{1}{2\omega}b_{s'}(\mathbf{p})u_{s'}(\mathbf{p}) + \tfrac{1}{2\omega}e^{2i\omega t}d_{s'}^\dagger(-\mathbf{p})v_{s'}(-\mathbf{p})\Big] . \tag{39.6}$$

Next, multiply on the left by $\overline{u}_s(\mathbf{p})\gamma^0$, and use $\overline{u}_s(\mathbf{p})\gamma^0 u_{s'}(\mathbf{p}) = 2\omega\delta_{ss'}$ and $\overline{u}_s(\mathbf{p})\gamma^0 v_{s'}(-\mathbf{p}) = 0$ from section 38. The result is

$$b_s(\mathbf{p}) = \int d^3x\ e^{-ipx}\,\overline{u}_s(\mathbf{p})\gamma^0\Psi(x) \ . \tag{39.7}$$

Note that $b_s(\mathbf{p})$ is time independent.

To get $b_s^\dagger(\mathbf{p})$, take the hermitian conjugate of eq. (39.7), using

$$\begin{aligned}\Big[\overline{u}_s(\mathbf{p})\gamma^0\Psi(x)\Big]^\dagger &= \overline{\overline{u}_s(\mathbf{p})\gamma^0\Psi(x)} \\ &= \overline{\Psi}(x)\overline{\gamma^0}u_s(\mathbf{p}) \\ &= \overline{\Psi}(x)\gamma^0 u_s(\mathbf{p})\,, \end{aligned} \tag{39.8}$$

where, for any general combination of gamma matrices A,

$$\overline{A} \equiv \beta A^\dagger \beta\,. \tag{39.9}$$

Thus we find

$$b_s^\dagger(\mathbf{p}) = \int d^3x\; e^{ipx}\,\overline{\Psi}(x)\gamma^0 u_s(\mathbf{p})\,. \tag{39.10}$$

To extract $d_s^\dagger(\mathbf{p})$ from $\Psi(x)$, we start with

$$\int d^3x\; e^{ipx}\Psi(x) = \sum_{s'=\pm}\Big[\tfrac{1}{2\omega}e^{-2i\omega t}b_{s'}(-\mathbf{p})u_{s'}(-\mathbf{p}) + \tfrac{1}{2\omega}d_{s'}^\dagger(\mathbf{p})v_{s'}(\mathbf{p})\Big]\,. \tag{39.11}$$

Next, multiply on the left by $\overline{v}_s(\mathbf{p})\gamma^0$, and use $\overline{v}_s(\mathbf{p})\gamma^0 v_{s'}(\mathbf{p}) = 2\omega\delta_{ss'}$ and $\overline{v}_s(\mathbf{p})\gamma^0 u_{s'}(-\mathbf{p}) = 0$ from section 38. The result is

$$d_s^\dagger(\mathbf{p}) = \int d^3x\; e^{ipx}\,\overline{v}_s(\mathbf{p})\gamma^0\Psi(x)\,. \tag{39.12}$$

To get $d_s(\mathbf{p})$, take the hermitian conjugate of eq. (39.12), which yields

$$d_s(\mathbf{p}) = \int d^3x\; e^{-ipx}\,\overline{\Psi}(x)\gamma^0 v_s(\mathbf{p})\,. \tag{39.13}$$

Next, let us work out the anticommutation relations of the b and d operators (and their hermitian conjugates). From eq. (39.2), it is immediately clear that

$$\begin{aligned}\{b_s(\mathbf{p}), b_{s'}(\mathbf{p}')\} &= 0\,, \\ \{d_s(\mathbf{p}), d_{s'}(\mathbf{p}')\} &= 0\,, \\ \{b_s(\mathbf{p}), d_{s'}^\dagger(\mathbf{p}')\} &= 0\,, \end{aligned} \tag{39.14}$$

because these involve only the anticommutator of Ψ with itself, and this vanishes. Of course, hermitian conjugation also yields

$$\begin{aligned}\{b_s^\dagger(\mathbf{p}), b_{s'}^\dagger(\mathbf{p}')\} &= 0\,, \\ \{d_s^\dagger(\mathbf{p}), d_{s'}^\dagger(\mathbf{p}')\} &= 0\,, \\ \{b_s^\dagger(\mathbf{p}), d_{s'}(\mathbf{p}')\} &= 0\,. \end{aligned} \tag{39.15}$$

Now consider

$$\begin{aligned}
\{b_s(\mathbf{p}), b^\dagger_{s'}(\mathbf{p}')\} &= \int d^3x\, d^3y\, e^{-ipx+ip'y}\, \overline{u}_s(\mathbf{p})\gamma^0\{\Psi(x), \overline{\Psi}(y)\}\gamma^0 u_{s'}(\mathbf{p}') \\
&= \int d^3x\, e^{-i(p-p')x}\, \overline{u}_s(\mathbf{p})\gamma^0\gamma^0\gamma^0 u_{s'}(\mathbf{p}') \\
&= (2\pi)^3\delta^3(\mathbf{p}-\mathbf{p}')\, \overline{u}_s(\mathbf{p})\gamma^0 u_{s'}(\mathbf{p}) \\
&= (2\pi)^3\delta^3(\mathbf{p}-\mathbf{p}')\, 2\omega\delta_{ss'}\ .
\end{aligned} \tag{39.16}$$

In the first line, we are free to set $x^0 = y^0$ because $b_s(\mathbf{p})$ and $b^\dagger_{s'}(\mathbf{p}')$ are actually time independent. In the third, we used $(\gamma^0)^2 = 1$, and in the fourth, $\overline{u}_s(\mathbf{p})\gamma^0 u_{s'}(\mathbf{p}) = 2\omega\delta_{ss'}$.

Similarly,

$$\begin{aligned}
\{d^\dagger_s(\mathbf{p}), d_{s'}(\mathbf{p}')\} &= \int d^3x\, d^3y\, e^{ipx-ip'y}\, \overline{v}_s(\mathbf{p})\gamma^0\{\Psi(x), \overline{\Psi}(y)\}\gamma^0 v_{s'}(\mathbf{p}') \\
&= \int d^3x\, e^{i(p-p')x}\, \overline{v}_s(\mathbf{p})\gamma^0\gamma^0\gamma^0 v_{s'}(\mathbf{p}') \\
&= (2\pi)^3\delta^3(\mathbf{p}-\mathbf{p}')\, \overline{v}_s(\mathbf{p})\gamma^0 v_{s'}(\mathbf{p}) \\
&= (2\pi)^3\delta^3(\mathbf{p}-\mathbf{p}')\, 2\omega\delta_{ss'}\ .
\end{aligned} \tag{39.17}$$

And finally,

$$\begin{aligned}
\{b_s(\mathbf{p}), d_{s'}(\mathbf{p}')\} &= \int d^3x\, d^3y\, e^{-ipx-ip'y}\, \overline{u}_s(\mathbf{p})\gamma^0\{\Psi(x), \overline{\Psi}(y)\}\gamma^0 v_{s'}(\mathbf{p}') \\
&= \int d^3x\, e^{-i(p+p')x}\, \overline{u}_s(\mathbf{p})\gamma^0\gamma^0\gamma^0 v_{s'}(\mathbf{p}') \\
&= (2\pi)^3\delta^3(\mathbf{p}+\mathbf{p}')\, \overline{u}_s(\mathbf{p})\gamma^0 v_{s'}(-\mathbf{p}) \\
&= 0\ .
\end{aligned} \tag{39.18}$$

According to the discussion in section 3, eqs. (39.14)–(39.18) are exactly what we need to describe the creation and annihilation of fermions. In this case, we have two different kinds: b-type and d-type, each with two possible spin states, $s = +$ and $s = -$.

Next, let us evaluate the hamiltonian

$$H = \int d^3x\, \overline{\Psi}(-i\gamma^i\partial_i + m)\Psi \tag{39.19}$$

in terms of the b and d operators. We have

$$
\begin{aligned}
(-i\gamma^i\partial_i + m)\Psi &= \sum_{s=\pm}\int \widetilde{dp}\,\Big(-i\gamma^i\partial_i + m\Big)\Big(b_s(\mathbf{p})u_s(\mathbf{p})e^{ipx} \\
&\qquad\qquad + d_s^\dagger(\mathbf{p})v_s(\mathbf{p})e^{-ipx}\Big) \\
&= \sum_{s=\pm}\int \widetilde{dp}\,\Big[\,b_s(\mathbf{p})(+\gamma^i p_i + m)u_s(\mathbf{p})e^{ipx} \\
&\qquad\qquad + d_s^\dagger(\mathbf{p})(-\gamma^i p_i + m)v_s(\mathbf{p})e^{-ipx}\Big] \\
&= \sum_{s=\pm}\int \widetilde{dp}\,\Big[\,b_s(\mathbf{p})(\gamma^0\omega)u_s(\mathbf{p})e^{ipx} \\
&\qquad\qquad + d_s^\dagger(\mathbf{p})(-\gamma^0\omega)v_s(\mathbf{p})e^{-ipx}\Big]\,.
\end{aligned}
\tag{39.20}
$$

Therefore

$$
\begin{aligned}
H = \sum_{s,s'}\int \widetilde{dp}\,\widetilde{dp}'\,d^3x\,&\Big(b_{s'}^\dagger(\mathbf{p}')\overline{u}_{s'}(\mathbf{p}')e^{-ip'x} + d_{s'}(\mathbf{p}')\overline{v}_{s'}(\mathbf{p}')e^{ip'x}\Big) \\
&\times\omega\,\Big(b_s(\mathbf{p})\gamma^0 u_s(\mathbf{p})e^{ipx} - d_s^\dagger(\mathbf{p})\gamma^0 v_s(\mathbf{p})e^{-ipx}\Big) \\
= \sum_{s,s'}\int \widetilde{dp}\,\widetilde{dp}'\,d^3x\,\omega\,\Big[\,&b_{s'}^\dagger(\mathbf{p}')b_s(\mathbf{p})\,\overline{u}_{s'}(\mathbf{p}')\gamma^0 u_s(\mathbf{p})\;e^{-i(p'-p)x} \\
&- b_{s'}^\dagger(\mathbf{p}')d_s^\dagger(\mathbf{p})\,\overline{u}_{s'}(\mathbf{p}')\gamma^0 v_s(\mathbf{p})\;e^{-i(p'+p)x} \\
&+ d_{s'}(\mathbf{p}')b_s(\mathbf{p})\,\overline{v}_{s'}(\mathbf{p}')\gamma^0 u_s(\mathbf{p})\;e^{+i(p'+p)x} \\
&- d_{s'}(\mathbf{p}')d_s^\dagger(\mathbf{p})\,\overline{v}_{s'}(\mathbf{p}')\gamma^0 v_s(\mathbf{p})\;e^{+i(p'-p)x}\,\Big] \\
= \sum_{s,s'}\int \widetilde{dp}\;\tfrac{1}{2}\Big[\,&b_{s'}^\dagger(\mathbf{p})b_s(\mathbf{p})\,\overline{u}_{s'}(\mathbf{p})\gamma^0 u_s(\mathbf{p}) \\
&- b_{s'}^\dagger(-\mathbf{p})d_s^\dagger(\mathbf{p})\,\overline{u}_{s'}(-\mathbf{p})\gamma^0 v_s(\mathbf{p})\;e^{+2i\omega t} \\
&+ d_{s'}(-\mathbf{p})b_s(\mathbf{p})\,\overline{v}_{s'}(-\mathbf{p})\gamma^0 u_s(\mathbf{p})\;e^{-2i\omega t} \\
&- d_{s'}(\mathbf{p})d_s^\dagger(\mathbf{p})\,\overline{v}_{s'}(\mathbf{p})\gamma^0 v_s(\mathbf{p})\,\Big] \\
= \sum_{s}\int \widetilde{dp}\;\omega\,\Big[\,&b_s^\dagger(\mathbf{p})b_s(\mathbf{p}) - d_s(\mathbf{p})d_s^\dagger(\mathbf{p})\,\Big]\,.
\end{aligned}
\tag{39.21}
$$

Using eq. (39.17), we can rewrite this as

$$
H = \sum_{s=\pm}\int \widetilde{dp}\;\omega\,\Big[\,b_s^\dagger(\mathbf{p})b_s(\mathbf{p}) + d_s^\dagger(\mathbf{p})d_s(\mathbf{p})\Big] - 4\mathcal{E}_0 V\,,
\tag{39.22}
$$

where $\mathcal{E}_0 = \frac{1}{2}(2\pi)^{-3}\int d^3k\ \omega$ is the zero-point energy per unit volume that we found for a real scalar field in section 3, and $V = (2\pi)^3\delta^3(\mathbf{0}) = \int d^3x$ is the volume of space. That the zero-point energy is negative rather than positive is characteristic of fermions; that it is larger in magnitude by a factor of four is due to the four types of particles that are associated with a Dirac field. We can cancel off this constant energy by including a constant term $\Omega_0 = -4\mathcal{E}_0$ in the original lagrangian density; from here on, we will assume that this has been done.

The ground state of the hamiltonian (39.22) is the *vacuum state* $|0\rangle$ that is annihilated by every $b_s(\mathbf{p})$ and $d_s(\mathbf{p})$,

$$b_s(\mathbf{p})|0\rangle = d_s(\mathbf{p})|0\rangle = 0\ . \tag{39.23}$$

Then, we can interpret the $b_s^\dagger(\mathbf{p})$ operator as creating a b-type particle with momentum $\mathbf{p}$, energy $\omega = (\mathbf{p}^2+m^2)^{1/2}$, and spin $S_z = \frac{1}{2}s$, and the $d_s^\dagger(\mathbf{p})$ operator as creating a d-type particle with the same properties. The b-type and d-type particles are distinguished by the value of the charge $Q = \int d^3x\, j^0$, where $j^\mu = \overline{\Psi}\gamma^\mu\Psi$ is the Noether current associated with the invariance of $\mathcal{L}$ under the U(1) transformation $\Psi \to e^{-i\alpha}\Psi$, $\overline{\Psi} \to e^{+i\alpha}\overline{\Psi}$. Following the same procedure that we used for the hamiltonian, we can show that

$$\begin{aligned} Q &= \int d^3x\ \overline{\Psi}\gamma^0\Psi \\ &= \sum_{s=\pm}\int \widetilde{dp}\ \Big[b_s^\dagger(\mathbf{p})b_s(\mathbf{p}) + d_s(\mathbf{p})d_s^\dagger(\mathbf{p})\Big] \\ &= \sum_{s=\pm}\int \widetilde{dp}\ \Big[b_s^\dagger(\mathbf{p})b_s(\mathbf{p}) - d_s^\dagger(\mathbf{p})d_s(\mathbf{p})\Big] + \text{constant}\ . \end{aligned} \tag{39.24}$$

Thus the conserved charge Q counts the total number of b-type particles minus the total number of d-type particles. (We are free to shift the overall value of Q to remove the constant term, and so we shall.) In quantum electrodynamics, we will identify the b-type particles as electrons and the d-type particles as positrons.

Now consider a Majorana field Ψ with lagrangian

$$\mathcal{L} = \tfrac{i}{2}\Psi^{\mathrm{T}}\mathcal{C}\partial\!\!\!/\Psi - \tfrac{1}{2}m\Psi^{\mathrm{T}}\mathcal{C}\Psi\ . \tag{39.25}$$

The equation of motion for Ψ is once again the Dirac equation, and so the general solution is once again given by eq. (39.5). However, Ψ must also obey the Majorana condition $\Psi = \mathcal{C}\overline{\Psi}^{\mathrm{T}}$. Starting from the barred form of eq. (39.5),

$$\overline{\Psi}(x) = \sum_{s=\pm} \int \widetilde{dp}\, \Big[b_s^\dagger(\mathbf{p})\overline{u}_s(\mathbf{p})e^{-ipx} + d_s(\mathbf{p})\overline{v}_s(\mathbf{p})e^{ipx} \Big] , \tag{39.26}$$

we have

$$\mathcal{C}\overline{\Psi}^{\mathrm{T}}(x) = \sum_{s=\pm} \int \widetilde{dp}\, \Big[b_s^\dagger(\mathbf{p})\, \mathcal{C}\overline{u}_s^{\mathrm{T}}(\mathbf{p})e^{-ipx} + d_s(\mathbf{p})\, \mathcal{C}\overline{v}_s^{\mathrm{T}}(\mathbf{p})e^{ipx} \Big] . \tag{39.27}$$

From section 38, we have

$$\begin{aligned} \mathcal{C}\overline{u}_s(\mathbf{p})^{\mathrm{T}} &= v_s(\mathbf{p}) \ , \\ \mathcal{C}\overline{v}_s(\mathbf{p})^{\mathrm{T}} &= u_s(\mathbf{p}) \ , \end{aligned} \tag{39.28}$$

and so

$$\mathcal{C}\overline{\Psi}^{\mathrm{T}}(x) = \sum_{s=\pm} \int \widetilde{dp}\, \Big[b_s^\dagger(\mathbf{p})v_s(\mathbf{p})e^{-ipx} + d_s(\mathbf{p})u_s(\mathbf{p})e^{ipx} \Big] . \tag{39.29}$$

Comparing eqs. (39.5) and (39.29), we see that we will have $\Psi = \mathcal{C}\overline{\Psi}^{\mathrm{T}}$ if

$$d_s(\mathbf{p}) = b_s(\mathbf{p}) \ . \tag{39.30}$$

Thus a free Majorana field can be written as

$$\Psi(x) = \sum_{s=\pm} \int \widetilde{dp}\, \Big[b_s(\mathbf{p})u_s(\mathbf{p})e^{ipx} + b_s^\dagger(\mathbf{p})v_s(\mathbf{p})e^{-ipx} \Big] . \tag{39.31}$$

The anticommutation relations for a Majorana field,

$$\{\Psi_\alpha(\mathbf{x},t), \Psi_\beta(\mathbf{y},t)\} = (\mathcal{C}\gamma^0)_{\alpha\beta}\, \delta^3(\mathbf{x}-\mathbf{y}) \ , \tag{39.32}$$

$$\{\Psi_\alpha(\mathbf{x},t), \overline{\Psi}_\beta(\mathbf{y},t)\} = (\gamma^0)_{\alpha\beta}\, \delta^3(\mathbf{x}-\mathbf{y}) \ , \tag{39.33}$$

can be used to show that

$$\begin{aligned} \{b_s(\mathbf{p}), b_{s'}(\mathbf{p}')\} &= 0 \ , \\ \{b_s(\mathbf{p}), b_{s'}^\dagger(\mathbf{p}')\} &= (2\pi)^3\delta^3(\mathbf{p}-\mathbf{p}')\, 2\omega\delta_{ss'} \ , \end{aligned} \tag{39.34}$$

as we would expect.

The hamiltonian for the Majorana field Ψ is

$$\begin{aligned} H &= \tfrac{1}{2}\int d^3x\ \Psi^{\mathrm{T}}\mathcal{C}(-i\gamma^i\partial_i + m)\Psi \\ &= \tfrac{1}{2}\int d^3x\ \overline{\Psi}(-i\gamma^i\partial_i + m)\Psi \ , \end{aligned} \tag{39.35}$$

and we can work through the same manipulations that led to eq. (39.21); the only differences are an extra overall factor of one-half, and $d_s(\mathbf{p}) = b_s(\mathbf{p})$. Thus we get

$$H = \tfrac{1}{2} \sum_{s=\pm} \int \widetilde{dp}\ \omega \left[b_s^\dagger(\mathbf{p}) b_s(\mathbf{p}) - b_s(\mathbf{p}) b_s^\dagger(\mathbf{p}) \right] . \tag{39.36}$$

Note that this would reduce to a constant if we tried to use commutators rather than anticommutators in eq. (39.34), a reflection of the spin-statistics theorem. Using eq. (39.34) as it is, we find

$$H = \sum_{s=\pm} \int \widetilde{dp}\ \omega\, b_s^\dagger(\mathbf{p}) b_s(\mathbf{p}) - 2\mathcal{E}_0 V . \tag{39.37}$$

Again, we can (and will) cancel off the zero-point energy by including a term $\Omega_0 = -2\mathcal{E}_0$ in the original lagrangian density.

The Majorana lagrangian has no U(1) symmetry. Thus there is no associated charge, and only one kind of particle (with two possible spin states).

Problems

39.1) Verify eq. (39.24).

39.2) Use $[\Psi(x), M^{\mu\nu}] = -i(x^\mu \partial^\nu - x^\nu \partial^\mu)\Psi(x) + S^{\mu\nu}\Psi(x)$, plus whatever spinor identities you need, to show that

$$\begin{aligned} J_z\, b_s^\dagger(p\hat{\mathbf{z}})|0\rangle &= \tfrac{1}{2} s\, b_s^\dagger(p\hat{\mathbf{z}})|0\rangle , \\ J_z\, d_s^\dagger(p\hat{\mathbf{z}})|0\rangle &= \tfrac{1}{2} s\, d_s^\dagger(p\hat{\mathbf{z}})|0\rangle , \end{aligned} \tag{39.38}$$

where $\mathbf{p} = p\hat{\mathbf{z}}$ is the three-momentum, and $\hat{\mathbf{z}}$ is a unit vector in the z direction.

39.3) a) Show that

$$U(\Lambda)^{-1} b_s(\mathbf{p}) U(\Lambda) = \textstyle\sum_{s'} R_{ss'}(\Lambda, \mathbf{p}) b_{s'}(\Lambda^{-1}\mathbf{p}) , \tag{39.39}$$

$$U(\Lambda)^{-1} d_s(\mathbf{p}) U(\Lambda) = \textstyle\sum_{s'} \widetilde{R}_{ss'}(\Lambda, \mathbf{p}) d_{s'}(\Lambda^{-1}\mathbf{p}) , \tag{39.40}$$

and find formulae for $R_{ss'}(\Lambda, \mathbf{p})$ and $\widetilde{R}_{ss'}(\Lambda, \mathbf{p})$ that involve matrix elements of $D(\Lambda)$ between appropriate u and v spinors.

b) Show that $\widetilde{R}_{ss'}(\Lambda, \mathbf{p}) = R_{ss'}(\Lambda, \mathbf{p})$.

c) Show that

$$U(\Lambda)|p, s, q\rangle = \textstyle\sum_{s'} R^*_{ss'}(\Lambda^{-1}, \mathbf{p})|\Lambda p, s', q\rangle , \tag{39.41}$$

where

$$\begin{aligned} |p, s, +\rangle &\equiv b_s^\dagger(\mathbf{p})|0\rangle , \\ |p, s, -\rangle &\equiv d_s^\dagger(\mathbf{p})|0\rangle \end{aligned} \tag{39.42}$$

are single-particle states.

39.4) *The spin-statistics theorem for spin-one-half particles.* We will follow the proof for spin-zero particles in section 4. We start with $b_s(\mathbf{p})$ and $b^\dagger_s(\mathbf{p})$ as the fundamental objects; we take them to have either commutation (−) or anticommutation (+) relations of the form

$$\begin{aligned}
[b_s(\mathbf{p}), b_{s'}(\mathbf{p}')]_\mp &= 0 \,, \\
[b^\dagger_s(\mathbf{p}), b^\dagger_{s'}(\mathbf{p}')]_\mp &= 0 \,, \\
[b_s(\mathbf{p}), b^\dagger_{s'}(\mathbf{p}')]_\mp &= (2\pi)^3 2\omega\delta^3(\mathbf{p}-\mathbf{p}')\delta_{ss'} \,.
\end{aligned} \tag{39.43}$$

Define

$$\begin{aligned}
\Psi^+(x) &\equiv \sum_{s=\pm} \int \widetilde{dp}\, b_s(\mathbf{p}) u_s(\mathbf{p}) e^{ipx} \,, \\
\Psi^-(x) &\equiv \sum_{s=\pm} \int \widetilde{dp}\, b^\dagger_s(\mathbf{p}) v_s(\mathbf{p}) e^{-ipx} \,.
\end{aligned} \tag{39.44}$$

a) Show that $U(\Lambda)^{-1}\Psi^\pm(x)U(\Lambda) = D(\Lambda)\Psi^\pm(\Lambda^{-1}x)$.
b) Show that $[\Psi^+(x)]^\dagger = [\Psi^-(x)]^\mathrm{T}\mathcal{C}\beta$. Thus a hermitian interaction term in the lagrangian must involve both $\Psi^+(x)$ and $\Psi^-(x)$.
c) Show that $[\Psi^+_\alpha(x), \Psi^-_\beta(y)]_\mp \neq 0$ for $(x-y)^2 > 0$.
d) Show that $[\Psi^+_\alpha(x), \Psi^-_\beta(y)]_\mp = -[\Psi^+_\beta(y), \Psi^-_\alpha(x)]_\mp$ for $(x-y)^2 > 0$.
e) Consider $\Psi(x) \equiv \Psi^+(x) + \lambda\Psi^-(x)$, where λ is an arbitrary complex number, and evaluate both $[\Psi_\alpha(x), \Psi_\beta(y)]_\mp$ and $[\Psi_\alpha(x), \overline{\Psi}_\beta(y)]_\mp$ for $(x-y)^2 > 0$. Show these can both vanish if and only if $|\lambda| = 1$ and we use anticommutators.

40

Parity, time reversal, and charge conjugation

Prerequisites: 23, 39

Recall that, under a Lorentz transformation Λ implemented by the unitary operator $U(\Lambda)$, a Dirac (or Majorana) field transforms as

$$U(\Lambda)^{-1}\Psi(x)U(\Lambda) = D(\Lambda)\Psi(\Lambda^{-1}x) \ . \tag{40.1}$$

For an infinitesimal transformation $\Lambda^\mu{}_\nu = \delta^\mu{}_\nu + \delta\omega^\mu{}_\nu$, the matrix $D(\Lambda)$ is given by

$$D(1+\delta\omega) = I + \tfrac{i}{2}\delta\omega_{\mu\nu}S^{\mu\nu} \ , \tag{40.2}$$

where the Lorentz generator matrices are

$$S^{\mu\nu} = \tfrac{i}{4}[\gamma^\mu, \gamma^\nu] \ . \tag{40.3}$$

In this section, we will consider the two Lorentz transformations that cannot be reached via a sequence of infinitesimal transformations away from the identity: parity and time reversal. We begin with parity.

Define the parity transformation

$$\mathcal{P}^\mu{}_\nu = (\mathcal{P}^{-1})^\mu{}_\nu = \begin{pmatrix} +1 & & & \\ & -1 & & \\ & & -1 & \\ & & & -1 \end{pmatrix} \tag{40.4}$$

and the corresponding unitary operator

$$P \equiv U(\mathcal{P}) \ . \tag{40.5}$$

Now we have

$$P^{-1}\Psi(x)P = D(\mathcal{P})\Psi(\mathcal{P}x) \ . \tag{40.6}$$

The question we wish to answer is, what is the matrix $D(\mathcal{P})$?

First of all, if we make a second parity transformation, we get

$$P^{-2}\Psi(x)P^2 = D(\mathcal{P})^2\Psi(x) \; , \tag{40.7}$$

and it is tempting to conclude that we should have $D(\mathcal{P})^2 = 1$, so that we return to the original field. This is correct for scalar fields, since they are themselves observable. With fermions, however, it takes an even number of fields to construct an observable. Therefore we need only require the weaker condition $D(\mathcal{P})^2 = \pm 1$.

We will also require the particle creation and annihilation operators to transform in a simple way. Because

$$P^{-1}\mathbf{P}P = -\mathbf{P} \; , \tag{40.8}$$

$$P^{-1}\mathbf{J}P = +\mathbf{J} \; , \tag{40.9}$$

where $\mathbf{P}$ is the total three-momentum operator and $\mathbf{J}$ is the total angular momentum operator, a parity transformation should reverse the three-momentum while leaving the spin direction unchanged. We therefore require

$$P^{-1}b_s^\dagger(\mathbf{p})P = \eta\, b_s^\dagger(-\mathbf{p}) \; ,$$

$$P^{-1}d_s^\dagger(\mathbf{p})P = \eta\, d_s^\dagger(-\mathbf{p}) \; , \tag{40.10}$$

where η is a possible phase factor that (by the previous argument about observables) should satisfy $\eta^2 = \pm 1$. We could, in principle, assign different phase factors to the b and d operators, but we choose them to be the same so that the parity transformation is compatible with the Majorana condition $d_s(\mathbf{p}) = b_s(\mathbf{p})$. Writing the mode expansion of the free field

$$\Psi(x) = \sum_{s=\pm} \int \widetilde{dp} \left[b_s(\mathbf{p})u_s(\mathbf{p})e^{ipx} + d_s^\dagger(\mathbf{p})v_s(\mathbf{p})e^{-ipx} \right] , \tag{40.11}$$

the parity transformation reads

$$\begin{aligned} P^{-1}\Psi(x)P \\ &= \sum_{s=\pm} \int \widetilde{dp} \left[\Big(P^{-1}b_s(\mathbf{p})P\Big)u_s(\mathbf{p})e^{ipx} + \Big(P^{-1}d_s^\dagger(\mathbf{p})P\Big)v_s(\mathbf{p})e^{-ipx} \right] \\ &= \sum_{s=\pm} \int \widetilde{dp} \left[\eta^* b_s(-\mathbf{p})u_s(\mathbf{p})e^{ipx} + \eta d_s^\dagger(-\mathbf{p})v_s(\mathbf{p})e^{-ipx} \right] \\ &= \sum_{s=\pm} \int \widetilde{dp} \left[\eta^* b_s(\mathbf{p})u_s(-\mathbf{p})e^{ip\mathcal{P}x} + \eta d_s^\dagger(\mathbf{p})v_s(-\mathbf{p})e^{-ip\mathcal{P}x} \right] . \end{aligned} \tag{40.12}$$

In the last line, we have changed the integration from $\mathbf{p}$ to $-\mathbf{p}$. We now use a result from section 38, namely that

$$\begin{aligned} u_s(-\mathbf{p}) &= +\beta u_s(\mathbf{p}) \, , \\ v_s(-\mathbf{p}) &= -\beta v_s(\mathbf{p}) \, , \end{aligned} \tag{40.13}$$

where

$$\beta = \begin{pmatrix} 0 & I \\ I & 0 \end{pmatrix} . \tag{40.14}$$

Then, if we choose $\eta = -i$, eq. (40.12) becomes

$$\begin{aligned} P^{-1}\Psi(x)P &= \sum_{s=\pm} \int \widetilde{dp} \left[i b_s(\mathbf{p}) \beta u_s(\mathbf{p}) e^{ip\mathcal{P}x} + i d_s^\dagger(\mathbf{p}) \beta v_s(\mathbf{p}) e^{-ip\mathcal{P}x} \right] \\ &= i\beta\, \Psi(\mathcal{P}x) \, . \end{aligned} \tag{40.15}$$

Thus we see that $D(\mathcal{P}) = i\beta$. (We could also have chosen $\eta = i$, resulting in $D(\mathcal{P}) = -i\beta$; either choice is acceptable.)

The factor of i has an interesting physical consequence. Consider a state of an electron and positron with zero center-of-mass momentum,

$$|\phi\rangle = \int \widetilde{dp}\; \phi(\mathbf{p}) b_s^\dagger(\mathbf{p}) d_{s'}^\dagger(-\mathbf{p}) |0\rangle \; ; \tag{40.16}$$

here $\phi(\mathbf{p})$ is the momentum-space wave function. Let us assume that the vacuum is parity invariant: $P|0\rangle = P^{-1}|0\rangle = |0\rangle$. Let us also assume that the wave function has definite parity: $\phi(-\mathbf{p}) = (-)^\ell \phi(\mathbf{p})$. Then, applying the inverse parity operator on $|\phi\rangle$, we get

$$\begin{aligned} P^{-1}|\phi\rangle &= \int \widetilde{dp}\; \phi(\mathbf{p}) \Big(P^{-1} b_s^\dagger(\mathbf{p}) P \Big) \Big(P^{-1} d_{s'}^\dagger(-\mathbf{p}) P \Big) P^{-1} |0\rangle \, . \\ &= (-i)^2 \int \widetilde{dp}\; \phi(\mathbf{p}) b_s^\dagger(-\mathbf{p}) d_{s'}^\dagger(\mathbf{p}) |0\rangle \\ &= (-i)^2 \int \widetilde{dp}\; \phi(-\mathbf{p}) b_s^\dagger(\mathbf{p}) d_{s'}^\dagger(-\mathbf{p}) |0\rangle \\ &= -(-)^\ell |\phi\rangle \, . \end{aligned} \tag{40.17}$$

Thus, the parity of this state is opposite to that of its wave function; an electron-positron pair has an *intrinsic parity* of -1. This also applies to a pair of Majorana fermions. This influences the selection rules for fermion pair annihilation in theories that conserve parity. (A pair of electrons also has negative intrinsic parity, but this is less interesting because the electrons are prevented from annihilating by charge conservation.)

Let us see what eq. (40.15) implies for the two Weyl fields that comprise the Dirac field. Recalling that

$$\Psi = \begin{pmatrix} \chi_a \\ \xi^{\dagger\dot{a}} \end{pmatrix} , \tag{40.18}$$

we see from eqs. (40.14) and (40.15) that

$$\begin{aligned} P^{-1}\chi_a(x)P &= i\xi^{\dagger\dot{a}}(\mathcal{P}x) , \\ P^{-1}\xi^{\dagger\dot{a}}(x)P &= i\chi_a(\mathcal{P}x) . \end{aligned} \tag{40.19}$$

Thus a parity transformation exchanges a left-handed field for a right-handed one.

If we take the hermitian conjugate of eq. (40.19), then raise the index on one side while lowering it on the other (and remember that this introduces a relative minus sign!), we get

$$\begin{aligned} P^{-1}\chi^{\dagger\dot{a}}(x)P &= i\xi_a(\mathcal{P}x) , \\ P^{-1}\xi_a(x)P &= i\chi^{\dagger\dot{a}}(\mathcal{P}x) . \end{aligned} \tag{40.20}$$

Comparing eqs. (40.19) and (40.20), we see that they are compatible with the Majorana condition $\chi_a(x) = \xi_a(x)$.

Next we take up time reversal. Define the time-reversal transformation

$$\mathcal{T}^\mu{}_\nu = (\mathcal{T}^{-1})^\mu{}_\nu = \begin{pmatrix} -1 & & & \\ & +1 & & \\ & & +1 & \\ & & & +1 \end{pmatrix} \tag{40.21}$$

and the corresponding operator

$$T \equiv U(\mathcal{T}) . \tag{40.22}$$

Now we have

$$T^{-1}\Psi(x)T = D(\mathcal{T})\Psi(\mathcal{T}x) . \tag{40.23}$$

The question we wish to answer is, what is the matrix $D(\mathcal{T})$?

As with parity, we can conclude that $D(\mathcal{T})^2 = \pm 1$, and we will require the particle creation and annihilation operators to transform in a simple way. Because

$$T^{-1}\mathbf{P}T = -\mathbf{P} , \tag{40.24}$$

$$T^{-1}\mathbf{J}T = -\mathbf{J} , \tag{40.25}$$

where $\mathbf{P}$ is the total three-momentum operator and $\mathbf{J}$ is the total angular momentum operator, a time-reversal transformation should reverse the direction of both the three-momentum and the spin. We therefore require

$$T^{-1}b_s^\dagger(\mathbf{p})T = \zeta_s\, b_{-s}^\dagger(-\mathbf{p}) \,,$$
$$T^{-1}d_s^\dagger(\mathbf{p})T = \zeta_s\, d_{-s}^\dagger(-\mathbf{p}) \,. \tag{40.26}$$

This time we allow for possible s-dependence of the phase factor. Also, we recall from section 23 that T must be an antiunitary operator, so that $T^{-1}iT = -i$. Then we have

$$\begin{aligned} T^{-1}\Psi(x)T & \\ &= \sum_{s=\pm}\int \widetilde{dp}\left[\Big(T^{-1}b_s(\mathbf{p})T\Big)u_s^*(\mathbf{p})e^{-ipx} + \Big(T^{-1}d_s^\dagger(\mathbf{p})T\Big)v_s^*(\mathbf{p})e^{ipx}\right] \\ &= \sum_{s=\pm}\int \widetilde{dp}\left[\zeta_s^* b_{-s}(-\mathbf{p})u_s^*(\mathbf{p})e^{-ipx} + \zeta_s d_{-s}^\dagger(-\mathbf{p})v_s^*(\mathbf{p})e^{ipx}\right] \\ &= \sum_{s=\pm}\int \widetilde{dp}\left[\zeta_{-s}^* b_s(\mathbf{p})u_{-s}^*(-\mathbf{p})e^{ip\mathcal{T}x} + \zeta_{-s} d_s^\dagger(\mathbf{p})v_{-s}^*(-\mathbf{p})e^{-ip\mathcal{T}x}\right] . \end{aligned} \tag{40.27}$$

In the last line, we have changed the integration variable from $\mathbf{p}$ to $-\mathbf{p}$, and the summation variable from s to $-s$. We now use a result from section 38, namely that

$$u_{-s}^*(-\mathbf{p}) = -s\,\mathcal{C}\gamma_5 u_s(\mathbf{p}) \,,$$
$$v_{-s}^*(-\mathbf{p}) = -s\,\mathcal{C}\gamma_5 v_s(\mathbf{p}) \,. \tag{40.28}$$

Then, if we choose $\zeta_s = s$, eq. (40.27) becomes

$$T^{-1}\Psi(x)T = \mathcal{C}\gamma_5\Psi(\mathcal{T}x) \,. \tag{40.29}$$

Thus we see that $D(\mathcal{T}) = \mathcal{C}\gamma_5$. (We could also have chosen $\zeta_s = -s$, resulting in $D(\mathcal{T}) = -\mathcal{C}\gamma_5$; either choice is acceptable.)

As with parity, we can consider the effect of time reversal on the Weyl fields. Using eqs. (40.18), (40.29),

$$\mathcal{C} = \begin{pmatrix} -\varepsilon^{ab} & 0 \\ 0 & -\varepsilon_{\dot{a}\dot{b}} \end{pmatrix}, \tag{40.30}$$

and

$$\gamma_5 = \begin{pmatrix} -\delta_a{}^c & 0 \\ 0 & +\delta^{\dot{a}}{}_{\dot{c}} \end{pmatrix}, \tag{40.31}$$

we see that

$$T^{-1}\chi_a(x)T = +\chi^a(\mathcal{T}x)\,,$$

$$T^{-1}\xi^{\dagger\dot{a}}(x)T = -\xi^{\dagger}_{\dot{a}}(\mathcal{T}x)\,. \qquad (40.32)$$

Thus left-handed Weyl fields transform into left-handed Weyl fields (and right-handed into right-handed) under time reversal.

If we take the hermitian conjugate of eq. (40.32), then raise the index on one side while lowering it on the other (and remember that this introduces a relative minus sign!), we get

$$T^{-1}\chi^{\dagger\dot{a}}(x)T = -\chi^{\dagger}_{\dot{a}}(\mathcal{T}x)\,,$$

$$T^{-1}\xi_a(x)T = +\xi^a(\mathcal{T}x)\,. \qquad (40.33)$$

Comparing eqs. (40.32) and (40.33), we see that they are compatible with the Majorana condition $\chi_a(x) = \xi_a(x)$.

It is interesting and important to evaluate the transformation properties of fermion bilinears of the form $\overline{\Psi}A\Psi$, where A is some combination of gamma matrices. We will consider As that satisfy $\overline{A} = A$, where $\overline{A} \equiv \beta A^{\dagger}\beta$; in this case, $\overline{\Psi}A\Psi$ is hermitian.

Let us begin with parity transformations. From $\overline{\Psi} = \Psi^{\dagger}\beta$ and eq. (40.15) we get

$$P^{-1}\overline{\Psi}(x)P = -i\overline{\Psi}(\mathcal{P}x)\beta \qquad (40.34)$$

Combining eqs. (40.15) and (40.34) we find

$$P^{-1}\Big(\overline{\Psi}A\Psi\Big)P = \overline{\Psi}\Big(\beta A\beta\Big)\Psi\,, \qquad (40.35)$$

where we have suppressed the spacetime arguments (which transform in the obvious way). For various particular choices of A we have

$$\begin{aligned}
\beta 1\beta &= +1\,,\\
\beta i\gamma_5\beta &= -i\gamma_5\,,\\
\beta\gamma^0\beta &= +\gamma^0\,,\\
\beta\gamma^i\beta &= -\gamma^i\,,\\
\beta\gamma^0\gamma_5\beta &= -\gamma^0\gamma_5\,,\\
\beta\gamma^i\gamma_5\beta &= +\gamma^i\gamma_5\,.
\end{aligned} \qquad (40.36)$$

Therefore, the corresponding hermitian bilinears transform as

$$\begin{aligned}
P^{-1}\Big(\overline{\Psi}\Psi\Big)P &= +\,\overline{\Psi}\Psi \;, \\
P^{-1}\Big(\overline{\Psi}i\gamma_5\Psi\Big)P &= -\,\overline{\Psi}i\gamma_5\Psi \;, \\
P^{-1}\Big(\overline{\Psi}\gamma^\mu\Psi\Big)P &= +\,\mathcal{P}^\mu{}_\nu\overline{\Psi}\gamma^\nu\Psi \;, \\
P^{-1}\Big(\overline{\Psi}\gamma^\mu\gamma_5\Psi\Big)P &= -\,\mathcal{P}^\mu{}_\nu\overline{\Psi}\gamma^\nu\gamma_5\Psi \;.
\end{aligned} \tag{40.37}$$

Thus we see that $\overline{\Psi}\Psi$ and $\overline{\Psi}\gamma^\mu\Psi$ are even under a parity transformation, while $\overline{\Psi}i\gamma_5\Psi$ and $\overline{\Psi}\gamma^\mu\gamma_5\Psi$ are odd. We say that $\overline{\Psi}\Psi$ is a scalar, $\overline{\Psi}\gamma^\mu\Psi$ is a vector or *polar vector*, $\overline{\Psi}i\gamma_5\Psi$ is a pseudoscalar, and $\overline{\Psi}\gamma^\mu\gamma_5\Psi$ is a *pseudovector* or *axial vector*.

Turning to time reversal, from eq. (40.29) we get

$$T^{-1}\overline{\Psi}(x)T = \overline{\Psi}(\mathcal{T}x)\gamma_5\mathcal{C}^{-1} \;. \tag{40.38}$$

Combining eqs. (40.29) and (40.38), along with $T^{-1}AT = A^*$, we find

$$T^{-1}\Big(\overline{\Psi}A\Psi\Big)T = \overline{\Psi}\Big(\gamma_5\mathcal{C}^{-1}A^*\mathcal{C}\gamma_5\Big)\Psi \;, \tag{40.39}$$

where we have suppressed the spacetime arguments (which transform in the obvious way). Recall that $\mathcal{C}^{-1}\gamma^\mu\mathcal{C} = -(\gamma^\mu)^{\mathrm{T}}$ and that $\mathcal{C}^{-1}\gamma_5\mathcal{C} = \gamma_5$. Also, γ^0 and γ_5 are real, hermitian, and square to one, while γ^i is antihermitian. Finally, γ_5 anticommutes with γ^μ. Using all of this information, we find

$$\begin{aligned}
\gamma_5\mathcal{C}^{-1}1^*\mathcal{C}\gamma_5 &= +1 \;, \\
\gamma_5\mathcal{C}^{-1}(i\gamma_5)^*\mathcal{C}\gamma_5 &= -i\gamma_5 \;, \\
\gamma_5\mathcal{C}^{-1}(\gamma^0)^*\mathcal{C}\gamma_5 &= +\gamma^0 \;, \\
\gamma_5\mathcal{C}^{-1}(\gamma^i)^*\mathcal{C}\gamma_5 &= -\gamma^i \;, \\
\gamma_5\mathcal{C}^{-1}(\gamma^0\gamma_5)^*\mathcal{C}\gamma_5 &= +\gamma^0\gamma_5 \;, \\
\gamma_5\mathcal{C}^{-1}(\gamma^i\gamma_5)^*\mathcal{C}\gamma_5 &= -\gamma^i\gamma_5 \;.
\end{aligned} \tag{40.40}$$

Therefore,

$$\begin{aligned}
T^{-1}\Big(\overline{\Psi}\Psi\Big)T &= +\,\overline{\Psi}\Psi \;, \\
T^{-1}\Big(\overline{\Psi}i\gamma_5\Psi\Big)T &= -\,\overline{\Psi}i\gamma_5\Psi \;, \\
T^{-1}\Big(\overline{\Psi}\gamma^\mu\Psi\Big)T &= -\,\mathcal{T}^\mu{}_\nu\overline{\Psi}\gamma^\nu\Psi \;, \\
T^{-1}\Big(\overline{\Psi}\gamma^\mu\gamma_5\Psi\Big)T &= -\,\mathcal{T}^\mu{}_\nu\overline{\Psi}\gamma^\nu\gamma_5\Psi \;.
\end{aligned} \tag{40.41}$$

Thus we see that $\overline{\Psi}\Psi$ is even under time reversal, while $\overline{\Psi}i\gamma_5\Psi$, $\overline{\Psi}\gamma^\mu\Psi$, and $\overline{\Psi}\gamma^\mu\gamma_5\Psi$ are odd.

For completeness we will also consider the transformation properties of bilinears under charge conjugation. Recall that

$$C^{-1}\Psi(x)C = \mathcal{C}\overline{\Psi}^{\mathrm{T}}(x) ,$$
$$C^{-1}\overline{\Psi}(x)C = \Psi^{\mathrm{T}}(x)\mathcal{C} . \tag{40.42}$$

The bilinear $\overline{\Psi}A\Psi$ therefore transforms as

$$C^{-1}\Big(\overline{\Psi}A\Psi\Big)C = \Psi^{\mathrm{T}}\mathcal{C}A\mathcal{C}\overline{\Psi}^{\mathrm{T}} . \tag{40.43}$$

Since all indices are contracted, we can rewrite the right-hand side as its transpose, with an extra minus sign for exchanging the order of the two fermion fields. We get

$$C^{-1}\Big(\overline{\Psi}A\Psi\Big)C = -\overline{\Psi}\mathcal{C}^{\mathrm{T}}A^{\mathrm{T}}\mathcal{C}^{\mathrm{T}}\Psi . \tag{40.44}$$

Recalling that $\mathcal{C}^{\mathrm{T}} = \mathcal{C}^{-1} = -\mathcal{C}$, we have

$$C^{-1}\Big(\overline{\Psi}A\Psi\Big)C = \overline{\Psi}\Big(\mathcal{C}^{-1}A^{\mathrm{T}}\mathcal{C}\Big)\Psi . \tag{40.45}$$

Once again we can go through the list:

$$\begin{aligned} \mathcal{C}^{-1}1^{\mathrm{T}}\mathcal{C} &= +1 , \\ \mathcal{C}^{-1}(i\gamma_5)^{\mathrm{T}}\mathcal{C} &= +i\gamma_5 , \\ \mathcal{C}^{-1}(\gamma^\mu)^{\mathrm{T}}\mathcal{C} &= -\gamma^\mu , \\ \mathcal{C}^{-1}(\gamma^\mu\gamma_5)^{\mathrm{T}}\mathcal{C} &= +\gamma^\mu\gamma_5 . \end{aligned} \tag{40.46}$$

Therefore,

$$\begin{aligned} C^{-1}\Big(\overline{\Psi}\Psi\Big)C &= +\overline{\Psi}\Psi , \\ C^{-1}\Big(\overline{\Psi}i\gamma_5\Psi\Big)C &= +\overline{\Psi}i\gamma_5\Psi , \\ C^{-1}\Big(\overline{\Psi}\gamma^\mu\Psi\Big)C &= -\overline{\Psi}\gamma^\mu\Psi , \\ C^{-1}\Big(\overline{\Psi}\gamma^\mu\gamma_5\Psi\Big)C &= +\overline{\Psi}\gamma^\mu\gamma_5\Psi . \end{aligned} \tag{40.47}$$

Thus we see that $\overline{\Psi}\Psi$, $\overline{\Psi}i\gamma_5\Psi$, and $\overline{\Psi}\gamma^\mu\gamma_5\Psi$ are even under charge conjugation, while $\overline{\Psi}\gamma^\mu\Psi$ is odd.

For a Majorana field, we have $C^{-1}\Psi C = \Psi$ and $C^{-1}\overline{\Psi}C = \overline{\Psi}$; this implies $C^{-1}(\overline{\Psi}A\Psi)C = \overline{\Psi}A\Psi$ for any combination of gamma matrices A. Since eq. (40.47) tells that $C^{-1}(\overline{\Psi}\gamma^\mu\Psi)C = -\overline{\Psi}\gamma^\mu\Psi$ for either a Dirac or Majorana field, it must be that $\overline{\Psi}\gamma^\mu\Psi = 0$ for a Majorana field.

Let us consider the combined effects of the three transformations (C, P, and T) on the bilinears. From eqs. (40.37), (40.41), and (40.47), we have

$$\begin{aligned}
(CPT)^{-1}\Big(\overline{\Psi}\Psi\Big)CPT &= +\,\overline{\Psi}\Psi\,, \\
(CPT)^{-1}\Big(\overline{\Psi}i\gamma_5\Psi\Big)CPT &= +\,\overline{\Psi}i\gamma_5\Psi\,, \\
(CPT)^{-1}\Big(\overline{\Psi}\gamma^\mu\Psi\Big)CPT &= -\,\overline{\Psi}\gamma^\mu\Psi\,, \\
(CPT)^{-1}\Big(\overline{\Psi}\gamma^\mu\gamma_5\Psi\Big)CPT &= -\,\overline{\Psi}\gamma^\mu\gamma_5\Psi\,,
\end{aligned} \tag{40.48}$$

where we have used $\mathcal{P}^\mu{}_\nu\mathcal{T}^\nu{}_\rho = -\delta^\mu{}_\rho$. We see that $\overline{\Psi}\Psi$ and $\overline{\Psi}i\gamma_5\Psi$ are both even under CPT, while $\overline{\Psi}\gamma^\mu\Psi$ and $\overline{\Psi}\gamma^\mu\gamma_5\Psi$ are both odd. These are (it turns out) examples of a more general rule: a fermion bilinear with n vector indices (and no uncontracted spinor indices) is even (odd) under CPT if n is even (odd). This also applies if we allow derivatives acting on the fields, since each component of ∂_μ is odd under the combination PT and even under C.

For scalar and vector fields, it is always possible to choose the phase factors in the C, P, and T transformations so that, overall, they obey the same rule: a hermitian combination of fields and derivatives is even or odd depending on the total number of uncontracted vector indices. Putting this together with our result for fermion bilinears, we see that any hermitian combination of any set of fields (scalar, vector, Dirac, Majorana) and their derivatives that is a Lorentz scalar (and so carries no indices) is even under CPT. Since the lagrangian must be formed out of such combinations, we have $\mathcal{L}(x) \to \mathcal{L}(-x)$ under CPT, and so the action $S = \int d^4x\,\mathcal{L}$ is invariant. This is the CPT theorem.

Reference notes

A detailed treatment of CPT for fields of any spin is given in *Weinberg I*.

Problems

40.1) Find the transformation properties of $\overline{\Psi}S^{\mu\nu}\Psi$ and $\overline{\Psi}iS^{\mu\nu}\gamma_5\Psi$ under P, T, and C. Verify that they are both even under CPT, as claimed. Do either or both vanish if Ψ is a Majorana field?

41

LSZ reduction for spin-one-half particles

Prerequisites: 5, 39

Let us now consider how to construct appropriate initial and final states for scattering experiments. We will first consider the case of a Dirac field Ψ, and assume that its interactions respect the U(1) symmetry that gives rise to the conserved current $j^\mu = \overline{\Psi}\gamma^\mu\Psi$ and its associated charge Q.

In the free theory, we can create a state of one particle by acting on the vacuum state with a creation operator:

$$|p, s, +\rangle = b_s^\dagger(\mathbf{p})|0\rangle \,, \tag{41.1}$$

$$|p, s, -\rangle = d_s^\dagger(\mathbf{p})|0\rangle \,, \tag{41.2}$$

where the label $\pm$ on the ket indicates the value of the U(1) charge Q, and

$$b_s^\dagger(\mathbf{p}) = \int d^3x \; e^{ipx}\, \overline{\Psi}(x)\gamma^0 u_s(\mathbf{p}) \,, \tag{41.3}$$

$$d_s^\dagger(\mathbf{p}) = \int d^3x \; e^{ipx}\, \overline{v}_s(\mathbf{p})\gamma^0 \Psi(x) \,. \tag{41.4}$$

Recall that $b_s^\dagger(\mathbf{p})$ and $d_s^\dagger(\mathbf{p})$ are time independent in the free theory. The states $|p, s, \pm\rangle$ have the Lorentz-invariant normalization

$$\langle p, s, q|p', s', q'\rangle = (2\pi)^3\, 2\omega\, \delta^3(\mathbf{p} - \mathbf{p}')\, \delta_{ss'}\, \delta_{qq'} \,, \tag{41.5}$$

where $\omega = (\mathbf{p}^2 + m^2)^{1/2}$.

Let us consider an operator that (in the free theory) creates a particle with definite spin and charge, localized in momentum space near $\mathbf{p}_1$, and localized in position space near the origin:

$$b_1^\dagger \equiv \int d^3p \; f_1(\mathbf{p}) b_{s_1}^\dagger(\mathbf{p}) \,, \tag{41.6}$$

where

$$f_1(\mathbf{p}) \propto \exp[-(\mathbf{p}-\mathbf{p}_1)^2/4\sigma^2] \tag{41.7}$$

is an appropriate wave packet, and σ is its width in momentum space. If we time evolve (in the Schrödinger picture) the state created by this time-independent operator, then the wave packet will propagate (and spread out). The particle will thus be localized far from the origin as $t \to \pm\infty$. If we consider instead an initial state of the form $|i\rangle = b_1^\dagger b_2^\dagger |0\rangle$, where $\mathbf{p}_1 \neq \mathbf{p}_2$, then we have two particles that are widely separated in the far past.

Let us guess that this still works in the interacting theory. One complication is that $b_s^\dagger(\mathbf{p})$ will no longer be time independent, and so $b_1^\dagger$, eq. (41.6), becomes time dependent as well. Our guess for a suitable initial state for a scattering experiment is then

$$|i\rangle = \lim_{t\to-\infty} b_1^\dagger(t) b_2^\dagger(t)|0\rangle\ . \tag{41.8}$$

By appropriately normalizing the wave packets, we can make $\langle i|i\rangle = 1$, and we will assume that this is the case. Similarly, we can consider a final state

$$|f\rangle = \lim_{t\to+\infty} b_{1'}^\dagger(t) b_{2'}^\dagger(t)|0\rangle\ , \tag{41.9}$$

where $\mathbf{p}_1' \neq \mathbf{p}_2'$, and $\langle f|f\rangle = 1$. This describes two widely separated particles in the far future. (We could also consider acting with more creation operators, if we are interested in the production of some extra particles in the collision of two particles, or using $d^\dagger$ operators instead of $b^\dagger$ operators for some or all of the initial and final particles.) Now the scattering amplitude is simply given by $\langle f|i\rangle$.

We need to find a more useful expression for $\langle f|i\rangle$. To this end, let us note that

$$\begin{aligned}
&b_1^\dagger(-\infty) - b_1^\dagger(+\infty)\\
&\quad = -\int_{-\infty}^{+\infty} dt\ \partial_0 b_1^\dagger(t)\\
&\quad = -\int d^3p\ f_1(\mathbf{p}) \int d^4x\ \partial_0\Big(e^{ipx}\,\overline{\Psi}(x)\gamma^0 u_{s_1}(\mathbf{p})\Big)\ .\\
&\quad = -\int d^3p\ f_1(\mathbf{p}) \int d^4x\ \overline{\Psi}(x)\Big(\gamma^0\overleftarrow{\partial}_0 - i\gamma^0 p^0\Big)u_{s_1}(\mathbf{p})e^{ipx}\\
&\quad = -\int d^3p\ f_1(\mathbf{p}) \int d^4x\ \overline{\Psi}(x)\Big(\gamma^0\overleftarrow{\partial}_0 - i\gamma^i p_i - im\Big)u_{s_1}(\mathbf{p})e^{ipx}\\
&\quad = -\int d^3p\ f_1(\mathbf{p}) \int d^4x\ \overline{\Psi}(x)\Big(\gamma^0\overleftarrow{\partial}_0 - \gamma^i \overrightarrow{\partial}_i - im\Big)u_{s_1}(\mathbf{p})e^{ipx}
\end{aligned}$$

$$= -\int d^3p\, f_1(\mathbf{p}) \int d^4x\, \overline{\Psi}(x)\Big(\gamma^0\overleftarrow{\partial}_0 + \gamma^i\overleftarrow{\partial}_i - im\Big) u_{s_1}(\mathbf{p}) e^{ipx}$$

$$= i\int d^3p\, f_1(\mathbf{p}) \int d^4x\, \overline{\Psi}(x)(+i\overleftarrow{\not\partial} + m) u_{s_1}(\mathbf{p}) e^{ipx} \; . \qquad (41.10)$$

The first equality is just the fundamental theorem of calculus. To obtain the second, we substituted the definition of $b_1^\dagger(t)$, and combined the d^3x from this definition with the dt to get d^4x. The third comes from straightforward evaluation of the time derivatives. The fourth uses $(\not p + m)u_s(\mathbf{p}) = 0$. The fifth writes ip_i as ∂_i acting on e^{ipx}. The sixth uses integration by parts to move the ∂_i onto the field $\overline{\Psi}(x)$; here the wave packet is needed to avoid a surface term. The seventh simply identifies $\gamma^0\partial_0 + \gamma^i\partial_i$ as $\not\partial$.

In free-field theory, the right-hand side of eq. (41.10) is zero, since $\Psi(x)$ obeys the Dirac equation, which, after barring it, reads

$$\overline{\Psi}(x)(+i\overleftarrow{\not\partial} + m) = 0 \; . \qquad (41.11)$$

In an interacting theory, however, the right-hand side of eq. (41.10) will not be zero.

We will also need the hermitian conjugate of eq. (41.10), which (after some slight rearranging) reads

$$b_1(+\infty) - b_1(-\infty)$$
$$= i\int d^3p\, f_1(\mathbf{p}) \int d^4x\, e^{-ipx}\, \overline{u}_{s_1}(\mathbf{p})(-i\not\partial + m)\Psi(x) \; , \qquad (41.12)$$

and the analogous formulae for the d operators,

$$d_1^\dagger(-\infty) - d_1^\dagger(+\infty)$$
$$= -i\int d^3p\, f_1(\mathbf{p}) \int d^4x\, e^{ipx}\, \overline{v}_{s_1}(\mathbf{p})(-i\not\partial + m)\Psi(x) \; , \qquad (41.13)$$

$$d_1(+\infty) - d_1(-\infty)$$
$$= -i\int d^3p\, f_1(\mathbf{p}) \int d^4x\, \overline{\Psi}(x)(+i\overleftarrow{\not\partial} + m) v_{s_1}(\mathbf{p}) e^{-ipx} \; . \qquad (41.14)$$

Let us now return to the scattering amplitude we were considering,

$$\langle f|i\rangle = \langle 0|b_{2'}(+\infty)b_{1'}(+\infty)b_1^\dagger(-\infty)b_2^\dagger(-\infty)|0\rangle \; . \qquad (41.15)$$

Note that the operators are in time order. Thus, if we feel like it, we can put in a *time-ordering symbol* without changing anything:

$$\langle f|i\rangle = \langle 0|\,\mathrm{T}\, b_{2'}(+\infty)b_{1'}(+\infty)b_1^\dagger(-\infty)b_2^\dagger(-\infty)|0\rangle \; . \qquad (41.16)$$

The symbol T means the product of operators to its right is to be ordered, not as written, but with operators at later times to the left of those at earlier times. However, *there is an extra minus sign if this rearrangement involves an odd number of exchanges of these anticommuting operators.*

Now let us use eqs. (41.10) and (41.12) in eq. (41.16). The time-ordering symbol automatically moves all $b_{i'}(-\infty)$s to the right, where they annihilate $|0\rangle$. Similarly, all $b_i^\dagger(+\infty)$s move to the left, where they annihilate $\langle 0|$.

The wave packets no longer play a key role, and we can take the $\sigma \to 0$ limit in eq. (41.7), so that $f_1(\mathbf{p}) = \delta^3(\mathbf{p} - \mathbf{p}_1)$. The initial and final states now have a delta-function normalization, the multiparticle generalization of eq. (41.5). We are left with the *Lehmann–Symanzik–Zimmermann reduction formula* for spin-one-half particles,

$$\begin{aligned}
\langle f|i\rangle = i^4 \int d^4x_1\, d^4x_2\, d^4x_{1'}\, d^4x_{2'} \\
&\times e^{-ip_1'x_1'}\, [\overline{u}_{s_{1'}}(\mathbf{p}_{1'})(-i\partial\!\!\!/_{1'} + m)]_{\alpha_{1'}} \\
&\times e^{-ip_2'x_2'}\, [\overline{u}_{s_{2'}}(\mathbf{p}_{2'})(-i\partial\!\!\!/_{2'} + m)]_{\alpha_{2'}} \\
&\times \langle 0|\, \mathrm{T}\, \Psi_{\alpha_{2'}}(x_{2'})\Psi_{\alpha_{1'}}(x_{1'})\overline{\Psi}_{\alpha_1}(x_1)\overline{\Psi}_{\alpha_2}(x_2)|0\rangle \\
&\times [(+i\overleftarrow{\partial\!\!\!/}_1 + m)u_{s_1}(\mathbf{p}_1)]_{\alpha_1}\, e^{ip_1x_1} \\
&\times [(+i\overleftarrow{\partial\!\!\!/}_2 + m)u_{s_2}(\mathbf{p}_2)]_{\alpha_2}\, e^{ip_2x_2}\ .
\end{aligned} \tag{41.17}$$

The generalization of the LSZ formula to other processes should be clear; insert a time-ordering symbol, and make the following replacements:

$$b_s^\dagger(\mathbf{p})_{\text{in}} \to +i \int d^4x\, \overline{\Psi}(x)(+i\overleftarrow{\partial\!\!\!/} + m)u_s(\mathbf{p})\, e^{+ipx}\ , \tag{41.18}$$

$$b_s(\mathbf{p})_{\text{out}} \to +i \int d^4x\, e^{-ipx}\, \overline{u}_s(\mathbf{p})(-i\partial\!\!\!/ + m)\Psi(x)\ , \tag{41.19}$$

$$d_s^\dagger(\mathbf{p})_{\text{in}} \to -i \int d^4x\, e^{+ipx}\, \overline{v}_s(\mathbf{p})(-i\partial\!\!\!/ + m)\Psi(x)\ , \tag{41.20}$$

$$d_s(\mathbf{p})_{\text{out}} \to -i \int d^4x\, \overline{\Psi}(x)(+i\overleftarrow{\partial\!\!\!/} + m)v_s(\mathbf{p})\, e^{-ipx}\ , \tag{41.21}$$

where we have used the subscripts "in" and "out" to denote $t \to -\infty$ and $t \to +\infty$, respectively.

All of this holds for a Majorana field as well. In that case, $d_s(\mathbf{p}) = b_s(\mathbf{p})$, and we can use *either* eq. (41.18) *or* eq. (41.20) for the incoming particles, and *either* eq. (41.19) *or* eq. (41.21) for the outgoing particles, whichever is more convenient. The Majorana condition $\overline{\Psi} = \Psi^{\mathrm{T}} \mathcal{C}$ guarantees that the results will be equivalent.

As in the case of a scalar field, we cheated a little in our derivation of the LSZ formula, because we assumed that the creation operators of *free-field* theory would work comparably in the *interacting* theory. After performing an analysis that is entirely analogous to what we did for the scalar in section 5, we come to the same conclusion: the LSZ formula holds provided the field is properly normalized. For a Dirac field, we must require

$$\langle 0|\Psi(x)|0\rangle = 0 \,, \tag{41.22}$$

$$\langle p,s,+|\Psi(x)|0\rangle = 0 \,, \tag{41.23}$$

$$\langle p,s,-|\Psi(x)|0\rangle = v_s(\mathbf{p})e^{-ipx} \,, \tag{41.24}$$

$$\langle p,s,+|\overline{\Psi}(x)|0\rangle = \overline{u}_s(\mathbf{p})e^{-ipx} \,, \tag{41.25}$$

$$\langle p,s,-|\overline{\Psi}(x)|0\rangle = 0 \,, \tag{41.26}$$

where $\langle 0|0\rangle = 1$, and the one-particle states are normalized according to eq. (41.5).

The zeros on the right-hand sides of eqs. (41.23) and (41.26) are required by charge conservation. To see this, start with $[\Psi(x), Q] = +\Psi(x)$, take the matrix elements indicated, and use $Q|0\rangle = 0$ and $Q|p,s,\pm\rangle = \pm|p,s,\pm\rangle$.

The zero on the right-hand side of eq. (41.22) is required by Lorentz invariance. To see this, start with $[\Psi(0), M^{\mu\nu}] = S^{\mu\nu}\Psi(0)$, and take the expectation value in the vacuum state $|0\rangle$. If $|0\rangle$ is Lorentz invariant (as we will assume), then it is annihilated by the Lorentz generators $M^{\mu\nu}$, which means that we must have $S^{\mu\nu}\langle 0|\Psi(0)|0\rangle = 0$; this is possible for all μ and ν only if $\langle 0|\Psi(0)|0\rangle = 0$, which (by translation invariance) is possible only if $\langle 0|\Psi(x)|0\rangle = 0$.

The right-hand sides of eqs. (41.24) and (41.25) are similarly fixed by Lorentz invariance: only the overall scale might be different in an interacting theory. However, the LSZ formula is correctly normalized if and only if eqs. (41.24) and (41.25) hold as written. We will enforce this by rescaling (or, one might say, *renormalizing*) $\Psi(x)$ by an overall constant. This is just a change of the name of the operator of interest, and does not affect the physics. However, the rescaled $\Psi(x)$ will obey eqs. (41.24) and (41.25). (These two equations are related by charge conjugation, and so actually constitute only one condition on Ψ.)

For a Majorana field, there is no conserved charge, and we have

$$\langle 0|\Psi(x)|0\rangle = 0 \,, \tag{41.27}$$

$$\langle p,s|\Psi(x)|0\rangle = v_s(p)e^{-ipx} \,, \tag{41.28}$$

$$\langle p,s|\overline{\Psi}(x)|0\rangle = \overline{u}_s(p)e^{-ipx} \,, \tag{41.29}$$

instead of eqs. (41.22)–(41.26).

The renormalization of Ψ necessitates including appropriate Z factors in the lagrangian. Consider, for example,

$$\mathcal{L} = iZ\overline{\Psi}\partial\!\!\!/\Psi - Z_m m\overline{\Psi}\Psi - \tfrac{1}{4}Z_g g(\overline{\Psi}\Psi)^2 \,, \tag{41.30}$$

where Ψ is a Dirac field, and g is a coupling constant. We choose the three constants Z, Z_m, and Z_g so that the following three conditions are satisfied: m is the mass of a single particle; g is fixed by some appropriate scattering cross section; and eq. (41.24) is obeyed. [Eq. (41.25) then follows by charge conjugation.]

Next, we must develop the tools needed to compute the correlation functions $\langle 0|\mathrm{T}\Psi_{\alpha_{1'}}(x_{1'})\ldots\overline{\Psi}_{\alpha_1}(x_1)\ldots|0\rangle$ in an interacting quantum field theory.

Problems

41.1) Assuming that eq. (39.40) holds for the exact single-particle states, verify eqs. (41.24) and (41.25), up to overall scale.

42

The free fermion propagator

Prerequisite: 39

Consider a free Dirac field

$$\Psi(x) = \sum_{s=\pm} \int \widetilde{dp} \left[b_s(\mathbf{p}) u_s(\mathbf{p}) e^{ipx} + d_s^\dagger(\mathbf{p}) v_s(\mathbf{p}) e^{-ipx} \right] , \tag{42.1}$$

$$\overline{\Psi}(y) = \sum_{s'=\pm} \int \widetilde{dp}' \left[b_{s'}^\dagger(\mathbf{p}') \overline{u}_{s'}(\mathbf{p}') e^{-ip'y} + d_{s'}(\mathbf{p}') \overline{v}_{s'}(\mathbf{p}') e^{ip'y} \right] , \tag{42.2}$$

where

$$b_s(\mathbf{p})|0\rangle = d_s(\mathbf{p})|0\rangle = 0 , \tag{42.3}$$

and

$$\{b_s(\mathbf{p}), b_{s'}^\dagger(\mathbf{p}')\} = (2\pi)^3 \delta^3(\mathbf{p} - \mathbf{p}')\, 2\omega \delta_{ss'} , \tag{42.4}$$

$$\{d_s(\mathbf{p}), d_{s'}^\dagger(\mathbf{p}')\} = (2\pi)^3 \delta^3(\mathbf{p} - \mathbf{p}')\, 2\omega \delta_{ss'} , \tag{42.5}$$

and all the other possible anticommutators between b and d operators (and their hermitian conjugates) vanish.

We wish to compute the Feynman propagator

$$S(x - y)_{\alpha\beta} \equiv i\langle 0|\mathrm{T}\Psi_\alpha(x)\overline{\Psi}_\beta(y)|0\rangle , \tag{42.6}$$

where T denotes the time-ordered product,

$$\mathrm{T}\Psi_\alpha(x)\overline{\Psi}_\beta(y) \equiv \theta(x^0 - y^0)\Psi_\alpha(x)\overline{\Psi}_\beta(y) - \theta(y^0 - x^0)\overline{\Psi}_\beta(y)\Psi_\alpha(x) , \tag{42.7}$$

and $\theta(t)$ is the unit step function. Note the minus sign in the second term; this is because the fields anticommute at spacelike separations.

We can now compute $\langle 0|\Psi_\alpha(x)\overline{\Psi}_\beta(y)|0\rangle$ and $\langle 0|\overline{\Psi}_\beta(y)\Psi_\alpha(x)|0\rangle$ by inserting eqs. (42.1) and (42.2), and then using eqs. (42.3)–(42.5). We get

$$\begin{aligned}
&\langle 0|\Psi_\alpha(x)\overline{\Psi}_\beta(y)|0\rangle \\
&\quad= \sum_{s,s'}\int \widetilde{dp}\,\widetilde{dp}'\,e^{ipx}\,e^{-ip'y}\,u_s(\mathbf{p})_\alpha\overline{u}_{s'}(\mathbf{p}')_\beta\,\langle 0|b_s(\mathbf{p})b^\dagger_{s'}(\mathbf{p}')|0\rangle \\
&\quad= \sum_{s,s'}\int \widetilde{dp}\,\widetilde{dp}'\,e^{ipx}\,e^{-ip'y}\,u_s(\mathbf{p})_\alpha\overline{u}_{s'}(\mathbf{p}')_\beta\,(2\pi)^3\delta^3(\mathbf{p}-\mathbf{p}')\,2\omega\delta_{ss'} \\
&\quad= \sum_{s}\int \widetilde{dp}\,e^{ip(x-y)}\,u_s(\mathbf{p})_\alpha\overline{u}_s(\mathbf{p})_\beta \\
&\quad= \int \widetilde{dp}\,e^{ip(x-y)}\,(-\not{p}+m)_{\alpha\beta}\,.
\end{aligned} \tag{42.8}$$

To get the last line, we used a result from section 38. Similarly,

$$\begin{aligned}
&\langle 0|\overline{\Psi}_\beta(y)\Psi_\alpha(x)|0\rangle \\
&\quad= \sum_{s,s'}\int \widetilde{dp}\,\widetilde{dp}'\,e^{-ipx}\,e^{ip'y}\,v_s(\mathbf{p})_\alpha\overline{v}_{s'}(\mathbf{p}')_\beta\,\langle 0|d_{s'}(\mathbf{p}')d^\dagger_s(\mathbf{p})|0\rangle \\
&\quad= \sum_{s,s'}\int \widetilde{dp}\,\widetilde{dp}'\,e^{-ipx}\,e^{ip'y}\,v_s(\mathbf{p})_\alpha\overline{v}_{s'}(\mathbf{p}')_\beta\,(2\pi)^3\delta^3(\mathbf{p}-\mathbf{p}')\,2\omega\delta_{ss'} \\
&\quad= \sum_{s}\int \widetilde{dp}\,e^{-ip(x-y)}\,v_s(\mathbf{p})_\alpha\overline{v}_s(\mathbf{p})_\beta \\
&\quad= \int \widetilde{dp}\,e^{-ip(x-y)}\,(-\not{p}-m)_{\alpha\beta}\,.
\end{aligned} \tag{42.9}$$

We can combine eqs. (42.8) and (42.9) into a compact formula for the time-ordered product by means of the identity

$$\begin{aligned}
\int \frac{d^4p}{(2\pi)^4}\,\frac{e^{ip(x-y)}f(p)}{p^2+m^2-i\epsilon} &= i\theta(x^0-y^0)\int \widetilde{dp}\,e^{ip(x-y)}\,f(p) \\
&\quad+ i\theta(y^0-x^0)\int \widetilde{dp}\,e^{-ip(x-y)}\,f(-p)\,,
\end{aligned} \tag{42.10}$$

where $f(p)$ is a polynomial in p; the derivation of eq. (42.10) was sketched in section 8. We get

$$\langle 0|\mathrm{T}\Psi_\alpha(x)\overline{\Psi}_\beta(y)|0\rangle = \frac{1}{i}\int \frac{d^4p}{(2\pi)^4}\,e^{ip(x-y)}\,\frac{(-\not{p}+m)_{\alpha\beta}}{p^2+m^2-i\epsilon}\,, \tag{42.11}$$

and so

$$S(x-y)_{\alpha\beta} = \int \frac{d^4p}{(2\pi)^4}\,e^{ip(x-y)}\,\frac{(-\not{p}+m)_{\alpha\beta}}{p^2+m^2-i\epsilon}\,. \tag{42.12}$$

Note that $S(x-y)$ is a Green's function for the Dirac wave operator:

$$(-i\partial\!\!\!/_x + m)_{\alpha\beta}S(x-y)_{\beta\gamma} = \int \frac{d^4p}{(2\pi)^4}\, e^{ip(x-y)}\, \frac{(p\!\!\!/ + m)_{\alpha\beta}(-p\!\!\!/ + m)_{\beta\gamma}}{p^2+m^2-i\epsilon}$$

$$= \int \frac{d^4p}{(2\pi)^4}\, e^{ip(x-y)}\, \frac{(p^2+m^2)\delta_{\alpha\gamma}}{p^2+m^2-i\epsilon}$$

$$= \delta^4(x-y)\delta_{\alpha\gamma}\,. \tag{42.13}$$

Similarly,

$$S(x-y)_{\alpha\beta}(+i\overleftarrow{\partial\!\!\!/}_y + m)_{\beta\gamma} = \int \frac{d^4p}{(2\pi)^4}\, e^{ip(x-y)}\, \frac{(-p\!\!\!/ + m)_{\alpha\beta}(p\!\!\!/ + m)_{\beta\gamma}}{p^2+m^2-i\epsilon}$$

$$= \int \frac{d^4p}{(2\pi)^4}\, e^{ip(x-y)}\, \frac{(p^2+m^2)\delta_{\alpha\gamma}}{p^2+m^2-i\epsilon}$$

$$= \delta^4(x-y)\delta_{\alpha\gamma}\,. \tag{42.14}$$

We can also consider $\langle 0|\mathrm{T}\Psi_\alpha(x)\Psi_\beta(y)|0\rangle$ and $\langle 0|\mathrm{T}\overline{\Psi}_\alpha(x)\overline{\Psi}_\beta(y)|0\rangle$, but it is easy to see that now there is no way to pair up a b with a $b^\dagger$ or a d with a $d^\dagger$, and so

$$\langle 0|\mathrm{T}\Psi_\alpha(x)\Psi_\beta(y)|0\rangle = 0\,, \tag{42.15}$$

$$\langle 0|\mathrm{T}\overline{\Psi}_\alpha(x)\overline{\Psi}_\beta(y)|0\rangle = 0\,. \tag{42.16}$$

Next, consider a Majorana field

$$\Psi(x) = \sum_{s=\pm}\int \widetilde{dp}\,\Big[b_s(\mathbf{p})u_s(\mathbf{p})e^{ipx} + b_s^\dagger(\mathbf{p})v_s(\mathbf{p})e^{-ipx}\Big]\,, \tag{42.17}$$

$$\overline{\Psi}(y) = \sum_{s'=\pm}\int \widetilde{dp}'\,\Big[b_{s'}^\dagger(\mathbf{p}')\overline{u}_{s'}(\mathbf{p}')e^{-ip'y} + b_{s'}(\mathbf{p}')\overline{v}_{s'}(\mathbf{p}')e^{ip'y}\Big]\,. \tag{42.18}$$

It is easy to see that $\langle 0|\mathrm{T}\Psi_\alpha(x)\overline{\Psi}_\beta(y)|0\rangle$ is the same as it is in the Dirac case; the only difference in the calculation is that we would have b and $b^\dagger$ in place of d and $d^\dagger$ in the second line of eq. (42.9), and this does not change the final result. Thus,

$$i\langle 0|\mathrm{T}\Psi_\alpha(x)\overline{\Psi}_\beta(y)|0\rangle = S(x-y)_{\alpha\beta}\,, \tag{42.19}$$

where $S(x-y)$ is given by eq. (42.12).

However, eqs. (42.15) and (42.16) no longer hold for a Majorana field. Instead, the Majorana condition $\overline{\Psi} = \Psi^{\mathrm{T}}\mathcal{C}$, which can be rewritten as

$\Psi^{\mathrm{T}} = \overline{\Psi}\mathcal{C}^{-1}$, implies

$$\begin{aligned} i\langle 0|\mathrm{T}\Psi_\alpha(x)\Psi_\beta(y)|0\rangle &= i\langle 0|\mathrm{T}\Psi_\alpha(x)\overline{\Psi}_\gamma(y)|0\rangle(\mathcal{C}^{-1})_{\gamma\beta} \\ &= [S(x-y)\mathcal{C}^{-1}]_{\alpha\beta}\,. \end{aligned} \tag{42.20}$$

Similarly, using $\mathcal{C}^{\mathrm{T}} = \mathcal{C}^{-1}$, we can write the Majorana condition as $\overline{\Psi}^{\mathrm{T}} = \mathcal{C}^{-1}\Psi$, and so

$$\begin{aligned} i\langle 0|\mathrm{T}\overline{\Psi}_\alpha(x)\overline{\Psi}_\beta(y)|0\rangle &= i(\mathcal{C}^{-1})_{\alpha\gamma}\langle 0|\mathrm{T}\Psi_\gamma(x)\overline{\Psi}_\beta(y)|0\rangle \\ &= [\mathcal{C}^{-1}S(x-y)]_{\alpha\beta}\,. \end{aligned} \tag{42.21}$$

Of course, $\mathcal{C}^{-1} = -\mathcal{C}$, but it will prove more convenient to leave eqs. (42.20) and (42.21) as they are.

We can also consider the vacuum expectation value of a time-ordered product of more than two fields. In the Dirac case, we must have an equal number of Ψs and $\overline{\Psi}$s to get a nonzero result; and then, the Ψs and $\overline{\Psi}$s must pair up to form propagators. There is an extra minus sign if the ordering of the fields in their pairs is an odd permutation of the original ordering. For example,

$$\begin{aligned} i^2\langle 0|\mathrm{T}\Psi_\alpha(x)\overline{\Psi}_\beta(y)\Psi_\gamma(z)\overline{\Psi}_\delta(w)|0\rangle = &+S(x-y)_{\alpha\beta}\,S(z-w)_{\gamma\delta} \\ &-S(x-w)_{\alpha\delta}\,S(z-y)_{\gamma\beta}\,. \end{aligned} \tag{42.22}$$

In the Majorana case, we may as well let all the fields be Ψs (since we can always replace a $\overline{\Psi}$ with $\Psi^{\mathrm{T}}\mathcal{C}$). Then we must pair them up in all possible ways. There is an extra minus sign if the ordering of the fields in their pairs is an odd permutation of the original ordering. For example,

$$\begin{aligned} i^2\langle 0|\mathrm{T}\Psi_\alpha(x)\Psi_\beta(y)\Psi_\gamma(z)\Psi_\delta(w)|0\rangle = &+[S(x-y)\mathcal{C}^{-1}]_{\alpha\beta}\,[S(z-w)\mathcal{C}^{-1}]_{\gamma\delta} \\ &-[S(x-z)\mathcal{C}^{-1}]_{\alpha\gamma}\,[S(y-w)\mathcal{C}^{-1}]_{\beta\delta} \\ &+[S(x-w)\mathcal{C}^{-1}]_{\alpha\delta}\,[S(y-z)\mathcal{C}^{-1}]_{\beta\gamma}\,. \end{aligned} \tag{42.23}$$

Note that the ordering within a pair does not matter, since

$$[S(x-y)\mathcal{C}^{-1}]_{\alpha\beta} = -[S(y-x)\mathcal{C}^{-1}]_{\beta\alpha}\,. \tag{42.24}$$

This follows from anticommutation of the fields and eq. (42.20).

Problems

42.1) Prove eq. (42.24) directly, using properties of the $\mathcal{C}$ matrix.

43

The path integral for fermion fields

Prerequisites: 9, 42

We would like to write down a path integral formula for the vacuum-expectation value of a time-ordered product of free Dirac or Majorana fields. Recall that for a real scalar field with

$$\begin{aligned} \mathcal{L}_0 &= -\tfrac{1}{2}\partial^\mu\varphi\partial_\mu\varphi - \tfrac{1}{2}m^2\varphi^2 \\ &= -\tfrac{1}{2}\varphi(-\partial^2 + m^2)\varphi - \tfrac{1}{2}\partial_\mu(\varphi\partial^\mu\varphi) \;, \end{aligned} \tag{43.1}$$

we have

$$\langle 0|\mathrm{T}\varphi(x_1)\ldots|0\rangle = \frac{1}{i}\frac{\delta}{\delta J(x_1)}\ldots Z_0(J)\Big|_{J=0} \;, \tag{43.2}$$

where

$$Z_0(J) = \int \mathcal{D}\varphi \, \exp\left[i\int d^4x\,(\mathcal{L}_0 + J\varphi)\right] . \tag{43.3}$$

In this formula, we use the epsilon trick (see section 6) of replacing m^2 with $m^2 - i\epsilon$ to construct the vacuum as the initial and final state. Then we get

$$Z_0(J) = \exp\left[\frac{i}{2}\int d^4x\,d^4y\,J(x)\Delta(x-y)J(y)\right] , \tag{43.4}$$

where the Feynman propagator

$$\Delta(x-y) = \int \frac{d^4k}{(2\pi)^4}\,\frac{e^{ik(x-y)}}{k^2+m^2-i\epsilon} \tag{43.5}$$

is the inverse of the Klein–Gordon wave operator:

$$(-\partial_x^2 + m^2)\Delta(x-y) = \delta^4(x-y) \,. \tag{43.6}$$

For a complex scalar field with

$$\begin{aligned}\mathcal{L}_0 &= -\partial^\mu \varphi^\dagger \partial_\mu \varphi - m^2 \varphi^\dagger \varphi \\ &= -\varphi^\dagger(-\partial^2 + m^2)\varphi - \partial_\mu(\varphi^\dagger \partial^\mu \varphi) ,\end{aligned} \tag{43.7}$$

we have instead

$$\langle 0|\mathrm{T}\varphi(x_1)\ldots\varphi^\dagger(y_1)\ldots|0\rangle = \frac{1}{i}\frac{\delta}{\delta J^\dagger(x_1)}\ldots\frac{1}{i}\frac{\delta}{\delta J(y_1)}\ldots Z_0(J^\dagger, J)\Big|_{J=J^\dagger=0} , \tag{43.8}$$

where

$$\begin{aligned}Z_0(J^\dagger, J) &= \int \mathcal{D}\varphi^\dagger \mathcal{D}\varphi \, \exp\left[i\int d^4x \, (\mathcal{L}_0 + J^\dagger\varphi + \varphi^\dagger J)\right] \\ &= \exp\left[i\int d^4x \, d^4y \, J^\dagger(x)\Delta(x-y)J(y)\right] .\end{aligned} \tag{43.9}$$

We treat J and $J^\dagger$ as independent variables when evaluating eq. (43.8).

In the case of a fermion field, we should have something similar, except that we need to account for the extra minus signs from anticommutation. For this to work out, a functional derivative with respect to an anticommuting variable must itself be treated as anticommuting. Thus if we define an anticommuting source $\eta(x)$ for a Dirac field, we can write

$$\frac{\delta}{\delta\eta(x)}\int d^4y \left[\overline{\eta}(y)\Psi(y) + \overline{\Psi}(y)\eta(y)\right] = -\overline{\Psi}(x) , \tag{43.10}$$

$$\frac{\delta}{\delta\overline{\eta}(x)}\int d^4y \left[\overline{\eta}(y)\Psi(y) + \overline{\Psi}(y)\eta(y)\right] = +\Psi(x) . \tag{43.11}$$

The minus sign in eq. (43.10) arises because the $\delta/\delta\eta$ must pass through $\overline{\Psi}$ before reaching η.

Thus, consider a free Dirac field with

$$\begin{aligned}\mathcal{L}_0 &= i\overline{\Psi}\partial\!\!\!/\Psi - m\overline{\Psi}\Psi \\ &= -\overline{\Psi}(-i\partial\!\!\!/ + m)\Psi .\end{aligned} \tag{43.12}$$

A natural guess for the appropriate path-integral formula, based on analogy with eq. (43.9), is

$$\begin{aligned}&\langle 0|\mathrm{T}\Psi_{\alpha_1}(x_1)\ldots\overline{\Psi}_{\beta_1}(y_1)\ldots|0\rangle \\ &\quad = \frac{1}{i}\frac{\delta}{\delta\overline{\eta}_{\alpha_1}(x_1)}\ \ldots\ i\frac{\delta}{\delta\eta_{\beta_1}(y_1)}\ \ldots\ Z_0(\overline{\eta}, \eta)\Big|_{\eta=\overline{\eta}=0} ,\end{aligned} \tag{43.13}$$

where

$$Z_0(\overline{\eta},\eta) = \int \mathcal{D}\Psi\,\mathcal{D}\overline{\Psi}\,\exp\left[i\int d^4x\,(\mathcal{L}_0 + \overline{\eta}\Psi + \overline{\Psi}\eta)\right]$$
$$= \exp\left[i\int d^4x\,d^4y\,\overline{\eta}(x)S(x-y)\eta(y)\right], \tag{43.14}$$

and the Feynman propagator

$$S(x-y) = \int \frac{d^4p}{(2\pi)^4}\,\frac{(-\not{p}+m)e^{ip(x-y)}}{p^2+m^2-i\epsilon} \tag{43.15}$$

is the inverse of the Dirac wave operator:

$$(-i\not{\partial}_x + m)S(x-y) = \delta^4(x-y)\;. \tag{43.16}$$

Note that each $\delta/\delta\eta$ in eq. (43.13) comes with a factor of i rather than the usual $1/i$; this reflects the extra minus sign of eq. (43.10). We treat η and $\overline{\eta}$ as independent variables when evaluating eq. (43.13). It is straightforward to check (by working out a few examples) that eqs. (43.13)–(43.16) do indeed reproduce the result of section 42 for the vacuum expectation value of a time-ordered product of Dirac fields.

This is really all we need to know. Recall that, for a complex scalar field with interactions specified by $\mathcal{L}_1(\varphi^\dagger,\varphi)$, we have

$$Z(J^\dagger,J) \propto \exp\left[i\int d^4x\,\mathcal{L}_1\!\left(\frac{1}{i}\frac{\delta}{\delta J(x)},\frac{1}{i}\frac{\delta}{\delta J^\dagger(x)}\right)\right] Z_0(J^\dagger,J)\;, \tag{43.17}$$

where the overall normalization is fixed by $Z(0,0)=1$. Thus, for a Dirac field with interactions specified by $\mathcal{L}_1(\overline{\Psi},\Psi)$, we have

$$Z(\overline{\eta},\eta) \propto \exp\left[i\int d^4x\,\mathcal{L}_1\!\left(i\frac{\delta}{\delta\eta(x)},\frac{1}{i}\frac{\delta}{\delta\overline{\eta}(x)}\right)\right] Z_0(\overline{\eta},\eta)\;, \tag{43.18}$$

where again the overall normalization is fixed by $Z(0,0)=1$. Vacuum expectation values of time-ordered products of Dirac fields in an interacting theory will now be given by eq. (43.13), but with $Z_0(\overline{\eta},\eta)$ replaced by $Z(\overline{\eta},\eta)$. Then, just as for a scalar field, this will lead to a Feynman-diagram expansion for $Z(\overline{\eta},\eta)$. There are two extra complications: we must keep track of the spinor indices, and we must keep track of the extra minus signs from anticommutation. Both tasks are straightforward; we will take them up in section 45.

Next, let us consider a Majorana field with

$$\mathcal{L}_0 = \tfrac{i}{2}\Psi^{\mathsf{T}}\mathcal{C}\not{\partial}\Psi - \tfrac{1}{2}m\Psi^{\mathsf{T}}\mathcal{C}\Psi$$
$$= -\tfrac{1}{2}\Psi^{\mathsf{T}}\mathcal{C}(-i\not{\partial}+m)\Psi\;. \tag{43.19}$$

A natural guess for the appropriate path-integral formula, based on analogy with eq. (43.2), is

$$\langle 0|\mathrm{T}\Psi_{\alpha_1}(x_1)\ldots|0\rangle = \frac{1}{i}\frac{\delta}{\delta\eta_{\alpha_1}(x_1)} \;\ldots\; Z_0(\eta)\bigg|_{\eta=0}\,, \tag{43.20}$$

where

$$\begin{aligned} Z_0(\eta) &= \int \mathcal{D}\Psi\, \exp\left[i\int d^4x\,(\mathcal{L}_0 + \eta^{\mathrm{T}}\Psi)\right] \\ &= \exp\left[-\frac{i}{2}\int d^4x\, d^4y\; \eta^{\mathrm{T}}(x)S(x-y)\mathcal{C}^{-1}\eta(y)\right] . \end{aligned} \tag{43.21}$$

The Feynman propagator $S(x-y)\mathcal{C}^{-1}$ is the inverse of the Majorana wave operator $\mathcal{C}(-i\partial\!\!\!/ + m)$:

$$\mathcal{C}(-i\partial\!\!\!/_x + m)S(x-y)\mathcal{C}^{-1} = \delta^4(x-y)\;. \tag{43.22}$$

The extra minus sign in eq. (43.21), as compared with eq. (43.14), arises because all functional derivatives in eq. (43.20) are accompanied by $1/i$, rather than half by $1/i$ and half by i, as in eq. (43.13). It is now straightforward to check (by working out a few examples) that eqs. (43.20)–(43.22) do indeed reproduce the result of section 42 for the vacuum expectation value of a time-ordered product of Majorana fields.

44

Formal development of fermionic path integrals

Prerequisite: 43

In section 43, we formally defined the fermionic path integral for a free Dirac field Ψ via

$$Z_0(\overline{\eta}, \eta) = \int \mathcal{D}\Psi\, \mathcal{D}\overline{\Psi}\; \exp\left[i \int d^4x\, \overline{\Psi}(i\partial\!\!\!/ - m)\Psi + \overline{\eta}\Psi + \overline{\Psi}\eta \right]$$

$$= \exp\left[i \int d^4x\, d^4y\; \overline{\eta}(x) S(x-y) \eta(y) \right], \tag{44.1}$$

where the Feynman propagator $S(x-y)$ is the inverse of the Dirac wave operator:

$$(-i\partial\!\!\!/_x + m) S(x-y) = \delta^4(x-y)\,. \tag{44.2}$$

We would like to find a mathematical framework that allows us to derive this formula, rather than postulating it by analogy.

Consider a set of *anticommuting numbers* or *Grassmann variables* ψ_i that obey

$$\{\psi_i, \psi_j\} = 0\,, \tag{44.3}$$

where $i = 1, \ldots, n$. Let us begin with the very simplest case of $n = 1$, and thus a single anticommuting number ψ that obeys $\psi^2 = 0$. We can define a function $f(\psi)$ of such an object via a Taylor expansion; because $\psi^2 = 0$, this expansion ends with the second term:

$$f(\psi) = a + \psi b\,. \tag{44.4}$$

The reason for writing the coefficient b to the right of the variable ψ will become clear in a moment.

Next we would like to define the derivative of $f(\psi)$ with respect to ψ. Before we can do so, we must decide if $f(\psi)$ itself is to be commuting or anticommuting; generally we will be interested in functions that are themselves commuting. In this case, a in eq. (44.4) should be treated as an ordinary commuting number, but b should be treated as an anticommuting number: $\{b,b\} = \{b,\psi\} = 0$. In this case, $f(\psi) = a + \psi b = a - b\psi$.

Now we can define two kinds of derivatives. The *left derivative* of $f(\psi)$ with respect to ψ is given by the coefficient of ψ when $f(\psi)$ is written with the ψ always on the far left:

$$\partial_\psi f(\psi) = +b \; . \tag{44.5}$$

Similarly, the *right derivative* of $f(\psi)$ with respect to ψ is given by the coefficient of ψ when $f(\psi)$ is written with the ψ always on the far right:

$$f(\psi)\overleftarrow{\partial}_\psi = -b \; . \tag{44.6}$$

Generally, when we write a derivative with respect to a Grassmann variable, we mean the left derivative. However, in section 37, when we wrote the canonical momentum for a fermionic field ψ as $\pi = \partial\mathcal{L}/\partial(\partial_0\psi)$, we actually meant the right derivative. (This is a standard, though rarely stated, convention.) Correspondingly, we wrote the hamiltonian density as $\mathcal{H} = \pi\partial_0\psi - \mathcal{L}$, with $\partial_0\psi$ to the right of π.

Finally, we would like to define a definite integral, analogous to integrating a real variable x from minus to plus infinity. The key features of such an integral over x (when it converges) are linearity,

$$\int_{-\infty}^{+\infty} dx\, cf(x) = c\int_{-\infty}^{+\infty} dx\, f(x) \; , \tag{44.7}$$

and invariance under shifts of the dependent variable x by a constant:

$$\int_{-\infty}^{+\infty} dx\, f(x+a) = \int_{-\infty}^{+\infty} dx\, f(x) \; . \tag{44.8}$$

Up to an overall numerical factor that is the same for every $f(\psi)$, the only possible nontrivial definition of $\int d\psi\, f(\psi)$ that is both linear and shift invariant is

$$\int d\psi\, f(\psi) = b \; . \tag{44.9}$$

Now let us generalize this to $n > 1$. We have

$$f(\psi) = a + \psi_i b_i + \tfrac{1}{2}\psi_{i_1}\psi_{i_2}c_{i_1 i_2} + \ldots + \tfrac{1}{n!}\psi_{i_1}\ldots\psi_{i_n}d_{i_1\ldots i_n} \; , \tag{44.10}$$

where the indices are implicitly summed. Here we have written the coefficients to the right of the variables to facilitate left-differentiation. These coefficients are completely antisymmetric on exchange of any two indices. The left derivative of $f(\psi)$ with respect to ψ_j is

$$\frac{\partial}{\partial\psi_j} f(\psi) = b_j + \psi_i c_{ji} + \ldots + \tfrac{1}{(n-1)!}\psi_{i_2} \ldots \psi_{i_n} d_{ji_2 \ldots i_n} \,. \tag{44.11}$$

Next we would like to find a linear, shift-invariant definition of the integral of $f(\psi)$. Note that the antisymmetry of the coefficients implies that

$$d_{i_1 \ldots i_n} = d\, \varepsilon_{i_1 \ldots i_n} \,, \tag{44.12}$$

where d is just a number (ordinary if f is commuting and n is even, Grassmann if f is commuting and n is odd, etc.), and $\varepsilon_{i_1 \ldots i_n}$ is the completely antisymmetric Levi-Civita symbol with $\varepsilon_{1 \ldots n} = +1$. This number d is a candidate (in fact, up to an overall numerical factor, the only candidate!) for the integral of $f(\psi)$:

$$\int d^n\psi\, f(\psi) = d \,. \tag{44.13}$$

Although eq. (44.13) really tells us everything we need to know about $\int d^n\psi$, we can, if we like, write $d^n\psi = d\psi_n \ldots d\psi_1$ (note the backwards ordering), and treat the individual differentials as anticommuting: $\{d\psi_i, d\psi_j\} = 0$, $\{d\psi_i, \psi_j\} = 0$. Then we take $\int d\psi_i = 0$ and $\int d\psi_i\, \psi_j = \delta_{ij}$ as our basic formulae, and use them to derive eq. (44.13).

Let us work out some consequences of eq. (44.13). Consider what happens if we make a linear change of variable,

$$\psi_i = J_{ij}\psi'_j \,, \tag{44.14}$$

where J_{ji} is a matrix of commuting numbers (and therefore can be written on either the left or right of ψ'_j). We now have

$$f(\psi) = a + \ldots + \tfrac{1}{n!}(J_{i_1 j_1}\psi'_{j_1}) \ldots (J_{i_n j_n}\psi'_{j_n})\varepsilon_{i_1 \ldots i_n} d \,. \tag{44.15}$$

Next we use

$$\varepsilon_{i_1 \ldots i_n} J_{i_1 j_1} \ldots J_{i_n j_n} = (\det J)\varepsilon_{j_1 \ldots j_n} \,, \tag{44.16}$$

which holds for any $n \times n$ matrix J, to get

$$f(\psi) = a + \ldots + \tfrac{1}{n!}\psi'_{i_1} \ldots \psi'_{i_n} \varepsilon_{i_1 \ldots i_n} (\det J) d \,. \tag{44.17}$$

If we now integrate $f(\psi)$ over $d^n\psi'$, eq. (44.13) tells us that the result is $(\det J)d$. Thus,

$$\int d^n\psi \, f(\psi) = (\det J)^{-1} \int d^n\psi' \, f(\psi) \,. \tag{44.18}$$

Recall that, for integrals over commuting real numbers x_i with $x_i = J_{ij}x'_j$, we have instead

$$\int d^n x \, f(x) = (\det J)^{+1} \int d^n x' \, f(x) \,. \tag{44.19}$$

Note the opposite sign on the power of the determinant.

Now consider a quadratic form $\psi^{\rm T}M\psi = \psi_i M_{ij}\psi_j$, where M is an antisymmetric matrix of commuting numbers (possibly complex). Let us evaluate the gaussian integral $\int d^n\psi \, \exp(\frac{1}{2}\psi^{\rm T}M\psi)$. For example, for $n = 2$, we have

$$M = \begin{pmatrix} 0 & +m \\ -m & 0 \end{pmatrix} , \tag{44.20}$$

and $\psi^{\rm T}M\psi = 2m\psi_1\psi_2$. Thus $\exp(\frac{1}{2}\psi^{\rm T}M\psi) = 1 + m\psi_1\psi_2$, and so

$$\int d^n\psi \, \exp(\tfrac{1}{2}\psi^{\rm T}M\psi) = m \,. \tag{44.21}$$

For larger n, we use the fact that a complex antisymmetric matrix can be brought to a block-diagonal form via

$$U^{\rm T}MU = \begin{pmatrix} 0 & +m_1 & & \\ -m_1 & 0 & & \\ & & & \ddots \end{pmatrix} , \tag{44.22}$$

where U is a unitary matrix, and each m_I is real and positive. (If n is odd there is a final row and column of all zeros; from here on, we assume n is even.) We can now let $\psi_i = U_{ij}\psi'_j$; then, dropping the primes, we have

$$\int d^n\psi \, \exp(\tfrac{1}{2}\psi^{\rm T}M\psi) = (\det U)^{-1} \prod_{I=1}^{n/2} \int d^2\psi_I \, \exp(\tfrac{1}{2}\psi^{\rm T}M_I\psi) \,, \tag{44.23}$$

where M_I represents one of the 2×2 blocks in eq. (44.22). Each of these two-dimensional integrals can be evaluated using eq. (44.21), and so

$$\int d^n\psi \, \exp(\tfrac{1}{2}\psi^{\rm T}M\psi) = (\det U)^{-1} \prod_{I=1}^{n/2} m_I \,. \tag{44.24}$$

Taking the determinant of eq. (44.22), we get

$$(\det U)^2(\det M) = \prod_{I=1}^{n/2} m_I^2 \,. \tag{44.25}$$

We can therefore rewrite the right-hand side of eq. (44.24) as

$$\int d^n\psi \, \exp(\tfrac{1}{2}\psi^{\mathrm{T}}M\psi) = (\det M)^{1/2} \,. \tag{44.26}$$

In this form, there is a sign ambiguity associated with the square root; it is resolved by eq. (44.24). However, the overall sign (more generally, any overall numerical factor) will never be of concern to us, so we can use eq. (44.26) without worrying about the correct branch of the square root.

It is instructive to compare eq. (44.26) with the corresponding gaussian integral for commuting real numbers,

$$\int d^n x \, \exp(-\tfrac{1}{2}x^{\mathrm{T}}Mx) = (2\pi)^{n/2}(\det M)^{-1/2} \,. \tag{44.27}$$

Here M is a complex symmetric matrix. Again, note the opposite sign on the power of the determinant.

Now let us introduce the notion of *complex* Grassmann variables via

$$\begin{aligned} \chi &\equiv \tfrac{1}{\sqrt{2}}(\psi_1 + i\psi_2) \,, \\ \bar{\chi} &\equiv \tfrac{1}{\sqrt{2}}(\psi_1 - i\psi_2) \,. \end{aligned} \tag{44.28}$$

We can invert this to get

$$\begin{pmatrix} \psi_1 \\ \psi_2 \end{pmatrix} = \frac{1}{\sqrt{2}} \begin{pmatrix} 1 & 1 \\ i & -i \end{pmatrix} \begin{pmatrix} \bar{\chi} \\ \chi \end{pmatrix} . \tag{44.29}$$

The determinant of this transformation matrix is $-i$, and so

$$d^2\psi = d\psi_2 d\psi_1 = (-i)^{-1} d\chi \, d\bar{\chi} \,. \tag{44.30}$$

Also, $\psi_1\psi_2 = -i\bar{\chi}\chi$. Thus we have

$$\int d\chi \, d\bar{\chi} \, \bar{\chi}\chi = (-i)(-i)^{-1} \int d\psi_2 d\psi_1 \, \psi_1\psi_2 = 1 \,. \tag{44.31}$$

Thus, if we have a function

$$f(\chi, \bar{\chi}) = a + \chi b + \bar{\chi}c + \bar{\chi}\chi d \,, \tag{44.32}$$

its integral is

$$\int d\chi \, d\bar{\chi} \, f(\chi, \bar{\chi}) = d \; . \tag{44.33}$$

In particular,

$$\int d\chi \, d\bar{\chi} \; \exp(m\bar{\chi}\chi) = m \; . \tag{44.34}$$

Let us now consider n complex Grassmann variables χ_i and their complex conjugates, $\bar{\chi}_i$. We define

$$d^n\chi \, d^n\bar{\chi} \equiv d\chi_n d\bar{\chi}_n \ldots d\chi_1 d\bar{\chi}_1 \; . \tag{44.35}$$

Then under a change of variable, $\chi_i = J_{ij}\chi'_j$ and $\bar{\chi}_i = K_{ij}\bar{\chi}'_j$, we have

$$d^n\chi \, d^n\bar{\chi} = (\det J)^{-1} (\det K)^{-1} \, d^n\chi' \, d^n\bar{\chi}' \; . \tag{44.36}$$

Note that we need not require $K_{ij} = J^*_{ij}$, because, as far as the integral is concerned, it does not matter whether or not $\bar{\chi}_i$ is the complex conjugate of χ_i.

We now have enough information to evaluate $\int d^n\chi \, d^n\bar{\chi} \; \exp(\chi^\dagger M \chi)$, where M is a general complex matrix. We make the change of variable $\chi = U\chi'$ and $\chi^\dagger = \chi'^\dagger V$, where U and V are unitary matrices with the property that VMU is diagonal with positive real entries m_i. Dropping the primes, we get

$$\begin{aligned} \int d^n\chi \, d^n\bar{\chi} \; \exp(\chi^\dagger M \chi) &= (\det U)^{-1} (\det V)^{-1} \prod_{i=1}^{n} \int d\chi_i d\bar{\chi}_i \; \exp(m_i \bar{\chi}_i \chi_i) \\ &= (\det U)^{-1} (\det V)^{-1} \prod_{i=1}^{n} m_i \\ &= \det M \; . \end{aligned} \tag{44.37}$$

This can be compared to the analogous integral for commuting complex variables $z_i = (x_i + iy_i)/\sqrt{2}$ and $\bar{z}_i = (x_i - iy_i)/\sqrt{2}$, with $d^n z \, d^n \bar{z} = d^n x \, d^n y$, namely

$$\int d^n z \, d^n \bar{z} \; \exp(-z^\dagger M z) = (2\pi)^n (\det M)^{-1} \; . \tag{44.38}$$

We can now generalize eqs. (44.26) and (44.37) by shifting the integration variables, and using shift invariance of the integrals. Thus, by making the replacement $\psi \to \psi - M^{-1}\eta$ in eq. (44.26), we get

$$\int d^n\psi \; \exp(\tfrac{1}{2}\psi^{\mathrm{T}} M \psi + \eta^{\mathrm{T}}\psi) = (\det M)^{1/2} \exp(\tfrac{1}{2}\eta^{\mathrm{T}} M^{-1} \eta) \; . \tag{44.39}$$

(In verifying this, remember that M and its inverse are both antisymmetric.) Similarly, by making the replacements $\chi \to \chi - M^{-1}\eta$ and $\chi^\dagger \to \chi^\dagger - \eta^\dagger M^{-1}$ in eq. (44.37), we get

$$\int d^n\chi\, d^n\bar{\chi}\, \exp(\chi^\dagger M \chi + \eta^\dagger \chi + \chi^\dagger \eta) = (\det M) \exp(-\eta^\dagger M^{-1} \eta)\ . \qquad (44.40)$$

We can now see that eq. (44.1) is simply a particular case of eq. (44.40), with the index on the complex Grassmann variable generalized to include both the ordinary spin index α and the continuous spacetime argument x of the field $\Psi_\alpha(x)$. Similarly, eq. (43.21) for the path integral for a free Majorana field is simply a particular case of eq. (44.39). In both cases, the determinant factors are constants (that is, independent of the fields and sources) that we simply absorb into the overall normalization of the path integral. We will meet determinants that cannot be so neatly absorbed in sections 53 and 71.

45

The Feynman rules for Dirac fields

Prerequisites: 10, 12, 41, 43

In this section we will derive the Feynman rules for *Yukawa theory*, a theory with a Dirac field Ψ (with mass m) and a real scalar field φ (with mass M), interacting via

$$\mathcal{L}_1 = g\varphi\overline{\Psi}\Psi \ , \tag{45.1}$$

where g is a coupling constant. In this section, we will be concerned with tree-level processes only, and so we omit renormalizing Z factors.

In four spacetime dimensions, φ has mass dimension $[\varphi] = 1$ and Ψ has mass dimension $[\Psi] = \frac{3}{2}$; thus the coupling constant g is dimensionless: $[g] = 0$. As discussed in section 12, this is generally the most interesting situation.

Note that $\mathcal{L}_1$ is invariant under the U(1) transformation $\Psi \to e^{-i\alpha}\Psi$, as is the free Dirac lagrangian. Thus, the corresponding Noether current $\overline{\Psi}\gamma^\mu\Psi$ is still conserved, and the associated charge Q (which counts the number of b-type particles minus the number of d-type particles) is constant in time.

We can think of Q as electric charge, and identify the b-type particle as the electron e^-, and the d-type particle as the positron e^+. The scalar particle is electrically neutral (and could, for example, be thought of as the Higgs boson; see section 88).

We now use the general result of sections 9 and 43 to write

$$Z(\overline{\eta},\eta,J) \propto \exp\left[ig\int d^4x\left(\frac{1}{i}\frac{\delta}{\delta J(x)}\right)\left(i\frac{\delta}{\delta\eta_\alpha(x)}\right)\left(\frac{1}{i}\frac{\delta}{\delta\overline{\eta}_\alpha(x)}\right)\right] Z_0(\overline{\eta},\eta,J) \ , \tag{45.2}$$

where

$$Z_0(\overline{\eta},\eta,J) = \exp\left[i\int d^4x\,d^4y\;\overline{\eta}(x)S(x-y)\eta(y)\right] \times \exp\left[\frac{i}{2}\int d^4x\,d^4y\;J(x)\Delta(x-y)J(y)\right], \tag{45.3}$$

and

$$S(x-y) = \int \frac{d^4p}{(2\pi)^4}\,\frac{(-\not{p}+m)e^{ip(x-y)}}{p^2+m^2-i\epsilon}\,, \tag{45.4}$$

$$\Delta(x-y) = \int \frac{d^4k}{(2\pi)^4}\,\frac{e^{ik(x-y)}}{k^2+M^2-i\epsilon} \tag{45.5}$$

are the appropriate Feynman propagators for the corresponding free fields. We impose the normalization $Z(0,0,0)=1$, and write

$$Z(\overline{\eta},\eta,J) = \exp[iW(\overline{\eta},\eta,J)]\,. \tag{45.6}$$

Then $iW(\overline{\eta},\eta,J)$ can be expressed as a series of connected Feynman diagrams with sources.

We use a dashed line to stand for the scalar propagator $\frac{1}{i}\Delta(x-y)$, and a solid line to stand for the fermion propagator $\frac{1}{i}S(x-y)$. The only allowed vertex joins two solid lines and one dashed line; the associated vertex factor is $ig\int d^4x$. The blob at the end of a dashed line stands for the φ source $i\int d^4x\,J(x)$, and the blob at the end of a solid line for either the Ψ source $i\int d^4x\,\overline{\eta}(x)$, or the $\overline{\Psi}$ source $i\int d^4x\,\eta(x)$. To tell which is which, we adopt the "arrow rule" of problem 9.3: the blob stands for $i\int d^4x\,\eta(x)$ if the arrow on the attached line points *away* from the blob, and the blob stands for $i\int d^4x\,\overline{\eta}(x)$ if the arrow on the attached line points *towards* the blob. Because $\mathcal{L}_1$ involves one Ψ and one $\overline{\Psi}$, we also have the rule that, at each vertex, *one arrow must point towards the vertex, and one away.* The first few tree diagrams that contribute to $iW(\overline{\eta},\eta,J)$ are shown in fig. 45.1. We omit tadpole diagrams; as in φ^3 theory, these can be canceled by shifting the φ field, or, equivalently, adding a term linear in φ to $\mathcal{L}$. The LSZ formula is valid only after all tadpole diagrams have been canceled in this way.

The spin indices on the fermionic sources and propagators are all contracted in the obvious way. For example, the complete expression corresponding to fig. 45.1(b) is

$$\text{Fig. 45.1(b)} = i^3\left(\tfrac{1}{i}\right)^3(ig)\int d^4x\,d^4y\,d^4z\,d^4w \times \left[\overline{\eta}(x)S(x-y)S(y-z)\eta(z)\right] \times \Delta(y-w)J(w)\,. \tag{45.7}$$

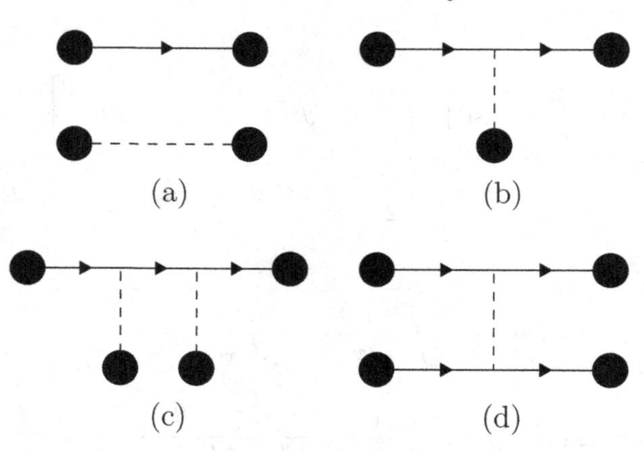

Figure 45.1. Tree contributions to $iW(\overline{\eta}, \eta, J)$ with four or fewer sources.

Our main purpose in this section is to compute the tree-level amplitudes for various two-body elastic scattering processes, such as $e^-\varphi \to e^-\varphi$ and $e^+e^- \to \varphi\varphi$; for these, we will need to evaluate the tree-level contributions to connected correlation functions of the form $\langle 0|\mathrm{T}\Psi\overline{\Psi}\varphi\varphi|0\rangle_{\mathrm{C}}$. Other processes of interest include $e^-e^- \to e^-e^-$ and $e^+e^- \to e^+e^-$; for these, we will need to evaluate the tree-level contributions to connected correlation functions of the form $\langle 0|\mathrm{T}\Psi\overline{\Psi}\Psi\overline{\Psi}|0\rangle_{\mathrm{C}}$.

For $\langle 0|\mathrm{T}\Psi\overline{\Psi}\varphi\varphi|0\rangle_{\mathrm{C}}$, the relevant tree-level contribution to $iW(\overline{\eta}, \eta, J)$ is given by fig. 45.1(c). We have

$$\begin{aligned}
&\langle 0|\mathrm{T}\Psi_\alpha(x)\overline{\Psi}_\beta(y)\varphi(z_1)\varphi(z_2)|0\rangle_{\mathrm{C}} \\
&= \frac{1}{i}\frac{\delta}{\delta\overline{\eta}_\alpha(x)}\, i\frac{\delta}{\delta\eta_\beta(y)}\,\frac{1}{i}\frac{\delta}{\delta J(z_1)}\,\frac{1}{i}\frac{\delta}{\delta J(z_2)}\, iW(\overline{\eta}, \eta, J)\bigg|_{\overline{\eta}=\eta=J=0} \\
&= \left(\tfrac{1}{i}\right)^5 (ig)^2 \int d^4w_1\, d^4w_2 \\
&\qquad \times \left[S(x{-}w_2)S(w_2{-}w_1)S(w_1{-}y)\right]_{\alpha\beta} \\
&\qquad \times \Delta(z_1{-}w_1)\Delta(z_2{-}w_2) \\
&\qquad + (z_1 \leftrightarrow z_2) + O(g^4)\,.
\end{aligned} \tag{45.8}$$

The corresponding diagrams, with sources removed, are shown in fig. 45.2.

For $\langle 0|\mathrm{T}\Psi\overline{\Psi}\Psi\overline{\Psi}|0\rangle_{\mathrm{C}}$, the relevant tree-level contribution to $iW(\overline{\eta}, \eta, J)$ is given by fig. 45.1(d), which has a symmetry factor $S = 2$. We have

$$\begin{aligned}
&\langle 0|\mathrm{T}\Psi_{\alpha_1}(x_1)\overline{\Psi}_{\beta_1}(y_1)\Psi_{\alpha_2}(x_2)\overline{\Psi}_{\beta_2}(y_2)|0\rangle_{\mathrm{C}} \\
&= \frac{1}{i}\frac{\delta}{\delta\overline{\eta}_{\alpha_1}(x_1)}\, i\frac{\delta}{\delta\eta_{\beta_1}(y_1)}\,\frac{1}{i}\frac{\delta}{\delta\overline{\eta}_{\alpha_2}(x_2)}\, i\frac{\delta}{\delta\eta_{\beta_2}(y_2)}\, iW(\overline{\eta}, \eta, J)\bigg|_{\overline{\eta}=\eta=J=0}\,.
\end{aligned} \tag{45.9}$$

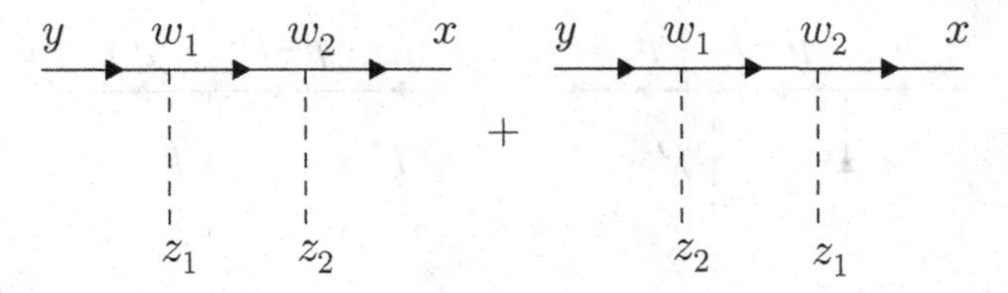

Figure 45.2. Diagrams corresponding to eq. (45.8).

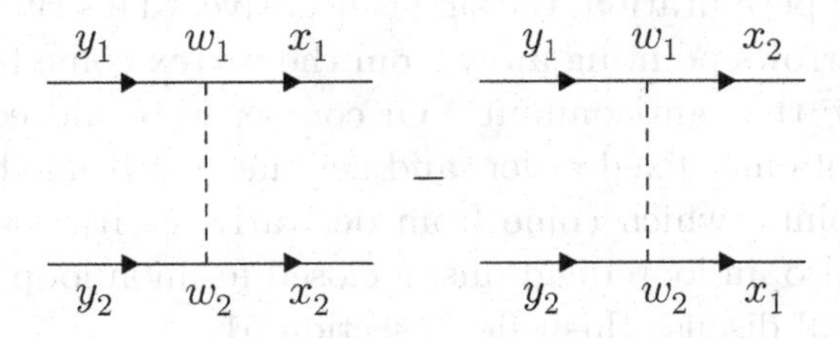

Figure 45.3. Diagrams corresponding to eq. (45.10).

The two η derivatives can act on the two ηs in the diagram in two different ways; ditto for the two $\overline{\eta}$ derivatives. This results in four different terms, but two of them are algebraic duplicates of the other two; this duplication cancels the symmetry factor (which is a general result for tree diagrams). We get

$$\begin{aligned}
&\langle 0|\mathrm{T}\Psi_{\alpha_1}(x_1)\overline{\Psi}_{\beta_1}(y_1)\Psi_{\alpha_2}(x_2)\overline{\Psi}_{\beta_2}(y_2)|0\rangle_{\mathrm{C}} \\
&\quad = \left(\tfrac{1}{i}\right)^5 (ig)^2 \int d^4w_1\, d^4w_2 \\
&\qquad\qquad \times [S(x_1{-}w_1)S(w_1{-}y_1)]_{\alpha_1\beta_1} \\
&\qquad\qquad \times \Delta(w_1{-}w_2) \\
&\qquad\qquad \times [S(x_2{-}w_2)S(w_2{-}y_2)]_{\alpha_2\beta_2} \\
&\qquad\qquad - ((y_1,\beta_1) \leftrightarrow (y_2,\beta_2)) + O(g^4)\ . \qquad (45.10)
\end{aligned}$$

The corresponding diagrams, with sources removed, are shown in fig. 45.3. Note that we now have a *relative minus sign* between the two diagrams, due to the anticommutation of the derivatives with respect to $\overline{\eta}$.

In general, the overall sign for a diagram can be determined by the following procedure. First, draw each diagram with all the fermion lines horizontal, with their arrows pointing from left to right, and with the left endpoints labeled in the same fixed order (from top to bottom). Next, in each diagram, note the ordering (from top to bottom) of the labels on the right endpoints of the fermion lines. If this ordering is an even permutation of an arbitrarily chosen fixed ordering, then the sign of that diagram is positive,

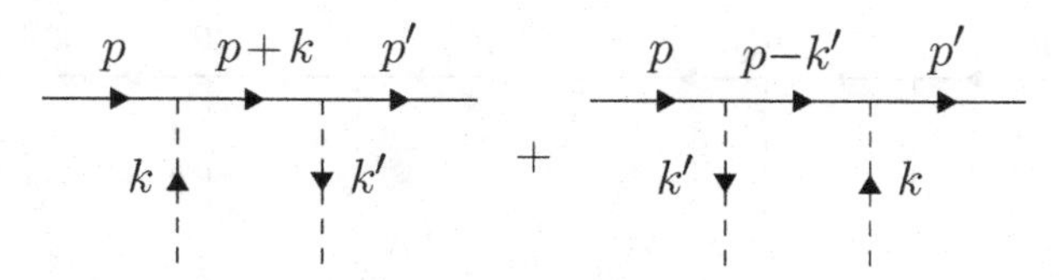

Figure 45.4. Diagrams for $e^-\varphi \to e^-\varphi$, corresponding to eq. (45.16).

and if it is an odd permutation, the sign is negative. (This rule arises because endpoints with arrows pointing away from the vertex come from derivatives with respect to $\overline{\eta}$ that anticommute. Of course, we could equally well put the right endpoints in a fixed order, and get the sign from the permutation of the left endpoints, which come from derivatives with respect to η that anticommute.) Also, in loop diagrams, a closed fermion loop yields an extra minus sign; we will discuss this rule in section 51.

Let us now consider a particular scattering process: $e^-\varphi \to e^-\varphi$. The scattering amplitude is

$$\langle f|i\rangle = \langle 0|\,\mathrm{T}\,a(\mathbf{k}')_{\rm out} b_{s'}(\mathbf{p}')_{\rm out} b^\dagger_s(\mathbf{p})_{\rm in} a^\dagger(\mathbf{k})_{\rm in}\,|0\rangle \;. \tag{45.11}$$

Next we make the replacements

$$b^\dagger_s(\mathbf{p})_{\rm in} \to i\int d^4y\;\overline{\Psi}(y)(+i\overleftarrow{\not\partial}+m)u_s(\mathbf{p})\,e^{+ipy}\;, \tag{45.12}$$

$$b_{s'}(\mathbf{p}')_{\rm out} \to i\int d^4x\; e^{-ip'x}\,\overline{u}_{s'}(\mathbf{p}')(-i\not\partial+m)\Psi(x)\;, \tag{45.13}$$

$$a^\dagger(\mathbf{k})_{\rm in} \to i\int d^4z_1\; e^{+ikz_1}(-\partial^2+m^2)\varphi(z_1)\;, \tag{45.14}$$

$$a(\mathbf{k}')_{\rm out} \to i\int d^4z_2\; e^{-ik'z_2}(-\partial^2+m^2)\varphi(z_2) \tag{45.15}$$

in eq. (45.11), and then use eq. (45.8). The wave operators (either Klein–Gordon or Dirac) act on the external propagators, and convert them to delta functions. After using eqs. (45.4) and (45.5) for the internal propagators, all dependence on the various spacetime coordinates is in the form of plane-wave factors, as in section 10. Integrating over the internal coordinates then generates delta functions that conserve four-momentum at each vertex. The only new feature arises from the spinor factors $u_s(\mathbf{p})$ and $\overline{u}_{s'}(\mathbf{p}')$. We find that $u_s(\mathbf{p})$ is associated with the external fermion line whose arrow points *towards* the vertex, and that $\overline{u}_{s'}(\mathbf{p}')$ is associated with the external fermion line whose arrow points *away* from the vertex. We can therefore draw the momentum-space diagrams of fig. 45.4. Since there is only one fermion line in each diagram, the relative sign is positive. The tree-level $e^-\varphi \to e^-\varphi$

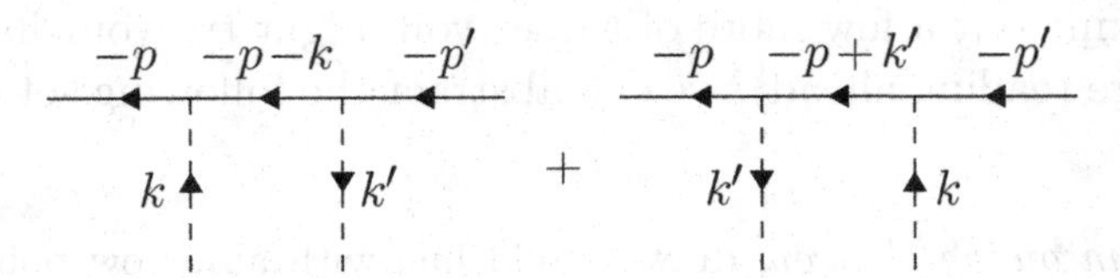

Figure 45.5. Diagrams for $e^+\varphi \to e^+\varphi$, corresponding to eq. (45.22).

scattering amplitude is then given by

$$i\mathcal{T}_{e^-\varphi\to e^-\varphi} = \tfrac{1}{i}(ig)^2\,\overline{u}_{s'}(\mathbf{p}')\left[\frac{-\not{p}-\not{k}+m}{-s+m^2}+\frac{-\not{p}+\not{k}'+m}{-u+m^2}\right]u_s(\mathbf{p})\,, \tag{45.16}$$

where $s = -(p+k)^2$ and $u = -(p-k')^2$. (We can safely ignore the $i\epsilon$s in the propagators, because their denominators cannot vanish for any physically allowed values of s and u.)

Next consider the process $e^+\varphi \to e^+\varphi$. We now have

$$\langle f|i\rangle = \langle 0|\,\mathrm{T}\,a(\mathbf{k}')_{\text{out}}d_{s'}(\mathbf{p}')_{\text{out}}d_s^\dagger(\mathbf{p})_{\text{in}}a^\dagger(\mathbf{k})_{\text{in}}\,|0\rangle\,. \tag{45.17}$$

The relevant replacements are

$$d_s^\dagger(\mathbf{p})_{\text{in}} \to -i\int d^4x\; e^{+ipx}\;\overline{v}_s(\mathbf{p})(-i\not{\partial}+m)\Psi(x)\,, \tag{45.18}$$

$$d_{s'}(\mathbf{p}')_{\text{out}} \to -i\int d^4y\;\overline{\Psi}(y)(+i\overleftarrow{\not{\partial}}+m)v_{s'}(\mathbf{p}')\,e^{-ip'y}\,, \tag{45.19}$$

$$a^\dagger(\mathbf{k})_{\text{in}} \to i\int d^4z_1\; e^{+ikz_1}(-\partial^2+m^2)\varphi(z_1)\,, \tag{45.20}$$

$$a(\mathbf{k}')_{\text{out}} \to i\int d^4z_2\; e^{-ik'z_2}(-\partial^2+m^2)\varphi(z_2)\,. \tag{45.21}$$

We substitute these into eq. (45.17), and then use eq. (45.8). This ultimately leads to the momentum-space Feynman diagrams of fig. 45.5. Note that we must now label the external fermion lines with *minus* their four-momenta; this is characteristic of d-type particles. (The same phenomenon occurs for a complex scalar field; see problem 10.2.) Regarding the spinor factors, we find that $-\overline{v}_s(\mathbf{p})$ is associated with the external fermion line whose arrow points *away* from the vertex, and $-v_{s'}(\mathbf{p}')$ with the external fermion line whose arrow points *towards* the vertex. The minus signs attached to each v and $\overline{v}$ can be consistently dropped, however, as they only affect the overall sign of the amplitude (and not the relative signs among contributing diagrams). The tree-level expression for the $e^+\varphi \to e^+\varphi$ amplitude is then

$$i\mathcal{T}_{e^+\varphi\to e^+\varphi} = \tfrac{1}{i}(ig)^2\,\overline{v}_s(\mathbf{p})\left[\frac{\not{p}-\not{k}'+m}{-u+m^2}+\frac{\not{p}+\not{k}+m}{-s+m^2}\right]v_{s'}(\mathbf{p}')\,, \tag{45.22}$$

where again $s = -(p+k)^2$ and $u = -(p-k')^2$.

After working out a few more of these (you might try your hand at some of them before reading ahead), we can abstract the following set of Feynman rules.

1) For each *incoming electron*, draw a solid line with an arrow pointed *towards* the vertex, and label it with the electron's four-momentum, p_i.
2) For each *outgoing electron*, draw a solid line with an arrow pointed *away* from the vertex, and label it with the electron's four-momentum, p'_i.
3) For each *incoming positron*, draw a solid line with an arrow pointed *away* from the vertex, and label it with *minus* the positron's four-momentum, $-p_i$.
4) For each *outgoing positron*, draw a solid line with an arrow pointed *towards* the vertex, and label it with *minus* the positron's four-momentum, $-p'_i$.
5) For each *incoming scalar*, draw a dashed line with an arrow pointed *towards* the vertex, and label it with the scalar's four-momentum, k_i.
6) For each *outgoing scalar*, draw a dashed line with an arrow pointed *away* from the vertex, and label it with the scalar's four-momentum, k'_i.
7) The only allowed vertex joins two solid lines, one with an arrow pointing towards it and one with an arrow pointing away from it, and one dashed line (whose arrow can point in either direction). Using this vertex, join up all the external lines, including extra internal lines as needed. In this way, draw all possible diagrams that are *topologically inequivalent*.
8) Assign each internal line its own four-momentum. Think of the four-momenta as flowing along the arrows, and conserve four-momentum at each vertex. For a tree diagram, this fixes the momenta on all the internal lines.
9) The value of a diagram consists of the following factors:
   for each incoming or outgoing scalar, 1;
   for each incoming electron, $u_{s_i}(\mathbf{p}_i)$;
   for each outgoing electron, $\overline{u}_{s'_i}(\mathbf{p}'_i)$;
   for each incoming positron, $\overline{v}_{s_i}(\mathbf{p}_i)$;
   for each outgoing positron, $v_{s'_i}(\mathbf{p}'_i)$;
   for each vertex, ig;
   for each internal scalar, $-i/(k^2 + M^2 - i\epsilon)$;
   for each internal fermion, $-i(-\not{p} + m)/(p^2 + m^2 - i\epsilon)$.
10) Spinor indices are contracted by starting at one end of a fermion line: specifically, the end that has the arrow pointing away from the vertex. The factor associated with the external line is either $\overline{u}$ or $\overline{v}$. Go along the complete fermion line, following the arrows backwards, and write down (in order from left to right) the factors associated with the vertices and propagators that you encounter. The last factor is either a u or v. Repeat this procedure for the other fermion lines, if any.
11) The overall sign of a tree diagram is determined by drawing all contributing diagrams in a standard form: all fermion lines horizontal, with their arrows pointing from left to right, and with the left endpoints labeled in the same

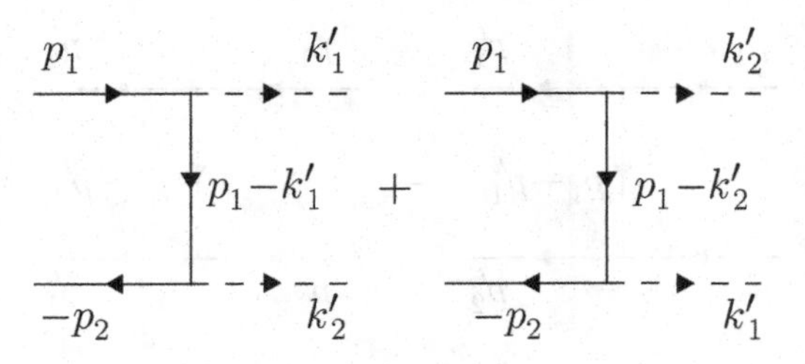

Figure 45.6. Diagrams for $e^+e^- \to \varphi\varphi$, corresponding to eq. (45.23).

fixed order (from top to bottom); if the ordering of the labels on the right endpoints of the fermion lines in a given diagram is an even (odd) permutation of an arbitrarily chosen fixed ordering, then the sign of that diagram is positive (negative).

12) The value of $i\mathcal{T}$ (at tree level) is given by a sum over the values of the contributing diagrams.

There are additional rules for counterterms and loops; in particular, each closed fermion loop contributes an extra minus sign. We will postpone discussion of loop corrections to section 51.

Let us apply these rules to $e^+e^- \to \varphi\varphi$. Let the initial electron and positron have four-momenta p_1 and p_2, respectively, and the two final scalars have four-momenta k_1' and k_2'. The relevant diagrams are shown in fig. 45.6; there is only one fermion line, and so the relative sign is positive. The result is

$$i\mathcal{T}_{e^+e^-\to\varphi\varphi} = \tfrac{1}{i}(ig)^2\,\overline{v}_{s_2}(\mathbf{p}_2)\left[\frac{-\not{p}_1+\not{k}_1'+m}{-t+m^2}+\frac{-\not{p}_1+\not{k}_2'+m}{-u+m^2}\right]u_{s_1}(\mathbf{p}_1)\,, \tag{45.23}$$

where $t = -(p_1-k_1')^2$ and $u = -(p_1-k_2')^2$.

Next, consider $e^-e^- \to e^-e^-$. Let the initial electrons have four-momenta p_1 and p_2, and the final electrons have four-momenta p_1' and p_2'. The relevant diagrams are shown in fig. 45.7, and according to rule no. 11 the relative sign is negative. Thus the result is

$$i\mathcal{T}_{e^-e^-\to e^-e^-} = \tfrac{1}{i}(ig)^2\left[\frac{(\overline{u}_1'u_1)(\overline{u}_2'u_2)}{-t+M^2}-\frac{(\overline{u}_2'u_1)(\overline{u}_1'u_2)}{-u+M^2}\right], \tag{45.24}$$

where u_1 is short for $u_{s_1}(\mathbf{p}_1)$, etc., and $t = -(p_1-p_1')^2$, $u = -(p_1-p_2')^2$.

One more: $e^+e^- \to e^+e^-$. Let the initial electron and positron have four-momenta p_1 and p_2, respectively, and the final electron and positron have four-momenta p_1' and p_2', respectively. The relevant diagrams are shown in fig. 45.8. If we redraw them in the standard form of rule no. 11, as shown in

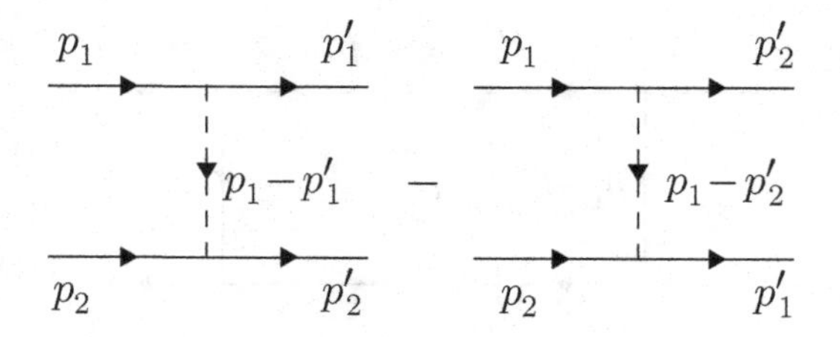

Figure 45.7. Diagrams for $e^-e^- \to e^-e^-$, corresponding to eq. (45.24).

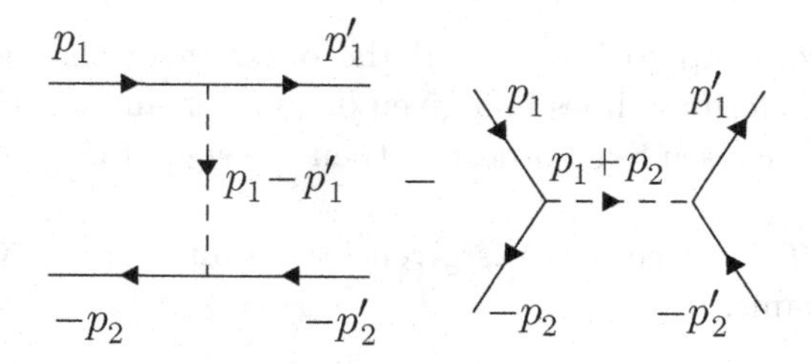

Figure 45.8. Diagrams for $e^+e^- \to e^+e^-$, corresponding to eq. (45.25).

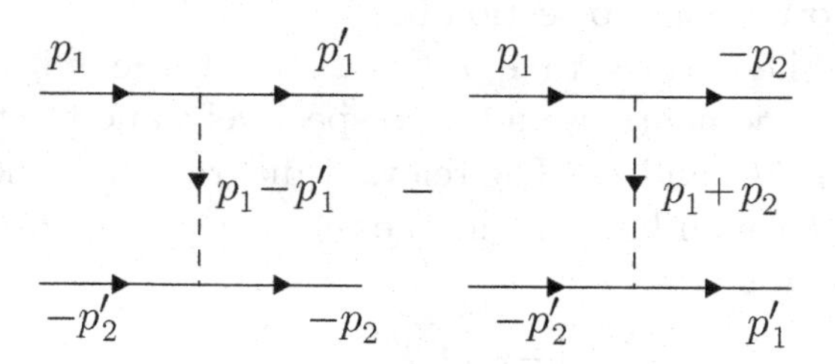

Figure 45.9. Same as fig. 45.8, but with the diagrams redrawn in the standard form given in rule no. 11.

fig. 45.9, we see that the relative sign is negative. Thus the result is

$$i\mathcal{T}_{e^+e^-\to e^+e^-} = \tfrac{1}{i}(ig)^2\left[\frac{(\overline{u}_1'u_1)(\overline{v}_2v_2')}{-t+M^2} - \frac{(\overline{v}_2u_1)(\overline{u}_1'v_2')}{-s+M^2}\right], \tag{45.25}$$

where $s = -(p_1+p_2)^2$ and $t = -(p_1-p_1')^2$.

Problems

45.1) a) Determine how $\varphi(x)$ must transform under parity, time reversal, and charge conjugation in order for these to all be symmetries of the theory. (Prerequisite: 40)

b) Same question, but with the interaction given by $\mathcal{L}_1 = ig\varphi\overline{\Psi}\gamma_5\Psi$ instead of eq. (45.1).

45.2) Use the Feynman rules to write down (at tree level) $i\mathcal{T}$ for the processes $e^+e^+ \to e^+e^+$ and $\varphi\varphi \to e^+e^-$.

46

Spin sums

Prerequisite: 45

In the last section, we calculated various tree-level scattering amplitudes in Yukawa theory. For example, for $e^-\varphi \to e^-\varphi$ we found

$$\mathcal{T} = g^2\,\overline{u}_{s'}(\mathbf{p}')\left[\frac{-\not{p} - \not{k} + m}{-s + m^2} + \frac{-\not{p} + \not{k}' + m}{-u + m^2}\right]u_s(\mathbf{p})\ , \tag{46.1}$$

where $s = -(p+k)^2$ and $u = -(p-k')^2$. In order to compute the corresponding cross section, we must evaluate $|\mathcal{T}|^2 = \mathcal{T}\mathcal{T}^*$. We begin by simplifying eq. (46.1) a little; we use $(\not{p} + m)u_s(\mathbf{p}) = 0$ to replace the $-\not{p}$ in each numerator with m. We then abbreviate eq. (46.1) as

$$\mathcal{T} = \overline{u}'Au\ , \tag{46.2}$$

where

$$A \equiv g^2\left[\frac{-\not{k} + 2m}{m^2 - s} + \frac{\not{k}' + 2m}{m^2 - u}\right] . \tag{46.3}$$

Then we have

$$\mathcal{T}^* = \overline{\mathcal{T}} = \overline{\overline{u}'Au} = \overline{u}\overline{A}u'\ , \tag{46.4}$$

where in general $\overline{A} \equiv \beta A^\dagger \beta$, and, for the particular A of eq. (46.3), $\overline{A} = A$. Thus we have

$$\begin{aligned}
|\mathcal{T}|^2 &= (\overline{u}'Au)(\overline{u}Au') \\
&= \sum_{\alpha\beta\gamma\delta} \overline{u}'_\alpha A_{\alpha\beta} u_\beta \overline{u}_\gamma A_{\gamma\delta} u'_\delta \\
&= \sum_{\alpha\beta\gamma\delta} u'_\delta \overline{u}'_\alpha A_{\alpha\beta} u_\beta \overline{u}_\gamma A_{\gamma\delta} \\
&= \mathrm{Tr}\Big[(u'\overline{u}')A(u\overline{u})A\Big] .
\end{aligned} \tag{46.5}$$

Next, we use a result from section 38:

$$u_s(\mathbf{p})\overline{u}_s(\mathbf{p}) = \tfrac{1}{2}(1 - s\gamma_5 \not{z})(-\not{p} + m)\ , \tag{46.6}$$

where $s = \pm$ tells us whether the spin is up or down along the spin quantization axis z. We then have

$$|\mathcal{T}|^2 = \tfrac{1}{4}\mathrm{Tr}\Big[(1 - s'\gamma_5 \not{z}')(-\not{p}' + m)A(1 - s\gamma_5 \not{z})(-\not{p} + m)A\Big]\ . \tag{46.7}$$

We now simply need to take traces of products of gamma matrices; we will work out the technology for this in the next section.

However, in practice, we are often not interested in (or are unable to easily measure or prepare) the spin states of the scattering particles. Thus, if we know that an electron with momentum p' landed in our detector, but know nothing about its spin, we should *sum* $|\mathcal{T}|^2$ over the two possible spin states of this outgoing electron. Similarly, if the spin state of the initial electron is not specially prepared for each scattering event, then we should *average* $|\mathcal{T}|^2$ over the two possible spin states of this initial electron. Then we can use

$$\sum_{s=\pm} u_s(\mathbf{p})\overline{u}_s(\mathbf{p}) = -\not{p} + m \tag{46.8}$$

in place of eq. (46.6).

Let us, then, take $|\mathcal{T}|^2$, sum over all final spins, and average over all initial spins, and call the result $\langle|\mathcal{T}|^2\rangle$. In the present case, we have

$$\begin{aligned}\langle|\mathcal{T}|^2\rangle &\equiv \tfrac{1}{2}\sum_{s,s'}|\mathcal{T}|^2 \\ &= \tfrac{1}{2}\mathrm{Tr}\Big[(-\not{p}' + m)A(-\not{p} + m)A\Big]\ ,\end{aligned} \tag{46.9}$$

which is much less cumbersome than eq. (46.7).

Next let us try something a little harder, namely $e^+e^- \to e^+e^-$. We found in section 45 that

$$\mathcal{T} = g^2\left[\frac{(\overline{u}'_1 u_1)(\overline{v}_2 v'_2)}{M^2 - t} - \frac{(\overline{v}_2 u_1)(\overline{u}'_1 v'_2)}{M^2 - s}\right]. \tag{46.10}$$

We then have

$$\overline{\mathcal{T}} = g^2\left[\frac{(\overline{u}_1 u'_1)(\overline{v}'_2 v_2)}{M^2 - t} - \frac{(\overline{u}_1 v_2)(\overline{v}'_2 u'_1)}{M^2 - s}\right]. \tag{46.11}$$

When we multiply $\mathcal{T}$ by $\overline{\mathcal{T}}$, we will get four terms. We want to arrange the factors in each of them so that every u and every v stands just to the left of the corresponding $\overline{u}$ and $\overline{v}$. In this way, we get

$$
\begin{aligned}
|\mathcal{T}|^2 = &+\frac{g^4}{(M^2-t)^2}\,\mathrm{Tr}\Big[u_1\overline{u}_1 u_1'\overline{u}_1'\Big]\,\mathrm{Tr}\Big[v_2'\overline{v}_2' v_2\overline{v}_2\Big] \\
&+\frac{g^4}{(M^2-s)^2}\,\mathrm{Tr}\Big[u_1\overline{u}_1 v_2\overline{v}_2\Big]\,\mathrm{Tr}\Big[v_2'\overline{v}_2' u_1'\overline{u}_1'\Big] \\
&-\frac{g^4}{(M^2-t)(M^2-s)}\,\mathrm{Tr}\Big[u_1\overline{u}_1 v_2\overline{v}_2 v_2'\overline{v}_2' u_1'\overline{u}_1'\Big] \\
&-\frac{g^4}{(M^2-s)(M^2-t)}\,\mathrm{Tr}\Big[u_1\overline{u}_1 u_1'\overline{u}_1' v_2'\overline{v}_2' v_2\overline{v}_2\Big]\,. \qquad (46.12)
\end{aligned}
$$

Then we average over initial spins and sum over final spins, and use eq. (46.8) and

$$
\sum_{s=\pm} v_s(\mathbf{p})\overline{v}_s(\mathbf{p}) = -\not{p} - m\,. \qquad (46.13)
$$

We then must evaluate traces of products of up to four gamma matrices.

47

Gamma matrix technology

Prerequisite: 36

In this section, we will learn some tricks for handling gamma matrices. We need the following information as a starting point:

$$\{\gamma^\mu, \gamma^\nu\} = -2g^{\mu\nu} , \tag{47.1}$$

$$\gamma_5^2 = 1 , \tag{47.2}$$

$$\{\gamma^\mu, \gamma_5\} = 0 , \tag{47.3}$$

$$\mathrm{Tr}\, 1 = 4 . \tag{47.4}$$

Now consider the trace of the product of n gamma matrices. We have

$$\begin{aligned}
\mathrm{Tr}[\gamma^{\mu_1} \ldots \gamma^{\mu_n}] &= \mathrm{Tr}[\gamma_5^2 \gamma^{\mu_1} \gamma_5^2 \ldots \gamma_5^2 \gamma^{\mu_n}] \\
&= \mathrm{Tr}[(\gamma_5 \gamma^{\mu_1} \gamma_5) \ldots (\gamma_5 \gamma^{\mu_n} \gamma_5)] \\
&= \mathrm{Tr}[(-\gamma_5^2 \gamma^{\mu_1}) \ldots (-\gamma_5^2 \gamma^{\mu_n})] \\
&= (-1)^n \,\mathrm{Tr}[\gamma^{\mu_1} \ldots \gamma^{\mu_n}] .
\end{aligned} \tag{47.5}$$

We used eq. (47.2) to get the first equality, the cyclic property of the trace for the second, eq. (47.3) for the third, and eq. (47.2) again for the fourth. If n is odd, eq. (47.5) tells us that this trace is equal to minus itself, and must therefore be zero:

$$\mathrm{Tr}[\,\text{odd no. of } \gamma^\mu\text{s}\,] = 0 . \tag{47.6}$$

Similarly,

$$\mathrm{Tr}[\,\gamma_5\,(\,\text{odd no. of } \gamma^\mu\text{s}\,)\,] = 0 . \tag{47.7}$$

Next, consider $\mathrm{Tr}[\gamma^\mu\gamma^\nu]$. We have

$$\begin{aligned}\mathrm{Tr}[\gamma^\mu\gamma^\nu] &= \mathrm{Tr}[\gamma^\nu\gamma^\mu] \\ &= \tfrac{1}{2}\mathrm{Tr}[\gamma^\mu\gamma^\nu + \gamma^\nu\gamma^\mu] \\ &= -g^{\mu\nu}\,\mathrm{Tr}\,1 \\ &= -4g^{\mu\nu}\ . \end{aligned} \tag{47.8}$$

The first equality follows from the cyclic property of the trace, the second averages the left- and right-hand sides of the first, the third uses eq. (47.1), and the fourth uses eq. (47.4).

A slightly nicer way of expressing eq. (47.8) is to introduce two arbitrary four-vectors a^μ and b^μ, and write

$$\mathrm{Tr}[\not{a}\not{b}] = -4(ab)\ , \tag{47.9}$$

where $\not{a} = a_\mu\gamma^\mu$, $\not{b} = b_\mu\gamma^\mu$, and $(ab) = a^\mu b_\mu$.

Next consider $\mathrm{Tr}[\not{a}\not{b}\not{c}\not{d}]$. We evaluate this by moving $\not{a}$ to the right, using eq. (47.1), which is now more usefully written as

$$\not{a}\not{b} = -\not{b}\not{a} - 2(ab)\ . \tag{47.10}$$

Using this repeatedly, we have

$$\begin{aligned}\mathrm{Tr}[\not{a}\not{b}\not{c}\not{d}] &= -\mathrm{Tr}[\not{b}\not{a}\not{c}\not{d}] - 2(ab)\mathrm{Tr}[\not{c}\not{d}] \\ &= +\mathrm{Tr}[\not{b}\not{c}\not{a}\not{d}] + 2(ac)\mathrm{Tr}[\not{b}\not{d}] - 2(ab)\mathrm{Tr}[\not{c}\not{d}] \\ &= -\mathrm{Tr}[\not{b}\not{c}\not{d}\not{a}] - 2(ad)\mathrm{Tr}[\not{b}\not{c}] + 2(ac)\mathrm{Tr}[\not{b}\not{d}] - 2(ab)\mathrm{Tr}[\not{c}\not{d}]\ . \end{aligned} \tag{47.11}$$

Now we note that the first term on the right-hand side of the last line is, by the cyclic property of the trace, equal to minus the left-hand side. We can then move this term to the left-hand side to get

$$2\,\mathrm{Tr}[\not{a}\not{b}\not{c}\not{d}] = -2(ad)\mathrm{Tr}[\not{b}\not{c}] + 2(ac)\mathrm{Tr}[\not{b}\not{d}] - 2(ab)\mathrm{Tr}[\not{c}\not{d}]\ . \tag{47.12}$$

Finally, we evaluate each $\mathrm{Tr}[\not{a}\not{b}]$ with eq. (47.9), and divide by two:

$$\mathrm{Tr}[\not{a}\not{b}\not{c}\not{d}] = 4\Big[(ad)(bc) - (ac)(bd) + (ab)(cd)\Big]\ . \tag{47.13}$$

This is our final result for this trace.

Clearly, we can use the same technique to evaluate the trace of the product of any even number of gamma matrices.

Next, let us consider traces that involve γ_5s and γ^μs. Since $\{\gamma_5, \gamma^\mu\} = 0$, we can always bring all the γ_5s together by moving them through the γ^μs (generating minus signs as we go). Then, since $\gamma_5^2 = 1$, we end up with either one γ_5 or none. So we need only consider $\mathrm{Tr}[\gamma_5\gamma^{\mu_1}\ldots\gamma^{\mu_n}]$. And, according to eq. (47.7), we need only be concerned with even n.

Recall that an explicit formula for γ_5 is

$$\gamma_5 = i\gamma^0\gamma^1\gamma^2\gamma^3 \; . \tag{47.14}$$

Eq. (47.13) then implies

$$\mathrm{Tr}\,\gamma_5 = 0 \; . \tag{47.15}$$

Similarly, we can show that

$$\mathrm{Tr}[\gamma_5\gamma^\mu\gamma^\nu] = 0 \; . \tag{47.16}$$

Finally, consider $\mathrm{Tr}[\gamma_5\gamma^\mu\gamma^\nu\gamma^\rho\gamma^\sigma]$. The only way to get a nonzero result is to have the four vector indices take on four different values. If we consider the special case $\mathrm{Tr}[\gamma_5\gamma^3\gamma^2\gamma^1\gamma^0]$, plug in eq. (47.14), and then use $(\gamma^i)^2 = -1$ and $(\gamma^0)^2 = 1$, we get $i(-1)^3\,\mathrm{Tr}\,1 = -4i$, or equivalently

$$\mathrm{Tr}[\gamma_5\gamma^\mu\gamma^\nu\gamma^\rho\gamma^\sigma] = -4i\varepsilon^{\mu\nu\rho\sigma} \; , \tag{47.17}$$

where $\varepsilon^{0123} = \varepsilon^{3210} = +1$.

Another category of gamma matrix combinations that we will eventually encounter is $\gamma^\mu \not{a} \ldots \gamma_\mu$. The simplest of these is

$$\begin{aligned}
\gamma^\mu\gamma_\mu &= g_{\mu\nu}\gamma^\mu\gamma^\nu \\
&= \tfrac{1}{2}g_{\mu\nu}\{\gamma^\mu,\gamma^\nu\} \\
&= -g_{\mu\nu}g^{\mu\nu} \\
&= -d \; .
\end{aligned} \tag{47.18}$$

To get the second equality, we used the fact that $g_{\mu\nu}$ is symmetric, and so only the symmetric part of $\gamma^\mu\gamma^\nu$ contributes. In the last line, d is the number of spacetime dimensions. Of course, our entire spinor formalism has been built around $d = 4$, but we will need formal results for $d = 4-\varepsilon$ when we dimensionally regulate loop diagrams involving fermions.

We move on to evaluate

$$\begin{aligned}
\gamma^\mu\not{a}\gamma_\mu &= \gamma^\mu(-\gamma_\mu\not{a} - 2a_\mu) \\
&= -\gamma^\mu\gamma_\mu\not{a} - 2\not{a} \\
&= (d-2)\not{a} \; .
\end{aligned} \tag{47.19}$$

We continue with

$$\gamma^\mu\not{a}\not{b}\gamma_\mu = 4(ab) - (d-4)\not{a}\not{b} \tag{47.20}$$

and

$$\gamma^\mu \not{a}\not{b}\not{c}\gamma_\mu = 2\not{c}\not{b}\not{a} + (d-4)\not{a}\not{b}\not{c} \; ; \tag{47.21}$$

the derivations are left as an exercise.

Problems

47.1) Verify eq. (47.16).

47.2) Verify eqs. (47.20) and (47.21).

47.3) Show that the most general 4×4 matrix can be written as a linear combination (with complex coefficients) of 1, γ^μ, $S^{\mu\nu}$, $\gamma^\mu\gamma_5$, and γ_5, where 1 is the identity matrix and $S^{\mu\nu} = \frac{i}{4}[\gamma^\mu, \gamma^\nu]$. Hint: if A and B are two different members of this set, prove linear independence by showing that $\mathrm{Tr}\, A^\dagger B = 0$. Then count.

48

Spin-averaged cross sections

Prerequisites: 46, 47

In section 46, we computed $|\mathcal{T}|^2$ for (among other processes) $e^+e^- \to e^+e^-$. We take the incoming and outgoing electrons to have momenta p_1 and p_1', respectively, and the incoming and outgoing positrons to have momenta p_2 and p_2', respectively. We have $p_i^2 = p_i'^2 = -m^2$, where m is the electron (and positron) mass. The Mandelstam variables are

$$\begin{aligned} s &= -(p_1+p_2)^2 = -(p_1'+p_2')^2 \; , \\ t &= -(p_1-p_1')^2 = -(p_2-p_2')^2 \; , \\ u &= -(p_1-p_2')^2 = -(p_2-p_1')^2 \; , \end{aligned} \tag{48.1}$$

and they obey $s+t+u = 4m^2$. Our result was

$$|\mathcal{T}|^2 = g^4 \left[\frac{\Phi_{ss}}{(M^2-s)^2} - \frac{\Phi_{st}+\Phi_{ts}}{(M^2-s)(M^2-t)} + \frac{\Phi_{tt}}{(M^2-t)^2} \right] , \tag{48.2}$$

where M is the scalar mass, and

$$\begin{aligned} \Phi_{ss} &= \mathrm{Tr}\Big[u_1\overline{u}_1 v_2\overline{v}_2\Big]\,\mathrm{Tr}\Big[v_2'\overline{v}_2' u_1'\overline{u}_1'\Big] \; , \\ \Phi_{tt} &= \mathrm{Tr}\Big[u_1\overline{u}_1 u_1'\overline{u}_1'\Big]\,\mathrm{Tr}\Big[v_2'\overline{v}_2' v_2\overline{v}_2\Big] \; , \\ \Phi_{st} &= \mathrm{Tr}\Big[u_1\overline{u}_1 u_1'\overline{u}_1' v_2'\overline{v}_2' v_2\overline{v}_2\Big] \; , \\ \Phi_{ts} &= \mathrm{Tr}\Big[u_1\overline{u}_1 v_2\overline{v}_2 v_2'\overline{v}_2' u_1'\overline{u}_1'\Big] \; . \end{aligned} \tag{48.3}$$

Next, we average over the two initial spins and sum over the two final spins to get

$$\langle|\mathcal{T}|^2\rangle = \tfrac{1}{4}\sum\nolimits_{s_1,s_2,s_1',s_2'} |\mathcal{T}|^2 \; . \tag{48.4}$$

Then we use

$$\sum_{s=\pm} u_s(\mathbf{p})\overline{u}_s(\mathbf{p}) = -\not{p} + m \,,$$

$$\sum_{s=\pm} v_s(\mathbf{p})\overline{v}_s(\mathbf{p}) = -\not{p} - m \,, \tag{48.5}$$

to get

$$\langle\Phi_{ss}\rangle = \tfrac{1}{4}\mathrm{Tr}\Big[(-\not{p}_1 + m)(-\not{p}_2 - m)\Big]\,\mathrm{Tr}\Big[(-\not{p}_2' - m)(-\not{p}_1' + m)\Big]\,, \tag{48.6}$$

$$\langle\Phi_{tt}\rangle = \tfrac{1}{4}\mathrm{Tr}\Big[(-\not{p}_1 + m)(-\not{p}_1' + m)\Big]\,\mathrm{Tr}\Big[(-\not{p}_2' - m)(-\not{p}_2 - m)\Big]\,, \tag{48.7}$$

$$\langle\Phi_{st}\rangle = \tfrac{1}{4}\mathrm{Tr}\Big[(-\not{p}_1 + m)(-\not{p}_1' + m)(-\not{p}_2' - m)(-\not{p}_2 - m)\Big]\,, \tag{48.8}$$

$$\langle\Phi_{ts}\rangle = \tfrac{1}{4}\mathrm{Tr}\Big[(-\not{p}_1 + m)(-\not{p}_2 - m)(-\not{p}_2' - m)(-\not{p}_1' + m)\Big]\,. \tag{48.9}$$

It is now merely tedious to evaluate these traces with the technology of section 47.

For example,

$$\begin{aligned}\mathrm{Tr}\Big[(-\not{p}_1 + m)(-\not{p}_2 - m)\Big] &= \mathrm{Tr}[\not{p}_1\not{p}_2] - m^2\,\mathrm{Tr}\,1 \\ &= -4(p_1p_2) - 4m^2 \,.\end{aligned} \tag{48.10}$$

It is convenient to write four-vector products in terms of the Mandelstam variables. We have

$$\begin{aligned}p_1p_2 &= p_1'p_2' = -\tfrac{1}{2}(s - 2m^2)\,, \\ p_1p_1' &= p_2p_2' = +\tfrac{1}{2}(t - 2m^2)\,, \\ p_1p_2' &= p_1'p_2 = +\tfrac{1}{2}(u - 2m^2)\,,\end{aligned} \tag{48.11}$$

and so

$$\mathrm{Tr}\Big[(-\not{p}_1 + m)(-\not{p}_2 - m)\Big] = 2s - 8m^2 \,. \tag{48.12}$$

Thus, we can easily work out eqs. (48.6) and (48.7):

$$\langle\Phi_{ss}\rangle = (s - 4m^2)^2 \,, \tag{48.13}$$

$$\langle\Phi_{tt}\rangle = (t - 4m^2)^2 \,. \tag{48.14}$$

Obviously, if we start with $\langle\Phi_{ss}\rangle$ and make the swap $s \leftrightarrow t$, we get $\langle\Phi_{tt}\rangle$. We could have anticipated this from eqs. (48.6) and (48.7): if we start with the right-hand side of eq. (48.6) and make the swap $p_2 \leftrightarrow -p'_1$, we get the right-hand side of eq. (48.7). But from eq. (48.11), we see that this momentum swap is equivalent to $s \leftrightarrow t$.

Let us move on to $\langle\Phi_{st}\rangle$ and $\langle\Phi_{ts}\rangle$. These two are also related by $p_2 \leftrightarrow -p'_1$, and so we only need to compute one of them. We have

$$\begin{aligned}\langle\Phi_{st}\rangle &= \tfrac{1}{4}\mathrm{Tr}[\not{p}_1\not{p}'_1\not{p}'_2\not{p}_2] \\ &\quad + \tfrac{1}{4}m^2\,\mathrm{Tr}[\not{p}_1\not{p}'_1 - \not{p}_1\not{p}'_2 - \not{p}_1\not{p}_2 - \not{p}'_1\not{p}'_2 - \not{p}'_1\not{p}_2 + \not{p}'_2\not{p}_2] + \tfrac{1}{4}m^4\,\mathrm{Tr}\,1 \\ &= (p_1p'_1)(p_2p'_2) - (p_1p'_2)(p_2p'_1) + (p_1p_2)(p'_1p'_2) \\ &\quad - m^2[p_1p'_1 - p_1p'_2 - p_1p_2 - p'_1p'_2 - p'_1p_2 + p_2p'_2] + m^4 \\ &= -\tfrac{1}{2}st + 2m^2u\ . \end{aligned} \tag{48.15}$$

To get the last line, we used eq. (48.11), and then simplified it as much as possible via $s+t+u = 4m^2$. Since our result is symmetric on $s \leftrightarrow t$, we have $\langle\Phi_{ts}\rangle = \langle\Phi_{st}\rangle$.

Putting all of this together, we get

$$\langle|\mathcal{T}|^2\rangle = g^4\left[\frac{(s-4m^2)^2}{(M^2-s)^2} + \frac{st-4m^2u}{(M^2-s)(M^2-t)} + \frac{(t-4m^2)^2}{(M^2-t)^2}\right]. \tag{48.16}$$

This can then be converted to a differential cross section (in any frame) via the formulae of section 11.

Let us do one more: $e^-\varphi \to e^-\varphi$. We take the incoming and outgoing electrons to have momenta p and p', respectively, and the incoming and outgoing scalars to have momenta k and k', respectively. We then have $p^2 = p'^2 = -m^2$ and $k^2 = k'^2 = -M^2$. The Mandelstam variables are

$$\begin{aligned} s &= -(p+k)^2 = -(p'+k')^2\ , \\ t &= -(p-p')^2 = -(k-k')^2\ , \\ u &= -(p-k')^2 = -(k-p')^2\ , \end{aligned} \tag{48.17}$$

and they obey $s+t+u = 2m^2+2M^2$. Our result in section 46 was

$$\langle|\mathcal{T}|^2\rangle = \tfrac{1}{2}\mathrm{Tr}\Big[A(-\not{p}+m)A(-\not{p}'+m)\Big]\ , \tag{48.18}$$

where

$$A = g^2\left[\frac{-\not{k}+2m}{m^2-s} + \frac{\not{k}'+2m}{m^2-u}\right]. \tag{48.19}$$

Thus we have

$$\langle|\mathcal{T}|^2\rangle = g^4\left[\frac{\langle\Phi_{ss}\rangle}{(m^2-s)^2} + \frac{\langle\Phi_{su}\rangle+\langle\Phi_{us}\rangle}{(m^2-s)(m^2-u)} + \frac{\langle\Phi_{uu}\rangle}{(m^2-u)^2}\right], \tag{48.20}$$

where now

$$\langle\Phi_{ss}\rangle = \tfrac{1}{2}\mathrm{Tr}\Big[(-\not{p}'+m)(-\not{k}+2m)(-\not{p}+m)(-\not{k}+2m)\Big]\,, \tag{48.21}$$

$$\langle\Phi_{uu}\rangle = \tfrac{1}{2}\mathrm{Tr}\Big[(-\not{p}'+m)(+\not{k}'+2m)(-\not{p}+m)(+\not{k}'+2m)\Big]\,, \tag{48.22}$$

$$\langle\Phi_{su}\rangle = \tfrac{1}{2}\mathrm{Tr}\Big[(-\not{p}'+m)(-\not{k}+2m)(-\not{p}+m)(+\not{k}'+2m)\Big]\,, \tag{48.23}$$

$$\langle\Phi_{us}\rangle = \tfrac{1}{2}\mathrm{Tr}\Big[(-\not{p}'+m)(+\not{k}'+2m)(-\not{p}+m)(-\not{k}+2m)\Big]\,. \tag{48.24}$$

We can evaluate these in terms of the Mandelstam variables by using our trace technology, along with

$$\begin{aligned} pk = p'k' &= -\tfrac{1}{2}(s-m^2-M^2)\,, \\ pp' &= +\tfrac{1}{2}(t-2m^2)\,, \\ kk' &= +\tfrac{1}{2}(t-2M^2)\,, \\ pk' = p'k &= +\tfrac{1}{2}(u-m^2-M^2)\,. \end{aligned} \tag{48.25}$$

Examining eqs. (48.21) and (48.22), we see that $\langle\Phi_{ss}\rangle$ and $\langle\Phi_{uu}\rangle$ are transformed into each other by $k \leftrightarrow -k'$. Examining eqs. (48.23) and (48.24), we see that $\langle\Phi_{su}\rangle$ and $\langle\Phi_{us}\rangle$ are also transformed into each other by $k \leftrightarrow -k'$. From eq. (48.25), we see that this is equivalent to $s \leftrightarrow u$. Thus we need only compute $\langle\Phi_{ss}\rangle$ and $\langle\Phi_{su}\rangle$, and then take $s \leftrightarrow u$ to get $\langle\Phi_{uu}\rangle$ and $\langle\Phi_{us}\rangle$. This is, again, merely tedious, and the results are

$$\langle\Phi_{ss}\rangle = -su + m^2(9s+u) + 7m^4 - 8m^2M^2 + M^4\,, \tag{48.26}$$

$$\langle\Phi_{uu}\rangle = -su + m^2(9u+s) + 7m^4 - 8m^2M^2 + M^4\,, \tag{48.27}$$

$$\langle\Phi_{su}\rangle = +su + 3m^2(s+u) + 9m^4 - 8m^2M^2 - M^4\,, \tag{48.28}$$

$$\langle\Phi_{us}\rangle = +su + 3m^2(s+u) + 9m^4 - 8m^2M^2 - M^4\,. \tag{48.29}$$

Problems

48.1) The tedium of these calculations is greatly alleviated by making use of a symbolic manipulation program like Mathematica or Maple. One approach is brute force: compute 4×4 matrices like $\not{p}$ in the CM frame, and take their products and traces. If you are familiar with a symbolic-manipulation program, write one that does this. See if you can verify eqs. (48.26)–(48.29).

48.2) Compute $\langle|\mathcal{T}|^2\rangle$ for $e^+e^- \to \varphi\varphi$. You should find that your result is the same as that for $e^-\varphi \to e^-\varphi$, but with $s \leftrightarrow t$, and an extra factor of minus one-half.

This relationship is known as *crossing symmetry*. There is an overall minus sign for each fermion that is moved from the initial to the final state.

48.3) Compute $\langle|\mathcal{T}|^2\rangle$ for $e^-e^- \to e^-e^-$. You should find that your result is the same as that for $e^+e^- \to e^+e^-$, but with $s \leftrightarrow u$. This is another example of crossing symmetry.

48.4) Suppose that $M > 2m$, so that the scalar can decay to an electron-positron pair.

a) Compute the decay rate, summed over final spins.

b) Compute $|\mathcal{T}|^2$ for decay into an electron with spin s_1 and a positron with spin s_2. Take the fermion three-momenta to be along the z axis, and let the x axis be the spin-quantization axis. You should find that $|\mathcal{T}|^2 = 0$ if $s_1 = -s_2$, or if $M = 2m$ (so that the outgoing three-momentum of each fermion is zero). Discuss this in light of conservation of angular momentum and of parity. (Prerequisite: 40.)

c) Compute $|\mathcal{T}|^2$ for decay into an electron with helicity s_1 and a positron with helicity s_2. (See section 38 for the definition of helicity.) You should find that the decay rate is zero if $s_1 = -s_2$. Discuss this in light of conservation of angular momentum and of parity.

d) Now consider changing the interaction to $\mathcal{L}_1 = ig\varphi\overline{\Psi}\gamma_5\Psi$, and compute the spin-summed decay rate. Explain (in light of conservation of angular momentum and of parity) why the decay rate is larger than it was without the $i\gamma_5$ in the interaction.

e) Repeat parts (b) and (c) for the new form of the interaction, and explain any differences in the results.

48.5) The *charged pion* π^- is represented by a complex scalar field φ, the *muon* μ^- by a Dirac field $\mathcal{M}$, and the *muon neutrino* ν_μ by a spin-projected Dirac field $P_{\rm L}\mathcal{N}$, where $P_{\rm L} = \frac{1}{2}(1-\gamma_5)$. The charged pion can decay to a muon and a muon antineutrino via the interaction

$$\mathcal{L}_1 = 2c_1 G_{\rm F} f_\pi \partial_\mu \varphi \overline{\mathcal{M}} \gamma^\mu P_{\rm L} \mathcal{N} + \text{h.c.}\,, \tag{48.30}$$

where c_1 is the cosine of the *Cabibbo angle*, $G_{\rm F}$ is the *Fermi constant*, and f_π is the *pion decay constant.*

a) Compute the charged pion decay rate Γ.

b) The charged pion mass is $m_\pi = 139.6\,\text{MeV}$, the muon mass is $m_\mu = 105.7\,\text{MeV}$, and the muon neutrino is massless. The Fermi constant is measured in muon decay to be $G_{\rm F} = 1.166 \times 10^{-5}\,\text{GeV}^{-2}$, and the cosine of the Cabibbo angle is measured in nuclear beta decays to be $c_1 = 0.974$. The measured value of the charged pion lifetime is $2.603 \times 10^{-8}\,$s. Determine the value of f_π in MeV. Your result is too large by 0.8%, due to neglect of electromagnetic loop corrections.

49

The Feynman rules for Majorana fields

Prerequisite: 45

In this section we will deduce the Feynman rules for Yukawa theory, but with a Majorana field instead of a Dirac field. We can think of the particles associated with the Majorana field as massive neutrinos.

We have

$$\begin{aligned}\mathcal{L}_1 &= \tfrac{1}{2} g \varphi \overline{\Psi} \Psi \\ &= \tfrac{1}{2} g \varphi \Psi^{\mathrm{T}} \mathcal{C} \Psi \,, \end{aligned} \tag{49.1}$$

where Ψ is a Majorana field (with mass m), φ is a real scalar field (with mass M), and g is a coupling constant. In this section, we will be concerned with tree-level processes only, and so we omit renormalizing Z factors.

From section 41, we have the LSZ rules appropriate for a Majorana field,

$$b^\dagger_s(\mathbf{p})_{\text{in}} \rightarrow -i \int d^4x \; e^{+ipx} \, \overline{v}_s(\mathbf{p})(-i\partial\!\!\!/ + m)\Psi(x) \tag{49.2}$$

$$= +i \int d^4x \; \Psi^{\mathrm{T}}(x) \mathcal{C}(+i\overleftarrow{\partial\!\!\!/} + m) u_s(\mathbf{p}) e^{+ipx} \,, \tag{49.3}$$

$$b_{s'}(\mathbf{p}')_{\text{out}} \rightarrow +i \int d^4x \; e^{-ip'x} \, \overline{u}_{s'}(\mathbf{p}')(-i\partial\!\!\!/ + m)\Psi(x) \,, \tag{49.4}$$

$$= -i \int d^4x \; e^{-ip'x} \, \Psi^{\mathrm{T}}(x) \mathcal{C}(+i\overleftarrow{\partial\!\!\!/} + m) v_{s'}(\mathbf{p}') \,. \tag{49.5}$$

Eq. (49.3) follows from eq. (49.2) by taking the transpose of the right-hand side, and using $\overline{v}_{s'}(\mathbf{p}')^{\mathrm{T}} = -\mathcal{C} u_{s'}(\mathbf{p}')$ and $(-i\partial\!\!\!/ + m)^{\mathrm{T}} = \mathcal{C}(+i\partial\!\!\!/ + m)\mathcal{C}^{-1}$; similarly, eq. (49.5) follows from eq. (49.4). Which form we use depends on convenience, and is best chosen on a diagram-by-diagram basis, as we will see shortly.

Eqs. (49.2)–(49.5) lead us to compute correlation functions containing Ψ's, but not $\overline{\Psi}$s. In position space, this leads to Feynman rules where the fermion propagator is $\frac{1}{i}S(x-y)\mathcal{C}^{-1}$, and the $\varphi\Psi\Psi$ vertex is $ig\mathcal{C}$; the factor of $\frac{1}{2}$ in $\mathcal{L}_1$ is canceled by a symmetry factor of 2! that arises from having two identical Ψ fields in $\mathcal{L}_1$. In a particular diagram, as we move along a fermion line, the $\mathcal{C}^{-1}$ in the propagator will cancel against the $\mathcal{C}$ in the vertex, leaving over a final $\mathcal{C}^{-1}$ at one end. This $\mathcal{C}^{-1}$ can be canceled by a $\mathcal{C}$ from eq. (49.3) (for an incoming particle) or eq. (49.5) (for an outgoing particle). On the other hand, for the other end of the same line, we should use either eq. (49.2) (for an incoming particle) or eq. (49.4) (for an outgoing particle) to avoid introducing an extra $\mathcal{C}$ at *that* end. In this way, we can avoid ever having explicit factors of $\mathcal{C}$ in our Feynman rules.

Using this approach, the Feynman rules for this theory are as follows.

1) The total number of incoming and outgoing neutrinos is always even; call this number $2n$. Draw n solid lines. Connect them with internal dashed lines, using a vertex that joins one dashed and two solid lines. Also, attach an external dashed line for each incoming or outgoing scalar. In this way, draw all possible diagrams that are *topologically inequivalent.*
2) Draw arrows on each segment of each solid line; keep the arrow direction continuous along each line.
3) Label each external dashed line with the momentum of an incoming or outgoing scalar. If the particle is incoming, draw an arrow on the dashed line that points *towards* the vertex; If the particle is outgoing, draw an arrow on the dashed line that points *away* from the vertex.
4) Label each external solid line with the momentum of an incoming or outgoing neutrino, but include a minus sign with the momentum if (a) the particle is incoming and the arrow points *away* from the vertex, or (b) the particle is outgoing and the arrow points *towards* the vertex. Do this labeling of external lines in all possible *inequivalent* ways. Two diagrams are considered *equivalent* if they can be transformed into each other by reversing all the arrows on one or more fermion lines, and correspondingly changing the signs of the external momenta on each arrow-reversed line.
5) Assign each internal line its own four-momentum. Think of the four-momenta as flowing along the arrows, and conserve four-momentum at each vertex. For a tree diagram, this fixes the momenta on all the internal lines.
6) The value of a diagram consists of the following factors:
   for each incoming or outgoing scalar, 1;
   for each incoming neutrino labeled with $+p_i$, $u_{s_i}(\mathbf{p}_i)$;
   for each incoming neutrino labeled with $-p_i$, $\overline{v}_{s_i}(\mathbf{p}_i)$;
   for each outgoing neutrino labeled with $+p'_i$, $\overline{u}_{s'_i}(\mathbf{p}'_i)$;
   for each outgoing neutrino labeled with $-p'_i$, $v_{s'_i}(\mathbf{p}'_i)$;
   for each vertex, ig;

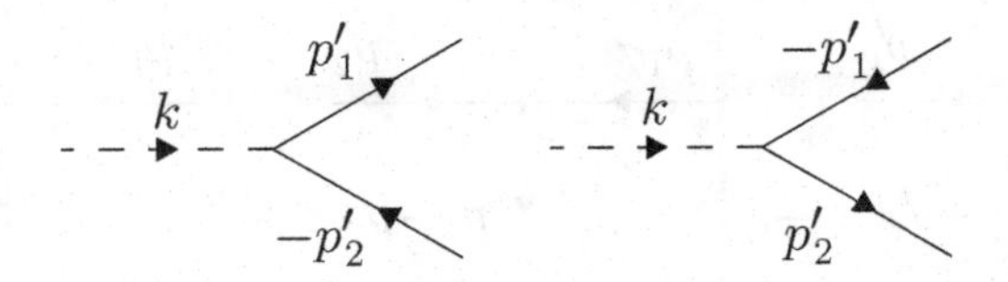

Figure 49.1. Two equivalent diagrams for $\varphi \to \nu\nu$.

for each internal scalar, $-i/(k^2 + M^2 - i\epsilon)$;
for each internal fermion, $-i(-\not{p} + m)/(p^2 + m^2 - i\epsilon)$.

7) Spinor indices are contracted by starting at one end of a fermion line: specifically, the end that has the arrow pointing away from the vertex. The factor associated with the external line is either $\overline{u}$ or $\overline{v}$. Go along the complete fermion line, following the arrows backwards, and writing down (in order from left to right) the factors associated with the vertices and propagators that you encounter. The last factor is either a u or a v. Repeat this procedure for the other fermion lines, if any.
8) The overall sign of a tree diagram is determined by drawing all contributing diagrams in a standard form: all fermion lines horizontal, with their arrows pointing from left to right, and with the left endpoints labeled in the same fixed order (from top to bottom); if the ordering of the labels on the right endpoints of the fermion lines in a given diagram is an even (odd) permutation of an arbitrarily chosen fixed ordering, then the sign of that diagram is positive (negative). To compare two diagrams, it may be necessary to use the arrow-reversing equivalence relation of rule no. 4; there is then an extra minus sign for each arrow-reversed line.
9) The value of $i\mathcal{T}$ is given by a sum over the values of all these diagrams.

There are additional rules for counterterms and loops, but we will postpone those to section 51.

Let us look at the simplest process, $\varphi \to \nu\nu$. There are two possible diagrams for this, shown in fig. 49.1. However, according to rule no. 4, these two diagrams are equivalent, and we should keep only one of them. The first diagram yields $i\mathcal{T}_1 = ig\,\overline{u}_1' v_2'$ and the second $i\mathcal{T}_2 = ig\,\overline{u}_2' v_1'$. Rule no. 8 then implies that we should have $\mathcal{T}_1 = -\mathcal{T}_2$. To check this, we note that (after dropping primes to simplify the notation)

$$\begin{aligned}\overline{u}_1 v_2 &= [\overline{u}_1 v_2]^{\mathrm{T}} \\ &= v_2^{\mathrm{T}}\,\overline{u}_1^{\mathrm{T}} \\ &= \overline{u}_2 \mathcal{C}^{-1}\mathcal{C}^{-1} v_1 \\ &= -\overline{u}_2 v_1\,, \end{aligned} \tag{49.6}$$

as required.

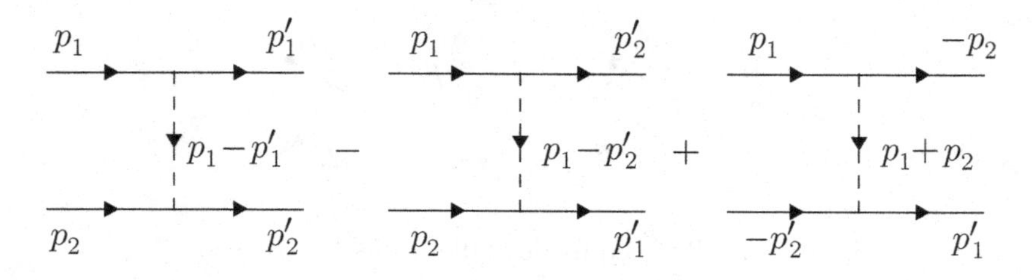

Figure 49.2. Diagrams for $\nu\nu \to \nu\nu$, corresponding to eq. (49.7).

In general, for processes with a total of just two incoming and outgoing neutrinos, such as $\nu\varphi \to \nu\varphi$ or $\nu\nu \to \varphi\varphi$, these rules give (up to an irrelevant overall sign) the same result for $i\mathcal{T}$ as we would get for the corresponding process in the Dirac case, $e^-\varphi \to e^-\varphi$ or $e^+e^- \to \varphi\varphi$. (Note, however, that in the Dirac case, we have $\mathcal{L}_1 = g\varphi\overline{\Psi}\Psi$, as compared with $\mathcal{L}_1 = \frac{1}{2}g\varphi\overline{\Psi}\Psi$ in the Majorana case.)

The differences between Dirac and Majorana fermions become more pronounced for $\nu\nu \to \nu\nu$. Now there are *three* inequivalent contributing diagrams, shown in fig. 49.2. The corresponding amplitude can be written as

$$i\mathcal{T} = \tfrac{1}{i}(ig)^2 \left[\frac{(\overline{u}_1'u_1)(\overline{u}_2'u_2)}{-t+M^2} - \frac{(\overline{u}_2'u_1)(\overline{u}_1'u_2)}{-u+M^2} + \frac{(\overline{v}_2u_1)(\overline{u}_1'v_2')}{-s+M^2} \right], \tag{49.7}$$

where $s = -(p_1+p_2)^2$, $t = -(p_1-p_1')^2$ and $u = -(p_1-p_2')^2$. After arbitrarily assigning the first diagram a plus sign, the minus sign of the second diagram follows from rule no. 8. To get the sign of the third diagram, we compare it with the first. To do so, we reverse the arrow direction on the lower line of the first diagram (which yields an extra minus sign), and then redraw it in standard form. Comparing this modified first diagram with the third diagram (and invoking rule no. 8) reveals a relative minus sign. Since the modified first diagram has a minus sign from the arrow reversal, we conclude that the third diagram has an overall plus sign.

After taking the absolute square of eq. (49.7), we can use relations like eq. (49.6) on a term-by-term basis to put everything into a form that allows the spin sums to be performed in the standard way. In fact, we have already done all the necessary work in the Dirac case. The s-s, s-t, and t-t terms in $\langle|\mathcal{T}|^2\rangle$ for $\nu\nu \to \nu\nu$ are the same as those for $e^+e^- \to e^+e^-$, while the t-t, t-u, and u-u terms are the same as those for the crossing-related process $e^-e^- \to e^-e^-$. Finally, the s-u terms can be obtained from the s-t terms via $t \leftrightarrow u$, or equivalently from the t-u terms via $t \leftrightarrow s$. Thus the result is

$$\langle|\mathcal{T}|^2\rangle = g^4\left[\frac{(s-4m^2)^2}{(M^2-s)^2}+\frac{st-4m^2u}{(M^2-s)(M^2-t)}\right.$$
$$+\frac{(t-4m^2)^2}{(M^2-t)^2}+\frac{tu-4m^2s}{(M^2-t)(M^2-u)}$$
$$\left.+\frac{(u-4m^2)^2}{(M^2-u)^2}+\frac{us-4m^2t}{(M^2-u)(M^2-s)}\right], \qquad (49.8)$$

which is neatly symmetric on permutations of s, t, and u.

Problems

49.1) Let Ψ be a Dirac field (representing the electron and positron), X be a Majorana field (representing the *photino*, the hypothetical supersymmetric partner of the photon, with mass $m_{\tilde{\gamma}}$), and $E_{\rm L}$ and $E_{\rm R}$ be two different complex scalar fields (representing the two *selectrons*, the hypothetical supersymmetric partners of the left-handed electron and the right-handed electron, with masses $M_{\rm L}$, and $M_{\rm R}$; note that the subscripts L and R are just part of their names, and do not signify anything about their Lorentz transformation properties). They interact via

$$\mathcal{L}_1 = \sqrt{2}eE^\dagger_{\rm L}\overline{X}P_{\rm L}\Psi + \sqrt{2}eE^\dagger_{\rm R}\overline{X}P_{\rm R}\Psi + \text{h.c.}\,, \qquad (49.9)$$

where $\alpha = e^2/4\pi \simeq 1/137$ is the fine-structure constant, and $P_{\rm L,R} = \frac{1}{2}(1\mp\gamma_5)$.

a) Write down the hermitian conjugate term explicitly.

b) Find the tree-level scattering amplitude for $e^+e^- \to \tilde{\gamma}\tilde{\gamma}$. Hint: there are four contributing diagrams, two each in the t and u channels, with exchange of either $E_{\rm L}$ or $E_{\rm R}$.

c) Compute the spin-averaged differential cross section for this process in the case that m_e (the electron mass) can be neglected, and $|t|,|u| \ll M_{\rm L} = M_{\rm R}$. Express it as a function of s and the center-of-mass scattering angle θ.

50

Massless particles and spinor helicity

Prerequisite: 48

Scattering amplitudes often simplify greatly if the particles are massless (or can be approximated as massless because the Mandelstam variables all have magnitudes much larger than the particle masses squared). In this section we will explore this phenomenon for spin-one-half (and spin-zero) particles. We will begin by developing the technology of *spinor helicity*, which will prove to be of indispensible utility in Part III.

Recall from section 38 that the u spinors for a massless spin-one-half particle obey

$$u_s(\mathbf{p})\overline{u}_s(\mathbf{p}) = \tfrac{1}{2}(1 + s\gamma_5)(-\not{p}) \, , \qquad (50.1)$$

where $s = \pm$ specifies the *helicity*, the component of the particle's spin measured along the axis specified by its three-momentum; in this notation the helicity is $\frac{1}{2}s$. The v spinors obey a similar relation,

$$v_s(\mathbf{p})\overline{v}_s(\mathbf{p}) = \tfrac{1}{2}(1 - s\gamma_5)(-\not{p}) \, . \qquad (50.2)$$

In fact, in the massless case, with the phase conventions of section 38, we have $v_s(\mathbf{p}) = u_{-s}(\mathbf{p})$. Thus we can confine our discussion to u-type spinors only, since we need merely change the sign of s to accomodate v-type spinors.

Consider a u spinor for a particle of negative helicity. We have

$$u_-(\mathbf{p})\overline{u}_-(\mathbf{p}) = \tfrac{1}{2}(1 - \gamma_5)(-\not{p}) \, . \qquad (50.3)$$

Let us define

$$p_{a\dot{a}} \equiv p_\mu \sigma^\mu_{a\dot{a}} \, . \qquad (50.4)$$

Then we also have

$$p^{\dot{a}a} = \varepsilon^{ac}\varepsilon^{\dot{a}\dot{c}} p_{c\dot{c}} = p_\mu \bar{\sigma}^{\mu\dot{a}a} \, . \qquad (50.5)$$

Then, using

$$\gamma^\mu = \begin{pmatrix} 0 & \sigma^\mu \\ \bar{\sigma}^\mu & 0 \end{pmatrix}, \qquad \tfrac{1}{2}(1-\gamma_5) = \begin{pmatrix} 1 & 0 \\ 0 & 0 \end{pmatrix} \tag{50.6}$$

in eq. (50.3), we find

$$u_-(\mathbf{p})\overline{u}_-(\mathbf{p}) = \begin{pmatrix} 0 & -p_{a\dot{a}} \\ 0 & 0 \end{pmatrix}. \tag{50.7}$$

On the other hand, we know that the lower two components of $u_-(\mathbf{p})$ vanish, and so we can write

$$u_-(\mathbf{p}) = \begin{pmatrix} \phi_a \\ 0 \end{pmatrix}. \tag{50.8}$$

Here ϕ_a is a two-component numerical spinor; it is not an anticommuting object. Such a commuting spinor is sometimes called a *twistor*. An explicit numerical formula for it (verified in problem 50.2) is

$$\phi_a = \sqrt{2\omega} \begin{pmatrix} -\sin(\frac{1}{2}\theta)e^{-i\phi} \\ +\cos(\frac{1}{2}\theta) \end{pmatrix}, \tag{50.9}$$

where θ and ϕ are the polar and azimuthal angles that specify the direction of the three-momentum $\mathbf{p}$, and $\omega = |\mathbf{p}|$. Barring eq. (50.8) yields

$$\overline{u}_-(\mathbf{p}) = (\,0,\;\; \phi^*_{\dot{a}}\,)\,, \tag{50.10}$$

where $\phi^*_{\dot{a}} = (\phi_a)^*$. Now, combining eqs. (50.8) and (50.10), we get

$$u_-(\mathbf{p})\overline{u}_-(\mathbf{p}) = \begin{pmatrix} 0 & \phi_a\phi^*_{\dot{a}} \\ 0 & 0 \end{pmatrix}. \tag{50.11}$$

Comparing with eq. (50.7), we see that

$$p_{a\dot{a}} = -\phi_a\phi^*_{\dot{a}}\,. \tag{50.12}$$

This expresses the four-momentum of the particle neatly in terms of the twistor that describes its spin state. The essence of the spinor helicity method is to treat ϕ_a as the fundamental object, and to express the particle's four-momentum in terms of it, via eq. (50.12).

Given eq. (50.8), and the phase conventions of section 38, the positive-helicity spinor is

$$u_+(\mathbf{p}) = \begin{pmatrix} 0 \\ \phi^{*\dot{a}} \end{pmatrix}, \tag{50.13}$$

where $\phi^{*\dot{a}} = \varepsilon^{\dot{a}\dot{c}}\phi^*_{\dot{c}}$. Barring eq. (50.13) yields

$$\overline{u}_+(\mathbf{p}) = (\,\phi^a, \quad 0\,)\,. \tag{50.14}$$

Computation of $u_+(\mathbf{p})\overline{u}_+(\mathbf{p})$ via eqs. (50.13) and (50.14), followed by comparison with eq. (50.1) with $s = +$, then reproduces eq. (50.12), but with the indices raised.

In fact, the decomposition of $p_{a\dot{a}}$ into the direct product of a twistor and its complex conjugate is unique (up to an overall phase for the twistor). To see this, use $\sigma^\mu = (I, \vec{\sigma})$ to write

$$p_{a\dot{a}} = \begin{pmatrix} -p^0 + p^3 & p^1 - ip^2 \\ p^1 + ip^2 & -p^0 - p^3 \end{pmatrix}. \tag{50.15}$$

The determinant of this matrix is $(p^0)^2 - \mathbf{p}^2$, and this vanishes because the particle is (by assumption) massless. Thus $p_{a\dot{a}}$ has a zero eigenvalue. Therefore, it can be written as a projection onto the eigenvector corresponding to the nonzero eigenvalue. That is what eq. (50.12) represents, with the nonzero eigenvalue absorbed into the normalization of the eigenvector ϕ_a.

Let us now introduce some useful notation. Let p and k be two four-momenta, and ϕ_a and κ_a the corresponding twistors. We define the twistor product

$$[p\,k] \equiv \phi^a \kappa_a\,. \tag{50.16}$$

Because $\phi^a\kappa_a = \varepsilon^{ac}\phi_c\kappa_a$, and the twistors commute, we have

$$[k\,p] = -[p\,k]\,. \tag{50.17}$$

From eqs. (50.8) and (50.14), we can see that

$$\overline{u}_+(\mathbf{p})u_-(\mathbf{k}) = [p\,k]\,. \tag{50.18}$$

Similarly, let us define

$$\langle p\,k\rangle \equiv \phi^*_{\dot{a}}\kappa^{*\dot{a}}\,. \tag{50.19}$$

Comparing with eq. (50.16) we see that

$$\langle p\,k\rangle = [k\,p]^*\,, \tag{50.20}$$

which implies that this product is also antisymmetric,

$$\langle k\,p\rangle = -\langle p\,k\rangle\,. \tag{50.21}$$

Also, from eqs. (50.10) and (50.13), we have

$$\overline{u}_-(\mathbf{p})u_+(\mathbf{k}) = \langle p\,k\rangle\,. \tag{50.22}$$

Note that the other two possible spinor products vanish:

$$\overline{u}_+(\mathbf{p})u_+(\mathbf{k}) = \overline{u}_-(\mathbf{p})u_-(\mathbf{k}) = 0\ . \tag{50.23}$$

The twistor products $\langle p\,k\rangle$ and $[p\,k]$ satisfy another important relation,

$$\begin{aligned}\langle p\,k\rangle[k\,p] &= (\phi^*_{\dot a}\kappa^{*\dot a})(\kappa^a\phi_a) \\ &= (\phi_a\phi^*_{\dot a})(\kappa^{*\dot a}\kappa^a) \\ &= p_{a\dot a}k^{\dot a a} \\ &= -2p^\mu k_\mu\ ,\end{aligned} \tag{50.24}$$

where the last line follows from $\bar\sigma^{\mu\dot a a}\sigma^\nu_{a\dot a} = -2g^{\mu\nu}$.

Let us apply this notation to the tree-level scattering amplitude for $e^-\varphi \to e^-\varphi$ in Yukawa theory, which we first computed in section 45, and which reads

$$\mathcal{T}_{s's} = g^2\,\overline{u}_{s'}(\mathbf{p}')\Big[\tilde S(p{+}k) + \tilde S(p{-}k')\Big]u_s(\mathbf{p})\ . \tag{50.25}$$

For a massless fermion, $\tilde S(p) = -\not{p}/p^2$. If the scalar is also massless, then $(p+k)^2 = 2p\cdot k$ and $(p-k')^2 = -2p\cdot k'$. Also, we can remove the $\not{p}$s in the propagator numerators in eq. (50.25), because $\not{p}u_s(\mathbf{p}) = 0$. Thus we have

$$\mathcal{T}_{s's} = g^2\,\overline{u}_{s'}(\mathbf{p}')\left[\frac{-\not{k}}{2p{\cdot}k} + \frac{-\not{k}'}{2p{\cdot}k'}\right]u_s(\mathbf{p})\ . \tag{50.26}$$

Now consider the case $s' = s = +$. From eqs. (50.13), (50.14), and

$$-\not{k} = \begin{pmatrix} 0 & \kappa_a\kappa^*_{\dot a} \\ \kappa^{*\dot a}\kappa^a & 0 \end{pmatrix}, \tag{50.27}$$

we get

$$\begin{aligned}\overline{u}_+(\mathbf{p}')(-\not{k})u_+(\mathbf{p}) &= \phi'^a\kappa_a\kappa^*_{\dot a}\phi^{*\dot a} \\ &= [p'\,k]\,\langle k\,p\rangle\ .\end{aligned} \tag{50.28}$$

Similarly, for $s' = s = -$, we find

$$\begin{aligned}\overline{u}_-(\mathbf{p}')(-\not{k})u_-(\mathbf{p}) &= \phi'^*_{\dot a}\kappa^{*\dot a}\kappa^a\phi_a \\ &= \langle p'\,k\rangle\,[k\,p]\ ,\end{aligned} \tag{50.29}$$

while for $s' \ne s$, the amplitude vanishes:

$$\overline{u}_-(\mathbf{p}')(-\not{k})u_+(\mathbf{p}) = \overline{u}_+(\mathbf{p}')(-\not{k})u_-(\mathbf{p}) = 0\ . \tag{50.30}$$

Then, using eq. (50.24) on the denominators in eq. (50.26), we find

$$\mathcal{T}_{++} = -g^2\left(\frac{[p'\,k]}{[p\,k]} + \frac{[p'\,k']}{[p\,k']}\right),$$

$$\mathcal{T}_{--} = -g^2\left(\frac{\langle p'\,k\rangle}{\langle p\,k\rangle} + \frac{\langle p'\,k'\rangle}{\langle p\,k'\rangle}\right), \tag{50.31}$$

while

$$\mathcal{T}_{+-} = \mathcal{T}_{-+} = 0\,. \tag{50.32}$$

Thus we have rather simple expressions for the fixed-helicity scattering amplitudes in terms of twistor products.

Reference notes

Spinor-helicity methods are discussed by *Siegel*.

Problems

50.1) Consider a bra-ket notation for twistors,

$$|p] = u_-(\mathbf{p}) = v_+(\mathbf{p})\,,$$

$$|p\rangle = u_+(\mathbf{p}) = v_-(\mathbf{p})\,,$$

$$[p| = \overline{u}_+(\mathbf{p}) = \overline{v}_-(\mathbf{p})\,,$$

$$\langle p| = \overline{u}_-(\mathbf{p}) = \overline{v}_+(\mathbf{p})\,. \tag{50.33}$$

We then have

$$\langle k|\,|p\rangle = \langle k\,p\rangle\,,$$

$$[k|\,|p] = [k\,p]\,,$$

$$\langle k|\,|p] = 0\,,$$

$$[k|\,|p\rangle = 0\,. \tag{50.34}$$

a) Show that

$$-\not{p} = |p\rangle[p| + |p]\langle p|\,, \tag{50.35}$$

where p is any massless four-momentum.

b) Use this notation to rederive eqs. (50.28)–(50.30).

50.2) a) Use eqs. (50.9) and (50.15) to verify eq. (50.12).

b) Let the three-momentum $\mathbf{p}$ be in the $+\hat{\mathbf{z}}$ direction. Use eq. (38.12) to compute $u_\pm(\mathbf{p})$ explicitly in the massless limit (corresponding to the limit $\eta \to \infty$, where $\sinh \eta = |\mathbf{p}|/m$). Verify that, when $\theta = 0$, your results agree with eqs. (50.8), (50.9), and (50.13).

50.3) Prove the *Schouten identity*,

$$\langle p\,q\rangle\langle r\,s\rangle + \langle p\,r\rangle\langle s\,q\rangle + \langle p\,s\rangle\langle q\,r\rangle = 0\,. \tag{50.36}$$

Hint: note that the left-hand side is completely antisymmetric in the three labels q, r, and s, and that each corresponding twistor has only two components.

50.4) Show that

$$\langle p\,q\rangle\,[q\,r]\,\langle r\,s\rangle\,[s\,p] = \mathrm{Tr}\,\tfrac{1}{2}(1-\gamma_5)\not{p}\not{q}\not{r}\not{s}\,, \tag{50.37}$$

and evaluate the right-hand side.

50.5) a) Prove the useful identities

$$\langle p|\gamma^\mu|k] = [k|\gamma^\mu|p\rangle\,, \tag{50.38}$$

$$\langle p|\gamma^\mu|k]^* = \langle k|\gamma^\mu|p]\,, \tag{50.39}$$

$$\langle p|\gamma^\mu|p] = 2p^\mu\,, \tag{50.40}$$

$$\langle p|\gamma^\mu|k\rangle = 0\,, \tag{50.41}$$

$$[p|\gamma^\mu|k] = 0\,. \tag{50.42}$$

b) Extend the last two identities of part (a): show that the product of an odd number of gamma matrices sandwiched between either $\langle p|$ and $|k\rangle$ or $[p|$ and $|k]$ vanishes. Also show that the product of an even number of gamma matrices between either $\langle p|$ and $|k]$ or $[p|$ and $|k\rangle$ vanishes.

c) Prove the Fierz identities,

$$-\tfrac{1}{2}\langle p|\gamma_\mu|q]\gamma^\mu = |q]\langle p| + |p\rangle[q|\,, \tag{50.43}$$

$$-\tfrac{1}{2}[p|\gamma_\mu|q\rangle\gamma^\mu = |q\rangle[p| + |p]\langle q|\,. \tag{50.44}$$

Now take the matrix element of eq. (50.44) between $\langle r|$ and $|s]$ to get another useful form of the Fierz identity,

$$[p|\gamma^\mu|q\rangle\,\langle r|\gamma_\mu|s] = 2\,[p\,s]\,\langle q\,r\rangle\,. \tag{50.45}$$

51

Loop corrections in Yukawa theory

Prerequisites: 19, 40, 48

In this section we will compute the one-loop corrections in Yukawa theory with a Dirac field. The basic concepts are all the same as for a scalar field, and so we will mainly be concerned with the extra technicalities arising from spin indices and anticommutation.

First let us note that the general discussion of sections 18 and 29 leads us to expect that we will need to add to the lagrangian all possible terms whose coefficients have positive or zero mass dimension, and that respect the symmetries of the original lagrangian. These include Lorentz symmetry, the U(1) phase symmetry of the Dirac field, and the discrete symmetries of parity, time reversal, and charge conjugation.

The mass dimensions of the fields (in four spacetime dimensions) are $[\varphi] = 1$ and $[\Psi] = \frac{3}{2}$. Thus any power of φ up to φ^4 is allowed. But there are no additional required terms involving Ψ: the only candidates contain either γ_5 (e.g., $i\overline{\Psi}\gamma_5\Psi$) and are forbidden by parity, or $\mathcal{C}$ (e.g., $\Psi^{\mathrm{T}}\mathcal{C}\Psi$) and are forbidden by the U(1) symmetry.

Nevertheless, having to deal with the addition of three new terms (φ, φ^3, φ^4) is annoying enough to prompt us to look for a simpler example. Consider, then, a modified form of the Yukawa interaction,

$$\mathcal{L}_{\text{Yuk}} = ig\varphi\overline{\Psi}\gamma_5\Psi \ . \tag{51.1}$$

This interaction will conserve parity if and only if φ is a pseudoscalar:

$$P^{-1}\varphi(\mathbf{x},t)P = -\varphi(-\mathbf{x},t) \ . \tag{51.2}$$

Then, φ and φ^3 are odd under parity, and so we will *not* need to add them to $\mathcal{L}$. The one term we will need to add is φ^4.

Therefore, the theory we will consider is

$$\mathcal{L} = \mathcal{L}_0 + \mathcal{L}_1 \,, \tag{51.3}$$

$$\mathcal{L}_0 = i\overline{\Psi}\partial\!\!\!/\Psi - m\overline{\Psi}\Psi - \tfrac{1}{2}\partial^\mu\varphi\partial_\mu\varphi - \tfrac{1}{2}M^2\varphi^2 \,, \tag{51.4}$$

$$\mathcal{L}_1 = iZ_g g\varphi\overline{\Psi}\gamma_5\Psi - \tfrac{1}{24}Z_\lambda\lambda\varphi^4 + \mathcal{L}_{\rm ct} \,, \tag{51.5}$$

$$\begin{aligned}\mathcal{L}_{\rm ct} = {}& i(Z_\Psi - 1)\overline{\Psi}\partial\!\!\!/\Psi - (Z_m - 1)m\overline{\Psi}\Psi \\ & - \tfrac{1}{2}(Z_\varphi - 1)\partial^\mu\varphi\partial_\mu\varphi - \tfrac{1}{2}(Z_M - 1)M^2\varphi^2 \,,\end{aligned} \tag{51.6}$$

where λ is a new coupling constant. We will use an on-shell renormalization scheme. The lagrangian parameter m is then the actual mass of the electron. We will define the couplings g and λ as the values of appropriate vertex functions when the external four-momenta vanish. Finally, the fields are normalized according to the requirements of the LSZ formula. In practice, this means that the scalar and fermion propagators must have appropriate poles with unit residue.

We will assume that $M < 2m$, so that the scalar is stable against decay into an electron-positron pair. The exact scalar propagator (in momentum space) can be then written in Lehmann–Källén form as

$$\tilde{\mathbf{\Delta}}(k^2) = \frac{1}{k^2 + M^2 - i\epsilon} + \int_{M_{\rm th}^2}^\infty ds \, \frac{\rho(s)}{k^2 + s - i\epsilon} \,, \tag{51.7}$$

where the spectral density $\rho(s)$ is real and nonnegative. The threshold mass $M_{\rm th}$ is either $2m$ (corresponding to the contribution of an electron-positron pair) or $3M$ (corresponding to the contribution of three scalars; by parity, there is no contribution from two scalars), whichever is less.

We can also write

$$\tilde{\mathbf{\Delta}}(k^2)^{-1} = k^2 + M^2 - i\epsilon - \Pi(k^2) \,, \tag{51.8}$$

where $i\Pi(k^2)$ is given by the sum of one-particle irreducible (1PI for short; see section 14) diagrams with two external scalar lines, and the external propagators removed. The fact that $\tilde{\mathbf{\Delta}}(k^2)$ has a pole at $k^2 = -M^2$ with residue one implies that $\Pi(-M^2) = 0$ and $\Pi'(-M^2) = 0$; this fixes the coefficients Z_φ and Z_M.

All of this is mimicked for the Dirac field. When parity is conserved, the exact propagator (in momentum space) can be written in Lehmann–Källén form as

$$\tilde{\mathbf{S}}(p\!\!\!/) = \frac{-p\!\!\!/ + m}{p^2 + m^2 - i\epsilon} + \int_{m_{\rm th}^2}^\infty ds \, \frac{-p\!\!\!/\rho_1(s) + \sqrt{s}\,\rho_2(s)}{p^2 + s - i\epsilon} \,, \tag{51.9}$$

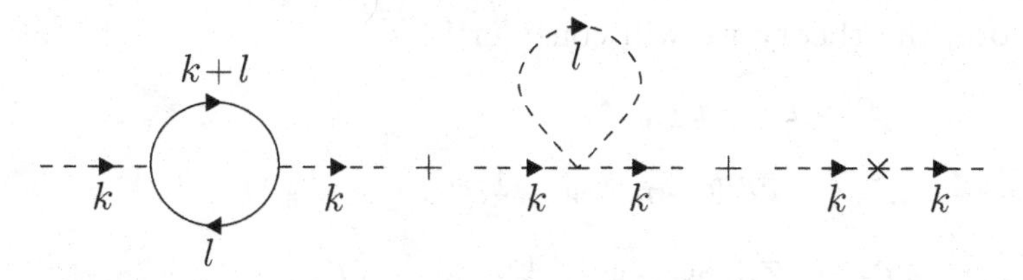

Figure 51.1. The one-loop and counterterm corrections to the scalar propagator in Yukawa theory.

where the spectral densities $\rho_1(s)$ and $\rho_2(s)$ are both real, and $\rho_1(s)$ is non-negative and greater than $\rho_2(s)$. The threshold mass $m_{\rm th}$ is $m+M$ (corresponding to the contribution of a fermion and a scalar), which, by assumption, is less than $3m$ (corresponding to the contribution of three fermions; by Lorentz invariance, there is no contribution from two fermions).

Since $p^2 = -\not{p}\not{p}$, we can rewrite eq. (51.9) as

$$\tilde{\mathbf{S}}(\not{p}) = \frac{1}{\not{p}+m-i\epsilon} + \int_{m_{\rm th}^2}^{\infty} ds\, \frac{-\not{p}\rho_1(s)+\sqrt{s}\,\rho_2(s)}{(-\not{p}+\sqrt{s}-i\epsilon)(\not{p}+\sqrt{s}-i\epsilon)}\,, \tag{51.10}$$

with the understanding that $1/(\ldots)$ refers to the matrix inverse. However, since $\not{p}$ is the only matrix involved, we can think of $\tilde{\mathbf{S}}(\not{p})$ as an analytic function of the single variable $\not{p}$. With this idea in mind, we see that $\tilde{\mathbf{S}}(\not{p})$ has an isolated pole at $\not{p} = -m$ with residue one. This residue corresponds to the field normalization that is needed for the validity of the LSZ formula.

We can also write the exact fermion propagator in the form

$$\tilde{\mathbf{S}}(\not{p})^{-1} = \not{p}+m-i\epsilon-\Sigma(\not{p})\,, \tag{51.11}$$

where $i\Sigma(\not{p})$ is given by the sum of 1PI diagrams with two external fermion lines, and the external propagators removed. The fact that $\tilde{\mathbf{S}}(\not{p})$ has a pole at $\not{p} = -m$ with residue one implies that $\Sigma(-m) = 0$ and $\Sigma'(-m) = 0$; this fixes the coefficients Z_Ψ and Z_m.

We proceed to the diagrams. The Yukawa vertex carries a factor of $i(iZ_g g)\gamma_5 = -Z_g g\gamma_5$. Since $Z_g = 1+O(g^2)$, we can set $Z_g = 1$ in the one-loop diagrams.

Consider first $\Pi(k^2)$, which receives the one-loop (and counterterm) corrections shown in fig. 51.1. The first diagram has a closed fermion loop. As we will see in problem 51.1 (and section 53), anticommutation of the fermion fields results in an extra factor of minus one for each closed fermion loop. The spin indices on the propagators and vertices are contracted in the usual way, following the arrows backwards. Since the loop closes on itself, we end up with a trace over the spin indices. Thus we have

$$i\Pi_{\Psi\,\text{loop}}(k^2) = (-1)(-g)^2\left(\tfrac{1}{i}\right)^2 \int \frac{d^4\ell}{(2\pi)^4}\,\text{Tr}\Big[\tilde{S}(\not\ell+\not k)\gamma_5\tilde{S}(\not\ell)\gamma_5\Big]\,, \tag{51.12}$$

where

$$\tilde{S}(\not p) = \frac{-\not p + m}{p^2 + m^2 - i\epsilon} \tag{51.13}$$

is the free fermion propagator in momentum space.

We now proceed to evaluate eq. (51.12). We have

$$\begin{aligned}\text{Tr}[(-\not\ell - \not k + m)\gamma_5(-\not\ell + m)\gamma_5] &= \text{Tr}[(-\not\ell - \not k + m)(+\not\ell + m)]\\ &= 4[(\ell + k)\ell + m^2]\\ &\equiv 4N\,.\end{aligned} \tag{51.14}$$

The first equality follows from $\gamma_5^2 = 1$ and $\gamma_5 \not p \gamma_5 = -\not p$.

Next we combine the denominators with Feynman's formula. Suppressing the $i\epsilon$s, we have

$$\frac{1}{(\ell{+}k)^2 + m^2}\,\frac{1}{\ell^2 + m^2} = \int_0^1 dx\,\frac{1}{(q^2 + D)^2}\,, \tag{51.15}$$

where $q = \ell + xk$ and $D = x(1{-}x)k^2 + m^2$.

We then change the integration variable in eq. (51.12) from ℓ to q; the result is

$$i\Pi_{\Psi\,\text{loop}}(k^2) = 4g^2 \int_0^1 dx \int \frac{d^4q}{(2\pi)^4}\,\frac{N}{(q^2 + D)^2}\,, \tag{51.16}$$

where now $N = (q + (1{-}x)k)(q - xk) + m^2$. The integral diverges, and so we analytically continue it to $d = 4 - \varepsilon$ spacetime dimensions. (Here we ignore a subtlety with the definition of γ_5 in d dimensions, and assume that $\gamma_5^2 = 1$ and $\gamma_5 \not p \gamma_5 = -\not p$ continue to hold.) We also make the replacement $g \to g\tilde{\mu}^{\varepsilon/2}$, where $\tilde{\mu}$ has dimensions of mass, so that g remains dimensionless.

Expanding out the numerator, we have

$$N = q^2 - x(1{-}x)k^2 + m^2 + (1{-}2x)kq\,. \tag{51.17}$$

The term linear in q integrates to zero. For the rest, we use the general result of section 14 to get

$$\tilde{\mu}^{\varepsilon}\int\frac{d^dq}{(2\pi)^d}\frac{1}{(q^2+D)^2}=\frac{i}{16\pi^2}\left[\frac{2}{\varepsilon}-\ln(D/\mu^2)\right], \tag{51.18}$$

$$\tilde{\mu}^{\varepsilon}\int\frac{d^dq}{(2\pi)^d}\frac{q^2}{(q^2+D)^2}=\frac{i}{16\pi^2}\left[\frac{2}{\varepsilon}+\tfrac{1}{2}-\ln(D/\mu^2)\right](-2D), \tag{51.19}$$

where $\mu^2=4\pi e^{-\gamma}\tilde{\mu}^2$, and we have dropped terms of order ε. Plugging eqs. (51.18) and (51.19) into eq. (51.16) yields

$$\Pi_{\Psi\,\text{loop}}(k^2)=-\frac{g^2}{4\pi^2}\left[\frac{1}{\varepsilon}(k^2+2m^2)+\tfrac{1}{6}k^2+m^2\right.$$
$$\left.-\int_0^1 dx\,(3x(1-x)k^2+m^2)\ln(D/\mu^2)\right]. \tag{51.20}$$

We see that the divergent term has (as expected) a form that permits cancellation by the counterterms.

We evaluated the second diagram of fig. 51.1 in section 31, with the result

$$\Pi_{\varphi\,\text{loop}}(k^2)=\frac{\lambda}{(4\pi)^2}\left[\frac{1}{\varepsilon}+\tfrac{1}{2}-\tfrac{1}{2}\ln(M^2/\mu^2)\right]M^2. \tag{51.21}$$

The third diagram gives the contribution of the counterterms,

$$\Pi_{\text{ct}}(k^2)=-(Z_\varphi-1)k^2-(Z_M-1)M^2. \tag{51.22}$$

Adding up eqs. (51.20)–(51.22), we see that finiteness of $\Pi(k^2)$ requires

$$Z_\varphi=1-\frac{g^2}{4\pi^2}\left(\frac{1}{\varepsilon}+\text{finite}\right), \tag{51.23}$$

$$Z_M=1+\left(\frac{\lambda}{16\pi^2}-\frac{g^2}{2\pi^2}\frac{m^2}{M^2}\right)\left(\frac{1}{\varepsilon}+\text{finite}\right), \tag{51.24}$$

plus higher-order (in g and/or λ) corrections. Note that, although there is an $O(\lambda)$ correction to Z_M, there is not an $O(\lambda)$ correction to Z_φ.

We can impose $\Pi(-M^2)=0$ by writing

$$\Pi(k^2)=\frac{g^2}{4\pi^2}\left[\int_0^1 dx\,(3x(1-x)k^2+m^2)\ln(D/D_0)+\kappa_\varphi(k^2+M^2)\right], \tag{51.25}$$

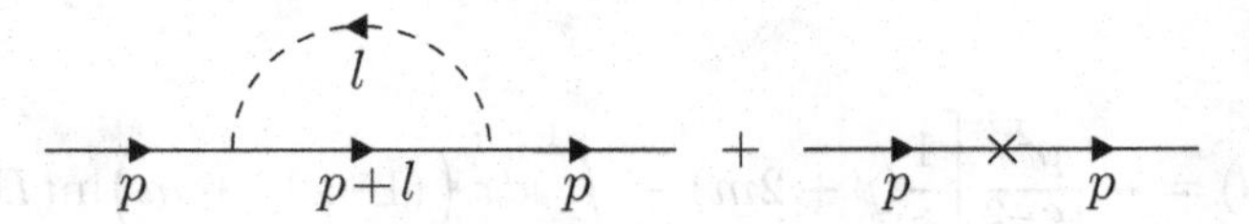

Figure 51.2. The one-loop and counterterm corrections to the fermion propagator in Yukawa theory.

where $D_0 = -x(1-x)M^2 + m^2$, and κ_φ is a constant to be determined. We fix κ_φ by imposing $\Pi'(-M^2) = 0$, which yields

$$\kappa_\varphi = \int_0^1 dx\; x(1-x)[3x(1-x)M^2 - m^2]/D_0 \; . \tag{51.26}$$

Note that, in this on-shell renormalization scheme, there is no $O(\lambda)$ correction to $\Pi(k^2)$.

Next we turn to the Ψ propagator, which receives the one-loop (and counterterm) corrections shown in fig. 51.2. The spin indices are contracted in the usual way, following the arrows backwards. We have

$$i\Sigma_{1\,\text{loop}}(\not{p}) = (-g)^2\left(\tfrac{1}{i}\right)^2 \int \frac{d^4\ell}{(2\pi)^4}\Big[\gamma_5 \tilde{S}(\not{p}+\not{\ell})\gamma_5\Big]\tilde{\Delta}(\ell^2)\; , \tag{51.27}$$

where $\tilde{S}(\not{p})$ is given by eq. (51.13), and

$$\tilde{\Delta}(\ell^2) = \frac{1}{\ell^2 + M^2 - i\epsilon} \tag{51.28}$$

is the free scalar propagator in momentum space.

We evaluate eq. (51.27) with the usual bag of tricks. The result is

$$i\Sigma_{1\,\text{loop}}(\not{p}) = -g^2 \int_0^1 dx \int \frac{d^4q}{(2\pi)^4}\frac{N}{(q^2+D)^2}\; , \tag{51.29}$$

where $q = \ell + xp$ and

$$N = \not{q} + (1-x)\not{p} + m\; , \tag{51.30}$$

$$D = x(1-x)p^2 + xm^2 + (1-x)M^2\; . \tag{51.31}$$

The integral diverges, and so we analytically continue it to $d = 4 - \varepsilon$ spacetime dimensions, make the replacement $g \to g\tilde{\mu}^{\varepsilon/2}$, and take the limit as $\varepsilon \to 0$. The term linear in q in eq. (51.30) integrates to zero. Using eq. (51.18), we get

$$\Sigma_{1\,\text{loop}}(\not{p}) = -\frac{g^2}{16\pi^2}\left[\frac{1}{\varepsilon}(\not{p}+2m) - \int_0^1 dx\,\Big((1-x)\not{p}+m\Big)\ln(D/\mu^2)\right]. \tag{51.32}$$

We see that the divergent term has (as expected) a form that permits cancellation by the counterterms, which give

$$\Sigma_{\text{ct}}(\not{p}) = -(Z_\Psi-1)\not{p} - (Z_m-1)m\,. \tag{51.33}$$

Adding up eqs. (51.32) and (51.33), we see that finiteness of $\Sigma(\not{p})$ requires

$$Z_\Psi = 1 - \frac{g^2}{16\pi^2}\left(\frac{1}{\varepsilon} + \text{finite}\right), \tag{51.34}$$

$$Z_m = 1 - \frac{g^2}{8\pi^2}\left(\frac{1}{\varepsilon} + \text{finite}\right), \tag{51.35}$$

plus higher-order corrections.

We can impose $\Sigma(-m) = 0$ by writing

$$\Sigma(\not{p}) = \frac{g^2}{16\pi^2}\left[\int_0^1 dx\,\Big((1-x)\not{p}+m\Big)\ln(D/D_0) + \kappa_\Psi(\not{p}+m)\right], \tag{51.36}$$

where D_0 is D evaluated at $p^2 = -m^2$, and κ_Ψ is a constant to be determined. We fix κ_Ψ by imposing $\Sigma'(-m) = 0$. In differentiating with respect to $\not{p}$, we take the p^2 in D, eq. (51.31), to be $-\not{p}^2$; we find

$$\kappa_\Psi = -2\int_0^1 dx\; x^2(1-x)m^2/D_0\,. \tag{51.37}$$

Next we turn to the correction to the Yukawa vertex. We define the vertex function $i\mathbf{V}_Y(p',p)$ as the sum of one-particle irreducible diagrams with one incoming fermion with momentum p, one outgoing fermion with momentum p', and one incoming scalar with momentum $k = p' - p$. The original vertex $-Z_g g\gamma_5$ is the first term in this sum, and the diagram of fig. 51.3 is the second. Thus we have

$$i\mathbf{V}_Y(p',p) = -Z_g g\gamma_5 + i\mathbf{V}_{Y,1\,\text{loop}}(p',p) + O(g^5)\,, \tag{51.38}$$

where

$$i\mathbf{V}_{Y,1\,\text{loop}}(p',p) = (-g)^3\left(\tfrac{1}{i}\right)^3\int\frac{d^4\ell}{(2\pi)^d}\Big[\gamma_5\tilde{S}(\not{p}'+\not{\ell})\gamma_5\tilde{S}(\not{p}+\not{\ell})\gamma_5\Big]\tilde{\Delta}(\ell^2)\,. \tag{51.39}$$

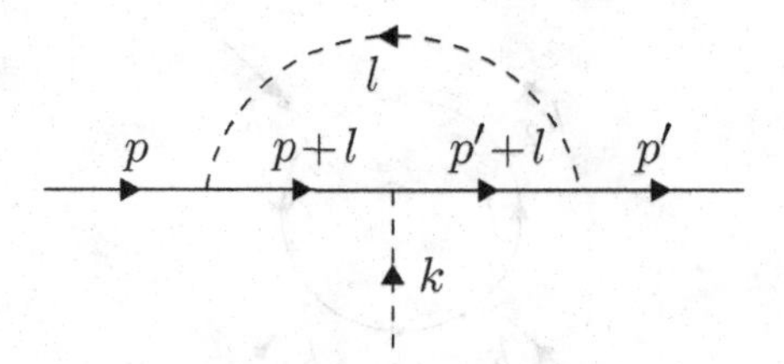

Figure 51.3. The one-loop correction to the scalar-fermion-fermion vertex in Yukawa theory.

The numerator can be written as

$$N = (\not{p}' + \not{\ell} + m)(-\not{p} - \not{\ell} + m)\gamma_5 \,, \tag{51.40}$$

and the denominators combined in the usual way. We then get

$$i\mathbf{V}_{Y,\,1\,\text{loop}}(p',p) = -ig^3 \int dF_3 \int \frac{d^4q}{(2\pi)^4} \frac{N}{(q^2+D)^3} \,, \tag{51.41}$$

where the integral over Feynman parameters was defined in section 16, and now

$$q = \ell + x_1 p + x_2 p' \,, \tag{51.42}$$

$$N = [\not{q} - x_1\not{p} + (1{-}x_2)\not{p}' + m][-\not{q} - (1{-}x_1)\not{p} + x_2\not{p}' + m]\gamma_5 \,, \tag{51.43}$$

$$D = x_1(1{-}x_1)p^2 + x_2(1{-}x_2)p'^2 - 2x_1x_2 p{\cdot}p' + (x_1{+}x_2)m^2 + x_3M^2 \,. \tag{51.44}$$

Using $\not{q}\not{q} = -q^2$, we can write N as

$$N = q^2\gamma_5 + \widetilde{N} + (\text{linear in } q) \,, \tag{51.45}$$

where

$$\widetilde{N} = [-x_1\not{p} + (1{-}x_2)\not{p}' + m][-(1{-}x_1)\not{p} + x_2\not{p}' + m]\gamma_5 \,. \tag{51.46}$$

The terms linear in q in eq. (51.45) integrate to zero, and only the first term is divergent. Performing the usual manipulations, we find

$$i\mathbf{V}_{Y,\,1\,\text{loop}}(p',p) = \frac{g^3}{8\pi^2}\left[\left(\frac{1}{\varepsilon} - \tfrac{1}{4} - \tfrac{1}{2}\int dF_3 \,\ln(D/\mu^2)\right)\gamma_5 + \tfrac{1}{4}\int dF_3 \,\frac{\widetilde{N}}{D}\right]. \tag{51.47}$$

From eq. (51.38), we see that finiteness of $\mathbf{V}_Y(p',p)$ requires

$$Z_g = 1 + \frac{g^2}{8\pi^2}\left(\frac{1}{\varepsilon} + \text{finite}\right) , \tag{51.48}$$

plus higher-order corrections.

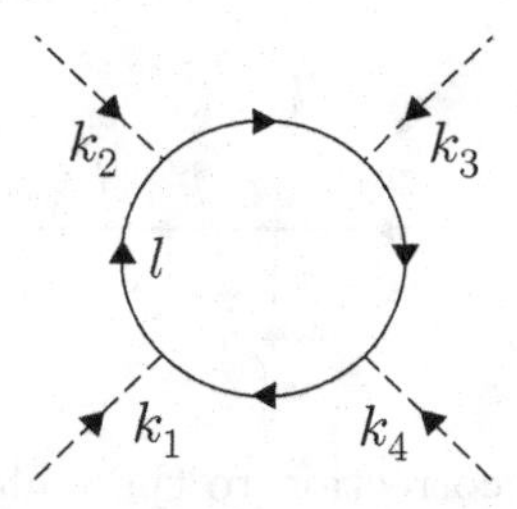

Figure 51.4. One of six diagrams with a closed fermion loop and four external scalar lines; the other five are obtained by permuting the external momenta in all possible inequivalent ways.

To fix the finite part of Z_g, we need a condition to impose on $\mathbf{V}_Y(p',p)$. We will mimic what we did in φ^3 theory in section 16, and require $\mathbf{V}_Y(0,0)$ to have the tree-level value $ig\gamma_5$. We leave the details to problem 51.2.

Next we turn to the corrections to the φ^4 vertex $i\mathbf{V}_4(k_1,k_2,k_3,k_4)$; the tree-level contribution is $-iZ_\lambda\lambda$. There are diagrams with a closed fermion loop, as shown in fig. 51.4, plus one-loop diagrams with φ particles only that we evaluated in section 31. We have

$$\begin{aligned} i\mathbf{V}_{4,\,\Psi\,\text{loop}} = (-1)(-g)^4\big(\tfrac{1}{i}\big)^4 \int \frac{d^4\ell}{(2\pi)^4}\,&\text{Tr}\Big[\tilde S(\not\ell)\gamma_5\tilde S(\not\ell-\not k_1)\gamma_5 \\ &\times \tilde S(\not\ell+\not k_2+\not k_3)\gamma_5\tilde S(\not\ell+\not k_2)\gamma_5\Big] \\ &+ 5 \text{ permutations of } (k_2,k_3,k_4)\ . \end{aligned} \tag{51.49}$$

Again we can employ the standard methods; there are no unfamiliar aspects. This being the case, let us concentrate on obtaining the divergent part; this will give us enough information to calculate the one-loop contributions to the beta functions for g and λ.

To obtain the divergent part of eq. (51.49), it is sufficient to set $k_i = 0$. The term in the numerator that contributes is $\text{Tr}\,(\not\ell\gamma_5)^4 = 4(\ell^2)^2$, and the denominator is $(\ell^2+m^2)^4$. Then we find, after including identical contributions from the other five permutations of the external momenta,

$$\mathbf{V}_{4,\,\Psi\,\text{loop}} = -\frac{3g^4}{\pi^2}\left(\frac{1}{\varepsilon} + \text{finite}\right). \tag{51.50}$$

From section 31, we have

$$\mathbf{V}_{4,\,\varphi\,\text{loop}} = \frac{3\lambda^2}{16\pi^2}\left(\frac{1}{\varepsilon} + \text{finite}\right). \tag{51.51}$$

Then, using

$$\mathbf{V}_4 = -Z_\lambda\lambda + \mathbf{V}_{4,\,\Psi\,\text{loop}} + \mathbf{V}_{4,\,\varphi\,\text{loop}} + \ldots\ , \tag{51.52}$$

we see that finiteness of $\mathbf{V}_4$ requires

$$Z_\lambda = 1 + \left(\frac{3\lambda}{16\pi^2} - \frac{3g^4}{\pi^2\lambda}\right)\left(\frac{1}{\varepsilon} + \text{finite}\right), \tag{51.53}$$

plus higher-order corrections.

Reference notes

A detailed derivation of the Lehmann–Källén form of the fermion propagator can be found in *Itzykson & Zuber*.

Problems

51.1) Derive the fermion-loop correction to the scalar propagator by working through eq. (45.2), and show that it has an extra minus sign relative to the case of a scalar loop.

51.2) Finish the computation of $\mathbf{V}_Y(p', p)$, imposing the condition

$$\mathbf{V}_Y(0,0) = ig\gamma_5 \,. \tag{51.54}$$

51.3) Consider making φ a scalar rather than a pseudoscalar, so that the Yukawa interaction is $\mathcal{L}_{\text{Yuk}} = g\varphi\overline{\Psi}\Psi$. In this case, renormalizability requires us to add a term $\mathcal{L}_{\varphi^3} = \frac{1}{6}Z_\kappa \kappa\varphi^3$, as well as term linear in φ to cancel tadpoles. Find the one-loop contributions to the renormalizing Z factors for this theory in the $\overline{\text{MS}}$ scheme.

52

Beta functions in Yukawa theory

Prerequisites: 28, 51

In this section we will compute the beta functions for the Yukawa coupling g and the φ^4 coupling λ in Yukawa theory, using the methods of section 28.

The relations between the bare and renormalized couplings are

$$g_0 = Z_\varphi^{-1/2} Z_\Psi^{-1} Z_g \tilde{\mu}^{\varepsilon/2} g \,, \tag{52.1}$$

$$\lambda_0 = Z_\varphi^{-2} Z_\lambda \tilde{\mu}^{\varepsilon} \lambda \,. \tag{52.2}$$

Let us define

$$\ln\Big(Z_\varphi^{-1/2} Z_\Psi^{-1} Z_g \Big) = \sum_{n=1}^{\infty} \frac{G_n(g,\lambda)}{\varepsilon^n} \,, \tag{52.3}$$

$$\ln\Big(Z_\varphi^{-2} Z_\lambda \Big) = \sum_{n=1}^{\infty} \frac{L_n(g,\lambda)}{\varepsilon^n} \,. \tag{52.4}$$

From our results in section 51, we have

$$G_1(g,\lambda) = \frac{5g^2}{16\pi^2} + \ldots \,, \tag{52.5}$$

$$L_1(g,\lambda) = \frac{3\lambda}{16\pi^2} + \frac{g^2}{2\pi^2} - \frac{3g^4}{\pi^2\lambda} + \ldots \,, \tag{52.6}$$

where the ellipses stand for higher-order (in g^2 and/or λ) corrections.

Taking the logarithm of eqs. (52.1) and (52.2), and using eqs. (52.3) and (52.4), we get

$$\ln g_0 = \sum_{n=1}^{\infty} \frac{G_n(g,\lambda)}{\varepsilon^n} + \ln g + \tfrac{1}{2}\varepsilon \ln \tilde{\mu} \,, \tag{52.7}$$

$$\ln \lambda_0 = \sum_{n=1}^{\infty} \frac{L_n(g,\lambda)}{\varepsilon^n} + \ln \lambda + \varepsilon \ln \tilde{\mu} \,. \tag{52.8}$$

We now use the fact that g_0 and λ_0 must be independent of μ. We differentiate eqs. (52.7) and (52.8) with respect to $\ln\mu$; the left-hand sides vanish, and we multiply the right-hand sides by g and λ, respectively. The result is

$$0 = \sum_{n=1}^{\infty}\left(g\frac{\partial G_n}{\partial g}\frac{dg}{d\ln\mu} + g\frac{\partial G_n}{\partial\lambda}\frac{d\lambda}{d\ln\mu}\right)\frac{1}{\varepsilon^n} + \frac{dg}{d\ln\mu} + \tfrac{1}{2}\varepsilon g\,, \tag{52.9}$$

$$0 = \sum_{n=1}^{\infty}\left(\lambda\frac{\partial L_n}{\partial g}\frac{dg}{d\ln\mu} + \lambda\frac{\partial L_n}{\partial\lambda}\frac{d\lambda}{d\ln\mu}\right)\frac{1}{\varepsilon^n} + \frac{d\lambda}{d\ln\mu} + \varepsilon\lambda\,. \tag{52.10}$$

In a renormalizable theory, $dg/d\ln\mu$ and $d\lambda/d\ln\mu$ must be finite in the $\varepsilon\to 0$ limit. Thus we can write

$$\frac{dg}{d\ln\mu} = -\tfrac{1}{2}\varepsilon g + \beta_g(g,\lambda)\,, \tag{52.11}$$

$$\frac{d\lambda}{d\ln\mu} = -\varepsilon\lambda + \beta_\lambda(g,\lambda)\,. \tag{52.12}$$

Substituting these into eqs. (52.9) and (52.10), and matching powers of ε, we find

$$\beta_g(g,\lambda) = g\left(\tfrac{1}{2}g\frac{\partial}{\partial g} + \lambda\frac{\partial}{\partial\lambda}\right)G_1\,, \tag{52.13}$$

$$\beta_\lambda(g,\lambda) = \lambda\left(\tfrac{1}{2}g\frac{\partial}{\partial g} + \lambda\frac{\partial}{\partial\lambda}\right)L_1\,. \tag{52.14}$$

The coefficients of all higher powers of $1/\varepsilon$ must also vanish, but this gives us no more information about the beta functions.

Using eqs. (52.5) and (52.6) in eqs. (52.13) and (52.14), we get

$$\beta_g(g,\lambda) = \frac{5g^3}{16\pi^2} + \ldots\,, \tag{52.15}$$

$$\beta_\lambda(g,\lambda) = \frac{1}{16\pi^2}\left(3\lambda^2 + 8\lambda g^2 - 48g^4\right) + \ldots\,. \tag{52.16}$$

The higher-order corrections have extra factors of g^2 and/or λ.

Problems

52.1) Compute the one-loop contributions to the anomalous dimensions of m, M, Ψ, and φ.

52.2) Consider the theory of problem 51.3. Compute the one-loop contributions to the beta functions for g, λ, and κ, and to the anomalous dimensions of m, M, Ψ, and φ.

52.3) Consider the beta functions of eqs. (52.15) and (52.16).

a) Let $\rho \equiv \lambda/g^2$, and compute $d\rho/d\ln\mu$. Express your answer in terms of g and ρ. Explain why it is better to work with g and ρ rather than g and λ. Hint: the answer is mathematical, not physical.

b) Show that there are two *fixed points*, ρ_+^* and ρ_-^*, where $d\rho/d\ln\mu = 0$, and find their values.

c) Suppose that, for some particular value of the renormalization scale μ, we have $\rho = 0$ and $g \lll 1$. What happens to ρ at much higher values of μ (but still low enough to keep $g \ll 1$)? At much lower values of μ?

d) Same question, but with an initial value of $\rho = 5$.

e) Same question, but with an initial value of $\rho = -5$.

f) Find the trajectory in the (ρ, g) plane that is followed for each of the three starting points as μ is varied up and down. Hint: you should find that the trajectories take the form

$$g = g_0 \left| \frac{\rho - \rho_+^*}{\rho - \rho_-^*} \right|^\nu$$

for some particular exponent ν. Put arrows on the trajectories that point in the direction of increasing μ.

g) Explain why ρ_-^* is called an *ultraviolet stable* fixed point, and why ρ_+^* is called an *infrared stable* fixed point.

53

Functional determinants

Prerequisites: 44, 45

In this section we will explore the meaning of the *functional determinants* that arise when doing gaussian path integrals, either bosonic or fermionic. We will be interested in situations where the path integral over one particular field is gaussian, but generates a functional determinant that depends on some other field. We will see how to relate this functional determinant to a certain infinite set of Feynman diagrams. We will need the technology we develop here to compute the path integral for nonabelian gauge theory in section 71.

We begin by considering a theory of a complex scalar field χ with

$$\mathcal{L} = -\partial^\mu \chi^\dagger \partial_\mu \chi - m^2 \chi^\dagger \chi + g\varphi\chi^\dagger\chi \,, \tag{53.1}$$

where φ is a real scalar *background field.* That is, $\varphi(x)$ is treated as a fixed function of spacetime. Next we define the path integral

$$Z(\varphi) = \int \mathcal{D}\chi^\dagger\, \mathcal{D}\chi\; e^{i\int d^4x\, \mathcal{L}} \,, \tag{53.2}$$

where we use the ϵ trick of section 6 to impose vacuum boundary conditions, and the normalization $Z(0) = 1$ is fixed by hand.

Recall from section 44 that if we have n complex variables z_i, then we can evaluate gaussian integrals by the general formula

$$\int d^n z\, d^n \bar{z}\, \exp\left(-i\bar{z}_i M_{ij} z_j\right) \propto (\det M)^{-1} \,. \tag{53.3}$$

In the case of the functional integral in eq. (53.2), the index i on the integration variable is replaced by the continuous spacetime label x, and the "matrix" M becomes

$$M(x,y) = [-\partial_x^2 + m^2 - g\varphi(x)]\delta^4(x-y) \ . \tag{53.4}$$

In order to apply eq. (53.3), we have to understand what it means to compute the determinant of this expression.

To this end, let us first note that we can write $M = M_0\widetilde{M}$, which is shorthand for

$$M(x,z) = \int d^4y\, M_0(x,y)\widetilde{M}(y,z) \ , \tag{53.5}$$

where

$$M_0(x,y) = (-\partial_x^2 + m^2)\delta^4(x-y) \ , \tag{53.6}$$

$$\widetilde{M}(y,z) = \delta^4(y-z) - g\Delta(y-z)\varphi(z) \ . \tag{53.7}$$

Here $\Delta(y-z)$ is the Feynman propagator, which obeys

$$(-\partial_y^2 + m^2)\Delta(y-z) = \delta^4(y-z) \ . \tag{53.8}$$

After various integrations by parts, it is easy to see that eqs. (53.5)–(53.7) reproduce eq. (53.4).

Now we can use the general matrix relation $\det AB = \det A \det B$ to conclude that

$$\det M = \det M_0 \det \widetilde{M} \ . \tag{53.9}$$

The advantage of this decomposition is that M_0 is independent of the background field φ, and so the resulting factor of $(\det M_0)^{-1}$ in $Z(\varphi)$ can simply be absorbed into the overall normalization. Furthermore, we have $\widetilde{M} = I - G$, where

$$I(x,y) = \delta^4(x-y) \tag{53.10}$$

is the identity matrix, and

$$G(x,y) = g\Delta(x-y)\varphi(y) \ . \tag{53.11}$$

Thus, for $\varphi(x) = 0$, we have $\widetilde{M} = I$ and so $\det\widetilde{M} = 1$. Then, using eq. (53.3) and the normalization condition $Z(0) = 1$, we see that for nonzero $\varphi(x)$ we must have simply

$$Z(\varphi) = (\det\widetilde{M})^{-1} \ . \tag{53.12}$$

Next, we need the general matrix relation $\det A = \exp \mathrm{Tr} \ln A$, which is most easily proved by working in a basis where A is in Jordan form (that

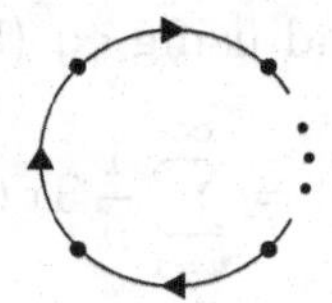

Figure 53.1. All connected diagrams with $\varphi(x)$ treated as an external field. Each of the n dots represents a factor of $ig\varphi(x)$, and each solid line is a χ or Ψ propagator.

is, all entries below the main diagonal are zero). Thus we can write

$$\begin{aligned}\det \widetilde{M} &= \exp \operatorname{Tr} \ln \widetilde{M} \\ &= \exp \operatorname{Tr} \ln(I - G) \\ &= \exp \operatorname{Tr} \left[-\sum_{n=1}^{\infty} \frac{1}{n} G^n \right] . \end{aligned} \tag{53.13}$$

Combining eqs. (53.12) and (53.13) we get

$$Z(\varphi) = \exp \sum_{n=1}^{\infty} \frac{1}{n} \operatorname{Tr} G^n , \tag{53.14}$$

where

$$\operatorname{Tr} G^n = g^n \int d^4x_1 \ldots d^4x_n \, \Delta(x_1 - x_2)\varphi(x_2) \ldots \Delta(x_n - x_1)\varphi(x_1) . \tag{53.15}$$

This is our final result for $Z(\varphi)$.

To better understand what it means, we will rederive it in a different way. Consider treating the $g\varphi\chi^\dagger\chi$ term in $\mathcal{L}$ as an interaction. This leads to a vertex that connects two χ propagators; the associated vertex factor is $ig\varphi(x)$. According to the general analysis of section 9, we have $Z(\varphi) = \exp i\Gamma(\varphi)$, where $i\Gamma(\varphi)$ is given by a sum of connected diagrams. (We have called the exponent Γ rather than W because it is naturally interpreted as a quantum action for φ after χ has been integrated out.) The only connected diagrams we can draw with these Feynman rules are those of fig. 53.1, with n insertions of the vertex, where $n \geq 1$. The diagram with n vertices has an n-fold cyclic symmetry, leading to a symmetry factor of $S = n$. The factor of i associated with each vertex is canceled by the factor of $1/i$ associated with each propagator. Thus the value of the n-vertex diagram is

$$\frac{1}{n} g^n \int d^4x_1 \ldots d^4x_n \, \Delta(x_1 - x_2)\varphi(x_2) \ldots \Delta(x_n - x_1)\varphi(x_1) . \tag{53.16}$$

Summing up these diagrams, and using eq. (53.15), we find

$$i\Gamma(\varphi) = \sum_{n=1}^{\infty} \frac{1}{n} \operatorname{Tr} G^n \; . \tag{53.17}$$

This neatly reproduces eq. (53.14). Thus we see that a functional determinant can be represented as an infinite sum of Feynman diagrams.

Next we consider a theory of a Dirac fermion Ψ with

$$\mathcal{L} = i\overline{\Psi}\partial\!\!\!/\Psi - m\overline{\Psi}\Psi + g\varphi\overline{\Psi}\Psi \; , \tag{53.18}$$

where φ is again a real scalar background field. We define the path integral

$$Z(\varphi) = \int \mathcal{D}\overline{\Psi}\,\mathcal{D}\Psi\; e^{i\int d^4x\,\mathcal{L}} \; , \tag{53.19}$$

where we again use the ϵ trick to impose vacuum boundary conditions, and the normalization $Z(0) = 1$ is fixed by hand.

Recall from section 44 that if we have n complex Grassmann variables ψ_i, then we can evaluate gaussian integrals by the general formula

$$\int d^n\bar{\psi}\, d^n\psi\; \exp\left(-i\bar{\psi}_i M_{ij}\psi_j\right) \propto \det M \; . \tag{53.20}$$

In the case of the functional integral in eq. (53.19), the index i on the integration variable is replaced by the continuous spacetime label x plus the spinor index α, and the "matrix" M becomes

$$M_{\alpha\beta}(x,y) = [-i\partial\!\!\!/_x + m - g\varphi(x)]_{\alpha\beta}\delta^4(x-y) \; . \tag{53.21}$$

In order to apply eq. (53.20), we have to understand what it means to compute the determinant of this expression.

To this end, let us first note that we can write $M = M_0\widetilde{M}$, which is shorthand for

$$M_{\alpha\gamma}(x,z) = \int d^4y\; M_{0\alpha\beta}(x,y)\widetilde{M}_{\beta\gamma}(y,z) \; , \tag{53.22}$$

where

$$M_{0\alpha\beta}(x,y) = (-i\partial\!\!\!/_x + m)_{\alpha\beta}\delta^4(x-y) \; , \tag{53.23}$$

$$\widetilde{M}_{\beta\gamma}(y,z) = \delta_{\beta\gamma}\delta^4(y-z) - gS_{\beta\gamma}(y-z)\varphi(z) \; . \tag{53.24}$$

Here $S_{\beta\gamma}(y-z)$ is the Feynman propagator, which obeys

$$(-i\partial\!\!\!/_y + m)_{\alpha\beta}S_{\beta\gamma}(y-z) = \delta_{\alpha\gamma}\delta^4(y-z) \; . \tag{53.25}$$

After various integrations by parts, it is easy to see that eqs. (53.22)–(53.24) reproduce eq. (53.21).

Now we can use eq. (53.9). The advantage of this decomposition is that M_0 is independent of the background field φ, and so the resulting factor of $\det M_0$ in $Z(\varphi)$ can simply be absorbed into the overall normalization. Furthermore, we have $\widetilde{M} = I - G$, where

$$I_{\alpha\beta}(x,y) = \delta_{\alpha\beta}\delta^4(x-y) \tag{53.26}$$

is the identity matrix, and

$$G_{\alpha\beta}(x,y) = gS_{\alpha\beta}(x-y)\varphi(y)\ . \tag{53.27}$$

Thus, for $\varphi(x) = 0$, we have $\widetilde{M} = I$ and so $\det \widetilde{M} = 1$. Then, using eq. (53.20) and the normalization condition $Z(0) = 1$, we see that for nonzero $\varphi(x)$ we must have simply

$$Z(\varphi) = \det \widetilde{M}\ . \tag{53.28}$$

Next, we use eqs. (53.13) and (53.28) to get

$$Z(\varphi) = \exp -\sum_{n=1}^{\infty} \frac{1}{n} \operatorname{Tr} G^n\ , \tag{53.29}$$

where now

$$\operatorname{Tr} G^n = g^n \int d^4x_1 \ldots d^4x_n \operatorname{tr} S(x_1{-}x_2)\varphi(x_2) \ldots S(x_n{-}x_1)\varphi(x_1)\ , \tag{53.30}$$

and "tr" denotes a trace over spinor indices. This is our final result for $Z(\varphi)$.

To better understand what it means, we will rederive it in a different way. Consider treating the $g\varphi\overline{\Psi}\Psi$ term in $\mathcal{L}$ as an interaction. This leads to a vertex that connects two Ψ propagators; the associated vertex factor is $ig\varphi(x)$. According to the general analysis of section 9, we have $Z(\varphi) = \exp i\Gamma(\varphi)$, where $i\Gamma(\varphi)$ is given by a sum of connected diagrams. (We have called the exponent Γ rather than W because it is naturally interpreted as a quantum action for φ after Ψ has been integrated out.) The only connected diagrams we can draw with these Feynman rules are those of fig. 53.1, with n insertions of the vertex, where $n \geq 1$. The diagram with n vertices has an n-fold cyclic symmetry, leading to a symmetry factor of $S = n$. The factor of i associated with each vertex is canceled by the factor of $1/i$ associated with each propagator. The closed fermion loop implies a trace over the spinor indices. Thus the value of the n-vertex diagram is

$$\frac{1}{n} g^n \int d^4x_1 \ldots d^4x_n \operatorname{tr} S(x_1 - x_2)\varphi(x_2) \ldots S(x_n - x_1)\varphi(x_1) \,. \tag{53.31}$$

Summing up these diagrams, we find that we are missing the overall minus sign in eq. (53.29). The appropriate conclusion is that we must associate an extra minus sign with each closed fermion loop.

Part III

Spin One

54

Maxwell's equations

Prerequisite: 3

The most common (and important) spin-one particle is the photon. Emission and absorption of photons by matter is an important phenomenon in many areas of physics, and so that is the context in which most physicists first encounter a serious treatment of photons. We will use a brief review of this subject (in this section and the next) as our entry point into the theory of quantum electrodynamics.

Let us begin with classical electrodynamics. Maxwell's equations are

$$\nabla\cdot\mathbf{E} = \rho \,, \tag{54.1}$$

$$\nabla\times\mathbf{B} - \dot{\mathbf{E}} = \mathbf{J} \,, \tag{54.2}$$

$$\nabla\times\mathbf{E} + \dot{\mathbf{B}} = 0 \,, \tag{54.3}$$

$$\nabla\cdot\mathbf{B} = 0 \,, \tag{54.4}$$

where $\mathbf{E}$ is the electric field, $\mathbf{B}$ is the magnetic field, ρ is the charge density, and $\mathbf{J}$ is the current density. We have written Maxwell's equations in *Heaviside–Lorentz units*, and also set $c = 1$. In these units, the magnitude of the force between two charges of magnitude Q is $Q^2/4\pi r^2$.

Maxwell's equations must be supplemented by formulae that give us the dynamics of the charges and currents (such as the Lorentz force law for point particles). For now, however, we will treat the charges and currents as specified sources, and focus on the dynamics of the electromagnetic fields.

The last two of Maxwell's equations, the ones with no sources on the right-hand side, can be solved by writing the $\mathbf{E}$ and $\mathbf{B}$ fields in terms of a scalar potential φ and a vector potential $\mathbf{A}$,

$$\mathbf{E} = -\nabla\varphi - \dot{\mathbf{A}} \,, \tag{54.5}$$

$$\mathbf{B} = \nabla \times \mathbf{A} \,. \tag{54.6}$$

The potentials uniquely determine the fields, but the fields do not uniquely determine the potentials. Given a particular φ and $\mathbf{A}$ that result in a particular $\mathbf{E}$ and $\mathbf{B}$, we will get the same $\mathbf{E}$ and $\mathbf{B}$ from any other potentials φ' and $\mathbf{A}'$ that are related by

$$\varphi' = \varphi + \dot{\Gamma} \,, \tag{54.7}$$

$$\mathbf{A}' = \mathbf{A} - \nabla\Gamma \,, \tag{54.8}$$

where Γ is an arbitrary function of spacetime. A change of potentials that does not change the fields is called a *gauge transformation.* The $\mathbf{E}$ and $\mathbf{B}$ fields are *gauge invariant.*

All this becomes more compact and elegant in a relativistic notation. We define the four-vector potential or *gauge field*

$$A^\mu \equiv (\varphi, \mathbf{A}) \,. \tag{54.9}$$

We also define the *field strength*

$$F^{\mu\nu} \equiv \partial^\mu A^\nu - \partial^\nu A^\mu \,. \tag{54.10}$$

Obviously, $F^{\mu\nu}$ is antisymmetric: $F^{\mu\nu} = -F^{\nu\mu}$. Comparing eqs. (54.5) and (54.6) with eqs. (54.9) and (54.10), we see that

$$F^{0i} = E^i \,, \tag{54.11}$$

$$F^{ij} = \varepsilon^{ijk} B_k \,. \tag{54.12}$$

The first two of Maxwell's equations can now be written as

$$\partial_\nu F^{\mu\nu} = J^\mu \,, \tag{54.13}$$

where

$$J^\mu \equiv (\rho, \mathbf{J}) \tag{54.14}$$

is the charge-current density four-vector.

If we take the four-divergence of eq. (54.13), we get $\partial_\mu\partial_\nu F^{\mu\nu} = \partial_\mu J^\mu$. The left-hand side of this equation vanishes, because $\partial_\mu\partial_\nu$ is symmetric on exchange of μ and ν, while $F^{\mu\nu}$ is antisymmetric. We conclude that we must have

$$\partial_\mu J^\mu = 0 \,, \tag{54.15}$$

or equivalently

$$\dot{\rho} + \nabla\cdot\mathbf{J} = 0 \,; \tag{54.16}$$

that is, the electromagnetic current must be conserved.

The last two of Maxwell's equations can be written as

$$\varepsilon_{\mu\nu\rho\sigma}\partial^\rho F^{\mu\nu} = 0 \,, \tag{54.17}$$

where $\varepsilon_{\mu\nu\rho\sigma}$ is the completely antisymmetric Levi-Civita tensor; see section 34. Plugging in eq. (54.10), we see that eq. (54.17) is automatically satisfied, since the antisymmetric combination of two derivatives vanishes.

Eqs. (54.7) and (54.8) can be combined into

$$A'^\mu = A^\mu - \partial^\mu \Gamma \,. \tag{54.18}$$

Setting $F'^{\mu\nu} = \partial^\mu A'^\nu - \partial^\nu A'^\mu$ and using eq. (54.18), we get

$$F'^{\mu\nu} = F^{\mu\nu} - (\partial^\mu\partial^\nu - \partial^\nu\partial^\mu)\Gamma \,. \tag{54.19}$$

The last term vanishes because derivatives commute; thus the field strength is gauge invariant,

$$F'^{\mu\nu} = F^{\mu\nu} \,. \tag{54.20}$$

Next we will find an action that results in Maxwell's equations as the equations of motion. We will treat the current as an external source. The action we seek should be Lorentz invariant, gauge invariant, parity and time-reversal invariant, and no more than second order in derivatives. The only candidate is $S = \int d^4x\, \mathcal{L}$, where

$$\mathcal{L} = -\tfrac{1}{4}F^{\mu\nu}F_{\mu\nu} + J^\mu A_\mu \,. \tag{54.21}$$

The first term is obviously gauge invariant, because $F^{\mu\nu}$ is. After a gauge transformation, eq. (54.18), the second term becomes $J^\mu A'_\mu$, and the difference is

$$\begin{aligned} J^\mu(A'_\mu - A_\mu) &= -J^\mu \partial_\mu \Gamma \\ &= (\partial_\mu J^\mu)\Gamma - \partial_\mu(J^\mu \Gamma) \,. \end{aligned} \tag{54.22}$$

The first term in eq. (54.22) vanishes because the current is conserved. The second term is a total divergence, and its integral over d^4x vanishes (assuming suitable boundary conditions at infinity). Thus the action specified by eq. (54.21) is gauge invariant.

Setting $F^{\mu\nu} = \partial^\mu A^\nu - \partial^\nu A^\mu$ and multiplying out the terms, eq. (54.21) becomes

$$\mathcal{L} = -\tfrac{1}{2}\partial^\mu A^\nu \partial_\mu A_\nu + \tfrac{1}{2}\partial^\mu A^\nu \partial_\nu A_\mu + J^\mu A_\mu \tag{54.23}$$

$$= +\tfrac{1}{2}A_\mu(g^{\mu\nu}\partial^2 - \partial^\mu\partial^\nu)A_\nu + J^\mu A_\mu - \partial^\mu K_\mu \,, \tag{54.24}$$

where $K_\mu = \frac{1}{2}A^\nu(\partial_\mu A_\nu - \partial_\nu A_\mu)$. The last term is a total divergence, and can be dropped. From eq. (54.24), we can see that varying A^μ while requiring S to be unchanged yields the equation of motion

$$(g^{\mu\nu}\partial^2 - \partial^\mu\partial^\nu)A_\nu + J^\mu = 0 \,. \tag{54.25}$$

Noting that $\partial_\nu F^{\mu\nu} = \partial_\nu(\partial^\mu A^\nu - \partial^\nu A^\mu) = (\partial^\mu\partial^\nu - g^{\mu\nu}\partial^2)A_\nu$, we see that eq. (54.25) is equivalent to eq. (54.13), and hence to Maxwell's equations.

55

Electrodynamics in Coulomb gauge

Prerequisite: 54

Next we would like to construct the hamiltonian, and quantize the electromagnetic field.

There is an immediate difficulty, caused by the gauge invariance: we have too many degrees of freedom. This problem manifests itself in several ways. For example, the lagrangian

$$\mathcal{L} = -\tfrac{1}{4}F^{\mu\nu}F_{\mu\nu} + J^\mu A_\mu \tag{55.1}$$

$$= -\tfrac{1}{2}\partial^\mu A^\nu \partial_\mu A_\nu + \tfrac{1}{2}\partial^\mu A^\nu \partial_\nu A_\mu + J^\mu A_\mu \tag{55.2}$$

does not contain the time derivative of A^0. Thus, this field has no canonically conjugate momentum and no dynamics.

To deal with this problem, we must eliminate the gauge freedom. We do this by *choosing a gauge.* We choose a gauge by imposing a *gauge condition.* This is a condition that we require $A^\mu(x)$ to satisfy. The idea is that there should be only one $A^\mu(x)$ that results in a given $F^{\mu\nu}(x)$ and that also satisfies the gauge condition.

One possible class of gauge conditions is $n^\mu A_\mu(x) = 0$, where n^μ is a constant four-vector. If n is spacelike ($n^2 > 0$), then we have chosen *axial gauge*; if n is lightlike ($n^2 = 0$), it is *lightcone gauge*; and if n is timelike ($n^2 < 0$), it is *temporal gauge.*

Another gauge is *Lorenz gauge*, where the condition is $\partial^\mu A_\mu = 0$. We will meet a family of closely related gauges in section 62.

In this section, we will work in *Coulomb gauge*, also known as *radiation gauge* or *transverse gauge.* The condition for Coulomb gauge is

$$\nabla\!\cdot\!\mathbf{A}(x) = 0\ . \tag{55.3}$$

We can impose eq. (55.3) by acting on $A_i(x)$ with a projection operator,

$$A_i(x) \to \left(\delta_{ij} - \frac{\nabla_i \nabla_j}{\nabla^2}\right) A_j(x) \,. \tag{55.4}$$

We construct the right-hand side of eq. (55.4) by Fourier-transforming $A_i(x)$ to $\widetilde{A}_i(k)$, multiplying $\widetilde{A}_i(k)$ by the matrix $\delta_{ij} - k_i k_j/\mathbf{k}^2$, and then Fourier-transforming back to position space. From now on, whenever we write A_i, we will implicitly mean the right-hand side of eq. (55.4).

Now let us write out the lagrangian in terms of the scalar and vector potentials, $\varphi = A^0$ and A_i, with A_i obeying the Coulomb gauge condition. Starting from eq. (55.2), we get

$$\begin{aligned}\mathcal{L} = {} & \tfrac{1}{2}\dot{A}_i\dot{A}_i - \tfrac{1}{2}\nabla_j A_i \nabla_j A_i + J_i A_i \\ & + \tfrac{1}{2}\nabla_i A_j \nabla_j A_i + \dot{A}_i \nabla_i \varphi \\ & + \tfrac{1}{2}\nabla_i \varphi \nabla_i \varphi - \rho\varphi \,. \end{aligned} \tag{55.5}$$

In the second line of eq. (55.5), the ∇_i in each term can be integrated by parts; in the first term, we will then get a factor of $\nabla_j(\nabla_i A_i)$, and in the second term, we will get a factor of $\nabla_i \dot{A}_i$. Both of these vanish by virtue of the gauge condition $\nabla_i A_i = 0$, and so both of these terms can simply be dropped.

If we now vary φ (and require $S = \int d^4x\, \mathcal{L}$ to be stationary), we find that φ obeys Poisson's equation,

$$-\nabla^2 \varphi = \rho \,. \tag{55.6}$$

The solution is

$$\varphi(\mathbf{x}, t) = \int d^3y \, \frac{\rho(\mathbf{y}, t)}{4\pi|\mathbf{x}-\mathbf{y}|} \,. \tag{55.7}$$

This solution is unique if we impose the boundary conditions that φ and ρ both vanish at spatial infinity.

Eq. (55.7) tells us that $\varphi(\mathbf{x}, t)$ is given entirely in terms of the charge density at the same time, and so has no dynamics of its own. It is therefore legitimate to plug eq. (55.7) back into the lagrangian. After an integration by parts to turn $\nabla_i \varphi \nabla_i \varphi$ into $-\varphi \nabla^2 \varphi = \varphi\rho$, the result is

$$\mathcal{L} = \tfrac{1}{2}\dot{A}_i\dot{A}_i - \tfrac{1}{2}\nabla_j A_i \nabla_j A_i + J_i A_i + \mathcal{L}_{\rm coul} \,, \tag{55.8}$$

where

$$\mathcal{L}_{\rm coul} = -\frac{1}{2} \int d^3y \, \frac{\rho(\mathbf{x}, t)\rho(\mathbf{y}, t)}{4\pi|\mathbf{x}-\mathbf{y}|} \,. \tag{55.9}$$

We can now vary A_i; keeping proper track of the implicit projection operator in eq. (55.4), we find that A_i obeys the massless Klein–Gordon equation with the projected current as a source,

$$-\partial^2 A_i(x) = \left(\delta_{ij} - \frac{\nabla_i \nabla_j}{\nabla^2}\right) J_j(x) \,. \tag{55.10}$$

For a free field ($J_i = 0$), the general solution is

$$\mathbf{A}(x) = \sum_{\lambda=\pm} \int \widetilde{dk} \left[\boldsymbol{\varepsilon}^*_\lambda(\mathbf{k}) a_\lambda(\mathbf{k}) e^{ikx} + \boldsymbol{\varepsilon}_\lambda(\mathbf{k}) a^\dagger_\lambda(\mathbf{k}) e^{-ikx} \right] , \tag{55.11}$$

where $k^0 = \omega = |\mathbf{k}|$, $\widetilde{dk} = d^3k/(2\pi)^3 2\omega$, and $\boldsymbol{\varepsilon}_+(\mathbf{k})$ and $\boldsymbol{\varepsilon}_-(\mathbf{k})$ are polarization vectors. In order to satisfy the Coulomb gauge condition, the polarization vectors must be orthogonal to the wave vector $\mathbf{k}$. We will choose them to correspond to right- and left-handed circular polarizations; for $\mathbf{k} = (0, 0, k)$, we then have

$$\begin{aligned} \boldsymbol{\varepsilon}_+(\mathbf{k}) &= \tfrac{1}{\sqrt{2}}(1, -i, 0) \,, \\ \boldsymbol{\varepsilon}_-(\mathbf{k}) &= \tfrac{1}{\sqrt{2}}(1, +i, 0) \,. \end{aligned} \tag{55.12}$$

More generally, the two polarization vectors along with the unit vector in the $\mathbf{k}$ direction form an orthonormal and complete set,

$$\mathbf{k} \cdot \boldsymbol{\varepsilon}_\lambda(\mathbf{k}) = 0 \,, \tag{55.13}$$

$$\boldsymbol{\varepsilon}_{\lambda'}(\mathbf{k}) \cdot \boldsymbol{\varepsilon}^*_\lambda(\mathbf{k}) = \delta_{\lambda'\lambda} \,, \tag{55.14}$$

$$\sum_{\lambda=\pm} \varepsilon^*_{i\lambda}(\mathbf{k}) \varepsilon_{j\lambda}(\mathbf{k}) = \delta_{ij} - \frac{k_i k_j}{\mathbf{k}^2} \,. \tag{55.15}$$

The coefficients $a_\lambda(\mathbf{k})$ and $a^\dagger_\lambda(\mathbf{k})$ will become operators after quantization, which is why we have used the dagger symbol for conjugation.

In complete analogy with the procedure used for a scalar field in section 3, we can invert eq. (55.11) and its time derivative to get

$$a_\lambda(\mathbf{k}) = +i\, \boldsymbol{\varepsilon}_\lambda(\mathbf{k}) \cdot \int d^3x \; e^{-ikx} \overleftrightarrow{\partial_0} \mathbf{A}(x) \,, \tag{55.16}$$

$$a^\dagger_\lambda(\mathbf{k}) = -i\, \boldsymbol{\varepsilon}^*_\lambda(\mathbf{k}) \cdot \int d^3x \; e^{+ikx} \overleftrightarrow{\partial_0} \mathbf{A}(x) \,, \tag{55.17}$$

where $f \overleftrightarrow{\partial_\mu} g = f(\partial_\mu g) - (\partial_\mu f) g$.

Now we can proceed to the hamiltonian formalism. First, we compute the canonically conjugate momentum to A_i,

$$\Pi_i = \frac{\partial \mathcal{L}}{\partial \dot{A}_i} = \dot{A}_i \; . \tag{55.18}$$

Note that $\nabla_i A_i = 0$ implies $\nabla_i \Pi_i = 0$. The hamiltonian density is then

$$\begin{aligned}\mathcal{H} &= \Pi_i \dot{A}_i - \mathcal{L} \\ &= \tfrac{1}{2}\Pi_i\Pi_i + \tfrac{1}{2}\nabla_j A_i \nabla_j A_i - J_i A_i + \mathcal{H}_{\text{coul}} \; ,\end{aligned} \tag{55.19}$$

where $\mathcal{H}_{\text{coul}} = -\mathcal{L}_{\text{coul}}$.

To quantize the field, we impose the canonical commutation relations. Keeping proper track of the implicit projection operator in eq. (55.4), we have

$$\begin{aligned}[A_i(\mathbf{x},t), \Pi_j(\mathbf{y},t)] &= i\left(\delta_{ij} - \frac{\nabla_i\nabla_j}{\nabla^2}\right)\delta^3(\mathbf{x}-\mathbf{y}) \\ &= i\int \frac{d^3k}{(2\pi)^3}\, e^{i\mathbf{k}\cdot(\mathbf{x}-\mathbf{y})}\left(\delta_{ij} - \frac{k_i k_j}{\mathbf{k}^2}\right) .\end{aligned} \tag{55.20}$$

The commutation relations of the $a_\lambda(\mathbf{k})$ and $a^\dagger_\lambda(\mathbf{k})$ operators follow from eq. (55.20) and $[A_i, A_j] = [\Pi_i, \Pi_j] = 0$ (at equal times). The result is

$$[a_\lambda(\mathbf{k}), a_{\lambda'}(\mathbf{k}')] = 0 \; , \tag{55.21}$$

$$[a^\dagger_\lambda(\mathbf{k}), a^\dagger_{\lambda'}(\mathbf{k}')] = 0 \; , \tag{55.22}$$

$$[a_\lambda(\mathbf{k}), a^\dagger_{\lambda'}(\mathbf{k}')] = (2\pi)^3 2\omega\, \delta^3(\mathbf{k}'-\mathbf{k})\delta_{\lambda\lambda'} \; . \tag{55.23}$$

We interpret $a^\dagger_\lambda(\mathbf{k})$ and $a_\lambda(\mathbf{k})$ as creation and annihilation operators for photons of definite helicity, with helicity $+1$ corresponding to right-circular polarization and helicity -1 to left-circular polarization.

It is now straightfoward to write the hamiltonian explicitly in terms of these operators. We find

$$H = \sum_{\lambda=\pm}\int \widetilde{dk}\; \omega\, a^\dagger_\lambda(\mathbf{k})a_\lambda(\mathbf{k}) + 2\mathcal{E}_0 V - \int d^3x\; \mathbf{J}(x)\cdot\mathbf{A}(x) + H_{\text{coul}} \; , \tag{55.24}$$

where $\mathcal{E}_0 = \frac{1}{2}(2\pi)^{-3}\int d^3k\; \omega$ is the zero-point energy per unit volume that we found for a real scalar field in section 3, V is the volume of space, the Coulomb hamiltonian is

$$H_{\text{coul}} = \frac{1}{2}\int d^3x\, d^3y\; \frac{\rho(\mathbf{x},t)\rho(\mathbf{y},t)}{4\pi|\mathbf{x}-\mathbf{y}|} \; , \tag{55.25}$$

and we use eq. (55.11) to express $A_i(x)$ in terms of $a_\lambda(\mathbf{k})$ and $a^\dagger_\lambda(\mathbf{k})$ at any one particular time (say, $t = 0$).

This form of the hamiltonian of electrodynamics is often used as the starting point for calculations of atomic transition rates, with the charges and currents treated via the nonrelativistic Schrödinger equation. The Coulomb interaction appears explicitly, and the $\mathbf{J}\cdot\mathbf{A}$ term allows for the creation and annihilation of photons of definite polarization.

Reference notes

A more rigorous treatment of quantization in Coulomb gauge can be found in *Weinberg I*.

Problems

55.1) Use eqs. (55.13)–(55.20) and $[A_i, A_j] = [\Pi_i, \Pi_j] = 0$ (at equal times) to verify eqs. (55.21)–(55.23).

55.2) Use eqs. (55.11), (55.14), (55.19), and (55.21)–(55.23) to verify eq. (55.24).

56

LSZ reduction for photons

Prerequisites: 5, 55

In section 55, we found that the creation and annihilation operators for free photons could be written as

$$a^\dagger_\lambda(\mathbf{k}) = -i\,\boldsymbol{\varepsilon}^*_\lambda(\mathbf{k})\cdot\int d^3x\; e^{+ikx}\,\overleftrightarrow{\partial_0}\,\mathbf{A}(x)\;, \tag{56.1}$$

$$a_\lambda(\mathbf{k}) = +i\,\boldsymbol{\varepsilon}_\lambda(\mathbf{k})\cdot\int d^3x\; e^{-ikx}\,\overleftrightarrow{\partial_0}\,\mathbf{A}(x)\;, \tag{56.2}$$

where $\boldsymbol{\varepsilon}_\lambda(\mathbf{k})$ is a polarization vector. From here, we can follow the analysis of section 5 line by line to deduce the LSZ reduction formula for photons. The result is that the creation operator for each incoming photon should be replaced by

$$a^\dagger_\lambda(\mathbf{k})_{\rm in} \to i\,\varepsilon^{\mu *}_\lambda(\mathbf{k})\int d^4x\; e^{+ikx}(-\partial^2)A_\mu(x)\;, \tag{56.3}$$

and the destruction operator for each outgoing photon should be replaced by

$$a_\lambda(\mathbf{k})_{\rm out} \to i\,\varepsilon^{\mu}_\lambda(\mathbf{k})\int d^4x\; e^{-ikx}(-\partial^2)A_\mu(x)\;, \tag{56.4}$$

and then we should take the vacuum expectation value of the time-ordered product. Note that, in writing eqs. (56.3) and (56.4), we have made them look nicer by introducing $\varepsilon^0_\lambda(\mathbf{k}) \equiv 0$, and then using four-vector dot products rather than three-vector dot products.

The LSZ formula is valid provided the field is normalized according to the free-field formulae

$$\langle 0|A^i(x)|0\rangle = 0\;, \tag{56.5}$$

$$\langle k,\lambda|A^i(x)|0\rangle = \varepsilon^i_\lambda(\mathbf{k})e^{-ikx}\;, \tag{56.6}$$

where $|k,\lambda\rangle$ is a single photon state, normalized according to

$$\langle k',\lambda'|k,\lambda\rangle = (2\pi)^3 2\omega\,\delta^3(\mathbf{k}'-\mathbf{k})\delta_{\lambda\lambda'}\ . \tag{56.7}$$

The zero on the right-hand side of eq. (56.5) is required by rotation invariance, and only the overall scale of the right-hand side of eq. (56.6) might be different in an interacting theory.

The renormalization of A_i necessitates including appropriate Z factors in the lagrangian,

$$\mathcal{L} = -\tfrac{1}{4}Z_3 F^{\mu\nu}F_{\mu\nu} + Z_1 J^\mu A_\mu\ . \tag{56.8}$$

Here Z_3 and Z_1 are the traditional names; we will meet Z_2 in section 62. We must choose Z_3 so that eq. (56.6) holds. We will fix Z_1 by requiring the corresponding vertex function to take on a certain value for a particular set of external momenta.

Next we must compute the correlation functions $\langle 0|\mathrm{T}A_i(x)\ldots|0\rangle$. As usual, we begin by working with free-field theory. The analysis is again almost identical to the case of a scalar field; see problem 8.4. We find that, in free-field theory,

$$\langle 0|\mathrm{T}A^i(x)A^j(y)|0\rangle = \tfrac{1}{i}\Delta^{ij}(x-y)\ , \tag{56.9}$$

where the propagator is

$$\Delta^{ij}(x-y) = \int \frac{d^4k}{(2\pi)^4}\,\frac{e^{ik(x-y)}}{k^2-i\epsilon}\,\textstyle\sum_{\lambda=\pm}\varepsilon^{i*}_\lambda(\mathbf{k})\varepsilon^j_\lambda(\mathbf{k})\ . \tag{56.10}$$

As with a free scalar field, correlations of an odd number of fields vanish, and correlations of an even number of fields are given in terms of the propagator by Wick's theorem; see section 8.

We would now like to evaluate the path integral for the free electromagnetic field

$$Z_0(J) \equiv \langle 0|0\rangle_J = \int \mathcal{D}A\; e^{i\int d^4x\,[-\frac{1}{4}F^{\mu\nu}F_{\mu\nu}+J^\mu A_\mu]}\ . \tag{56.11}$$

Here we treat the current $J^\mu(x)$ as an external source.

We will evaluate $Z_0(J)$ in Coulomb gauge. This means that we will integrate over only those field configurations that satisfy $\nabla\!\cdot\!\mathbf{A}=0$.

We begin by integrating over A^0. Because the action is quadratic in A^μ, this is equivalent to solving the variational equation for A^0, and then substituting the solution back into the lagrangian. The result is that we have the Coulomb term in the action,

$$S_{\rm coul} = -\frac{1}{2}\int d^4x\, d^4y\, \delta(x^0{-}y^0)\,\frac{J^0(x)J^0(y)}{4\pi|\mathbf{x}{-}\mathbf{y}|}\,. \tag{56.12}$$

Since this term does not depend on the vector potential, we simply get a factor of $\exp(iS_{\rm coul})$ in front of the remaining path integral over A_i. We will perform this integral formally (as we did for fermion fields in section 43) by requiring it to yield the correct results for the correlation functions of A_i when we take functional derivatives with respect to J_i. In this way we find that

$$Z_0(J) = \exp\left[iS_{\rm coul} + \frac{i}{2}\int d^4x\, d^4y\, J_i(x)\Delta^{ij}(x-y)J_j(y)\right]. \tag{56.13}$$

We can make $Z_0(J)$ look prettier by writing it as

$$Z_0(J) = \exp\left[\frac{i}{2}\int d^4x\, d^4y\, J_\mu(x)\Delta^{\mu\nu}(x-y)J_\nu(y)\right], \tag{56.14}$$

where we have defined

$$\Delta^{\mu\nu}(x-y) \equiv \int \frac{d^4k}{(2\pi)^4}\, e^{ik(x-y)}\, \tilde{\Delta}^{\mu\nu}(k)\,, \tag{56.15}$$

$$\tilde{\Delta}^{\mu\nu}(k) \equiv -\frac{1}{\mathbf{k}^2}\,\delta^{\mu 0}\delta^{\nu 0} + \frac{1}{k^2 - i\epsilon}\sum\nolimits_{\lambda=\pm}\varepsilon_\lambda^{\mu *}(\mathbf{k})\varepsilon_\lambda^{\nu}(\mathbf{k})\,. \tag{56.16}$$

The first term on the right-hand side of eq. (56.16) reproduces the Coulomb term in eq. (56.13) by virtue of the facts that

$$\int_{-\infty}^{+\infty}\frac{dk^0}{2\pi}\, e^{-ik^0(x^0-y^0)} = \delta(x^0-y^0)\,, \tag{56.17}$$

$$\int \frac{d^3k}{(2\pi)^3}\,\frac{e^{i\mathbf{k}\cdot(\mathbf{x}-\mathbf{y})}}{\mathbf{k}^2} = \frac{1}{4\pi|\mathbf{x}-\mathbf{y}|}\,. \tag{56.18}$$

The second term on the right-hand side of eq. (56.16) reproduces the second term in eq. (56.13) by virtue of the fact that $\varepsilon_\lambda^0(\mathbf{k}) = 0$.

Next we will simplify eq. (56.16). We begin by introducing a unit vector in the time direction,

$$\hat{t}^\mu = (1, \mathbf{0})\,. \tag{56.19}$$

Next we need a unit vector in the $\mathbf{k}$ direction, which we will call $\hat{z}^\mu$. We first note that $\hat{t}\cdot k = -k^0$, and so we can write

$$(0, \mathbf{k}) = k^\mu + (\hat{t}\cdot k)\hat{t}^\mu\,. \tag{56.20}$$

The square of this four-vector is

$$\mathbf{k}^2 = k^2 + (\hat{t}\cdot k)^2 \,, \tag{56.21}$$

where we have used $\hat{t}^2 = -1$. Thus the unit vector that we want is

$$\hat{z}^\mu = \frac{k^\mu + (\hat{t}\cdot k)\hat{t}^\mu}{[k^2 + (\hat{t}\cdot k)^2]^{1/2}} \,. \tag{56.22}$$

Now we recall from section 55 that

$$\sum_{\lambda=\pm} \varepsilon_\lambda^{i*}(\mathbf{k})\varepsilon_\lambda^{j}(\mathbf{k}) = \delta_{ij} - \frac{k_i k_j}{\mathbf{k}^2} \,. \tag{56.23}$$

This can be extended to $i \to \mu$ and $j \to \nu$ by writing

$$\sum_{\lambda=\pm} \varepsilon_\lambda^{\mu*}(\mathbf{k})\varepsilon_\lambda^{\nu}(\mathbf{k}) = g^{\mu\nu} + \hat{t}^\mu\hat{t}^\nu - \hat{z}^\mu\hat{z}^\nu \,. \tag{56.24}$$

It is not hard to check that the right-hand side of eq. (56.24) vanishes if $\mu = 0$ or $\nu = 0$, and agrees with eq. (56.23) for $\mu = i$ and $\nu = j$. Putting all this together, we can now write eq. (56.16) as

$$\tilde{\Delta}^{\mu\nu}(k) = -\frac{\hat{t}^\mu\hat{t}^\nu}{k^2 + (\hat{t}\cdot k)^2} + \frac{g^{\mu\nu} + \hat{t}^\mu\hat{t}^\nu - \hat{z}^\mu\hat{z}^\nu}{k^2 - i\epsilon} \,. \tag{56.25}$$

The next step is to consider the terms in this expression that contain factors of k^μ or k^ν; from eq. (56.22), we see that these will arise from the $\hat{z}^\mu\hat{z}^\nu$ term. In eq. (56.15), a factor of k^μ can be written as a derivative with respect to x^μ acting on $e^{ik(x-y)}$. This derivative can then be integrated by parts in eq. (56.14) to give a factor of $\partial^\mu J_\mu(x)$. But $\partial^\mu J_\mu(x)$ vanishes, because the current must be conserved. Similarly, a factor of k^ν can be turned into $\partial^\nu J_\nu(y)$, and also leads to a vanishing contribution. Therefore, *we can ignore any terms in* $\tilde{\Delta}^{\mu\nu}(k)$ *that contain factors of* k^μ *or* k^ν.

From eq. (56.22), we see that this means we can make the substitution

$$\hat{z}^\mu \to \frac{(\hat{t}\cdot k)\hat{t}^\mu}{[k^2 + (\hat{t}\cdot k)^2]^{1/2}} \,. \tag{56.26}$$

Then eq. (56.25) becomes

$$\tilde{\Delta}^{\mu\nu}(k) = \frac{1}{k^2 - i\epsilon}\left[g^{\mu\nu} + \left(-\frac{k^2}{k^2 + (\hat{t}\cdot k)^2} + 1 - \frac{(\hat{t}\cdot k)^2}{k^2 + (\hat{t}\cdot k)^2}\right)\hat{t}^\mu\hat{t}^\nu\right], \tag{56.27}$$

where the three coefficients of $\hat{t}^\mu\hat{t}^\nu$ come from the Coulomb term, the $\hat{t}^\mu\hat{t}^\nu$ term in the polarization sum, and the $\hat{z}^\mu\hat{z}^\nu$ term, respectively. A bit of algebra now reveals that the net coefficient of $\hat{t}^\mu \hat{t}^\nu$ vanishes, leaving us with the elegant expression

$$\tilde{\Delta}^{\mu\nu}(k) = \frac{g^{\mu\nu}}{k^2 - i\epsilon} \, . \tag{56.28}$$

Written in this way, the photon propagator is said to be in *Feynman gauge.* (It would still be in Coulomb gauge if we had retained the k^μ and k^ν terms that we previously dropped.)

In the next section, we will rederive eq. (56.28) from a more explicit path-integral point of view.

Problems

56.1) Use eqs. (55.11) and (55.21)–(55.23) to verify eqs. (56.9) and (56.10).

57

The path integral for photons

Prerequisites: 8, 56

In this section, in order to get a better understanding of the photon path integral, we will evaluate it directly, using the methods of section 8. We begin with

$$Z_0(J) = \int \mathcal{D}A\, e^{iS_0}, \tag{57.1}$$

$$S_0 = \int d^4x \left[-\tfrac{1}{4}F^{\mu\nu}F_{\mu\nu} + J^\mu A_\mu \right]. \tag{57.2}$$

Following section 8, we Fourier-transform to momentum space, where we find

$$S_0 = \frac{1}{2}\int \frac{d^4k}{(2\pi)^4} \Big[-\widetilde{A}_\mu(k)\Big(k^2 g^{\mu\nu} - k^\mu k^\nu\Big)\widetilde{A}_\nu(-k) + \widetilde{J}^\mu(k)\widetilde{A}_\mu(-k) + \widetilde{J}^\mu(-k)\widetilde{A}_\mu(k)\Big]. \tag{57.3}$$

The next step is to shift the integration variable $\widetilde{A}$ so as to "complete the square". This involves inverting the 4×4 matrix $k^2 g^{\mu\nu} - k^\mu k^\nu$. However, this matrix has a zero eigenvalue, and cannot be inverted.

To see this, let us write

$$k^2 g^{\mu\nu} - k^\mu k^\nu = k^2 P^{\mu\nu}(k) , \tag{57.4}$$

where we have defined

$$P^{\mu\nu}(k) \equiv g^{\mu\nu} - k^\mu k^\nu / k^2 . \tag{57.5}$$

This is a projection matrix because, as is easily checked,

$$P^{\mu\nu}(k) P_\nu{}^\lambda(k) = P^{\mu\lambda}(k) . \tag{57.6}$$

Thus the only allowed eigenvalues of P are zero and one. There is at least one zero eigenvalue, because

$$P^{\mu\nu}(k)k_\nu = 0 \; . \tag{57.7}$$

On the other hand, the sum of the eigenvalues is given by the trace

$$g_{\mu\nu}P^{\mu\nu}(k) = 3 \; . \tag{57.8}$$

Thus the remaining three eigenvalues must all be one.

Now let us imagine carrying out the path integral of eq. (57.1), with S_0 given by eq. (57.3). Let us decompose the field $\widetilde{A}_\mu(k)$ into components aligned along a set of linearly independent four-vectors, one of which is k_μ. (It will not matter whether or not this basis set is orthonormal.) Because the term quadratic in $\widetilde{A}_\mu$ involves the matrix $k^2P^{\mu\nu}(k)$, and $P^{\mu\nu}(k)k_\nu = 0$, the component of $\widetilde{A}_\mu(k)$ that lies along k_μ does not contribute to this quadratic term. Furthermore, it does not contribute to the linear term either, because $\partial^\mu J_\mu(x) = 0$ implies $k^\mu \widetilde{J}_\mu(k) = 0$. Thus this component does not appear in the path integral at all! It then makes no sense to integrate over it. We therefore define $\int \mathcal{D}A$ to mean integration over only those components that are spanned by the remaining three basis vectors, and therefore satisfy $k^\mu \widetilde{A}_\mu(k) = 0$. This is equivalent to imposing Lorenz gauge, $\partial^\mu A_\mu(x) = 0$.

The matrix $P^{\mu\nu}(k)$ is simply the matrix that projects a four-vector into the subspace orthogonal to k^μ. Within the subspace, $P^{\mu\nu}(k)$ is equivalent to the identity matrix. Therefore, within the subspace, the inverse of $k^2P^{\mu\nu}(k)$ is $(1/k^2)P^{\mu\nu}(k)$. Employing the ϵ trick to pick out vacuum boundary conditions replaces k^2 with $k^2 - i\epsilon$.

We can now continue following the procedure of section 8, with the result that

$$\begin{aligned} Z_0(J) &= \exp\left[\frac{i}{2}\int \frac{d^4k}{(2\pi)^4}\,\widetilde{J}_\mu(k)\,\frac{P^{\mu\nu}(k)}{k^2 - i\epsilon}\,\widetilde{J}_\nu(-k)\right] \\ &= \exp\left[\frac{i}{2}\int d^4x\,d^4y\;J_\mu(x)\Delta^{\mu\nu}(x-y)J_\nu(y)\right] , \end{aligned} \tag{57.9}$$

where

$$\Delta^{\mu\nu}(x-y) = \int \frac{d^4k}{(2\pi)^4}\,e^{ik(x-y)}\,\frac{P^{\mu\nu}(k)}{k^2 - i\epsilon} \tag{57.10}$$

is the photon propagator in *Lorenz gauge* (also known as *Landau gauge*). Of course, because the current is conserved, the $k^\mu k^\nu$ term in $P^{\mu\nu}(k)$ does not contribute, and so the result is equivalent to that of Feynman gauge, where $P^{\mu\nu}(k)$ is replaced by $g^{\mu\nu}$.

58

Spinor electrodynamics

Prerequisites: 45, 57

In this section, we will study *spinor electrodynamics*: the theory of photons interacting with the electrons and positrons of a Dirac field. (We will use the term *quantum electrodynamics* to denote any theory of photons, irrespective of the kinds of particles with which they interact.)

We construct spinor electrodynamics by taking the electromagnetic current $j^\mu(x)$ to be proportional to the Noether current corresponding to the U(1) symmetry of a Dirac field; see section 36. Specifically,

$$j^\mu(x) = e\overline{\Psi}(x)\gamma^\mu\Psi(x) \ . \tag{58.1}$$

Here $e = -0.302\,822$ is the charge of the electron in Heaviside–Lorentz units, with $\hbar = c = 1$. (We will rely on context to distinguish this e from the base of natural logarithms.) In these units, the fine-structure constant is $\alpha = e^2/4\pi = 1/137.036$. With the normalization of eq. (58.1), $Q = \int d^3x\, j^0(x)$ is the electric charge operator.

Of course, when we specify a number in quantum field theory, we must always have a renormalization scheme in mind; $e = -0.302\,822$ corresponds to a specific version of on-shell renormalization that we will explore in sections 62 and 63. The value of e is different in other renormalization schemes, such as $\overline{\text{MS}}$, as we will see in section 66.

The complete lagrangian of our theory is thus

$$\mathcal{L} = -\tfrac{1}{4}F^{\mu\nu}F_{\mu\nu} + i\overline{\Psi}\partial\!\!\!/\Psi - m\overline{\Psi}\Psi + e\overline{\Psi}\gamma^\mu\Psi A_\mu \ . \tag{58.2}$$

In this section, we will be concerned with tree-level processes only, and so we omit renormalizing Z factors.

We have a problem, though. A Noether current is conserved only when the fields obey the equations of motion, or, equivalently, only at points in field space where the action is stationary. On the other hand, in our development of photon path integrals in sections 56 and 57, we assumed that the current was *always* conserved.

This issue is resolved by enlarging the definition of a gauge transformation to include a transformation on the Dirac field as well as the electromagnetic field. Specifically, we define a gauge transformation to consist of

$$A^\mu(x) \to A^\mu(x) - \partial^\mu\Gamma(x) \,, \tag{58.3}$$

$$\Psi(x) \to \exp[-ie\Gamma(x)]\Psi(x) \,, \tag{58.4}$$

$$\overline{\Psi}(x) \to \exp[+ie\Gamma(x)]\overline{\Psi}(x) \,. \tag{58.5}$$

It is not hard to check that $\mathcal{L}(x)$ is *invariant* under this transformation, whether or not the fields obey their equations of motion. To perform this check most easily, we first rewrite $\mathcal{L}$ as

$$\mathcal{L} = -\tfrac{1}{4}F^{\mu\nu}F_{\mu\nu} + i\overline{\Psi}\not{D}\Psi - m\overline{\Psi}\Psi \,, \tag{58.6}$$

where we have defined the *gauge covariant derivative* (or just *covariant derivative* for short)

$$D_\mu \equiv \partial_\mu - ieA_\mu \,. \tag{58.7}$$

In the last section, we found that $F^{\mu\nu}$ is invariant under eq. (58.3), and so the FF term in $\mathcal{L}$ is obviously invariant as well. It is also obvious that the $m\overline{\Psi}\Psi$ term in $\mathcal{L}$ is invariant under eqs. (58.4) and (58.5). This leaves the $\overline{\Psi}\not{D}\Psi$ term. This term will also be invariant if, under the gauge transformation, the covariant derivative of Ψ transforms as

$$D_\mu\Psi(x) \to \exp[-ie\Gamma(x)]D_\mu\Psi(x) \,. \tag{58.8}$$

To see if this is true, we note that

$$\begin{aligned}
D_\mu\Psi \to{}& \Big(\partial_\mu - ie[A_\mu - \partial_\mu\Gamma]\Big)\Big(\exp[-ie\Gamma]\Psi\Big) \\
={}& \exp[-ie\Gamma]\Big(\partial_\mu\Psi - ie(\partial_\mu\Gamma)\Psi - ie[A_\mu - \partial_\mu\Gamma]\Psi\Big) \\
={}& \exp[-ie\Gamma]\Big(\partial_\mu - ieA_\mu\Big)\Psi \\
={}& \exp[-ie\Gamma]D_\mu\Psi \,.
\end{aligned} \tag{58.9}$$

So eq. (58.8) holds, and $\overline{\Psi}\not{D}\Psi$ is gauge invariant.

We can also write the transformation rule for D_μ a little more abstractly as

$$D_\mu \to e^{-ie\Gamma} D_\mu \, e^{+ie\Gamma} \, , \tag{58.10}$$

where the ordinary derivative in D_μ is defined to act on anything to its right, including any fields that are left unwritten in eq. (58.10). Thus we have

$$\begin{aligned} D_\mu \Psi &\to \Big(e^{-ie\Gamma} D_\mu e^{+ie\Gamma}\Big)\Big(e^{-ie\Gamma}\Psi\Big) \\ &= e^{-ie\Gamma} D_\mu \Psi \, , \end{aligned} \tag{58.11}$$

which is, of course, the same as eq. (58.9). We can also express the field strength in terms of the covariant derivative by noting that

$$[D^\mu, D^\nu]\Psi(x) = -ieF^{\mu\nu}(x)\Psi(x) \, . \tag{58.12}$$

We can write this more abstractly as

$$F^{\mu\nu} = \tfrac{i}{e}[D^\mu, D^\nu] \, , \tag{58.13}$$

where, again, the ordinary derivative in each covariant derivative acts on anything to its right. From eqs. (58.10) and (58.13), we see that, under a gauge transformation,

$$\begin{aligned} F^{\mu\nu} &\to \tfrac{i}{e}\Big[e^{-ie\Gamma} D^\mu \, e^{+ie\Gamma} \, , e^{-ie\Gamma} D^\nu \, e^{+ie\Gamma}\Big] \\ &= e^{-ie\Gamma}\Big(\tfrac{i}{e}[D^\mu, D^\nu]\Big)e^{+ie\Gamma} \\ &= e^{-ie\Gamma} F^{\mu\nu} e^{+ie\Gamma} \\ &= F^{\mu\nu} \, . \end{aligned} \tag{58.14}$$

In the last line, we are able to cancel the $e^{\pm ie\Gamma}$ factors against each other because no derivatives act on them. Eq. (58.14) shows us that (as we already knew) $F^{\mu\nu}$ is gauge invariant.

It is interesting to note that the gauge transformation on the fermion fields, eqs. (58.4) and (58.5), is a generalization of the U(1) transformation

$$\Psi \to e^{-i\alpha}\Psi \, , \tag{58.15}$$

$$\overline{\Psi} \to e^{+i\alpha}\overline{\Psi} \, , \tag{58.16}$$

that is a symmetry of the free Dirac lagrangian. The difference is that, in the gauge transformation, the phase factor is allowed to be a function of spacetime, rather than a constant that is the same everywhere. Thus, the gauge transformation is also called a *local* U(1) transformation, while eqs. (58.15) and (58.16) correspond to a *global* U(1) transformation. We say that, in a gauge theory, the global U(1) symmetry is promoted to a local U(1) symmetry, or that we have *gauged* the U(1) symmetry.

In section 57, we argued that the path integral over A_μ should be restricted to those components of $\widetilde{A}_\mu(k)$ that are orthogonal to k_μ, because the component parallel to k_μ did not appear in the integrand. Now we must make a slightly more subtle argument. We argue that the path integral over the parallel component is redundant, because the fermionic path integral over Ψ and $\overline{\Psi}$ already includes all possible values of $\Gamma(x)$. Therefore, as in section 57, we should not integrate over the parallel component. (We will make a more precise and careful version of this argument when we discuss the quantization of nonabelian gauge theories in section 71.)

By the standard procedure, this leads us to the following form of the path integral for spinor electrodynamics:

$$Z(\overline{\eta},\eta,J) \propto \exp\left[ie\int d^4x\,\left(\frac{1}{i}\frac{\delta}{\delta J^\mu(x)}\right)\left(i\frac{\delta}{\delta\eta_\alpha(x)}\right)(\gamma^\mu)_{\alpha\beta}\left(\frac{1}{i}\frac{\delta}{\delta\overline{\eta}_\beta(x)}\right)\right] \times Z_0(\overline{\eta},\eta,J)\,, \tag{58.17}$$

where

$$Z_0(\overline{\eta},\eta,J) = \exp\left[i\int d^4x\,d^4y\;\overline{\eta}(x)S(x-y)\eta(y)\right] \times \exp\left[\frac{i}{2}\int d^4x\,d^4y\;J^\mu(x)\Delta_{\mu\nu}(x-y)J^\nu(y)\right], \tag{58.18}$$

and

$$S(x-y) = \int\frac{d^4p}{(2\pi)^4}\,\frac{(-\not{p}+m)}{p^2+m^2-i\epsilon}\,e^{ip(x-y)}\,, \tag{58.19}$$

$$\Delta_{\mu\nu}(x-y) = \int\frac{d^4k}{(2\pi)^4}\,\frac{g_{\mu\nu}}{k^2-i\epsilon}\,e^{ik(x-y)} \tag{58.20}$$

are the appropriate Feynman propagators for the corresponding free fields, with the photon propagator in Feynman gauge. We impose the normalization $Z(0,0,0) = 1$, and write

$$Z(\overline{\eta},\eta,J) = \exp[iW(\overline{\eta},\eta,J)]\,. \tag{58.21}$$

Then $iW(\overline{\eta},\eta,J)$ can be expressed as a series of connected Feynman diagrams with sources.

The rules for internal and external Dirac fermions were worked out in the context of Yukawa theory in section 45, and they follow here with no change. For external photons, the LSZ analysis of section 56 implies that each external photon line carries a factor of the polarization vector $\varepsilon^\mu(\mathbf{k})$.

Putting everything together, we get the following set of Feynman rules for tree-level processes in spinor electrodynamics.

1) For each *incoming electron*, draw a solid line with an arrow pointed *towards* the vertex, and label it with the electron's four-momentum, p_i.
2) For each *outgoing electron*, draw a solid line with an arrow pointed *away* from the vertex, and label it with the electron's four-momentum, p'_i.
3) For each *incoming positron*, draw a solid line with an arrow pointed *away* from the vertex, and label it with *minus* the positron's four-momentum, $-p_i$.
4) For each *outgoing positron*, draw a solid line with an arrow pointed *towards* the vertex, and label it with *minus* the positron's four-momentum, $-p'_i$.
5) For each *incoming photon*, draw a wavy line with an arrow pointed *towards* the vertex, and label it with the photon's four-momentum, k_i. (Wavy lines for photons is a standard convention.)
6) For each *outgoing photon*, draw a wavy line with an arrow pointed *away* from the vertex, and label it with the photon's four-momentum, k'_i.
7) The only allowed vertex joins two solid lines, one with an arrow pointing towards it and one with an arrow pointing away from it, and one wavy line (whose arrow can point in either direction). Using this vertex, join up all the external lines, including extra internal lines as needed. In this way, draw all possible diagrams that are *topologically inequivalent.*
8) Assign each internal line its own four-momentum. Think of the four-momenta as flowing along the arrows, and conserve four-momentum at each vertex. For a tree diagram, this fixes the momenta on all the internal lines.
9) The value of a diagram consists of the following factors:
   for each incoming photon, $\varepsilon^{\mu *}_{\lambda_i}(\mathbf{k}_i)$;
   for each outgoing photon, $\varepsilon^{\mu}_{\lambda'_i}(\mathbf{k}'_i)$;
   for each incoming electron, $u_{s_i}(\mathbf{p}_i)$;
   for each outgoing electron, $\overline{u}_{s'_i}(\mathbf{p}'_i)$;
   for each incoming positron, $\overline{v}_{s_i}(\mathbf{p}_i)$;
   for each outgoing positron, $v_{s'_i}(\mathbf{p}'_i)$;
   for each vertex, $ie\gamma^\mu$;
   for each internal photon, $-ig^{\mu\nu}/(k^2 - i\epsilon)$;
   for each internal fermion, $-i(-\not{p} + m)/(p^2 + m^2 - i\epsilon)$.
10) Spinor indices are contracted by starting at one end of a fermion line: specifically, the end that has the arrow pointing away from the vertex. The factor associated with the external line is either $\overline{u}$ or $\overline{v}$. Go along the complete fermion line, following the arrows backwards, and write down (in order from left to right) the factors associated with the vertices and propagators that you encounter.

The last factor is either a u or v. Repeat this procedure for the other fermion lines, if any. The vector index on each vertex is contracted with the vector index on either the photon propagator (if the attached photon line is internal) or the photon polarization vector (if the attached photon line is external).

11) The overall sign of a tree diagram is determined by drawing all contributing diagrams in a standard form: all fermion lines horizontal, with their arrows pointing from left to right, and with the left endpoints labeled in the same fixed order (from top to bottom); if the ordering of the labels on the right endpoints of the fermion lines in a given diagram is an even (odd) permutation of an arbitrarily chosen fixed ordering, then the sign of that diagram is positive (negative).
12) The value of $i\mathcal{T}$ (at tree level) is given by a sum over the values of all the contributing diagrams.

In the next section, we will do a sample calculation.

Problems

58.1) Compute $P^{-1}A^{\mu}(\mathbf{x},t)P$, $T^{-1}A^{\mu}(\mathbf{x},t)T$, and $C^{-1}A^{\mu}(\mathbf{x},t)C$, assuming that P, T, and C are symmetries of the lagrangian. (Prerequisite: 40.)

58.2) *Furry's theorem.* Show that any scattering amplitude with no external fermions, and an odd number of external photons, is zero.

59

Scattering in spinor electrodynamics

Prerequisites: 48, 58

In the last section, we wrote down the Feynman rules for spinor electrodynamics. In this section, we will compute the scattering amplitude (and its spin-averaged square) at tree level for the process of electron-positron annihilation into a pair of photons, $e^+e^- \to \gamma\gamma$.

The contributing diagrams are shown in fig. 59.1, and the associated expression for the scattering amplitude is

$$\mathcal{T} = e^2\, \varepsilon^{\mu}_{1'}\varepsilon^{\nu}_{2'}\, \overline{v}_2 \left[\gamma_\nu \left(\frac{-\not{p}_1 + \not{k}'_1 + m}{-t + m^2} \right) \gamma_\mu + \gamma_\mu \left(\frac{-\not{p}_1 + \not{k}'_2 + m}{-u + m^2} \right) \gamma_\nu \right] u_1 \,, \tag{59.1}$$

where $\varepsilon^{\mu}_{1'}$ is shorthand for $\varepsilon^{\mu}_{\lambda'_1}(\mathbf{k}'_1)$, $\overline{v}_2$ is shorthand for $\overline{v}_{s_2}(\mathbf{p}_2)$, and so on. The Mandelstam variables are

$$\begin{aligned} s &= -(p_1 + p_2)^2 = -(k'_1 + k'_2)^2 \,, \\ t &= -(p_1 - k'_1)^2 = -(p_2 - k'_2)^2 \,, \\ u &= -(p_1 - k'_2)^2 = -(p_2 - k'_1)^2 \,, \end{aligned} \tag{59.2}$$

and they obey $s + t + u = 2m^2$.

Following the procedure of section 46, we write eq. (59.1) as

$$\mathcal{T} = \varepsilon^{\mu}_{1'}\varepsilon^{\nu}_{2'}\, \overline{v}_2 A_{\mu\nu} u_1 \,, \tag{59.3}$$

where

$$A_{\mu\nu} \equiv e^2 \left[\gamma_\nu \left(\frac{-\not{p}_1 + \not{k}'_1 + m}{-t + m^2} \right) \gamma_\mu + \gamma_\mu \left(\frac{-\not{p}_1 + \not{k}'_2 + m}{-u + m^2} \right) \gamma_\nu \right] . \tag{59.4}$$

We also have

$$\mathcal{T}^* = \overline{\mathcal{T}} = \varepsilon^{\rho *}_{1'}\varepsilon^{\sigma *}_{2'}\, \overline{u}_1\, \overline{A_{\rho\sigma}}\, v_2 \,. \tag{59.5}$$

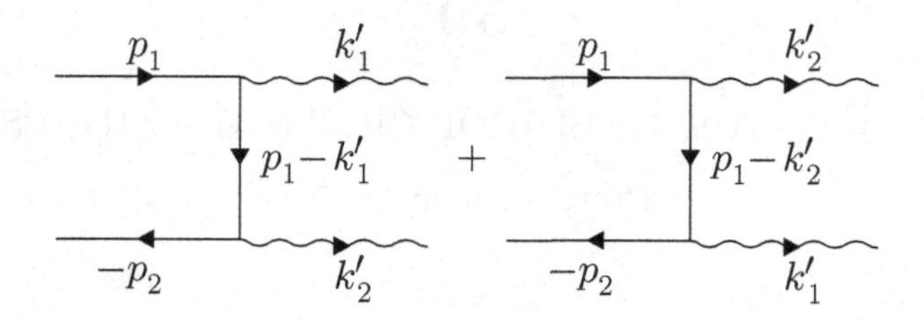

Figure 59.1. Diagrams for $e^+e^- \to \gamma\gamma$, corresponding to eq. (59.1).

Using $\overline{\not{a}\not{b}\ldots} = \ldots\not{b}\not{a}$, we see from eq. (59.4) that

$$\overline{A_{\rho\sigma}} = A_{\sigma\rho} \ . \tag{59.6}$$

Thus we have

$$|\mathcal{T}|^2 = \varepsilon^{\mu}_{1'}\varepsilon^{\nu}_{2'}\,\varepsilon^{\rho *}_{1'}\varepsilon^{\sigma *}_{2'}\,(\overline{v}_2 A_{\mu\nu} u_1)(\overline{u}_1 A_{\sigma\rho} v_2) \ . \tag{59.7}$$

Next, we will average over the initial electron and positron spins, using the technology of section 46; the result is

$$\tfrac{1}{4}\sum_{s_1,s_2} |\mathcal{T}|^2 = \tfrac{1}{4}\varepsilon^{\mu}_{1'}\varepsilon^{\nu}_{2'}\varepsilon^{\rho *}_{1'}\varepsilon^{\sigma *}_{2'}\ \mathrm{Tr}\Big[A_{\mu\nu}(-\not{p}_1 + m)A_{\sigma\rho}(-\not{p}_2 - m)\Big] \ . \tag{59.8}$$

We would also like to sum over the final photon polarizations. From eq. (59.8), we see that we must evaluate

$$\sum_{\lambda=\pm} \varepsilon^{\mu}_{\lambda}(\mathbf{k})\varepsilon^{\rho *}_{\lambda}(\mathbf{k}) \ . \tag{59.9}$$

We did this polarization sum in Coulomb gauge in section 56, with the result that

$$\sum_{\lambda=\pm} \varepsilon^{\mu}_{\lambda}(\mathbf{k})\varepsilon^{\rho *}_{\lambda}(\mathbf{k}) = g^{\mu\rho} + \hat{t}^{\mu}\hat{t}^{\rho} - \hat{z}^{\mu}\hat{z}^{\rho} \ , \tag{59.10}$$

where $\hat{t}^{\mu}$ is a unit vector in the time direction, and $\hat{z}^{\mu}$ is a unit vector in the $\mathbf{k}$ direction that can be expressed as

$$\hat{z}^{\mu} = \frac{k^{\mu} + (\hat{t}\cdot k)\hat{t}^{\mu}}{[k^2 + (\hat{t}\cdot k)^2]^{1/2}} \ . \tag{59.11}$$

It is tempting to drop the k^{μ} and k^{ρ} terms in eq. (59.10), on the grounds that the photons couple to a conserved current, and so these terms should not contribute. (We indeed used this argument to drop the analogous terms in the photon propagator.) This also follows from the notion that the scattering amplitude should be invariant under a gauge transformation, as represented by a transformation of the external polarization vectors of the form

$$\varepsilon^{\mu}_{\lambda}(\mathbf{k}) \to \varepsilon^{\mu}_{\lambda}(\mathbf{k}) - i\tilde{\Gamma}(k)k^{\mu} \,. \tag{59.12}$$

Thus, if we write a scattering amplitude $\mathcal{T}$ for a process that includes a particular outgoing photon with four-momentum k^{μ} as

$$\mathcal{T} = \varepsilon^{\mu}_{\lambda}(\mathbf{k})\mathcal{M}_{\mu} \,, \tag{59.13}$$

or a particular incoming photon with four-momentum k^{μ} as

$$\mathcal{T} = \varepsilon^{\mu*}_{\lambda}(\mathbf{k})\mathcal{M}_{\mu} \,, \tag{59.14}$$

then in either case we should have

$$k^{\mu}\mathcal{M}_{\mu} = 0 \,. \tag{59.15}$$

Eq. (59.15) is in fact valid; we will give a proof of it, based on the Ward identity for the electromagnetic current, in section 67. For now, we will take eq. (59.15) as given, and so drop the k^{μ} and k^{ρ} terms in eq. (59.10).

This leaves us with

$$\sum_{\lambda=\pm} \varepsilon^{\mu}_{\lambda}(\mathbf{k})\varepsilon^{\rho*}_{\lambda}(\mathbf{k}) \to g^{\mu\rho} + \hat{t}^{\mu}\hat{t}^{\rho} - \frac{(\hat{t}\cdot k)^2}{k^2 + (\hat{t}\cdot k)^2}\,\hat{t}^{\mu}\hat{t}^{\rho} \,. \tag{59.16}$$

But, for an external photon, $k^2 = 0$. Thus the second and third terms in eq. (59.16) cancel, leaving us with the beautifully simple substitution rule

$$\sum_{\lambda=\pm} \varepsilon^{\mu}_{\lambda}(\mathbf{k})\varepsilon^{\rho*}_{\lambda}(\mathbf{k}) \to g^{\mu\rho} \,. \tag{59.17}$$

Using eq. (59.17), we can sum $|\mathcal{T}|^2$ over the polarizations of the outgoing photons, in addition to averaging over the spins of the incoming fermions; the result is

$$\begin{aligned}
\langle|\mathcal{T}|^2\rangle &\equiv \tfrac{1}{4}\sum_{\lambda'_1,\lambda'_2}\sum_{s_1,s_2}|\mathcal{T}|^2 \\
&= \tfrac{1}{4}\mathrm{Tr}\Big[A_{\mu\nu}(-\not p_1+m)A^{\nu\mu}(-\not p_2-m)\Big] \\
&= e^4\left[\frac{\langle\Phi_{tt}\rangle}{(m^2-t)^2} + \frac{\langle\Phi_{tu}\rangle+\langle\Phi_{ut}\rangle}{(m^2-t)(m^2-u)} + \frac{\langle\Phi_{uu}\rangle}{(m^2-u)^2}\right],
\end{aligned} \tag{59.18}$$

where

$$\langle\Phi_{tt}\rangle = \tfrac{1}{4}\mathrm{Tr}\Big[\gamma_\nu(-\not{p}_1+\not{k}'_1+m)\gamma_\mu(-\not{p}_1+m)\gamma^\mu(-\not{p}_1+\not{k}'_1+m)\gamma^\nu(-\not{p}_2-m)\Big]\,,$$

$$\langle\Phi_{uu}\rangle = \tfrac{1}{4}\mathrm{Tr}\Big[\gamma_\mu(-\not{p}_1+\not{k}'_2+m)\gamma_\nu(-\not{p}_1+m)\gamma^\nu(-\not{p}_1+\not{k}'_2+m)\gamma^\mu(-\not{p}_2-m)\Big]\,,$$

$$\langle\Phi_{tu}\rangle = \tfrac{1}{4}\mathrm{Tr}\Big[\gamma_\nu(-\not{p}_1+\not{k}'_1+m)\gamma_\mu(-\not{p}_1+m)\gamma^\nu(-\not{p}_1+\not{k}'_2+m)\gamma^\mu(-\not{p}_2-m)\Big]\,,$$

$$\langle\Phi_{ut}\rangle = \tfrac{1}{4}\mathrm{Tr}\Big[\gamma_\mu(-\not{p}_1+\not{k}'_2+m)\gamma_\nu(-\not{p}_1+m)\gamma^\mu(-\not{p}_1+\not{k}'_1+m)\gamma^\nu(-\not{p}_2-m)\Big]\,. \tag{59.19}$$

Examining $\langle\Phi_{tt}\rangle$ and $\langle\Phi_{uu}\rangle$, we see that they are transformed into each other by $k'_1 \leftrightarrow k'_2$, which is equivalent to $t \leftrightarrow u$. The same is true of $\langle\Phi_{tu}\rangle$ and $\langle\Phi_{ut}\rangle$. Thus we need only compute $\langle\Phi_{tt}\rangle$ and $\langle\Phi_{tu}\rangle$, and then take $t \leftrightarrow u$ to get $\langle\Phi_{uu}\rangle$ and $\langle\Phi_{ut}\rangle$.

Now we can apply the gamma-matrix technology of section 47. In particular, we will need the $d = 4$ relations

$$\begin{aligned}
\gamma^\mu\gamma_\mu &= -4\,,\\
\gamma^\mu\not{a}\gamma_\mu &= 2\not{a}\,,\\
\gamma^\mu\not{a}\not{b}\gamma_\mu &= 4(ab)\,,\\
\gamma^\mu\not{a}\not{b}\not{c}\gamma_\mu &= 2\not{c}\not{b}\not{a}\,,
\end{aligned} \tag{59.20}$$

in addition to the trace formulae. We also need

$$\begin{aligned}
p_1p_2 &= -\tfrac{1}{2}(s-2m^2)\,,\\
k'_1k'_2 &= -\tfrac{1}{2}s\,,\\
p_1k'_1 = p_2k'_2 &= +\tfrac{1}{2}(t-m^2)\,,\\
p_1k'_2 = p_2k'_1 &= +\tfrac{1}{2}(u-m^2)\,,
\end{aligned} \tag{59.21}$$

which follow from eq. (59.2) plus the mass-shell conditions $p_1^2 = p_2^2 = -m^2$ and $k_1'^2 = k_2'^2 = 0$. After a lengthy and tedious calculation, we find

$$\langle\Phi_{tt}\rangle = 2[tu - m^2(3t+u) - m^4]\,, \tag{59.22}$$

$$\langle\Phi_{tu}\rangle = 2m^2(s-4m^2)\,, \tag{59.23}$$

which then implies

$$\langle\Phi_{uu}\rangle = 2[tu - m^2(3u+t) - m^4]\,, \tag{59.24}$$

$$\langle\Phi_{ut}\rangle = 2m^2(s-4m^2)\,. \tag{59.25}$$

This completes our calculation.

Other tree-level scattering processes in spinor electrodynamics pose no new calculational difficulties, and are left to the problems.

In the high-energy limit, where the electron can be treated as massless, we can reduce our labor with the method of *spinor helicity*, which was introduced in section 50. We take this up in the next section.

Problems

59.1) Compute $\langle|\mathcal{T}|^2\rangle$ for *Compton scattering*, $e^-\gamma \to e^-\gamma$. You should find that your result is the same as that for $e^+e^- \to \gamma\gamma$, but with $s \leftrightarrow t$, and an extra overall minus sign. This is an example of *crossing symmetry*; there is an overall minus sign for each fermion that is moved from the initial to the final state.

59.2) Compute $\langle|\mathcal{T}|^2\rangle$ for *Bhabha scattering*, $e^+e^- \to e^+e^-$.

59.3) Compute $\langle|\mathcal{T}|^2\rangle$ for *Møller scattering*, $e^-e^- \to e^-e^-$. You should find that your result is the same as that for $e^+e^- \to e^+e^-$, but with $s \leftrightarrow u$. This is another example of crossing symmetry.

60

Spinor helicity for spinor electrodynamics

Prerequisites: 50, 59

In section 50, we introduced a special notation for u and v spinors of definite helicity for *massless* electrons and positrons. This notation greatly simplifies calculations in the high-energy limit (s, $|t|$, and $|u|$ all much greater than m^2).

We define the *twistors*

$$
\begin{aligned}
|p] &\equiv u_-(p) = v_+(p) \, , \\
|p\rangle &\equiv u_+(p) = v_-(p) \, , \\
[p| &\equiv \overline{u}_+(p) = \overline{v}_-(p) \, , \\
\langle p| &\equiv \overline{u}_-(p) = \overline{v}_+(p) \, .
\end{aligned}
\tag{60.1}
$$

We then have

$$
\begin{aligned}
[k| \, |p] &= [k \, p] \, , \\
\langle k| \, |p\rangle &= \langle k \, p\rangle \, , \\
[k| \, |p\rangle &= 0 \, , \\
\langle k| \, |p] &= 0 \, ,
\end{aligned}
\tag{60.2}
$$

where the *twistor products* $[k \, p]$ and $\langle k \, p\rangle$ are antisymmetric,

$$
\begin{aligned}
[k \, p] &= -[p \, k] \, , \\
\langle k \, p\rangle &= -\langle p \, k\rangle \, ,
\end{aligned}
\tag{60.3}
$$

and related by complex conjugation, $\langle p \, k\rangle^* = [k \, p]$. They can be expressed explicitly in terms of the components of the massless four-momenta k and

p. However, more useful are the relations

$$\begin{aligned}\langle k\,p\rangle\,[p\,k] &= \mathrm{Tr}\,\tfrac{1}{2}(1-\gamma_5)\not{k}\not{p} \\ &= -2k\cdot p \\ &= -(k+p)^2 \end{aligned} \tag{60.4}$$

and

$$\begin{aligned}\langle p\,q\rangle\,[q\,r]\,\langle r\,s\rangle\,[s\,p] &= \mathrm{Tr}\,\tfrac{1}{2}(1-\gamma_5)\not{p}\not{q}\not{r}\not{s} \\ &= 2[(p\cdot q)(r\cdot s) - (p\cdot r)(q\cdot s) + (p\cdot s)(q\cdot r) \\ &\qquad + i\varepsilon^{\mu\nu\rho\sigma}p_\mu q_\nu r_\rho s_\sigma]\;. \end{aligned} \tag{60.5}$$

Finally, for any massless four-momentum p we can write

$$-\not{p} = |p\rangle[p| + |p]\langle p|\;. \tag{60.6}$$

We will quote other results from section 50 as we need them.

To apply this formalism to spinor electrodynamics, we need to write photon polarization vectors in terms of twistors. The formulae we need are

$$\varepsilon^\mu_+(k) = -\,\frac{\langle q|\gamma^\mu|k]}{\sqrt{2}\,\langle q\,k\rangle}\;, \tag{60.7}$$

$$\varepsilon^\mu_-(k) = -\,\frac{[q|\gamma^\mu|k\rangle}{\sqrt{2}\,[q\,k]}\;, \tag{60.8}$$

where q is an arbitrary massless *reference momentum.*

We will verify eqs. (60.7) and (60.8) for a specific choice of k, and then rely on the Lorentz transformation properties of twistors to conclude that the result must hold in any frame (and therefore for any massless four-momentum k).

We will choose $k^\mu = (\omega, \omega\hat{\mathbf{z}}) = \omega(1,0,0,1)$. Then, the most general form of $\varepsilon^\mu_+(k)$ is

$$\varepsilon^\mu_+(k) = e^{i\phi}\tfrac{1}{\sqrt{2}}(0,1,-i,0) + Ck^\mu\;. \tag{60.9}$$

Here $e^{i\phi}$ is an arbitrary phase factor, and C is an arbitrary complex number; the freedom to add a multiple of k comes from the underlying gauge invariance.

To verify that eq. (60.7) reproduces eq. (60.9), we need the explicit form of the twistors $|k]$ and $|k\rangle$ when the three-momentum is in the z direction.

Using results in section 50 we find

$$|k] = \sqrt{2\omega}\begin{pmatrix}0\\1\\0\\0\end{pmatrix}, \qquad |k\rangle = \sqrt{2\omega}\begin{pmatrix}0\\0\\1\\0\end{pmatrix}. \tag{60.10}$$

For any value of q, the twistor $\langle q|$ takes the form

$$\langle q| = (0,\, 0,\, \alpha,\, \beta)\,, \tag{60.11}$$

where α and β are complex numbers. Plugging eqs. (60.10) and (60.11) into eq. (60.7), and using

$$\gamma^\mu = \begin{pmatrix}0 & \sigma^\mu\\ \bar\sigma^\mu & 0\end{pmatrix} \tag{60.12}$$

along with $\sigma^\mu = (I, \vec\sigma)$ and $\bar\sigma^\mu = (I, -\vec\sigma)$, we find that we reproduce eq. (60.9) with $e^{i\phi} = 1$ and $C = -\beta/(\sqrt{2}\alpha\omega)$. There is now no need to check eq. (60.8), because $\varepsilon_-^\mu(k) = -[\varepsilon_+^\mu(k)]^*$, as can be seen by using $\langle q\, k\rangle^* = -[q\, k]$ along with another result from section 50, $\langle q|\gamma^\mu|k]^* = \langle k|\gamma^\mu|q]$.

In spinor electrodynamics, the vector index on a photon polarization vector is always contracted with the vector index on a gamma matrix. We can get a convenient formula for $\not\varepsilon_\pm(k)$ by using the Fierz identities

$$-\tfrac{1}{2}\gamma^\mu\langle q|\gamma_\mu|k] = |k]\langle q| + |q\rangle[k|\,, \tag{60.13}$$

$$-\tfrac{1}{2}\gamma^\mu[q|\gamma_\mu|k\rangle = |k\rangle[q| + |q]\langle k|\,. \tag{60.14}$$

We then have

$$\not\varepsilon_+(k;q) = \frac{\sqrt{2}}{\langle q\, k\rangle}\Big(|k]\langle q| + |q\rangle[k|\Big)\,, \tag{60.15}$$

$$\not\varepsilon_-(k;q) = \frac{\sqrt{2}}{[q\, k]}\Big(|k\rangle[q| + |q]\langle k|\Big)\,, \tag{60.16}$$

where we have added the reference momentum as an explicit argument on the left-hand sides.

Now we have all the tools we need for doing calculations. However, we can simplify things even further by making maximal use of crossing symmetry.

Note from eq. (60.1) that u_- (which is the factor associated with an incoming electron) and v_+ (an outgoing positron) are both represented by the twistor $|p]$, while $\overline{u}_+$ (an outgoing electron) and $\overline{v}_-$ (an incoming positron)

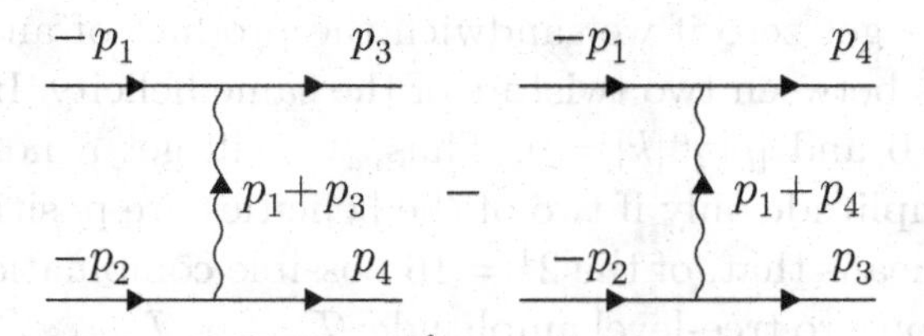

Figure 60.1. Diagrams for fermion–fermion scattering, with all momenta treated as outgoing.

are both represented by $[p|$. Thus the square-bracket twistors correspond to outgoing fermions with positive helicity, and incoming fermions with negative helicity. Similarly, the angle-bracket twistors correspond to outgoing fermions with negative helicity, and incoming fermions with positive helicity.

Let us adopt a convention in which all particles are assigned four-momenta that are treated as outgoing. A particle that has an assigned four-momentum p then has physical four-momentum $\epsilon_p p$, where $\epsilon_p = \text{sign}(p^0) = +1$ if the particle is physically outgoing, and $\epsilon_p = \text{sign}(p^0) = -1$ if the particle is physically incoming.

Since the physical three-momentum of an incoming particle is opposite to its assigned three-momentum, a particle with negative helicity relative to its physical three-momentum has positive helicity relative to its assigned three-momentum. From now on, we will refer to the helicity of a particle relative to its assigned momentum. Thus a particle that we say has "positive helicity" actually has negative physical helicity if it is incoming, and positive physical helicity if it is outgoing.

With this convention, the square-bracket twistors $|p]$ and $[p|$ represent positive-helicity fermions, and the angle-bracket twistors $|p\rangle$ and $\langle p|$ represent negative-helicity fermions. When $\epsilon_p = \text{sign}(p^0) = -1$, we analytically continue the twistors by replacing each $\omega^{1/2}$ in eq. (60.10) with $i|\omega|^{1/2}$. Then all of our formulae for twistors and polarizations hold without change, with the exception of the rule for complex conjugation of a twistor product, which becomes

$$\langle p\, k\rangle^* = \epsilon_p \epsilon_k [k\, p]\,. \tag{60.17}$$

Now we are ready to calculate some amplitudes. Consider first the process of fermion-fermion scattering. The contributing tree-level diagrams are shown in fig. 60.1.

The first thing to notice is that a diagram is zero if two external fermion lines that meet at a vertex have the same helicity. This is because (as shown in section 50) we get zero if we sandwich the product of an odd number of gamma matrices between two twistors of the same helicity. In particular, we have $\langle p|\gamma^\mu|k\rangle = 0$ and $[p|\gamma^\mu|k] = 0$. Thus, we will get a nonzero result for the tree-level amplitude only if two of the helicities are positive, and two are negative. This means that, of the $2^4 = 16$ possible combinations of helicities, only six give a nonzero tree-level amplitude: $\mathcal{T}_{++--}$, $\mathcal{T}_{+-+-}$, $\mathcal{T}_{+--+}$, $\mathcal{T}_{--++}$, $\mathcal{T}_{-+-+}$, and $\mathcal{T}_{-++-}$, where the notation is $\mathcal{T}_{s_1 s_2 s_3 s_4}$. Furthermore, the last three of these are related to the first three by complex conjugation, so we only have three amplitudes to compute.

Let us begin with $\mathcal{T}_{+--+}$. Only the first diagram of fig. 60.1 contributes, because the second has two positive-helicity lines meeting at a vertex. To evaluate the first diagram, we note that the two vertices contribute a factor of $(ie)^2 = -e^2$, and the internal photon line contributes a factor of $ig_{\mu\nu}/s_{13}$, where we have defined the Mandelstam variable

$$s_{ij} \equiv -(p_i + p_j)^2 \, . \tag{60.18}$$

Following the charge arrows backwards on each fermion line, and dividing by i to get $\mathcal{T}$ (rather than $i\mathcal{T}$), we find

$$\begin{aligned}\mathcal{T}_{+--+} &= -e^2 \langle 3|\gamma^\mu|1]\,[4|\gamma_\mu|2\rangle\,/s_{13} \\ &= +2e^2\,[1\,4]\,\langle 2\,3\rangle\,/s_{13}\,,\end{aligned} \tag{60.19}$$

where $\langle 3|$ is short for $\langle p_3|$, etc., and we have used yet another form of the Fierz identity to get the second line.

The computation of $\mathcal{T}_{+-+-}$ is exactly analogous, except that now it is only the second diagram of fig. 60.1 that contributes. According to the Feynman rules, this diagram comes with a relative minus sign, and so we have

$$\mathcal{T}_{+-+-} = -2e^2\,[1\,3]\,\langle 2\,4\rangle\,/s_{14}\,. \tag{60.20}$$

Finally, we turn to $\mathcal{T}_{++--}$. Now both diagrams contribute, and we have

$$\begin{aligned}\mathcal{T}_{++--} &= -e^2 \left(\frac{\langle 3|\gamma^\mu|1]\,\langle 4|\gamma_\mu|2]}{s_{13}} - \frac{\langle 4|\gamma^\mu|1]\,\langle 3|\gamma_\mu|2]}{s_{14}} \right) \\ &= -2e^2\,[1\,2]\,\langle 3\,4\rangle \left(\frac{1}{s_{13}} + \frac{1}{s_{14}} \right) \\ &= +2e^2\,[1\,2]\,\langle 3\,4\rangle \left(\frac{s_{12}}{s_{13}s_{14}} \right),\end{aligned} \tag{60.21}$$

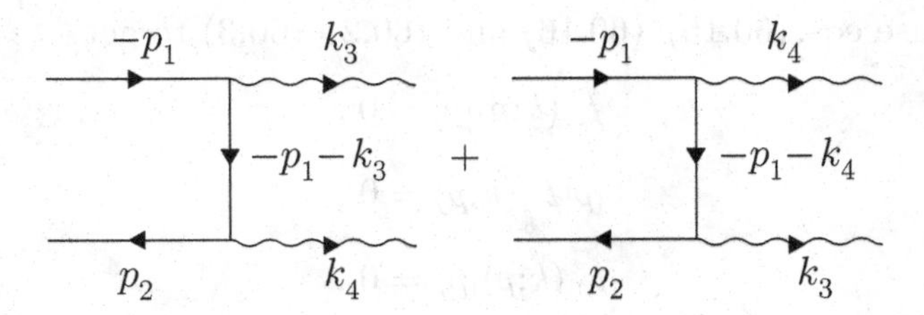

Figure 60.2. Diagrams for fermion–photon scattering, with all momenta treated as outgoing.

where we used the Mandelstam relation $s_{12}+s_{13}+s_{14}=0$ to get the last line.

To get the cross section for a particular set of helicities, we must take the absolute squares of the amplitudes. These follow from eqs. (60.4), (60.17), and (60.18):

$$|\langle 1\,2\rangle|^2 = |[1\,2]|^2 = \epsilon_1\epsilon_2 s_{12} = |s_{12}|\ . \tag{60.22}$$

We can then compute the spin-averaged cross section by summing the absolute squares of eqs. (60.19)–(60.21), multiplying by two to account for the processes in which all helicities are opposite (and which have amplitudes that are related by complex conjugation), and then dividing by four to average over the initial helicities. Making use of $s_{34}=s_{12}$ and its permutations, we find

$$\begin{aligned}\langle|\mathcal{T}|^2\rangle &= 2e^4\left(\frac{s_{14}^2}{s_{13}^2}+\frac{s_{13}^2}{s_{14}^2}+\frac{s_{12}^4}{s_{13}^2 s_{14}^2}\right)\\ &= 2e^4\left(\frac{s_{12}^4+s_{13}^4+s_{14}^4}{s_{13}^2 s_{14}^2}\right).\end{aligned} \tag{60.23}$$

For the processes of $e^-e^-\to e^-e^-$ and $e^+e^+\to e^+e^+$, we have $s_{12}=s$, $s_{13}=t$, and $s_{14}=u$; for $e^+e^-\to e^+e^-$, we have $s_{13}=s$, $s_{14}=t$, and $s_{12}=u$.

Now we turn to processes with two external fermions and two external photons, as shown in fig. 60.2. The first thing to notice is that a diagram is zero if the two external fermion lines have the same helicity. This is because the corresponding twistors sandwich an odd number of gamma matrices: one from each vertex, and one from the massless fermion propagator $\tilde{S}(p) = -\not{p}/p^2$. Thus we need only compute $\mathcal{T}_{+-\lambda_3\lambda_4}$ since $\mathcal{T}_{-+\lambda_3\lambda_4}$ is related by complex conjugation.

Next we use eqs. (60.15)–(60.16) and (60.2)–(60.3) to get

$$\not{\varepsilon}_-(k;p)|p] = 0 \,, \tag{60.24}$$

$$[p|\not{\varepsilon}_-(k;p) = 0 \,, \tag{60.25}$$

$$\not{\varepsilon}_+(k;p)|p\rangle = 0 \,, \tag{60.26}$$

$$\langle p|\not{\varepsilon}_+(k;p) = 0 \,. \tag{60.27}$$

Thus we can get some amplitudes to vanish with appropriate choices of the reference momenta in the photon polarizations.

So, let us consider

$$\begin{aligned} \mathcal{T}_{+-\lambda_3\lambda_4} = &-e^2 \, \langle 2|\not{\varepsilon}_{\lambda_4}(k_4;q_4)(\not{p}_1 + \not{k}_3)\not{\varepsilon}_{\lambda_3}(k_3;q_3)|1] \,/\, s_{13} \\ &-e^2 \, \langle 2|\not{\varepsilon}_{\lambda_3}(k_3;q_3)(\not{p}_1 + \not{k}_4)\not{\varepsilon}_{\lambda_4}(k_4;q_4)|1] \,/\, s_{14} \,. \end{aligned} \tag{60.28}$$

If we take $\lambda_3 = \lambda_4 = -$, then we can get both terms in eq. (60.28) to vanish by choosing $q_3 = q_4 = p_1$, and using eq. (60.24). If we take $\lambda_3 = \lambda_4 = +$, then we can get both terms in eq. (60.28) to vanish by choosing $q_3 = q_4 = p_2$, and using eq. (60.27).

Thus, we need only compute $\mathcal{T}_{+--+}$ and $\mathcal{T}_{+-+-}$. For $\mathcal{T}_{+-+-}$, we can get the second term in eq. (60.28) to vanish by choosing $q_3 = p_2$, and using eq. (60.27). Then we have

$$\begin{aligned} \mathcal{T}_{+-+-} &= -e^2 \, \langle 2|\not{\varepsilon}_-(k_4;q_4)(\not{p}_1 + \not{k}_3)\not{\varepsilon}_+(k_3;p_2)|1] \,/\, s_{13} \\ &= -e^2 \, \frac{\sqrt{2}}{[q_4 4]} \, \langle 2\,4\rangle \, [q_4|(\not{p}_1 + \not{k}_3)|2\rangle \, [3\,1] \, \frac{\sqrt{2}}{\langle 2\,3\rangle} \, \frac{1}{s_{13}} \,. \end{aligned} \tag{60.29}$$

Next we note that $[p|\not{p} = 0$, and so it is useful to choose either $q_4 = p_1$ or $q_4 = k_3$. There is no obvious advantage in one choice over the other, and they must give equivalent results, so let us take $q_4 = k_3$. Then, using eq. (60.6) for $\not{p}_1$, we get

$$\mathcal{T}_{+-+-} = 2e^2 \, \frac{\langle 2\,4\rangle \, [3\,1] \, \langle 1\,2\rangle \, [3\,1]}{[3\,4] \, \langle 2\,3\rangle \, s_{13}} \,. \tag{60.30}$$

Now we use $[3\,1]\,\langle 1\,2\rangle = -[3\,4]\,\langle 4\,2\rangle$ in the numerator (see problem 60.2b), and set $s_{13} = \langle 1\,3\rangle\,[3\,1]$ in the denominator. Canceling common factors and using antisymmetry of the twistor product then yields

$$\mathcal{T}_{+-+-} = 2e^2 \, \frac{\langle 2\,4\rangle^2}{\langle 1\,3\rangle\langle 2\,3\rangle} \,. \tag{60.31}$$

We can now get $\mathcal{T}_{+--+}$ simply by exchanging the labels 3 and 4,

$$\mathcal{T}_{+--+} = 2e^2 \frac{\langle 2\,3\rangle^2}{\langle 1\,4\rangle\langle 2\,4\rangle} . \tag{60.32}$$

We can compute the spin-averaged cross section by summing the absolute squares of eqs. (60.31) and (60.32), multiplying by two to account for the processes in which all helicities are opposite (and which have amplitudes that are related by complex conjugation), and then dividing by four to average over the initial helicities. The result is

$$\langle|\mathcal{T}|^2\rangle = 2e^4 \left(\left|\frac{s_{13}}{s_{14}}\right| + \left|\frac{s_{14}}{s_{13}}\right| \right) . \tag{60.33}$$

For the processes of $e^-\gamma \to e^-\gamma$ and $e^+\gamma \to e^+\gamma$, we have $s_{13} = s$, $s_{12} = t$, and $s_{14} = u$; for $e^+e^- \to \gamma\gamma$ and $\gamma\gamma \to e^+e^-$ we have $s_{12} = s$, $s_{13} = t$, and $s_{14} = u$.

Problems

60.1) a) Show that

$$p\cdot\varepsilon_+(k;q) = \frac{\langle q\,p\rangle\,[p\,k]}{\sqrt{2}\,\langle q\,k\rangle} , \tag{60.34}$$

$$p\cdot\varepsilon_-(k;q) = \frac{[q\,p]\,\langle p\,k\rangle}{\sqrt{2}\,[q\,k]} . \tag{60.35}$$

Use this result to show that

$$k\cdot\varepsilon_\pm(k;q) = 0 , \tag{60.36}$$

which is required by gauge invariance, and also that

$$q\cdot\varepsilon_\pm(k;q) = 0 . \tag{60.37}$$

b) Show that

$$\varepsilon_+(k;q)\cdot\varepsilon_+(k';q') = \frac{\langle q\,q'\rangle\,[k\,k']}{\langle q\,k\rangle\,\langle q'\,k'\rangle} , \tag{60.38}$$

$$\varepsilon_-(k;q)\cdot\varepsilon_-(k';q') = \frac{[q\,q']\,\langle k\,k'\rangle}{[q\,k]\,[q'\,k']} , \tag{60.39}$$

$$\varepsilon_+(k;q)\cdot\varepsilon_-(k';q') = \frac{\langle q\,k'\rangle\,[k\,q']}{\langle q\,k\rangle\,[q'\,k']} . \tag{60.40}$$

Note that the right-hand sides of eqs. (60.38) and (60.39) vanish if $q' = q$, and that the right-hand side of eq. (60.40) vanishes if $q = k'$ or $q' = k$.

60.2) a) For a process with n external particles, and all momenta treated as outgoing, show that

$$\sum_{j=1}^{n} \langle i\, j\rangle\, [j\, k] = 0 \quad \text{and} \quad \sum_{j=1}^{n} [i\, j]\, \langle j\, k\rangle = 0\,. \tag{60.41}$$

Hint: make use of eq. (60.6).

b) For $n = 4$, show that $[3\,1]\,\langle 1\,2\rangle = -\,[3\,4]\,\langle 4\,2\rangle$.

60.3) Use various identities to show that eq. (60.31) can also be written as

$$\mathcal{T}_{+-+-} = -2e^2\,\frac{[1\,3]^2}{[1\,4]\,[2\,4]}\,. \tag{60.42}$$

60.4) a) Show explicitly that you would get the same result as eq. (60.31) if you set $q_4 = p_1$ in eq. (60.29).

b) Show explicitly that you would get the same result as eq. (60.31) if you set $q_4 = p_2$ in eq. (60.29).

61

Scalar electrodynamics

Prerequisite: 58

In this section, we will consider how charged spin-zero particles interact with photons. We begin with the lagrangian for a complex scalar field with a quartic interaction,

$$\mathcal{L} = -\partial^\mu\varphi^\dagger\partial_\mu\varphi - m^2\varphi^\dagger\varphi - \tfrac{1}{4}\lambda(\varphi^\dagger\varphi)^2 \; . \tag{61.1}$$

This lagrangian is obviously invariant under the global U(1) symmetry

$$\begin{aligned} \varphi(x) &\to e^{-i\alpha}\varphi(x) \, , \\ \varphi^\dagger(x) &\to e^{+i\alpha}\varphi^\dagger(x) \; . \end{aligned} \tag{61.2}$$

We would like to promote this global symmetry to a local symmetry,

$$\varphi(x) \to \exp[-ie\Gamma(x)]\varphi(x) \, , \tag{61.3}$$

$$\varphi^\dagger(x) \to \exp[+ie\Gamma(x)]\varphi^\dagger(x) \; . \tag{61.4}$$

To do so, we must replace each ordinary derivative in eq. (61.1) with a covariant derivative

$$D_\mu \equiv \partial_\mu - ieA_\mu \, , \tag{61.5}$$

where A_μ transforms as

$$A^\mu(x) \to A^\mu(x) - \partial^\mu\Gamma(x) \, , \tag{61.6}$$

which implies that D_μ transforms as

$$D_\mu \to \exp[-ie\Gamma(x)]D_\mu \exp[+ie\Gamma(x)] \; . \tag{61.7}$$

Our complete lagrangian for *scalar electrodynamics* is then

$$\mathcal{L} = -(D^\mu\varphi)^\dagger D_\mu\varphi - m^2\varphi^\dagger\varphi - \tfrac{1}{4}\lambda(\varphi^\dagger\varphi)^2 - \tfrac{1}{4}F^{\mu\nu}F_{\mu\nu}\,. \tag{61.8}$$

We have added the usual gauge-invariant kinetic term for the gauge field. The quartic interaction term has a dimensionless coefficient, and so is necessary for renormalizability. For now, we omit the renormalizing Z factors.

Of course, eq. (61.8) is invariant under a global U(1) transformation as well as a local U(1) transformation: we simply set $\Gamma(x)$ to a constant. Then we can find the conserved Noether current corresponding to this symmetry, following the procedure of section 22. In the case of spinor electrodynamics, this current is same as it is for a free Dirac field, $j^\mu = \overline{\Psi}\gamma^\mu\Psi$. In the case of a complex scalar field, we find

$$j^\mu = -i[\varphi^\dagger D^\mu\varphi - (D^\mu\varphi)^\dagger\varphi]\,. \tag{61.9}$$

With a factor of e, this current should be identified as the electromagnetic current. Because the covariant derivative appears in eq. (61.9), the electromagnetic current depends explicitly on the gauge field. We had not previously contemplated this possibility, but in scalar electrodynamics it arises naturally, and is essential for gauge invariance.

It also poses no special problem in the quantum theory. We will make the same assumption that we did for spinor electrodynamics: namely, that the correct procedure is to omit integration over the component of $\widetilde{A}_\mu(k)$ that is parallel to k_μ, on the grounds that this integration is redundant. This leads to the same Feynman rules for internal and external photons as in section 58. The Feyman rules for internal and external scalars are the same as those of problem 10.2. We will call the spin-zero particle with electric charge $+e$ a *scalar electron* or *selectron* (recall that our convention is that e is negative), and the spin-zero particle with electric charge $-e$ a *scalar positron* or *spositron.* Scalar lines (traditionally drawn as dashed in scalar electrodynamics) carry a charge arrow whose direction must be preserved when lines are joined by vertices.

To determine the kinds of vertices we have, we first write out the interaction terms in the lagrangian of eq. (61.8):

$$\mathcal{L}_1 = ieA^\mu[(\partial_\mu\varphi^\dagger)\varphi - \varphi^\dagger\partial_\mu\varphi] - e^2A^\mu A_\mu\varphi^\dagger\varphi - \tfrac{1}{4}\lambda(\varphi^\dagger\varphi)^2\,. \tag{61.10}$$

This leads to the vertices shown in fig. 61.1. The vertex factors associated with the last two terms are $-2ie^2g_{\mu\nu}$ and $-i\lambda$. To get the vertex factor for the first term, we note that if $|k\rangle$ is an incoming selectron state, then $\langle 0|\varphi(x)|k\rangle = e^{ikx}$ and $\langle 0|\varphi^\dagger(x)|k\rangle = 0$; and if $\langle k'|$ is an outgoing selectron state, then $\langle k'|\varphi^\dagger(x)|0\rangle = e^{-ik'x}$ and $\langle k'|\varphi(x)|0\rangle = 0$. Therefore, in free-field

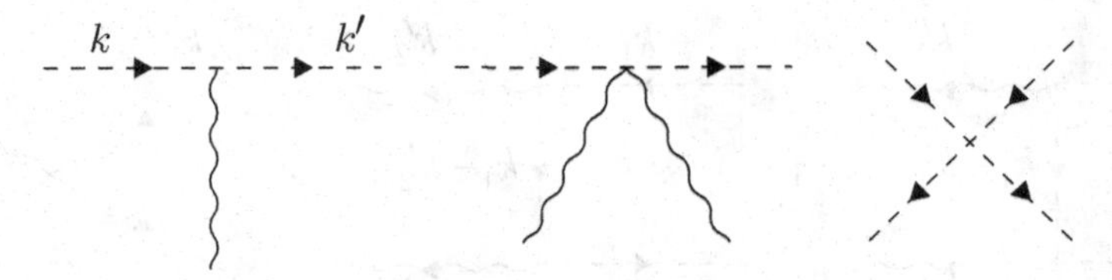

Figure 61.1. The three vertices of scalar electrodynamics; the corresponding vertex factors are $ie(k+k')_\mu$, $-2ie^2 g_{\mu\nu}$, and $-i\lambda$.

theory,

$$\langle k'|(\partial_\mu \varphi^\dagger)\varphi|k\rangle = -ik'_\mu e^{-i(k'-k)x} \,, \tag{61.11}$$

$$\langle k'|\varphi^\dagger \partial_\mu \varphi|k\rangle = +ik_\mu e^{-i(k'-k)x} . \tag{61.12}$$

This implies that the vertex factor for the first term in eq. (61.10) is given by $i(ie)[(-ik'_\mu)-(ik_\mu)] = ie(k+k')_\mu$.

Putting everything together, we get the following set of Feynman rules for tree-level processes in scalar electrodynamics.

1) For each *incoming selectron*, draw a dashed line with an arrow pointed *towards* the vertex, and label it with the selectron's four-momentum, k_i.
2) For each *outgoing selectron*, draw a dashed line with an arrow pointed *away* from the vertex, and label it with the selectron's four-momentum, k'_i.
3) For each *incoming spositron*, draw a dashed line with an arrow pointed *away* from the vertex, and label it with *minus* the spositron's four-momentum, $-k_i$.
4) For each *outgoing spositron*, draw a dashed line with an arrow pointed *towards* the vertex, and label it with *minus* the spositron's four-momentum, $-k'_i$.
5) For each *incoming photon*, draw a wavy line with an arrow pointed *towards* the vertex, and label it with the photon's four-momentum, k_i.
6) For each *outgoing photon*, draw a wavy line with an arrow pointed *away* from the vertex, and label it with the photon's four-momentum, k'_i.
7) There are three allowed vertices, shown in fig. 61.1. Using these vertices, join up all the external lines, including extra internal lines as needed. In this way, draw all possible diagrams that are *topologically inequivalent.*
8) Assign each internal line its own four-momentum. Think of the four-momenta as flowing along the arrows, and conserve four-momentum at each vertex. For a tree diagram, this fixes the momenta on all the internal lines.
9) The value of a diagram consists of the following factors:
for each incoming photon, $\varepsilon^{\mu *}_{\lambda_i}(\mathbf{k}_i)$;
for each outgoing photon, $\varepsilon^{\mu}_{\lambda_i}(\mathbf{k}_i)$;
for each incoming or outgoing selectron or spositron, 1;
for each scalar-scalar-photon vertex, $ie(k+k')_\mu$;

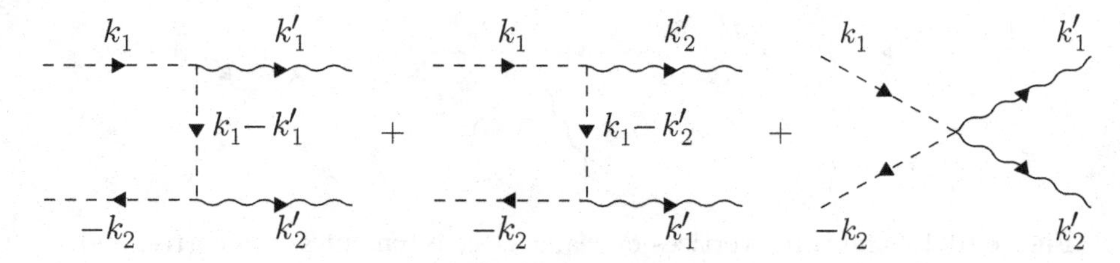

Figure 61.2. Diagrams for $\widetilde{e}^+\widetilde{e}^- \to \gamma\gamma$.

for each scalar-scalar-photon-photon vertex, $-2ie^2 g_{\mu\nu}$;
for each four-scalar vertex, $-i\lambda$;
for each internal photon, $-ig^{\mu\nu}/(k^2 - i\epsilon)$;
for each internal scalar, $-i/(k^2 + m^2 - i\epsilon)$.

10) The vector index on each vertex is contracted with the vector index on either the photon propagator (if the attached photon line is internal) or the photon polarization vector (if the attached photon line is external).
11) The value of $i\mathcal{T}$ (at tree level) is given by a sum over the values of all the contributing diagrams.

Let us compute the scattering amplitude for a particular process, $\widetilde{e}^+\widetilde{e}^- \to \gamma\gamma$, where $\widetilde{e}^-$ denotes a selectron. We have the diagrams of fig. 61.2. The amplitude is

$$i\mathcal{T} = (ie)^2 \, \frac{1}{i} \, \frac{(2k_1 - k'_1)_\mu \varepsilon^\mu_{1'} (k_1 - k'_1 - k_2)_\nu \varepsilon^\nu_{2'}}{m^2 - t} + (1' \leftrightarrow 2')$$
$$- 2ie^2 g_{\mu\nu} \varepsilon^\mu_{1'} \varepsilon^\nu_{2'} , \tag{61.13}$$

where $t = -(k_1 - k'_1)^2$ and $u = -(k_1 - k'_2)^2$. This expression can be simplified by noting that $k_1 - k'_1 - k_2 = k'_2 - 2k_2$, and that $k'_i \cdot \varepsilon'_i = 0$. Then we have

$$\mathcal{T} = -e^2 \left[\frac{4(k_1 \cdot \varepsilon_{1'})(k_2 \cdot \varepsilon_{2'})}{m^2 - t} + \frac{4(k_1 \cdot \varepsilon_{2'})(k_2 \cdot \varepsilon_{1'})}{m^2 - u} + 2(\varepsilon_{1'} \cdot \varepsilon_{2'}) \right] . \tag{61.14}$$

To get the polarization-summed cross section, we take the absolute square of eq. (61.14), and use the substitution rule

$$\sum_{\lambda = \pm} \varepsilon^\mu_\lambda(\mathbf{k}) \varepsilon^{\rho *}_\lambda(\mathbf{k}) \to g^{\mu\rho} \; . \tag{61.15}$$

This is a straightforward calculation, which we leave to the problems.

Problems

61.1) Compute $\langle|\mathcal{T}|^2\rangle$ for $\widetilde{e}^+\widetilde{e}^- \to \gamma\gamma$, and express your answer in terms of the Mandelstam variables.

61.2) Compute $\langle|\mathcal{T}|^2\rangle$ for the process $\widetilde{e}^-\gamma \to \widetilde{e}^-\gamma$. You should find that your result is the same as that for $\widetilde{e}^+\widetilde{e}^- \to \gamma\gamma$, but with $s \leftrightarrow t$, an example of crossing symmetry.

62

Loop corrections in spinor electrodynamics

Prerequisites: 51, 59

In this section we will compute the one-loop corrections in spinor electrodynamics.

First let us note that the general discussion of sections 18 and 29 leads us to expect that we will need to add to the free lagrangian

$$\mathcal{L}_0 = i\overline{\Psi}\partial\!\!\!/\Psi - m\overline{\Psi}\Psi - \tfrac{1}{4}F^{\mu\nu}F_{\mu\nu} \tag{62.1}$$

all possible terms whose coefficients have positive or zero mass dimension, and that respect the symmetries of the original lagrangian. These include Lorentz symmetry, the U(1) gauge symmetry, and the discrete symmetries of parity, time reversal, and charge conjugation.

The mass dimensions of the fields (in four spacetime dimensions) are $[A^\mu] = 1$ and $[\Psi] = \frac{3}{2}$. Gauge invariance requires that A^μ appear only in the form of a covariant derivative D^μ. (Recall that the field strength $F^{\mu\nu}$ can be expressed as the commutator of two covariant derivatives.) Thus, the only possible term we could add to $\mathcal{L}_0$ that does not involve the Ψ field, and that has mass dimension four or less, is $\varepsilon_{\mu\nu\rho\sigma}F^{\mu\nu}F^{\rho\sigma}$. This term, however, is odd under parity and time reversal. Similarly, there are no terms meeting all the requirements that involve Ψ: the only candidates contain either γ_5 (e.g., $i\overline{\Psi}\gamma_5\Psi$) and are forbidden by parity, or $\mathcal{C}$ (e.g., $\Psi^{\mathrm{T}}\mathcal{C}\Psi$) and are forbidden by the U(1) symmetry.

Therefore, the theory we will consider is specified by $\mathcal{L} = \mathcal{L}_0 + \mathcal{L}_1$, where $\mathcal{L}_0$ is given by eq. (62.1), and

$$\mathcal{L}_1 = Z_1 e\overline{\Psi}A\!\!\!/\Psi + \mathcal{L}_{\rm ct}\,, \tag{62.2}$$

$$\mathcal{L}_{\rm ct} = i(Z_2-1)\overline{\Psi}\partial\!\!\!/\Psi - (Z_m-1)m\overline{\Psi}\Psi - \tfrac{1}{4}(Z_3-1)F^{\mu\nu}F_{\mu\nu}\,. \tag{62.3}$$

We will use an on-shell renormalization scheme.

We can write the exact photon propagator (in momentum space) as a geometric series of the form

$$\boldsymbol{\tilde{\Delta}}_{\mu\nu}(k) = \tilde{\Delta}_{\mu\nu}(k) + \tilde{\Delta}_{\mu\rho}(k)\Pi^{\rho\sigma}(k)\tilde{\Delta}_{\sigma\nu}(k) + \ldots\ , \tag{62.4}$$

where $i\Pi^{\mu\nu}(k)$ is given by a sum of one-particle irreducible (1PI for short; see section 14) diagrams with two external photon lines (and the external propagators removed), and $\tilde{\Delta}_{\mu\nu}(k)$ is the free photon propagator,

$$\tilde{\Delta}_{\mu\nu}(k) = \frac{1}{k^2 - i\epsilon}\left(g_{\mu\nu} - (1{-}\xi)\frac{k_\mu k_\nu}{k^2}\right). \tag{62.5}$$

Here we have used the freedom to add k_μ or k_ν terms to put the propagator into *generalized Feynman gauge* or R_ξ *gauge*. (The name R_ξ *gauge* has historically been used only in the context of spontaneous symmetry breaking – see section 85 – but we will use it here as well. R stands for *renormalizable* and ξ stands for ξ.) Setting $\xi = 1$ gives Feynman gauge, and setting $\xi = 0$ gives Lorenz gauge (also known as Landau gauge).

Observable squared amplitudes should not depend on the value of ξ. This suggests that $\Pi^{\mu\nu}(k)$ should be *transverse*,

$$k_\mu \Pi^{\mu\nu}(k) = k_\nu \Pi^{\mu\nu}(k) = 0\ , \tag{62.6}$$

so that the ξ dependent term in $\tilde{\Delta}_{\mu\nu}(k)$ vanishes when an internal photon line is attached to $\Pi^{\mu\nu}(k)$. Eq. (62.6) is in fact valid; we will give a proof of it, based on the Ward identity for the electromagnetic current, in problem 68.1. For now, we will take eq. (62.6) as given. This implies that we can write

$$\Pi^{\mu\nu}(k) = \Pi(k^2)\left(k^2 g^{\mu\nu} - k^\mu k^\nu\right) \tag{62.7}$$

$$= k^2\,\Pi(k^2)P^{\mu\nu}(k)\ , \tag{62.8}$$

where $\Pi(k^2)$ is a scalar function, and $P^{\mu\nu}(k) = g^{\mu\nu} - k^\mu k^\nu/k^2$ is the projection matrix introduced in section 57.

Note that we can also write

$$\tilde{\Delta}_{\mu\nu}(k) = \frac{1}{k^2 - i\epsilon}\left(P_{\mu\nu}(k) + \xi\,\frac{k_\mu k_\nu}{k^2}\right). \tag{62.9}$$

Then, using eqs. (62.8) and (62.9) in eq. (62.4), and summing the geometric series, we find

$$\boldsymbol{\tilde{\Delta}}_{\mu\nu}(k) = \frac{P_{\mu\nu}(k)}{k^2[1 - \Pi(k^2)] - i\epsilon} + \xi\,\frac{k_\mu k_\nu/k^2}{k^2 - i\epsilon}\,. \tag{62.10}$$

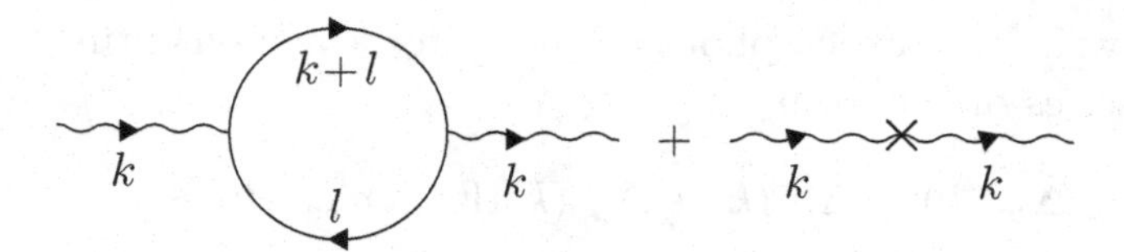

Figure 62.1. The one-loop and counterterm corrections to the photon propagator in spinor electrodynamics.

The ξ dependent term should be physically irrelevant (and can be set to zero by the gauge choice $\xi = 0$, corresponding to Lorenz gauge). The remaining term has a pole at $k^2 = 0$ with residue $P_{\mu\nu}(k)/[1 - \Pi(0)]$. In our on-shell renormalization scheme, we should have

$$\Pi(0) = 0 \,. \tag{62.11}$$

This corresponds to the field normalization that is needed for validity of the LSZ formula.

Let us now turn to the calculation of $\Pi^{\mu\nu}(k)$. The one-loop and counterterm contributions are shown in fig. 62.1. We have

$$\begin{aligned} i\Pi^{\mu\nu}(k) = (-1)(iZ_1 e)^2(\tfrac{1}{i})^2 \int \frac{d^4\ell}{(2\pi)^4} \,\mathrm{Tr}\Big[\tilde{S}(\not{\ell}+\not{k})\gamma^\mu \tilde{S}(\not{\ell})\gamma^\nu\Big] \\ - i(Z_3-1)(k^2 g^{\mu\nu} - k^\mu k^\nu) + O(e^4) \,, \end{aligned} \tag{62.12}$$

where the factor of minus one is for the closed fermion loop, and $\tilde{S}(\not{p}) = (-\not{p}+m)/(p^2+m^2-i\epsilon)$ is the free fermion propagator in momentum space. Anticipating that $Z_1 = 1 + O(e^2)$, we set $Z_1 = 1$ in the first term.

We can write

$$\mathrm{Tr}\Big[\tilde{S}(\not{\ell}+\not{k})\gamma^\mu \tilde{S}(\not{\ell})\gamma^\nu\Big] = \int_0^1 dx\, \frac{4N^{\mu\nu}}{(q^2+D)^2} \,, \tag{62.13}$$

where we have combined denominators in the usual way: $q = \ell + xk$ and

$$D = x(1-x)k^2 + m^2 - i\epsilon \,. \tag{62.14}$$

The numerator is

$$4N^{\mu\nu} = \mathrm{Tr}\Big[(-\not{\ell}-\not{k}+m)\gamma^\mu(-\not{\ell}+m)\gamma^\nu\Big] \,. \tag{62.15}$$

Completing the trace, we get

$$N^{\mu\nu} = (\ell+k)^\mu \ell^\nu + \ell^\mu(\ell+k)^\nu - [\ell(\ell+k) + m^2]g^{\mu\nu} \,. \tag{62.16}$$

Setting $\ell = q - xk$ and dropping terms linear in q (because they integrate to zero), we find

$$N^{\mu\nu} \to 2q^\mu q^\nu - 2x(1-x)k^\mu k^\nu - [q^2 - x(1-x)k^2 + m^2]g^{\mu\nu} \,. \tag{62.17}$$

The integrals diverge, and so we analytically continue to $d = 4 - \varepsilon$ dimensions, and replace e with $e\tilde{\mu}^{\varepsilon/2}$ (so that e remains dimensionless for any d).

Next we recall a result from problem 14.3,

$$\int d^d q \, q^\mu q^\nu f(q^2) = \frac{1}{d} \, g^{\mu\nu} \int d^d q \, q^2 f(q^2) \,. \tag{62.18}$$

This allows the replacement

$$N^{\mu\nu} \to -2x(1-x)k^\mu k^\nu + \Big[\Big(\tfrac{2}{d} - 1\Big)q^2 + x(1-x)k^2 - m^2\Big]g^{\mu\nu} \,. \tag{62.19}$$

Using the results of section 14, along with a little manipulation of gamma functions, we can show that

$$\Big(\tfrac{2}{d} - 1\Big) \int \frac{d^d q}{(2\pi)^d} \, \frac{q^2}{(q^2 + D)^2} = D \int \frac{d^d q}{(2\pi)^d} \, \frac{1}{(q^2 + D)^2} \,. \tag{62.20}$$

Thus we can make the replacement $(\frac{2}{d} - 1)q^2 \to D$ in eq. (62.19), and we find

$$N^{\mu\nu} \to 2x(1-x)(k^2 g^{\mu\nu} - k^\mu k^\nu) \,. \tag{62.21}$$

This guarantees that the one-loop contribution to $\Pi^{\mu\nu}(k)$ is transverse (as we expected) in any number of spacetime dimensions.

Now we evaluate the integral over q, using

$$\begin{aligned} \tilde{\mu}^\varepsilon \int \frac{d^d q}{(2\pi)^d} \, \frac{1}{(q^2 + D)^2} &= \frac{i}{16\pi^2} \, \Gamma(\tfrac{\varepsilon}{2}) \Big(4\pi\tilde{\mu}^2/D\Big)^{\varepsilon/2} \\ &= \frac{i}{8\pi^2} \left[\frac{1}{\varepsilon} - \tfrac{1}{2}\ln(D/\mu^2)\right], \end{aligned} \tag{62.22}$$

where $\mu^2 = 4\pi e^{-\gamma}\tilde{\mu}^2$, and we have dropped terms of order ε in the last line. Combining eqs. (62.7), (62.12), (62.13), (62.21), and (62.22), we get

$$\Pi(k^2) = -\frac{e^2}{\pi^2} \int_0^1 dx \, x(1-x) \left[\frac{1}{\varepsilon} - \tfrac{1}{2}\ln(D/\mu^2)\right] - (Z_3 - 1) + O(e^4) \,. \tag{62.23}$$

Imposing $\Pi(0) = 0$ fixes

$$Z_3 = 1 - \frac{e^2}{6\pi^2} \left[\frac{1}{\varepsilon} - \ln(m/\mu)\right] + O(e^4) \tag{62.24}$$

and

$$\Pi(k^2) = \frac{e^2}{2\pi^2} \int_0^1 dx\, x(1-x) \ln(D/m^2) + O(e^4) . \tag{62.25}$$

Next we turn to the fermion propagator. The exact propagator can be written in Lehmann–Källén form as

$$\tilde{\mathbf{S}}(\not{p}) = \frac{1}{\not{p} + m - i\epsilon} + \int_{m_{\rm th}^2}^{\infty} ds\, \frac{\rho_\Psi(s)}{\not{p} + \sqrt{s} - i\epsilon} . \tag{62.26}$$

We see that the first term has a pole at $\not{p} = -m$ with residue one. This residue corresponds to the field normalization that is needed for the validity of the LSZ formula.

There is a problem, however: in quantum electrodynamics, the threshold mass $m_{\rm th}$ is m, corresponding to the contribution of a fermion and a zero-energy photon. Thus the second term has a branch point at $\not{p} = -m$. The pole in the first term is therefore not isolated, and its residue is ill defined.

This is a reflection of an underlying infrared divergence, associated with the massless photon. To deal with it, we must impose an infrared cutoff that moves the branch point away from the pole. The most direct method is to change the denominator of the photon propagator from k^2 to $k^2 + m_\gamma^2$, where m_γ is a fictitious photon mass. Ultimately, as in section 26, we must deal with this issue by computing cross sections that take into account detector inefficiencies. In quantum electrodynamics, we must specify the lowest photon energy $\omega_{\rm min}$ that can be detected. Only after computing cross sections with extra undetectable photons, and then summing over them, is it safe to take the limit $m_\gamma \to 0$. It turns out that it is not also necessary to abandon the on-shell renormalization scheme (as we were forced to do in massless φ^3 theory in section 27), as long as the electron is massive.

An alternative is to use dimensional regularization for the infrared divergences as well as the ultraviolet ones. As discussed in section 25, there are no soft-particle infrared divergences for $d > 4$ (and no co-linear divergences at all in quantum electrodynamics with massive charged particles). In practice, infrared-divergent integrals are finite away from even-integer dimensions, just like ultraviolet-divergent integrals. Thus we simply keep $d = 4 - \varepsilon$ all the way through to the very end, taking the $\varepsilon \to 0$ limit only after summing over cross sections with extra undetectable photons, all computed in $4 - \varepsilon$ dimensions. This method is calculationally the simplest, but requires careful bookkeeping to segregate the infrared and ultraviolet singularities. For that reason, we will not pursue it further.

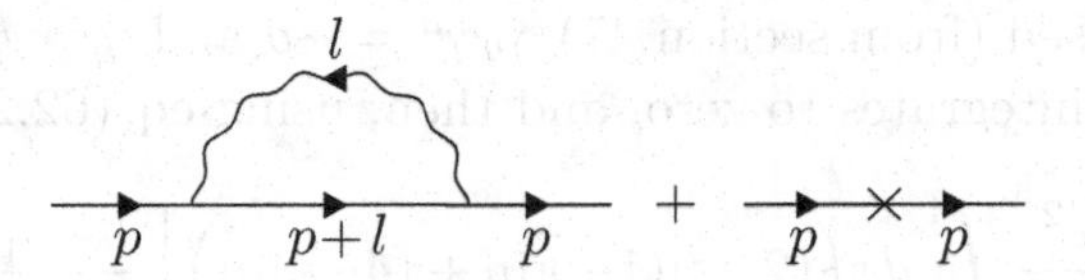

Figure 62.2. The one-loop and counterterm corrections to the fermion propagator in spinor electrodynamics.

We can write the exact fermion propagator in the form

$$\tilde{\mathbf{S}}(\not{p})^{-1} = \not{p} + m - i\epsilon - \Sigma(\not{p}) \ , \tag{62.27}$$

where $i\Sigma(\not{p})$ is given by the sum of 1PI diagrams with two external fermion lines (and the external propagators removed). The fact that $\tilde{\mathbf{S}}(\not{p})$ has a pole at $\not{p} = -m$ with residue one implies that $\Sigma(-m) = 0$ and $\Sigma'(-m) = 0$; this fixes the coefficients Z_2 and Z_m. As we will see, we must have an infrared cutoff in place in order to have a finite value for $\Sigma'(-m)$.

Let us now turn to the calculation of $\Sigma(\not{p})$. The one-loop and counterterm contributions are shown in fig. 62.2. We have

$$\begin{aligned} i\Sigma(\not{p}) = (iZ_1 e)^2 \left(\tfrac{1}{i}\right)^2 \int \frac{d^4\ell}{(2\pi)^4} \Big[\gamma^\nu \tilde{S}(\not{p} + \not{\ell})\gamma^\mu\Big] \tilde{\Delta}_{\mu\nu}(\ell) \\ - i(Z_2 - 1)\not{p} - i(Z_m - 1)m + O(e^4) \ . \end{aligned} \tag{62.28}$$

It is simplest to work in Feynman gauge, where we take

$$\tilde{\Delta}_{\mu\nu}(\ell) = \frac{g_{\mu\nu}}{\ell^2 + m_\gamma^2 - i\epsilon} \ ; \tag{62.29}$$

here we have included the fictitious photon mass m_γ as an infrared cutoff.

We now apply the usual bag of tricks to get

$$\begin{aligned} i\Sigma(\not{p}) = e^2 \tilde{\mu}^\varepsilon \int_0^1 dx \int \frac{d^d q}{(2\pi)^d} \frac{N}{(q^2 + D)^2} \\ - i(Z_2 - 1)\not{p} - i(Z_m - 1)m + O(e^4) \ , \end{aligned} \tag{62.30}$$

where $q = \ell + xp$ and

$$D = x(1-x)p^2 + xm^2 + (1-x)m_\gamma^2 \ , \tag{62.31}$$

$$\begin{aligned} N &= \gamma_\mu(-\not{p} - \not{\ell} + m)\gamma^\mu \\ &= -(d-2)(\not{p} + \not{\ell}) - dm \\ &= -(d-2)[\not{q} + (1-x)\not{p}] - dm \ , \end{aligned} \tag{62.32}$$

where we have used (from section 47) $\gamma_\mu\gamma^\mu = -d$ and $\gamma_\mu \not{p} \gamma^\mu = (d-2)\not{p}$. The term linear in q integrates to zero, and then, using eq. (62.22), we get

$$\Sigma(\not{p}) = -\frac{e^2}{8\pi^2}\int_0^1 dx\Big((2-\varepsilon)(1-x)\not{p} + (4-\varepsilon)m\Big)\Big[\frac{1}{\varepsilon} - \tfrac{1}{2}\ln(D/\mu^2)\Big]$$

$$- (Z_2-1)\not{p} - (Z_m-1)m + O(e^4)\ . \tag{62.33}$$

We see that finiteness of $\Sigma(\not{p})$ requires

$$Z_2 = 1 - \frac{e^2}{8\pi^2}\Big(\frac{1}{\varepsilon} + \text{finite}\Big) + O(e^4)\ , \tag{62.34}$$

$$Z_m = 1 - \frac{e^2}{2\pi^2}\Big(\frac{1}{\varepsilon} + \text{finite}\Big) + O(e^4)\ . \tag{62.35}$$

We can impose $\Sigma(-m) = 0$ by writing

$$\Sigma(\not{p}) = \frac{e^2}{8\pi^2}\Big[\int_0^1 dx\,\Big((1-x)\not{p} + 2m\Big)\ln(D/D_0) + \kappa_2(\not{p}+m)\Big] + O(e^4)\ , \tag{62.36}$$

where D_0 is D evaluated at $p^2 = -m^2$,

$$D_0 = x^2m^2 + (1-x)m_\gamma^2\ , \tag{62.37}$$

and κ_2 is a constant to be determined. We fix κ_2 by imposing $\Sigma'(-m) = 0$. In differentiating with respect to $\not{p}$, we take the p^2 in D, eq. (62.31), to be $-\not{p}^2$; we find

$$\kappa_2 = -2\int_0^1 dx\, x(1-x^2)m^2/D_0$$

$$= -2\ln(m/m_\gamma) + 1\ , \tag{62.38}$$

where we have dropped terms that go to zero with the infrared cutoff m_γ.

Next we turn to the loop correction to the vertex. We define the vertex function $i\mathbf{V}^\mu(p',p)$ as the sum of one-particle irreducible diagrams with one incoming fermion with momentum p, one outgoing fermion with momentum p', and one incoming photon with momentum $k = p' - p$. The original vertex $iZ_1e\gamma^\mu$ is the first term in this sum, and the diagram of fig. 62.3 is the second. Thus we have

$$i\mathbf{V}^\mu(p',p) = iZ_1e\gamma^\mu + i\mathbf{V}^\mu_{1\,\text{loop}}(p',p) + O(e^5)\ , \tag{62.39}$$

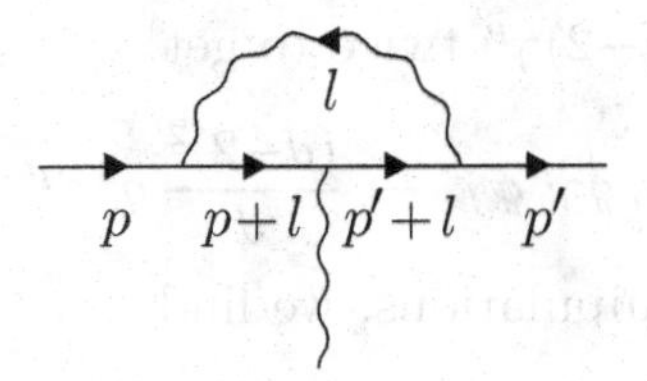

Figure 62.3. The one-loop correction to the photon-fermion-fermion vertex in spinor electrodynamics.

where

$$i\mathbf{V}^{\mu}_{1\,\text{loop}}(p',p) = (ie)^3\left(\tfrac{1}{i}\right)^3 \int \frac{d^4\ell}{(2\pi)^4} \left[\gamma^\rho \tilde{S}(\not{p}'+\not{\ell})\gamma^\mu \tilde{S}(\not{p}+\not{\ell})\gamma^\nu\right]\tilde{\Delta}_{\nu\rho}(\ell)\,. \tag{62.40}$$

We again use eq. (62.29) for the photon propagator, and combine denominators in the usual way. We then get

$$i\mathbf{V}^{\mu}_{1\,\text{loop}}(p',p) = e^3 \int dF_3 \int \frac{d^4q}{(2\pi)^4}\,\frac{N^\mu}{(q^2+D)^3}\,, \tag{62.41}$$

where the integral over Feynman parameters is

$$\int dF_3 \equiv 2\int_0^1 dx_1\,dx_2\,dx_3\,\delta(x_1{+}x_2{+}x_3{-}1)\,, \tag{62.42}$$

and

$$q = \ell + x_1 p + x_2 p'\,, \tag{62.43}$$

$$\begin{aligned} D = {} & x_1(1{-}x_1)p^2 + x_2(1{-}x_2)p'^2 - 2x_1x_2\,p{\cdot}p' \\ & + (x_1{+}x_2)m^2 + x_3 m_\gamma^2\,, \end{aligned} \tag{62.44}$$

$$\begin{aligned} N^\mu &= \gamma_\nu(-\not{p}' - \not{\ell} + m)\gamma^\mu(-\not{p} - \not{\ell} + m)\gamma^\nu \\ &= \gamma_\nu[-\not{q} + x_1\not{p} - (1{-}x_2)\not{p}' + m]\gamma^\mu[-\not{q} - (1{-}x_1)\not{p} + x_2\not{p}' + m]\gamma^\nu \\ &= \gamma_\nu\not{q}\gamma^\mu\not{q}\gamma^\nu + \widetilde{N}^\mu + (\text{linear in } q)\,, \end{aligned} \tag{62.45}$$

where

$$\widetilde{N}^\mu = \gamma_\nu[x_1\not{p} - (1{-}x_2)\not{p}' + m]\gamma^\mu[-(1{-}x_1)\not{p} + x_2\not{p}' + m]\gamma^\nu\,. \tag{62.46}$$

The terms linear in q in eq. (62.45) integrate to zero, and only the first term is divergent. After continuing to d dimensions, we can use eq. (62.18) to make the replacement

$$\gamma_\nu\not{q}\gamma^\mu\not{q}\gamma^\nu \to \frac{1}{d}\,q^2\,\gamma_\nu\gamma_\rho\gamma^\mu\gamma^\rho\gamma^\nu\,. \tag{62.47}$$

Then we use $\gamma_\rho\gamma^\mu\gamma^\rho = (d-2)\gamma^\mu$ twice to get

$$\gamma_\nu \not{q} \gamma^\mu \not{q} \gamma^\nu \to \frac{(d-2)^2}{d}\, q^2\, \gamma^\mu \,. \tag{62.48}$$

Performing the usual manipulations, we find

$$\mathbf{V}^\mu_{1\,\text{loop}}(p',p) = \frac{e^3}{8\pi^2}\left[\left(\frac{1}{\varepsilon} - 1 - \tfrac{1}{2}\int dF_3\, \ln(D/\mu^2)\right)\gamma^\mu + \tfrac{1}{4}\int dF_3\, \frac{\widetilde{N}^\mu}{D}\right] . \tag{62.49}$$

From eq. (62.39), we see that finiteness of $\mathbf{V}^\mu(p',p)$ requires

$$Z_1 = 1 - \frac{e^2}{8\pi^2}\left(\frac{1}{\varepsilon} + \text{finite}\right) + O(e^4) \,. \tag{62.50}$$

To completely fix $\mathbf{V}^\mu(p',p)$, we need a suitable condition to impose on it. We take this up in the next section.

Problems

62.1) Show that adding a *gauge fixing term* $-\frac{1}{2}\xi^{-1}(\partial^\mu A_\mu)^2$ to $\mathcal{L}$ results in eq. (62.9) as the photon propagator. Explain why $\xi = 0$ corresponds to Lorenz gauge, $\partial^\mu A_\mu = 0$.

62.2) Find the coefficients of e^2/ε in $Z_{1,2,3,m}$ in R_ξ gauge. In particular, show that $Z_1 = Z_2 = 1 + O(e^4)$ in Lorenz gauge.

62.3) Consider the six one-loop diagrams with four external photons (and no external fermions). Show that, even though each diagram is logarithmically divergent, their sum is finite. Use gauge invariance to explain why this must be the case.

63

The vertex function in spinor electrodynamics

Prerequisite: 62

In the last section, we computed the one-loop contribution to the vertex function $\mathbf{V}^{\mu}(p',p)$ in spinor electrodynamics, where p is the four-momentum of an incoming electron (or outgoing positron), and p' is the four-momentum of an outgoing electron (or incoming positron). We left open the issue of the renormalization condition we wish to impose on $\mathbf{V}^{\mu}(p',p)$.

For the theories we have studied previously, we have usually made the mathematically convenient (but physically obscure) choice to define the coupling constant as the value of the vertex function when all external four-momenta are set to zero. However, in the case of spinor electrodynamics, the masslessness of the photon gives us the opportunity to do something more physically meaningful: we can define the coupling constant as the value of the vertex function when all three particles are on shell: $p^2 = p'^2 = -m^2$, and $q^2 = 0$, where $q \equiv p' - p$ is the photon four-momentum. Because the photon is massless, these three on-shell conditions are compatible with momentum conservation.

To be more precise, let us sandwich $\mathbf{V}^{\mu}(p',p)$ between the spinor factors that are appropriate for an incoming electron with momentum p and an outgoing electron with momentum p', impose the on-shell conditions, and define the electron charge e via

$$\overline{u}_{s'}(\mathbf{p}')\mathbf{V}^{\mu}(p',p)u_s(\mathbf{p})\bigg|_{\substack{p^2=p'^2=-m^2\\(p'-p)^2=0}} = e\,\overline{u}_{s'}(\mathbf{p}')\gamma^{\mu}u_s(\mathbf{p})\bigg|_{\substack{p^2=p'^2=-m^2\\(p'-p)^2=0}} . \tag{63.1}$$

This definition is in accord with the usual one provided by Coulomb's law. To see why, consider the process of electron-electron scattering. According to the general discussion in section 19, we compute the exact amplitude for this process by using tree diagrams with exact internal propagators and vertices, as shown in fig. 63.1. In the last section, we renormalized the photon

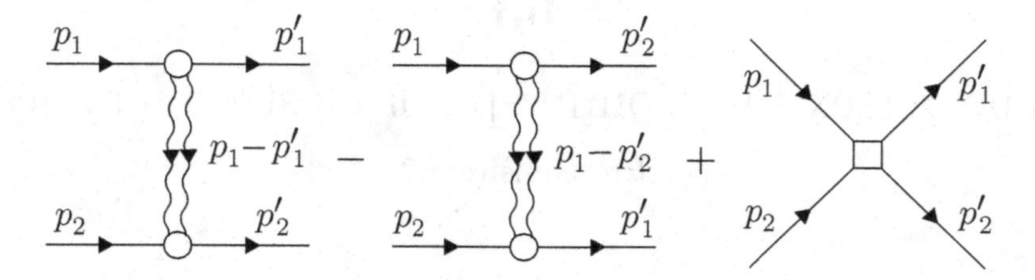

Figure 63.1. Diagrams for the exact electron-electron scattering amplitude. The vertices and photon propagator are exact; external lines stand for the usual u and $\overline{u}$ spinor factors, times the unit residue of the pole at $p^2 = -m^2$.

propagator so that it approaches its tree-level value $\tilde{\Delta}_{\mu\nu}(q)$ when $q^2 \to 0$. And we have just chosen to renormalize the electron-photon vertex function by requiring it to approach its tree-level value $e\gamma^\mu$ when $q^2 \to 0$, and when sandwiched between external spinors for on-shell incoming and outgoing electrons. Therefore, as $q^2 \to 0$, the first two diagrams in fig. 63.1 approach the tree-level scattering amplitude, with the electron charge equal to e. Furthermore, the third diagram does not have a pole at $q^2 = 0$, and so can be neglected in this limit. Physically, $q^2 \to 0$ means that the electron's momentum changes very little during the scattering. Measuring a slight deflection in the trajectory of one charged particle (due to the presence of another) is how we measure the coefficient in Coulomb's law. Thus, eq. (63.1) corresponds to this traditional definition of the charge of the electron.

We can simplify eq. (63.1) by noting that the on-shell conditions actually enforce $p' = p$. So we can rewrite eq. (63.1) as

$$\begin{aligned} \overline{u}_s(\mathbf{p})\mathbf{V}^\mu(p,p)u_s(\mathbf{p}) &= e\,\overline{u}_s(\mathbf{p})\gamma^\mu u_s(\mathbf{p}) \\ &= 2ep^\mu\ , \end{aligned} \tag{63.2}$$

where $p^2 = -m^2$ is implicit. We have taken $s' = s$, because otherwise the right-hand side vanishes (and hence does not specify a value for e).

Now we can use eq. (63.2) to completely determine $\mathbf{V}^\mu(p',p)$. Using the freedom to choose the finite part of Z_1, we write

$$\mathbf{V}^\mu(p',p) = e\gamma^\mu - \frac{e^3}{16\pi^2}\int dF_3 \left[\Big(\ln(D/D_0) + 2\kappa_1\Big)\gamma^\mu - \frac{N^\mu}{2D}\right] + O(e^5)\ , \tag{63.3}$$

where

$$\begin{aligned} D = {} & x_1(1{-}x_1)p^2 + x_2(1{-}x_2)p'^2 - 2x_1x_2 p{\cdot}p' \\ & + (x_1{+}x_2)m^2 + x_3 m_\gamma^2\ , \end{aligned} \tag{63.4}$$

D_0 is D evaluated at $p' = p$ and $p^2 = -m^2$,

$$\begin{aligned} D_0 &= (x_1{+}x_2)^2 m^2 + x_3 m_\gamma^2 \\ &= (1{-}x_3)^2 m^2 + x_3 m_\gamma^2 \,, \end{aligned} \tag{63.5}$$

and

$$N^\mu = \gamma_\nu [x_1 \not{p} - (1{-}x_2)\not{p}' + m]\gamma^\mu[-(1{-}x_1)\not{p} + x_2\not{p}' + m]\gamma^\nu \,; \tag{63.6}$$

N^μ was called $\widetilde{N}^\mu$ in section 62, but we have dropped the tilde for notational convenience.

We fix the constant κ_1 in eq. (63.3) by imposing eq. (63.2). This yields

$$4\kappa_1 p^\mu = \int dF_3 \, \frac{\overline{u}_s(\mathbf{p}) N_0^\mu u_s(\mathbf{p})}{2D_0} \,, \tag{63.7}$$

where N_0^μ is N^μ with $p' = p$ and $p^2 = p'^2 = -m^2$.

So now we must evaluate $\overline{u} N_0^\mu u$. To do so, we first write

$$N^\mu = \gamma_\nu (\not{a}_1 {+} m)\gamma^\mu (\not{a}_2 {+} m)\gamma^\nu \,, \tag{63.8}$$

where

$$\begin{aligned} a_1 &= x_1 p - (1{-}x_2)p' \,, \\ a_2 &= x_2 p' - (1{-}x_1)p \,. \end{aligned} \tag{63.9}$$

Now we use the gamma matrix contraction identities to get

$$N^\mu = 2\not{a}_2 \gamma^\mu \not{a}_1 + 4m(a_1{+}a_2)^\mu + 2m^2\gamma^\mu \,. \tag{63.10}$$

Here we have set $d = 4$, because we have already removed the divergence and taken the limit $\varepsilon \to 0$. Setting $p' = p$, and using $\not{p}u = -mu$ and $\overline{u}\not{p} = -m\overline{u}$, along with $\overline{u}\gamma^\mu u = 2p^\mu$ and $\overline{u}u = 2m$, and recalling that $x_1{+}x_2{+}x_3 = 1$, we find

$$\overline{u} N_0^\mu u = 4(1{-}4x_3{+}x_3^2)m^2 p^\mu \,. \tag{63.11}$$

Using eqs. (63.5), (63.7), and (63.11), we get

$$\begin{aligned} \kappa_1 &= \frac{1}{2}\int dF_3 \, \frac{1{-}4x_3{+}x_3^2}{(1{-}x_3)^2 + x_3 m_\gamma^2/m^2} \\ &= \int_0^1 dx_3 \, (1{-}x_3) \, \frac{1{-}4x_3{+}x_3^2}{(1{-}x_3)^2 + x_3 m_\gamma^2/m^2} \\ &= -2\ln(m/m_\gamma) + \tfrac{5}{2} \end{aligned} \tag{63.12}$$

in the limit of $m_\gamma \to 0$. We see that an infrared regulator is necessary for the vertex function as well as the fermion propagator.

Now that we have $\mathbf{V}^\mu(p',p)$, we can extract some physics from it. Consider again the process of electron-electron scattering, shown in fig. 63.1. In order to compute the contributions of these diagrams, we must evaluate $\overline{u}_{s'}(\mathbf{p}')\mathbf{V}^\mu(p',p)u_s(\mathbf{p})$ with $p^2 = p'^2 = -m^2$, but with $q^2 = (p'-p)^2$ arbitrary.

To evaluate $\overline{u}'N^\mu u$, we start with eq. (63.10), and use the anticommutation relations of the gamma matrices to move all the $\not{p}$s in N^μ to the far right (where we can use $\not{p}u = -mu$) and all the $\not{p}'$s to the far left (where we can use $\overline{u}'\not{p}' = -m\overline{u}'$). This results in

$$N^\mu \to [4(1-x_1-x_2+x_1x_2)p\cdot p' + 2(2x_1-x_1^2+2x_2-x_2^2)m^2]\gamma^\mu$$
$$+ 4m(x_1^2-x_2+x_1x_2)p^\mu + 4m(x_2^2-x_1+x_1x_2)p'^\mu \,. \qquad (63.13)$$

Next, replace $p\cdot p'$ with $-\frac{1}{2}q^2-m^2$, group the p^μ and p'^μ terms into $p'+p$ and $p'-p$ combinations, and make use of $x_1+x_2+x_3=1$ to simplify some coefficients. The result is

$$N^\mu \to 2[(1-2x_3-x_3^2)m^2 - (x_3+x_1x_2)q^2]\gamma^\mu$$
$$- 2m(x_3-x_3^2)(p'+p)^\mu$$
$$- 2m[(x_1+x_1^2)-(x_2+x_2^2)](p'-p)^\mu \,. \qquad (63.14)$$

In the denominator, set $p^2 = p'^2 = -m^2$ and $p\cdot p' = -\frac{1}{2}q^2-m^2$ to get

$$D \to x_1x_2q^2 + (1-x_3)^2m^2 + x_3m_\gamma^2 \,. \qquad (63.15)$$

Note that the right-hand side of eq. (63.15) is symmetric under $x_1 \leftrightarrow x_2$. Thus the last line of eq. (63.14) will vanish when we integrate $\overline{u}'N^\mu u/D$ over the Feynman parameters. Finally, we use the Gordon identity from section 38,

$$\overline{u}'(p'+p)^\mu u = \overline{u}'[2m\gamma^\mu + 2iS^{\mu\nu}q_\nu]u \,, \qquad (63.16)$$

where $S^{\mu\nu} = \frac{i}{4}[\gamma^\mu,\gamma^\nu]$, to get

$$N^\mu \to 2[(1-4x_3+x_3^2)m^2 - (x_3+x_1x_2)q^2]\gamma^\mu$$
$$- 4im(x_3-x_3^2)S^{\mu\nu}q_\nu \,. \qquad (63.17)$$

So now we have

$$\overline{u}_{s'}(\mathbf{p}')\mathbf{V}^\mu(p',p)u_s(\mathbf{p}) = e\overline{u}'\Big[F_1(q^2)\gamma^\mu - \tfrac{i}{m}F_2(q^2)S^{\mu\nu}q_\nu\Big]u \,, \qquad (63.18)$$

where we have defined the *form factors*

$$F_1(q^2) = 1 - \frac{e^2}{16\pi^2}\int dF_3 \left[\ln\left(1 + \frac{x_1 x_2 q^2/m^2}{(1-x_3)^2}\right) + \frac{1-4x_3+x_3^2}{(1-x_3)^2 + x_3 m_\gamma^2/m^2} \right.$$
$$\left. + \frac{(x_3+x_1x_2)q^2/m^2 - (1-4x_3+x_3^2)}{x_1x_2q^2/m^2 + (1-x_3)^2 + x_3 m_\gamma^2/m^2} \right] + O(e^4)\,, \quad (63.19)$$

$$F_2(q^2) = \frac{e^2}{8\pi^2}\int dF_3 \, \frac{x_3 - x_3^2}{x_1x_2q^2/m^2 + (1-x_3)^2} + O(e^4)\,. \quad (63.20)$$

We have set $m_\gamma = 0$ in eq. (63.20), and in the logarithm term in eq. (63.19), because these terms do not suffer from infrared divergences.

We can simplify $F_2(q^2)$ by using the delta function in dF_3 to do the integral over x_2 (which replaces x_2 with $1-x_3-x_1$), making the change of variable $x_1 = (1-x_3)y$, and performing the integral over x_3 from zero to one; the result is

$$F_2(q^2) = \frac{e^2}{8\pi^2}\int_0^1 \frac{dy}{1 + y(1-y)q^2/m^2} + O(e^4)\,. \quad (63.21)$$

This last integral can also be done in closed form, but we will be mostly interested in its value at $q^2 = 0$, corresponding to an on-shell photon:

$$F_2(0) = \frac{\alpha}{2\pi} + O(\alpha^2)\,, \quad (63.22)$$

where $\alpha = e^2/4\pi = 1/137.036$ is the fine-structure constant. We will explore the physical consequences of eq. (63.22) in the next section.

Problems

63.1) The most general possible form of $\overline{u}'V^\mu(p',p)u$ is a linear combination of γ^μ, p^μ, and p'^μ sandwiched between $\overline{u}'$ and u, with coefficients that depend on q^2. (The only other possibility is to include terms with γ_5, but γ_5 does not appear in the tree-level propagators or vertex, and so it cannot be generated in any Feynman diagram; this is a consequence of parity conservation.) Thus we can write

$$\overline{u}_{s'}(\mathbf{p}')\mathbf{V}^\mu(p',p)u_s(\mathbf{p}) = e\overline{u}'[A(q^2)\gamma^\mu + B(q^2)(p'+p)^\mu + C(q^2)(p'-p)^\mu]u\,. \quad (63.23)$$

a) Use gauge invariance to show that $q_\mu\overline{u}'\mathbf{V}^\mu(p',p)u = 0$, and determine the consequences for A, B, and C.

b) Express F_1 and F_2 in terms of A, B, and C.

64

The magnetic moment of the electron

Prerequisite: 63

In the last section, we computed the one-loop contribution to the vertex function $\mathbf{V}^\mu(p',p)$ in spinor electrodynamics, where p is the four-momentum of an incoming electron, and p' is the four-momentum of an outgoing electron. We found

$$\overline{u}_{s'}(\mathbf{p}')\mathbf{V}^\mu(p',p)u_s(\mathbf{p}) = e\overline{u}'\Big[F_1(q^2)\gamma^\mu - \tfrac{i}{m}F_2(q^2)S^{\mu\nu}q_\nu\Big]u\,, \tag{64.1}$$

where $q = p' - p$ is the four-momentum of the photon (treated as incoming), and with complicated expressions for the *form factors* $F_1(q^2)$ and $F_2(q^2)$. For our purposes in this section, all we will need to know is that

$$\begin{aligned} F_1(0) &= 1 \quad \text{exactly,} \\ F_2(0) &= \tfrac{\alpha}{2\pi} + O(\alpha^2)\,. \end{aligned} \tag{64.2}$$

Eq. (64.1) follows from a quantum action of the form

$$\Gamma = \int d^4x\,\Big[eF_1(0)\overline{\Psi}A\!\!\!/\Psi + \tfrac{e}{2m}F_2(0)F_{\mu\nu}\overline{\Psi}S^{\mu\nu}\Psi + \ldots\Big]\,, \tag{64.3}$$

where the ellipses stand for terms with more derivatives. The displayed terms yield the vertex factor of eq. (64.1) with $q^2 = 0$. To see this, recall that an incoming photon translates into a factor of $A_\mu \sim \varepsilon^*_\mu e^{iqx}$, and therefore of $F_{\mu\nu} \sim i(q_\mu\varepsilon^*_\nu - q_\nu\varepsilon^*_\mu)e^{iqx}$; the two terms in $F_{\mu\nu}$ cancel the extra factor of one half in the second term in eq. (64.3).

Now we will see what eq. (64.3) predicts for the *magnetic moment* of the electron. We define the magnetic moment by the following procedure. We take the photon field A^μ to be a classical field that corresponds to a constant magnetic field in the z direction: $A^0 = 0$ and $\mathbf{A} = (0, Bx, 0)$. This yields $F_{12} = -F_{21} = B$, with all other components of $F_{\mu\nu}$ vanishing. Then we define a normalized state of an electron at rest, with spin up along the

z axis:

$$|e\rangle \equiv \int \widetilde{dp}\, f(\mathbf{p}) b_+^\dagger(\mathbf{p})|0\rangle\ , \tag{64.4}$$

where the wave packet is rotationally invariant (so that there is no orbital angular momentum) and sharply peaked at $\mathbf{p} = 0$, something like

$$f(\mathbf{p}) \sim \exp(-a^2\mathbf{p}^2/2)\ , \tag{64.5}$$

with $a \ll 1/m$. We normalize the wave packet by $\int \widetilde{dp}\, |f(\mathbf{p})|^2 = 1$; then we have $\langle e|e\rangle = 1$.

Now we define the interaction hamiltonian as what we get from the two displayed terms in eq. (64.3), using our specified field A^μ, and with the form-factor values of eq. (64.2):

$$H_1 \equiv -eB \int d^3x\, \overline{\Psi} \left[x\gamma^2 + \frac{\alpha}{2\pi m} S^{12} \right] \Psi\ . \tag{64.6}$$

Then the electron's magnetic moment μ is specified by

$$\mu B \equiv -\langle e|H_1|e\rangle\ . \tag{64.7}$$

In quantum mechanics in general, if we identify H_1 as the piece of the hamiltonian that is linear in the external magnetic field, then eq. (64.7) defines the magnetic moment of a normalized quantum state with definite angular momentum in the $\mathbf{B}$ direction.

Now we turn to the computation. We need to evaluate $\langle e|\overline{\Psi}_\alpha(x)\Psi_\beta(x)|e\rangle$. Using the usual plane-wave expansions, we have

$$\langle 0|b_+(\mathbf{p}')\overline{\Psi}_\alpha(x)\Psi_\beta(x)b_+^\dagger(\mathbf{p})|0\rangle = \overline{u}_+(\mathbf{p}')_\alpha u_+(\mathbf{p})_\beta\, e^{i(p-p')x}\ . \tag{64.8}$$

Thus we get

$$\begin{aligned} \langle e|H_1|e\rangle = -eB \int \widetilde{dp}\, \widetilde{dp}'\, d^3x\, e^{i(p-p')x} \\ \times f^*(\mathbf{p}')\overline{u}_+(\mathbf{p}')\Big[x\gamma^2 + \frac{\alpha}{2\pi m} S^{12} \Big] u_+(\mathbf{p}) f(\mathbf{p})\ . \end{aligned} \tag{64.9}$$

We can write the factor of x as $-i\partial_{p_1}$ acting on $e^{i(p-p')x}$, and integrate by parts to put this derivative onto $u_+(\mathbf{p})f(\mathbf{p})$; the wave packets kill any surface terms. Then we can complete the integral over d^3x to get a factor of $(2\pi)^3\delta^3(\mathbf{p}' - \mathbf{p})$, and do the integral over $\widetilde{dp}'$. The result is

$$\langle e|H_1|e\rangle = -eB \int \frac{\widetilde{dp}}{2\omega}\, f^*(\mathbf{p})\overline{u}_+(\mathbf{p}) \Big[i\gamma^2\partial_{p_1} + \frac{\alpha}{2\pi m} S^{12} \Big] u_+(\mathbf{p}) f(\mathbf{p})\ . \tag{64.10}$$

Suppose the ∂_{p_1} acts on $f(\mathbf{p})$. Since $f(\mathbf{p})$ is rotationally invariant, the result is odd in p_1. We then use $\overline{u}_+(\mathbf{p})\gamma^i u_+(\mathbf{p}) = 2p^i$ to conclude that this term is odd in both p_1 and p_2, and hence integrates to zero.

The remaining contribution from the first term has the ∂_{p_1} acting on $u_+(\mathbf{p})$. Recall from section 38 that

$$u_s(\mathbf{p}) = \exp(i\eta\,\hat{\mathbf{p}}\cdot\mathbf{K})u_s(\mathbf{0}) \; , \tag{64.11}$$

where $K^j = S^{j0} = \frac{i}{2}\gamma^j\gamma^0$ is the boost matrix, $\hat{\mathbf{p}}$ is a unit vector in the $\mathbf{p}$ direction, and $\eta = \sinh^{-1}(|\mathbf{p}|/m)$ is the rapidity. Since the wave packet is sharply peaked at $\mathbf{p} = 0$, we can expand eq. (64.11) to linear order in $\mathbf{p}$, take the derivative with respect to p_1, and then set $\mathbf{p} = 0$; the result is

$$\begin{aligned} \partial_{p_1} u_+(\mathbf{p})\Big|_{\mathbf{p}=0} &= \tfrac{i}{m}K^1 u_+(\mathbf{0}) \\ &= -\frac{1}{2m}\gamma^1\gamma^0 u_+(\mathbf{0}) \\ &= -\frac{1}{2m}\gamma^1 u_+(\mathbf{0}) \; , \end{aligned} \tag{64.12}$$

where we used $\gamma^0 u_s(\mathbf{0}) = u_s(\mathbf{0})$ to get the last line. Then we have

$$\begin{aligned} \overline{u}_+(\mathbf{p})i\gamma^2\partial_{p_1}u_+(\mathbf{p})\Big|_{\mathbf{p}=0} &= \overline{u}_+(\mathbf{0})\frac{-i}{2m}\gamma^2\gamma^1 u_+(\mathbf{0}) \\ &= \frac{1}{m}\,\overline{u}_+(\mathbf{0})S^{12}u_+(\mathbf{0}) \; . \end{aligned} \tag{64.13}$$

Plugging this into eq. (64.10) yields

$$\begin{aligned} \langle e|H_1|e\rangle &= -eB\int \widetilde{dp}\,|f(\mathbf{p})|^2\left(1+\tfrac{\alpha}{2\pi}\right)\tfrac{1}{m}\,\overline{u}_+(\mathbf{0})S^{12}u_+(\mathbf{0}) \\ &= -\frac{eB}{2m^2}\left(1+\tfrac{\alpha}{2\pi}\right)\overline{u}_+(\mathbf{0})S^{12}u_+(\mathbf{0}) \; . \end{aligned} \tag{64.14}$$

Next we use $S^{12}u_\pm(\mathbf{0}) = \pm\frac{1}{2}u_\pm(\mathbf{0})$ and $\overline{u}_\pm(\mathbf{0})u_\pm(\mathbf{0}) = 2m$ to get

$$\langle e|H_1|e\rangle = -\frac{eB}{2m}\left(1+\tfrac{\alpha}{2\pi}\right) . \tag{64.15}$$

Comparing with eq. (64.7), we see that the magnetic moment of the electron is

$$\mu = g\,\frac{1}{2}\,\frac{e}{2m} \; , \tag{64.16}$$

where $e/2m$ is the *Bohr magneton*, the extra factor of one-half is for the electron's spin (a classical spinning ball of charge would have a magnetic moment equal to the Bohr magneton times its angular momemtum), and g

is the *Landé g factor*, given by

$$g = 2\left(1 + \frac{\alpha}{2\pi} + O(\alpha^2)\right). \tag{64.17}$$

Since g can be measured to high precision, calculations of μ provide a stringent test of spinor electrodynamics. Corrections up through the α^4 term have been computed; the result is currently in good agreement with experiment.

Problems

64.1) Let the wave packet be $f(\mathbf{p}) \sim \exp(-a^2\mathbf{p}^2/2)Y_{\ell m}(\hat{\mathbf{p}})$, where $Y_{\ell m}(\hat{\mathbf{p}})$ is a spherical harmonic. Find the contribution of the orbital angular momentum to the magnetic moment.

65

Loop corrections in scalar electrodynamics

Prerequisites: 61, 62

In this section we will compute the one-loop corrections in scalar electrodynamics. We will concentrate on the divergent parts of the diagrams, enabling us to compute the renormalizing Z factors in the $\overline{\rm MS}$ scheme, and hence the beta functions. This gives us the most important qualitative information about the theory: whether it becomes strongly coupled at high or low energies.

Our lagrangian for scalar electrodynamics in section 61 already includes all possible terms whose coefficients have positive or zero mass dimension, and that respect Lorentz symmetry, the U(1) gauge symmetry, parity, time reversal, and charge conjugation. Therefore, the theory we will consider is

$$\mathcal{L} = \mathcal{L}_0 + \mathcal{L}_1 \,, \tag{65.1}$$

$$\mathcal{L}_0 = -\partial^\mu\varphi^\dagger\partial_\mu\varphi - m^2\varphi^\dagger\varphi - \tfrac{1}{4}F^{\mu\nu}F_{\mu\nu} \,, \tag{65.2}$$

$$\begin{aligned}\mathcal{L}_1 = {}& iZ_1 e[\varphi^\dagger\partial^\mu\varphi - (\partial^\mu\varphi^\dagger)\varphi]A_\mu - Z_4 e^2\varphi^\dagger\varphi A^\mu A_\mu \\ &- \tfrac{1}{4}Z_\lambda\lambda(\varphi^\dagger\varphi)^2 + \mathcal{L}_{\rm ct} \,,\end{aligned} \tag{65.3}$$

$$\mathcal{L}_{\rm ct} = -(Z_2{-}1)\partial^\mu\varphi^\dagger\partial_\mu\varphi - (Z_m{-}1)m^2\varphi^\dagger\varphi - \tfrac{1}{4}(Z_3{-}1)F^{\mu\nu}F_{\mu\nu} \,. \tag{65.4}$$

We will use the $\overline{\rm MS}$ renormalization scheme to fix the values of the Zs.

We begin with the photon self-energy, $\Pi^{\mu\nu}(k)$. The one-loop and counterterm contributions are shown in fig. 65.1. We have

$$\begin{aligned} i\Pi^{\mu\nu}(k) = {}& (iZ_1 e)^2\big(\tfrac{1}{i}\big)^2 \int \frac{d^4\ell}{(2\pi)^4}\,\frac{(2\ell+k)^\mu(2\ell+k)^\nu}{((\ell{+}k)^2+m^2)(\ell^2+m^2)} \\ &+(-2iZ_4e^2g^{\mu\nu})\big(\tfrac{1}{i}\big)\int\frac{d^4\ell}{(2\pi)^4}\,\frac{1}{\ell^2+m^2} \\ &- i(Z_3{-}1)(k^2g^{\mu\nu} - k^\mu k^\mu) + \ldots \,, \end{aligned} \tag{65.5}$$

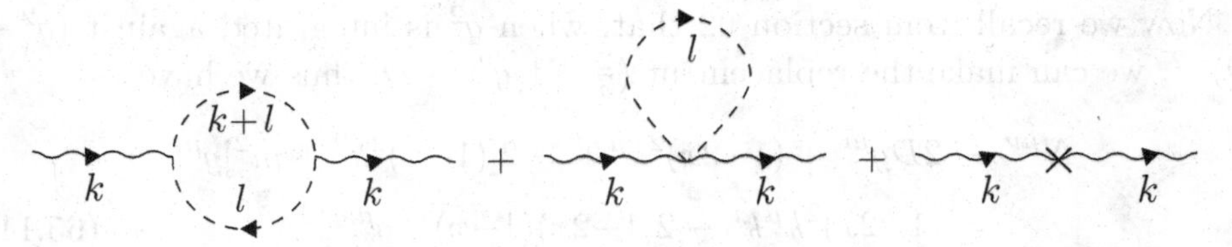

Figure 65.1. The one-loop and counterterm corrections to the photon propagator in scalar electrodynamics.

where the ellipses stand for higher-order (in e^2 and/or λ) terms. We can set $Z_i = 1 + O(e^2,\lambda)$ in the first two terms.

It will prove convenient to combine these first two terms into

$$i\Pi^{\mu\nu}(k) = e^2 \int \frac{d^4\ell}{(2\pi)^4} \frac{N^{\mu\nu}}{((\ell{+}k)^2 + m^2)(\ell^2 + m^2)} - i(Z_3{-}1)(k^2 g^{\mu\nu} - k^\mu k^\mu) + \ldots \,, \tag{65.6}$$

where

$$N^{\mu\nu} = (2\ell + k)^\mu (2\ell + k)^\nu - 2[(\ell{+}k)^2 + m^2]g^{\mu\nu} \,. \tag{65.7}$$

Then we continue to d dimensions, replace e with $e\tilde{\mu}^{\varepsilon/2}$, and combine the denominators with Feynman's formula; the result is

$$i\Pi^{\mu\nu}(k) = e^2 \tilde{\mu}^\varepsilon \int_0^1 dx \int \frac{d^dq}{(2\pi)^d} \frac{N^{\mu\nu}}{(q^2 + D)^2} - i(Z_3{-}1)(k^2 g^{\mu\nu} - k^\mu k^\mu) + \ldots \,, \tag{65.8}$$

where $q = \ell + xk$ and $D = x(1{-}x)k^2 + m^2$. The numerator is

$$\begin{aligned} N^{\mu\nu} &= (2q + (1{-}2x)k)^\mu (2q + (1{-}2x)k)^\nu - 2[(q + (1{-}x)k)^2 + m^2]g^{\mu\nu} \\ &= 4q^\mu q^\nu + (1{-}2x)^2 k^\mu k^\nu - 2[q^2 + (1{-}x)^2 k^2 + m^2]g^{\mu\nu} \\ &\quad + (\text{linear in } q) \\ &\to 4d^{-1} g^{\mu\nu} q^2 + (1{-}2x)^2 k^\mu k^\nu - 2[q^2 + (1{-}x)^2 k^2 + m^2]g^{\mu\nu} \,, \end{aligned} \tag{65.9}$$

where we used the symmetric-integration identity from problem 14.3 to get the last line. We can rearrange eq. (65.9) into

$$N^{\mu\nu} = 2\Big(\tfrac{2}{d} - 1\Big) g^{\mu\nu} q^2 + (1{-}2x)^2 k^\mu k^\nu - 2[(1{-}x)^2 k^2 + m^2]g^{\mu\nu} \,. \tag{65.10}$$

Now we recall from section 62 that, when q^2 is integrated against $(q^2 + D)^{-2}$, we can make the replacement $(\frac{2}{d} - 1)q^2 \to D$; thus we have

$$\begin{aligned} N^{\mu\nu} &\to 2Dg^{\mu\nu} + (1-2x)^2 k^\mu k^\nu - 2[(1-x)^2 k^2 + m^2] g^{\mu\nu} \\ &= (1-2x)^2 k^\mu k^\nu - 2(1-2x)(1-x) k^2 g^{\mu\nu} \,. \end{aligned} \tag{65.11}$$

Next we note that if we make the change of variable $x = y + \frac{1}{2}$, then we have $D = (\frac{1}{4} - y^2)k^2 + m^2$, and y is integrated from $-\frac{1}{2}$ to $+\frac{1}{2}$. Therefore, any term in $N^{\mu\nu}$ that is odd in y will integrate to zero. We then get

$$\begin{aligned} N^{\mu\nu} &= 4y^2 k^\mu k^\nu - 2(2y^2 - y) k^2 g^{\mu\nu} \\ &\to -4y^2 (k^2 g^{\mu\nu} - k^\mu k^\nu) \,. \end{aligned} \tag{65.12}$$

Thus we see that $\Pi^{\mu\nu}(k)$ is transverse, as expected.

Performing the integral over q in eq. (65.8), and focusing on the divergent part, we get

$$\tilde{\mu}^\varepsilon \int \frac{d^d q}{(2\pi)^d} \frac{1}{(q^2 + D)^2} = \frac{i}{8\pi^2} \frac{1}{\varepsilon} + O(\varepsilon^0) \,. \tag{65.13}$$

Then performing the integral over y yields

$$\int_{-1/2}^{1/2} dy \, N^{\mu\nu} = -\tfrac{1}{3}(k^2 g^{\mu\nu} - k^\mu k^\nu) \,. \tag{65.14}$$

Combining eqs. (65.8), (65.13), and (65.14), we get

$$\Pi^{\mu\nu}(k) = \Pi(k^2)(k^2 g^{\mu\nu} - k^\mu k^\nu) \,, \tag{65.15}$$

where

$$\Pi(k^2) = -\frac{e^2}{24\pi^2} \frac{1}{\varepsilon} + \text{finite} - (Z_3 - 1) + \ldots \,. \tag{65.16}$$

Thus we find, in the $\overline{\text{MS}}$ scheme,

$$Z_3 = 1 - \frac{e^2}{24\pi^2} \frac{1}{\varepsilon} + \ldots \,. \tag{65.17}$$

Now we turn to the one-loop corrections to the scalar propagator, shown in fig. 65.2. It will prove very convenient to work in Lorenz gauge, where the photon propagator is

$$\tilde{\Delta}_{\mu\nu}(\ell) = \frac{P_{\mu\nu}(\ell)}{\ell^2 - i\epsilon} \,, \tag{65.18}$$

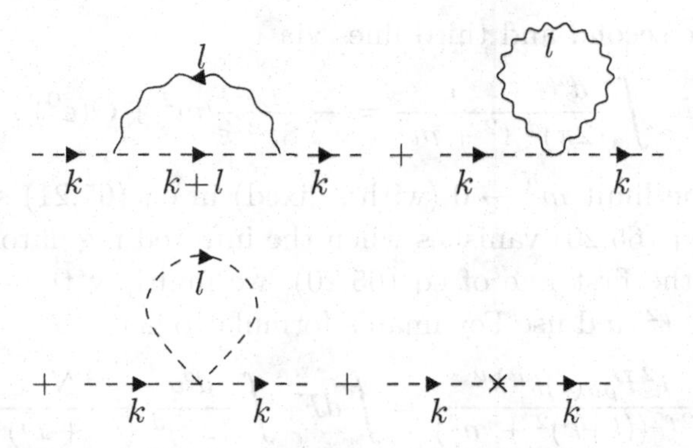

Figure 65.2. The one-loop and counterterm corrections to the scalar propagator in scalar electrodynamics.

with $P_{\mu\nu}(\ell) = g_{\mu\nu} - \ell_\mu \ell_\nu / \ell^2$. The diagrams in fig. 65.2 then yield

$$
\begin{aligned}
i\Pi_\varphi(k^2) = {} & (iZ_1 e)^2 (\tfrac{1}{i})^2 \int \frac{d^4\ell}{(2\pi)^4} \frac{P_{\mu\nu}(\ell)(\ell+2k)^\mu(\ell+2k)^\nu}{\ell^2((\ell{+}k)^2+m^2)} \\
& + (-2iZ_4 e^2 g^{\mu\nu})(\tfrac{1}{i}) \int \frac{d^4\ell}{(2\pi)^4} \frac{P_{\mu\nu}(\ell)}{\ell^2+m_\gamma^2} \\
& + (-iZ_\lambda \lambda)(\tfrac{1}{i}) \int \frac{d^4\ell}{(2\pi)^4} \frac{1}{\ell^2+m^2} \\
& - i(Z_2{-}1)k^2 - i(Z_m{-}1)m^2 + \ldots . \qquad (65.19)
\end{aligned}
$$

In the second line, m_γ is a fictitious photon mass; it appears here as an infrared regulator.

We can set $Z_i = 1 + O(e^2,\lambda)$ in the first three lines. Continuing to d dimensions, making the replacements $e \to e\tilde{\mu}^{\varepsilon/2}$ and $\lambda \to \lambda\tilde{\mu}^{\varepsilon}$, and using the relations $\ell^\mu P_{\mu\nu}(\ell) = \ell^\nu P_{\mu\nu}(\ell) = 0$ and $g^{\mu\nu}P_{\mu\nu}(\ell) = d-1$, we get

$$
\begin{aligned}
i\Pi_\varphi(k^2) = {} & 4e^2\tilde{\mu}^\varepsilon \int \frac{d^d\ell}{(2\pi)^d} \frac{P_{\mu\nu}(\ell)k^\mu k^\nu}{\ell^2((\ell{+}k)^2+m^2)} \\
& - 2(d{-}1)e^2\tilde{\mu}^\varepsilon \int \frac{d^d\ell}{(2\pi)^d} \frac{1}{\ell^2+m_\gamma^2} \\
& - \lambda\tilde{\mu}^\varepsilon \int \frac{d^d\ell}{(2\pi)^4} \frac{1}{\ell^2+m^2} \\
& - i(Z_2{-}1)k^2 - i(Z_m{-}1)m^2 + \ldots . \qquad (65.20)
\end{aligned}
$$

We evaluate the second and third lines via

$$\tilde{\mu}^{\varepsilon} \int \frac{d^d\ell}{(2\pi)^d} \frac{1}{\ell^2 + m^2} = -\frac{i}{8\pi^2} \frac{1}{\varepsilon} m^2 + O(\varepsilon^0) . \tag{65.21}$$

Then, taking the limit $m^2 \to 0$ (with ε fixed) in eq. (65.21) shows that the second line of eq. (65.20) vanishes when the infrared regulator is removed.

To evaluate the first line of eq. (65.20), we multiply the numerator and denominator by ℓ^2 and use Feynman's formula to get

$$\int \frac{d^d\ell}{(2\pi)^d} \frac{\ell^2 P_{\mu\nu}(\ell) k^\mu k^\nu}{\ell^2 \ell^2((\ell+k)^2 + m^2)} = \int dF_3 \int \frac{d^dq}{(2\pi)^d} \frac{N}{(q^2 + D)^3} , \tag{65.22}$$

where $q = \ell + x_3 k$, $D = x_3(1-x_3)k^2 + x_3 m^2$, and

$$\begin{aligned} N &= \ell^2 k^2 - (\ell\cdot k)^2 \\ &= (q - x_3 k)^2 k^2 - (q\cdot k - x_3 k^2)^2 \\ &= q^2 k^2 - (q\cdot k)^2 + (\text{linear in } q) \\ &\to q^2 k^2 - d^{-1} q^2 k^2 . \end{aligned} \tag{65.23}$$

Now we use

$$\tilde{\mu}^{\varepsilon} \int \frac{d^dq}{(2\pi)^d} \frac{q^2}{(q^2 + D)^3} = \frac{i}{8\pi^2} \frac{1}{\varepsilon} + O(\varepsilon^0) . \tag{65.24}$$

Combining eqs. (65.20)–(65.24), and requiring $\Pi_\varphi(k^2)$ to be finite, we find

$$Z_2 = 1 + \frac{3e^2}{8\pi^2} \frac{1}{\varepsilon} + \ldots , \tag{65.25}$$

$$Z_m = 1 + \frac{\lambda}{8\pi^2} \frac{1}{\varepsilon} + \ldots \tag{65.26}$$

in the $\overline{\text{MS}}$ scheme.

Now we turn to the one-loop corrections to the three-point (scalar-scalar-photon) vertex, shown in fig. 65.3. In order to simplify the calculation of the divergent terms as much as possible, we have chosen a special set of external momenta. (If we wanted the complete vertex function, including the finite terms, we would need to use a general set of external momenta.) We take the incoming scalar to have zero four-momentum, and the photon (treated as incoming) to have four-momentum k; then, by momentum conservation, the outgoing scalar also has four-momentum k. We take the internal photon to have four-momentum ℓ.

Now comes the magic of Lorenz gauge: in the second and third diagrams of fig. 65.3, the vertex factor for the leftmost vertex is $ie\ell^\mu$, and this is zero

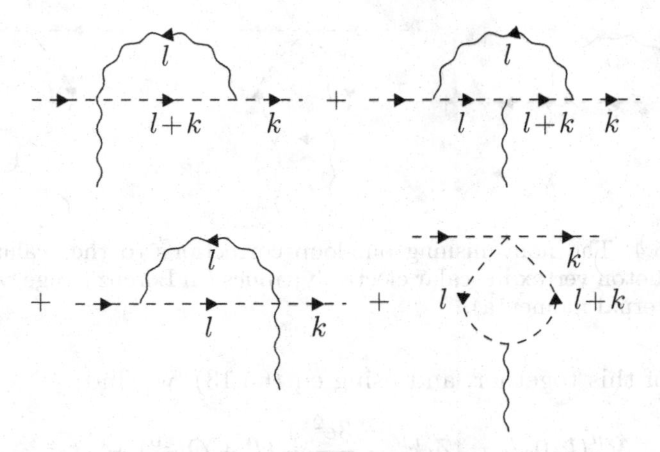

Figure 65.3. The one-loop corrections to the three-point vertex in scalar electrodynamics.

when contracted with the $P_{\mu\nu}(\ell)$ of the internal photon propagator. Thus the second and third diagrams vanish.

Alas, we will have to do some more work to evaluate the first and fourth diagrams. We have

$$
\begin{aligned}
i\mathbf{V}_3^\mu(k,0) &= ieZ_1 k^\mu \\
&\quad + (iZ_1 e)(-2iZ_4 e^2 g^{\mu\nu})\left(\tfrac{1}{i}\right)^2 \int \frac{d^4\ell}{(2\pi)^4}\,\frac{P_{\nu\rho}(\ell)(\ell+2k)^\rho}{\ell^2((\ell{+}k)^2+m^2)} \\
&\quad + (-iZ_\lambda\lambda)(iZ_1 e)\left(\tfrac{1}{i}\right)^2 \int \frac{d^4\ell}{(2\pi)^4}\,\frac{(2\ell+k)^\mu}{(\ell^2+m^2)((\ell{+}k)^2+m^2)} \\
&\quad + \ldots\,.
\end{aligned}
\tag{65.27}
$$

We can set $Z_i = 1 + O(e^2,\lambda)$ in the second and third lines. We then do the usual manipulations; the integral in the third line becomes

$$
\int_0^1 dx \int \frac{d^dq}{(2\pi)^d}\,\frac{2q^\mu + (1{-}2x)k^\mu}{(q^2+D)^2}\;, \tag{65.28}
$$

where $q = \ell + xk$ and $D = x(1{-}x)k^2 + m^2$. The term linear in q vanishes upon integration over q, and the term linear in k vanishes upon integration over x. Thus the third line of eq. (65.27) evaluates to zero.

To evaluate the second line of eq. (65.27), we note that, since $P_{\nu\rho}(\ell)\ell^\rho = 0$, it already has an overall factor of k. We can then treat k as infinitesimal, and set $k = 0$ in the denominator. We can then use the symmetric-integration identity to make the replacement $\ell_\nu\ell_\rho/\ell^2 \to d^{-1}g_{\nu\rho}$ in the numerator.

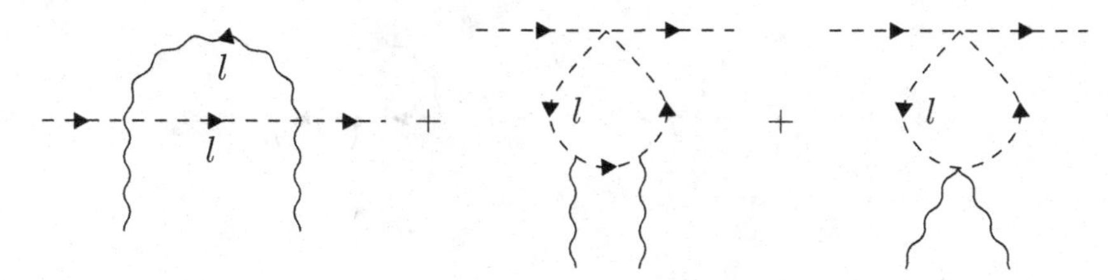

Figure 65.4. The nonvanishing one-loop corrections to the scalar-scalar-photon-photon vertex in scalar electrodynamics (in Lorenz gauge with vanishing external momenta).

Putting all of this together, and using eq. (65.13), we find

$$\mathbf{V}_3^\mu(k,0)/e = Z_1 k^\mu - \frac{3e^2}{8\pi^2}\frac{1}{\varepsilon}\,k^\mu + O(\varepsilon^0) + \ldots\,, \tag{65.29}$$

and so

$$Z_1 = 1 + \frac{3e^2}{8\pi^2}\frac{1}{\varepsilon} + \ldots\,, \tag{65.30}$$

in the $\overline{\text{MS}}$ scheme.

Next up is the four-point, scalar-scalar-photon-photon vertex. Because the tree-level vertex factor, $-2iZ_4 e^2 g^{\mu\nu}$, does not depend on the external four-momenta, we can simply set them all to zero. Then, whenever an internal photon line attaches to an external scalar with a three-point vertex, the diagram is zero, for the same reason that the second and third diagrams of fig. 65.3 were zero. This kills a lot of diagrams; the survivors are shown in fig. 65.4. We have

$$\begin{aligned} i\mathbf{V}_4^{\mu\nu}(0,0,0) = {}& -2iZ_4 e^2 g^{\mu\nu} \\ & + (-2iZ_4e^2)^2\left(\tfrac{1}{i}\right)^2 \int \frac{d^4\ell}{(2\pi)^4}\,\frac{g^{\mu\rho}P_{\rho\sigma}(\ell)g^{\sigma\nu}}{\ell^2(\ell^2+m^2)} + (\mu\!\leftrightarrow\!\nu) \\ & + (iZ_1e)^2(-iZ_\lambda\lambda)\left(\tfrac{1}{i}\right)^3 \int \frac{d^4\ell}{(2\pi)^4}\,\frac{(2\ell)^\mu(2\ell)^\nu}{(\ell^2+m^2)^3} + (\mu\!\leftrightarrow\!\nu) \\ & + (-iZ_\lambda\lambda)(-2iZ_4e^2g^{\mu\nu})\left(\tfrac{1}{i}\right)^2 \int \frac{d^4\ell}{(2\pi)^4}\,\frac{1}{(\ell^2+m^2)^2} \\ & + \ldots\,. \end{aligned} \tag{65.31}$$

The notation $+(\mu\!\leftrightarrow\!\nu)$ in the second and third lines means that we must add the same expression with these indices swapped; this is because the original and swapped versions of each diagram are topologically distinct, and contribute separately to the vertex function.

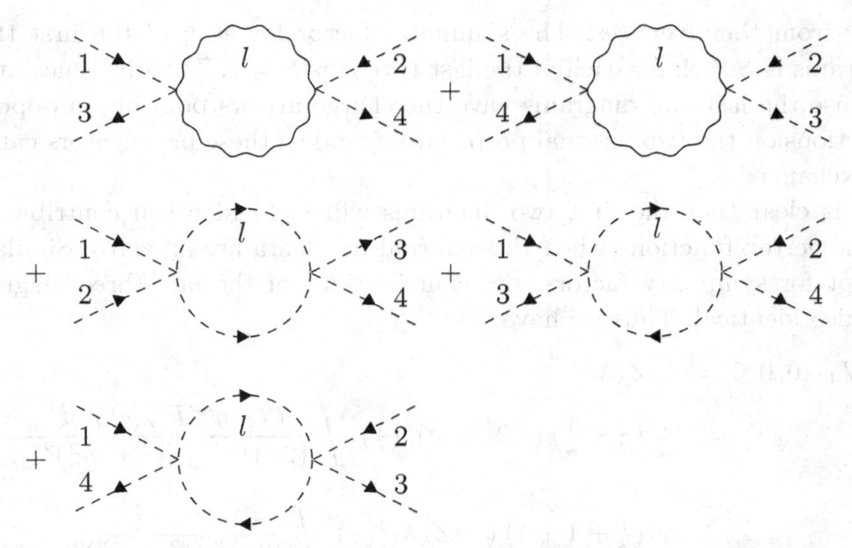

Figure 65.5. The nonvanishing one-loop corrections to the four-scalar vertex in scalar electrodynamics (in Lorenz gauge with vanishing external momenta).

As usual, we set $Z_i = 1 + O(e^2,\lambda)$ in the second through fourth lines. After using the symmetric-integration identity, along with eqs. (65.13) and (65.24), we can see that the divergent parts of the third and fourth lines cancel each other. The first line is easily evaluated with symmetric integration and eq. (65.13). Then we have

$$\mathbf{V}_4^{\mu\nu}(0,0,0)/e^2 = -2Z_4 g^{\mu\nu} + \frac{3e^2}{4\pi^2}\frac{1}{\varepsilon}\, g^{\mu\nu} + O(\varepsilon^0) + \ldots \,, \tag{65.32}$$

and so

$$Z_4 = 1 + \frac{3e^2}{8\pi^2}\frac{1}{\varepsilon} + \ldots \tag{65.33}$$

in the $\overline{\text{MS}}$ scheme.

Finally, we have the one-loop corrections to the four-scalar vertex. Once again, because the tree-level vertex factor, $-iZ_\lambda\lambda$, does not depend on the external four-momenta, we can set them all to zero. Then, whenever an internal photon line attaches to an external scalar with a three-point vertex, the diagram is zero. The remaining diagrams are shown in fig. 65.5.

Even though we have set the external momenta to zero, we still have to keep track of which particle is which, in order to count the diagrams correctly; thus the external lines are labeled 1 through 4. Lines 1 and 2 have arrows pointing towards their vertices, and 3 and 4 have arrows pointing away from their vertices. The symmetry factor for each of the first three diagrams is $S = 2$; for each of the last two, it is $S = 1$. The difference arises because the last two diagrams have the charge arrows pointing in opposite directions on the two internal propagators, and so these propagators cannot be exchanged.

It is clear that the first two diagrams will yield identical contributions to the vertex function (when the external momenta are all zero). Similarly, except for symmetry factors, the contributions of the last three diagrams are also identical. Thus we have

$$\begin{aligned} i\mathbf{V}_{4\varphi}(0,0,0) &= -iZ_\lambda\lambda \\ &\quad + \left(\tfrac{1}{2}+\tfrac{1}{2}\right)(-2iZ_4e^2)^2\left(\tfrac{1}{i}\right)^2 \int \frac{d^4\ell}{(2\pi)^4}\,\frac{g^{\mu\nu}P_{\nu\rho}(\ell)g^{\rho\sigma}P_{\sigma\mu}(\ell)}{(\ell^2+m_\gamma^2)^2} \\ &\quad + \left(\tfrac{1}{2}+1+1\right)(-iZ_\lambda\lambda)^2\left(\tfrac{1}{i}\right)^2 \int \frac{d^4\ell}{(2\pi)^4}\,\frac{1}{(\ell^2+m^2)^2} \\ &\quad + \ldots\,. \end{aligned} \tag{65.34}$$

Using the familiar techniques, we find

$$\mathbf{V}_{4\varphi}(0,0,0) = -iZ_\lambda\lambda + \frac{3e^4}{2\pi^2}\,\frac{1}{\varepsilon} + \frac{5\lambda^2}{16\pi^2}\,\frac{1}{\varepsilon} + O(\varepsilon^0) + \ldots\,, \tag{65.35}$$

and so

$$Z_\lambda = 1 + \left(\frac{3e^4}{2\pi^2\lambda} + \frac{5\lambda}{16\pi^2}\right)\frac{1}{\varepsilon} + \ldots \tag{65.36}$$

in the $\overline{\text{MS}}$ scheme.

Problems

65.1) What conditions should be imposed on $\mathbf{V}_3^\mu(p',p)$ and $\mathbf{V}_4^{\mu\nu}(k,p',p)$ in the OS scheme? (Here k is the incoming four-momentum of the photon at the μ vertex, and p' and p are the four-momenta of the outgoing and incoming scalars, respectively.)

65.2) Consider a gauge transformaton $A^\mu \to A^\mu - \partial^\mu\Gamma$. Show that there is a transformation of φ that leaves the lagrangian of eqs. (65.1)–(65.4) invariant if and only if $Z_4 = Z_1^2/Z_2$.

66

Beta functions in quantum electrodynamics

Prerequisites: 52, 62

In this section we will compute the beta function for the electromagnetic coupling e in spinor electrodynamics and scalar electrodynamics. We will also compute the beta function for the φ^4 coupling λ in scalar electrodynamics.

In spinor electrodynamics, the relation between the bare and renormalized couplings is

$$e_0 = Z_3^{-1/2} Z_2^{-1} Z_1 \,\tilde{\mu}^{\varepsilon/2} e\,. \tag{66.1}$$

It is convenient to recast this formula in terms of the fine-structure constant $\alpha = e^2/4\pi$ and its bare counterpart $\alpha_0 = e_0^2/4\pi$,

$$\alpha_0 = Z_3^{-1} Z_2^{-2} Z_1^2 \,\tilde{\mu}^{\varepsilon} \alpha\,. \tag{66.2}$$

From section 62, we have

$$Z_1 = 1 - \frac{\alpha}{2\pi}\frac{1}{\varepsilon} + O(\alpha^2)\,, \tag{66.3}$$

$$Z_2 = 1 - \frac{\alpha}{2\pi}\frac{1}{\varepsilon} + O(\alpha^2)\,, \tag{66.4}$$

$$Z_3 = 1 - \frac{2\alpha}{3\pi}\frac{1}{\varepsilon} + O(\alpha^2)\,, \tag{66.5}$$

in the $\overline{\text{MS}}$ scheme. Let us write

$$\ln\Bigl(Z_3^{-1} Z_2^{-2} Z_1^2\Bigr) = \sum_{n=1}^{\infty} \frac{E_n(\alpha)}{\varepsilon^n}\,. \tag{66.6}$$

Then we have

$$\ln \alpha_0 = \sum_{n=1}^{\infty} \frac{E_n(\alpha)}{\varepsilon^n} + \ln \alpha + \varepsilon \ln \tilde{\mu} \ . \tag{66.7}$$

From eqs. (66.3)–(66.5), we get

$$E_1(\alpha) = \frac{2\alpha}{3\pi} + O(\alpha^2) \ . \tag{66.8}$$

Then, the general analysis of section 28 yields

$$\beta(\alpha) = \alpha^2 E_1'(\alpha) \ , \tag{66.9}$$

where the prime denotes differentiation with respect to α. Thus we find

$$\beta(\alpha) = \frac{2\alpha^2}{3\pi} + O(\alpha^3) \tag{66.10}$$

in spinor electrodynamics, We can, if we like, restate this in terms of e as

$$\beta(e) = \frac{e^3}{12\pi^2} + O(e^5) \ . \tag{66.11}$$

To go from eq. (66.10) to eq. (66.11), we use $\alpha = e^2/4\pi$ and $\dot{\alpha} = e\dot{e}/2\pi$, where the dot denotes $d/d\ln\mu$.

The most important feature of either eq. (66.10) or eq. (66.11) is that the beta function is positive: the electromagnetic coupling in spinor electrodynamics gets stronger at high energies, and weaker at low energies.

It is easy to generalize eqs. (66.10) and (66.11) to the case of N Dirac fields with electric charges $Q_i e$. There is now a factor of Z_{2i} for each field, and of Z_{1i} for each interaction. These are found by replacing α in eqs. (66.3) and (66.4) with $Q_i^2\alpha$. Then we find $Z_{1i}/Z_{2i} = 1 + O(\alpha^2)$, so that this ratio is universal, at least through $O(\alpha)$. In fact, as we will see in section 67, Z_{1i}/Z_{2i} is always exactly equal to one, and so it always cancels in eq. (66.6). As for Z_3, now each Dirac field contributes separately to the fermion loop in the photon self-energy, and so we should replace α in eq. (66.5) with $\sum_i Q_i^2\alpha$. Thus we find that the generalization of eq. (66.11) is

$$\beta(e) = \frac{\sum_{i=1}^N Q_i^2}{12\pi^2}\, e^3 + O(e^5) \ . \tag{66.12}$$

Now we turn to scalar electrodynamics. (Prerequisite: 65.) The relations between the bare and renormalized couplings are

$$e_0 = Z_3^{-1/2} Z_2^{-1} Z_1 \,\tilde{\mu}^{\varepsilon/2} e \,, \tag{66.13}$$

$$e_0^2 = Z_3^{-1} Z_2^{-1} Z_4 \,\tilde{\mu}^{\varepsilon} e^2 \,, \tag{66.14}$$

$$\lambda_0 = Z_2^{-2} Z_\lambda \tilde{\mu}^{\varepsilon} \lambda \,. \tag{66.15}$$

We have two different relations between e and e_0, coming from the two types of vertices. We can guess (and will demonstrate in section 67) that these two renormalizations must work out to give the same answer. Indeed, from section 65, we have

$$Z_1 = 1 + \frac{3e^2}{8\pi^2}\frac{1}{\varepsilon} + \ldots \,, \tag{66.16}$$

$$Z_2 = 1 + \frac{3e^2}{8\pi^2}\frac{1}{\varepsilon} + \ldots \,, \tag{66.17}$$

$$Z_3 = 1 - \frac{e^2}{24\pi^2}\frac{1}{\varepsilon} + \ldots \,, \tag{66.18}$$

$$Z_4 = 1 + \frac{3e^2}{8\pi^2}\frac{1}{\varepsilon} + \ldots \,, \tag{66.19}$$

$$Z_\lambda = 1 + \left(\frac{3e^4}{2\pi^2\lambda} + \frac{5\lambda}{16\pi^2}\right)\frac{1}{\varepsilon} + \ldots \,, \tag{66.20}$$

in Lorenz gauge in the $\overline{\text{MS}}$ scheme; the ellipses stand for higher powers of e^2 and/or λ. We see that $Z_1 = Z_2 = Z_4$, at least through $O(e^2,\lambda)$. The correct guess is that this is true exactly. Thus eqs. (66.13) and (66.14) both collapse to $e_0 = Z_3^{-1/2} e$, just as in spinor electrodynamics.

Thus we can write

$$\ln\Bigl(Z_3^{-1/2}\Bigr) = \sum_{n=1}^{\infty} \frac{E_n(e,\lambda)}{\varepsilon^n} \,, \tag{66.21}$$

$$\ln\Bigl(Z_2^{-2} Z_\lambda\Bigr) = \sum_{n=1}^{\infty} \frac{L_n(e,\lambda)}{\varepsilon^n} \,. \tag{66.22}$$

Then we have

$$\ln e_0 = \sum_{n=1}^{\infty} \frac{E_n(e,\lambda)}{\varepsilon^n} + \ln e + \tfrac{1}{2}\varepsilon \ln \tilde{\mu} \,, \tag{66.23}$$

$$\ln \lambda_0 = \sum_{n=1}^{\infty} \frac{L_n(e,\lambda)}{\varepsilon^n} + \ln \lambda + \varepsilon \ln \tilde{\mu} \,. \tag{66.24}$$

Using eqs. (66.17), (66.18), and (66.20), we have

$$E_1(e,\lambda) = \frac{e^2}{48\pi^2} + \ldots \,, \tag{66.25}$$

$$L_1(e,\lambda) = \frac{1}{16\pi^2}\Big(5\lambda + 24e^4/\lambda - 12e^2\Big) + \ldots \,. \tag{66.26}$$

Now applying the general analysis of section 52 yields

$$\beta_e(e,\lambda) = \frac{e^3}{48\pi^2} + \ldots \,, \tag{66.27}$$

$$\beta_\lambda(e,\lambda) = \frac{1}{16\pi^2}\Big(5\lambda^2 - 12\lambda e^2 + 24e^4\Big) + \ldots \,. \tag{66.28}$$

Both right-hand sides are strictly positive, and so both e and λ become large at high energies, and small at low energies.

Generalizing eq. (66.25) to the case of several complex scalar fields with charges $Q_i e$ works in the same way as it does in spinor electrodynamics. For a theory with both Dirac fields and complex scalar fields, the one-loop contributions to Z_3 simply add, and so the beta function for e is

$$\beta_e(e,\lambda) = \frac{1}{12\pi^2}\Big(\textstyle\sum_\Psi Q_\Psi^2 + \frac{1}{4}\sum_\varphi Q_\varphi^2\Big)e^3 + \ldots \,. \tag{66.29}$$

Problems

66.1) Compute the one-loop contributions to the anomalous dimensions of m, Ψ, and A^μ in spinor electrodynamics in Feynman gauge.

66.2) Compute the one-loop contributions to the anomalous dimensions of m, φ, and A^μ in scalar electrodynamics in Lorenz gauge.

66.3) Use the results of problem 62.2 to compute the anomalous dimension of m and the beta function for e in spinor electrodynamics in R_ξ gauge. You should find that the results are independent of ξ.

66.4) *The value of* $\alpha(M_{\rm W})$. The solution of eq. (66.12) is

$$\frac{1}{\alpha(M_{\rm W})} = \frac{1}{\alpha(\mu)} - \frac{2}{3\pi}\sum_i Q_i^2 \ln(M_{\rm W}/\mu) \,, \tag{66.30}$$

where the sum is over all quarks and leptons (each color of quark counts separately), and we have chosen the $W^{\pm}$ boson mass $M_{\rm W}$ as a reference scale. We can define a different renormalization scheme, *modified decoupling subtraction* or $\overline{\rm DS}$, where we imagine integrating out a field when μ is below its mass. In this scheme, eq. (66.30) becomes

$$\frac{1}{\alpha(M_{\rm W})} = \frac{1}{\alpha(\mu)} - \frac{2}{3\pi}\sum_i Q_i^2 \ln[M_{\rm W}/\max(m_i,\mu)] \, , \tag{66.31}$$

where the sum is now over all quarks and leptons with mass less than $M_{\rm W}$. For $\mu < m_e$, the $\overline{\rm DS}$ scheme coincides with the OS scheme, and we have

$$\frac{1}{\alpha(M_{\rm W})} = \frac{1}{\alpha} - \frac{2}{3\pi}\sum_i Q_i^2 \ln(M_{\rm W}/m_i) \, , \tag{66.32}$$

where $\alpha = 1/137.036$ is the fine-structure constant in the OS scheme. Using $m_u = m_d = m_s \sim 300$ MeV for the light quark masses (because quarks should be replaced by hadrons at lower energies), and other quark and lepton masses from sections 83 and 88, compute $\alpha(M_{\rm W})$.

67

Ward identities in quantum electrodynamics I

Prerequisites: 22, 59

In section 59, we assumed that scattering amplitudes would be gauge invariant, in the sense that they would be unchanged if we replaced any photon polarization vector ε^μ with $\varepsilon^\mu + ck^\mu$, where k^μ is the photon's four-momentum and c is an arbitrary constant. Thus, if we write a scattering amplitude $\mathcal{T}$ for a process that includes an external photon with four-momentum k^μ as

$$\mathcal{T} = \varepsilon^\mu \mathcal{M}_\mu \; , \tag{67.1}$$

then we should have

$$k^\mu \mathcal{M}_\mu = 0 \; . \tag{67.2}$$

In this section, we will use the Ward identity for the electromagnetic current to prove eq. (67.2).

We begin by recalling the LSZ formula for scalar fields,

$$\langle f|i\rangle = i\int d^4x_1\, e^{-ik_1x_1}(-\partial_1^2 + m^2)\ldots\langle 0|\mathrm{T}\varphi(x_1)\ldots|0\rangle \; . \tag{67.3}$$

We have treated all external particles as outgoing; an incoming particle has $k_i^0 < 0$. We can rewrite eq. (67.3) as

$$\langle f|i\rangle = \lim_{k_i^2\to -m^2}(k_1^2 + m^2)\ldots\langle 0|\mathrm{T}\tilde\varphi(k_1)\ldots|0\rangle \; . \tag{67.4}$$

Here $\tilde\varphi(k) = i\int d^4x\, e^{-ikx}\varphi(x)$ is the field in momentum space (with an extra factor of i), and we do not fix $k^2 = -m^2$.

We know that the right-hand side of eq. (67.4) must include an overall energy-momentum delta function, so let us write

$$\langle 0|\mathrm{T}\tilde\varphi(k_1)\ldots|0\rangle = (2\pi)^4\delta^4(\textstyle\sum_i k_i)\mathcal{F}(k_i^2, k_i\!\cdot\! k_j) \; , \tag{67.5}$$

where $\mathcal{F}(k_i^2, k_i \cdot k_j)$ is a function of the Lorentz scalars k_i^2 and $k_i \cdot k_j$. Then, since

$$\langle f|i\rangle = i(2\pi)^4\delta^4(\textstyle\sum_i k_i)\mathcal{T} , \tag{67.6}$$

eq. (67.4) tells us that $\mathcal{F}$ should have a multivariable pole as each k_i^2 approaches $-m^2$, and that $i\mathcal{T}$ is the residue of this pole. That is, near $k_i^2 = -m^2$, $\mathcal{F}$ takes the form

$$\mathcal{F}(k_i^2, k_i \cdot k_j) = \frac{i\mathcal{T}}{(k_1^2+m^2)\ldots(k_n^2+m^2)} + \text{nonsingular} . \tag{67.7}$$

The key point is this: contributions to $\mathcal{F}$ that do *not* have this multivariable pole do *not* contribute to $\mathcal{T}$.

We have framed this discussion in terms of scalar fields in order to keep the notation as simple as possible, but the general point holds for fields of any spin.

In section 22, we analyzed how various classical field equations apply to quantum correlation functions. For example, we derived the *Schwinger–Dyson equations*

$$\begin{aligned}\langle 0|\mathrm{T}\frac{\delta S}{\delta\phi_a(x)}\phi_{a_1}(x_1)\ldots\phi_{a_n}(x_n)|0\rangle \\ = i\sum_{j=1}^{n}\langle 0|\mathrm{T}\phi_{a_1}(x_1)\ldots\delta_{aa_j}\delta^4(x{-}x_j)\ldots\phi_{a_n}(x_n)|0\rangle .\end{aligned} \tag{67.8}$$

Here we have used $\phi_a(x)$ to denote any kind of field, not necessarily a scalar field, carrying any kind of index or indices. The classical equation of motion for the field $\phi_a(x)$ is $\delta S/\delta\phi_a(x) = 0$. Thus, eq. (67.8) tells us that the classical equation of motion holds for a field inside a quantum correlation function, as long as its spacetime argument and indices do not match up exactly with those of any other field in the correlation function. These matches, which constitute the right-hand side of eq. (67.8), are called *contact terms.*

Suppose we have a correlation function that, for whatever reason, includes a contact term with a factor of, say, $\delta^4(x_1{-}x_2)$. After Fourier-transforming to momentum space, this contact term is a function of $k_1 + k_2$, but is independent of $k_1 - k_2$; hence it cannot take the form of the singular term in eq. (67.7). Therefore, *contact terms in a correlation function $\mathcal{F}$ do not contribute to the scattering amplitude $\mathcal{T}$.*

Now let us consider a scattering process in quantum electrodyamics that involves an external photon with four-momentum k. In Lorenz gauge (the simplest for this analysis), the LSZ formula reads

$$\langle f|i\rangle = i\varepsilon^{\mu}\int d^4x\, e^{-ikx}(-\partial^2)\ldots\langle 0|\mathrm{T}A_{\mu}(x)\ldots|0\rangle\ , \tag{67.9}$$

and the classical equation of motion for A^{μ} is

$$-Z_3\,\partial^2 A_{\mu} = \frac{\partial\mathcal{L}}{\partial A^{\mu}}\ . \tag{67.10}$$

In spinor electrodynamics, the right-hand side of eq. (67.10) is $Z_1 j^{\mu}$, where j^{μ} is the electromagnetic current. (This is also true for scalar electrodynamics if $Z_4 = Z_1^2/Z_2$; we saw in problem 65.2 that this condition is necessary for gauge invariance.) We therefore have

$$\langle f|i\rangle = iZ_3^{-1}Z_1\varepsilon^{\mu}\int d^4x\, e^{-ikx}\ldots\Big[\langle 0|\mathrm{T}j_{\mu}(x)\ldots|0\rangle + \text{contact terms}\Big]. \tag{67.11}$$

The contact terms arise because, as we saw in eq. (67.8), the classical equations of motion hold inside quantum correlation functions only up to contact terms. However, the contact terms cannot generate singularities in the k^2s of the other particles, and so they do not contribute to the left-hand side. (Remember that, for each of the other particles, there is still an appropriate wave operator, such as the Klein–Gordon wave operator for a scalar, acting on the correlation function. These wave operators kill any term that does not have an appropriate singularity.)

Now let us try replacing ε^{μ} in eq. (67.11) with k^{μ}. We are attempting to prove that the result is zero, and we are almost there. We can write the factor of ik^{μ} as $-\partial^{\mu}$ acting on the e^{-ikx}, and then we can integrate by parts to get ∂^{μ} acting on the correlation function. (Strictly speaking, we need a wave packet for the external photon to kill surface terms.) Then we have $\partial^{\mu}\langle 0|\mathrm{T}j_{\mu}(x)\ldots|0\rangle$ on the right-hand side. Now we use another result from section 22, namely that a Noether current for an exact symmetry obeys $\partial^{\mu}j_{\mu} = 0$ classically, and

$$\partial^{\mu}\langle 0|\mathrm{T}j_{\mu}(x)\ldots|0\rangle = \text{contact terms} \tag{67.12}$$

quantum mechanically; this is the *Ward* (or *Ward–Takahashi*) *identity.* But once again, the contact terms do not have the right singularities to contribute to $\langle f|i\rangle$. Thus we conclude that $\langle f|i\rangle$ vanishes if we replace an external photon's polarization vector ε^{μ} with its four-momentum k^{μ}, quod erat demonstrandum.

Reference notes

Diagrammatic proofs of the Ward identity in spinor electrodynamics can be found in *Peskin & Schroeder*, and *Zee.*

Problems

67.1) Show explicitly that the tree-level $\widetilde{e}^+\widetilde{e}^- \to \gamma\gamma$ scattering amplitude in scalar electrodynamics,

$$\mathcal{T} = -e^2\left[\frac{4(k_1\cdot\varepsilon_{1'})(k_2\cdot\varepsilon_{2'})}{m^2-t} + \frac{4(k_1\cdot\varepsilon_{2'})(k_2\cdot\varepsilon_{1'})}{m^2-u} + 2(\varepsilon_{1'}\cdot\varepsilon_{2'})\right],$$

vanishes if $\varepsilon^{\mu}_{1'}$ is replaced with k'^{μ}_1.

67.2) Show explicitly that the tree-level $e^+e^- \to \gamma\gamma$ scattering amplitude in spinor electrodynamics,

$$\mathcal{T} = e^2\,\overline{v}_2\left[\not\varepsilon_{2'}\left(\frac{-\not p_1 + \not k'_1 + m}{m^2-t}\right)\not\varepsilon_{1'} + \not\varepsilon_{1'}\left(\frac{-\not p_1 + \not k'_2 + m}{m^2-u}\right)\not\varepsilon_{2'}\right]u_1\,,$$

vanishes if $\varepsilon^{\mu}_{1'}$ is replaced with k'^{μ}_1.

68

Ward identities in quantum electrodynamics II

Prerequisites: 63, 67

In this section, we will show that $Z_1 = Z_2$ in spinor electrodynamics, and that $Z_1 = Z_2 = Z_4$ in scalar electrodynamics (in the OS and $\overline{\rm MS}$ renormalization schemes).

Let us specialize to the case of spinor electrodynamics with a single Dirac field, and consider the correlation function

$$C^{\mu}_{\alpha\beta}(k,p',p) \equiv iZ_1 \int d^4x\, d^4y\, d^4z\, e^{ikx-ip'y+ipz}\, \langle 0|\mathrm{T}j^{\mu}(x)\Psi_{\alpha}(y)\overline{\Psi}_{\beta}(z)|0\rangle\ , \tag{68.1}$$

where $j^{\mu} = e\overline{\Psi}\gamma^{\mu}\Psi$ is the electromagnetic current. As we saw in section 67, including $Z_1 j^{\mu}(x)$ inside a correlation function adds a vertex for an external photon; the factor of Z_1 provides the necessary renormalization of this vertex. The explicit fermion fields on the right-hand side of eq. (68.1) combine with the fermion fields in the current to generate propagators. Thus we have

$$C^{\mu}_{\alpha\beta}(k,p',p) = (2\pi)^4\delta^4(k+p-p')\Big[\tfrac{1}{i}\tilde{\mathbf{S}}(p')i\mathbf{V}^{\mu}(p',p)\tfrac{1}{i}\tilde{\mathbf{S}}(p)\Big]_{\alpha\beta}\ , \tag{68.2}$$

where $\tilde{\mathbf{S}}(p)$ is the exact fermion propagator, and $\mathbf{V}^{\mu}(p',p)$ is the exact 1PI photon-fermion-fermion vertex function.

Now let us consider $k_{\mu}C^{\mu}_{\alpha\beta}(k,p',p)$. Using eq. (68.1), we can write the factor of ik_{μ} on the right-hand side as ∂_{μ} acting on e^{ikx}, and then integrate by parts to get $-\partial_{\mu}$ acting on $j^{\mu}(x)$. (Strictly speaking, we need a wave packet for the external photon to kill surface terms.) Thus we have

$$k_{\mu}C^{\mu}_{\alpha\beta}(k,p',p) = -Z_1 \int d^4x\, d^4y\, d^4z\, e^{ikx-ip'y+ipz}\, \partial_{\mu}\langle 0|\mathrm{T}j^{\mu}(x)\Psi_{\alpha}(y)\overline{\Psi}_{\beta}(z)|0\rangle . \tag{68.3}$$

Now we use the Ward identity from section 22, which in general reads

$$-\partial_\mu \langle 0|\mathrm{T}J^\mu(x)\phi_{a_1}(x_1)\ldots\phi_{a_n}(x_n)|0\rangle$$
$$= i\sum_{j=1}^{n}\langle 0|\mathrm{T}\phi_{a_1}(x_1)\ldots\delta\phi_{a_j}(x)\delta^4(x{-}x_j)\ldots\phi_{a_n}(x_n)|0\rangle\,. \quad (68.4)$$

Here $\delta\phi_a(x)$ is the change in a field $\phi_a(x)$ under an infinitesimal transformation that leaves the action invariant, and

$$J^\mu = \frac{\partial\mathcal{L}}{\partial(\partial_\mu\phi_a)}\delta\phi_a \quad (68.5)$$

is the corresponding Noether current. In the case of spinor electrodynamics,

$$\delta\Psi(x) = -ie\Psi(x)\,,$$
$$\delta\overline{\Psi}(x) = +ie\overline{\Psi}(x)\,, \quad (68.6)$$

where we have dropped an infinitesimal parameter on the right-hand sides, but included a factor of the electron charge e. Using $\partial\mathcal{L}/\partial(\partial_\mu\Psi) = iZ_2\overline{\Psi}\gamma^\mu$ and $\partial\mathcal{L}/\partial(\partial_\mu\overline{\Psi}) = 0$ we find that, with these conventions, the Noether current is $J^\mu = Z_2 e\overline{\Psi}\gamma^\mu\Psi = Z_2 j^\mu$. Thus the Ward identity becomes

$$-Z_2\partial_\mu\langle 0|\mathrm{T}j^\mu(x)\Psi_\alpha(y)\overline{\Psi}_\beta(z)|0\rangle = +e\,\delta^4(x{-}y)\langle 0|\mathrm{T}\Psi_\alpha(y)\overline{\Psi}_\beta(z)|0\rangle$$
$$-e\,\delta^4(x{-}z)\langle 0|\mathrm{T}\Psi_\alpha(y)\overline{\Psi}_\beta(z)|0\rangle\,. \quad (68.7)$$

Recall that

$$\langle 0|\mathrm{T}\Psi_\alpha(y)\overline{\Psi}_\beta(z)|0\rangle = \frac{1}{i}\int\frac{d^4q}{(2\pi)^4}\,e^{iq(y-z)}\,\tilde{\mathbf{S}}(q)_{\alpha\beta}\,. \quad (68.8)$$

Using eqs. (68.7) and (68.8) in eq. (68.3), and carrying out the coordinate integrals, we get

$$k_\mu C^\mu_{\alpha\beta}(k,p',p) = -iZ_2^{-1}Z_1(2\pi)^4\delta^4(k{+}p{-}p')\Big[e\tilde{\mathbf{S}}(p) - e\tilde{\mathbf{S}}(p')\Big]_{\alpha\beta}\,. \quad (68.9)$$

On the other hand, from eq. (68.2) we have

$$k_\mu C^\mu_{\alpha\beta}(k,p',p) = -i(2\pi)^4\delta^4(k{+}p{-}p')\Big[\tilde{\mathbf{S}}(p')k_\mu\mathbf{V}^\mu(p',p)\tilde{\mathbf{S}}(p)\Big]_{\alpha\beta}\,. \quad (68.10)$$

Comparing eqs. (68.9) and (68.10) shows that

$$(p'{-}p)_\mu\tilde{\mathbf{S}}(p')\mathbf{V}^\mu(p',p)\tilde{\mathbf{S}}(p) = Z_2^{-1}Z_1 e\Big[\tilde{\mathbf{S}}(p) - \tilde{\mathbf{S}}(p')\Big]\,, \quad (68.11)$$

where we have dropped the spin indices. We can simplify eq. (68.11) by multiplying on the left by $\tilde{\mathbf{S}}(p')^{-1}$, and on the right by $\tilde{\mathbf{S}}(p)^{-1}$, to get

$$(p'-p)_\mu \mathbf{V}^\mu(p',p) = Z_2^{-1} Z_1 e\Big[\tilde{\mathbf{S}}(p')^{-1} - \tilde{\mathbf{S}}(p)^{-1}\Big] \,. \tag{68.12}$$

Thus we find a relation between the exact photon-fermion-fermion vertex function $\mathbf{V}^\mu(p',p)$ and the exact fermion propagator $\tilde{\mathbf{S}}(p)$.

Since both $\tilde{\mathbf{S}}(p)$ and $\mathbf{V}^\mu(p',p)$ are finite, eq. (68.12) implies that Z_1/Z_2 must be finite as well. In the $\overline{\text{MS}}$ scheme (where all corrections to $Z_i = 1$ are divergent), this immediately implies that

$$Z_1 = Z_2 \,. \tag{68.13}$$

In the OS scheme, we recall that near the on-shell point $p^2 = p'^2 = -m^2$ and $(p'-p)^2 = 0$ we have $\tilde{\mathbf{S}}(p)^{-1} = \not{p} + m$ and $\mathbf{V}^\mu(p',p) = e\gamma^\mu$. Plugging these expressions into eq. (68.12) then yields $Z_1 = Z_2$ for the OS scheme.

To better understand this result, we note that when $Z_1 = Z_2$, we can combine the fermion kinetic term $iZ_2\overline{\Psi}\not{\partial}\Psi$ and the interaction term $Z_1 e\overline{\Psi}\not{A}\Psi$ into $iZ_2\overline{\Psi}\not{D}\Psi$, where $D^\mu = \partial^\mu - ieA^\mu$ is the covariant derivative. Recall that it is D^μ that has a simple gauge transformation, and so we might expect the lagrangian, written in terms of renormalized fields, to include ∂^μ and A^μ only in the combination D^μ. It is still necessary to go through the analysis that led to eq. (68.12), however, because quantization requires fixing a gauge, and this renders suspect any naive arguments based on gauge invariance. Still, in this case, those arguments yield the correct result.

We can make a similar analysis in scalar electrodynamics. We leave the details to the problems.

Reference notes

BRST symmetry (see section 74) can be used to derive the Ward identities; see *Ramond I*.

Problems

68.1) Consider the current correlation function $\langle 0|\mathrm{T}j^\mu(x)j^\nu(y)|0\rangle$ in spinor electrodynamics.

a) Show that its Fourier transform is proportional to

$$\Pi^{\mu\nu}(k) + \Pi^{\mu\rho}(k)\tilde{\Delta}_{\rho\sigma}(k)\Pi^{\sigma\nu}(k) + \ldots \,. \tag{68.14}$$

b) Use this to prove that $\Pi^{\mu\nu}(k)$ is transverse: $k_\mu\Pi^{\mu\nu}(k) = 0$.

68.2) Verify that eq. (68.12) holds at the one-loop level in any renormalization scheme with $Z_1 = Z_2$.

68.3) *Scalar electrodynamics.* (Prerequisite: 65.)

a) Consider the Fourier transform of $\langle 0|\mathrm{T}J^\mu(x)\varphi(y)\varphi^\dagger(z)|0\rangle$, where

$$J^\mu = -ieZ_2[\varphi^\dagger\partial^\mu\varphi - (\partial^\mu\varphi^\dagger)\varphi] - 2Z_1e^2A^\mu\varphi^\dagger\varphi \tag{68.15}$$

is the Noether current. You may assume that $Z_4 = Z_1^2/Z_2$ (which is necessary for gauge invariance). Show that

$$(p'-p)_\mu \mathbf{V}_3^\mu(p',p) = Z_2^{-1}Z_1e\Big[\tilde{\mathbf{\Delta}}(p')^{-1} - \tilde{\mathbf{\Delta}}(p)^{-1}\Big]\,, \tag{68.16}$$

where $\mathbf{V}_3^\mu(p',p)$ is the exact scalar-scalar-photon vertex function, and $\tilde{\mathbf{\Delta}}(p)$ is the exact scalar propagator.

b) Use this result to show that $Z_1 = Z_2$ in both the $\overline{\text{MS}}$ and OS renormalization schemes.

c) Consider the Fourier transform of $\langle 0|\mathrm{T}J^\mu(x)A^\nu(w)\varphi(y)\varphi^\dagger(z)|0\rangle$. Show that

$$k_\mu \mathbf{V}_4^{\mu\nu}(k,p',p) = Z_1^{-1}Z_4e\Big[\mathbf{V}_3^\nu(p'-k,p) - \mathbf{V}_3^\nu(p',p+k)\Big]\,, \tag{68.17}$$

where $\mathbf{V}_4^{\mu\nu}(k,p',p)$ is the exact scalar–scalar–photon–photon vertex function, with k the incoming momentum of the photon at the μ vertex.

69

Nonabelian gauge theory

Prerequisites: 24, 58

Consider a lagrangian with N scalar or spinor fields $\phi_i(x)$ that is invariant under a continuous SU(N) or SO(N) symmetry,

$$\phi_i(x) \to U_{ij}\phi_j(x) , \tag{69.1}$$

where U_{ij} is an $N \times N$ special unitary matrix in the case of SU(N), or an $N \times N$ special orthogonal matrix in the case of SO(N). (*Special* means that the determinant of U is one.) Eq. (69.1) is called a *global* symmetry transformation, because the matrix U does not depend on the spacetime label x.

In section 58, we saw that quantum electrodynamics could be understood as having a *local* U(1) symmetry,

$$\phi(x) \to U(x)\phi(x) , \tag{69.2}$$

where $U(x) = \exp[-ie\Gamma(x)]$ can be thought of as a 1×1 unitary matrix that *does* depend on the spacetime label x. Eq. (69.2) can be a symmetry of the lagrangian only if we include a *U(1) gauge field* $A_\mu(x)$, and promote ordinary derivatives ∂_μ of $\phi(x)$ to covariant derivatives $D_\mu = \partial_\mu - ieA_\mu$. Under the transformation of eq. (69.2), we have

$$D_\mu \to U(x)D_\mu U^\dagger(x) . \tag{69.3}$$

With this transformation rule, a scalar kinetic term like $-(D_\mu\varphi)^\dagger D^\mu\varphi$, or a fermion kinetic term like $i\overline{\Psi}\not{D}\Psi$, is invariant, as are mass terms like $m^2\varphi^\dagger\varphi$ and $m\overline{\Psi}\Psi$. We call eq. (69.3) a *gauge transformation*, and say that the lagrangian is *gauge invariant*.

Eq. (69.3) implies that the gauge field transforms as

$$A_\mu(x) \to U(x)A_\mu(x)U^\dagger(x) + \tfrac{i}{e}U(x)\partial_\mu U^\dagger(x) . \tag{69.4}$$

If we use $U(x) = \exp[-ie\Gamma(x)]$, then eq. (69.4) simplifies to

$$A_\mu(x) \to A_\mu(x) - \partial_\mu \Gamma(x) \,, \tag{69.5}$$

which is what we originally had in section 54.

We can now easily generalize this construction of U(1) gauge theory to SU(N) or SO(N). (We will consider other possibilities later.) To be concrete, let us consider SU(N). Recall from section 24 that we can write an infinitesimal SU(N) transformation as

$$U_{jk}(x) = \delta_{jk} - ig\theta^a(x)(T^a)_{jk} + O(\theta^2) \,, \tag{69.6}$$

where we have inserted a coupling constant g for later convenience. The indices j and k run from 1 to N, the index a runs from 1 to N^2-1 (and is implicitly summed), and the *generator matrices* T^a are hermitian and traceless. (These properties of T^a follow immediately from the special unitarity of U.) The generator matrices obey commutation relations of the form

$$[T^a, T^b] = if^{abc}T^c \,, \tag{69.7}$$

where the real numerical factors f^{abc} are called the *structure coefficients* of the group. If f^{abc} does not vanish, the group is *nonabelian.*

We can choose the generator matrices so that they obey the normalization condition

$$\mathrm{Tr}(T^aT^b) = \tfrac{1}{2}\delta^{ab} \,; \tag{69.8}$$

then eqs. (69.7) and (69.8) can be used to show that f^{abc} is completely antisymmetric. For SU(2), we have $T^a = \frac{1}{2}\sigma^a$, where σ^a is a Pauli matrix, and $f^{abc} = \varepsilon^{abc}$, where ε^{abc} is the completely antisymmetric Levi-Civita symbol.

Now we define an SU(N) gauge field $A_\mu(x)$ as a traceless hermitian $N \times N$ matrix of fields with the gauge transformation property

$$A_\mu(x) \to U(x)A_\mu(x)U^\dagger(x) + \frac{i}{g}U(x)\partial_\mu U^\dagger(x) \,. \tag{69.9}$$

Note that this is identical to eq. (69.4), except that now $U(x)$ is a special unitary matrix (rather than a phase factor), and $A_\mu(x)$ is a traceless hermitian matrix (rather than a real number). (Also, the electromagnetic coupling e has been replaced by g.) We can write $U(x)$ in terms of the generator matrices as

$$U(x) = \exp[-ig\Gamma^a(x)T^a] \,, \tag{69.10}$$

where the real parameters $\Gamma^a(x)$ are no longer infinitesimal.

The covariant derivative is

$$D_\mu = \partial_\mu - igA_\mu(x) \,, \tag{69.11}$$

where there is an understood $N \times N$ identity matrix multiplying ∂_μ. Acting on the set of N fields $\phi_i(x)$ that transform according to eq. (69.2), the covariant derivative can be written more explicitly as

$$(D_\mu\phi)_j(x) = \partial_\mu\phi_j(x) - igA_\mu(x)_{jk}\phi_k(x) \,, \tag{69.12}$$

with an understood sum over k. The covariant derivative transforms according to eq. (69.3). Replacing all ordinary derivatives in $\mathcal{L}$ with covariant derivatives renders $\mathcal{L}$ gauge invariant (assuming, of course, that $\mathcal{L}$ originally had a global SU(N) symmetry).

We still need a kinetic term for $A_\mu(x)$. Let us define the *field strength*

$$F_{\mu\nu}(x) \equiv \tfrac{i}{g}[D_\mu, D_\nu] \tag{69.13}$$

$$= \partial_\mu A_\nu - \partial_\nu A_\mu - ig[A_\mu, A_\nu] \,. \tag{69.14}$$

Because A_μ is a matrix, the final term in eq. (69.14) does not vanish, as it does in U(1) gauge theory. Eqs. (69.3) and (69.13) imply that, under a gauge transformation,

$$F_{\mu\nu}(x) \to U(x)F_{\mu\nu}(x)U^\dagger(x) \,. \tag{69.15}$$

Therefore,

$$\mathcal{L}_{\text{kin}} = -\tfrac{1}{2}\text{Tr}(F^{\mu\nu}F_{\mu\nu}) \tag{69.16}$$

is gauge invariant, and can serve as a kinetic term for the SU(N) gauge field. (Note, however, that the field strength itself is *not* gauge invariant, in contrast to the situation in U(1) gauge theory.)

Since we have taken $A_\mu(x)$ to be hermitian and traceless, we can expand it in terms of the generator matrices:

$$A_\mu(x) = A^a_\mu(x)T^a \,. \tag{69.17}$$

Then we can use eq. (69.8) to invert eq. (69.17):

$$A^a_\mu(x) = 2\,\text{Tr}\,A_\mu(x)T^a \,. \tag{69.18}$$

Similarly, we have

$$F_{\mu\nu}(x) = F^a_{\mu\nu}T^a \,, \tag{69.19}$$

$$F^a_{\mu\nu}(x) = 2\,\text{Tr}\,F_{\mu\nu}T^a \,. \tag{69.20}$$

Using eq. (69.19) in eq. (69.14), we get

$$\begin{aligned} F^c_{\mu\nu}T^c &= (\partial_\mu A^c_\nu - \partial_\nu A^c_\mu)T^c - igA^a_\mu A^b_\nu[T^a,T^b] \\ &= (\partial_\mu A^c_\nu - \partial_\nu A^c_\mu + gf^{abc}A^a_\mu A^b_\nu)T^c \ . \end{aligned} \tag{69.21}$$

Then using eqs. (69.20) and (69.8) yields

$$F^c_{\mu\nu} = \partial_\mu A^c_\nu - \partial_\nu A^c_\mu + gf^{abc}A^a_\mu A^b_\nu \ . \tag{69.22}$$

Also, using eqs. (69.19) and (69.8) in eq. (69.16), we get

$$\mathcal{L}_{\text{kin}} = -\tfrac{1}{4}F^{c\mu\nu}F^c_{\mu\nu} \ . \tag{69.23}$$

From eq. (69.22), we see that $\mathcal{L}_{\text{kin}}$ includes interactions among the gauge fields. A theory of this type, with nonzero f^{abc}, is called *nonabelian gauge theory* or *Yang–Mills theory*.

Everything we have just said about SU(N) also goes through for SO(N), with *unitary* replaced by *orthogonal*, and *traceless* replaced by *antisymmetric*. There is also another class of compact nonabelian groups called Sp($2N$), and five exceptional compact groups: G(2), F(4), E(6), E(7), and E(8). *Compact* means that $\text{Tr}(T^aT^b)$ is a positive definite matrix. Nonabelian gauge theory must be based on a compact group, because otherwise some of the terms in $\mathcal{L}_{\text{kin}}$ would have the wrong sign, leading to a hamiltonian that is unbounded below.

As a specific example, let us consider *quantum chromodynamics*, or QCD, which is based on the gauge group SU(3). There are several Dirac fields corresponding to *quarks*. Each quark comes in three *colors*; these are the values of the SU(3) index. (These colors have nothing to do with ordinary color.) There are also six *flavors*: up, down, strange, charm, bottom (or beauty), and top (or truth). Thus we consider the Dirac field $\Psi_{iI}(x)$, where i is the color index and I is the flavor index. The lagrangian is

$$\mathcal{L} = i\overline{\Psi}_{iI}\not{D}_{ij}\Psi_{jI} - m_I\overline{\Psi}_{iI}\Psi_{iI} - \tfrac{1}{2}\text{Tr}(F^{\mu\nu}F_{\mu\nu}) \ , \tag{69.24}$$

where all indices are summed. The different quark flavors have different masses, ranging from a few MeV for the up and down quarks to 174 GeV for the top quark. (The quarks also have electric charges: $+\frac{2}{3}|e|$ for the u, c, and t quarks, and $-\frac{1}{3}|e|$ for the d, s, and b quarks. For now, however, we omit the appropriate coupling to the electromagnetic field.) The covariant derivative in eq. (69.24) is

$$(D_\mu)_{ij} = \delta_{ij}\partial_\mu - igA^a_\mu T^a_{ij} \ . \tag{69.25}$$

The index a on A^a_μ runs from 1 to 8, and the corresponding massless spin-one particles are the eight *gluons*.

In a nonabelian gauge theory in general, we can consider scalar or spinor fields in different *representations* of the group. A representation of a compact nonabelian group is a set of finite-dimensional hermitian matrices $T^a_{\rm R}$ (the R is part of the name, not an index) that obey the same commutation relations as the original generator matrices T^a. Given such a set of $D(\mathrm{R}) \times D(\mathrm{R})$ matrices (where $D(\mathrm{R})$ is the *dimension* of the representation), and a field $\phi(x)$ with $D(\mathrm{R})$ components, we can write its covariant derivative as $D_\mu = \partial_\mu - igA^a_\mu T^a_{\rm R}$, with an understood $D(\mathrm{R}) \times D(\mathrm{R})$ identity matrix multiplying ∂_μ. Under a gauge transformation, $\phi(x) \to U_{\rm R}(x)\phi(x)$, where $U_{\rm R}(x)$ is given by eq. (69.10) with T^a replaced by $T^a_{\rm R}$. The theory will be gauge invariant provided that the transformation rule for A^a_μ is indepedent of the representation used in eq. (69.9); we show in problem 69.1 that it is.

We will not need to know a lot of representation theory, but we collect some useful facts in the next section.

Problems

69.1) Show that eq. (69.9) implies a transformation rule for A^a_μ that is independent of the representation used in eq. (69.9). Hint: consider an infinitesimal transformation.

69.2) Show that $[T^a T^a, T^b] = 0$.

70

Group representations

Prerequisite: 69

Given the structure coefficients f^{abc} of a compact nonabelian group, a *representation* of that group is specified by a set of $D(\mathrm{R}) \times D(\mathrm{R})$ traceless hermitian matrices T^a_R (the R is part of the name, not an index) that obey the same commutation relations as the original generators matrices T^a, namely

$$[T^a_\mathrm{R}, T^b_\mathrm{R}] = i f^{abc} T^c_\mathrm{R} . \tag{70.1}$$

The number $D(\mathrm{R})$ is the *dimension* of the representation. The original T^as correspond to the *fundamental* or *defining* representation.

Consider taking the complex conjugate of the commutation relations, eq. (70.1). Since the structure coefficients are real, we see that the matrices $-(T^a_\mathrm{R})^*$ also obey these commutation relations. If $-(T^a_\mathrm{R})^* = T^a_\mathrm{R}$, or if we can find a unitary transformation $T^a_\mathrm{R} \to U^{-1} T^a_\mathrm{R} U$ that makes $-(T^a_\mathrm{R})^* = T^a_\mathrm{R}$ for every a, then the representation R is *real.* If such a unitary transformation does not exist, but we can find a unitary matrix $V \neq I$ such that $-(T^a_\mathrm{R})^* = V^{-1} T^a_\mathrm{R} V$ for every a, then the representation R is *pseudoreal.* If such a unitary matrix also does not exist, then the representation R is *complex.* In this case, the *complex conjugate representation* $\overline{\mathrm{R}}$ is specified by

$$T^a_{\overline{\mathrm{R}}} = -(T^a_\mathrm{R})^* . \tag{70.2}$$

One way to prove that a representation is complex is to show that at least one generator matrix T^a_R (or a real linear combination of them) has eigenvalues that do not come in plus-minus pairs. This is the case for the fundamental representation of SU(N) with $N \geq 3$. For SU(2), the fundamental representation is pseudoreal, because $-(\frac{1}{2}\sigma^a)^* \neq \frac{1}{2}\sigma^a$, but $-(\frac{1}{2}\sigma^a)^* = V^{-1}(\frac{1}{2}\sigma^a)V$ with $V = \sigma_2$. For SO(N), the fundamental representation is real, because the generator matrices are antisymmetric, and every antisymmetric hermitian matrix is equal to minus its complex conjugate.

An important representation for any compact nonabelian group is the *adjoint representation* A. This is given by

$$(T^a_{\rm A})^{bc} = -if^{abc}\,. \tag{70.3}$$

Because f^{abc} is real and completely antisymmetric, $T^a_{\rm A}$ is manifestly hermitian, and also satisfies $T^a_{\rm A} = -(T^a_{\rm A})^*$; thus the adjoint representation is real. The dimension of the adjoint representation $D({\rm A})$ is equal to the number of generators of the group; this number is also called the dimension of the group.

To see that the $T^a_{\rm A}$s satisfy the commutation relations, we use the *Jacobi identity*

$$f^{abd}f^{dce} + f^{bcd}f^{dae} + f^{cad}f^{dbe} = 0\,, \tag{70.4}$$

which holds for the structure coefficients of any group. To prove the Jacobi identity, we note that

$$\operatorname{Tr} T^e\Big([[T^a,T^b],T^c] + [[T^b,T^c],T^a] + [[T^c,T^a],T^b]\Big) = 0\,, \tag{70.5}$$

where the T^as are the original generator matrices. That the left-hand side of eq. (70.5) vanishes can be seen by writing out all the commutators as matrix products, and noting that they cancel in pairs. Employing the commutation relations twice in each term, followed by

$$\operatorname{Tr}(T^aT^b) = \tfrac{1}{2}\delta^{ab}\,, \tag{70.6}$$

ultimately yields eq. (70.4). Then, using the antisymmetry of the structure coefficients, inserting some judicious factors of i, and moving the last term of eq. (70.4) to the right-hand side, we can rewrite it as

$$(-if^{abd})(-if^{cde}) - (-if^{cbd})(-if^{ade}) = if^{acd}(-if^{dbe})\,. \tag{70.7}$$

Now we use eq. (70.3) in eq. (70.7) to get

$$(T^a_{\rm A})^{bd}(T^c_{\rm A})^{de} - (T^c_{\rm A})^{bd}(T^a_{\rm A})^{de} = if^{acd}(T^d_{\rm A})^{be}\,, \tag{70.8}$$

or equivalently $[T^a_{\rm A},T^c_{\rm A}] = if^{acd}T^d_{\rm A}$. Thus the $T^a_{\rm A}$s satisfy the appropriate commutation relations.

Two related numbers usefully characterize a representation: the *index* $T({\rm R})$ and the *quadratic Casimir* $C({\rm R})$. The index is defined via

$$\operatorname{Tr}(T^a_{\rm R}T^b_{\rm R}) = T({\rm R})\delta^{ab}\,. \tag{70.9}$$

Next we recall from problem 69.2 that the matrix $T^a_{\rm R}T^a_{\rm R}$ commutes with every generator, and so must be a number times the identity matrix; this number is the quadratic Casimir $C(\rm R)$. It is easy to show that

$$T({\rm R})D({\rm A}) = C({\rm R})D({\rm R}) . \tag{70.10}$$

With the standard normalization conventions for the generators, we have $T({\rm N}) = \frac{1}{2}$ for the fundamental representation of SU(N) and $T({\rm N}) = 2$ for the fundamental representation of SO(N). We show in problem 70.2 that $T({\rm A}) = N$ for the adjoint representation of SU(N), and in problem 70.3 that $T({\rm A}) = 2N - 4$ for the adjoint representation of SO(N).

A representation R is *reducible* if there is a unitary transformation $T^a_{\rm R} \to U^{-1}T^a_{\rm R}U$ that puts all the nonzero entries into the same diagonal blocks for each a; otherwise it is *irreducible.* Consider a reducible representation R whose generators can be put into (for example) two blocks, with the blocks forming the generators of the irreducible representations $\rm R_1$ and $\rm R_2$. Then R is the *direct sum* representation ${\rm R} = {\rm R_1} \oplus {\rm R_2}$, and we have

$$D({\rm R_1}{\oplus}{\rm R_2}) = D({\rm R_1}) + D({\rm R_2}) , \tag{70.11}$$

$$T({\rm R_1}{\oplus}{\rm R_2}) = T({\rm R_1}) + T({\rm R_2}) . \tag{70.12}$$

Suppose we have a field $\varphi_{iI}(x)$ that carries two group indices, one for the representation $\rm R_1$ and one for the representation $\rm R_2$, denoted by i and I respectively. This field is in the *direct product* representation ${\rm R_1} \otimes {\rm R_2}$. The corresponding generator matrix is

$$(T^a_{{\rm R_1}\otimes{\rm R_2}})_{iI,jJ} = (T^a_{\rm R_1})_{ij}\delta_{IJ} + \delta_{ij}(T^a_{\rm R_2})_{IJ} , \tag{70.13}$$

where i and I together constitute the row index, and j and J together constitute the column index. We then have

$$D({\rm R_1}{\otimes}{\rm R_2}) = D({\rm R_1})D({\rm R_2}) , \tag{70.14}$$

$$T({\rm R_1}{\otimes}{\rm R_2}) = T({\rm R_1})D({\rm R_2}) + D({\rm R_1})T({\rm R_2}) . \tag{70.15}$$

To get eq. (70.15), we need to use the fact that the generator matrices are traceless, $(T^a_{\rm R})_{ii} = 0$.

At this point it is helpful to introduce a slightly more refined notation for the indices of a complex representation. Consider a field φ in the complex representation R. We will adopt the convention that such a field carries a "down" index: φ_i, where $i = 1, 2, \ldots, D({\rm R})$. Hermitian conjugation changes the representation from R to $\overline{\rm R}$, and we will adopt the convention that this also raises the index on the field,

$$(\varphi_i)^\dagger = \varphi^{\dagger i} \,. \tag{70.16}$$

Thus a down index corresponds to the representation R, and an up index to $\overline{\rm R}$. Indices can be contracted only if one is up and one is down. Generator matrices for R are then written with the first index down and the second index up: $(T^a_{\rm R})_i{}^j$. An infinitesimal group transformation of φ_i takes the form

$$\begin{aligned}\varphi_i &\to (1 - i\theta^a T^a_{\rm R})_i{}^j \varphi_j \\ &= \varphi_i - i\theta^a (T^a_{\rm R})_i{}^j \varphi_j \,.\end{aligned} \tag{70.17}$$

The generator matrices for $\overline{\rm R}$ are then given by

$$(T^a_{\overline{\rm R}})^i{}_j = -(T^a_{\rm R})_j{}^i \,, \tag{70.18}$$

where we have used the hermiticity to trade complex conjugation for transposition of the indices. An infinitesimal group transformation of $\varphi^{\dagger i}$ takes the form

$$\begin{aligned}\varphi^{\dagger i} &\to (1 - i\theta^a T^a_{\overline{\rm R}})^i{}_j \varphi^{\dagger j} \\ &= \varphi^{\dagger i} - i\theta^a (T^a_{\overline{\rm R}})^i{}_j \varphi^{\dagger j} \\ &= \varphi^{\dagger i} + i\theta^a (T^a_{\rm R})_j{}^i \varphi^{\dagger j} \,,\end{aligned} \tag{70.19}$$

where we used eq. (70.18) to get the last line. Note that eqs. (70.17) and (70.19) together imply that $\varphi^{\dagger i}\varphi_i$ is invariant, as expected.

Consider the Kronecker delta symbol with one index down and one up: $\delta_i{}^j$. Under a group transformation, we have

$$\begin{aligned}\delta_i{}^j &\to (1 - i\theta^a T^a_{\rm R})_i{}^k (1 - i\theta^a T^a_{\overline{\rm R}})^j{}_l \delta_k{}^l \\ &= (1 - i\theta^a T^a_{\rm R})_i{}^k \delta_k{}^l (1 + i\theta^a T^a_{\rm R})_l{}^j \\ &= \delta_i{}^j + O(\theta^2) \,.\end{aligned} \tag{70.20}$$

Eq. (70.20) shows that $\delta_i{}^j$ is an *invariant symbol* of the group. The existence of this invariant symbol, which carries one index for R and one for $\overline{\rm R}$, tells us that the product of the representations R and $\overline{\rm R}$ must contain the *singlet* representation 1, specified by $T^a_1 = 0$. We therefore can write

$${\rm R} \otimes \overline{\rm R} = 1 \oplus \ldots \,. \tag{70.21}$$

The generator matrix $(T^a_{\rm R})_i{}^j$, which carries one index for R, one for $\overline{\rm R}$, and one for the adjoint representation A, is also an invariant symbol. To see this, we make a simultaneous infinitesimal group transformation on each of these indices,

$$
\begin{aligned}
(T^b_{\rm R})_i{}^j \to {} & (1 - i\theta^a T^a_{\rm R})_i{}^k (1 - i\theta^a T^a_{\overline{\rm R}})^j{}_l (1 - i\theta^a T^a_{\rm A})^{bc} (T^c_{\rm R})_k{}^l \\
= {} & (T^b_{\rm R})_i{}^j - i\theta^a [(T^a_{\rm R})_i{}^k (T^b_{\rm R})_k{}^j + (T^a_{\overline{\rm R}})^j{}_l (T^b_{\rm R})_i{}^l + (T^a_{\rm A})^{bc} (T^c_{\rm R})_i{}^j] \\
& + O(\theta^2) \,. \qquad (70.22)
\end{aligned}
$$

The factor in square brackets should vanish if (as we claim) the generator matrix is an invariant symbol. Using eqs. (70.3) and (70.18), we have

$$
\begin{aligned}
[\ldots] &= (T^a_{\rm R})_i{}^k (T^b_{\rm R})_k{}^j - (T^a_{\rm R})_l{}^j (T^b_{\rm R})_i{}^l - i f^{abc} (T^c_{\rm R})_i{}^j \\
&= (T^a_{\rm R} T^b_{\rm R})_i{}^j - (T^b_{\rm R} T^a_{\rm R})_i{}^j - i f^{abc} (T^c_{\rm R})_i{}^j \\
&= 0 \,, \qquad (70.23)
\end{aligned}
$$

where the last line follows from the commutation relations. The fact that $(T^a_{\rm R})_i{}^j$ is an invariant symbol implies that

$$
{\rm R} \otimes \overline{\rm R} \otimes {\rm A} = 1 \oplus \ldots \,. \qquad (70.24)
$$

If we now multiply both sides of eq. (70.24) by A, and use $\rm A \otimes A = 1 \oplus \ldots$ [which follows from eq. (70.21) and the reality of A], we find $\rm R \otimes \overline{R} = A \oplus \ldots$. Combining this with eq. (70.21), we have

$$
{\rm R} \otimes \overline{\rm R} = 1 \oplus {\rm A} \oplus \ldots \,. \qquad (70.25)
$$

That is, the product of a representation with its complex conjugate is always reducible into a sum that includes (at least) the singlet and adjoint representations.

For the fundamental representation N of SU(N), we have

$$
{\rm N} \otimes \overline{\rm N} = 1 \oplus {\rm A} \,, \qquad (70.26)
$$

with no other representations on the right-hand side. To see this, recall that $D(1) = 1$, $D({\rm N}) = D(\overline{\rm N}) = N$, and, as shown in section 24, $D({\rm A}) = N^2 - 1$. From eq. (70.14), we see that there is no room for anything else on the right-hand side of eq. (70.26).

Consider now a real representation R. From eq. (70.25), with $\overline{\rm R} = {\rm R}$, we have

$$
{\rm R} \otimes {\rm R} = 1 \oplus {\rm A} \oplus \ldots \,. \qquad (70.27)
$$

The singlet on the right-hand side implies the existence of an invariant symbol with two R indices; this symbol is the Kronecker delta δ_{ij}. It is invariant because

$$\begin{aligned}\delta_{ij} &\to (1 - i\theta^a T^a_{\rm R})_i{}^k (1 - i\theta^a T^a_{\rm R})_j{}^l \delta_{kl} \\ &= \delta_{ij} - i\theta^a [(T^a_{\rm R})_{ij} + (T^a_{\rm R})_{ji}] + O(\theta^2)\ . \end{aligned} \qquad (70.28)$$

The term in square brackets vanishes by hermiticity and eq. (70.18). The fact that $\delta_{ij} = \delta_{ji}$ implies that the singlet on the right-hand side of eq. (70.28) appears in the symmetric part of this product of two identical representations.

The fundamental representation N of SO(N) is real, and we have

$$\mathrm{N} \otimes \mathrm{N} = 1_{\rm S} \oplus \mathrm{A}_{\rm A} \oplus \mathrm{S}_{\rm S}\ . \qquad (70.29)$$

The subscripts tell whether the representation appears in the symmetric or antisymmetric part of the product. The representation S corresponds to a field with a symmetric traceless pair of fundamental indices: $\varphi_{ij} = \varphi_{ji}$, $\varphi_{ii} = 0$, where the repeated index is summed. We have $D(1) = 1$, $D(\mathrm{N}) = N$, and, as shown in section 24, $D(\mathrm{A}) = \frac{1}{2}N(N{-}1)$. Also, a traceless symmetric tensor has $D(\mathrm{S}) = \frac{1}{2}N(N{+}1){-}1$ independent components; thus eq. (70.14) is fulfilled.

Consider now a pseudoreal representation R. Since R is equivalent to its complex conjugate, up to a change of basis, eq. (70.27) still holds. However, we cannot identify δ_{ij} as the corresponding invariant symbol, because then eq. (70.28) shows that R would have to be real, rather than pseudoreal. From the perspective of the direct product, the only alternative is to have the singlet appear in the antisymmetric part of the product, rather than the symmetric part. The corresponding invariant symbol must then be antisymmetric on exchange of its two R indices.

An example (the only one that will be of interest to us) is the fundamental representation of SU(2). For SU(N) in general, another invariant symbol is the Levi-Civita tensor $\varepsilon_{i_1 \ldots i_N}$, which carries N fundamental indices and is completely antisymmetric. It is invariant because, under an SU(N) transformation,

$$\begin{aligned}\varepsilon_{i_1 \ldots i_N} &\to U_{i_1}{}^{j_1} \ldots U_{i_N}{}^{j_N} \varepsilon_{j_1 \ldots j_N} \\ &= (\det U) \varepsilon_{i_1 \ldots i_N}\ . \end{aligned} \qquad (70.30)$$

Since $\det U = 1$ for SU(N), we see that the Levi-Civita symbol is invariant. We can similarly consider $\varepsilon^{i_1 \ldots i_N}$, which carries N completely antisymmetric antifundamental indices. For SU(2), the Levi-Civita symbol is $\varepsilon_{ij} = -\varepsilon_{ji}$; this is the two-index invariant symbol that corresponds to the singlet in the product

$$2 \otimes 2 = 1_{\rm A} \oplus 3_{\rm S} \,, \tag{70.31}$$

where 3 is the adjoint representation.

We can use ε^{ij} and ε_{ij} to raise and lower SU(2) indices. This is another way to see that there is no distinction between the fundamental representation 2 and its complex conjugate $\overline{2}$. That is, if we have a field φ_i in the representation 2, we can get a field in the representation $\overline{2}$ by raising the index: $\varphi^i = \varepsilon^{ij}\varphi_j$.

The structure constants f^{abc} are another invariant symbol. This follows from $(T^a_{\rm A})^{bc} = -if^{abc}$, since we have seen that generator matrices (in any representation) are invariant. Alternatively, given the generator matrices in a representation R, we can write

$$T({\rm R})f^{abc} = -i\,{\rm Tr}(T^a_{\rm R}[T^b_{\rm R}, T^c_{\rm R}]) \,. \tag{70.32}$$

Since the right-hand side is invariant, the left-hand side must be as well.

If we use an anticommutator in place of the commutator in eq. (70.32), we get another invariant symbol,

$$A({\rm R})d^{abc} \equiv \tfrac{1}{2}{\rm Tr}(T^a_{\rm R}\{T^b_{\rm R}, T^c_{\rm R}\}) \,, \tag{70.33}$$

where $A({\rm R})$ is the *anomaly coefficient* of the representation. The cyclic property of the trace implies that $A({\rm R})d^{abc}$ is symmetric on exchange of any pair of indices. Using eq. (70.18), we can see that

$$A(\overline{\rm R}) = -A({\rm R}) \,. \tag{70.34}$$

Thus, if R is real or pseudoreal, $A({\rm R}) = 0$. We also have

$$A({\rm R}_1 \oplus {\rm R}_2) = A({\rm R}_1) + A({\rm R}_2) \,, \tag{70.35}$$

$$A({\rm R}_1 \otimes {\rm R}_2) = A({\rm R}_1)D({\rm R}_2) + D({\rm R}_1)A({\rm R}_2) \,. \tag{70.36}$$

We normalize the anomaly coefficient so that it equals one for the smallest complex representation. In particular, for SU(N) with $N \geq 3$, the smallest complex representation is the fundamental, and $A({\rm N}) = 1$. For SU(2), all representations are real or pseudoreal, and $A({\rm R}) = 0$ for all of them.

Reference notes

More group and representation theory can be found in *Ramond II*.

Problems

70.1) Verify eq. (70.10).

70.2) a) Use eqs. (70.12) and (70.26) to compute $T(\mathrm{A})$ for SU(N).

b) For SU(2), the adjoint representation is specified by $(T^a_{\mathrm{A}})^{bc} = -i\varepsilon^{abc}$. Use this to compute $T(\mathrm{A})$ explicitly for SU(2). Does your result agree with part (a)?

c) Consider the SU(2) *subgroup* of SU(N) that acts on the first two components of the fundamental representation of SU(N). Under this SU(2) subgroup, the N of SU(N) transforms as $2 \oplus (N-2)1$s. Using eq. (70.26), figure out how the adjoint representation of SU(N) transforms under this SU(2) subgroup.

d) Use your results from parts (b) and (c) to compute $T(\mathrm{A})$ for SU(N). Does your result agree with part (a)?

70.3) a) Consider the SO(3) subgroup of SO(N) that acts on the first three components of the fundamental representation of SO(N). Under this SO(3) subgroup, the N of SO(N) transforms as $3 \oplus (N-3)1$s. Using eq. (70.29), work out how the adjoint representation of SO(N) transforms under this SO(3) subgroup.

b) Use your results from part (a) and from problem 70.2 to compute $T(\mathrm{A})$ for SO(N).

70.4) a) For SU(N), we have

$$\mathrm{N} \otimes \mathrm{N} = \mathcal{A}_{\mathrm{A}} \oplus \mathcal{S}_{\mathrm{S}} \,, \tag{70.37}$$

where $\mathcal{A}$ corresponds to a field with two antisymmetric fundamental SU(N) indices, $\varphi_{ij} = -\varphi_{ji}$, and $\mathcal{S}$ corresponds to a field with two symmetric fundamental SU(N) indices, $\varphi_{ij} = +\varphi_{ji}$. Compute $D(\mathcal{A})$ and $D(\mathcal{S})$.

b) By considering an SU(2) subgroup of SU(N), compute $T(\mathcal{A})$ and $T(\mathcal{S})$.

c) For SU(3), show that $\mathcal{A} = \overline{3}$.

d) By considering an SU(3) subgroup of SU(N), compute $A(\mathcal{A})$ and $A(\mathcal{S})$.

70.5) Consider a field φ_i in the representation R_1 and a field χ_I in the representation R_2. Their product $\varphi_i\chi_I$ is then in the direct product representation $\mathrm{R}_1 \otimes \mathrm{R}_2$, with generator matrices given by eq. (70.13).

a) Prove the distribution rule for the covariant derivative,

$$[D_\mu(\varphi\chi)]_{iI} = (D_\mu\varphi)_i\chi_I + \varphi_i(D_\mu\chi)_I \,. \tag{70.38}$$

b) Consider a field φ_i in the complex representation R. Show that

$$\partial_\mu(\varphi^{\dagger i}\varphi_i) = (D_\mu\varphi^\dagger)^i\varphi_i + \varphi^{\dagger i}(D_\mu\varphi)_i \,. \tag{70.39}$$

Explain why this is a special case of eq. (70.38).

70.6) The field strength in Yang–Mills theory is in the adjoint representation, and so its covariant derivative is

$$(D_\rho F_{\mu\nu})^a = \partial_\rho F^a_{\mu\nu} - igA^c_\rho (T^c_{\rm A})^{ab} F^b_{\mu\nu} \,. \tag{70.40}$$

Prove the *Bianchi identity*,

$$(D_\mu F_{\nu\rho})^a + (D_\nu F_{\rho\mu})^a + (D_\rho F_{\mu\nu})^a = 0 \,. \tag{70.41}$$

71

The path integral for nonabelian gauge theory

Prerequisites: 53, 69

We wish to evaluate the path integral for nonabelian gauge theory (also known as Yang–Mills theory),

$$Z(J) \propto \int \mathcal{D}A \; e^{iS_{\text{YM}}(A,J)} \;, \tag{71.1}$$

$$S_{\text{YM}}(A,J) = \int d^4x \left[-\tfrac{1}{4} F^{a\mu\nu} F^a_{\mu\nu} + J^{a\mu} A^a_\mu \right] . \tag{71.2}$$

In section 57, we evaluated the path integral for U(1) gauge theory by arguing that, in momentum space, the component of the U(1) gauge field parallel to the four-momentum k^μ did not appear in the action, and hence should not be integrated over. This argument relied on the form of the U(1) gauge transformation,

$$A_\mu(x) \to A_\mu(x) - \partial_\mu \Gamma(x) \;. \tag{71.3}$$

In the nonabelian case, however, the gauge transformation is nonlinear,

$$A_\mu(x) \to U(x) A_\mu(x) U^\dagger(x) + \tfrac{i}{g} U(x) \partial_\mu U^\dagger(x) \;, \tag{71.4}$$

where $A_\mu(x) = A^a_\mu(x) T^a$. For an infinitesimal transformation,

$$\begin{aligned} U(x) &= I - ig\theta(x) + O(\theta^2) \\ &= I - ig\theta^a(x) T^a + O(\theta^2) \;, \end{aligned} \tag{71.5}$$

we have

$$A_\mu(x) \to A_\mu(x) + ig[A_\mu(x), \theta(x)] - \partial_\mu \theta(x) \;, \tag{71.6}$$

or equivalently

$$\begin{aligned}
A^a_\mu(x) &\to A^a_\mu(x) - gf^{abc}A^b_\mu(x)\theta^c(x) - \partial_\mu\theta^a(x) \\
&= A^a_\mu(x) - [\delta^{ac}\partial_\mu + gf^{abc}A^b_\mu(x)]\theta^c(x) \\
&= A^a_\mu(x) - [\delta^{ac}\partial_\mu - igA^b_\mu(-if^{bac})]\theta^c(x) \\
&= A^a_\mu(x) - [\delta^{ac}\partial_\mu - igA^b_\mu(T^b_{\rm A})^{ac}]\theta^c(x) \\
&= A^a_\mu(x) - D^{ac}_\mu\theta^c(x) \,, \qquad (71.7)
\end{aligned}$$

where D^{ac}_μ is the covariant derivative in the adjoint representation. We see the similarity with the abelian case, eq. (71.3). However, the fact that it is D_μ that appears in eq. (71.7), rather than ∂_μ, means that we cannot account for gauge redundancy in the path integral by simply excluding the components of A^a_μ that are parallel to k_μ. We will have to do something more clever.

Consider an ordinary integral of the form

$$Z \propto \int dx\, dy\, e^{iS(x)} \,, \qquad (71.8)$$

where both x and y are integrated from minus to plus infinity. Because y does not appear in $S(x)$, the integral over y is redundant. We can then define Z by simply dropping the integral over y,

$$Z \equiv \int dx\, e^{iS(x)} \,. \qquad (71.9)$$

This is how we dealt with gauge redundancy in the abelian case.

We could get the same answer by inserting a delta function, rather than by dropping the y integral:

$$Z = \int dx\, dy\, \delta(y)\, e^{iS(x)} \,. \qquad (71.10)$$

Furthermore, the argument of the delta function can be shifted by an arbitrary function of x, without changing the result:

$$Z = \int dx\, dy\, \delta(y - f(x))\, e^{iS(x)} \,. \qquad (71.11)$$

Suppose we are not given $f(x)$ explicitly, but rather are told that $y = f(x)$ is the unique solution, for fixed x, of $G(x,y) = 0$. Then we can write

$$\delta(G(x,y)) = \frac{\delta(y - f(x))}{|\partial G/\partial y|} \,, \qquad (71.12)$$

where we have used a standard rule for delta functions. We can drop the absolute-value signs if we assume that $\partial G/\partial y$ is positive when evaluated at $y = f(x)$. Then we have

$$Z = \int dx\, dy\, \frac{\partial G}{\partial y}\, \delta(G)\, e^{iS}\,. \tag{71.13}$$

Now let us generalize this result to an integral over $d^n x\, d^n y$. We will need n functions $G_i(x, y)$ to fix all n components of y. The generalization of eq. (71.13) is

$$Z = \int d^n x\, d^n y\, \det\left(\frac{\partial G_i}{\partial y_j}\right) \textstyle\prod_i \delta(G_i)\, e^{iS}\,. \tag{71.14}$$

Now we are ready to translate these results to path integrals over non-abelian gauge fields. The role of the redundant integration variable y is played by the set of all gauge transformations $\theta^a(x)$. The role of the integration variables x and y together is played by the gauge field $A^a_\mu(x)$. The role of G is played by a *gauge-fixing function.* We will use the gauge-fixing function appropriate for R_ξ gauge, which is

$$G^a(x) \equiv \partial^\mu A^a_\mu(x) - \omega^a(x)\,, \tag{71.15}$$

where $\omega^a(x)$ is a fixed, arbitrarily chosen function of x. (We will see how the parameter ξ enters later.) In eq. (71.15), the spacetime argument x and the index a play the role of the index i in eq. (71.14). Our path integral becomes

$$Z(J) \propto \int \mathcal{D}A\, \det\left(\frac{\delta G}{\delta \theta}\right) \textstyle\prod_{x,a} \delta(G)\, e^{iS_{\text{YM}}}\,, \tag{71.16}$$

where S_{YM} is given by eq. (71.2).

Now we have to evaluate the functional derivative $\delta G^a(x)/\delta\theta^b(y)$, and then its functional determinant. From eqs. (71.7) and (71.15), we find that, under an infinitesimal gauge transformation,

$$G^a(x) \to G^a(x) - \partial^\mu D^{ab}_\mu \theta^b(x)\,. \tag{71.17}$$

Thus we have

$$\frac{\delta G^a(x)}{\delta \theta^b(y)} = -\partial^\mu D^{ab}_\mu \delta^4(x-y)\,, \tag{71.18}$$

where the derivatives are with respect to x.

Now we need to compute the functional determinant of eq. (71.18). Luckily, we learned how to do this in section 53. A functional determinant can be written as a path integral over complex Grassmann variables. So let us introduce the complex Grassmann field $c^a(x)$, and its hermitian conjugate

$\bar{c}^a(x)$. (We use a bar rather than a dagger to keep the notation a little less cluttered.) These fields are called *Faddeev–Popov ghosts*. Then we can write

$$\det \frac{\delta G^a(x)}{\delta \theta^b(y)} \propto \int \mathcal{D}c\, \mathcal{D}\bar{c}\; e^{iS_{\rm gh}} \,, \qquad (71.19)$$

where the ghost action is $S_{\rm gh} = \int d^4x\, \mathcal{L}_{\rm gh}$, and the ghost lagrangian is

$$\begin{aligned}
\mathcal{L}_{\rm gh} &= \bar{c}^a \partial^\mu D^{ab}_\mu c^b \\
&= -\partial^\mu \bar{c}^a D^{ab}_\mu c^b \\
&= -\partial^\mu \bar{c}^a \partial_\mu c^a + ig \partial^\mu \bar{c}^a A^c_\mu (T^c_{\rm A})^{ab} c^b \\
&= -\partial^\mu \bar{c}^a \partial_\mu c^a + g f^{abc} A^c_\mu \partial^\mu \bar{c}^a c^b \,. \qquad (71.20)
\end{aligned}$$

We dropped a total divergence in the second line. We see that $c^a(x)$ has the standard kinetic term for a complex scalar field. (We need the factor of i in front of $S_{\rm gh}$ in eq. (71.19) for this to work out; this factor affects only the overall phase of $Z(J)$, and so we can choose it at will.) The ghost field is also a Grassmann field, and so a closed loop of ghost lines in a Feynman diagram carries an extra factor of minus one. We see from eq. (71.20) that the ghost field interacts with the gauge field, and so we will have such loops.

Since the particles associated with the ghost field do not in fact exist (and would violate the spin-statistics theorem if they did), it must be that the amplitude to produce them in any scattering process is zero. This is indeed the case, as we will discuss in section 74.

We note that in abelian gauge theory, where $f^{abc} = 0$, there is no interaction term for the ghost field. In that case, it is simply an extra free field, and we can absorb its path integral into the overall normalization.

We have one final trick to perform. Our gauge-fixing function, $G^a(x)$, contains an arbitrary function $\omega^a(x)$. The path integral $Z(J)$ is, however, independent of $\omega^a(x)$. So, we can multiply $Z(J)$ by an arbitrary functional of ω, and then perform a path integral over ω; the result can change only the overall normalization of $Z(J)$. In particular, let us multiply $Z(J)$ by

$$\exp\left[-\frac{i}{2\xi} \int d^4x\, \omega^a \omega^a \right] .$$

Because of the delta-functional in eq. (71.16), it is easy to integrate over ω. The final result for $Z(J)$ is

$$Z(J) \propto \int \mathcal{D}A\, \mathcal{D}\bar{c}\, \mathcal{D}c\; \exp\Bigl(iS_{\rm YM} + iS_{\rm gh} + iS_{\rm gf} \Bigr) \,, \qquad (71.21)$$

where $S_{\rm YM}$ is given by eq. (71.2), $S_{\rm gh}$ is given by the integral over d^4x of eq. (71.20), and $S_{\rm gf}$ (gf stands for *gauge fixing*) is given by the integral over d^4x of

$$\mathcal{L}_{\rm gf} = -\tfrac{1}{2}\xi^{-1}\partial^\mu A^a_\mu \partial^\nu A^a_\nu \,. \tag{71.22}$$

In the next section, we will derive the Feynman rules that follow from this path integral.

72

The Feynman rules for nonabelian gauge theory

Prerequisite: 71

Let us begin by considering nonabelian gauge theory without any scalar or spinor fields. The lagrangian is

$$\begin{aligned}\mathcal{L}_{\text{YM}} &= -\tfrac{1}{4}F^{e\mu\nu}F^{e}_{\mu\nu} \\ &= -\tfrac{1}{4}(\partial^\mu A^{e\nu} - \partial^\nu A^{e\mu} + gf^{abe}A^{a\mu}A^{b\nu})(\partial_\mu A^e_\nu - \partial_\nu A^e_\mu + gf^{cde}A^c_\mu A^d_\nu) \\ &= -\tfrac{1}{2}\partial^\mu A^{e\nu}\partial_\mu A^e_\nu + \tfrac{1}{2}\partial^\mu A^{e\nu}\partial_\nu A^e_\mu \\ &\quad - gf^{abe}A^{a\mu}A^{b\nu}\partial_\mu A^e_\nu - \tfrac{1}{4}g^2 f^{abe}f^{cde}A^{a\mu}A^{b\nu}A^c_\mu A^d_\nu \,. \end{aligned} \tag{72.1}$$

To this we should add the gauge-fixing term for R_ξ gauge,

$$\mathcal{L}_{\text{gf}} = -\tfrac{1}{2}\xi^{-1}\partial^\mu A^e_\mu \partial^\nu A^e_\nu \,. \tag{72.2}$$

Adding eqs. (72.1) and (72.2), and doing some integrations-by-parts in the quadratic terms, we find

$$\begin{aligned}\mathcal{L}_{\text{YM}} + \mathcal{L}_{\text{gf}} = \tfrac{1}{2}A^{e\mu}(g_{\mu\nu}\partial^2 - \partial_\mu\partial_\nu)A^{e\nu} + \tfrac{1}{2}\xi^{-1}A^{e\mu}\partial_\mu\partial_\nu A^{e\nu} \\ - gf^{abc}A^{a\mu}A^{b\nu}\partial_\mu A^c_\nu - \tfrac{1}{4}g^2 f^{abe}f^{cde}A^{a\mu}A^{b\nu}A^c_\mu A^d_\nu \,. \end{aligned} \tag{72.3}$$

The first line of eq. (72.3) yields the gluon propagator in R_ξ gauge,

$$\tilde{\Delta}^{ab}_{\mu\nu}(k) = \frac{\delta^{ab}}{k^2 - i\epsilon}\left(g_{\mu\nu} - \frac{k_\mu k_\nu}{k^2} + \xi\frac{k_\mu k_\nu}{k^2}\right). \tag{72.4}$$

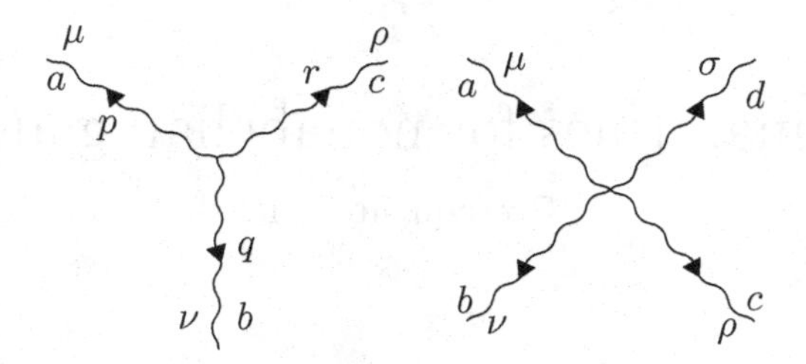

Figure 72.1. The three-gluon and four-gluon vertices in nonabelian gauge theory.

The second line of eq. (72.3) yields three- and four-gluon vertices, shown in fig. 72.1. The three-gluon vertex factor is

$$\begin{aligned} i\mathbf{V}^{abc}_{\mu\nu\rho}(p,q,r) &= i(-gf^{abc})(-ir_\mu g_{\nu\rho}) \\ &\qquad + [\,5 \text{ permutations of } (a,\mu,p), (b,\nu,q), (c,\rho,r)\,] \\ &= gf^{abc}[(q-r)_\mu g_{\nu\rho} + (r-p)_\nu g_{\rho\mu} + (p-q)_\rho g_{\mu\nu}]\,. \end{aligned} \tag{72.5}$$

The four-gluon vertex factor is

$$\begin{aligned} i\mathbf{V}^{abcd}_{\mu\nu\rho\sigma} &= -ig^2 f^{abe} f^{cde} g_{\mu\rho} g_{\nu\sigma} \\ &\qquad + [\,5 \text{ permutations of } (b,\nu), (c,\rho), (d,\sigma)\,] \\ &= -ig^2 \,[\, f^{abe} f^{cde} (g_{\mu\rho} g_{\nu\sigma} - g_{\mu\sigma} g_{\nu\rho}) \\ &\qquad + f^{ace} f^{dbe} (g_{\mu\sigma} g_{\rho\nu} - g_{\mu\nu} g_{\rho\sigma}) \\ &\qquad + f^{ade} f^{bce} (g_{\mu\nu} g_{\sigma\rho} - g_{\mu\rho} g_{\sigma\nu})\,]\,. \end{aligned} \tag{72.6}$$

These vertex factors are quite a bit more complicated that the ones we are used to, and they lead to rather involved formulae for scattering cross sections. For example, the tree-level $gg \to gg$ cross section (where g is a gluon), averaged over initial spins and colors and summed over final spins and colors, has 12 996 terms! Of course, many are identical and the final result can be expressed much more simply, but this is no help to us at the initial stages of computation. For this reason, we postpone any attempt at tree-level calculations until section 81, where we will make use of some techniques (color ordering and Gervais–Neveu gauge) that, combined with spinor-helicity methods, greatly reduce the necessary labor.

For loop calculations, we need to include the ghosts. The ghost lagrangian is

$$\begin{aligned} \mathcal{L}_{\text{gh}} &= -\partial^\mu \bar{c}^b D^{bc}_\mu c^c \\ &= -\partial^\mu \bar{c}^c \partial_\mu c^c + ig\partial^\mu \bar{c}^b A^a_\mu (T^a_{\text{A}})^{bc} c^c \\ &= -\partial^\mu \bar{c}^c \partial_\mu c^c + gf^{abc} A^a_\mu \partial^\mu \bar{c}^b c^c\,. \end{aligned} \tag{72.7}$$

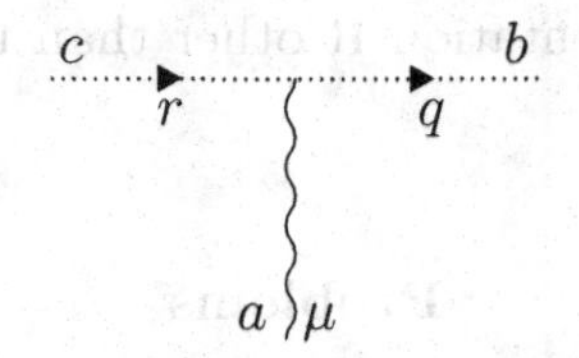

Figure 72.2. The ghost-ghost-gluon vertex in nonabelian gauge theory.

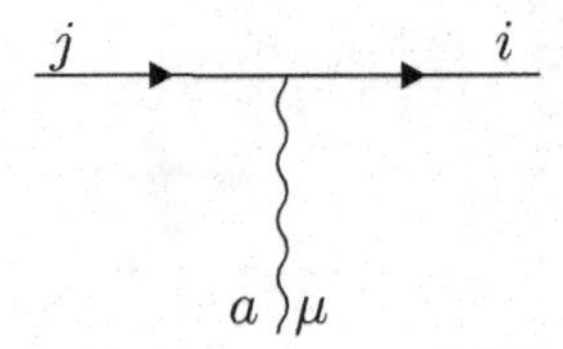

Figure 72.3. The quark-quark-gluon vertex in nonabelian gauge theory.

The ghost propagator is

$$\tilde{\Delta}^{ab}(k^2) = \frac{\delta^{ab}}{k^2 - i\epsilon} \,. \tag{72.8}$$

Because the ghosts are complex scalars, their propagators carry a charge arrow. The ghost–ghost–gluon vertex is shown in fig. 72.2; the associated vertex factor is

$$\begin{aligned} i\mathbf{V}^{abc}_{\mu}(q,r) &= i(gf^{abc})(-iq_\mu) \\ &= gf^{abc}q_\mu \,. \end{aligned} \tag{72.9}$$

If we include a quark coupled to the gluons, we have the quark lagrangian

$$\begin{aligned} \mathcal{L}_{\rm q} &= i\overline{\Psi}_i \not{D}_{ij} \Psi_j - m\overline{\Psi}_i\Psi_i \\ &= i\overline{\Psi}_i \not{\partial}\Psi_i - m\overline{\Psi}_i\Psi_i + gA^a_\mu \overline{\Psi}_i\gamma^\mu T^a_{ij}\Psi_j \,. \end{aligned} \tag{72.10}$$

The quark propagator is

$$\tilde{S}_{ij}(p) = \frac{(-\not{p} + m)\delta_{ij}}{p^2 + m^2 - i\epsilon} \,. \tag{72.11}$$

The quark–quark–gluon vertex is shown in fig. 72.3; the associated vertex factor is

$$i\mathbf{V}^{\mu a}_{ij} = ig\gamma^\mu T^a_{ij} \,. \tag{72.12}$$

If the quark is in a representation R other than the fundamental, then T^a_{ij} becomes $(T^a_{\rm R})_{ij}$.

Problems

72.1) Consider a complex scalar field φ_i in a representation R of the gauge group. Find the vertices that involve this field, and the associated vertex factors.

73

The beta function in nonabelian gauge theory

Prerequisites: 70, 72

In this section, we will do enough loop calculations to compute the beta function for the Yang–Mills coupling g.

We can write the complete lagrangian, including Z factors, as

$$\begin{aligned}\mathcal{L} = {} & \tfrac{1}{2}Z_3 A^{a\mu}(g_{\mu\nu}\partial^2 - \partial_\mu\partial_\nu)A^{a\nu} + \tfrac{1}{2}\xi^{-1}A^{a\mu}\partial_\mu\partial_\nu A^{a\nu} \\ & - Z_{3g}\, g f^{abc}A^{a\mu}A^{b\nu}\partial_\mu A^c_\nu - \tfrac{1}{4}Z_{4g}\, g^2 f^{abe}f^{cde}A^{a\mu}A^{b\nu}A^c_\mu A^d_\nu \\ & - Z_{2'}\partial^\mu \bar{c}^a\partial_\mu c^a + Z_{1'}\, g f^{abc}A^c_\mu \partial^\mu \bar{c}^a c^b \\ & + iZ_2 \overline{\Psi}_i \partial\!\!\!/ \Psi_i - Z_m m\overline{\Psi}_i\Psi_i + Z_1 g A^a_\mu \overline{\Psi}_i\gamma^\mu T^a_{ij}\Psi_j\,. \end{aligned} \tag{73.1}$$

Note that the gauge-fixing term in the first line does not need a Z factor; we saw in section 62 that the ξ-dependent term in the propagator is not renormalized.

We see that g appears in several places in $\mathcal{L}$, and gauge invariance leads us to expect that it will renormalize in the same way in each place. If we rewrite $\mathcal{L}$ in terms of bare fields and parameters, and compare with eq. (73.1), we find that

$$g_0^2 = \frac{Z_1^2}{Z_2^2 Z_3}g^2\tilde{\mu}^\varepsilon = \frac{Z_{1'}^2}{Z_{2'}^2 Z_3}g^2\tilde{\mu}^\varepsilon = \frac{Z_{3g}^2}{Z_3^3}g^2\tilde{\mu}^\varepsilon = \frac{Z_{4g}}{Z_3^2}g^2\tilde{\mu}^\varepsilon\,, \tag{73.2}$$

where $d = 4 - \varepsilon$ is the number of spacetime dimensions. To prove eq. (73.2), we have to derive the nonabelian analogs of the Ward identities, known as *Slavnov–Taylor identities*. For now, we simply assume that eq. (73.2) holds; we will return to this issue in section 74.

The simplest computation to perform is the renormalization of the quark-quark-gluon vertex. This is partly because much of the calculation is the

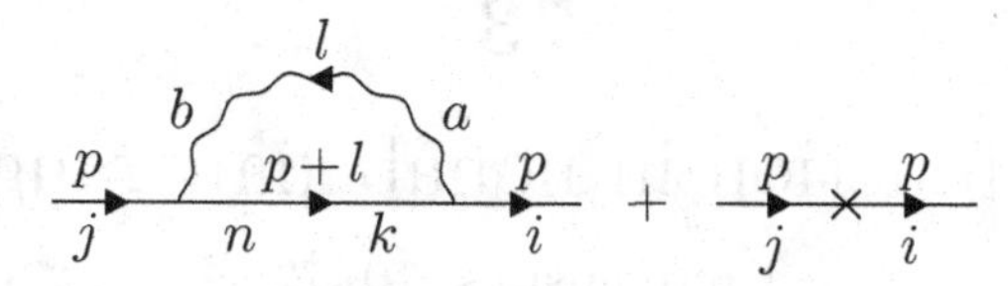

Figure 73.1. The one-loop and counterterm corrections to the quark propagator in quantum chromodynamics.

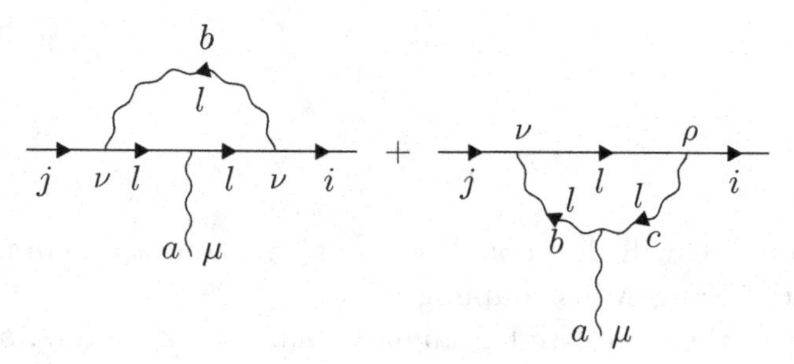

Figure 73.2. The one-loop corrections to the quark–quark–gluon vertex in quantum chromodynamics.

same as it is in spinor electrodynamics, and so we can make use of our results in section 62. We then must compute Z_1, Z_2, and Z_3. We will work in Feynman gauge, and use the $\overline{\text{MS}}$ renormalization scheme.

We begin with Z_2. The $O(g^2)$ corrections to the fermion propagator are shown in fig. 73.1. These diagrams are the same as in spinor electrodynamics, except for the factors related to the color indices. The loop diagram has a color-factor of $(T^aT^a)_{ij} = C(\text{R})\delta_{ij}$. (Here we have allowed the quark to be in an arbitrary representation R; for notational simplicity, we will continue to omit the label R on the generator matrices.) In section 62, we found that, in spinor electrodynamics, the divergent part of this diagram contributes $-(e^2/8\pi^2\varepsilon)\not{p}$ to the electron self-energy $\Sigma(\not{p})$. Thus in Yang–Mills gauge theory, the divergent part of this diagram contributes $-(g^2/8\pi^2\varepsilon)C(\text{R})\delta_{ij}\not{p}$ to the quark self-energy $\Sigma_{ij}(\not{p})$. This divergent term must be canceled by the counterterm contribution of $-(Z_2{-}1)\delta_{ij}\not{p}$. Therefore, in Yang–Mills theory, with a quark in the representation R, using Feynman gauge and the $\overline{\text{MS}}$ renormalization scheme, we have

$$Z_2 = 1 - C(\text{R})\,\frac{g^2}{8\pi^2}\,\frac{1}{\varepsilon} + O(g^4)\,. \tag{73.3}$$

Moving on to the quark–quark–gluon vertex, the contributing one-loop diagrams are shown in fig. 73.2. The first diagram is again the same as it is in spinor electrodynamics, except for the color factor of $(T^bT^aT^b)_{ij}$. We can simplify this via

$$\begin{aligned} T^bT^aT^b &= T^b\Big(T^bT^a + if^{abc}T^c\Big) \\ &= C(\mathrm{R})T^a + \tfrac{1}{2}if^{abc}[T^b,T^c] \\ &= C(\mathrm{R})T^a + \tfrac{1}{2}(if^{abc})(if^{bcd})T^d \\ &= C(\mathrm{R})T^a - \tfrac{1}{2}(T^a_{\mathrm{A}})^{bc}(T^d_{\mathrm{A}})^{cb}T^d \\ &= \Big[C(\mathrm{R}) - \tfrac{1}{2}T(\mathrm{A})\Big]T^a \,. \end{aligned} \tag{73.4}$$

In the second line, we used the complete antisymmetry of f^{abc} to replace T^bT^c with $\frac{1}{2}[T^b,T^c]$. To get the last line, we used $\mathrm{Tr}(T^a_{\mathrm{A}}T^d_{\mathrm{A}}) = T(\mathrm{A})\delta^{ad}$. In section 62, we found that, in spinor electrodynamics, the divergent part of this diagram contributes $(e^2/8\pi^2\varepsilon)ie\gamma^\mu$ to the vertex function $i\mathbf{V}^\mu(p',p)$. Thus in Yang–Mills theory, the divergent part of this diagram contributes

$$\Big[C(\mathrm{R}) - \tfrac{1}{2}T(\mathrm{A})\Big]\frac{g^2}{8\pi^2\varepsilon}\,igT^a_{ij}\gamma^\mu \tag{73.5}$$

to the quark–quark–gluon vertex function $i\mathbf{V}^{a\mu}_{ij}(p',p)$. This divergent term, along with any divergent term from the second diagram of fig. 73.2, must be canceled by the tree-level vertex $iZ_1g\gamma^\mu T^a_{ij}$.

Now we must evaluate the second diagram of fig. 73.2. The divergent part is independent of the external momenta, and so we can set them to zero. Then we get a contribution to $i\mathbf{V}^{a\mu}_{ij}(0,0)$ of

$$\begin{aligned} (ig)^2 g f^{abc}(T^cT^b)_{ij}\left(\tfrac{1}{i}\right)^3 &\int \frac{d^4\ell}{(2\pi)^4}\,\frac{\gamma_\rho(-\not{\ell}+m)\gamma_\nu}{\ell^2\ell^2(\ell^2+m^2)} \\ &\times \left[(\ell-(-\ell))^\mu g^{\nu\rho} + (-\ell-0)^\nu g^{\rho\mu} + (0-\ell)^\rho g^{\mu\nu}\right] . \end{aligned} \tag{73.6}$$

We can simplify the color factor with the manipulations of eq. (73.4),

$$\begin{aligned} f^{abc}T^cT^b &= \tfrac{1}{2}f^{abc}[T^c,T^b] \\ &= \tfrac{1}{2}if^{abc}f^{cbd}T^d \\ &= -\tfrac{1}{2}iT(\mathrm{A})T^a \,. \end{aligned} \tag{73.7}$$

The numerator in function (73.6) is

$$N^\mu = \gamma_\rho(-\gamma_\sigma\ell^\sigma + m)\gamma_\nu(2\ell^\mu g^{\nu\rho} - \ell^\nu g^{\rho\mu} - \ell^\rho g^{\mu\nu}) \,. \tag{73.8}$$

We can drop the terms linear in ℓ, and make the replacement $\ell^\sigma \ell^\mu \to d^{-1}\ell^2 g^{\sigma\mu}$. Thus we have

$$\begin{aligned}
N^\mu &\to -d^{-1}\ell^2\,(\gamma_\rho\gamma_\sigma\gamma_\nu)(2g^{\sigma\mu}g^{\nu\rho} - g^{\sigma\nu}g^{\rho\mu} - g^{\sigma\rho}g^{\mu\nu}) \\
&\to -d^{-1}\ell^2\,(2\gamma^\nu\gamma^\mu\gamma_\nu - \gamma^\mu\gamma^\nu\gamma_\nu - \gamma_\rho\gamma^\rho\gamma^\mu) \\
&\to -d^{-1}\ell^2\,(2(d{-}2) + d + d)\gamma^\mu\,. \qquad (73.9)
\end{aligned}$$

Because we are only keeping track of the divergent term, we are free to set $d = 4$, which yields

$$N^\mu \to -3\ell^2\gamma^\mu\,. \qquad (73.10)$$

Using eqs. (73.7) and (73.10) in function (73.6), we get

$$\tfrac{3}{2}T(\mathrm{A})\,g^3\,T^a_{ij}\gamma^\mu \int \frac{d^4\ell}{(2\pi)^4}\,\frac{1}{\ell^2(\ell^2{+}m^2)}\,. \qquad (73.11)$$

After continuing to d dimensions, the integral becomes $i/8\pi^2\varepsilon + O(\varepsilon^0)$. Combining eqs. (73.5) and (73.6), we find that the divergent part of the quark-quark-gluon vertex function is

$$\mathbf{V}^{a\mu}_{ij}(0,0)_{\text{div}} = \left(Z_1 + \Big[C(\mathrm{R}){-}\tfrac{1}{2}T(\mathrm{A})\Big]\frac{g^2}{8\pi^2\varepsilon} + \tfrac{3}{2}T(\mathrm{A})\,\frac{g^2}{8\pi^2\varepsilon}\right) gT^a_{ij}\gamma^\mu\,. \qquad (73.12)$$

Requiring $\mathbf{V}^{a\mu}_{ij}(0,0)$ to be finite yields

$$Z_1 = 1 - \Big[C(\mathrm{R}) + T(\mathrm{A})\Big]\frac{g^2}{8\pi^2}\,\frac{1}{\varepsilon} + O(g^4) \qquad (73.13)$$

in Feynman gauge and the $\overline{\text{MS}}$ renormalization scheme.

Note that we have found that Z_1 does not equal Z_2. In electrodynamics, we argued that gauge invariance requires all derivatives in the lagrangian to be covariant derivatives, and that both pieces of $D_\mu = \partial_\mu - ieA_\mu$ should therefore be renormalized by the same factor; this then implies that Z_1 must equal Z_2. In Yang–Mills theory, however, this argument fails. This failure is due to the introduction of the ordinary derivative in the gauge-fixing function for R_ξ gauge: once we have added $\mathcal{L}_{\text{gf}}$ and $\mathcal{L}_{\text{gh}}$ to $\mathcal{L}_{\text{YM}}$, we find that both ordinary and covariant derivatives appear. (This is especially obvious for $\mathcal{L}_{\text{gh}}$.) Therefore, to be certain of what gauge invariance does and does not imply, we must derive the appropriate Slavnov–Taylor identities, a subject we will take up in section 74.

Next we turn to the calculation of Z_3. The $O(g^2)$ corrections to the gluon propagator are shown in fig. 73.3. The first diagram is proportional

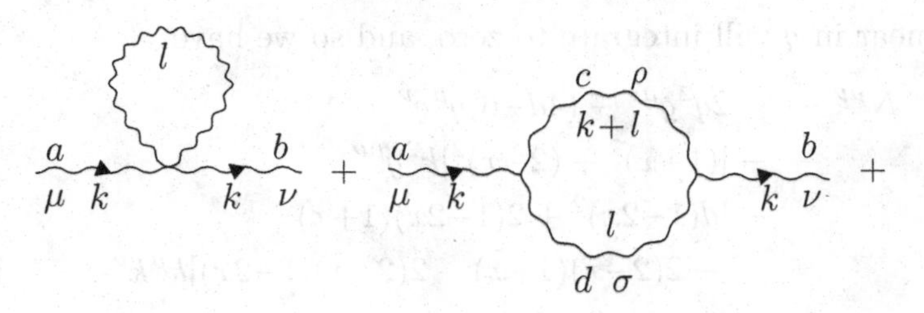

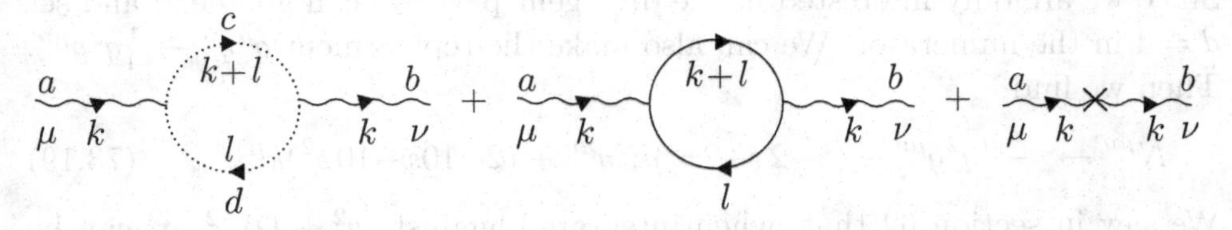

Figure 73.3. The one-loop and counterterm corrections to the gluon propagator in quantum chromodynamics.

to $\int d^4\ell/\ell^2$; as we saw in section 65, this integral vanishes after dimensional regularization.

The second diagram yields a contribution to $i\Pi^{\mu\nu ab}(k)$ of

$$\tfrac{1}{2}g^2 f^{acd} f^{bcd} \left(\tfrac{1}{i}\right)^2 \int \frac{d^4\ell}{(2\pi)^4} \frac{N^{\mu\nu}}{\ell^2(\ell+k)^2} \,, \tag{73.14}$$

where the one-half is a symmetry factor, and

$$\begin{aligned} N^{\mu\nu} &= [(k+\ell)-(-\ell))^\mu g^{\rho\sigma} + (-\ell-(-k))^\rho g^{\sigma\mu} + ((-k)-(k+\ell))^\sigma g^{\mu\rho}] \\ &\quad\times[(-k-\ell)-\ell)^\nu g_{\rho\sigma} + (\ell-k)_\rho \delta_\sigma{}^\nu + (k-(-k-\ell))_\sigma \delta^\nu{}_\rho] \\ &= -[(2\ell+k)^\mu g^{\rho\sigma} - (\ell-k)^\rho g^{\sigma\mu} - (\ell+2k)^\sigma g^{\mu\rho}] \\ &\quad\times[(2\ell+k)^\nu g_{\rho\sigma} - (\ell-k)_\rho \delta_\sigma{}^\nu - (\ell+2k)_\sigma \delta^\nu{}_\rho] \,. \end{aligned} \tag{73.15}$$

The color factor can be simplified via $f^{acd}f^{bcd} = T(\mathrm{A})\delta^{ab}$. We combine denominators with Feynman's formula, and continue to $d = 4 - \varepsilon$ dimensions; we now have

$$-\tfrac{1}{2}g^2\, T(\mathrm{A})\delta^{ab}\, \tilde{\mu}^\varepsilon \int_0^1 dx \int \frac{d^dq}{(2\pi)^d} \frac{N^{\mu\nu}}{(q^2+D)^2} \,, \tag{73.16}$$

where $D = x(1-x)k^2$ and $q = \ell + xk$. The numerator is

$$\begin{aligned} N^{\mu\nu} = -&[(2q+(1-2x)k)^\mu g^{\rho\sigma} - (q-(1+x)k)^\rho g^{\sigma\mu} - (q+(2-x)k)^\sigma g^{\mu\rho}] \\ \times&[(2q+(1-2x)k)^\nu g_{\rho\sigma} - (q-(1+x)k)_\rho \delta_\sigma{}^\nu - (q+(2-x)k)_\sigma \delta^\nu{}_\rho] \,. \end{aligned} \tag{73.17}$$

Terms linear in q will integrate to zero, and so we have

$$\begin{aligned} N^{\mu\nu} \to & - 2q^2 g^{\mu\nu} - (4d{-}6)q^\mu q^\nu \\ & - [(1{+}x)^2 + (2{-}x)^2]k^2 g^{\mu\nu} \\ & - [d(1{-}2x)^2 + 2(1{-}2x)(1{+}x) \\ & \quad - 2(2{-}x)(1{+}x) - 2(2{-}x)(1{-}2x)]k^\mu k^\nu \ . \end{aligned} \tag{73.18}$$

Since we are only interested in the divergent part, we can go ahead and set $d = 4$ in the numerator. We can also make the replacement $q^\mu q^\nu \to \frac{1}{4}q^2 g^{\mu\nu}$. Then we find

$$N^{\mu\nu} \to - \tfrac{9}{2}q^2 g^{\mu\nu} - (5{-}2x{+}2x^2)k^2 g^{\mu\nu} + (2{+}10x{-}10x^2)k^\mu k^\nu \ . \tag{73.19}$$

We saw in section 62 that, when integrated against $(q^2 + D)^{-2}$, q^2 can be replaced with $(\frac{2}{d}{-}1)^{-1}D$; in our case this is $-2x(1{-}x)k^2$. This yields

$$N^{\mu\nu} \to - (5{-}11x{+}11x^2)k^2 g^{\mu\nu} + (2{+}10x{-}10x^2)k^\mu k^\nu \ . \tag{73.20}$$

We now use

$$\tilde{\mu}^\varepsilon \int \frac{d^d q}{(2\pi)^d} \frac{1}{(q^2 + D)^2} = \frac{i}{8\pi^2 \varepsilon} + O(\varepsilon^0) \tag{73.21}$$

in eq. (73.16) to get

$$- \frac{ig^2}{16\pi^2} T(\mathrm{A})\delta^{ab} \frac{1}{\varepsilon} \int_0^1 dx \, N^{\mu\nu} + O(\varepsilon^0) \ . \tag{73.22}$$

Performing the integral over x yields

$$- \frac{ig^2}{16\pi^2} T(\mathrm{A})\delta^{ab} \frac{1}{\varepsilon} \Big(- \tfrac{19}{6}k^2 g^{\mu\nu} + \tfrac{11}{3}k^\mu k^\nu \Big) \tag{73.23}$$

as the divergent contribution of the second diagram to $i\Pi^{\mu\nu ab}(k)$.

Next we have the third diagram of fig. 73.3, which makes a contribution to $i\Pi^{\mu\nu ab}(k)$ of

$$(-1)g^2 f^{acd} f^{bdc} \big(\tfrac{1}{i}\big)^2 \int \frac{d^4\ell}{(2\pi)^4} \frac{(\ell{+}k)^\mu \ell^\nu}{\ell^2(\ell{+}k)^2} \ , \tag{73.24}$$

where the factor of minus one is from the closed ghost loop. The color factor is $f^{acd} f^{bdc} = -T(\mathrm{A})\delta^{ab}$. After combining denominators, the numerator becomes

$$\begin{aligned} (\ell{+}k)^\mu \ell^\nu &= (q + (1{-}x)k)^\mu (q - xk)^\nu \\ &\to \tfrac{1}{4}q^2 g^{\mu\nu} - x(1{-}x)k^\mu k^\nu \\ &\to -\tfrac{1}{2}x(1{-}x)k^2 g^{\mu\nu} - x(1{-}x)k^\mu k^\nu \ . \end{aligned} \tag{73.25}$$

We then use eq. (73.21) in eq. (73.24); performing the integral over x yields

$$-\frac{ig^2}{8\pi^2}\,T(\mathrm{A})\delta^{ab}\,\frac{1}{\varepsilon}\left(-\tfrac{1}{12}k^2g^{\mu\nu}-\tfrac{1}{6}k^\mu k^\nu\right) \tag{73.26}$$

as the divergent contribution of the third diagram to $i\Pi^{\mu\nu ab}(k)$.

Finally, we have the fourth diagram. This is the same as it is in spinor electrodynamics, except for the color factor of $\mathrm{Tr}(T^aT^b) = T(\mathrm{R})\delta^{ab}$. If there is more than one flavor of quark, each contributes separately, leading to a factor of the number of flavors n_{F}. Then, using our results in section 62, we find

$$-\frac{ig^2}{6\pi^2}\,n_{\mathrm{F}}T(\mathrm{R})\delta^{ab}\,\frac{1}{\varepsilon}\left(k^2g^{\mu\nu}-k^\mu k^\nu\right) \tag{73.27}$$

as the divergent contribution of the fourth diagram to $i\Pi^{\mu\nu ab}(k)$.

Adding up eqs. (73.23), (73.26), and (73.27), as well as the counterterm contribution, we find that the gluon self-energy is transverse,

$$\Pi^{\mu\nu ab}(k) = \Pi(k^2)(k^2g^{\mu\nu}-k^\mu k^\nu)\delta^{ab}\,, \tag{73.28}$$

and that

$$\Pi(k^2)_{\mathrm{div}} = -\,(Z_3-1)+\left[\tfrac{5}{3}T(\mathrm{A})-\tfrac{4}{3}n_{\mathrm{F}}T(\mathrm{R})\right]\frac{g^2}{8\pi^2}\frac{1}{\varepsilon}+O(g^4)\,. \tag{73.29}$$

Thus we find

$$Z_3 = 1+\left[\tfrac{5}{3}T(\mathrm{A})-\tfrac{4}{3}n_{\mathrm{F}}T(\mathrm{R})\right]\frac{g^2}{8\pi^2}\frac{1}{\varepsilon}+O(g^4) \tag{73.30}$$

in Feynman gauge and the $\overline{\mathrm{MS}}$ renormalization scheme.

Let us collect our results:

$$Z_1 = 1-\left[C(\mathrm{R})+T(\mathrm{A})\right]\frac{g^2}{8\pi^2}\frac{1}{\varepsilon}+O(g^4)\,, \tag{73.31}$$

$$Z_2 = 1-C(\mathrm{R})\,\frac{g^2}{8\pi^2}\frac{1}{\varepsilon}+O(g^4)\,, \tag{73.32}$$

$$Z_3 = 1+\left[\tfrac{5}{3}T(\mathrm{A})-\tfrac{4}{3}n_{\mathrm{F}}T(\mathrm{R})\right]\frac{g^2}{8\pi^2}\frac{1}{\varepsilon}+O(g^4)\,, \tag{73.33}$$

in Feynman gauge and the $\overline{\mathrm{MS}}$ renormalization scheme. We define

$$\alpha \equiv \frac{g^2}{4\pi}\,. \tag{73.34}$$

Then we have

$$\alpha_0 = \frac{Z_1^2}{Z_2^2 Z_3} \alpha \tilde{\mu}^{\varepsilon} \, . \tag{73.35}$$

Let us write

$$\ln\Big(Z_3^{-1} Z_2^{-2} Z_1^2\Big) = \sum_{n=1}^{\infty} \frac{G_n(\alpha)}{\varepsilon^n} \, . \tag{73.36}$$

Then we have

$$\ln \alpha_0 = \sum_{n=1}^{\infty} \frac{G_n(\alpha)}{\varepsilon^n} + \ln \alpha + \varepsilon \ln \tilde{\mu} \, . \tag{73.37}$$

From eqs. (73.31)–(73.33), we get

$$G_1(\alpha) = -\Big[\tfrac{11}{3} T(\mathrm{A}) - \tfrac{4}{3} n_{\mathrm{F}} T(\mathrm{R})\Big] \frac{\alpha}{2\pi} + O(\alpha^2) \, . \tag{73.38}$$

Then, the general analysis of section 28 yields

$$\beta(\alpha) = \alpha^2 G_1'(\alpha) \, , \tag{73.39}$$

where the prime denotes differentiation with respect to α. Thus we find

$$\beta(\alpha) = -\Big[\tfrac{11}{3} T(\mathrm{A}) - \tfrac{4}{3} n_{\mathrm{F}} T(\mathrm{R})\Big] \frac{\alpha^2}{2\pi} + O(\alpha^3) \tag{73.40}$$

in nonabelian gauge theory with n_{F} Dirac fermions in the representation R of the gauge group.

We can, if we like, restate eq. (73.40) in terms of g as

$$\beta(g) = -\Big[\tfrac{11}{3} T(\mathrm{A}) - \tfrac{4}{3} n_{\mathrm{F}} T(\mathrm{R})\Big] \frac{g^3}{16\pi^2} + O(g^5) \, . \tag{73.41}$$

To go from eq. (73.40) to eq. (73.41), we use $\alpha = g^2/4\pi$ and $\dot{\alpha} = g\dot{g}/2\pi$, where the dot denotes $d/d\ln\mu$.

In quantum chromodynamics, the gauge group is SU(3), and the quarks are in the fundamental representation. Thus $T(\mathrm{A}) = 3$ and $T(\mathrm{R}) = \frac{1}{2}$, and the factor in square brackets in eq. (73.41) is $11 - \frac{2}{3} n_{\mathrm{F}}$. So for $n_{\mathrm{F}} \le 16$, the beta function is negative: the gauge coupling in quantum chromodynamics gets weaker at high energies, and stronger at low energies.

This has dramatic physical consequences. Perturbation theory cannot serve as a reliable guide to the low-energy physics. And indeed, in nature we do not see isolated quarks or gluons. (Quarks, in particular, have fractional electric charges and would be easy to discover.) The appropriate conclusion is that color is *confined*: all finite-energy states are invariant under a global

SU(3) transformation. This has not yet been rigorously proven, but it is the only hypothesis that is consistent with all of the available theoretical and experimental information.

Problems

73.1) Compute the beta function for g in Yang–Mills theory with a complex scalar field in the representation R of the gauge group. Hint: all the real work has been done already in this section, problem 72.1, and section 66.

73.2) Write down the beta function for the gauge coupling in Yang–Mills theory with several Dirac fermions in the representations R_i, and several complex scalars in the representations R'_j.

73.3) Compute the one-loop contributions to the anomalous dimensions of m, Ψ, and A^μ.

74

BRST symmetry

Prerequisites: 70, 71

In this section we will rederive the gauge-fixed path integral for nonabelian gauge theory from a different point of view. We will discover that the complete gauge-fixed lagrangian, $\mathcal{L} \equiv \mathcal{L}_{\rm YM} + \mathcal{L}_{\rm gf} + \mathcal{L}_{\rm gh}$, still has a residual form of the gauge symmetry, known as *Becchi–Rouet–Stora–Tyutin symmetry*, or *BRST symmetry* for short. BRST symmetry can be used to derive the Slavnov–Taylor identities that, among other useful things, show that the coupling constant is renormalized by the same factor at each of its appearances in $\mathcal{L}$. Also, we can use BRST symmetry to show that gluons whose polarizations are not both spacelike and transverse (perpendicular to the four-momentum) decouple from physical scattering amplitudes (as do particles that are created by the ghost field).

Consider a nonabelian gauge theory with a gauge field $A^a_\mu(x)$, and a scalar or spinor field $\phi_i(x)$ in the representation R. Then, under an infinitesimal gauge transformation parameterized by $\theta^a(x)$, we have

$$\delta A^a_\mu(x) = -D^{ab}_\mu \theta^b(x) \,, \tag{74.1}$$

$$\delta\phi_i(x) = -ig\theta^a(x)(T^a_{\rm R})_{ij}\phi_j(x) \,. \tag{74.2}$$

We now introduce a scalar Grassmann field $c^a(x)$ in the adjoint representation; this field will turn out to be the ghost field that we introduced in section 71. We define an infinitesimal BRST transformation via

$$\delta_{\rm B} A^a_\mu(x) \equiv D^{ab}_\mu c^b(x) \tag{74.3}$$

$$= \partial_\mu c^a(x) - g f^{abc} A^c_\mu(x) c^b(x) \,, \tag{74.4}$$

$$\delta_{\rm B}\phi_i(x) \equiv igc^a(x)(T^a_{\rm R})_{ij}\phi_j(x) \,. \tag{74.5}$$

This is simply an infinitesimal gauge transformation, with the ghost field $c^a(x)$ in place of the infinitesimal parameter $-\theta^a(x)$. Therefore, any combination of fields that is gauge invariant is also BRST invariant. In particular, the Yang–Mills lagrangian $\mathcal{L}_{\rm YM}$ (including the appropriate lagrangian for the scalar or spinor field ϕ_i) is BRST invariant,

$$\delta_{\rm B}\mathcal{L}_{\rm YM} = 0 \,. \tag{74.6}$$

We now place a further restriction on the BRST transformation: we require a BRST variation of a BRST variation to be zero. This requirement will determine the BRST transformation of the ghost field. Consider

$$\delta_{\rm B}(\delta_{\rm B}\phi_i) = ig(\delta_{\rm B}c^a)(T^a_{\rm R})_{ij}\phi_j - igc^a(T^a_{\rm R})_{ij}\delta_{\rm B}\phi_j \,. \tag{74.7}$$

There is a minus sign in front of the second term because $\delta_{\rm B}$ acts as an anticommuting object, and it generates a minus sign when it passes through another anticommuting object, in this case c^a. Using eq. (74.5), we have

$$\delta_{\rm B}(\delta_{\rm B}\phi_i) = ig(\delta_{\rm B}c^a)(T^a_{\rm R})_{ij}\phi_j + g^2c^ac^b(T^a_{\rm R}T^b_{\rm R})_{ik}\phi_k \,. \tag{74.8}$$

We now use $c^bc^a = -c^ac^b$ in the second term to replace $T^a_{\rm R}T^b_{\rm R}$ with its antisymmetric part, $\frac{1}{2}[T^a_{\rm R}, T^b_{\rm R}] = \frac{i}{2}f^{abc}T^c_{\rm R}$. Then, after relabeling some dummy indices, we have

$$\delta_{\rm B}(\delta_{\rm B}\phi_i) = ig(\delta_{\rm B}c^c + \tfrac{1}{2}gf^{abc}c^ac^b)(T^c_{\rm R})_{ij}\phi_j \,. \tag{74.9}$$

The right-hand side of eq. (74.9) will vanish for all $\phi_j(x)$ if and only if

$$\delta_{\rm B}c^c(x) = -\tfrac{1}{2}gf^{abc}c^a(x)c^b(x) \,. \tag{74.10}$$

We therefore adopt eq. (74.10) as the BRST variation of the ghost field.

Let us now check to see that the BRST variation of the BRST variation of the gauge field also vanishes. From eq. (74.4) we have

$$\begin{aligned}
\delta_{\rm B}(\delta_{\rm B}A^a_\mu) &= (\delta^{ab}\partial_\mu - gf^{abc}A^c_\mu)(\delta_{\rm B}c^b) - gf^{abc}(\delta_{\rm B}A^c_\mu)c^b \\
&= D^{ab}_\mu(\delta_{\rm B}c^b) - gf^{abc}(D^{cd}_\mu c^d)c^b \\
&= D^{ab}_\mu(\delta_{\rm B}c^b) - gf^{abc}(\partial_\mu c^c)c^b + g^2f^{abc}f^{cde}A^e_\mu c^dc^b \,.
\end{aligned} \tag{74.11}$$

We now use the antisymmetry of f^{abc} in the second term to replace $(\partial_\mu c^c)c^b$ with its antisymmetric part,

$$\begin{aligned}
(\partial_\mu c^{[c})c^{b]} &\equiv \tfrac{1}{2}(\partial_\mu c^c)c^b - \tfrac{1}{2}(\partial_\mu c^b)c^c \\
&= \tfrac{1}{2}(\partial_\mu c^c)c^b + \tfrac{1}{2}c^c(\partial_\mu c^b) \\
&= \tfrac{1}{2}\partial_\mu(c^cc^b) \,.
\end{aligned} \tag{74.12}$$

Similarly, we use the antisymmetry of $c^d c^b$ in the third term to replace $f^{abc}f^{cde}$ with its antisymmetric part,

$$\begin{aligned}\tfrac{1}{2}(f^{abc}f^{cde} - f^{adc}f^{cbe}) &= -\tfrac{1}{2}[(T^b_{\rm A})^{ac}(T^d_{\rm A})^{ce} - (T^d_{\rm A})^{ac}(T^b_{\rm A})^{ce}] \\ &= -\tfrac{1}{2}if^{bdh}(T^h_{\rm A})^{ae} \\ &= -\tfrac{1}{2}f^{bdh}f^{hae} , \end{aligned} \tag{74.13}$$

which is just the Jacobi identity. Now we have

$$\begin{aligned}\delta_{\rm B}(\delta_{\rm B}A^a_\mu) &= D^{ab}_\mu(\delta_{\rm B}c^b) - \tfrac{1}{2}gf^{abc}(\partial_\mu c^c c^b) - \tfrac{1}{2}g^2 f^{bdh}f^{hae}A^e_\mu c^d c^b \\ &= D^{ah}_\mu(\delta_{\rm B}c^h) - (\delta^{ah}\partial_\mu - gf^{ahe}A^e_\mu)\tfrac{1}{2}gf^{bch}c^c c^b \\ &= D^{ah}_\mu(\delta_{\rm B}c^h + \tfrac{1}{2}gf^{bch}c^b c^c) . \end{aligned} \tag{74.14}$$

We see that this vanishes if the BRST variation of the ghost field is given by eq. (74.10).

Now we introduce the *antighost* field $\bar{c}^a(x)$. We take its BRST transformation to be

$$\delta_{\rm B}\bar{c}^a(x) = B^a(x) , \tag{74.15}$$

where $B^a(x)$ is a commuting (as opposed to Grassmann) scalar field, the *Lautrup–Nakanishi auxiliary field.* Because $B^a(x)$ is itself a BRST variation, we have

$$\delta_{\rm B}B^a(x) = 0 . \tag{74.16}$$

Note that eq. (74.15) is in apparent contradiction with eq. (74.10). However, there is actually no need to identify $\bar{c}^a(x)$ as the hermitian conjugate of $c^a(x)$. The role of these fields (in producing the functional determinant that must accompany the gauge-fixing delta functional) is fulfilled as long as $c^a(x)$ and $\bar{c}^a(x)$ are treated as independent when we integrate them; whether or not they are hermitian conjugates of each other is irrelevant. We identified them as hermitian conjugates in section 71 only for the sake of familiarity in deriving the associated Feynman rules. Now, however, we must abandon this notion. In fact, it will be most convenient to treat $c^a(x)$ and $\bar{c}^a(x)$ as two real Grassmann fields.

Now that we have introduced a collection of new fields—$c^a(x)$, $\bar{c}^a(x)$, and $B^a(x)$—what are we to do with them?

Consider adding to $\mathcal{L}_{\rm YM}$ a new term that is the BRST variation of some object $\mathcal{O}$,

$$\mathcal{L} = \mathcal{L}_{\rm YM} + \delta_{\rm B}\mathcal{O} . \tag{74.17}$$

Clearly $\mathcal{L}$ is BRST invariant, because $\mathcal{L}_{\rm YM}$ is, and because $\delta_{\rm B}(\delta_{\rm B}\mathcal{O}) = 0$. We will see that adding $\delta_{\rm B}\mathcal{O}$ corresponds to fixing a gauge; which gauge we get depends on $\mathcal{O}$.

We will choose

$$\mathcal{O}(x) = \bar{c}^a(x)\Big[\tfrac{1}{2}\xi B^a(x) - G^a(x)\Big] \,, \tag{74.18}$$

where $G^a(x)$ is a gauge-fixing function, and ξ is a parameter. If we further choose

$$G^a(x) = \partial^\mu A^a_\mu(x) \,, \tag{74.19}$$

then we end up with R_ξ gauge.

Let us see how this works. We have

$$\delta_{\rm B}\mathcal{O} = (\delta_{\rm B}\bar{c}^a)\Big[\tfrac{1}{2}\xi B^a - \partial^\mu A^a_\mu\Big] - \bar{c}^a\Big[\tfrac{1}{2}\xi(\delta_{\rm B}B^a) - \partial^\mu(\delta_{\rm B}A^a_\mu)\Big] \,. \tag{74.20}$$

There is a minus sign in front of the second set of terms because $\delta_{\rm B}$ acts as an anticommuting object, and so it generates a minus sign when it passes through another anticommuting object, in this case $\bar{c}^a$. Now using eqs. (74.3), (74.15), and (74.16), we get

$$\delta_{\rm B}\mathcal{O} = \tfrac{1}{2}\xi B^a B^a - B^a\partial^\mu A^a_\mu + \bar{c}^a\partial^\mu D^{ab}_\mu c^b \,. \tag{74.21}$$

We see that the last term is the ghost lagrangian $\mathcal{L}_{\rm gh}$ that we found in section 71. If we like, we can integrate the ordinary derivative by parts, so that it acts on the antighost field,

$$\delta_{\rm B}\mathcal{O} \to \tfrac{1}{2}\xi B^a B^a - B^a\partial^\mu A^a_\mu - \partial^\mu\bar{c}^a D^{ab}_\mu c^b \,. \tag{74.22}$$

Examining the first two terms in eq. (74.22), we see that no derivatives act on the auxiliary field $B^a(x)$. Furthermore, it appears only quadratically and linearly in $\delta_{\rm B}\mathcal{O}$. We can, therefore, perform the path integral over it; the result is equivalent to solving the classical equation of motion

$$\frac{\partial(\delta_{\rm B}\mathcal{O})}{\partial B^a(x)} = \xi B^a(x) - \partial^\mu A^a_\mu(x) = 0 \,, \tag{74.23}$$

and substituting the result back into $\delta_{\rm B}\mathcal{O}$. This yields

$$\delta_{\rm B}\mathcal{O} \to -\tfrac{1}{2}\xi^{-1}\partial^\mu A^a_\mu\partial^\nu A^a_\nu - \partial^\mu\bar{c}^a D^{ab}_\mu c^b \,. \tag{74.24}$$

We see that the first term is the gauge-fixing lagrangian $\mathcal{L}_{\rm gf}$ that we found in section 71.

We now take note of all the symmetries of the action $S = \int d^4x\,\mathcal{L}$, where $\mathcal{L} = \mathcal{L}_{\rm YM} + \delta_{\rm B}\mathcal{O}$. With our choice of $\mathcal{O}$, they are: (1) Lorentz invariance; (2) the discrete symmetries of parity, time reversal, and charge conjugation;

(3) global gauge invariance (that is, invariance under a gauge transformation with a spacetime-independent parameter θ^a); (4) BRST invariance; (5) ghost number conservation; and (6) antighost translation invariance.

Global gauge invariance simply requires every term in $\mathcal{L}$ to have all the group indices contracted in a group-invariant manner. Ghost number conservation corresponds to assigning ghost number $+1$ to c^a, -1 to $\bar{c}^a$, and zero to all other fields, and requiring every term in $\mathcal{L}$ to have ghost number zero. Antighost translation invariance corresponds to $\bar{c}^a(x) \to \bar{c}^a(x) + \chi$, where χ is a Grassmann constant. This leaves $\mathcal{L}$ invariant because, in the form of eq. (74.22), $\mathcal{L}$ contains only a derivative of $\bar{c}^b(x)$.

We now claim that $\mathcal{L}$ already includes all terms consistent with these symmetries that have coefficients with positive or zero mass dimension. This means that we will not encounter any divergences in perturbation theory that cannot be absorbed by including a Z factor for each term in $\mathcal{L}$. Furthermore, loop corrections should respect the symmetries, and BRST symmetry requires that g renormalize in the same way at each of its appearances. (Filling in the mathematical details of these claims is a lengthy project that we will not undertake.)

We can regard a BRST transformation as infinitesimal, and hence construct the associated Noether current via the standard formula

$$j^\mu_{\rm B}(x) = \sum_I \frac{\partial \mathcal{L}}{\partial(\partial_\mu \Phi_I(x))} \delta_{\rm B} \Phi_I(x) \,, \tag{74.25}$$

where $\Phi_I(x)$ stands for all the fields, including the matter (scalar and/or spinor), gauge, ghost, antighost, and auxiliary fields. We can then define the BRST charge

$$Q_{\rm B} = \int d^3x \, j^0_{\rm B}(x) \,. \tag{74.26}$$

If we think of $c^a(x)$ and $\bar{c}^a(x)$ as independent hermitian fields, then $Q_{\rm B}$ is hermitian. The BRST charge generates a BRST transformation,

$$i[Q_{\rm B}, A^a_\mu(x)] = D^{ab}_\mu c^b(x) \,, \tag{74.27}$$

$$i\{Q_{\rm B}, c^a(x)\} = -\tfrac{1}{2} g f^{abc} c^b(x) c^c(x) \,, \tag{74.28}$$

$$i\{Q_{\rm B}, \bar{c}^a(x)\} = B^a(x) \,, \tag{74.29}$$

$$i[Q_{\rm B}, B^a(x)] = 0 \,, \tag{74.30}$$

$$i[Q_{\rm B}, \phi_i(x)]_\pm = igc^a(x)(T^a_{\rm R})_{ij}\phi_j(x) \,, \tag{74.31}$$

where $\{A, B\} = AB + BA$ is the anticommutator, and $[\ ,\ ]_\pm$ is the commutator if ϕ_i is a scalar field, and the anticommutator if ϕ_i is a spinor field.

Also, since the BRST transformation of a BRST transformation is zero, Q_{B} must be *nilpotent*,

$$Q_{\text{B}}^2 = 0 \; . \tag{74.32}$$

Eq. (74.32) has far-reaching consequences. In order for it to be satisfied, many states must be annihilated by Q_{B}; such states are said to be in the *kernel* of Q_{B}. A state $|\psi\rangle$ which is annihilated by Q_{B} may take the form of Q_{B} acting on some other state; such states are said to be in the *image* of Q_{B}. There may be some states in the kernel of Q_{B} that are not in the image; such states are said to be in the *cohomology* of Q_{B}. Two states in the cohomology of Q_{B} are identified if their difference is in the image; that is, if $Q_{\text{B}}|\psi\rangle = 0$ but $|\psi\rangle \neq Q_{\text{B}}|\chi\rangle$ for any state $|\chi\rangle$, and if $|\psi'\rangle = |\psi\rangle + Q_{\text{B}}|\zeta\rangle$ for some state $|\zeta\rangle$, then we identify $|\psi\rangle$ and $|\psi'\rangle$ as a single element of the cohomology of Q_{B}.

Note any state in the image of Q_{B} has zero norm, since if $|\psi\rangle = Q_{\text{B}}|\chi\rangle$, then $\langle\psi|\psi\rangle = \langle\psi|Q_{\text{B}}|\chi\rangle = 0$. (Here we have used the hermiticity of Q_{B} to conclude that $Q_{\text{B}}|\psi\rangle = 0$ implies $\langle\psi|Q_{\text{B}} = 0$.)

Now consider starting at some initial time with a normalized state $|\psi\rangle$ in the cohomology: $\langle\psi|\psi\rangle = 1$, $Q_{\text{B}}|\psi\rangle = 0$, $|\psi\rangle \neq Q_{\text{B}}|\chi\rangle$. (This last equation is actually redundant, because if $|\psi\rangle = Q_{\text{B}}|\chi\rangle$ for some state $|\chi\rangle$, then $|\psi\rangle$ has zero norm.) Since $\mathcal{L}$ is BRST invariant, the hamiltonian that we derive from it must commute with the BRST charge: $[H, Q_{\text{B}}] = 0$. Thus, an initial state $|\psi\rangle$ that is annihilated by Q_{B} must still be annihilated by it at later times, since $Q_{\text{B}}e^{-iHt}|\psi\rangle = e^{-iHt}Q_{\text{B}}|\psi\rangle = 0$. Also, since unitary time evolution does not change the norm of a state, the time-evolved state must still be in the cohomology.

We now claim that *the physical states of the theory correspond to the cohomology of* Q_{B}. We have already shown that if we start with a state in the cohomology, it remains in the cohomology under time evolution. Consider, then, an initial state of widely separated wave packets of incoming particles. According to our discussion in section 5, we can treat these states as being created by the appropriate Fourier modes of the fields, and ignore interactions. We will suppress the group index (because it plays no essential role when interactions can be neglected) and write the mode expansions

$$A^\mu(x) = \sum_{\substack{\lambda=>,<,\\+,-}} \int \widetilde{dk} \left[\varepsilon_\lambda^{\mu*}(\mathbf{k})a_\lambda(\mathbf{k})e^{ikx} + \varepsilon_\lambda^\mu(\mathbf{k})a_\lambda^\dagger(\mathbf{k})e^{-ikx}\right] \, , \tag{74.33}$$

$$c(x) = \int \widetilde{dk} \left[c(\mathbf{k})e^{ikx} + c^\dagger(\mathbf{k})e^{-ikx}\right] \, , \tag{74.34}$$

$$\bar{c}(x) = \int \widetilde{dk} \left[b(\mathbf{k}) e^{ikx} + b^\dagger(\mathbf{k}) e^{-ikx} \right] , \tag{74.35}$$

$$\phi(x) = \int \widetilde{dk} \left[a_\phi(\mathbf{k}) e^{ikx} + a_\phi^\dagger(\mathbf{k}) e^{-ikx} \right] . \tag{74.36}$$

Here, for maximum simplicity, we have taken $\phi(x)$ to be a real scalar field. (This is possible if R is a real representation.) In eq. (74.33), we have included four polarization vectors that span four-dimensional spacetime. For $k^\mu = (\omega, \mathbf{k}) = \omega(1,0,0,1)$, we choose these four polarization vectors to be

$$\begin{aligned}
\varepsilon^\mu_>(\mathbf{k}) &= \tfrac{1}{\sqrt{2}}(1,0,0,1) , \\
\varepsilon^\mu_<(\mathbf{k}) &= \tfrac{1}{\sqrt{2}}(1,0,0,-1) , \\
\varepsilon^\mu_+(\mathbf{k}) &= \tfrac{1}{\sqrt{2}}(0,1,-i,0) , \\
\varepsilon^\mu_-(\mathbf{k}) &= \tfrac{1}{\sqrt{2}}(0,1,+i,0) .
\end{aligned} \tag{74.37}$$

The first two of these, $>$ and $<$, are lightlike vectors; $\varepsilon^\mu_>(\mathbf{k})$ is parallel to k^μ, and $\varepsilon^\mu_<(\mathbf{k})$ is spatially opposite. The latter two, $+$ and $-$, are spacelike and transverse: they correspond to physical photon polarizations of definite helicity.

We set $g = 0$, plug eqs. (74.33)–(74.36) into eqs. (74.27)–(74.31), and use eq. (74.23) to eliminate the auxiliary field. Matching coefficients of e^{-ikx}, we find

$$[Q_{\rm B}, a^\dagger_\lambda(\mathbf{k})] = \sqrt{2}\omega\, \delta_{\lambda>}\, c^\dagger(\mathbf{k}) , \tag{74.38}$$

$$\{Q_{\rm B}, c^\dagger(\mathbf{k})\} = 0 , \tag{74.39}$$

$$\{Q_{\rm B}, b^\dagger(\mathbf{k})\} = \xi^{-1}\sqrt{2}\omega\, a^\dagger_<(\mathbf{k}) , \tag{74.40}$$

$$[Q_{\rm B}, a^\dagger_\phi(\mathbf{k})] = 0 . \tag{74.41}$$

Consider a normalized state $|\psi\rangle$ in the cohomology: $\langle\psi|\psi\rangle = 1$, $Q_{\rm B}|\psi\rangle = 0$. Eq. (74.38) tells us that if we add a photon with the unphysical polarization $>$ by acting on $|\psi\rangle$ with $a^\dagger_>(\mathbf{k})$, then this state is not annihilated by $Q_{\rm B}$; hence the state $a^\dagger_>(\mathbf{k})|\psi\rangle$ is not in the cohomology. Eq. (74.40) tells us that the state $a^\dagger_<(\mathbf{k})|\psi\rangle$ is proportional to $Q_{\rm B} b^\dagger(\mathbf{k})|\psi\rangle$; hence the state $a^\dagger_<(\mathbf{k})|\psi\rangle$ is also not in the cohomology. On the other hand, the states $a^\dagger_+(\mathbf{k})|\psi\rangle$ and $a^\dagger_-(\mathbf{k})|\psi\rangle$ are annihilated by $Q_{\rm B}$, but they cannot be written as $Q_{\rm B}$ acting on some other state; hence these states *are* in the cohomology. Also, by similar reasoning, the state with one extra ϕ particle, $a^\dagger_\phi(\mathbf{k})|\psi\rangle$, is in the cohomology.

Eq. (74.38) tells us that if we add a ghost particle by acting on $|\psi\rangle$ with $c^\dagger(\mathbf{k})$, then this state is proportional to $Q_{\rm B} a^\dagger_{>}(\mathbf{k})|\psi\rangle$; hence the state $c^\dagger(\mathbf{k})|\psi\rangle$ is not in the cohomology. Eq. (74.40) tells us that if we add an antighost particle by acting on $|\psi\rangle$ with $b^\dagger(\mathbf{k})$, then this state is not annihilated by $Q_{\rm B}$; hence the state $b^\dagger(\mathbf{k})|\psi\rangle$ is also not in the cohomology.

We conclude that the only particle creation operators that do not take a state out of the cohomology are $a^\dagger_\phi(\mathbf{k})$, $a^\dagger_+(\mathbf{k})$, and $a^\dagger_-(\mathbf{k})$. Of course, it is precisely these operators that create the expected physical particles.

Finally, we note that the vacuum $|0\rangle$ must be in the cohomology, because it is the unique state with zero energy and positive norm.

Thus we can conclude that we can build an initial state of widely separated particles that is in the cohomology only if we do not include any ghost or antighost particles, or photons with polarizations other than $+$ and $-$. Since a state in the cohomology must evolve to another state in the cohomology, no ghosts, antighosts, or unphysically polarized photons can be produced in the scattering process.

Reference notes

A detailed treatment of BRST symmetry can be found in *Weinberg II*.

Problems

74.1) The creation operator for a photon of positive helicity can be written as

$$a^\dagger_+(\mathbf{k}) = -i\,\varepsilon^{\mu *}_+(\mathbf{k}) \int d^3x\; e^{+ikx}\overleftrightarrow{\partial_0} A_\mu(x)\,. \tag{74.42}$$

Consider the state $a^\dagger_+(\mathbf{k})|\psi\rangle$, where $|\psi\rangle$ is in the BRST cohomology. Define a gauge-transformed polarization vector

$$\tilde{\varepsilon}^\mu_+(\mathbf{k}) = \varepsilon^\mu_+(\mathbf{k}) + ck^\mu\;, \tag{74.43}$$

where c is a constant, and a corresponding creation operator $\tilde{a}^\dagger_+(\mathbf{k})$. Show that

$$\tilde{a}^\dagger_+(\mathbf{k})|\psi\rangle = a^\dagger_+(\mathbf{k})|\psi\rangle + Q_{\rm B}|\chi\rangle\;, \tag{74.44}$$

which implies that $\tilde{a}^\dagger_+(\mathbf{k})|\psi\rangle$ and $a^\dagger_+(\mathbf{k})|\psi\rangle$ represent the same element of the cohomology, and hence are physically equivalent. Find the state $|\chi\rangle$.

75

Chiral gauge theories and anomalies

Prerequisites: 70, 72

So far, we have only discussed gauge theories with Dirac fermion fields. Recall that a Dirac field Ψ can be written in terms of two left-handed Weyl fields χ and ξ as

$$\Psi = \begin{pmatrix} \chi \\ \xi^\dagger \end{pmatrix} . \tag{75.1}$$

If Ψ is in a representation R of the gauge group, then χ and $\xi^\dagger$ must be as well. Equivalently, χ must be in the representation R, and ξ must be in the complex conjugate representation $\overline{\mathrm{R}}$. (For an abelian theory, this means that if Ψ has charge $+Q$, then χ has charge $+Q$ and ξ has charge $-Q$.) Thus a Dirac field in a representation R is equivalent to two left-handed Weyl fields, one in R and one in $\overline{\mathrm{R}}$.

If the representation R is real, then we can have a Majorana field

$$\Psi = \begin{pmatrix} \psi \\ \psi^\dagger \end{pmatrix} \tag{75.2}$$

instead of a Dirac field; the left-handed Weyl field ψ and and its hermitian conjugate $\psi^\dagger$ are both in the representation R. Thus a Majorana field in a real representation R is equivalent to a single left-handed Weyl field in R.

Now suppose that we have a single left-handed Weyl field ψ in a *complex* representation R. Such a gauge theory is automatically parity violating (because the right-handed hermitian conjugate of the left-handed Weyl field is in an inequivalent representation of the gauge group), and is said to be *chiral*. The lagrangian is

$$\mathcal{L} = i\psi^\dagger \bar{\sigma}^\mu D_\mu \psi - \tfrac{1}{4} F^{a\mu\nu} F^a_{\mu\nu} , \tag{75.3}$$

where $D_\mu = \partial_\mu - igA^a_\mu T^a_{\rm R}$. Since $T^a_{\rm R}$ is a hermitian matrix (even when R is a complex representation), $i\psi^\dagger\bar\sigma^\mu D_\mu\psi$ is hermitian (up to a total divergence, as usual). We cannot include a mass term for ψ, though, because $\psi\psi$ transforms as $\rm R\otimes R$, and $\rm R\otimes R$ does not contain a singlet if R is complex. Thus, $\psi\psi$ is not gauge invariant. But without a mass term, this lagrangian would appear to possess all the required properties: Lorentz invariance, gauge invariance, and no terms with coefficients with negative mass dimension.

However, it turns out that most chiral gauge theories do not exist as quantum field theories; they are *anomalous.* The problem can ultimately be traced back to the functional measure for the fermion field; it turns out that this measure is, in general, not gauge invariant. We will explore this surprising fact in section 77.

For now we will content ourselves with analyzing Feynman diagrams. We will find an insuperable problem with gauge invariance at the one-loop level that afflicts most chiral gauge theories.

We will work with the simplest possible example: a U(1) theory with a single Weyl field ψ with charge $+1$. The lagrangian is

$$\mathcal{L} = i\psi^\dagger\bar\sigma^\mu(\partial_\mu - igA_\mu)\psi - \tfrac14 F^{\mu\nu}F_{\mu\nu}\,. \tag{75.4}$$

We can use the following trick to write this theory in terms of a Dirac field Ψ. We note that

$$P_{\rm L}\Psi = \begin{pmatrix}\psi\\ 0\end{pmatrix}, \tag{75.5}$$

where $P_{\rm L} = \frac12(1-\gamma_5)$ is the left-handed projection matrix, does not involve the right-handed components of Ψ. Then we can write eq. (75.4) as

$$\mathcal{L} = i\overline{\Psi}\gamma^\mu(\partial_\mu - igA_\mu)P_{\rm L}\Psi - \tfrac14 F^{\mu\nu}F_{\mu\nu}\,, \tag{75.6}$$

and treat Ψ as a Dirac field when we derive the Feynman rules.

To better understand the physical consequences of eq. (75.5), consider the case of a free field. The mode expansion is

$$P_{\rm L}\Psi(x) = \sum_{s=\pm}\int\widetilde{dp}\,\Big[b_s(\mathbf{p})P_{\rm L}u_s(\mathbf{p})e^{ipx} + d^\dagger_s(\mathbf{p})P_{\rm L}v_s(\mathbf{p})e^{-ipx}\Big]\,. \tag{75.7}$$

For a massless field, we learned in section 38 that $P_{\rm L}u_+(\mathbf{p}) = 0$ and $P_{\rm L}v_-(\mathbf{p}) = 0$. Thus we can write eq. (75.7) as

$$P_{\rm L}\Psi(x) = \int\widetilde{dp}\,\Big[b_-(\mathbf{p})u_-(\mathbf{p})e^{ipx} + d^\dagger_+(\mathbf{p})v_+(\mathbf{p})e^{-ipx}\Big]\,. \tag{75.8}$$

Eq. (75.8) shows us that there are only two kinds of particles associated with this field (as opposed to four with a Dirac field): $b_-^\dagger(\mathbf{p})$ creates a particle with charge +1 and helicity −1/2, and $d_+^\dagger(\mathbf{p})$ creates a particle with charge −1 and helicity +1/2. In this theory, charge and spin are correlated.

We can easily read the Feynman rules off of eq. (75.6). In particular, the fermion propagator in momentum space is $-P_{\rm L}\,\not{p}/p^2$, and the fermion-fermion-photon vertex is $ig\gamma^\mu P_{\rm L}$.

When we go to evaluate loop diagrams, we need a method of regulating the divergent integrals. However, our usual choice, dimensional regularization, is problematic, due to the close connection between γ_5 and four-dimensional spacetime. In particular, in four dimensions we have

$$\mathrm{Tr}[\gamma_5\gamma^\mu\gamma^\nu\gamma^\rho\gamma^\sigma] = -4i\varepsilon^{\mu\nu\rho\sigma}\,, \tag{75.9}$$

where $\varepsilon^{0123} = +1$. It is not obvious what should be done with this formula in d dimensions. One possibility is to take $d > 4$ and define $\gamma_5 \equiv i\gamma^0\gamma^1\gamma^2\gamma^3$. Then eq. (75.9) holds, but with each of the four vector indices restricted to span 0, 1, 2, 3. We also have $\{\gamma^\mu, \gamma_5\} = 0$ for $\mu = 0, 1, 2, 3$, but $[\gamma^\mu, \gamma_5] = 0$ for $\mu > 3$. This approach is workable, but cumbersome in practice.

It is therefore tempting to abandon dimensional regularization in favor of, say, Pauli–Villars regularization, which involves the replacement

$$P_{\rm L}\frac{-\not{p}}{p^2} \to P_{\rm L}\left(\frac{-\not{p}}{p^2} - \frac{-\not{p}+\Lambda}{p^2+\Lambda^2}\right). \tag{75.10}$$

Pauli–Villars regularization is equivalent to adding an extra fermion field with mass Λ, and a propagator with the wrong sign (corresponding to changing the signs of the kinetic and mass terms in the lagrangian). But, a Dirac field with a chiral coupling to the gauge field cannot have a mass, since the mass term would not be gauge invariant. So, in a chiral gauge theory, Pauli–Villars regularization violates gauge invariance, and hence is unacceptable.

Given the difficulty with regulating chiral gauge theories (which is a hint that they may not make sense), we will sidestep the issue for now, and see what we can deduce about loop diagrams without a regulator in place.

Consider the correction to the photon propagator, shown in fig. 75.1. We have

$$\begin{aligned} i\Pi^{\mu\nu}(k) = {}& (-1)(ig)^2\left(\tfrac{1}{i}\right)^2\int\frac{d^4\ell}{(2\pi)^4}\,\frac{N^{\mu\nu}}{(\ell+k)^2\ell^2} \\ & - i(Z_3-1)(k^2g^{\mu\nu} - k^\mu k^\nu) + O(g^4)\,, \end{aligned} \tag{75.11}$$

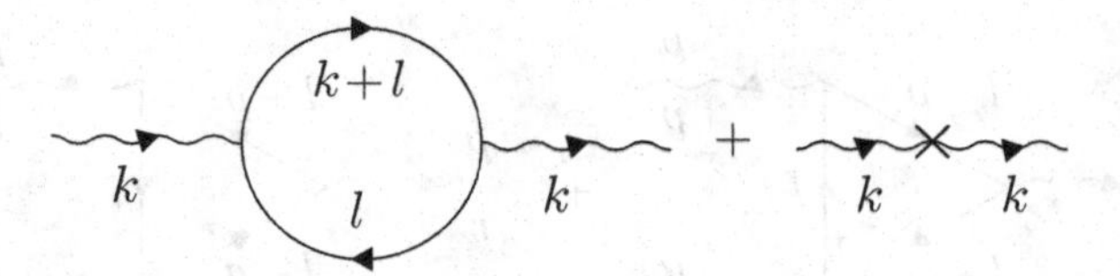

Figure 75.1. The one-loop and counterterm corrections to the photon propagator.

where the numerator is

$$N^{\mu\nu} = \mathrm{Tr}[P_{\rm L}(\not\ell+\not k)\gamma^\mu P_{\rm L}P_{\rm L}\not\ell\gamma^\nu P_{\rm L}] \,. \tag{75.12}$$

We have $P_{\rm L}^2 = P_{\rm L}$ and $P_{\rm L}\gamma^\mu = \gamma^\mu P_{\rm R}$ (and hence $P_{\rm L}\gamma^\mu\gamma^\nu = \gamma^\mu\gamma^\nu P_{\rm L}$), and so all the $P_{\rm L}$s in eq. (75.12) can be collapsed into just one; this is generically true along any fermion line. Thus we have

$$N^{\mu\nu} = \mathrm{Tr}[(\not\ell+\not k)\gamma^\mu\not\ell\gamma^\nu P_{\rm L}] \,. \tag{75.13}$$

The term in eq. (75.13) with $P_{\rm L} \to \frac{1}{2}$ simply yields half the result that we get in spinor electrodynamics with a Dirac field.

The term in eq. (75.13) with $P_{\rm L} \to -\frac{1}{2}\gamma_5$, on the other hand, yields a vanishing contribution to $\Pi^{\mu\nu}(k)$. To see this, first note that

$$\begin{aligned} N^{\mu\nu} &\to -\tfrac{1}{2}\mathrm{Tr}[(\not\ell+\not k)\gamma^\mu\not\ell\gamma^\nu\gamma_5] \\ &= 2i\varepsilon^{\alpha\mu\beta\nu}(\ell+k)_\alpha\ell_\beta \\ &= 2i\varepsilon^{\alpha\mu\beta\nu}k_\alpha\ell_\beta \,. \end{aligned} \tag{75.14}$$

Thus we have

$$\begin{aligned} \Pi^{\mu\nu}(k) = \tfrac{1}{2}\,\Pi^{\mu\nu}(k)_{\rm Dirac} - 2g^2\varepsilon^{\alpha\mu\beta\nu}k_\alpha \int \frac{d^4\ell}{(2\pi)^4}\,\frac{\ell_\beta}{(\ell+k)^2\ell^2} \\ - (Z_3{-}1)(k^2g^{\mu\nu} - k^\mu k^\nu) + O(g^4) \,. \end{aligned} \tag{75.15}$$

The integral is logarithmically divergent. But, it carries a single vector index β, and the only vector it depends on is k. Therefore, any Lorentz-invariant regularization must yield a result that is proportional to k_β. This then vanishes when contracted with $\varepsilon^{\alpha\mu\beta\nu}k_\alpha$. We therefore conclude that, at the one-loop level, the contribution to $\Pi^{\mu\nu}(k)$ of a single charged Weyl field is half that of a Dirac field. This is physically reasonable, since a Dirac field is equivalent to two charged Weyl fields.

Nothing interesting happens in the one-loop corrections to the fermion propagator, or the fermion-fermion-photon vertex. There is simply an extra factor of $P_{\rm L}$ along the fermion line, which can be moved to the far right.

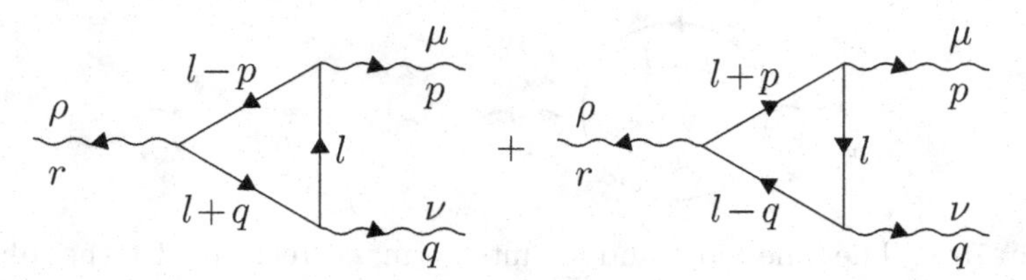

Figure 75.2. One-loop contributions to the three-photon vertex.

Except for this factor, the results exactly duplicate those of spinor electrodynamics.

All of this implies that a single Weyl field makes half the contribution of a Dirac field to the leading term in the beta function for the gauge coupling.

Next we turn to diagrams with three external photons, and no external fermions, shown in fig. 75.2. In spinor electrodynamics, the fact that the vector potential is odd under charge conjugation implies that the sum of these diagrams must vanish; see problem 58.2. For the present case of a single Weyl field, there is no charge-conjugation symmetry, and so we must evaluate these diagrams.

The second diagram in fig. 75.2 is the same as the first, with $p \leftrightarrow q$ and $\mu \leftrightarrow \nu$. Thus we have

$$i\mathbf{V}^{\mu\nu\rho}(p,q,r) = (-1)(ig)^3\left(\tfrac{1}{i}\right)^3 \int \frac{d^4\ell}{(2\pi)^4}\,\frac{N^{\mu\nu\rho}}{(\ell-p)^2\ell^2(\ell+q)^2}$$
$$+ (p,\mu \leftrightarrow q,\nu) + O(g^5)\,, \tag{75.16}$$

where

$$N^{\mu\nu\rho} = \mathrm{Tr}[(-\not\ell+\not p)\gamma^\mu(-\not\ell)\gamma^\nu(-\not\ell-\not q)\gamma^\rho P_{\mathrm L}]\,. \tag{75.17}$$

The term in eq. (75.17) with $P_{\mathrm L} \to \frac{1}{2}$ simply yields half the result that we get in spinor electrodynamics with a Dirac field, which gives a vanishing contribution to $\mathbf{V}^{\mu\nu\rho}(p,q,r)$. Hence, we can make the replacement $P_{\mathrm L} \to -\frac{1}{2}\gamma_5$ in eq. (75.17). Then, after canceling some minus signs, we have

$$N^{\mu\nu\rho} \to \tfrac{1}{2}\mathrm{Tr}[(\not\ell-\not p)\gamma^\mu\not\ell\gamma^\nu(\not\ell+\not q)\gamma^\rho\gamma_5]\,. \tag{75.18}$$

We would now like to verify that $\mathbf{V}^{\mu\nu\rho}(p,q,r)$ is gauge invariant. We should have

$$p_\mu\mathbf{V}^{\mu\nu\rho}(p,q,r) = 0\,, \tag{75.19}$$

$$q_\nu\mathbf{V}^{\mu\nu\rho}(p,q,r) = 0\,, \tag{75.20}$$

$$r_\rho\mathbf{V}^{\mu\nu\rho}(p,q,r) = 0\,. \tag{75.21}$$

Let us first check the last of these. From eq. (75.16) we find

$$r_\rho \mathbf{V}^{\mu\nu\rho}(p,q,r) = ig^3 \int \frac{d^4\ell}{(2\pi)^4} \frac{r_\rho N^{\mu\nu\rho}}{(\ell-p)^2\ell^2(\ell+q)^2} + (p,\mu \leftrightarrow q,\nu) + O(g^5)\,, \tag{75.22}$$

where

$$r_\rho N^{\mu\nu\rho} = \tfrac{1}{2}\mathrm{Tr}[(\not{\ell}-\not{p})\gamma^\mu \not{\ell}\gamma^\nu(\not{\ell}+\not{q})r_\rho\gamma^\rho\gamma_5]\,. \tag{75.23}$$

It will be convenient to use the cyclic property of the trace to rewrite eq. (75.23) as

$$r_\rho N^{\mu\nu\rho} = \tfrac{1}{2}\mathrm{Tr}[\not{\ell}\gamma^\nu(\not{\ell}+\not{q})r_\rho\gamma^\rho(\not{\ell}-\not{p})\gamma^\mu\gamma_5]\,. \tag{75.24}$$

To simplify eq. (75.24), we write $r_\rho\gamma^\rho = \not{r} = -(\not{q}+\not{p}) = -(\not{\ell}+\not{q}) + (\not{\ell}-\not{p})$. Then

$$\begin{aligned}(\not{\ell}+\not{q})r_\rho\gamma^\rho(\not{\ell}-\not{p}) &= (\not{\ell}+\not{q})[-(\not{\ell}+\not{q}) + (\not{\ell}-\not{p})](\not{\ell}-\not{p}) \\ &= (\ell+q)^2(\not{\ell}-\not{p}) - (\ell-p)^2(\not{\ell}+\not{q})\,.\end{aligned} \tag{75.25}$$

Now we have

$$\begin{aligned} r_\rho N^{\mu\nu\rho} &= \tfrac{1}{2}(\ell+q)^2\,\mathrm{Tr}[\not{\ell}\gamma^\nu(\not{\ell}-\not{p})\gamma^\mu\gamma_5] - \tfrac{1}{2}(\ell-p)^2\,\mathrm{Tr}[\not{\ell}\gamma^\nu(\not{\ell}+\not{q})\gamma^\mu\gamma_5] \\ &= -2i\varepsilon^{\alpha\nu\beta\mu}\Big[(\ell+q)^2\ell_\alpha(\ell-p)_\beta - (\ell-p)^2\ell_\alpha(\ell+q)_\beta\Big] \\ &= +2i\varepsilon^{\alpha\nu\beta\mu}\Big[(\ell+q)^2\ell_\alpha p_\beta + (\ell-p)^2\ell_\alpha q_\beta\Big]\,. \end{aligned} \tag{75.26}$$

Putting eq. (75.26) into eq. (75.22), we get

$$r_\rho \mathbf{V}^{\mu\nu\rho}(p,q,r) = -2g^3\,\varepsilon^{\alpha\nu\beta\mu} \int \frac{d^4\ell}{(2\pi)^4} \left[\frac{\ell_\alpha p_\beta}{\ell^2(\ell-p)^2} + \frac{\ell_\alpha q_\beta}{\ell^2(\ell+q)^2}\right] + (p,\mu \leftrightarrow q,\nu) + O(g^5)\,. \tag{75.27}$$

Consider the first term in the integrand. Because the only four-vector that it depends on is p, any Lorentz-invariant regularization of its integral must yield a result proportional to $p_\alpha p_\beta$. Similarly, any Lorentz-invariant regularization of the integral of the second term must yield a result proportional to $q_\alpha q_\beta$. Both $p_\alpha p_\beta$ and $q_\alpha q_\beta$ vanish when contracted with $\varepsilon^{\mu\alpha\nu\beta}$. Therefore, we have shown that

$$r_\rho \mathbf{V}^{\mu\nu\rho}(p,q,r) = 0\,, \tag{75.28}$$

as required by gauge invariance.

It might seem that now we are done: we can invoke symmetry among the external lines to conclude that we must also have $p_\mu \mathbf{V}^{\mu\nu\rho}(p,q,r) = 0$ and $q_\nu \mathbf{V}^{\mu\nu\rho}(p,q,r) = 0$. However, eq. (75.16) is not *manifestly* symmetric on the exchanges $(p,\mu \leftrightarrow r,\rho)$ and $(q,\nu \leftrightarrow r,\rho)$. So it still behooves us to compute either $p_\mu \mathbf{V}^{\mu\nu\rho}(p,q,r)$ or $q_\nu \mathbf{V}^{\mu\nu\rho}(p,q,r)$.

From eq. (75.16) we find

$$p_\mu \mathbf{V}^{\mu\nu\rho}(p,q,r) = ig^3 \int \frac{d^4\ell}{(2\pi)^4} \left[\frac{p_\mu N^{\mu\nu\rho}}{(\ell-p)^2 \ell^2 (\ell+q)^2} + \frac{p_\mu (N^{\nu\mu\rho}|_{p\leftrightarrow q})}{(\ell-q)^2 \ell^2 (\ell+p)^2} \right] + O(g^5)\,, \tag{75.29}$$

where

$$p_\mu N^{\mu\nu\rho} = \tfrac{1}{2} \mathrm{Tr}[(\not{\ell}-\not{p}) p_\mu \gamma^\mu \not{\ell} \gamma^\nu (\not{\ell}+\not{q}) \gamma^\rho \gamma_5]\,. \tag{75.30}$$

To simplify eq. (75.30), we write $p_\mu \gamma^\mu = \not{p} = -(\not{\ell}-\not{p}) + \not{\ell}$. Then

$$\begin{aligned}(\not{\ell}-\not{p}) p_\mu \gamma^\mu \not{\ell} &= (\not{\ell}-\not{p})[-(\not{\ell}-\not{p}) + \not{\ell}]\not{\ell} \\ &= (\ell-p)^2 \not{\ell} - \ell^2 (\not{\ell}-\not{p})\,. \end{aligned} \tag{75.31}$$

Now we have

$$\begin{aligned} p_\mu N^{\mu\nu\rho} &= \tfrac{1}{2}(\ell-p)^2\, \mathrm{Tr}[\not{\ell}\gamma^\nu(\not{\ell}+\not{q})\gamma^\rho\gamma_5] - \tfrac{1}{2}\ell^2\, \mathrm{Tr}[(\not{\ell}-\not{p})\gamma^\nu(\not{\ell}+\not{q})\gamma^\rho\gamma_5] \\ &= -2i\varepsilon^{\alpha\nu\beta\rho}\Big[(\ell-p)^2 \ell_\alpha q_\beta - \ell^2 (\ell-p)_\alpha (\ell+q)_\beta\Big] \\ &= -2i\varepsilon^{\alpha\nu\beta\rho}\Big[(\ell-p)^2 \ell_\alpha q_\beta - \ell^2 (\ell-p)_\alpha (\ell-p+p+q)_\beta\Big] \\ &= -2i\varepsilon^{\alpha\nu\beta\rho}\Big[(\ell-p)^2 \ell_\alpha q_\beta - \ell^2 (\ell-p)_\alpha (p+q)_\beta\Big]\,. \end{aligned} \tag{75.32}$$

With a similar evaluation of $p_\mu (N^{\nu\mu\rho}|_{p\leftrightarrow q})$, we get

$$p_\mu \mathbf{V}^{\mu\nu\rho}(p,q,r) = 2g^3\, \varepsilon^{\alpha\nu\beta\rho} \int \frac{d^4\ell}{(2\pi)^4} \left[\frac{\ell_\alpha q_\beta}{\ell^2 (\ell+q)^2} - \frac{(\ell-p)_\alpha (p+q)_\beta}{(\ell-p)^2 (\ell+q)^2} \right] - (\nu \leftrightarrow \rho, \ell \leftrightarrow -\ell) + O(g^5)\,, \tag{75.33}$$

The first term on the right-hand side of eq. (75.33) must vanish, because any Lorentz-invariant regularization of the integral must yield a result proportional to $q_\alpha q_\beta$, and this vanishes when contracted with $\varepsilon^{\alpha\nu\beta\rho}$.

As for the second term, we can shift the loop momentum from ℓ to $\ell + p$, which results in

$$\frac{(\ell-p)_\alpha (p+q)_\beta}{(\ell-p)^2 (\ell+q)^2} \to \frac{\ell_\alpha (p+q)_\beta}{\ell^2 (\ell+p+q)^2}\,. \tag{75.34}$$

We can now use Lorentz invariance to argue that the integral of the right-hand side of eq. (75.34) must yield something proportional to $(p+q)_\alpha(p+q)_\beta$; this vanishes when contracted with $\varepsilon^{\alpha\nu\beta\rho}$. Thus, we have shown that $p_\mu \mathbf{V}^{\mu\nu\rho}(p,q,r) = 0$, as required by gauge invariance, *provided* that the shift of the loop momentum did not change the value of the integral. This would of course be true if the integral were convergent. Instead, however, the integral is linearly divergent, and so we must be more careful.

Consider a one-dimensional example of a linearly divergent integral: let

$$I(a) \equiv \int_{-\infty}^{+\infty} dx\, f(x+a)\,, \tag{75.35}$$

where $f(\pm\infty) = c_\pm$, with c_+ and c_- two finite constants. If the integral converged, then $I(a)$ would be independent of a. In the present case, however, we can Taylor expand $f(x+a)$ in powers of a, and note that $f(\pm\infty) = c_\pm$ implies that every derivative of $f(x)$ vanishes at $x = \pm\infty$. Thus we have

$$\begin{aligned} I(a) &= \int_{-\infty}^{+\infty} dx \left[f(x) + af'(x) + \tfrac{1}{2}a^2 f''(x) + \ldots \right] \\ &= I(0) + a(c_+ - c_-)\,. \end{aligned} \tag{75.36}$$

We see that $I(a)$ is *not* independent of a. Furthermore, even if we cannot assign a definite value to $I(0)$ (because the integral is divergent), we *can* assign a definite value to the difference

$$I(a) - I(0) = a(c_+ - c_-)\,. \tag{75.37}$$

Now let us return to eqs. (75.33) and (75.34). Define

$$f_\alpha(\ell) \equiv \frac{\ell_\alpha}{\ell^2(\ell+p+q)^2}\,. \tag{75.38}$$

Using Lorentz invariance, we can argue that

$$\int \frac{d^4\ell}{(2\pi)^4}\, f_\alpha(\ell) = A(p+q)_\alpha\,, \tag{75.39}$$

where A is a scalar that will depend on the regularization scheme. Now consider

$$f_\alpha(\ell-p) = f_\alpha(\ell) - p^\beta \frac{\partial}{\partial \ell^\beta} f_\alpha(\ell) + \ldots\,. \tag{75.40}$$

The integral of the first term on the right-hand side of eq. (75.40) is given by eq. (75.39). The integrals of the remaining terms can be converted to surface integrals at infinity. Only the second term in eq. (75.40) falls off slowly enough to contribute. To determine the value of its integral, we make a Wick rotation to euclidean space, which yields a factor of i as usual; then we have

$$\int \frac{d^4\ell}{(2\pi)^4} \frac{\partial}{\partial \ell^\beta} f_\alpha(\ell) = i \lim_{\ell\to\infty} \int \frac{dS_\beta}{(2\pi)^4} f_\alpha(\ell) \,, \tag{75.41}$$

where $dS_\beta = \ell^2 \ell_\beta d\Omega$ is a surface-area element, and $d\Omega$ is the differential solid angle in four dimensions. We thus find

$$\begin{aligned} \int \frac{d^4\ell}{(2\pi)^4} \frac{\partial}{\partial \ell^\beta} f_\alpha(\ell) &= i \lim_{\ell\to\infty} \int \frac{d\Omega}{(2\pi)^4} \frac{\ell_\beta \ell_\alpha}{(\ell+p+q)^2} \\ &= i \frac{\Omega_4}{(2\pi)^4} \frac{1}{4} g_{\alpha\beta} \\ &= \frac{i}{32\pi^2} g_{\alpha\beta} \,, \end{aligned} \tag{75.42}$$

where we used $\Omega_4 = 2\pi^2$. Combining eqs. (75.38)–(75.42), we find

$$\int \frac{d^4\ell}{(2\pi)^4} \frac{(\ell-p)_\alpha}{(\ell-p)^2(\ell+q)^2} = A(p+q)_\alpha - \frac{i}{32\pi^2} p_\alpha \,. \tag{75.43}$$

Using this in eq. (75.33), we find

$$\begin{aligned} p_\mu \mathbf{V}^{\mu\nu\rho}(p,q,r) &= \frac{ig^3}{16\pi^2} \varepsilon^{\alpha\nu\beta\rho} p_\alpha (p+q)_\beta - (\nu \leftrightarrow \rho) + O(g^5) \\ &= \frac{ig^3}{8\pi^2} \varepsilon^{\alpha\nu\beta\rho} p_\alpha q_\beta + O(g^5) \,. \end{aligned} \tag{75.44}$$

An exactly analogous calculation results in

$$q_\nu \mathbf{V}^{\mu\nu\rho}(p,q,r) = \frac{ig^3}{8\pi^2} \varepsilon^{\alpha\rho\beta\mu} p_\alpha q_\beta + O(g^5) \,. \tag{75.45}$$

Eqs. (75.44) and (75.45) show that the three-photon vertex is *not* gauge invariant. Since $r_\rho \mathbf{V}^{\mu\nu\rho}(p,q,r) = 0$, eqs. (75.44) and (75.45) also show that the three-photon vertex does not exhibit the expected symmetry among the external lines.

This is a puzzle, because the only asymmetric aspects of the diagrams in fig. 75.2 are the momentum labels on the *internal* lines. The resolution of the puzzle lies in the fact that the integral in eq. (75.16) is linearly divergent, and so shifting the loop momentum changes its value. To account for this, let us write $\mathbf{V}^{\mu\nu\rho}(p,q,r)$ with ℓ replaced with $\ell + a$, where a is an arbitrary linear combination of p and q. We define

$$\mathbf{V}^{\mu\nu\rho}(p,q,r;a) \equiv \tfrac{1}{2}ig^3 \int \frac{d^4\ell}{(2\pi)^4} \frac{\mathrm{Tr}[(\not\ell+\not a-\not p)\gamma^\mu(\not\ell+\not a)\gamma^\nu(\not\ell+\not a+\not q)\gamma^\rho\gamma_5]}{(\ell+a-p)^2(\ell+a)^2(\ell+a+q)^2}$$
$$+ (p,\mu \leftrightarrow q,\nu) + O(g^5)\,. \tag{75.46}$$

Our previous expression, eq. (75.16), corresponds to $a = 0$. The integral in eq. (75.46) is linearly divergent, and so we can express the difference between $\mathbf{V}^{\mu\nu\rho}(p,q,r;a)$ and $\mathbf{V}^{\mu\nu\rho}(p,q,r;0)$ as a surface integral. Let us write

$$\mathbf{V}^{\mu\nu\rho}(p,q,r;a) = \tfrac{1}{2}ig^3 I_{\alpha\beta\gamma}(a)\,\mathrm{Tr}[\gamma^\alpha\gamma^\mu\gamma^\beta\gamma^\nu\gamma^\gamma\gamma^\rho\gamma_5]$$
$$+ (p,\mu \leftrightarrow q,\nu) + O(g^5)\,, \tag{75.47}$$

where

$$I^{\alpha\beta\gamma}(a) \equiv \int \frac{d^4\ell}{(2\pi)^4} \frac{(\ell+a-p)^\alpha(\ell+a)^\beta(\ell+a+q)^\gamma}{(\ell+a-p)^2(\ell+a)^2(\ell+a+q)^2}\,. \tag{75.48}$$

Then we have

$$\begin{aligned}
I_{\alpha\beta\gamma}(a) - I_{\alpha\beta\gamma}(0) &= a^\delta \int \frac{d^4\ell}{(2\pi)^4} \frac{\partial}{\partial\ell^\delta}\left[\frac{(\ell-p)_\alpha\ell_\beta(\ell+q)_\gamma}{(\ell-p)^2\ell^2(\ell+q)^2}\right] \\
&= ia^\delta \lim_{\ell\to\infty} \int \frac{d\Omega}{(2\pi)^4} \frac{\ell_\delta(\ell-p)_\alpha\ell_\beta(\ell+q)_\gamma}{(\ell-p)^2(\ell+q)^2} \\
&= ia^\delta \frac{\Omega_4}{(2\pi)^4}\frac{1}{24}\Big(g_{\delta\alpha}g_{\beta\gamma} + g_{\delta\beta}g_{\gamma\alpha} + g_{\delta\gamma}g_{\alpha\beta}\Big) \\
&= \frac{i}{192\pi^2}\Big(a_\alpha g_{\beta\gamma} + a_\beta g_{\gamma\alpha} + a_\gamma g_{\alpha\beta}\Big)\,.
\end{aligned} \tag{75.49}$$

Using this in eq. (75.47), we get contractions of the form $g_{\alpha\beta}\gamma^\alpha\gamma^\mu\gamma^\beta = 2\gamma^\mu$. The three terms in eq. (75.49) all end up contributing equally, and after using eq. (75.9) to compute the trace, we find

$$\mathbf{V}^{\mu\nu\rho}(p,q,r;a) - \mathbf{V}^{\mu\nu\rho}(p,q,r;0) = -\frac{ig^3}{16\pi^2}\,\varepsilon^{\mu\nu\rho\beta}a_\beta$$
$$+ (p,\mu \leftrightarrow q,\nu) + O(g^5)\,. \tag{75.50}$$

Since the Levi-Civita symbol is antisymmetric on $\mu \leftrightarrow \nu$, only the part of a that is antisymmetric on $p \leftrightarrow q$ contributes to $\mathbf{V}^{\mu\nu\rho}(p,q,r;a)$. Therefore we will set $a = c(p-q)$, where c is a numerical constant. Then we have

$$\mathbf{V}^{\mu\nu\rho}(p,q,r;a) - \mathbf{V}^{\mu\nu\rho}(p,q,r;0) = -\frac{ig^3}{8\pi^2}\,c\,\varepsilon^{\mu\nu\rho\beta}(p-q)_\beta + O(g^5)\,. \tag{75.51}$$

Using this, along with eqs. (75.28), (75.44), and (75.45), and making some simplifying rearrangements of the indices and momenta on the right-hand sides (using $p+q+r=0$), we find

$$p_\mu \mathbf{V}^{\mu\nu\rho}(p,q,r;a) = -\frac{ig^3}{8\pi^2}(1-c)\varepsilon^{\nu\rho\alpha\beta}q_\alpha r_\beta + O(g^5)\ , \tag{75.52}$$

$$q_\nu \mathbf{V}^{\mu\nu\rho}(p,q,r;a) = -\frac{ig^3}{8\pi^2}(1-c)\varepsilon^{\rho\mu\alpha\beta}r_\alpha p_\beta + O(g^5)\ , \tag{75.53}$$

$$r_\rho \mathbf{V}^{\mu\nu\rho}(p,q,r;a) = -\frac{ig^3}{8\pi^2}(2c)\varepsilon^{\mu\nu\alpha\beta}p_\alpha q_\beta + O(g^5)\ . \tag{75.54}$$

We see that choosing $c=1$ removes the anomalous right-hand side from eqs. (75.52) and (75.53), but it then necessarily appears in eq. (75.54). Choosing $c=\frac{1}{3}$ restores symmetry among the external lines, but now all three right-hand sides are anomalous. (This is what results from dimensional regularization of this theory with $\gamma_5 = i\gamma^0\gamma^1\gamma^2\gamma^3$.) We have therefore failed to construct a gauge-invariant U(1) theory with a single charged Weyl field.

Consider now a U(1) gauge theory with several left-handed Weyl fields ψ_i, with charges Q_i, so that the covariant derivative of ψ_i is $(\partial_\mu - igQ_iA_\mu)\psi_i$. Then each of these fields circulates in the loop in fig. 75.2, and each vertex has an extra factor of Q_i. The right-hand sides of eqs. (75.52)–(75.54) are now multiplied by $\sum_i Q_i^3$. And if $\sum_i Q_i^3$ happens to be zero, then gauge invariance is restored! The simplest possibility is to have the ψs come in pairs with equal and opposite charges. (In this case, they can be assembled into Dirac fields.) But there are other possibilities as well: for example, one field with charge $+2$ and eight with charge -1. Such a gauge theory is still chiral, but it is *anomaly free.* (It could be that further obstacles to gauge invariance arise with more external photons and/or more loops, but this turns out not to be the case. We will discuss this in section 77.)

All of this has a straightforward generalization to nonabelian gauge theories. Suppose we have a single Weyl field in a (possibly reducible) representation R of the gauge group. Then we must attach an extra factor of $\mathrm{Tr}(T^a_{\mathrm{R}}T^b_{\mathrm{R}}T^c_{\mathrm{R}})$ to the first diagram in fig. 75.2, and a factor of $\mathrm{Tr}(T^a_{\mathrm{R}}T^c_{\mathrm{R}}T^b_{\mathrm{R}})$ to the second; here the group indices a, b, c go along with the momenta p, q, r, respectively. Repeating our analysis shows that the diagrams with $P_{\mathrm{L}} \to \frac{1}{2}$ come with an extra factor of $\frac{1}{2}\mathrm{Tr}([T^a_{\mathrm{R}}, T^b_{\mathrm{R}}]T^c_{\mathrm{R}}) = \frac{i}{2}T(\mathrm{R})f^{abc}$; these contribute to the renormalization of the tree-level three-gluon vertex. Diagrams with $P_{\mathrm{L}} \to -\frac{1}{2}\gamma_5$ come with an extra factor of

$$\tfrac{1}{2}\mathrm{Tr}(\{T^a_{\mathrm{R}}, T^b_{\mathrm{R}}\}T^c_{\mathrm{R}}) = A(\mathrm{R})d^{abc}\ . \tag{75.55}$$

Here d^{abc} is a completely symmetric tensor that is independent of the representation, and $A(\mathrm{R})$ is the *anomaly coefficient* of R, introduced in section 70. In order for this theory to exist, we must have $A(\mathrm{R}) = 0$. As shown in section 70, $A(\overline{\mathrm{R}}) = -A(\mathrm{R})$; thus a theory whose left-handed Weyl fields come in $\mathrm{R} \oplus \overline{\mathrm{R}}$ pairs is automatically anomaly free (as is one whose Weyl fields are all in real representations). Otherwise, we must arrange the cancellation by hand. For SU(2) and SO(N), $N \neq 2, 6$, all representations have $A(\mathrm{R}) = 0$. For SU(N) with $N \geq 3$, the fundamental representation has $A(\mathrm{N}) = 1$, and most complex SU(N) representations R have $A(\mathrm{R}) \neq 0$. So the cancellation is nontrivial.

We mention in passing two other kinds of anomalies: if we couple our theory to gravity, we can draw a triangle diagram with two gravitons and one gauge boson. This diagram violates general coordinate invariance (the gauge symmetry of gravity). If the gauge boson is from a nonabelian group, the diagram is accompanied by a factor of $\mathrm{Tr}\, T^a_{\mathrm{R}} = 0$, and so there is no anomaly. If the gauge boson is from a U(1) group, the diagram is accompanied by a factor of $\sum_i Q_i$, and this must vanish to cancel the anomaly.

There is also a *global anomaly* that afflicts theories with an odd number of Weyl fermions in a pseudoreal representation, such as the fundamental representation of SU(2). The global anomaly cannot be seen in perturbation theory; we will discuss it briefly in section 77.

Reference notes

Discussions of anomalies emphasizing different aspects can be found in *Georgi*, *Peskin & Schroeder*, and *Weinberg II*.

Problems

75.1) Consider a theory with a nonabelian gauge symmetry, and also a U(1) gauge symmetry. The theory contains left-handed Weyl fields in the representations (R_i, Q_i), where R_i is the representation of the nonabelian group, and Q_i is the U(1) charge. Find the conditions for this theory to be anomaly free.

76

Anomalies in global symmetries

Prerequisite: 75

In this section we will study anomalies in *global* symmetries that can arise in gauge theories that are free of anomalies in the *local* symmetries (and are therefore consistent quantum field theories). A phenomenological application will be discussed in section 90.

The simplest example is electrodynamics with a massless Dirac field Ψ with charge $Q = +1$. The lagrangian is

$$\mathcal{L} = i\overline{\Psi}\not{D}\Psi - \tfrac{1}{4}F^{\mu\nu}F_{\mu\nu} \, , \tag{76.1}$$

where $\not{D} = \gamma^\mu D_\mu$ and $D_\mu = \partial_\mu - igA_\mu$. (We call the coupling constant g rather than e because we are using this theory as a formal example rather than a physical model.) We can write Ψ in terms of two left-handed Weyl fields χ and ξ via

$$\Psi = \begin{pmatrix} \chi \\ \xi^\dagger \end{pmatrix} , \tag{76.2}$$

where χ has charge $Q = +1$ and ξ has charge $Q = -1$. In terms of χ and ξ, the lagrangian is

$$\mathcal{L} = i\chi^\dagger\bar{\sigma}^\mu(\partial_\mu - igA_\mu)\chi + i\xi^\dagger\bar{\sigma}^\mu(\partial_\mu + igA_\mu)\xi - \tfrac{1}{4}F^{\mu\nu}F_{\mu\nu} \, . \tag{76.3}$$

The lagrangian is invariant under a U(1) gauge transformation

$$\Psi(x) \to e^{-ig\Gamma(x)}\Psi(x) \, , \tag{76.4}$$

$$\overline{\Psi}(x) \to e^{+ig\Gamma(x)}\overline{\Psi}(x) \, , \tag{76.5}$$

$$A^\mu(x) \to A^\mu(x) - \partial^\mu\Gamma(x) \, . \tag{76.6}$$

In terms of the Weyl fields, eqs. (76.4) and (76.5) become

$$\chi(x) \to e^{-ig\Gamma(x)}\chi(x)\,, \tag{76.7}$$

$$\xi(x) \to e^{+ig\Gamma(x)}\xi(x)\,. \tag{76.8}$$

Because the fermion field is massless, the lagrangian is also invariant under a global symmetry in which χ and ξ transform with the same phase,

$$\chi(x) \to e^{+i\alpha}\chi(x)\,, \tag{76.9}$$

$$\xi(x) \to e^{+i\alpha}\xi(x)\,. \tag{76.10}$$

In terms of Ψ, this is

$$\Psi(x) \to e^{-i\alpha\gamma_5}\Psi(x)\,, \tag{76.11}$$

$$\overline{\Psi}(x) \to \overline{\Psi}(x)e^{-i\alpha\gamma_5}\,. \tag{76.12}$$

This is called *axial U(1) symmetry*, because the associated Noether current

$$j^\mu_{\rm A}(x) \equiv \overline{\Psi}(x)\gamma^\mu\gamma_5\Psi(x) \tag{76.13}$$

is an axial vector (that is, its spatial part is odd under parity). Noether's theorem leads us to expect that this current is conserved: $\partial_\mu j^\mu_{\rm A} = 0$. However, in this section we will show that the axial current actually has an *anomalous divergence*,

$$\partial_\mu j^\mu_{\rm A} = -\frac{g^2}{16\pi^2}\,\varepsilon^{\mu\nu\rho\sigma}F_{\mu\nu}F_{\rho\sigma}\,. \tag{76.14}$$

We will see in section 77 that eq. (76.14) is exact; there are no higher-order corrections.

We will demonstrate eq. (76.14) by making use of our results in section 75. Consider the matrix element $\langle p,q|j^\rho_{\rm A}(z)|0\rangle$, where $\langle p,q|$ is a state of two outgoing photons with four-momenta p and q, and polarization vectors ε_μ and ε'_ν, respectively. (We omit the helicity label, which will not play an essential role.) Using the LSZ formula for photons (see section 67), we have

$$\langle p,q|j^\rho_{\rm A}(z)|0\rangle = (ig)^2\,\varepsilon_\mu\varepsilon'_\nu\int d^4x\,d^4y\,e^{-i(px+qy)}\langle 0|{\rm T}j^\mu(x)j^\nu(y)j^\rho_{\rm A}(z)|0\rangle\,, \tag{76.15}$$

where

$$j^\mu(x) \equiv \overline{\Psi}(x)\gamma^\mu\Psi(x) \tag{76.16}$$

is the Noether current corresponding to the U(1) gauge symmetry. Since both $j^\mu(x)$ and $j^\mu_{\rm A}(x)$ are Noether currents, we expect the Ward identities

$$\frac{\partial}{\partial x^\mu}\langle 0|\mathrm{T}j^\mu(x)j^\nu(y)j^\rho_{\rm A}(z)|0\rangle = 0\;, \tag{76.17}$$

$$\frac{\partial}{\partial y^\nu}\langle 0|\mathrm{T}j^\mu(x)j^\nu(y)j^\rho_{\rm A}(z)|0\rangle = 0\;, \tag{76.18}$$

$$\frac{\partial}{\partial z^\rho}\langle 0|\mathrm{T}j^\mu(x)j^\nu(y)j^\rho_{\rm A}(z)|0\rangle = 0\;. \tag{76.19}$$

to be satisfied. Note that there are no contact terms in eqs. (76.17)–(76.19), because both $j^\mu(x)$ and $j^\mu_{\rm A}(x)$ are invariant under both U(1) transformations. If we use eq. (76.19) in eq. (76.15), we see that we expect

$$\frac{\partial}{\partial z^\rho}\langle p,q|j^\rho_{\rm A}(z)|0\rangle = 0\;. \tag{76.20}$$

However, our experience in section 75 leads us to proceed more cautiously.

Let us define $C^{\mu\nu\rho}(p,q,r)$ via

$$\begin{aligned}(2\pi)^4\delta^4(p{+}q{+}r)&C^{\mu\nu\rho}(p,q,r)\\ &\equiv \int d^4x\,d^4y\,d^4z\,e^{-i(px+qy+rz)}\langle 0|\mathrm{T}j^\mu(x)j^\nu(y)j^\rho_{\rm A}(z)|0\rangle\;.\end{aligned} \tag{76.21}$$

Then we can rewrite eq. (76.15) as

$$\langle p,q|j^\rho_{\rm A}(z)|0\rangle = -g^2\varepsilon_\mu\varepsilon'_\nu C^{\mu\nu\rho}(p,q,r)e^{irz}\Big|_{r=-p-q}\;. \tag{76.22}$$

Taking the divergence of the current yields

$$\langle p,q|\partial_\rho j^\rho_{\rm A}(z)|0\rangle = -ig^2\varepsilon_\mu\varepsilon'_\nu r_\rho C^{\mu\nu\rho}(p,q,r)e^{irz}\Big|_{r=-p-q}\;. \tag{76.23}$$

The expected Ward identities become

$$p_\mu C^{\mu\nu\rho}(p,q,r) = 0\;, \tag{76.24}$$

$$q_\nu C^{\mu\nu\rho}(p,q,r) = 0\;, \tag{76.25}$$

$$r_\rho C^{\mu\nu\rho}(p,q,r) = 0\;. \tag{76.26}$$

To check eqs. (76.24)–(76.26), we compute $C^{\mu\nu\rho}(p,q,r)$ with Feynman diagrams. At the one-loop level, the contributing diagrams are exactly those we computed in section 75, except that the three vertex factors are now γ^μ, γ^ν, and $\gamma^\rho\gamma_5$, instead of $ig\gamma^\mu P_{\rm L}$, $ig\gamma^\nu P_{\rm L}$, and $ig\gamma^\rho P_{\rm L}$. But, as we saw, the three $P_{\rm L}$s can be combined into just one at the last vertex, and then this one can be replaced by $-\frac{1}{2}\gamma_5$. Thus, the vertex function $i\mathbf{V}^{\mu\nu\rho}(p,q,r)$ of section 75 is related to $C^{\mu\nu\rho}(p,q,r)$ by

$$i\mathbf{V}^{\mu\nu\rho}(p,q,r) = -\tfrac{1}{2}(ig)^3 C^{\mu\nu\rho}(p,q,r) + O(g^5)\ . \tag{76.27}$$

In section 75, we saw that we could choose a regularization scheme that preserved eqs. (76.24) and (76.25), but not also (76.26). For the theory of this section, we definitely want to preserve eqs. (76.24) and (76.25), because these imply conservation of the current coupled to the gauge field, which is necessary for gauge invariance. On the other hand, we are less enamored of eq. (76.26), because it implies conservation of the current for a mere global symmetry.

Using eq. (76.27) and our results from section 75, we find that preserving eqs. (76.24) and (76.25) results in

$$r_\rho C^{\mu\nu\rho}(p,q,r) = -\frac{i}{2\pi^2}\varepsilon^{\mu\nu\alpha\beta}p_\alpha q_\beta + O(g^2) \tag{76.28}$$

in place of eq. (76.26). Using this in eq. (76.23), we find

$$\langle p,q|\partial_\rho j^\rho_{\rm A}(z)|0\rangle = -\frac{g^2}{2\pi^2}\varepsilon^{\mu\nu\alpha\beta}p_\alpha q_\beta \varepsilon_\mu \varepsilon'_\nu e^{-i(p+q)z} + O(g^4)\ . \tag{76.29}$$

Now we come to the point. The right-hand side of eq. (76.29) is exactly what we get in free-field theory for the matrix element of the right-hand side of eq. (76.14). We conclude that eq. (76.14) is correct, up to possible higher-order corrections.

In the next section, we will see that eq. (76.14) is exact.

Problems

76.1) Verify that the right-hand side of eq. (76.29) is exactly what we get in free-field theory for the matrix element of the right-hand side of eq. (76.14).

77

Anomalies and the path integral for fermions

Prerequisite: 76

In the last section, we saw that in a U(1) gauge theory with a massless Dirac field Ψ with charge $Q = +1$, the axial vector current

$$j_{\rm A}^{\mu} = \overline{\Psi}\gamma^{\mu}\gamma_5\Psi\ , \tag{77.1}$$

which should (according to Noether's theorem) be conserved, actually has an anomalous divergence,

$$\partial_\mu j_{\rm A}^{\mu} = -\frac{g^2}{16\pi^2}\,\varepsilon^{\mu\nu\rho\sigma}F_{\mu\nu}F_{\rho\sigma}\ . \tag{77.2}$$

In this section, we will derive eq. (77.2) directly from the path integral, using the *Fujikawa method.* We will see that eq. (77.2) is exact; there are no higher-order corrections.

We can also consider a nonabelian gauge theory with a massless Dirac field Ψ in a (possibly reducible) representation R of the gauge group. In this case, the triangle diagrams that we analyzed in the last section carry an extra factor of $\mathrm{Tr}(T^a_{\rm R}T^b_{\rm R}) = T({\rm R})\delta^{ab}$, and we have

$$\partial_\mu j_{\rm A}^{\mu} = -\frac{g^2}{16\pi^2}T({\rm R})\varepsilon^{\mu\nu\rho\sigma}\partial_{[\mu}A^a_{\nu]}\partial_{[\rho}A^a_{\sigma]} + O(g^3)\ , \tag{77.3}$$

where $\partial_{[\mu}A^a_{\nu]} \equiv \partial_\mu A^a_\nu - \partial_\nu A^a_\mu$. We expect the right-hand side of eq. (77.3) to be gauge invariant (since this theory is free of anomalies in the currents coupled to the gauge fields); this suggests that we should have

$$\partial_\mu j_{\rm A}^{\mu} = -\frac{g^2}{16\pi^2}T({\rm R})\varepsilon^{\mu\nu\rho\sigma}F^a_{\mu\nu}F^a_{\rho\sigma}\ , \tag{77.4}$$

where $F^a_{\mu\nu} = \partial_\mu A^a_\nu - \partial_\nu A^a_\mu + gf^{abc}A^b_\mu A^c_\nu$ is the nonabelian field strength. We will see that eq. (77.4) is correct, and that there are no higher-order corrections.

We can write eq. (77.4) more compactly by using the matrix-valued gauge field

$$A_\mu \equiv T^a_{\text{R}} A^a_\mu \tag{77.5}$$

and field strength

$$F_{\mu\nu} = \partial_\mu A_\nu - \partial_\nu A_\mu - ig[A_\mu, A_\nu] \ . \tag{77.6}$$

Then eq. (77.4) can be written as

$$\partial_\mu j^\mu_{\text{A}} = -\frac{g^2}{16\pi^2}\,\varepsilon^{\mu\nu\rho\sigma}\,\text{Tr}\,F_{\mu\nu}F_{\rho\sigma} \ . \tag{77.7}$$

We now turn to the derivation of eqs. (77.2) and (77.7). We begin with the path integral over the Dirac field, with the gauge field treated as a fixed background, to be integrated later. We have

$$Z(A) \equiv \int \mathcal{D}\Psi\,\mathcal{D}\overline{\Psi}\; e^{iS(A)} \ , \tag{77.8}$$

where

$$S(A) \equiv \int d^4x\; \overline{\Psi} i\rlap{\,/}D\Psi \tag{77.9}$$

is the Dirac action, $i\rlap{\,/}D = i\gamma^\mu D_\mu$ is the Dirac wave operator, and

$$D_\mu = \partial_\mu - igA_\mu \tag{77.10}$$

is the covariant derivative. Here A_μ is either the U(1) gauge field, or the matrix-valued nonabelian gauge field of eq. (77.5), depending on the theory under consideration. Our notation allows us to treat both cases simultaneously.

We can formally evaluate eq. (77.8) as a functional determinant,

$$Z(A) = \det(i\rlap{\,/}D) \ . \tag{77.11}$$

However, this expression is not useful without some form of regularization. We will take up this issue shortly.

Now consider an axial U(1) transformation of the Dirac field, but with a spacetime dependent parameter $\alpha(x)$:

$$\Psi(x) \to e^{-i\alpha(x)\gamma_5}\,\Psi(x) \ , \tag{77.12}$$

$$\overline{\Psi}(x) \to \overline{\Psi}(x) e^{-i\alpha(x)\gamma_5} \ . \tag{77.13}$$

We can think of eqs. (77.12) and (77.13) as a change of integration variable in eq. (77.8); then $Z(A)$ should be independent of $\alpha(x)$. The corresponding change in the action is

$$S(A) \to S(A) + \int d^4x \, j^\mu_{\rm A}(x)\partial_\mu \alpha(x) \,. \tag{77.14}$$

We can integrate by parts to write this as

$$S(A) \to S(A) - \int d^4x \, \alpha(x)\partial_\mu j^\mu_{\rm A}(x) \,. \tag{77.15}$$

If we assume that the measure $\mathcal{D}\Psi\,\mathcal{D}\overline{\Psi}$ is invariant under the axial U(1) transformation, then we have

$$Z(A) \to \int \mathcal{D}\Psi\,\mathcal{D}\overline{\Psi}\; e^{iS(A)} e^{-i\int d^4x\,\alpha(x)\partial_\mu j^\mu_{\rm A}(x)} \,. \tag{77.16}$$

This must be equal to the original expression for $Z(A)$, eq. (77.8). This implies that $\partial_\mu j^\mu_{\rm A}(x) = 0$ holds inside quantum correlation functions, up to contact terms, as discussed in section 22.

However, the assumption that the measure $\mathcal{D}\Psi\,\mathcal{D}\overline{\Psi}$ is invariant under the axial U(1) transformation must be examined more closely. The change of variable in eqs. (77.12) and (77.13) is implemented by the functional matrix

$$J(x,y) = \delta^4(x-y)e^{-i\alpha(x)\gamma_5} \,. \tag{77.17}$$

Because the path integral is over fermionic variables (rather than bosonic), we get a jacobian factor of $(\det J)^{-1}$ (rather than $\det J$) for each of the transformations in eqs. (77.12) and (77.13), so that we have

$$\mathcal{D}\Psi\,\mathcal{D}\overline{\Psi} \to (\det J)^{-2}\,\mathcal{D}\Psi\,\mathcal{D}\overline{\Psi} \,. \tag{77.18}$$

Using $\log\det J = \mathrm{Tr}\log J$, we can write

$$(\det J)^{-2} = \exp\left[\, 2i \int d^4x \, \alpha(x) \,\mathrm{Tr}\, \delta^4(x-x)\gamma_5 \right] , \tag{77.19}$$

where the explicit trace is over spin and group indices. Like eq. (77.11), this expression is not useful without some form of regularization.

We could try to replace the delta function with a gaussian; this is equivalent to

$$\delta^4(x-y) \to e^{\partial_x^2/M^2}\delta^4(x-y) \,, \tag{77.20}$$

where M is a regulator mass that we would take to infinity at the end of the calculation. However, the appearance of the ordinary derivative ∂, rather than the covariant derivative D, implies that eq. (77.20) is not properly gauge invariant. So, another possibility is

$$\delta^4(x-y) \to e^{D_x^2/M^2}\delta^4(x-y) . \tag{77.21}$$

However, eq. (77.21) presents us with a more subtle problem. Our regularization scheme for eq. (77.19) should be compatible with our regularization scheme for eq. (77.11). It is not obvious whether or not eq. (77.21) meets this criterion, because D^2 has no simple relation to $i\not{D}$. To resolve this issue, we use

$$\delta^4(x-y) \to e^{(i\not{D}_x)^2/M^2}\delta^4(x-y) \tag{77.22}$$

to regulate the delta function in eq. (77.19).

To evaluate eq. (77.22), we write the delta function on the right-hand side of eq. (77.22) as a Fourier integral,

$$\delta^4(x-y) \to \int \frac{d^4k}{(2\pi)^4}\, e^{(i\not{D}_x)^2/M^2} e^{ik(x-y)} . \tag{77.23}$$

Then we use $f(\partial)e^{ikx} = e^{ikx}f(\partial + ik)$; eq. (77.23) becomes

$$\delta^4(x-y) \to \int \frac{d^4k}{(2\pi)^4}\, e^{ik(x-y)}\, e^{(i\not{D}-\not{k})^2/M^2} , \tag{77.24}$$

where a derivative acting on the far right now yields zero. We have

$$\begin{aligned}(i\not{D} - \not{k})^2 &= \not{k}^2 - i\{\not{k}, \not{D}\} - \not{D}^2 \\ &= -k^2 - i\{\gamma^\mu, \gamma^\nu\}k_\mu D_\nu - \gamma^\mu\gamma^\nu D_\mu D_\nu .\end{aligned} \tag{77.25}$$

Next we use $\gamma^\mu\gamma^\nu = \frac{1}{2}(\{\gamma^\mu,\gamma^\nu\} + [\gamma^\mu,\gamma^\nu]) = -g^{\mu\nu} - 2iS^{\mu\nu}$ to get

$$(i\not{D} - \not{k})^2 = -k^2 + 2ik\cdot D + D^2 + 2iS^{\mu\nu}D_\mu D_\nu . \tag{77.26}$$

In the last term, we can use the antisymmetry of $S^{\mu\nu}$ to replace $D_\mu D_\nu$ with $\frac{1}{2}[D_\mu, D_\nu] = -\frac{1}{2}igF_{\mu\nu}$, which yields

$$(i\not{D} - \not{k})^2 = -k^2 + 2ik\cdot D + D^2 + gS^{\mu\nu}F_{\mu\nu} . \tag{77.27}$$

We use eq. (77.27) in eq. (77.24), and then rescale k by M; the result is

$$\delta^4(x-y) \to M^4 \int \frac{d^4k}{(2\pi)^4}\, e^{iMk(x-y)}\, e^{-k^2} e^{2ik\cdot D/M + D^2/M^2 + gS^{\mu\nu}F_{\mu\nu}/M^2} . \tag{77.28}$$

Thus we have

$$\operatorname{Tr}\delta^4(x-x)\gamma_5 \to M^4 \int \frac{d^4k}{(2\pi)^4}\, e^{-k^2}\, \operatorname{Tr} e^{2ik\cdot D/M + D^2/M^2 + gS^{\mu\nu}F_{\mu\nu}/M^2}\gamma_5 \;. \tag{77.29}$$

We can now expand the exponential in inverse powers of M; only terms up to M^{-4} will survive the $M\to\infty$ limit. Furthermore, the trace over spin indices will vanish unless there are four or more gamma matrices multiplying γ_5. Together, these considerations imply that the only term that can make a nonzero contribution is $\frac{1}{2}(gS^{\mu\nu}F_{\mu\nu})^2/M^4$. Thus we find

$$\operatorname{Tr}\delta^4(x-x)\gamma_5 \to \tfrac{1}{2}g^2 \int \frac{d^4k}{(2\pi)^4}\, e^{-k^2}\, (\operatorname{Tr} F_{\mu\nu}F_{\rho\sigma})(\operatorname{Tr} S^{\mu\nu}S^{\rho\sigma}\gamma_5)\;, \tag{77.30}$$

where the first trace is over group indices (in the nonabelian case), and the second trace is over spin indices. The spin trace is

$$\begin{aligned}\operatorname{Tr} S^{\mu\nu}S^{\rho\sigma}\gamma_5 &= \operatorname{Tr}\,(\tfrac{i}{2}\gamma^\mu\gamma^\nu)(\tfrac{i}{2}\gamma^\rho\gamma^\sigma)\gamma_5 \\ &= -\tfrac{1}{4}\operatorname{Tr}\gamma^\mu\gamma^\nu\gamma^\rho\gamma^\sigma\gamma_5 \\ &= i\varepsilon^{\mu\nu\rho\sigma}\;. \end{aligned}\tag{77.31}$$

To evaluate the integral over k in eq. (77.30), we analytically continue to euclidean spacetime; this results in an overall factor of i, as usual. Then each of the four gaussian integrals gives a factor of $\pi^{1/2}$. So we find

$$\operatorname{Tr}\delta^4(x-x)\gamma_5 \to -\frac{g^2}{32\pi^2}\,\varepsilon^{\mu\nu\rho\sigma}\operatorname{Tr} F_{\mu\nu}F_{\rho\sigma}\;. \tag{77.32}$$

Using this in eq. (77.19), we get

$$(\det J)^{-2} = \exp\!\left[-\frac{ig^2}{16\pi^2}\int d^4x\,\alpha(x)\,\varepsilon^{\mu\nu\rho\sigma}\operatorname{Tr} F_{\mu\nu}(x)F_{\rho\sigma}(x)\right]. \tag{77.33}$$

Including the transformation of the measure, eq. (77.18), in the transformation of the path integral, eq. (77.16), then yields

$$Z(A) \to \int \mathcal{D}\Psi\,\mathcal{D}\overline{\Psi}\; e^{iS(A)} e^{-i\int d^4x\,\alpha(x)[(g^2/16\pi^2)\varepsilon^{\mu\nu\rho\sigma}\operatorname{Tr} F_{\mu\nu}(x)F_{\rho\sigma}(x)+\partial_\mu j^\mu_{\rm A}(x)]} \tag{77.34}$$

in place of eq. (77.16). This must be equal to the original expression for $Z(A)$, eq. (77.8). This implies that eq. (77.7) holds inside quantum correlation functions, up to possible contact terms.

Note that this derivation of eq. (77.7) did not rely on an expansion in powers of g, and so eq. (77.7) is exact; there are no higher-order corrections.

This result is known as the *Adler–Bardeen theorem.* It can also be (and originally was) established by a careful study of Feynman diagrams.

The Fujikawa method can be used to find the anomaly in the chiral gauge theories that we studied in section 75, but the analysis is more involved. Here we will quote only the final result.

Consider a left-handed Weyl field in a (possibly reducible) representation R of the gauge group. We define the chiral gauge current $j^{a\mu} \equiv \overline{\Psi} T^a_{\rm R} \gamma^\mu P_{\rm L} \Psi$. Its covariant divergence (which should be zero, according to Noether's theorem) is given by

$$D^{ab}_\mu j^{b\mu} = \frac{g^2}{24\pi^2}\,\varepsilon^{\mu\nu\rho\sigma}\partial_\mu {\rm Tr}\Big[T^a_{\rm R}(A_\nu\partial_\rho A_\sigma - \tfrac{1}{2}igA_\nu A_\rho A_\sigma)\Big]\;. \qquad (77.35)$$

Note that the right-hand side of eq. (77.35) is *not* gauge invariant. The anomaly spoils gauge invariance in chiral gauge theories, unless this right-hand side happens to vanish for group-theoretic reasons. We show in problem 77.1 that this occurs if and only if $A({\rm R}) = 0$, where $A({\rm R})$ is the anomaly coefficient of the representation R.

For comparison, note that eq. (77.7) can be written as

$$\partial_\mu j^\mu_{\rm A} = -\frac{g^2}{4\pi^2}\,\varepsilon^{\mu\nu\rho\sigma}\partial_\mu {\rm Tr}\Big[A_\nu\partial_\rho A_\sigma - \tfrac{2}{3}igA_\nu A_\rho A_\sigma\Big]\;. \qquad (77.36)$$

The relative value of the overall numerical prefactor in eqs. (77.35) and (77.36) is easy to understand: there is a minus one-half in eq. (77.35) from $P_{\rm L} \to -\frac{1}{2}\gamma_5$, and a one-third from regularizing to preserve symmetry among the three external lines in the triangle diagram. (The relative coefficients of the second terms have no comparably simple explanation.)

Finally, a related but more subtle problem, known as a *global anomaly*, arises for theories with an odd number of Weyl fields in a pseudoreal representation, such as the fundamental representation of SU(2). In this case, every gauge field configuration A_μ can be smoothly deformed into another gauge field configuration A'_μ that has the same action, but has $Z(A') = -Z(A)$. Thus, when we integrate over A, the contribution from A' cancels the contribution from A, and the result is zero. Since its path integral is trivial, this theory does not exist.

Problems

77.1) Show that the right-hand side of eq. (77.35) vanishes if and only if $A({\rm R}) = 0$.

77.2) Show that the right-hand side of eq. (77.36) equals the right-hand side of eq. (77.7).

78

Background field gauge

Prerequisite: 73

In this section, we will introduce a clever choice of gauge, *background field gauge*, that greatly simplifies the calculation of the beta function for Yang–Mills theory, especially at the one-loop level.

We begin with the lagrangian for Yang–Mills theory,

$$\mathcal{L}_{\rm YM} = -\tfrac{1}{4}F^{a\mu\nu}F^a_{\mu\nu}\;, \tag{78.1}$$

where the field strength is

$$F^a_{\mu\nu} = \partial_\mu A^a_\nu - \partial_\nu A^a_\mu + gf^{abc}A^b_\mu A^c_\nu\;. \tag{78.2}$$

To evaluate the path integral, we must choose a gauge. As we saw in section 71, one large class of gauges corresponds to choosing a gauge-fixing function $G^a(x)$, and adding $\mathcal{L}_{\rm gf} + \mathcal{L}_{\rm gh}$ to $\mathcal{L}_{\rm YM}$, where

$$\mathcal{L}_{\rm gf} = -\tfrac{1}{2}\xi^{-1}G^aG^a\;, \tag{78.3}$$

$$\mathcal{L}_{\rm gh} = \bar{c}^a\frac{\partial G^a}{\partial A^b_\mu}D^{bc}_\mu c^c\;. \tag{78.4}$$

Here $D^{bc}_\mu = \delta^{bc}\partial_\mu - ig(T^a_{\rm A})^{bc}A^a_\mu = \delta^{bc}\partial_\mu + gf^{bac}A^a_\mu$ is the covariant derivative in the adjoint representation, and c and $\bar{c}$ are the ghost and antighost fields. The notation $\partial G^a/\partial A^b_\mu$ means that any derivatives that act on A^b_μ in G^a now act to the right in eq. (78.4).

We get R_ξ gauge by choosing $G^a = \partial^\mu A^a_\mu$. To get *background field gauge*, we first introduce a fixed, classical *background field* $\bar{A}^a_\mu(x)$, and the corresponding background covariant derivative,

$$\bar{D}_\mu \equiv \partial_\mu - igT^a_{\rm A}\bar{A}^a_\mu\;. \tag{78.5}$$

Then we choose

$$G^a = (\bar{D}^\mu)^{ab}(A - \bar{A})^b_\mu \,. \tag{78.6}$$

The ghost lagrangian becomes $\mathcal{L}_{\rm gh} = \bar{c}^a \bar{D}^{\mu ab} D^{bc}_\mu c^c$, or, after an integration by parts,

$$\mathcal{L}_{\rm gh} = -(\bar{D}^\mu \bar{c})^a (D_\mu c)^a \,, \tag{78.7}$$

where $(\bar{D}^\mu \bar{c})^a = \bar{D}^{\mu ab}\bar{c}^b$ and $(D_\mu c)^a = D^{ac}_\mu c^c$.

Under an infinitesimal gauge transformation, the change in the fields is

$$\delta_{\rm G} A^a_\mu(x) = -D^{ac}_\mu \theta^c(x) \,, \tag{78.8}$$

$$\delta_{\rm G} c^b(x) = -ig\theta^a(x)(T^a_{\rm A})^{bc} c^c(x) \,, \tag{78.9}$$

$$\delta_{\rm G} \bar{A}^a_\mu(x) = 0 \,. \tag{78.10}$$

The antighost $\bar{c}$ transforms in the same way as c (since the adjoint representation is real). The background field $\bar{A}$ is fixed, and so does not change under a gauge transformation. Of course, this means that $\mathcal{L}_{\rm gf}$ and $\mathcal{L}_{\rm gh}$ are *not* gauge invariant; their role is to fix the gauge.

We can, however, define a *background field gauge transformation*, under which only the background field transforms,

$$\delta_{\rm BG} \bar{A}^a_\mu(x) = -\bar{D}^{ac}_\mu \theta^c(x) \,, \tag{78.11}$$

$$\delta_{\rm BG} A^a_\mu(x) = 0 \,, \tag{78.12}$$

$$\delta_{\rm BG} c^b(x) = 0 \,. \tag{78.13}$$

Obviously, $\mathcal{L}_{\rm YM}$ is invariant under this transformation (since it does not involve the background field at all), but $\mathcal{L}_{\rm gf}$ and $\mathcal{L}_{\rm gh}$ are not. However, $\mathcal{L}_{\rm gf}$ and $\mathcal{L}_{\rm gh}$ *are* invariant under the *combined* transformation $\delta_{\rm G+BG}$. For $\mathcal{L}_{\rm gh}$, as given by eq. (78.7), this follows immediately from the fact that D_μ and $\bar{D}_\mu$ have the same transformation property under the combined transformation, and that using covariant derivatives with all group indices contracted always yields a gauge-invariant expression.

To use this argument on $\mathcal{L}_{\rm gf}$, as given by eqs. (78.3) and (78.6), we need to show that $(A - \bar{A})^a_\mu$ transforms under the combined transformation in the same way as does an ordinary field in the adjoint representation, such as the ghost field in eq. (78.9). To show this, we write

$$\begin{aligned} \delta_{\rm G+BG}(A - \bar{A})^b_\mu &= -(D - \bar{D})^{ba}_\mu \theta^a \\ &= +ig(A - \bar{A})^c_\mu (T^c_{\rm A})^{ba}\theta^a \\ &= -ig\theta^a (T^a_{\rm A})^{bc}(A - \bar{A})^c_\mu \,. \end{aligned} \tag{78.14}$$

We used the complete antisymmetry of $(T^c_{\text{A}})^{ba} = -if^{cba}$ to get the last line. We see that $(A - \bar{A})^a_\mu$ transforms like an ordinary field in the adjoint representation, and so any expression that involves only covariant derivatives (either $\bar{D}_\mu$ or D_μ) acting on this field, with all group indices contracted, is invariant under the combined transformation.

Therefore, *the complete lagrangian, $\mathcal{L} = \mathcal{L}_{\text{YM}} + \mathcal{L}_{\text{gf}} + \mathcal{L}_{\text{gh}}$, is invariant under the combined transformation.*

Now consider constructing the quantum action $\Gamma(A, c, \bar{c}; \bar{A})$. Recall from section 21 that the quantum action can be expressed as the sum of all 1PI diagrams, with the external propagators replaced by the corresponding fields. In a gauge theory, the quantum action is in general *not* gauge invariant, because we had to fix a gauge in order to carry out the path integral. The quantum action thus depends on the choice of gauge, and hence (in the case of background field gauge) on the background field $\bar{A}$. This is why we have written $\bar{A}$ as an argument of Γ, but separated by a semicolon to indicate its special role.

An important property of the quantum action is that it inherits all linear symmetries of the classical action; see problem 21.2. In the present case, these symmetries include the combined gauge transformation $\delta_{\text{G+BG}}$. Therefore, *the quantum action is also invariant under the combined transformation.* The quantum action takes its simplest form if we set the external field A equal to the background field $\bar{A}$. Then, $\Gamma(\bar{A}, c, \bar{c}; \bar{A})$ is invariant under a gauge transformation of the form

$$\delta_{\text{G+BG}} \bar{A}^a_\mu(x) = -\bar{D}^{ac}_\mu \theta^c(x) \,, \tag{78.15}$$

$$\delta_{\text{G+BG}} c^b(x) = -ig\theta^a(x)(T^a_{\text{A}})^{bc} c^c(x) \,. \tag{78.16}$$

This is now simply an ordinary gauge transformation, with $\bar{A}$ as the gauge field.

The quantum action can be expressed as the classical action, plus loop corrections. For $A = \bar{A}$, we have

$$\Gamma(\bar{A}, c, \bar{c}; \bar{A}) = \int d^4x \left[-\tfrac{1}{4}\bar{F}^{a\mu\nu}\bar{F}^a_{\mu\nu} - (\bar{D}^\mu \bar{c})^a (\bar{D}_\mu c)^a \right] + \ldots \,, \tag{78.17}$$

where the ellipses stand for the loop corrections. Note that $\mathcal{L}_{\text{gf}}$ has disappeared [because we set $A = \bar{A}$ in eq. (78.6)], and $\mathcal{L}_{\text{gh}}$ has the form of a kinetic term for a complex scalar field in the adjoint representation. This term is therefore manifestly gauge invariant, as is the $\bar{F}\bar{F}$ term.

The gauge invariance of the quantum action has an important consequence for the loop corrections. In background field gauge, the renormalizing

Z factors must respect the gauge invariance of the quantum action. Therefore, using the notation of section 73, we must have

$$Z_1 = Z_2 \,, \tag{78.18}$$

$$Z_{1'} = Z_{2'} \,, \tag{78.19}$$

$$Z_3 = Z_{3g} = Z_{4g} \,. \tag{78.20}$$

Thus the relation between the bare and renormalized gauge couplings becomes

$$g_0^2 = Z_3^{-1} g^2 \tilde{\mu}^{\varepsilon} \,. \tag{78.21}$$

This relation now involves only Z_3. We can therefore compute the beta function from Z_3 alone. This is the major advantage of background field gauge.

To compute the loop corrections, we need to evaluate 1PI diagrams in background field gauge with the external propagators removed and replaced with external fields; the external gauge field should be set equal to the background field. The easiest way to do this is to set

$$A = \bar{A} + \mathcal{A} \tag{78.22}$$

at the beginning, and to write the path integral in terms of $\mathcal{A}$. Then the $\mathcal{A}$ field appears only on internal lines, and the $\bar{A}$ field only on external lines. The gauge-fixing term now reads

$$\mathcal{L}_{\text{gf}} = -\tfrac{1}{2}\xi^{-1}(\bar{D}^{\mu}\mathcal{A}_{\mu})^a(\bar{D}^{\nu}\mathcal{A}_{\nu})^a \,, \tag{78.23}$$

and the ghost term is given by eq. (78.7).

The Feynman rules that follow from $\mathcal{L}_{\text{YM}} + \mathcal{L}_{\text{gf}} + \mathcal{L}_{\text{gh}}$ are closely related to those we found in R_ξ gauge in section 72. The ghost and gluon propagators are the same, and vertices involving all *internal* lines are also the same. But if one or more gluon lines are *external*, then there are additional contributions to the vertices from $\mathcal{L}_{\text{gf}}$ and $\mathcal{L}_{\text{gh}}$. We leave the details to problem 78.1.

Further simplifications arise at the one-loop level. Using eq. (78.22) in eq. (78.2), we find

$$\begin{aligned} F^a_{\mu\nu} &= \partial_\mu \bar{A}^a_\nu - \partial_\nu \bar{A}^a_\mu + g f^{abc}\bar{A}^b_\mu \bar{A}^c_\nu \\ &\quad + \partial_\mu \mathcal{A}^a_\nu - \partial_\nu \mathcal{A}^a_\mu + g f^{abc}(\bar{A}^b_\mu \mathcal{A}^c_\nu + \mathcal{A}^b_\mu \bar{A}^c_\nu) + g f^{abc}\mathcal{A}^b_\mu \mathcal{A}^c_\nu \\ &= \bar{F}^a_{\mu\nu} + (\bar{D}_\mu \mathcal{A}_\nu)^a - (\bar{D}_\nu \mathcal{A}_\mu)^a + g f^{abc}\mathcal{A}^b_\mu \mathcal{A}^c_\nu \,. \end{aligned} \tag{78.24}$$

We then have

$$\mathcal{L}_{\rm YM} = -\tfrac{1}{4}\bar{F}^{a\mu\nu}\bar{F}^a_{\mu\nu} - \tfrac{1}{2}(\bar{D}^\mu\mathcal{A}^\nu)^a(\bar{D}_\mu\mathcal{A}_\nu)^a + \tfrac{1}{2}(\bar{D}^\mu\mathcal{A}^\nu)^a(\bar{D}_\nu\mathcal{A}_\mu)^a$$
$$- \tfrac{1}{2}gf^{abc}\bar{F}^{a\mu\nu}\mathcal{A}^b_\mu\mathcal{A}^c_\nu + \ldots\;, \qquad (78.25)$$

where the ellipses stand for terms that are linear, cubic, or quartic in $\mathcal{A}$. Vertices arising from terms linear in $\mathcal{A}$ cannot appear in a 1PI diagram, and the cubic and quartic vertices do not appear in the one-loop contribution to the $\bar{A}$ propagator.

The last term on the first line of eq. (78.25) can be usefully manipulated with some dummy-index relabelings and integrations by parts; we have

$$\begin{aligned}(\bar{D}^\mu\mathcal{A}^\nu)^a(\bar{D}_\nu\mathcal{A}_\mu)^a &= (\bar{D}^\nu\mathcal{A}_\mu)^c(\bar{D}^\mu\mathcal{A}_\nu)^c \\ &= -\mathcal{A}^b_\mu(\bar{D}^\nu\bar{D}^\mu)^{bc}\mathcal{A}^c_\nu \\ &= -\mathcal{A}^b_\mu(\bar{D}^\mu\bar{D}^\nu - [\bar{D}^\mu,\bar{D}^\nu])^{bc}\mathcal{A}^c_\nu \\ &= -\mathcal{A}^b_\mu(\bar{D}^\mu\bar{D}^\nu)^{bc}\mathcal{A}^c_\nu - ig(T^a_{\rm A})^{bc}\bar{F}^{a\mu\nu}\mathcal{A}^b_\mu\mathcal{A}^c_\nu \\ &= +(\bar{D}^\mu\mathcal{A}_\mu)^c(\bar{D}^\nu\mathcal{A}_\nu)^c - gf^{abc}\bar{F}^{a\mu\nu}\mathcal{A}^b_\mu\mathcal{A}^c_\nu\;. \qquad (78.26)\end{aligned}$$

Now the first term on the right-hand side of eq. (78.26) has the same form as the gauge-fixing term. If we choose $\xi = 1$, these two terms will cancel.

Setting $\xi = 1$, and including a renormalizing factor of Z_3, the terms of interest in the complete lagrangian become

$$\mathcal{L} = -\tfrac{1}{4}Z_3\bar{F}^{a\mu\nu}\bar{F}^a_{\mu\nu} - \tfrac{1}{2}Z_3(\bar{D}^\mu\mathcal{A}^\nu)^a(\bar{D}_\mu\mathcal{A}_\nu)^a - (\bar{D}^\mu\bar{c})^a(\bar{D}_\mu c)^a$$
$$- Z_3 gf^{abc}\bar{F}^{a\mu\nu}\mathcal{A}^b_\mu\mathcal{A}^c_\nu\;. \qquad (78.27)$$

In the ghost term, we have replaced D_μ with $\bar{D}_\mu$; the vertex corresponding to the dropped $\mathcal{A}$ term does not appear in the one-loop contribution to the $\bar{A}$ propagator. Also, we can rescale $\mathcal{A}$ to absorb Z_3 in all terms except the first; since $\mathcal{A}$ never appears on an external line, its normalization is irrelevant, and always cancels among propagators and vertices. (The same is true of the ghost field.)

The one-loop diagrams that contribute to the $\bar{A}$ propagator are shown in fig. 78.1. The dashed lines in the first two diagrams represent either the $\mathcal{A}$ field or the ghost fields. In either case, the second diagram vanishes, because it is proportional to $\int d^4\ell/\ell^2$, which is zero after dimensional regularization.

Note that the ghost term in eq. (78.27) has the form of a kinetic term for a complex scalar field in the adjoint representation. In problem 73.1, we found the contribution of a complex scalar field in a representation $\rm R_{CS}$ to

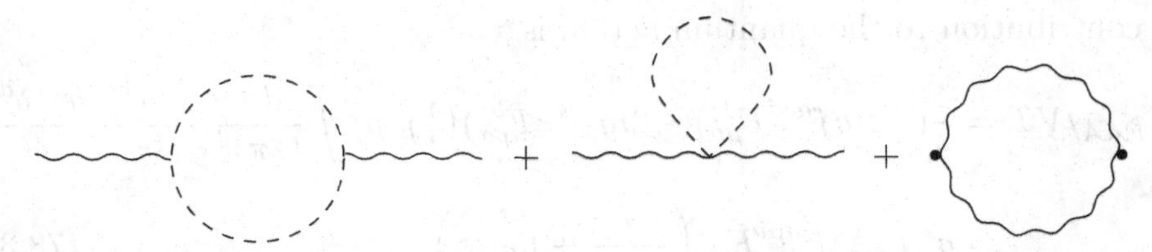

Figure 78.1. The one-loop contributions to the $\bar{A}$ propagator in background field gauge; the dashed lines can be either ghosts or internal $\mathcal{A}$ gauge fields. The dots denote the $\bar{F}\mathcal{A}\mathcal{A}$ vertex.

$\Pi(k^2)$ is

$$\Pi_{\rm CS}(k^2) = -\frac{g^2}{24\pi^2}\,T(\mathrm{R_{CS}})\,\frac{1}{\varepsilon} + \text{finite} \,. \tag{78.28}$$

The ghost contribution is minus this, with $\mathrm{R_{CS}} \to \mathrm{A}$. (The minus sign is from the closed ghost loop.) Thus we have

$$\Pi_{\rm gh}(k^2) = +\frac{g^2}{24\pi^2}\,T(\mathrm{A})\,\frac{1}{\varepsilon} + \text{finite} + O(g^4) \,. \tag{78.29}$$

For reference we recall that the counterterm contribution is

$$\Pi_{\rm ct}(k^2) = -(Z_3 - 1) \,. \tag{78.30}$$

Next we consider the diagrams with $\mathcal{A}$ fields in the loop. If the $\bar{F}\mathcal{A}\mathcal{A}$ interaction term was absent, the calculation would again be a familiar one; the $\bar{D}\mathcal{A}\bar{D}\mathcal{A}$ term in eq. (78.27) has the form of a kinetic term for a real scalar field that carries an extra index ν. That this index is a Lorentz vector index is immaterial for the diagrammatic calculation; the index is simply summed around the loop, yielding an extra factor of $d = 4$. There is also an extra factor of one-half (relative to the case of a complex scalar) because $\mathcal{A}$ is real rather than complex. (Equivalently, the diagram has a symmetry factor of $S = 2$ from exchange of the top and bottom internal propagators when they do not carry charge arrows.) We thus have

$$\Pi_{\bar{D}\mathcal{A}\bar{D}\mathcal{A}}(k^2) = -\frac{g^2}{12\pi^2}\,T(\mathrm{A})\,\frac{1}{\varepsilon} + \text{finite} + O(g^4) \,. \tag{78.31}$$

If we now include the $\bar{F}\mathcal{A}\mathcal{A}$ interaction, we can think of $\bar{F}^a_{\mu\nu}$ as a constant external field. We can then draw the third diagram of fig. 78.1, where each dot denotes a vertex factor of $-2igf^{abc}\bar{F}^a_{\mu\nu}$. This vacuum diagram has a symmetry factor of $S = 2 \times 2$: one factor of two for exchanging the top and bottom propagators, and one for exchanging the left- and right-hand sources.

Its contribution to the quantum action is

$$i\Gamma_{\bar{F}\mathcal{AA}}/VT = \frac{1}{4}(-2igf^{acd}\bar{F}^{a}_{\mu\nu})(-2igf^{beg}\bar{F}^{b}_{\rho\sigma})\left(\tfrac{1}{i}\right)^2 \tilde{\mu}^{\varepsilon}\int\frac{d^d\ell}{(2\pi)^d}\,\frac{g^{\mu\rho}\delta^{ce}}{\ell^2}\,\frac{g^{\nu\sigma}\delta^{dg}}{\ell^2}$$

$$= g^2\,T(\mathrm{A})\bar{F}^{a\mu\nu}\bar{F}^{a}_{\mu\nu}\left(\frac{i}{8\pi^2\varepsilon}+\text{finite}\right), \tag{78.32}$$

where VT is the volume of spacetime. Comparing this with the tree-level lagrangian $-\frac{1}{4}Z_3\bar{F}\bar{F}$, and recalling eq. (78.30), we see that eq. (78.32) is equivalent to a contribution to $\Pi(k^2)$ of

$$\Pi_{\bar{F}\mathcal{AA}}(k^2) = +\frac{g^2}{2\pi^2}\,T(\mathrm{A})\,\frac{1}{\varepsilon}+\text{finite}\,. \tag{78.33}$$

There is also a one-loop diagram with one $\bar{D}\mathcal{A}\bar{D}\mathcal{A}$ vertex and one $\bar{F}\mathcal{AA}$ vertex; however, contracting the vector indices on the $\mathcal{A}$ fields around the loop leads to a factor of $\bar{F}^{\mu\nu}g_{\mu\nu}=0$. Similarly, a one-loop diagram with a single $\bar{F}\mathcal{AA}$ vertex vanishes.

We could also couple the gauge field to a Dirac fermion in the representation R_{DF}, and a complex scalar in the representation R_{CS}. The corresponding contributions to $\Pi(k^2)$ were computed in section 73, and are given by eq. (78.28) and

$$\Pi_{\mathrm{DF}}(k^2) = -\frac{g^2}{6\pi^2}\,T(\mathrm{R}_{\mathrm{DF}})\,\frac{1}{\varepsilon}+\text{finite}\,. \tag{78.34}$$

Adding up eqs. (78.28)–(78.31), (78.33), and (78.34), we find that finiteness of $\Pi(k^2)$ requires

$$Z_3 = 1+\frac{g^2}{24\pi^2}\Big[\Big(+1-2+12\Big)T(\mathrm{A})-4T(\mathrm{R}_{\mathrm{DF}})-T(\mathrm{R}_{\mathrm{CS}})\Big]\frac{1}{\varepsilon}+O(g^4) \tag{78.35}$$

in the $\overline{\mathrm{MS}}$ renormalization scheme.

The analysis of section 28 now results in a beta function of

$$\beta(g) = -\frac{g^3}{48\pi^2}\Big[11T(\mathrm{A})-4T(\mathrm{R}_{\mathrm{DF}})-T(\mathrm{R}_{\mathrm{CS}})\Big]+O(g^5)\,. \tag{78.36}$$

A Majorana fermion or a Weyl fermion makes half the contribution of a Dirac fermion in the same representation; a real scalar field makes half the contribution of a complex scalar field. (Majorana fermions and real scalars must be in real representations of the gauge group.)

In quantum chromodynamics, the gauge group is SU(3), and there are $n_{\mathrm{F}}=6$ flavors of quarks (which are Dirac fermions) in the fundamental representation. We therefore have $T(\mathrm{A})=3$, $T(\mathrm{R}_{\mathrm{DF}})=\frac{1}{2}n_{\mathrm{F}}$, and $T(\mathrm{R}_{\mathrm{CS}})=0$;

therefore

$$\beta(g) = -\frac{g^3}{16\pi^2}\left(11 - \tfrac{2}{3}n_{\rm F}\right) + O(g^5)\ . \tag{78.37}$$

We see that the beta function is negative for $n_{\rm F} \le 16$, and so QCD is asymptotically free.

Problems

78.1) Compute the tree-level vertex factors in background field gauge for all vertices that connect one or more external gluons with two or more internal lines (ghost or gluon).

78.2) Our one-loop corrections can be interpreted as functional determinants. Define

$$\Box_{{\rm R},(a,b)} \equiv \bar{D}^2 + gT^a_{\rm R}\bar{F}^a_{\mu\nu}S^{\mu\nu}_{(a,b)}\ , \tag{78.38}$$

where $\bar{D}_\mu = \partial_\mu - ig(T^a_{\rm R})\bar{A}^a_\mu$ is the background-covariant derivative in the representation R, implicitly multiplied by the identity matrix for the (a,b) representation of the Lorentz group, and $S^{\mu\nu}_{(a,b)}$ are the Lorentz generators for that representation; in particular,

$$S^{\mu\nu}_{(1,1)} = 0\ , \tag{78.39}$$

$$S^{\mu\nu}_{(2,1)\oplus(1,2)} = \tfrac{i}{4}[\gamma^\mu,\gamma^\nu]\ , \tag{78.40}$$

$$(S^{\mu\nu}_{(2,2)})_{\alpha\beta} = -i(\delta^\mu{}_\alpha\delta^\nu{}_\beta - \delta^\nu{}_\alpha\delta^\mu{}_\beta)\ . \tag{78.41}$$

Show that the one-loop contribution to the terms in the quantum action that do not depend on the ghost fields is given by

$$\begin{aligned}\exp i\Gamma_{1-{\rm loop}}(\bar{A},0,0;\bar{A}) \propto\ & (\det \Box_{{\rm A},(1,1)})^{+1} \\ & \times(\det \Box_{{\rm A},(2,2)})^{-1/2} \\ & \times(\det \Box_{{\rm R_{DF}},(2,1)\oplus(1,2)})^{+1/2} \\ & \times(\det \Box_{{\rm R_{CS}},(1,1)})^{-1}\ .\end{aligned} \tag{78.42}$$

Verify that this expression agrees with the diagrammatic analysis in this section.

79

Gervais–Neveu gauge

Prerequisite: 78

In section 78, we used background field gauge to set up the computation of a quantum action that is gauge invariant. Given this quantum action, we can use it to compute scattering amplitudes via the corresponding tree diagrams, as discussed in section 19. Since the ghost fields in the quantum action do not contribute to tree diagrams, we can simply drop all the ghost terms in the quantum action.

Because the quantum action computed in background field gauge is itself gauge invariant, it requires further gauge fixing to specify the gluon propagator and vertices. We can choose whatever gauge is most convenient; for example, R_ξ gauge. In principle, this gauge fixing involves introducing *new* ghost fields, but, once again, these do not contribute to tree diagrams, and so we can ignore them.

If we start with the tree-level approximation to the quantum action, then, in R_ξ gauge, we simply get the gluon propagator and vertices of section 72. As we noted there, the complexity of the three- and four-gluon vertices in R_ξ gauge leads to long, involved computations of even simple processes like gluon-gluon scattering.

In this section, we will introduce another gauge, *Gervais–Neveu gauge*, that simplifies these tree-level computations.

We begin by specializing to the gauge group SU(N), and working with the matrix-valued field $A_\mu = A^a_\mu T^a$. For later convenience, we will normalize the generators via

$$\operatorname{Tr} T^a T^b = \delta^{ab} \; . \tag{79.1}$$

With this choice, their commutation relations become

$$[T^a, T^b] = i\sqrt{2} f^{abc} T^c \; . \tag{79.2}$$

The tree-level action is specified by the Yang–Mills lagrangian,

$$\mathcal{L}_{\text{YM}} = -\tfrac{1}{4}\text{Tr}\, F^{\mu\nu}F_{\mu\nu}\ , \tag{79.3}$$

where the matrix-valued field strength is

$$F_{\mu\nu} = \partial_\mu A_\nu - \partial_\nu A_\mu - \tfrac{ig}{\sqrt{2}}[A_\mu, A_\nu]\ . \tag{79.4}$$

Let us introduce the matrix-valued complex tensor

$$H_{\mu\nu} \equiv \partial_\mu A_\nu - \tfrac{ig}{\sqrt{2}} A_\mu A_\nu\ . \tag{79.5}$$

Then $F_{\mu\nu}$ is the antisymmetric part of $H_{\mu\nu}$,

$$F_{\mu\nu} = H_{\mu\nu} - H_{\nu\mu}\ . \tag{79.6}$$

The Yang–Mills lagrangian can now be written as

$$\mathcal{L}_{\text{YM}} = -\tfrac{1}{2}\text{Tr}\Big(H^{\mu\nu}H_{\mu\nu} - H^{\mu\nu}H_{\nu\mu}\Big)\ . \tag{79.7}$$

To fix the gauge, we choose a matrix-valued gauge-fixing function $G(x)$, and add

$$\mathcal{L}_{\text{gf}} = -\tfrac{1}{2}\text{Tr}\, GG \tag{79.8}$$

to $\mathcal{L}_{\text{YM}}$. Here we have set the gauge parameter ξ to one, and ignored the ghost lagrangian (since, as we have already discussed, it does not affect tree diagrams). The choice of G that yields Gervais–Neveu gauge is

$$G = H^\mu{}_\mu\ . \tag{79.9}$$

At first glance, this choice seems untenable, because we see from eq. (79.5) that this G (and hence $\mathcal{L}_{\text{gf}}$) is not hermitian. However, because the role of $\mathcal{L}_{\text{gf}}$ is merely to fix the gauge, it is acceptable for $\mathcal{L}_{\text{gf}}$ to be nonhermitian.

Combining eqs. (79.7) and (79.8), we get a total, gauge-fixed lagrangian

$$\mathcal{L} = -\tfrac{1}{2}\text{Tr}\Big(H^{\mu\nu}H_{\mu\nu} - H^{\mu\nu}H_{\nu\mu} + H^\mu{}_\mu H^\nu{}_\nu\Big)\ . \tag{79.10}$$

Consider the terms in $\mathcal{L}$ with two derivatives. After some integrations by parts, those from the third term in eq. (79.10) cancel those from the second, leading to

$$\mathcal{L}_{2\partial} = -\tfrac{1}{2}\text{Tr}\,\partial^\mu A^\nu \partial_\mu A_\nu\ , \tag{79.11}$$

just as in R_ξ gauge with $\xi = 1$. Now consider the terms with no derivatives. Once again, those from the third term in eq. (79.10) cancel those from the second (after using the cyclic property of the trace), leading to

$$\mathcal{L}_{0\partial} = +\tfrac{1}{4}g^2 \,\mathrm{Tr}\, A^\mu A^\nu A_\mu A_\nu \,. \tag{79.12}$$

Finally, we have the terms with one derivative,

$$\mathcal{L}_{1\partial} = +\frac{ig}{\sqrt{2}}\,\mathrm{Tr}\Big(\partial^\mu A^\nu A_\mu A_\nu - \partial^\mu A^\nu A_\nu A_\mu + \partial^\mu A_\mu A^\nu A_\nu\Big)\,. \tag{79.13}$$

Each derivative acts only on the field to its immediate right. If we integrate by parts in the last term in eq. (79.13), we generate two terms; one of these cancels the first term in eq. (79.13), and the other duplicates the second. Thus we have

$$\mathcal{L}_{1\partial} = -i\sqrt{2}g\,\mathrm{Tr}\,\partial^\mu A^\nu A_\nu A_\mu \,. \tag{79.14}$$

Combining eqs. (79.11), (79.12), and (79.14), we find

$$\mathcal{L} = \mathrm{Tr}\Big(-\tfrac{1}{2}\partial^\mu A^\nu \partial_\mu A_\nu - i\sqrt{2}g\,\partial^\mu A^\nu A_\nu A_\mu + \tfrac{1}{4}g^2 A^\mu A^\nu A_\mu A_\nu\Big)\,. \tag{79.15}$$

Because this lagrangian has a rather simple structure in terms of the matrix-valued field A_μ, it is helpful to stick with this notation, rather than trying to reexpress $\mathcal{L}$ in terms of $A^a_\mu = \mathrm{Tr}(T^a A_\mu)$. In the next section, we explore the Feynman rules for a matrix-valued field in a simplified context.

Reference notes

Gervais–Neveu gauge and some interesting variations are discussed in *Siegel*.

80

The Feynman rules for $N \times N$ matrix fields

Prerequisite: 10

In section 79, we found that the lagrangian for SU(N) Yang–Mills theory in Gervais–Neveu gauge is

$$\mathcal{L} = \mathrm{Tr}\Big(-\tfrac{1}{2}\partial^\mu A^\nu \partial_\mu A_\nu - i\sqrt{2}g\, \partial^\mu A^\nu A_\nu A_\mu + \tfrac{1}{4}g^2 A^\mu A^\nu A_\mu A_\nu\Big)\,, \qquad (80.1)$$

where $A_\mu(x)$ is a traceless hermitian $N \times N$ matrix. In this section, we will work out the Feynman rules for a simplified model of a scalar field that keeps the essence of the matrix structure.

Let $B(x)$ be a hermitian $N \times N$ matrix that is *not* traceless. Let T^a be a complete set of N^2 hermitian $N \times N$ matrices normalized according to

$$\mathrm{Tr}\, T^a T^b = \delta^{ab}\,. \qquad (80.2)$$

We will take one of these matrices, T^{N^2}, to be proportional to the identity matrix; then eq. (80.2) requires the rest of the T^as to be traceless. We can expand $B(x)$ in the T^as, with coefficient fields $B^a(x)$,

$$B(x) = B^a(x) T^a\,, \qquad (80.3)$$

$$B^a(x) = \mathrm{Tr}\, T^a B(x)\,, \qquad (80.4)$$

where the repeated index in eq. (80.3) is implicitly summed over $a = 1$ to N^2.

Consider a lagrangian for $B(x)$ of the form

$$\mathcal{L} = \mathrm{Tr}\Big(-\tfrac{1}{2}\partial^\mu B \partial_\mu B + \tfrac{1}{3}gB^3 - \tfrac{1}{4}\lambda B^4\Big)\,. \qquad (80.5)$$

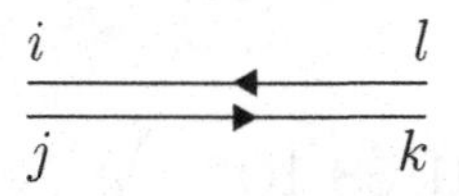

Figure 80.1. The double-line notation for the propagator of a hermitian matrix field.

Using eqs. (80.2) and (80.3), we find an expression for $\mathcal{L}$ in terms of the coefficient fields,

$$\mathcal{L} = -\tfrac{1}{2}\partial^\mu B^a \partial_\mu B^a + \tfrac{1}{3}g \,\mathrm{Tr}(T^a T^b T^c) B^a B^b B^c \\ -\tfrac{1}{4}\lambda \,\mathrm{Tr}(T^a T^b T^c T^d) B^a B^b B^c B^d \,. \tag{80.6}$$

It is easy to read off the Feynman rules from this form of $\mathcal{L}$. The propagator for the coefficient field B^a is

$$\tilde{\Delta}^{ab}(k^2) = \frac{\delta^{ab}}{k^2 - i\epsilon} \,. \tag{80.7}$$

There is a three-point vertex with vertex factor $2ig\,\mathrm{Tr}(T^a T^b T^c)$, and a four-point vertex with vertex factor $-6i\lambda\,\mathrm{Tr}(T^a T^b T^c T^d)$. This clearly leads to messy and complicated formulae for scattering amplitudes.

Instead, let us work with $\mathcal{L}$ in the form of eq. (80.5). Writing the matrix indices explicitly, with one up and one down (and employing the rule that two indices can be contracted only if one is up and one is down), we have $B(x)_i{}^j = B^a(x)(T^a)_i{}^j$. This implies that the propagator for $B_i{}^j$ is

$$\tilde{\Delta}_i{}^j{}_k{}^l(k^2) = \frac{(T^a)_i{}^j (T^a)_k{}^l}{k^2 - i\epsilon} \,. \tag{80.8}$$

Since the T^a matrices form a complete set, there is a completeness relation of the form $(T^a)_i{}^j (T^a)_k{}^l \propto \delta_i{}^l \delta_k{}^j$. To get the constant of proportionality, set $j{=}k$ and $l{=}i$ to turn the left-hand side into $(T^a)_i{}^k (T^a)_k{}^i = \mathrm{Tr}(T^a T^a)$, and the right-hand side into $\delta_i{}^i \delta_k{}^k = N^2$. From eq. (80.2) we have $\mathrm{Tr}(T^a T^a) = \delta^{aa} = N^2$. So the constant of proportionality is one, and

$$(T^a)_i{}^j (T^a)_k{}^l = \delta_i{}^l \delta_k{}^j \,. \tag{80.9}$$

We can represent the B propagator with a double-line notation, as shown in fig. 80.1. The arrow on each line points from an up index to a down index. Since the interactions are simple matrix products, with an up index from one field contracted with a down index from an adjacent field, the vertices follow the pattern shown in fig. 80.2. Since an n-point vertex of this type has only an n-fold cyclic symmetry (rather than an $n!$-fold permutation symmetry), the vertex factor is i times the coefficient of $\mathrm{Tr}(B^n)$ in $\mathcal{L}$ times n (rather

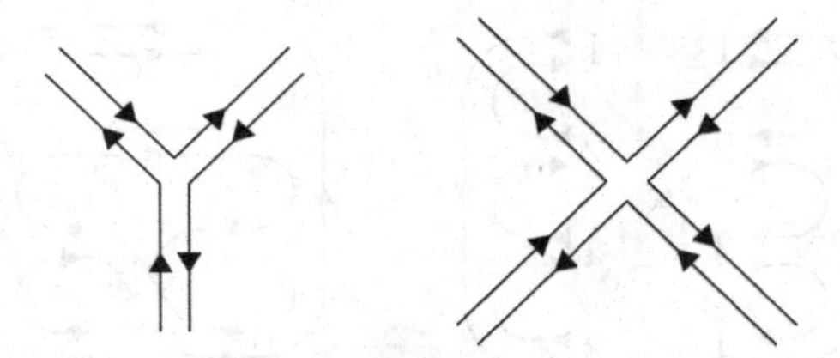

Figure 80.2. 3- and 4-point vertices in the double-line notation.

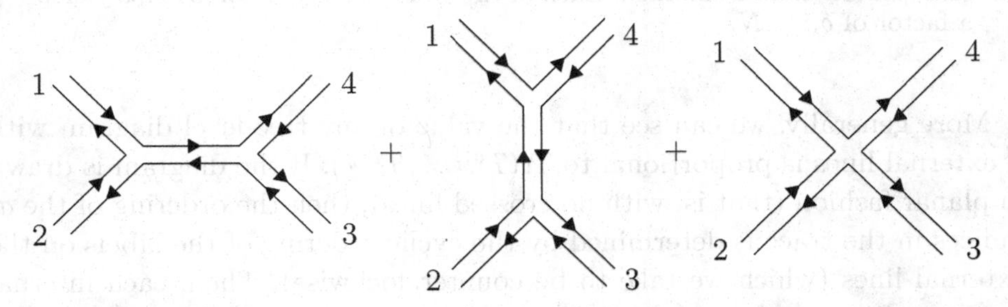

Figure 80.3. Tree diagrams with four external lines. Five more diagrams of each of these three types, with the external labels 2, 3, and 4 permuted, also contribute.

than $n!$). Thus, for the lagrangian of eq. (80.5), the 3- and 4-point vertex factors are ig and $-i\lambda$.

Now consider a scattering process. Particles corresponding to the coefficient fields labeled by the indices a_1 and a_2 (and with four-momenta k_1 and k_2) scatter into particles corresponding to the coefficient fields labeled by the indices a_3 and a_4 (and with four-momenta k_3 and k_4). We wish to compute the scattering amplitude for this process, at tree level.

There are 18 contributing Feynman diagrams. Three are shown in fig. 80.3; the remaining 15 are obtained by making *noncylic* permutations of the labels $1, 2, 3, 4$ (equivalent to making unrestricted permutations of $2, 3, 4$). For simplicity, we will treat all external momenta as outgoing; then k_1^0 and k_2^0 are negative, and $k_1 + k_2 + k_3 + k_4 = 0$. Each external line carries a factor of T^{a_i}, with its matrix indices contracted by following the arrows backward through the diagrams. Omitting the $i\epsilon$s in the propagators (which are not relevant for tree diagrams), the resulting tree-level amplitude is

$$i\mathcal{T} = \mathrm{Tr}(T^{a_1}T^{a_2}T^{a_3}T^{a_4})\left(\frac{(ig)^2(-i)}{(k_1{+}k_2)^2} + \frac{(ig)^2(-i)}{(k_1{+}k_4)^2} - i\lambda\right)$$

$$+ \Big((234) \to (342), (423), (243), (432), (324)\Big). \qquad (80.10)$$

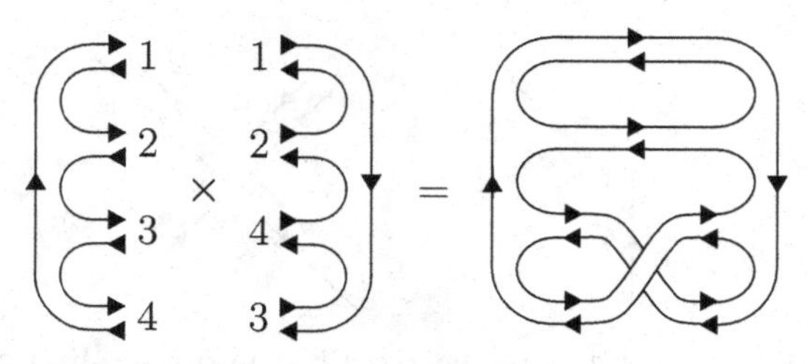

Figure 80.4. Evaluation of $\mathrm{Tr}(T^{a_1}T^{a_2}T^{a_3}T^{a_4})[\mathrm{Tr}(T^{a_1}T^{a_2}T^{a_4}T^{a_3})]^*$, with all repeated indices summed. Each of the two closed single-line loops yields a factor of $\delta_i{}^i = N$.

More generally, we can see that the value of any tree-level diagram with n external lines is proportional to $\mathrm{Tr}(T^{a_{i_1}} \ldots T^{a_{i_n}})$. If the diagram is drawn in planar fashion (that is, with no crossed lines), then the ordering of the a_i indices in the trace is determined by the cyclic ordering of the labels on the external lines (which we take to be counterclockwise). Then, each internal line contributes a factor of $-i/k^2$, each 3-point vertex a factor of ig, and each four-point vertex a factor of $-i\lambda$. These are the *color-ordered* Feynman rules for this theory.

Return now to $i\mathcal{T}$ as given by eq. (80.10). Suppose that we wish to square this amplitude, and sum over all possible particle types for each incoming or outgoing particle. We then have to evaluate expressions like

$$\mathrm{Tr}(T^{a_1}T^{a_2}T^{a_3}T^{a_4})[\mathrm{Tr}(T^{a_1}T^{a_2}T^{a_4}T^{a_3})]^* , \tag{80.11}$$

with all repeated indices summed. Using the hermiticity of the T^a matrices, we have

$$[\mathrm{Tr}(T^{a_1} \ldots T^{a_n})]^* = \mathrm{Tr}(T^{a_n} \ldots T^{a_1}) . \tag{80.12}$$

It is then easiest to evaluate eq. (80.11) diagrammatically, as shown in fig. 80.4. Each closed single-line loop yields a factor of $\delta_i{}^i = N$. The result is that the absolute square of any particular trace yields a factor of N^4, and the product of any trace times the complex conjugate of any other different trace yields a factor of N^2.

The coefficient of both $\mathrm{Tr}(T^{a_1}T^{a_2}T^{a_3}T^{a_4})$ and $\mathrm{Tr}(T^{a_1}T^{a_4}T^{a_3}T^{a_2})$ in eq. (80.10) is

$$A_3 \equiv \frac{g^2}{(k_1+k_2)^2} + \frac{g^2}{(k_1+k_4)^2} - \lambda . \tag{80.13}$$

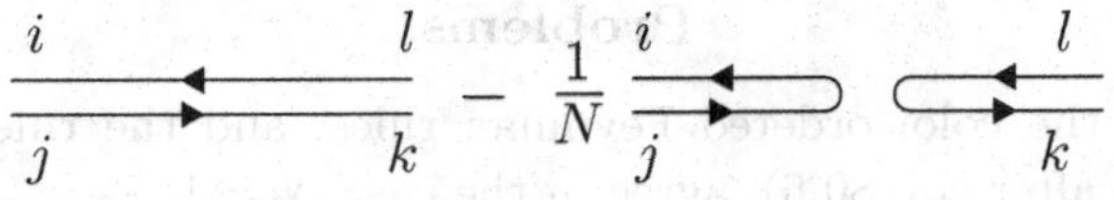

Figure 80.5. The propagator for a traceless hermitian field.

Similarly, the coefficient of both $\mathrm{Tr}(T^{a_1}T^{a_3}T^{a_4}T^{a_2})$ and $\mathrm{Tr}(T^{a_1}T^{a_2}T^{a_4}T^{a_3})$ is

$$A_4 \equiv \frac{g^2}{(k_1{+}k_3)^2} + \frac{g^2}{(k_1{+}k_2)^2} - \lambda \,, \tag{80.14}$$

and of both $\mathrm{Tr}(T^{a_1}T^{a_4}T^{a_2}T^{a_3})$ and $\mathrm{Tr}(T^{a_1}T^{a_3}T^{a_2}T^{a_4})$ is

$$A_2 \equiv \frac{g^2}{(k_1{+}k_4)^2} + \frac{g^2}{(k_1{+}k_3)^2} - \lambda \,. \tag{80.15}$$

Thus we have

$$\begin{aligned}\sum_{a_1,a_2,a_3,a_4} |\mathcal{T}|^2 &= (2N^4 + 2N^2)\sum_j |A_j|^2 + 4N^2 \sum_{j\neq k} A_j^* A_k \\ &= (2N^4 - 2N^2)\sum_j |A_j|^2 + 4N^2 \Big(\sum_j A_j^*\Big)\Big(\sum_k A_k\Big) \,,\end{aligned} \tag{80.16}$$

where j and k are summed over $2, 3, 4$.

Now suppose we wish to impose the condition that the matrix field B is *traceless*: $\mathrm{Tr}\, B = 0$. This means that we eliminate the component field with $a{=}N^2$, corresponding to the matrix $T^{N^2} = N^{-1/2} I$. We must also eliminate T^{N^2} from the sum in eq. (80.9), leading to

$$(T^a)_i{}^j (T^a)_k{}^l = \delta_i{}^l \delta_k{}^j - \tfrac{1}{N}\delta_i{}^j \delta_k{}^l \,. \tag{80.17}$$

This can all be done diagrammatically by replacing the propagator in fig. 80.1 with the one in fig. 80.5. (The kinematic factor, $-i/k^2$, is unchanged.) Fig. 80.5 must now be used as the internal propagator in the diagrams of fig. 80.3. Also, when we multiply one diagram by the complex conjugate of another in the computation of $\sum_{a_1 \ldots a_n} |\mathcal{T}|^2$, we must use the propagator of fig. 80.5 to connect the external line of one diagram with the matching external line of the complex conjugate diagram. Although these computations are still straightforward, they can become considerably more involved.

Problems

80.1) Show that the color-ordered Feynman rules, and the rules for component fields given after eq. (80.6), agree in the case $N = 1$.

80.2) Verify the results quoted after eq. (80.12).

80.3) Compute $\sum_{a_1,a_2,a_3,a_4} |\mathrm{Tr}(T^{a_1}T^{a_2}T^{a_3}T^{a_4})|^2$ for the case of traceless T^as.

80.4) *The large-N limit.* Let $\lambda = cg^2$, where c is a number of order one. Now consider evaluating the path integral, without sources, as a function of g and N,

$$Z(g,N) = e^{iW(g,N)} = \int \mathcal{D}B\, e^{i\int d^dx\, \mathcal{L}} , \tag{80.18}$$

where $W(g,N)$ is normalized by $W(0,N) = 0$. As usual, W can be expressed as a sum of connected vacuum diagrams, which we draw in the double-line notation. Consider a diagram with V_3 three-point vertices, V_4 four-point vertices, E propagators or *edges*, and F closed single-line loops or *faces*.

a) Find the dependence on g and N of a diagram specified by the values of V_3, V_4, E, and F.

b) Express E for a vacuum diagram in terms of V_3 and V_4.

c) Recall, derive, or look up the formula for the *Euler character* χ of the two-dimensional surface of a polyhedron in terms of the values of $V \equiv V_3 + V_4$, E, and F. The Euler character is related to the *genus* $\mathcal{G}$ of the surface by $\chi = 2 - 2\mathcal{G}$; $\mathcal{G}$ counts the number of handles, so that a sphere has genus zero, a torus has genus one, etc.

d) Consider the limit $g \to 0$ and $N \to \infty$ with the *'t Hooft coupling* $\bar{\lambda} = g^2N$ held fixed (and not necessarily small). Show that $W(\bar{\lambda}, N)$ has a *topological expansion* of the form

$$W(\bar{\lambda}, N) = \sum_{\mathcal{G}=0}^{\infty} N^{2-2\mathcal{G}}\, W_{\mathcal{G}}(\bar{\lambda}) , \tag{80.19}$$

where $W_{\mathcal{G}}(\bar{\lambda})$ is given by a sum over diagrams that form polyhedra with genus $\mathcal{G}$. In particular, the leading term, $W_0(\bar{\lambda})$, is given by a sum over diagrams with spherical topology, also known as *planar diagrams*.

81

Scattering in quantum chromodynamics

Prerequisites: 60, 79, 80

In section 79, we found that the lagrangian for SU(N) Yang–Mills theory in Gervais–Neveu gauge is

$$\mathcal{L} = \text{Tr}\Big(-\tfrac{1}{2}\partial^\mu A^\nu \partial_\mu A_\nu - i\sqrt{2}g\,\partial^\mu A^\nu A_\nu A_\mu + \tfrac{1}{4}g^2 A^\mu A^\nu A_\mu A_\nu\Big)\,, \qquad (81.1)$$

where $A_\mu(x)$ is a traceless hermitian $N \times N$ matrix. For quantum chromodynamics, $N = 3$, but we will leave N unspecified in our calculations. In section 80, we worked out the *color-ordered Feynman rules* for a scalar matrix field; the same technology applies here as well. In particular, we draw each tree diagram in planar fashion (that is, with no crossed lines). Then the cyclic, counterclockwise ordering $i_1 \dots i_n$ of the external lines fixes the color factor as $\text{Tr}(T^{a_{i_1}} \dots T^{a_{i_n}})$, where the generator matrices are normalized via $\text{Tr}(T^a T^b) = \delta^{ab}$. The tree-level n-gluon scattering amplitude is then written as

$$\mathcal{T} = g^{n-2} \sum_{\substack{\text{noncyclic}\\ \text{perms}}} \text{Tr}(T^{a_1} \dots T^{a_n}) A(1, \dots, n)\,, \qquad (81.2)$$

where we have pulled out the coupling constant dependence, and $A(1, \dots, n)$ is a *partial amplitude* that we compute with the color-ordered Feynman rules. The partial amplitudes are cyclically symmetric,

$$A(2, \dots, n, 1) = A(1, 2, \dots, n)\,. \qquad (81.3)$$

The sum in eq. (81.2) is over all *noncyclic* permutations of $1 \dots n$, which is equivalent to a sum over *all* permutations of $2 \dots n$.

From the first term in eq. (81.1), we see that the gluon propagator is simply

$$\tilde{\Delta}_{\mu\nu}(k) = \frac{g_{\mu\nu}}{k^2 - i\epsilon} \ . \tag{81.4}$$

Here we have left out the matrix indices since we have already accounted for them with the color factor in eq. (81.2). The second and third terms in eq. (81.1) yield three- and four-gluon vertices. The three-gluon vertex factor (again without the matrix indices) is

$$\begin{aligned} i\mathbf{V}_{\mu\nu\rho}(p,q,r) &= i(-i\sqrt{2}g)(-ip_\rho g_{\mu\nu}) \\ &\qquad + [\,2 \text{ cyclic permutations of } (\mu,p), (\nu,q), (\rho,r)\,] \\ &= -i\sqrt{2}g(p_\rho g_{\mu\nu} + q_\mu g_{\nu\rho} + r_\nu g_{\rho\mu}) \ , \end{aligned} \tag{81.5}$$

where the four-momenta p, q, and r are all taken to be outgoing. The four-gluon vertex factor is simply

$$i\mathbf{V}_{\mu\nu\rho\sigma} = ig^2 g_{\mu\rho} g_{\nu\sigma} \ . \tag{81.6}$$

However, in the context of the color-ordered rules, it is simpler to designate the outgoing four-momentum on each external line as k_i, and contract the vector index with the corresponding polarization vector ε_i. (For now we suppress the helicity label $\lambda = \pm$.) In this notation, the vertex factors become

$$i\mathbf{V}_{123} = -i\sqrt{2}g\Big[(\varepsilon_1\varepsilon_2)(k_1\varepsilon_3) + (\varepsilon_2\varepsilon_3)(k_2\varepsilon_1) + (\varepsilon_3\varepsilon_1)(k_3\varepsilon_2)\Big] \ , \tag{81.7}$$

$$i\mathbf{V}_{1234} = +ig^2(\varepsilon_1\varepsilon_3)(\varepsilon_2\varepsilon_4) \ , \tag{81.8}$$

where the external lines are numbered sequentially, counterclockwise around the vertex. (Of course, if an attached line is internal, the corresponding polarization vector is simply a placeholder for an internal propagator.)

The color-ordered three-point vertex, eq. (81.7), is antisymmetric on the reversal $123 \leftrightarrow 321$, while the four-point vertex, eq. (81.8), is symmetric on the reversal $1234 \leftrightarrow 4321$. This implies the *reflection identity*,

$$A(n, \ldots, 2, 1) = (-1)^n A(1, 2, \ldots, n) \ , \tag{81.9}$$

which will be useful later.

It is clear from eqs. (81.7) and (81.8) that every term in any tree-level scattering amplitude is proportional to products of polarization vectors with each other, or with external momenta. (Actually, this follows directly from Lorentz invariance, and the fact that the scattering amplitude is linear in each polarization.) We get one momentum factor from each three-point vertex. Since every tree diagram with n external lines has no more than $n-2$

vertices, there are no more than $n-2$ momenta to contract with the n polarizations. Therefore, every term in every tree-level amplitude must include at least one product of two polarization vectors. Then, if the product of every possible pair of polarization vectors vanishes, the tree-level amplitude for that process is zero.

We will now show that this is indeed the case if all, or all but one, of the external gluons have the same helicity. (Here we are using the semantic convention of section 60: the helicity of an external gluon is specified relative to the outgoing four-momentum k_i that labels the corresponding external line. If that gluon is actually incoming—as indicated by a negative value of k_i^0—then its physical helicity is opposite to its labeled helicity.)

To proceed, we recall from section 60 some formulae for products of polarization vectors in the spinor-helicity formalism,

$$\varepsilon_+(k;q)\cdot\varepsilon_+(k';q') = \frac{\langle q\,q'\rangle\,[k\,k']}{\langle q\,k\rangle\,\langle q'\,k'\rangle}\,, \tag{81.10}$$

$$\varepsilon_-(k;q)\cdot\varepsilon_-(k';q') = \frac{[q\,q']\,\langle k\,k'\rangle}{[q\,k]\,[q'\,k']}\,, \tag{81.11}$$

$$\varepsilon_+(k;q)\cdot\varepsilon_-(k';q') = \frac{\langle q\,k'\rangle\,[k\,q']}{\langle q\,k\rangle\,[q'\,k']}\,. \tag{81.12}$$

The first argument of each ε is the momentum of the corresponding line; the second argument is an arbitrary reference momentum. Recall that the twistor producs $\langle q\,k\rangle$ and $[q\,k]$ are antisymmetric, and hence $\langle q\,q\rangle = [q\,q] = 0$. Using this fact in eq. (81.10), we see that choosing the same reference momentum q for all positive-helicity polarizations results in a vanishing product for any pair of them. Furthermore, if we choose this q equal to the momentum k' of a negative-helicity gluon, eq. (81.12) tells us that the product of its polarization with that of any positive-helicity gluon also vanishes. Thus, if all, or all but one, of the external gluons have positive helicity, all possible polarization products are zero, and hence the tree-level scattering amplitude is also zero. Thus we have shown that

$$A(1^\pm,2^+,\ldots,n^+) = 0\,, \tag{81.13}$$

where the superscripts are the helicities. Of course, the same is true if all, or all but one, of the helicities are negative,

$$A(1^\pm,2^-,\ldots,n^-) = 0\,. \tag{81.14}$$

Now we turn to the calculation of some nonzero tree-level partial amplitudes, beginning with $A(1^-,2^-,3^+,4^+)$. The contributing color-ordered

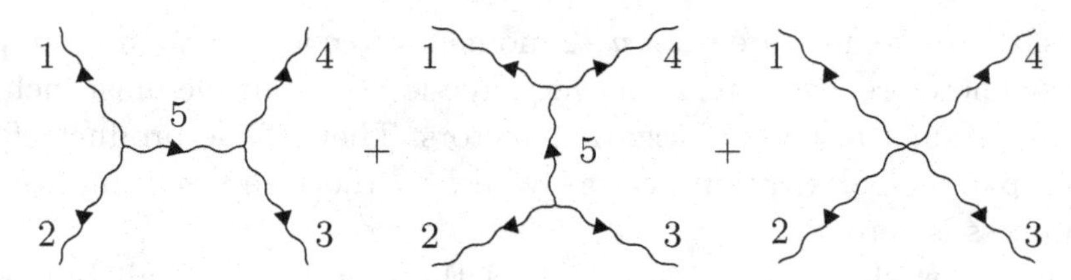

Figure 81.1. Diagrams for the partial amplitude $A(1, 2, 3, 4)$.

Feynman diagrams are shown in fig. 81.1. We choose the reference momenta to be $q_1 = q_2 = k_3$ and $q_3 = q_4 = k_2$. Then all polarization products vanish, with the exception of

$$\begin{aligned} \varepsilon_1 \cdot \varepsilon_4 &= \varepsilon_-(k_1, q_1) \cdot \varepsilon_+(k_4, q_4) \\ &= \varepsilon_-(k_1, k_3) \cdot \varepsilon_+(k_4, k_2) \\ &= \frac{\langle 2\,1\rangle\,[4\,3]}{\langle 2\,4\rangle\,[3\,1]} \,. \end{aligned} \tag{81.15}$$

With this choice of the reference momenta, the third diagram in fig. 81.1 obviously vanishes, because it has a factor of $\varepsilon_1 \cdot \varepsilon_3 = 0$ (and also, for good measure, $\varepsilon_2 \cdot \varepsilon_4 = 0$). Now consider the 235 vertex in the second diagram; we have

$$\mathbf{V}_{235} \propto (\varepsilon_2 \varepsilon_3)(k_2 \varepsilon_5) + (\varepsilon_3 \varepsilon_5)(k_3 \varepsilon_2) + (\varepsilon_5 \varepsilon_2)(k_5 \varepsilon_3) \,, \tag{81.16}$$

where ε_5 is a placeholder for an internal propagator. The first term in eq. (81.16) vanishes because $\varepsilon_2 \cdot \varepsilon_3 = 0$. The second term vanishes because $k_3 = q_2$ and $q_2 \cdot \varepsilon_2 = 0$. Finally, the third term vanishes because $k_5 = -k_2 - k_3 = -q_3 - k_3$, and $q_3 \cdot \varepsilon_3 = k_3 \cdot \varepsilon_3 = 0$. Hence the 235 vertex vanishes, and therefore so does the second diagram.

That leaves only the first diagram. We then have

$$ig^2 A(1^-, 2^-, 3^+, 4^+) = (i\mathbf{V}_{125})(i\mathbf{V}_{345'})\Big|_{\varepsilon_5^\mu \varepsilon_5^\nu \to\, i g^{\mu\nu}/s_{12}} \,, \tag{81.17}$$

where $5'$ means the momentum is $-k_5$ rather than k_5, and

$$s_{12} \equiv -(k_1 + k_2)^2 = \langle 1\,2\rangle\,[2\,1] \,. \tag{81.18}$$

We have

$$i\mathbf{V}_{125} = -i\sqrt{2}g\Big[(\varepsilon_1 \varepsilon_2)(k_1 \varepsilon_5) + (\varepsilon_2 \varepsilon_5)(k_2 \varepsilon_1) + (\varepsilon_5 \varepsilon_1)(k_5 \varepsilon_2)\Big] \,, \tag{81.19}$$

but the first term vanishes because $\varepsilon_1{\cdot}\varepsilon_2 = 0$. Similarly, the first term of

$$i\mathbf{V}_{345'} = -i\sqrt{2}g\Big[(\varepsilon_3\varepsilon_4)(k_3\varepsilon_5) + (\varepsilon_4\varepsilon_5)(k_4\varepsilon_3) + (\varepsilon_5\varepsilon_3)(-k_5\varepsilon_4)\Big] \tag{81.20}$$

also vanishes. When we take the product of these two vertices, and replace the internal polarizations with the propagator, as indicated in eq. (81.17), only the product of the third term of eq. (81.19) with the second term of eq. (81.20) is nonzero; all other terms include a vanishing product of polarizations. We get

$$ig^2 A(1^-,2^-,3^+,4^+) = (-i\sqrt{2}g)^2(i/s_{12})(\varepsilon_1\varepsilon_4)(k_5\varepsilon_2)(k_4\varepsilon_3)\,. \tag{81.21}$$

Since $k_5 = -k_1 - k_2$, and $k_2{\cdot}\varepsilon_2 = 0$, we have $k_5{\cdot}\varepsilon_2 = -k_1{\cdot}\varepsilon_2$. We evaluate $k_1{\cdot}\varepsilon_2$ and $k_4{\cdot}\varepsilon_3$ via the general formulae

$$p{\cdot}\varepsilon_+(k;q) = \frac{\langle q\,p\rangle\,[p\,k]}{\sqrt{2}\,\langle q\,k\rangle}\,, \tag{81.22}$$

$$p{\cdot}\varepsilon_-(k;q) = \frac{[q\,p]\,\langle p\,k\rangle}{\sqrt{2}\,[q\,k]}\,. \tag{81.23}$$

Setting $q_2 = k_3$ and $q_3 = k_2$, we get

$$k_1{\cdot}\varepsilon_2 = \frac{[3\,1]\,\langle 1\,2\rangle}{\sqrt{2}\,[3\,2]}\,, \tag{81.24}$$

$$k_4{\cdot}\varepsilon_3 = \frac{\langle 2\,4\rangle\,[4\,3]}{\sqrt{2}\,\langle 2\,3\rangle}\,. \tag{81.25}$$

Using eqs. (81.15), (81.18), (81.24), and (81.25) in eq. (81.21), and using antisymmetry of the twistor products to cancel common factors, we get

$$A(1^-,2^-,3^+,4^+) = \frac{\langle 2\,1\rangle\,[4\,3]^2}{[2\,1]\,[3\,2]\,\langle 2\,3\rangle}\,. \tag{81.26}$$

We can make our result look nicer by multiplying the numerator and denominator by $\langle 3\,4\rangle$. In the numerator, we use

$$\langle 3\,4\rangle\,[4\,3] = s_{34} = s_{12} = \langle 1\,2\rangle\,[2\,1]\,, \tag{81.27}$$

and cancel the $[2\,1]$ with the one in the denominator. Now multiply the numerator and denominator by $\langle 4\,1\rangle$, and use the momentum-conservation identity (see problem 60.2) to replace $\langle 4\,1\rangle\,[4\,3]$ in the numerator with $-\langle 2\,1\rangle\,[2\,3]$, and cancel the $[2\,3]$ with the $[3\,2]$ in the denominator (which yields a minus sign). Finally, multiply the numerator and denominator by $\langle 1\,2\rangle$ to get

$$A(1^-,2^-,3^+,4^+) = \frac{\langle 1\,2\rangle^4}{\langle 1\,2\rangle\langle 2\,3\rangle\langle 3\,4\rangle\langle 4\,1\rangle} \,. \tag{81.28}$$

This is our final result for $A(1^-,2^-,3^+,4^+)$.

Now, using cyclic symmetry, we can get any partial amplitude where the two negative helicities are adjacent; for example,

$$A(1^+,2^-,3^-,4^+) = \frac{\langle 2\,3\rangle^4}{\langle 1\,2\rangle\langle 2\,3\rangle\langle 3\,4\rangle\langle 4\,1\rangle} \,. \tag{81.29}$$

We must still calculate one partial amplitude where the negative helicities are not adjacent, such as $A(1^-,2^+,3^-,4^+)$. Once we have it, we can use cyclic symmetry to get all the remaining partial amplitudes.

Before turning to this calculation, let us consider the problem of squaring the total amplitude and summing over colors. Because the generator matrices are traceless, we should (as we discussed in section 80) use the completeness relation

$$(T^a)_i{}^j(T^a)_k{}^l = \delta_i{}^l\delta_k{}^j - \tfrac{1}{N}\delta_i{}^j\delta_k{}^l \,. \tag{81.30}$$

However, recall that the Yang–Mills field strength is

$$F_{\mu\nu} = \partial_\mu A_\nu - \partial_\nu A_\mu - \tfrac{ig}{\sqrt{2}}[A_\mu,A_\nu] \,. \tag{81.31}$$

If we allow a generator matrix proportional to the identity, which corresponds to a gauge group of U(N) rather than SU(N), then this extra U(1) generator commutes with every other generator. Thus the U(1) field does not appear in the commutator term in eq. (81.31). Since it is this commutator term that is responsible for the interaction of the gluons, the U(1) field is a free field. Therefore, any scattering amplitude involving the associated particle (which we will call the *fictitious photon*) must be zero. Thus, if we write a scattering amplitude in the form of eq. (81.2), and replace one of the T^as with the identity matrix, the result must be zero.

This *decoupling of the fictitious photon* allows us to use the much simpler completeness relation

$$(T^a)_i{}^j(T^a)_k{}^l = \delta_i{}^l\delta_k{}^j \tag{81.32}$$

in place of eq. (81.30). There is no need to subtract the U(1) generator from the sum over the generators, as we did in eq. (81.30), because the terms involving it vanish anyway.

The decoupling of the fictitious photon is useful in another way. Let us apply it to the case of $n = 4$, and set $T^{a_4} \propto I$ in eq. (81.2). Then we have

$$\begin{aligned} 0 = {} & \text{Tr}(T^{a_1}T^{a_2}T^{a_3})\Big[A(1,2,3,4) + A(1,2,4,3) + A(1,4,2,3)\Big] \\ & + \text{Tr}(T^{a_1}T^{a_3}T^{a_2})\Big[A(1,3,2,4) + A(1,3,4,2) + A(1,4,3,2)\Big] \,. \end{aligned} \tag{81.33}$$

The contents of each square bracket must vanish. Requiring this of the first term yields

$$A(1,2,3,4) = \, - A(1,2,4,3) - A(1,4,2,3) \,. \tag{81.34}$$

Assigning some helicities, this reads

$$A(1^-,2^+,3^-,4^+) = \, - A(1^-,2^+,4^+,3^-) - A(1^-,4^+,2^+,3^-) \,. \tag{81.35}$$

Note that we have now expressed a partial amplitude with nonadjacent negative helicities in terms of partial amplitudes with adjacent negative helicities, which we have already calculated. Thus we have

$$\begin{aligned} A(1^-,2^+,3^-,4^+) &= -\left[\frac{\langle 3\,1\rangle^4}{\langle 3\,1\rangle\langle 1\,2\rangle\langle 2\,4\rangle\langle 4\,3\rangle} + \frac{\langle 3\,1\rangle^4}{\langle 3\,1\rangle\langle 1\,4\rangle\langle 4\,2\rangle\langle 2\,3\rangle}\right] \\ &= -\frac{\langle 1\,3\rangle^3}{\langle 2\,4\rangle}\left[\frac{1}{\langle 1\,2\rangle\langle 3\,4\rangle} + \frac{1}{\langle 1\,4\rangle\langle 2\,3\rangle}\right] \\ &= -\frac{\langle 1\,3\rangle^3}{\langle 2\,4\rangle}\left[\frac{\langle 1\,4\rangle\langle 2\,3\rangle + \langle 1\,2\rangle\langle 3\,4\rangle}{\langle 1\,2\rangle\langle 3\,4\rangle\langle 1\,4\rangle\langle 2\,3\rangle}\right] \\ &= -\frac{\langle 1\,3\rangle^3}{\langle 2\,4\rangle}\left[\frac{-\,\langle 1\,3\rangle\langle 4\,2\rangle}{\langle 1\,2\rangle\langle 3\,4\rangle\langle 1\,4\rangle\langle 2\,3\rangle}\right], \end{aligned} \tag{81.36}$$

where the last line follows from the Schouten identity (see problem 50.3). A final clean-up yields

$$A(1^-,2^+,3^-,4^+) = \frac{\langle 1\,3\rangle^4}{\langle 1\,2\rangle\langle 2\,3\rangle\langle 3\,4\rangle\langle 4\,1\rangle} \,. \tag{81.37}$$

Now that we have all the partial amplitudes, we can compute the color-summed $|\mathcal{T}|^2$. There are only three partial amplitudes that are not related by either cyclic permutations, eq. (81.3), or reflections, eq. (81.9); we can take these to be

$$A_3 \equiv A(1,2,3,4) \,, \tag{81.38}$$

$$A_4 \equiv A(1,3,4,2) \,, \tag{81.39}$$

$$A_2 \equiv A(1,4,2,3) \,, \tag{81.40}$$

where the subscript on the left-hand side is the third argument on the right-hand side. (Switching the second and fourth arguments is equivalent to a reflection and a cyclic permutation, and so leaves the partial amplitude unchanged.) This mimics the notation we used at the end of section 80, and we can apply our result from there to the color sum,

$$\sum_{\text{colors}} |\mathcal{T}|^2 = 2N^2(N^2{-}1)g^4 \sum_j |A_j|^2 + 4N^2 g^4 (\textstyle\sum_j A_j^*)(\sum_k A_k) \,, \tag{81.41}$$

where j and k are summed over $2, 3, 4$. In the present case, however, eq. (81.34) is equivalent to $\sum_j A_j = 0$, so the second term in eq. (81.41) vanishes. Our result for the color-summed squared amplitude is then

$$\sum_{\text{colors}} |\mathcal{T}|^2 = 2N^2(N^2{-}1)g^4 \Big(|A_2|^2 + |A_3|^2 + |A_4|^2\Big) \,. \tag{81.42}$$

For the case where 1 and 2 are the incoming gluons, and 3 and 4 are the outgoing gluons, we can write this in terms of the usual Mandelstam variables $s = s_{12} = s_{34}$, $t = s_{13} = s_{24}$, and $u = s_{14} = s_{23}$ by recalling that $|\langle 1\,2\rangle|^2 = |[1\,2]|^2 = |s_{12}|$, etc. Let us take the case where gluons 1 and 2 have negative helicity, and 3 and 4 have positive helicity. In this case, we see from eqs. (81.28) and (81.37) that the numerator in every nonvanishing partial amplitude is $\langle 1\,2\rangle^4$. Then we get

$$\sum_{\text{colors}} |\mathcal{T}|^2_{1^-2^-3^+4^+} = 2N^2(N^2{-}1)g^4 s^4 \left(\frac{1}{s^2t^2} + \frac{1}{t^2u^2} + \frac{1}{u^2s^2}\right) . \tag{81.43}$$

We can also sum over helicities. There are six patterns of two positive and two negative helicities; $--++$ and $++--$ yield a factor of s^4, $-+-+$ and $+-+-$ yield t^4, and $-++-$ and $+--+$ yield u^4. The helicity sum is therefore

$$\sum_{\substack{\text{colors}\\ \text{helicities}}} |\mathcal{T}|^2 = 4N^2(N^2{-}1)g^4 (s^4 + t^4 + u^4) \left(\frac{1}{s^2t^2} + \frac{1}{t^2u^2} + \frac{1}{u^2s^2}\right) . \tag{81.44}$$

Of course, we really want to *average* (rather than sum) over the *initial* colors and helicities; to do so we must divide eq. (81.44) by $4(N^2{-}1)^2$.

Next we turn to scattering of quarks and gluons. We consider a single type of massless quark: a Dirac field in the N representation of SU(N). The lagrangian for this field is $\mathcal{L} = i\overline{\Psi}\not{D}\Psi$, where the covariant derivative is $D_\mu = \partial_\mu - (ig/\sqrt{2})A_\mu$. Thus the color-ordered vertex factor is

$$i\mathbf{V}^\mu = (ig/\sqrt{2})\gamma^\mu \,. \tag{81.45}$$

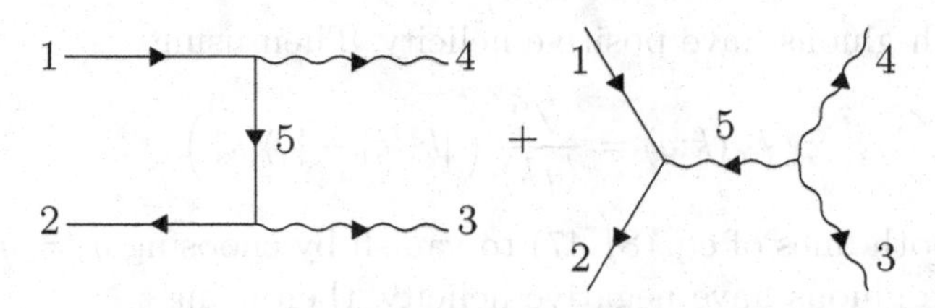

Figure 81.2. Color-ordered diagrams for $\bar{q}qgg$ scattering.

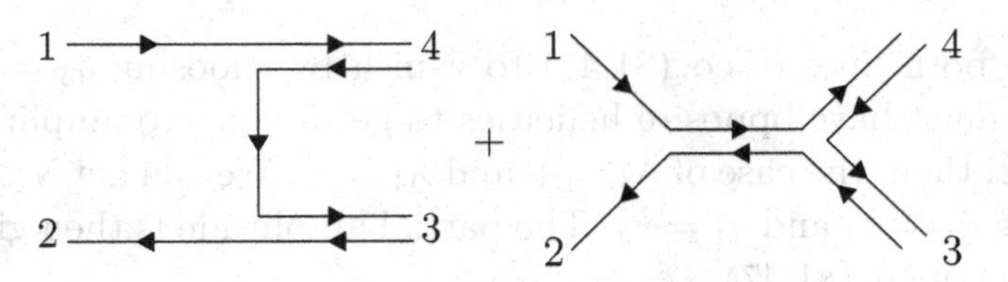

Figure 81.3. The double-line version of fig. 81.2; the associated color factor is $(T^{a_3}T^{a_4})_{i_2}{}^{i_1}$.

To get the color factor, we use the double-line notation, with a single line for the quark. As an example, consider the process of $\bar{q}q \to gg$ (and its crossing-related cousins). The contributing color-ordered tree diagrams are shown in fig. 81.2. The corresponding double-line diagrams are shown in fig. 81.3; the quark is represented by a single line, with an arrow direction that matches its charge arrow. To get the color factor, we start with line 2, and follow the arrows backwards; the result is $(T^{a_3}T^{a_4})_{i_2}{}^{i_1}$. The complete amplitude can then be written as

$$\mathcal{T} = g^2\Big[(T^{a_3}T^{a_4})_{i_2}{}^{i_1}A(1_{\bar{q}}, 2_q, 3, 4) + (T^{a_4}T^{a_3})_{i_2}{}^{i_1}A(1_{\bar{q}}, 2_q, 4, 3)\Big] , \quad (81.46)$$

where $A(1_{\bar{q}}, 2_q, 3, 4)$ is the appropriate partial amplitude. The subscripts q and $\bar{q}$ indicate the labels that correspond to an outgoing quark and outgoing antiquark, respectively.

From our results for spinor electrodynamics in section 60, we know that a nonzero amplitude requires opposite helicities on the two ends of any fermion line. Consider, then, the case of $\mathcal{T}_{-+\lambda_3\lambda_4}$. The partial amplitude corresponding to the diagrams of fig. 81.2 is

$$ig^2A(1^-_{\bar{q}}, 2^+_q, 3, 4) = (ig/\sqrt{2})^2(1/i)[2|\not{\varepsilon}_3(-\not{p}_5/p_5^2)\not{\varepsilon}_4|1\rangle$$
$$+ (ig/\sqrt{2})[2|\not{\varepsilon}_5|1\rangle i\mathbf{V}_{345}\Big|_{\varepsilon_5^\mu\varepsilon_5^\nu \to ig^{\mu\nu}/s_{12}} . \quad (81.47)$$

Suppose both gluons have positive helicity. Then using

$$\not{\varepsilon}_+(k;q) = \frac{\sqrt{2}}{\langle q\, k\rangle}\Big(\,|k]\langle q| + |q\rangle [k|\,\Big)\,, \tag{81.48}$$

we can get both lines of eq. (81.47) to vanish by choosing $q_3 = q_4 = p_1$. Similarly, if both gluons have negative helicity, then using

$$\not{\varepsilon}_-(k;q) = \frac{\sqrt{2}}{[q\, k]}\Big(\,|k\rangle [q| + |q]\langle k|\,\Big)\,, \tag{81.49}$$

we can get both lines of eq. (81.47) to vanish by choosing $q_3 = q_4 = p_2$. So the gluons must have opposite helicities to get a nonzero amplitude.

Consider, then, the case of $\lambda_3 = +$ and $\lambda_4 = -$. We can get $\mathbf{V}_{345}$ to vanish by choosing $q_3 = k_4$ and $q_4 = k_3$. The partial amplitude is then given by just the first line of eq. (81.47),

$$A(1_{\bar{q}}^-\,, 2_q^+\,, 3^+, 4^-) = \tfrac{1}{2}\,[2|\not{\varepsilon}_{3+}(\not{p}_1 + \not{k}_4)\not{\varepsilon}_{4-}|1\rangle/(-s_{14})\;. \tag{81.50}$$

With $q_3 = k_4$ and $q_4 = k_3$, we have

$$\not{\varepsilon}_{3+} = \frac{\sqrt{2}}{\langle 4\, 3\rangle}\Big(|4\rangle [3| + |3]\langle 4|\Big)\,, \tag{81.51}$$

$$\not{\varepsilon}_{4-} = \frac{\sqrt{2}}{[3\, 4]}\Big(|4\rangle [3| + |3]\langle 4|\Big)\,. \tag{81.52}$$

Using the identity

$$\not{p} = -\,|p\rangle [p| - |p]\langle p|\,, \tag{81.53}$$

eq. (81.50) becomes

$$A(1_{\bar{q}}^-\,, 2_q^+\,, 3^+, 4^-) = \frac{[2\, 3]\,\langle 4\, 1\rangle\,[1\, 3]\,\langle 4\, 1\rangle}{\langle 4\, 3\rangle\,[3\, 4]\,s_{14}}\;. \tag{81.54}$$

In the numerator, we use $[1\,3]\,\langle 4\,1\rangle = -[2\,3]\,\langle 4\,2\rangle$. In the denominator, we set $s_{14} = s_{23} = \langle 2\,3\rangle\,[3\,2]$. Then we multiply the numerator and denominator by $\langle 1\,2\rangle$, and use $[2\,3]\,\langle 1\,2\rangle = -[4\,3]\,\langle 1\,4\rangle$ in the numerator. Finally we multiply both by $\langle 1\,4\rangle$, and rearrange to get

$$A(1_{\bar{q}}^-\,, 2_q^+\,, 3^+, 4^-) = \frac{\langle 1\, 4\rangle^3\,\langle 2\, 4\rangle}{\langle 1\, 2\rangle\langle 2\, 3\rangle\langle 3\, 4\rangle\langle 4\, 1\rangle}\;. \tag{81.55}$$

An analogous calculation yields

$$A(1_{\bar{q}}^-\,, 2_q^+\,, 3^-, 4^+) = \frac{\langle 1\, 3\rangle^3\,\langle 2\, 3\rangle}{\langle 1\, 2\rangle\langle 2\, 3\rangle\langle 3\, 4\rangle\langle 4\, 1\rangle}\;. \tag{81.56}$$

The remaining nonzero amplitudes are related by complex conjugation.

Now that we have all the partial amplitudes, we can compute the color-summed $|\mathcal{T}|^2$. To do so, we multiply eq. (81.46) by its complex conjugate, and use hermiticity of the generator matrices to get

$$\sum_{\text{colors}} |\mathcal{T}|^2 = g^4 \Big[\mathrm{Tr}(T^a T^b T^b T^a) \Big(|A_3|^2 + |A_4|^2 \Big) + \mathrm{Tr}(T^a T^b T^a T^b) \Big(A_3^* A_4 + A_4^* A_3 \Big) \Big] , \qquad (81.57)$$

where $A_3 \equiv A(1_{\bar{q}}, 2_q, 3, 4)$ and $A_4 \equiv A(1_{\bar{q}}, 2_q, 4, 3)$. The traces are easily evaluated with the double-line technique of section 80; because the fictitious photon couples to the quark, we must use eq. (81.30) to project it out. The traces in eq. (81.57) are also easily evaluated with the group-theoretic methods of section 70, with the normalization that the index of the fundamental representation is one: $T(\mathrm{N}) = 1$. Either way, the results are

$$\mathrm{Tr}(T^a T^b T^b T^a) = +(N^2-1)^2/N , \qquad (81.58)$$

$$\mathrm{Tr}(T^a T^b T^a T^b) = -(N^2-1)/N . \qquad (81.59)$$

The sum over the four possible helicity patterns $(-{+}{+}-, -{+}-{+}, +{-}-{+}, +{-}{+}-)$ is left as an exercise.

Now that we have calculated these scattering amplitudes for quarks and gluons, an important question arises: why did we bother to do it? Quarks and gluons are confined inside colorless bound states, the *hadrons*, and so apparently cannot appear as incoming and outgoing particles in a scattering event.

To answer this question, suppose we collide two hadrons with a center-of-mass energy $E = \sqrt{s}$ large enough so that the QCD coupling g is small when renormalized in the $\overline{\mathrm{MS}}$ scheme with $\mu = E$. (In the real world, we have $\alpha \equiv g^2/4\pi = 0.12$ for $\mu = M_{\mathrm{Z}} = 91\,\mathrm{GeV}$.) Then we can think of each hadron as being made up of a loose collection of quarks and gluons, and these parts of a hadron, or *partons*, can be treated as independent participants in scattering processes. In order to extract quantitative results for hadron scattering (a project beyond the scope of this book), we need to know how each hadron's energy and momentum is shared among its partons. This is described by *parton distribution functions.* At present, these cannot be calculated from first principles, but they have to satisfy a variety of consistency conditions that *can* be derived from perturbation theory, and that relate their values at different energies. These conditions are well satisfied by current experimental data.

Reference notes

More detail on how hadron scattering experiments can be compared with parton scattering amplitudes can be found in *Peskin & Schroeder*, *Muta*, *Quigg*, and *Sterman*.

Problems

81.1) Compute the four-gluon partial amplitude $A(1^-,2^+,3^-,4^+)$ directly from the Feynman diagrams, and verify eq. (81.37).

81.2) Compute the $\bar{q}qgg$ partial amplitude $A(1_{\bar{q}}^-,2_q^+,3^+,4^-)$ with $q_4 = p_1$ and $q_3 = k_4$. Show that, with this choice of the reference momenta, the first line of eq. (81.47) vanishes. Evaluate the second line, and verify eq. (81.55).

81.3) Compute $A(1_{\bar{q}}^-,2_q^+,3^-,4^+)$, and verify eq. (81.56).

81.4) a) Verify eqs. (81.58) and (81.59) using the double-line notation of section 80.

b) Compute $\mathrm{Tr}(T_{\mathrm{R}}^a T_{\mathrm{R}}^b T_{\mathrm{R}}^b T_{\mathrm{R}}^a)$ and $\mathrm{Tr}(T_{\mathrm{R}}^a T_{\mathrm{R}}^b T_{\mathrm{R}}^a T_{\mathrm{R}}^b)$ in terms of the index $T(\mathrm{R})$ and dimension $D(\mathrm{R})$ of the representation R, and the index $T(\mathrm{A})$ and dimension $D(\mathrm{A})$ of the adjoint representation. Verify that your results reproduce eqs. (81.58) and (81.59).

81.5) Compute the sum over helicities of eq. (81.57). Express your answer in terms of s, t, and u for the process $\bar{q}q \to gg$.

81.6) Consider the partial amplitude $A(1_{\bar{q}}^-,2_q^+,3^-,4^+,5^+)$. Show that, with the choice $q_3 = k_2$ and $q_4 = q_5 = k_1$, there are just two contributing diagrams. Evaluate them. After some manipulations, you should be able to put your result in the form

$$A(1_{\bar{q}}^-,2_q^+,3^-,4^+,5^+) = \frac{\langle 1\,3\rangle^3\,\langle 2\,3\rangle}{\langle 1\,2\rangle\langle 2\,3\rangle\langle 3\,4\rangle\langle 4\,5\rangle\langle 5\,1\rangle}\,. \tag{81.60}$$

82

Wilson loops, lattice theory, and confinement

Prerequisites: 29, 73

In this section, we will contruct a gauge-invariant operator, the *Wilson loop*, whose vacuum expectation value (VEV for short) can diagnose whether or not a gauge theory exhibits *confinement.* A theory is *confining* if all finite-energy states are invariant under a global gauge transformation. U(1) gauge theory—quantum electrodynamics—is *not* confining, because there are finite-energy states (such as the state of a single electron) that have nonzero electric charge, and hence change by a phase under a global gauge transformation.

Confinement is a *nonperturbative* phenomenon; it cannot be seen at any finite order in the kind of weak-coupling perturbation theory that we have been doing. (This is why we had no trouble calculating quark and gluon scattering amplitudes.) In this section, we will introduce *lattice gauge theory,* in which spacetime is replaced by a discrete set of points; the inverse lattice spacing $1/a$ then acts as an ultraviolet cutoff (see section 29). This cutoff theory can be analyzed at *strong* coupling, and, as we will see, in this regime the VEV of the Wilson loop is indicative of confinement. The outstanding question is whether this phenomenon persists as we simultaneously lower the coupling and increase the ultraviolet cutoff (with the relationship between the two governed by the beta function), or whether we encounter a *phase transition,* signaled by a sudden change in the behavior of the Wilson loop VEV.

We take the gauge group to be SU(N). Consider two spacetime points x^μ and $x^\mu + \varepsilon^\mu$, where ε^μ is infinitesimal. Define the *Wilson link*

$$W(x+\varepsilon, x) \equiv \exp[ig\varepsilon^\mu A_\mu(x)]\ , \tag{82.1}$$

where $A_\mu(x)$ is an $N \times N$ matrix-valued traceless hermitian gauge field. Since ε is infinitesimal, we also have

$$W(x+\varepsilon, x) = I + ig\varepsilon^\mu A_\mu(x) + O(\varepsilon^2) . \tag{82.2}$$

Let us determine the behavior of the Wilson link under a gauge transformation. Using the gauge transformation of $A_\mu(x)$ from section 69, we find

$$W(x+\varepsilon, x) \to 1 + ig\varepsilon^\mu U(x)A_\mu(x)U^\dagger(x) - \varepsilon^\mu U(x)\partial_\mu U^\dagger(x) , \tag{82.3}$$

where $U(x)$ is a spacetime-dependent special unitary matrix. Since $UU^\dagger = 1$, we have $-U\partial_\mu U^\dagger = +(\partial_\mu U)U^\dagger$; thus we can rewrite eq. (82.3) as

$$W(x+\varepsilon, x) \to \Big((1 + \varepsilon^\mu\partial_\mu)U(x)\Big)U^\dagger(x) + ig\varepsilon^\mu U(x)A_\mu(x)U^\dagger(x) . \tag{82.4}$$

In the first term, we can use $(1 + \varepsilon^\mu\partial_\mu)U(x) = U(x+\varepsilon) + O(\varepsilon^2)$. In the second term, which already contains an explicit factor of ε^μ, we can replace $U(x)$ with $U(x+\varepsilon)$ at the cost of an $O(\varepsilon^2)$ error. Then we get

$$W(x+\varepsilon, x) \to U(x+\varepsilon)\Big(1 + ig\varepsilon^\mu A_\mu(x)\Big)U^\dagger(x) , \tag{82.5}$$

which is equivalent to

$$W(x+\varepsilon, x) \to U(x+\varepsilon)W(x+\varepsilon, x)U^\dagger(x) . \tag{82.6}$$

Note also that eq. (82.1) implies $W^\dagger(x+\varepsilon, x) = W(x-\varepsilon, x)$. We can shift x to $x+\varepsilon$ at the cost of an $O(\varepsilon^2)$ error, and so

$$W^\dagger(x+\varepsilon, x) = W(x, x+\varepsilon) , \tag{82.7}$$

which is consistent with eq. (82.6).

Now consider mutiplying together a string of Wilson links, specified by a starting point x and n sequential infinitesimal displacement vectors ε_j. The ordered set of εs defines a *path* P through spacetime that starts at x and ends at $y = x + \varepsilon_1 + \ldots + \varepsilon_n$. The *Wilson line* for this path is

$$W_P(y, x) \equiv W(y, y-\varepsilon_n) \ldots W(x+\varepsilon_1+\varepsilon_2, x+\varepsilon_1)W(x+\varepsilon_1, x) . \tag{82.8}$$

Using eq. (82.6) and the unitarity of $U(x)$, we see that, under a gauge transformation, the Wilson line transforms as

$$W_P(y, x) \to U(y)W_P(y, x)U^\dagger(x) . \tag{82.9}$$

Also, since hermitian conjugation reverses the order of the product in eq. (82.8), using eq. (82.7) yields

$$W_P^\dagger(y,x) = W_{-P}(x,y)\ , \tag{82.10}$$

where $-P$ denotes the reverse of the path P.

Now consider a path that returns to its starting point, forming a closed, oriented curve C in spacetime. The *Wilson loop* is the trace of the Wilson line for this path,

$$W_C \equiv \mathrm{Tr}\, W_C(x,x)\ . \tag{82.11}$$

Using eq. (82.9), we see that the Wilson loop is *gauge invariant*,

$$W_C \to W_C\ . \tag{82.12}$$

Also, eq. (82.10) implies

$$W_C^\dagger = W_{-C}\ , \tag{82.13}$$

where $-C$ denotes the curve C traversed in the opposite direction.

To gain some intuition, we will calculate $\langle 0|W_C|0\rangle$ for U(1) gauge theory, without charged fields. This is simply a free-field theory, and the calculation can be done exactly.

In order to avoid dealing with $i\epsilon$ issues, it is convenient to make a Wick rotation to euclidean spacetime (see section 29). The action is then

$$S = \int d^4x\ \tfrac{1}{4}F_{\mu\nu}F_{\mu\nu}\ , \tag{82.14}$$

where $F_{\mu\nu} = \partial_\mu A_\nu - \partial_\nu A_\mu$. The VEV of the Wilson loop is now given by the path integral

$$\langle 0|W_C|0\rangle = \int \mathcal{D}A\ e^{ig\oint_C dx_\mu A_\mu} e^{-S}\ . \tag{82.15}$$

If we formally identify $g\oint_C dx_\mu$ as a current $J_\mu(x)$, we can apply our results for the path integral from section 57. After including a factor of i from the Wick rotation, we get

$$\langle 0|W_C|0\rangle = \exp\!\left[-\tfrac{1}{2}g^2\oint_C dx_\mu \oint_C dy_\nu\ \Delta_{\mu\nu}(x{-}y)\right]\ , \tag{82.16}$$

where $\Delta_{\mu\nu}(x-y)$ is the photon propagator in euclidean spacetime. In Feynman gauge, we have

$$\begin{aligned}
\Delta_{\mu\nu}(x-y) &= \delta_{\mu\nu} \int \frac{d^4k}{(2\pi)^4} \frac{e^{ik\cdot(x-y)}}{k^2} \\
&= \delta_{\mu\nu} \frac{4\pi}{(2\pi)^4} \int_0^\infty \frac{k^3\,dk}{k^2} \int_0^\pi d\theta\, \sin^2\theta\, e^{ik|x-y|\cos\theta} \\
&= \delta_{\mu\nu} \frac{4\pi}{(2\pi)^4} \int_0^\infty \frac{k^3\,dk}{k^2} \frac{\pi J_1(k|x-y|)}{k|x-y|} \\
&= \frac{\delta_{\mu\nu}}{4\pi^2(x-y)^2} \int_0^\infty du\, J_1(u) \\
&= \frac{\delta_{\mu\nu}}{4\pi^2(x-y)^2} \,,
\end{aligned} \tag{82.17}$$

where $J_1(u)$ is a Bessel function. Since $\Delta_{\mu\nu}(x-y)$ depends only on $x-y$, the double line integral in eq. (82.16) will yield a factor of the perimeter P of the curve C. There is also an ultraviolet divergence as x approaches y; we will cut this off at a length scale a. The result is then

$$\langle 0|W_C|0\rangle = \exp[\,-(\tilde{c}g^2/a)P\,]\,, \tag{82.18}$$

where $\tilde{c}$ is a numerical constant that depends on the shape of C and the details of the cutoff procedure. This behavior of the Wilson loop in euclidean spacetime—exponential decay with the length of the perimeter—is called the *perimeter law*. It is indicative of unconfined charges.

We can gain more insight into the meaning of $\langle 0|W_C|0\rangle$ by taking C to be a rectangle, with length T in the time direction and R in a space direction, where $a \ll R \ll T$. (Of course, in euclidean spacetime, the choice of the time direction is an arbitrary convention.) The reason for this particular shape is that the current $g \oint_C dx_\mu$ corresponds to a point charge moving along the curve C. When the particle is moving backwards in time, the associated minus sign is equivalent to a change in the sign of its charge. So when we compute $\langle 0|W_C|0\rangle$, we are doing the path integral in the presence of a pair of point charges with opposite sign, separated by a distance R, that exists for a time T. On general principles (see section 6), this path integral is proportional to $\exp(-E_{\text{pair}}T)$, where E_{pair} is the ground-state energy of the quantum electrodynamic field in the presence of the charged particle pair.

We now turn to the calculation. If both x and y are on the same side of the rectangle, we find

$$\int_0^L\!\!\int_0^L \frac{dx\,dy}{(x-y)^2} = 2L/a - 2\ln(L/a) + O(1)\ , \tag{82.19}$$

where L is the length of the side (either R or T), and the $O(1)$ term is a numerical constant that depends on the details of the short-distance cutoff. If x and y are on perpendicular sides, the double line integral is zero, because then $dx \cdot dy = 0$. If x is on one short side and y on the other, the integral evaluates to R^2/T^2, and this we can neglect. Finally, if x is on one long side and y is on the other, we have

$$\int_0^T\!\!\int_0^T \frac{dx\,dy}{(x-y)^2+R^2} = \pi T/R - 2\ln(T/R) - 2 + O(R^2/T^2)\ . \tag{82.20}$$

Adding up all these contributions, we find in the limit of large T that

$$\oint_C\oint_C \frac{dx\cdot dy}{(x-y)^2} = \Big(4/a - 2\pi/R\Big)T + O(\ln T)\ . \tag{82.21}$$

Combining this with eqs. (82.16) and (82.17), and setting $\alpha = g^2/4\pi$, we find

$$\langle 0|W_C|0\rangle = \exp\!\left[-\left(\frac{2\alpha}{\pi a} - \frac{\alpha}{R}\right)T\right] . \tag{82.22}$$

Comparing this with the general expectation $\langle 0|W_C|0\rangle \propto \exp(-E_{\rm pair}T)$, we find a cutoff-dependent contribution to $E_{\rm pair}$ that represents a divergent self-energy for each point particle, plus the Coulomb potential energy for the pair, $V(R) = -\alpha/R$.

In the nonabelian case, where there are interactions among the gluons, we must expand everything in powers of g. Then we find

$$\langle 0|W_C|0\rangle = {\rm Tr}\!\left[1 - \tfrac{1}{2}g^2 T^a T^a \oint_C dx_\mu \oint_C dy_\nu\ \Delta_{\mu\nu}(x{-}y) + O(g^4)\right] . \tag{82.23}$$

Since T^aT^a equals the quadratic Casimir $C({\rm N})$ for the fundamental representation (times an identity matrix), we see that to leading order in g^2 we simply reproduce the results of the abelian case, but with $g^2 \to g^2C({\rm N})$. We can also consider a Wilson loop in a different representation by setting $A_\mu(x) = A^a_\mu(x)T^a_{\rm R}$. Then, at leading order, we get a factor of $C({\rm R})$ instead of $C({\rm N})$. Perturbative corrections can be computed via standard Feynman diagrams with gluon lines that, in position space, have at least one end on the curve C.

Next we turn to a strong coupling analysis. We begin by constructing a *lattice action* for nonabelian gauge theory. Consider a hypercubic lattice of points in four-dimensional euclidean spacetime, with a *lattice spacing* a

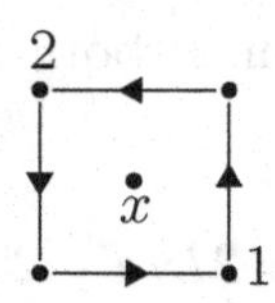

Figure 82.1. The minimal Wilson loop on a hypercubic lattice goes around an elementary plaquette; this one lies in the 1-2 plane.

between nearest-neighbor points. The smallest Wilson loop we can make on this lattice goes around an elementary square or *plaquette*, as shown in fig. 82.1. Let ε_1 and ε_2 be vectors of length a in the 1 and 2 directions, and let x be the point at the center of the plaquette. Using the center of each link as the argument of the gauge field, and using the lower-left corner as the starting point, we have (multiplying the Wilson links from right to left along the path)

$$W_{\text{plaq}} = \text{Tr}\, e^{-igaA_2(x-\varepsilon_1/2)} e^{-igaA_1(x+\varepsilon_2/2)} e^{+igaA_2(x+\varepsilon_1/2)} e^{+igaA_1(x-\varepsilon_2/2)} \,. \tag{82.24}$$

If we now treat the gauge field as smooth and expand in a, we get

$$\begin{aligned} W_{\text{plaq}} = \text{Tr}\, & e^{-igaA_2(x)+iga^2\partial_1 A_2(x)/2+\ldots} e^{-igaA_1(x)-iga^2\partial_2 A_1(x)/2+\ldots} \\ & \times e^{+igaA_2(x)+iga^2\partial_1 A_2(x)/2+\ldots} e^{+igaA_1(x)-iga^2\partial_2 A_1(x)/2+\ldots} \,. \end{aligned} \tag{82.25}$$

Next we use $e^A e^B = e^{A+B+[A,B]/2+\ldots}$ to combine the two exponential factors on the first line of eq. (82.25), and also the two exponential factors on the second line. Then we use this formula once again to combine the two results. We get

$$W_{\text{plaq}} = \text{Tr}\, e^{+iga^2(\partial_1 A_2 - \partial_2 A_1 - ig[A_1,A_2])+\ldots} \,, \tag{82.26}$$

where all fields are evaluated at x. If we now take $W_{\text{plaq}} + W_{-\text{plaq}}$ and expand the exponentials, we find

$$W_{\text{plaq}} + W_{-\text{plaq}} = 2N - g^2 a^4\, \text{Tr} F_{12}^2 + \ldots \,, \tag{82.27}$$

where $F_{12} = \partial_1 A_2 - \partial_2 A_1 - ig[A_1, A_2]$ is the Yang–Mills field strength. From eq. (82.27), we conclude that an appropriate action for Yang–Mills theory on a euclidean spacetime lattice is

$$S = -\frac{1}{2g^2} \sum_{\text{plaq}} W_{\text{plaq}} \,, \tag{82.28}$$

where the sum includes both orientations of all plaquettes. Each W_{plaq} is expressed as the trace of the product of four special unitary $N \times N$ matrices, one for each oriented link in the plaquette. If U is the matrix associated with one orientation of a particular link, then $U^\dagger$ is the matrix associated with the opposite orientation of that link. The path integral for this *lattice gauge theory* is

$$Z = \int \mathcal{D}U \; e^{-S} \,, \tag{82.29}$$

where

$$\mathcal{D}U = \prod_{\text{links}} dU_{\text{link}} \,, \tag{82.30}$$

and dU is the *Haar measure* for a special unitary matrix. The Haar measure is invariant under the transformation $U \to VU$, where V is a constant special unitary matrix, and is normalized via $\int dU = 1$; this fixes it uniquely. For $N \geq 3$, it obeys

$$\int dU \; U_{ij} = 0 \,, \tag{82.31}$$

$$\int dU \; U_{ij} U_{kl} = 0 \,, \tag{82.32}$$

$$\int dU \; U_{ij} U^*_{kl} = \tfrac{1}{N}\delta_{ik}\delta_{jl} \,, \tag{82.33}$$

which is all we will need to know.

Now consider a Wilson loop, expressed as the trace of the product of the Us associated with the oriented links that form a closed curve C. For simplicity, we take this curve to lie in a plane. We have

$$\langle 0|W_C|0\rangle = Z^{-1} \int \mathcal{D}U \; W_C \, e^{-S} \,. \tag{82.34}$$

We will evaluate eq. (82.34) in the *strong coupling limit* by expanding e^{-S} in powers of $1/g^2$. At zeroth order, $e^{-S} \to 1$; then eq. (82.31) tells us that the integral over every link in C vanishes. To get a nonzero result, we need to have a corresponding $U^\dagger$ from the expansion of e^{-S}. This can only come from a plaquette containing that link. But then the integral over the other links of this plaquette will vanish, unless there is a compensating $U^\dagger$ for each of them. We conclude that a nonzero result for $\langle 0|W_C|0\rangle$ requires us to fill the interior of C with plaquettes from the expansion of e^{-S}. Since each plaquette is accompanied by a factor of $1/g^2$, we have

$$\langle 0|W_C|0\rangle \sim (1/g^2)^{A/a^2} \,, \tag{82.35}$$

where A is the *area* of the surface bounded by C, and A/a^2 is the number of plaquettes in this surface. Eq. (82.35) yields the *area law* for a Wilson loop,

$$\langle 0|W_C|0\rangle \propto e^{-\tau A} , \tag{82.36}$$

where

$$\tau = c(g)/a^2 \tag{82.37}$$

is the *string tension.* In the strong coupling limit, $c(g) = \ln(g^2) + O(1)$.

The area law for the Wilson loop implies confinement. To see why, let us again consider a rectangular loop with area $A = RT$. Comparing eq. (82.36) with the general expectation $\langle 0|W_C|0\rangle \propto \exp(-E_{\text{pair}}T)$, we see that $E_{\text{pair}} = V(R) = \tau R$. This corresponds to a *linear potential* between nonabelian point charges in the fundamental representation. It takes an infinite amount of energy to separate these charges by an infinite distance; the charges are therefore *confined.* The coefficient τ of R in $V(R)$ is called the string tension because a linear potential is what we get from two points joined by a string with a fixed energy per unit length; the energy per unit length of a string is its tension.

The string tension τ is a physical quantity that should remain fixed as we remove the cutoff by lowering a. Thus lowering a requires us to lower g. The outstanding question is whether $c(g)$ reaches zero at a finite, nonzero value of g. If so, at this point there is a *phase transition* to an unconfined phase with zero string tension. This has been proven to be the case for abelian gauge theory (which also exhibits an area law at strong coupling, by the identical argument). In nonabelian gauge theory, on the other hand, analytic and numerical evidence strongly suggests that $c(g)$ remains nonzero for all nonzero values of g.

At small a and small g, the behavior of g as a function of a is governed by the beta function, $\beta(g) = -a\,dg/da$. (The minus sign arises because the ultraviolet cutoff is a^{-1}.) Requiring τ to be independent of a yields

$$c'(g)/c(g) = -2/\beta(g) . \tag{82.38}$$

At small g, we have

$$\beta(g) = -b_1 g^3 + O(g^5) , \tag{82.39}$$

where $b_1 = 11N/48\pi^2$ for SU(N) gauge theory without quarks. Solving for $c(g)$ yields

$$c(g) = C\exp(-1/b_1 g^2) , \tag{82.40}$$

where C is an integration constant, which is nonzero if there is no phase transition. In this case, at small g, the string tension has the form

$$\tau = C \exp(-1/b_1 g^2)/a^2 \,. \tag{82.41}$$

We then want to take the *continuum limit* of $a \to 0$ and $g \to 0$ with τ held fixed.

Note that eq. (82.41) shows that the string tension, at weak coupling, is not analytic in g, and so cannot be computed via the Taylor expansion in g that is provided by conventional weak-coupling perturbation theory. Instead, the path integrals of eqs. (82.29) and (82.34) can be performed on a finite-size lattice via numerical integration. The limiting factor in such a calculation is computer resources.

Reference notes

An introduction to lattice theory is given in *Smit*.

Problems

82.1) Let C be a circle of radius R. Evaluate the constant $\tilde{c}$ in eq. (82.18), where $P = 2\pi R$ is the circumference of the circle. Replace $1/(x-y)^2$ with zero when $|x-y| < a$. Assume $a \ll R$.

83

Chiral symmetry breaking

Prerequisites: 76, 82

In the previous section, we discussed confinement in Yang–Mills theory without quarks. In the real world, there are six different *flavors* of quark; see Table 83.1. Each flavor has a different mass, and is represented by a Dirac field in the fundamental or 3 representation of the color group SU(3). Such a Dirac field is equivalent to two left-handed Weyl fields, one in the fundamental representation, and one in the antifundamental or $\overline{3}$ representation.

The lightest quarks are the up and down quarks, with masses of a few MeV. These masses are small, in the following sense. The gauge coupling g of QCD becomes large at low energies. If we truncate the beta function after some number of terms (in practice, four or fewer), and integrate it, we find that g becomes infinite at some finite, nonzero value of the $\overline{\text{MS}}$ parameter μ; this value is called Λ_{QCD}. Measurements of the strength of the gauge coupling at high energies imply $\Lambda_{\text{QCD}} \sim 0.2\,\text{GeV}$. The up and down quark masses are much less than Λ_{QCD}. We can therefore begin with the approximation that the up and down quarks are massless. The mass of the strange quark is also somewhat less than Λ_{QCD}. It is sometimes useful (though clearly less justified) to treat the strange quark as massless as well.

If we are interested in hadron physics at energies below $\sim$1 GeV, we can ignore the charm, bottom, and top quarks entirely; we will also ignore the strange quark for now. Let us, then, consider QCD with $n_{\text{F}} = 2$ flavors of massless quarks. We then have left-handed Weyl fields $\chi_{\alpha i}$, where $\alpha = 1, 2, 3$ is a color index for the 3 representation, and $i = 1, 2$ is a flavor index, and left-handed Weyl fields $\xi^{\alpha\bar{\imath}}$, where $\alpha = 1, 2, 3$ is a color index for the $\overline{3}$ representation, and $\bar{\imath} = 1, 2$ is a flavor index; we distinguish this flavor index from the one for the χs by putting a bar over it, and we write it as a superscript for later notational convenience. We suppress the undotted spinor index carried

Table 83.1. The six flavors of quark. Each flavor is represented by a Dirac field in the 3 representation of the color group SU(3). Q is the electric charge in units of the proton charge. Masses are approximate, and are $\overline{\text{MS}}$ parameters. For the u, d, and s quarks, the $\overline{\text{MS}}$ scale μ is taken to be 2 GeV. For the c, b, and t quarks, μ is taken to be equal to the corresponding mass; e.g., the bottom quark mass is 4.3 GeV when $\mu = 4.3$ GeV.

Name	Symbol	Mass (GeV)	Q
up	u	0.0017	+2/3
down	d	0.0039	−1/3
strange	s	0.076	−1/3
charm	c	1.3	+2/3
bottom	b	4.3	−1/3
top	t	174	+2/3

by both χ and ξ. The lagrangian is

$$\mathcal{L} = i\chi^{\dagger\alpha i}\,\bar{\sigma}^{\mu}(D_{\mu})_{\alpha}{}^{\beta}\,\chi_{\beta i} + i\xi^{\dagger}_{\bar{\imath}\alpha}\,\bar{\sigma}^{\mu}(\bar{D}_{\mu})^{\alpha}{}_{\beta}\,\xi^{\beta\bar{\imath}} - \tfrac{1}{4}F^{a\mu\nu}F^{a}_{\mu\nu}\,, \tag{83.1}$$

where $D_{\mu} = \partial_{\mu} - igT^{a}_{3}A^{a}_{\mu}$ and $\bar{D}_{\mu} = \partial_{\mu} - igT^{a}_{\bar{3}}A^{a}_{\mu}$, with $(T^{a}_{\bar{3}})^{\alpha}{}_{\beta} = -(T^{a}_{3})_{\beta}{}^{\alpha}$. In addition to the SU(3) color gauge symmetry, this lagrangian has a global U(2) × U(2) flavor symmetry: $\mathcal{L}$ is invariant under

$$\chi_{\alpha i} \to L_{i}{}^{j}\chi_{\alpha j}\,, \tag{83.2}$$

$$\xi^{\alpha\bar{\imath}} \to (R^{*})^{\bar{\imath}}{}_{\bar{\jmath}}\,\xi^{\alpha\bar{\jmath}}\,, \tag{83.3}$$

where L and R^{*} are independent 2×2 constant unitary matrices. (The complex conjugation of R is a notational convention that turns out to be convenient.) In terms of the Dirac field

$$\Psi_{\alpha i} = \begin{pmatrix} \chi_{\alpha i} \\ \xi^{\dagger}_{\alpha\bar{\imath}} \end{pmatrix}, \tag{83.4}$$

eqs. (83.2) and (83.3) read

$$P_{\rm L}\Psi_{\alpha i} \to L_{i}{}^{j}P_{\rm L}\Psi_{\alpha j}\,, \tag{83.5}$$

$$P_{\rm R}\Psi_{\alpha\bar{\imath}} \to R_{\bar{\imath}}{}^{\bar{\jmath}}P_{\rm R}\Psi_{\alpha\bar{\jmath}}\,, \tag{83.6}$$

where $P_{\mathrm{L,R}} = \frac{1}{2}(1 \mp \gamma_5)$. Thus the global flavor symmetry is often called $\mathrm{U(2)_L} \times \mathrm{U(2)_R}$. A symmetry that treats the left- and right-handed parts of a Dirac field differently is said to be *chiral.*

However, there is an anomaly in the axial U(1) symmetry corresponding to $L = R^* = e^{i\alpha}I$ (which is equivalent to $\Psi \to e^{-i\alpha\gamma_5}\Psi$ for the Dirac field). Thus the nonanomalous global flavor symmetry is $\mathrm{SU(2)_L} \times \mathrm{SU(2)_R} \times \mathrm{U(1)_V}$, where V stands for vector. The $\mathrm{U(1)_V}$ transformation corresponds to $L = R = e^{-i\alpha}I$, or equivalently $\Psi \to e^{-i\alpha}\Psi$. The corresponding conserved charge is the *quark number*, the number of quarks minus the number of antiquarks; this is three times the *baryon number*, the number of baryons minus the number of antibaryons. (*Baryons* are color-singlet bound states of three quarks; the proton and neutron are baryons. *Mesons* are color-singlet bound states of a quark and an antiquark; pions are mesons.)

Thus, $\mathrm{U(1)_V}$ results in classification of hadrons by their baryon number. How is the $\mathrm{SU(2)_L} \times \mathrm{SU(2)_R}$ symmetry realized in nature? The vector subgroup $\mathrm{SU(2)_V}$, obtained by setting $R = L$ in eq. (83.3), is known as *isotopic spin* or *isospin* symmetry. Hadrons clearly come in representations of $\mathrm{SU(2)_V}$: the lightest spin-one-half hadrons (the proton, mass 0.938 GeV, and the neutron, mass 0.940 GeV) form a doublet or 2 representation, while the lightest spin-zero hadrons (the π^0, mass 0.135 GeV, and the $\pi^\pm$, mass 0.140 GeV) form a triplet or 3 representation. Isospin is not an exact symmetry; it is violated by the small mass difference between the up and down quarks, and by electromagnetism. Thus we see small differences in the masses of the hadrons assigned to a particular isotopic multiplet.

The role of the axial part of the $\mathrm{SU(2)_L} \times \mathrm{SU(2)_R}$ symmetry, obtained by setting $R = L^\dagger$ in eq. (83.3), is harder to identify. The hadrons do not appear to be classified into multiplets by a second SU(2) symmetry group. In particular, there is no evidence for a classification that distinguishes the left- and right-handed components of spin-one-half hadrons like the proton and neutron.

Reconciliation of these observations with the $\mathrm{SU(2)_L} \times \mathrm{SU(2)_R}$ symmetry of the underlying lagrangian is only possible if the axial generators are *spontaneously broken.* The three pions (which have spin zero, odd parity, and are by far the lightest hadrons) are then identified as the corresponding Goldstone bosons. They are not exactly massless (and hence are sometimes called *pseudogoldstone bosons*) because the $\mathrm{SU(2)_L} \times \mathrm{SU(2)_R}$ symmetry is, as we just discussed, not exact.

To spontaneously break the axial part of the $\mathrm{SU(2)_L} \times \mathrm{SU(2)_R}$, some operator that transforms nontrivially under it must acquire a nonzero vacuum expectation value, or VEV for short. To avoid spontaneous breakdown of Lorentz invariance, this operator must be a Lorentz scalar, and to avoid spontaneous breakdown of the SU(3) gauge symmetry, it must be a color singlet. Since we have no fundamental scalar fields that could acquire a nonzero VEV, we must turn to *composite fields* instead. The simplest candidate is $\chi^a_{\alpha i}\xi^{\alpha\bar{\jmath}}_a = \overline{\Psi}^{\alpha\bar{\jmath}} P_{\rm L}\Psi_{\alpha i}$, where a is an undotted spinor index. (The product of two fields is generically singular, and a renormalization scheme must be specified to define it.) We assume that

$$\langle 0|\chi^a_{\alpha i}\xi^{\alpha\bar{\jmath}}_a|0\rangle = -v^3\delta_i{}^{\bar{\jmath}}\,, \tag{83.7}$$

where v is a parameter with dimensions of mass. Its numerical value depends on the renormalization scheme; for $\overline{\rm MS}$ with $\mu = 2\,{\rm GeV}$, $v \simeq 0.23\,{\rm GeV}$.

To see that this *fermion condensate* does the job of breaking the axial generators of $\mathrm{SU}(2)_{\rm L}\times\mathrm{SU}(2)_{\rm R}$ while preserving the vector generators, we note that, under the transformation of eqs. (83.2) and (83.3),

$$\begin{aligned}\langle 0|\chi^a_{\alpha i}\xi^{\alpha\bar{\jmath}}_a|0\rangle &\to L_i{}^k (R^*)^{\bar{\jmath}}{}_{\bar{n}}\langle 0|\chi^a_{\alpha k}\xi^{\alpha\bar{n}}_a|0\rangle \\ &\to -v^3(LR^\dagger)_i{}^{\bar{\jmath}}\,,\end{aligned} \tag{83.8}$$

where we used eq. (83.7) to get the second line. If we take $R = L$, corresponding to an $\mathrm{SU}(2)_{\rm V}$ transformation, the right-hand side of eq. (83.8) is unchanged from its value in eq. (83.7). This signifies that $\mathrm{SU}(2)_{\rm V}$ [and also $\mathrm{U}(1)_{\rm V}$] is unbroken. However, for a more general transformation with $R \neq L$, the right-hand side of eq. (83.8) does not match that of eq. (83.7), signifying the spontaneous breakdown of the axial generators.

Eq. (83.7) is *nonperturbative*: $\langle 0|\chi^a_{\alpha i}\xi^{\alpha\bar{\jmath}}_a|0\rangle$ vanishes at tree level. Perturbative corrections then also vanish, because of the chiral flavor symmetry of the lagrangian. Thus the value of v is not accessible in perturbation theory. On general grounds, we expect $v \sim \Lambda_{\rm QCD}$, since $\Lambda_{\rm QCD}$ is the only mass scale in the theory when the quarks are massless. Similarly, $\Lambda_{\rm QCD}$ sets the scale for the masses of all the hadrons that are not pseudogoldstone bosons, including the proton and neutron.

We can construct a low-energy effective lagrangian for the three pseudogoldstone bosons (to be identified as the pions) in the following way. Let $|U\rangle$ be a low-energy state for which the expectation value of $\chi^a_{\alpha i}\xi^{\alpha\bar{\jmath}}_a$ varies slowly in orientation (in flavor space) as a function of spacetime:

$$\langle U|\chi^a_{\alpha i}(x)\xi^{\alpha\bar{\jmath}}_a(x)|U\rangle = -v^3 U_i{}^{\bar{\jmath}}(x)\,, \tag{83.9}$$

where $U(x)$ is a spacetime-dependent unitary matrix. We can write it as

$$U(x) = \exp[2i\pi^a(x)T^a/f_\pi]\,, \tag{83.10}$$

where $T^a = \frac{1}{2}\sigma^a$ with $a = 1, 2, 3$ are the generator matrices of SU(2), $\pi^a(x)$ are three real scalar fields to be identified with the pions, and f_π is a parameter with dimensions of mass, the *pion decay constant.* We do not include a fourth generator matrix proportional to the identity, since the corresponding field would be the Goldstone boson for the $\mathrm{U}(1)_\mathrm{A}$ symmetry that is eliminated by the anomaly. Equivalently, we require $\det U(x) = 1$.

We will think of $U(x)$ as an effective, low energy field. Its lagrangian should be the most general one that is consistent with the underlying $\mathrm{SU}(2)_\mathrm{L} \times \mathrm{SU}(2)_\mathrm{R}$ symmetry.[1] Under a general $\mathrm{SU}(2)_\mathrm{L} \times \mathrm{SU}(2)_\mathrm{R}$ transformation, we have

$$U(x) \to LU(x)R^\dagger \; , \tag{83.11}$$

where L and R are independent special unitary matrices. We can organize the terms in the effective lagrangian for $U(x)$ (also known as the *chiral lagrangian*) by the number of derivatives they contain. Because $U^\dagger U = 1$, there are no terms with no derivatives. There is one term with two (all others being equivalent after integration by parts),

$$\mathcal{L} = -\tfrac{1}{4} f_\pi^2 \, \mathrm{Tr}\, \partial^\mu U^\dagger \partial_\mu U \; . \tag{83.12}$$

If we substitute in eq. (83.10) for U, and expand in inverse powers of f_π, we find

$$\mathcal{L} = -\tfrac{1}{2}\partial^\mu \pi^a \partial_\mu \pi^a + \tfrac{1}{6} f_\pi^{-2} (\pi^a \pi^a \partial^\mu \pi^b \partial_\mu \pi^b - \pi^a \pi^b \partial^\mu \pi^b \partial_\mu \pi^a) + \ldots \; . \tag{83.13}$$

Thus the pion fields are conventionally normalized, and they have interactions that are dictated by the general form of eq. (83.12). These interactions lead to Feynman vertices that contain factors of momenta p divided by f_π. Therefore, we can think of p/f_π as an expansion parameter. Of course, we should also add to $\mathcal{L}$ all possible inequivalent terms with four or more derivatives, with coefficients that include inverse powers of f_π. These will lead to more vertices, but their effects will be suppressed by additional powers of p/f_π. Comparison with experiment then yields $f_\pi = 92.4\,\mathrm{MeV}$. (In practice, the value of f_π is more readily determined from the decay rate of the pion via the weak interaction; see section 90 and problem 48.5.)

This value for f_π may seem low; it is, for example, less than the mass of the "almost massless" pions. However, it turns out that tree and loop diagrams contribute roughly equally to any particular process if each extra derivative in $\mathcal{L}$ is accompanied by a factor of $(4\pi f_\pi)^{-1}$ rather that f_π^{-1}, and

[1] $\mathrm{U}(1)_\mathrm{V}$ acts trivially on $U(x)$, and so we need not be concerned with it.

each loop momentum is cut off at $4\pi f_\pi$. Thus it is $4\pi f_\pi \sim 1\,\text{GeV}$ that sets the scale of the interactions, rather than $f_\pi \sim 100\,\text{MeV}$.

Now let us consider the effect of including the small masses for the up and down quarks. The most general mass term we can add to the lagrangian is

$$\begin{aligned}\mathcal{L}_{\text{mass}} &= -\xi^{\alpha\bar{\jmath}} M_{\bar{\jmath}}{}^{i} \chi_{\alpha i} + \text{h.c.} \\ &= -M_{\bar{\jmath}}{}^{i} \chi_{\alpha i} \xi^{\alpha\bar{\jmath}} + \text{h.c.} \\ &= -\text{Tr}\, M \chi_\alpha \xi^\alpha + \text{h.c.}\,, \end{aligned} \tag{83.14}$$

where M is a complex 2×2 matrix. By making an $\text{SU}(2)_\text{L} \times \text{SU}(2)_\text{R}$ transformation, we can bring M to the form

$$M = \begin{pmatrix} m_u & 0 \\ 0 & m_d \end{pmatrix} e^{-i\theta/2}\,, \tag{83.15}$$

where m_u and m_d are real and positive. We cannot remove the overall phase θ, however, without making a forbidden $\text{U}(1)_\text{A}$ transformation. A nonzero value of θ has physical consequences, as we will discuss in section 94. For now, we note that experimental observations fix $|\theta| < 10^{-9}$, and so we will set $\theta = 0$ on this phenomenological basis.

Next, we replace $\chi_\alpha \xi^\alpha$ in eq. (83.14) with its spacetime-dependent VEV, eq. (83.9). The result is a term in the chiral lagrangian that incorporates the leading effect of the quark masses,

$$\mathcal{L}_{\text{mass}} = v^3\, \text{Tr}(MU + M^\dagger U^\dagger)\,. \tag{83.16}$$

Here we continue to distinguish M and $M^\dagger$, even though, with $\theta = 0$, they are the same matrix. If we think of M as transforming as $M \to RML^\dagger$, while U transforms as $U \to LUR^\dagger$, then $\text{Tr}\, MU$ is formally invariant. We then require all terms in the chiral lagrangian to exhibit this formal invariance.

If we expand $\mathcal{L}_{\text{mass}}$ in inverse powers of f_π, and use $M^\dagger = M$, we find

$$\begin{aligned}\mathcal{L}_{\text{mass}} &= -4(v^3/f_\pi^2)\,\text{Tr}(MT^aT^b)\pi^a\pi^b + \ldots \\ &= -2(v^3/f_\pi^2)\,\text{Tr}(M\{T^a,T^b\})\pi^a\pi^b + \ldots \\ &= -(v^3/f_\pi^2)(\text{Tr}\, M)\pi^a\pi^a + \ldots\,. \end{aligned} \tag{83.17}$$

We used the SU(2) relation $\{T^a, T^b\} = \frac{1}{2}\delta^{ab}$ to get the last line. From eq. (83.17), we see that all three pions have the same mass, given by the *Gell-Mann–Oakes–Renner relation*,

$$m_\pi^2 = 2(m_u + m_d)v^3/f_\pi^2\,. \tag{83.18}$$

On the right-hand side, the quark masses and v^3 depend on the renormalization scheme, but their product does not. In the real world, electromagnetic interactions raise the mass of the $\pi^\pm$ slightly above that of the π^0.

This framework is easily expanded to include the strange quark. The three pions (π^+, π^-, mass 0.140 GeV; π^0, mass 0.135 GeV), the four kaons (K^+, K^-, mass 0.494 GeV; K^0, $\overline{K}^0$, mass 0.498 GeV), and the eta (η, mass 0.548 GeV) are identified as the eight expected Goldstone bosons. We can assemble them into the hermitian matrix

$$\Pi \equiv \pi^a T^a/f_\pi = \frac{1}{2f_\pi}\begin{pmatrix} \pi^0 + \frac{1}{\sqrt{3}}\eta & \sqrt{2}\pi^+ & \sqrt{2}K^+ \\ \sqrt{2}\pi^- & -\pi^0 + \frac{1}{\sqrt{3}}\eta & \sqrt{2}K^0 \\ \sqrt{2}K^- & \sqrt{2}\,\overline{K}^0 & -\frac{2}{\sqrt{3}}\eta \end{pmatrix} . \tag{83.19}$$

The second line of eq. (83.17) still applies, but now the T^as are the generators of SU(3), and M includes a third diagonal entry for the strange quark mass. We leave the details to the problems.

Next we turn to the coupling of the pions to the nucleons (the proton and neutron). We define a Dirac field N_i, where $N_1 = p$ (the proton) and $N_2 = n$ (the neutron). We assume that, under an $\mathrm{SU(2)_L \times SU(2)_R}$ transformation,

$$P_{\rm L} N_i \to L_i{}^j P_{\rm L} N_j \,, \tag{83.20}$$

$$P_{\rm R} N_{\bar\imath} \to R_{\bar\imath}{}^{\bar\jmath} P_{\rm R} N_{\bar\jmath} \,. \tag{83.21}$$

The standard Dirac kinetic term $i\overline{N}\partial\!\!\!/ N$ is then $\mathrm{SU(2)_L \times SU(2)_R}$ invariant, but the standard mass term $m_N\overline{N}N$ is not. (Here m_N is the value of the nucleon mass in the limit of zero up and down quark masses.) However, we can construct an invariant mass term by including appropriate factors of U and $U^\dagger$,

$$\mathcal{L}_{\rm mass} = -m_N\overline{N}(U^\dagger P_{\rm L} + UP_{\rm R})N \,. \tag{83.22}$$

There is one other parity, time-reversal, and $\mathrm{SU(2)_L \times SU(2)_R}$ invariant term with one derivative. Including this term, we have

$$\begin{aligned}\mathcal{L} = {}& i\overline{N}\partial\!\!\!/ N - m_N\overline{N}(U^\dagger P_{\rm L} + UP_{\rm R})N \\ & - \tfrac{1}{2}(g_{\rm A}{-}1)i\overline{N}\gamma^\mu(U\partial_\mu U^\dagger P_{\rm L} + U^\dagger\partial_\mu U P_{\rm R})N \,,\end{aligned} \tag{83.23}$$

where $g_{\rm A} = 1.27$ is the *axial vector coupling*. Its value is determined from the decay rate of the neutron via the weak interaction; see section 90.

The form of the lagrangian in eq. (83.23) is somewhat awkward. It can be simplified by first defining

$$u(x) \equiv \exp[\,i\pi^a(x)T^a/f_\pi]\,, \tag{83.24}$$

so that $U(x) \equiv u^2(x)$. Then we define a new nucleon field

$$\mathcal{N} \equiv (u^\dagger P_{\rm L} + uP_{\rm R})N \, . \tag{83.25}$$

(This is a *field redefinition* in the sense of problem 10.5.) Equivalently, using the unitarity of u, we have

$$N = (uP_{\rm L} + u^\dagger P_{\rm R})\mathcal{N} \, . \tag{83.26}$$

Using eq. (83.26) in eq. (83.23), along with the identities $\partial_\mu U = (\partial_\mu u)u + u(\partial_\mu u)$, $(\partial_\mu u^\dagger)u = -u^\dagger(\partial_\mu u)$, etc., we ultimately find

$$\mathcal{L} = i\overline{\mathcal{N}}\partial\!\!\!/\mathcal{N} - m_N\overline{\mathcal{N}}\mathcal{N} + \overline{\mathcal{N}}v\!\!\!/\mathcal{N} - g_{\rm A}\overline{\mathcal{N}}a\!\!\!/\gamma_5\mathcal{N} \, , \tag{83.27}$$

where we have defined the hermitian vector fields

$$v_\mu \equiv \tfrac{1}{2}i[u^\dagger(\partial_\mu u) + u(\partial_\mu u^\dagger)] \, , \tag{83.28}$$

$$a_\mu \equiv \tfrac{1}{2}i[u^\dagger(\partial_\mu u) - u(\partial_\mu u^\dagger)] \, . \tag{83.29}$$

If we now expand u and $u^\dagger$ in inverse powers of f_π, we get

$$\mathcal{L} = i\overline{\mathcal{N}}\partial\!\!\!/\mathcal{N} - m_N\overline{\mathcal{N}}\mathcal{N} + (g_{\rm A}/f_\pi)\partial_\mu\pi^a\overline{\mathcal{N}}T^a\gamma^\mu\gamma_5\mathcal{N} + \ldots \, . \tag{83.30}$$

We can integrate by parts in the interaction term to put the derivative on the $\overline{\mathcal{N}}$ and $\mathcal{N}$ fields. Then, if we consider a process where an off-shell pion is scattered by an on-shell nucleon, we can use the Dirac equation to replace the derivatives of $\overline{\mathcal{N}}$ and $\mathcal{N}$ with factors of m_N. We then find a coupling of the pion to an on-shell nucleon of the form

$$\mathcal{L}_{\pi\overline{N}N} = -ig_{\pi\overline{N}N}\pi^a\overline{\mathcal{N}}\sigma^a\gamma_5\mathcal{N} \, , \tag{83.31}$$

where we have set $T^a = \frac{1}{2}\sigma^a$, and identified the *pion-nucleon coupling constant*,

$$g_{\pi\overline{N}N} = g_{\rm A}m_N/f_\pi \, . \tag{83.32}$$

The value of $g_{\pi\overline{N}N}$ can be determined from measurements of the neutron-proton scattering cross section, assuming that it is dominated by pion exchange; the result is $g_{\pi\overline{N}N} = 13.5$. Eq. (83.32), known as the *Goldberger–Treiman relation*, is then satisfied to within about 5%.

Reference notes

The chiral lagrangian is treated in *Georgi*, *Ramond II*, and *Weinberg II*. Light quark masses are taken from *MILC*.

Problems

83.1) Suppose that the color group is SO(3) rather than SU(3), and that each quark flavor is represented by a Dirac field in the 3 representation of SO(3).

a) With n_F flavors of massless quarks, what is the nonanomalous flavor symmetry group?

b) Assume the formation of a color-singlet, Lorentz scalar, fermion condensate. Assume that it preserves the largest possible unbroken subgroup of the flavor symmetry. What is this unbroken subgroup?

c) For the case $n_F = 2$, how many massless Goldstone bosons are there?

d) Now suppose that the color group is SU(2) rather than SU(3), and that each quark flavor is represented by a Dirac field in the 2 representation of SU(2). Repeat parts (a), (b), and (c) for this case. Hint: at least one of the answers is different!

83.2) Why is there a minus sign on the right-hand side of eq. (83.7)?

83.3) Verify that eq. (83.13) follows from eq. (83.12).

83.4) Use eqs. (83.12) and (83.16) to compute the tree-level contribution to the scattering amplitude for $\pi^a\pi^b \to \pi^c\pi^d$. Work in the isospin limit, $m_u = m_d \equiv m$. Express your answer in terms of the Mandelstam variables and the pion mass m_π.

83.5) Verify that eq. (83.27) follows from eqs. (83.23) and (83.26).

83.6) Consider the case of three light quark flavors, with masses m_u, m_d, and m_s.

a) Find the masses-squared of the eight pseudogoldstone bosons. Take the limit $m_{u,d} \ll m_s$, and drop terms that are of order $m_{u,d}^2/m_s$.

b) Assume that $m_{\pi^\pm}^2$ and $m_{K^\pm}^2$ each receive an electromagnetic contribution; to zeroth order in the quark masses, this contribution is the same for both, but the comparatively large strange quark mass results in an electromagnetic contribution to $m_{K^\pm}^2$ that is roughly twice as large as the electromagnetic contribution $\Delta m_{\rm EM}^2$ to $m_{\pi^\pm}^2$. Use the observed masses of the $\pi^\pm$, π^0, $K^\pm$, and K^0 to compute $m_u v^3/f_\pi^2$, $m_d v^3/f_\pi^2$, $m_s v^3/f_\pi^2$, and $\Delta m_{\rm EM}^2$.

c) Compute the quark mass ratios m_u/m_d and m_s/m_d.

d) Use your results from part (b) to predict the η mass. How good is your prediction?

83.7) Suppose that the $\mathrm{U}(1)_\mathrm{A}$ symmetry is not anomalous, so that we must include a ninth Goldstone boson. We can write

$$U(x) = \exp[\,2i\pi^a(x)T^a/f_\pi + i\pi^9(x)/f_9]\,. \tag{83.33}$$

The ninth Goldstone boson is given its own decay constant f_9, since there is no symmetry that forces it to be equal to f_π. We write the two-derivative terms in the lagrangian as

$$\mathcal{L} = -\tfrac{1}{4}f_\pi^2 \operatorname{Tr} \partial^\mu U^\dagger \partial_\mu U - \tfrac{1}{4}F^2 \partial^\mu(\det U^\dagger)\partial_\mu(\det U)\,. \tag{83.34}$$

a) By requiring all nine Goldstone fields to have canonical kinetic terms, determine F in terms of f_π and f_9.

b) To simplify the analysis, let $m_u = m_d \equiv m \ll m_s$. Find the masses of the nine pseudogoldstone bosons. Identify the three lightest as the pions, and call their mass m_π. Show that another one of the nine has a mass less than or equal to $\sqrt{3}\,m_\pi$. (The nonexistence of such a particle in nature is *the U(1) problem*; the axial anomaly solves this problem.)

83.8) a) Write down all possible parity and time-reversal invariant terms with no derivatives that are bilinear in the nucleon field N and that have one factor of the quark mass matrix M.

b) Reexpress your result in terms of the nucleon field $\mathcal{N}$.

c) Use the observed neutron-proton mass difference, $m_n - m_p = 1.293\,\text{MeV}$, and the m_u/m_d ratio you found in problem 83.6, to detemine as much as you can about the coefficients of the terms you wrote down. (Ignore the mass difference due to electromagnetism.)

84

Spontaneous breaking of gauge symmetries

Prerequisites: 32, 70

Consider scalar electrodynamics, specified by the lagrangian

$$\mathcal{L} = -(D^\mu\varphi)^\dagger D_\mu\varphi - V(\varphi) - \tfrac{1}{4}F^{\mu\nu}F_{\mu\nu} \ , \tag{84.1}$$

where φ is a complex scalar field, $D_\mu = \partial_\mu - igA_\mu$, and

$$V(\varphi) = m^2\varphi^\dagger\varphi + \tfrac{1}{4}\lambda(\varphi^\dagger\varphi)^2 \ . \tag{84.2}$$

(We call the gauge coupling constant g rather than e because we are using this theory as a formal example rather than a physical model.) So far we have always taken $m^2 > 0$, but now let us consider $m^2 < 0$. We analyzed this model in the absence of the gauge field in section 32. Classically, the field has a nonzero vacuum expectation value (VEV for short), given by

$$\langle 0|\varphi(x)|0\rangle = \tfrac{1}{\sqrt{2}}v \ , \tag{84.3}$$

where we have made a global U(1) transformation to set the phase of the VEV to zero, and

$$v = (4|m^2|/\lambda)^{1/2} \ . \tag{84.4}$$

We therefore write

$$\varphi(x) = \tfrac{1}{\sqrt{2}}(v + \rho(x))e^{-i\chi(x)/v} \ , \tag{84.5}$$

where $\rho(x)$ and $\chi(x)$ are real scalar fields. The scalar potential depends only on ρ, and is given by

$$V(\varphi) = \tfrac{1}{4}\lambda v^2\rho^2 + \tfrac{1}{4}\lambda v\rho^3 + \tfrac{1}{16}\lambda\rho^4 \ . \tag{84.6}$$

Since χ does not appear in the potential, it is massless; it is the Goldstone boson for the spontaneously broken U(1) symmetry.

The big difference in the gauge theory is that we can make a gauge transformation that shifts the phase of $\varphi(x)$ by an arbitrary spacetime function. We can use this gauge freedom to set $\chi(x) = 0$; this choice is called *unitary gauge*. Using eq. (84.5) with $\chi(x) = 0$ in eq. (84.1), we have

$$\begin{aligned} -(D^\mu\varphi)^\dagger D_\mu\varphi &= -\tfrac{1}{2}(\partial^\mu\rho + ig(v+\rho)A^\mu)(\partial_\mu\rho - ig(v+\rho)A_\mu) \\ &= -\tfrac{1}{2}\partial^\mu\rho\partial_\mu\rho - \tfrac{1}{2}g^2(v+\rho)^2A^\mu A_\mu \,. \end{aligned} \tag{84.7}$$

Expanding out the last term, we see that the gauge field now has a mass

$$M = gv \,. \tag{84.8}$$

This is the *Higgs mechanism*: the Goldstone boson disappears, and the gauge field acquires a mass. Note that this leaves the counting of particle spin states unchanged: a massless spin-one particle has two spin states, but a massive one has three. The Goldstone boson has become the third or *longitudinal* state of the now-massive gauge field. A scalar field whose VEV breaks a gauge symmetry is generically called a *Higgs field.*

This generalizes in a straightforward way to a nonabelian gauge theory. Consider a complex scalar field φ in a representation R of the gauge group. The kinetic term for φ is $-(D^\mu\varphi)^\dagger D_\mu\varphi$, where the covariant derivative is $(D_\mu\varphi)_i = \partial_\mu\varphi_i - igA^a_\mu(T^a_{\rm R})_i{}^j\varphi_j$, and the indices i and j run from 1 to D(R). We assume that φ acquires a VEV

$$\langle 0|\varphi_i(x)|0\rangle = \tfrac{1}{\sqrt{2}}v_i \,, \tag{84.9}$$

where the value of v_i is determined (up to a global gauge transformation) by minimizing the potential. If we replace φ by its VEV in $-(D^\mu\varphi)^\dagger D_\mu\varphi$, we find a mass term for the gauge fields,

$$\mathcal{L}_{\rm mass} = -\tfrac{1}{2}(M^2)^{ab}A^{a\mu}A^b_\mu \,, \tag{84.10}$$

where the mass-squared matrix is

$$(M^2)^{ab} = \tfrac{1}{2}g^2 v_i^*\{T^a_{\rm R}, T^b_{\rm R}\}_{ij}v_j \,. \tag{84.11}$$

The anticommutator appears because $A^{a\mu}A^b_\mu$ is symmetric on $a \leftrightarrow b$, and so we replaced $T^a_{\rm R}T^b_{\rm R}$ with $\tfrac{1}{2}\{T^a_{\rm R}, T^b_{\rm R}\}$.

If the field φ is real rather than complex (which is possible only if R is a real representation), then we remove the factor of root-two from the right-hand side of eq. (84.9), but this is compensated by an extra factor of one-half from the kinetic term for a real scalar field; thus eq. (84.11) holds as written. If there is more than one gauge group, then the g^2 in eq. (84.11) is replaced by $g_a g_b$, where g_a is the coupling constant that goes along with the generator T^a, and all generators of all gauge groups are included in the mass-squared matrix.

Recall from section 32 that a generator T^a is spontaneously broken if $(T^a_{\rm R})_{ij}v_j \neq 0$. From eq. (84.11), we see that gauge fields corresponding to broken generators get a mass, while those corresponding to unbroken generators do not. The unbroken generators (if any) form a gauge group with massless gauge fields. The massive gauge fields (and all other fields) form representations of this unbroken group.

Let us work out some simple examples.

Consider the gauge group SU(N), with a complex scalar field φ in the fundamental representation. We can make a global SU(N) transformation to bring the VEV entirely into the last component, and furthermore make it real. Any generator $(T^a)_i{}^j$ that does not have a nonzero entry in the last column will remain unbroken. These generators form an unbroken SU($N-1$) gauge group. There are three classes of broken generators: those with $(T^a)_i{}^N = \frac{1}{2}$ for $i \neq N$ (there are $N-1$ of these); those with $(T^a)_i{}^N = -\frac{1}{2}i$ for $i \neq N$ (there are also $N-1$ of these), and finally the single generator $T^{N^2-1} = [2N(N-1)]^{-1/2}\,{\rm diag}(1,\ldots,1,-(N-1))$. The gauge fields corresponding to the generators in the first two classes get a mass $M = \frac{1}{2}gv$; we can group them into a complex vector field that transforms in the fundamental representation of the unbroken SU($N-1$) subgroup. The gauge field corresponding to T^{N^2-1} gets a mass $M = [(N-1)/2N]^{1/2}gv$; it is a singlet of SU($N-1$).

Consider the gauge group SO(N), with a real scalar field in the fundamental representation. We can make a global SO(N) transformation to bring the VEV entirely into the last component. Any generator $(T^a)_i{}^j$ that does not have a nonzero entry in the last column will remain unbroken. These generators form an unbroken SO($N-1$) subgroup. There are $N-1$ broken generators, those with $(T^a)_i{}^N = -i$ for $i \neq N$. The corresponding gauge fields get a mass $M = gv$; they form a fundamental representation of the unbroken SO($N-1$) subgroup. In the case $N = 3$, this subgroup is SO(2), which is equivalent to U(1).

Consider the gauge group SU(N), with a real scalar field Φ^a in the adjoint representation. It will prove more convenient to work with the matrix-valued field $\Phi = \Phi^a T^a$; the covariant derivative of Φ is $D_\mu\Phi = \partial_\mu\Phi - igA^a_\mu[T^a,\Phi]$, and the VEV of Φ is a traceless hermitian $N \times N$ matrix V. Thus the mass-squared matrix for the gauge fields is $(M^2)^{ab} = -\frac{1}{2}g^2\,{\rm Tr}\{[T^a,V],[T^b,V]\}$. We can make a global SU(N) transformation to bring V into diagonal form.

Suppose the diagonal entries consist of N_1 v_1s, followed by N_2 v_2s, etc., where $v_1 < v_2 < \ldots$, and $\sum_i N_i v_i = 0$. Then all generators whose nonzero entries lie entirely within the ith block commute with V, and hence form an unbroken SU(N_i) subgroup. Furthermore, the linear combination of diagonal generators that is proportional to V also commutes with V, and forms a U(1) subgroup. Thus the unbroken gauge group is SU(N_1) × SU(N_2) × ... × U(1). The gauge coupling constants for the different groups are all the same, and equal to the original SU(N) gauge coupling constant.

As a specific example, consider the case of SU(5), which has 24 generators. Let the diagonal entries of V be given by $(-\frac{1}{3}, -\frac{1}{3}, -\frac{1}{3}, +\frac{1}{2}, +\frac{1}{2})v$. The unbroken subgroup is then SU(3) × SU(2) × U(1). The number of broken generators is $24 - 8 - 3 - 1 = 12$. The generator of the U(1) subgroup is $T^{24} = \text{diag}(-\frac{1}{3}c, -\frac{1}{3}c, -\frac{1}{3}c, +\frac{1}{2}c, +\frac{1}{2}c)$, where $c^2 = 3/5$. Under the unbroken SU(3) × SU(2) × U(1) subgroup, the 5 representation of SU(5) transforms as

$$5 \to (3, 1, -\tfrac{1}{3}) \oplus (1, 2, +\tfrac{1}{2}) \,. \tag{84.12}$$

Here the last entry is the value of T^{24}/c. The $\overline{5}$ of SU(5) then transforms as

$$\overline{5} \to (\overline{3}, 1, +\tfrac{1}{3}) \oplus (1, 2, -\tfrac{1}{2}) \,. \tag{84.13}$$

To find out how the adjoint or 24 representation of SU(5) transforms under the SU(3) × SU(2) × U(1) subgroup, we use the SU(5) relation

$$5 \otimes \overline{5} = 24 \oplus 1 \,. \tag{84.14}$$

From eqs. (84.12) and (84.13), we have

$$5 \otimes \overline{5} \to [(3, 1, -\tfrac{1}{3}) \oplus (1, 2, +\tfrac{1}{2})] \otimes [(\overline{3}, 1, +\tfrac{1}{3}) \oplus (1, 2, -\tfrac{1}{2})] \,. \tag{84.15}$$

If we expand this out, and compare with eq. (84.14), we see that

$$\begin{aligned} 24 \to\ & (8, 1, 0) \oplus (1, 3, 0) \oplus (1, 1, 0) \\ & \oplus (3, 2, -\tfrac{5}{6}) \oplus (\overline{3}, 2, +\tfrac{5}{6}) \,. \end{aligned} \tag{84.16}$$

The first line on the right-hand side of eq. (84.16) is the adjoint representation of SU(3) × SU(2) × U(1); the corresponding gauge fields remain massless. The second line shows us that the gauge fields corresponding to the twelve broken generators can be grouped into a complex vector field in the representation $(3, 2, -\frac{5}{6})$. Since it is an irreducible representation of the unbroken subgroup, all twelve vector fields must have the same mass. This mass is most easily computed from $(M^2)^{44} = -g^2\,\text{Tr}([T^4, V][T^4, V])$, where we have defined $(T^4)_i{}^j = \frac{1}{2}(\delta_i{}^1\delta^j{}_4 + \delta_i{}^4\delta^j{}_1)$; the result is $M = \frac{5}{6\sqrt{2}}gv$.

Problems

84.1) Consider a theory with gauge group SU(N), with a real scalar field Φ in the adjoint representation, and potential

$$V(\Phi) = \tfrac{1}{2}m^2 \operatorname{Tr}\Phi^2 + \tfrac{1}{4}\lambda_1 \operatorname{Tr}\Phi^4 + \tfrac{1}{4}\lambda_2 (\operatorname{Tr}\Phi^2)^2\,. \tag{84.17}$$

This is the most general potential consistent with SU(N) symmetry and a Z_2 symmetry $\Phi \leftrightarrow -\Phi$, which we impose to keep things simple. We assume $m^2 < 0$. We can work in a basis in which $\Phi = v\,\mathrm{diag}(\alpha_1, \ldots, \alpha_N)$, with the constraints $\sum_i \alpha_i = 0$ and $\sum_i \alpha_i^2 = 1$.

a) Extremize $V(\Phi)$ with respect to v. Solve for v, and plug your result back into $V(\Phi)$. You should find

$$V(\Phi) = \frac{-\frac{1}{4}(m^2)^2}{\lambda_1 A(\alpha) + \lambda_2 B(\alpha)}\,, \tag{84.18}$$

where $A(\alpha)$ and $B(\alpha)$ are functions of α_i.

b) Show that $\lambda_1 A(\alpha) + \lambda_2 B(\alpha)$ must be everywhere positive in order for the potential to be bounded below.

c) Show that the absolute minimum of the potential (assuming that it is bounded below) occurs at the absolute minimum of $\lambda_1 A(\alpha) + \lambda_2 B(\alpha)$.

d) Show that, at any extremum of the potential, the α_i take on at most three different values, and that these three values sum to zero. Hint: impose the constraints with Lagrange multipliers.

e) Show that, for $\lambda_1 > 0$ and $\lambda_2 > 0$, at the absolute minimum of $V(\Phi)$ the unbroken symmetry group is SU(N_+) × SU(N_-) × U(1), where $N_+ = N_- = \frac{1}{2}N$ if N is even, and $N_\pm = \frac{1}{2}(N\pm1)$ if N is odd.

85

Spontaneously broken abelian gauge theory

Prerequisites: 61, 84

Consider scalar electrodynamics, specified by the lagrangian

$$\mathcal{L} = -(D^\mu\varphi)^\dagger D_\mu\varphi - V(\varphi) - \tfrac{1}{4}F^{\mu\nu}F_{\mu\nu}\,, \tag{85.1}$$

where φ is a complex scalar field and $D_\mu = \partial_\mu - igA_\mu$. We choose

$$V(\varphi) = \tfrac{1}{4}\lambda(\varphi^\dagger\varphi - \tfrac{1}{2}v^2)^2\,, \tag{85.2}$$

which yields a nonzero VEV for φ. We therefore write

$$\varphi(x) = \tfrac{1}{\sqrt{2}}(v+\rho(x))e^{-i\chi(x)/v}\,, \tag{85.3}$$

where $\rho(x)$ and $\chi(x)$ are real scalar fields. The scalar potential depends only on ρ, and is given by

$$V(\varphi) = \tfrac{1}{4}\lambda v^2\rho^2 + \tfrac{1}{4}\lambda v\rho^3 + \tfrac{1}{16}\lambda\rho^4\,. \tag{85.4}$$

We can now make a gauge transformation to set

$$\chi(x) = 0\,. \tag{85.5}$$

This is *unitary gauge*. The kinetic term for φ becomes

$$-(D^\mu\varphi)^\dagger D_\mu\varphi = -\tfrac{1}{2}\partial^\mu\rho\partial_\mu\rho - \tfrac{1}{2}g^2(v+\rho)^2A^\mu A_\mu\,. \tag{85.6}$$

We see that the gauge field has acquired a mass

$$M = gv\,. \tag{85.7}$$

The terms in $\mathcal{L}$ that are quadratic in A_μ are

$$\mathcal{L}_0 = -\tfrac{1}{4}F^{\mu\nu}F_{\mu\nu} - \tfrac{1}{2}M^2A^\mu A_\mu\,. \tag{85.8}$$

The equation of motion that follows from eq. (85.8) is

$$[(-\partial^2 + M^2)g^{\mu\nu} + \partial^\mu\partial^\nu]A_\nu = 0 \,. \tag{85.9}$$

If we act with ∂_μ on this equation, we get

$$M^2\partial^\nu A_\nu = 0 \,. \tag{85.10}$$

If we now use eq. (85.10) in eq. (85.9), we find that each component of A_ν obeys the Klein–Gordon equation,

$$(-\partial^2 + M^2)A_\nu = 0 \,. \tag{85.11}$$

The general solution of eqs. (85.9) and (85.10) is

$$A^\mu(x) = \sum_{\lambda=-,0,+} \int \widetilde{dk} \left[\varepsilon_\lambda^{\mu *}(k)a_\lambda(k)e^{ikx} + \varepsilon_\lambda^{\mu}(k)a_\lambda^\dagger(k)e^{-ikx}\right] , \tag{85.12}$$

where the polarization vectors must satisfy $k_\mu\varepsilon_\lambda^\mu(k) = 0$. In the rest frame, where $k = (M, 0, 0, 0)$, we choose the polarization vectors to correspond to definite spin along the $\hat{z}$ axis,

$$\begin{aligned}
\varepsilon_+(0) &= \tfrac{1}{\sqrt{2}}(0, 1, -i, 0) \,, \\
\varepsilon_-(0) &= \tfrac{1}{\sqrt{2}}(0, 1, +i, 0) \,, \\
\varepsilon_0(0) &= (0, 0, 0, 1) \,.
\end{aligned} \tag{85.13}$$

More generally, the three polarization vectors along with the timelike unit vector k^μ/M form an orthonormal and complete set,

$$k\cdot\varepsilon_\lambda(k) = 0 \,, \tag{85.14}$$

$$\varepsilon_{\lambda'}(k)\cdot\varepsilon_\lambda^*(k) = \delta_{\lambda'\lambda} \,, \tag{85.15}$$

$$\sum_{\lambda=-,0,+} \varepsilon_\lambda^{\mu *}(k)\varepsilon_\lambda^\nu(k) = g^{\mu\nu} + \frac{k^\mu k^\nu}{M^2} \,. \tag{85.16}$$

Since the lagrangian of eq. (85.8) has no manifest gauge invariance, quantization is straightforward. The coefficients $a_\lambda^\dagger(k)$ and $a_\lambda(k)$ become particle creation and annihilation operators in the usual way, and the propagator of the A_μ field is given by

$$\begin{aligned}
i\langle 0|\mathrm{T}A^\mu(x)A^\nu(y)|0\rangle &= \int \frac{d^4k}{(2\pi)^4}\,\frac{e^{ik(x-y)}}{k^2 + M^2 - i\epsilon} \textstyle\sum_\lambda \varepsilon_\lambda^{\mu *}(k)\varepsilon_\lambda^\nu(k) \\
&= \int \frac{d^4k}{(2\pi)^4}\,\frac{e^{ik(x-y)}}{k^2 + M^2 - i\epsilon}\left(g^{\mu\nu} + \frac{k^\mu k^\nu}{M^2}\right) .
\end{aligned} \tag{85.17}$$

The interactions of the massive vector field A_μ with the real scalar field ρ can be read off of eq. (85.6). The self-interactions of the ρ field can be read off of eq. (85.4). The resulting Feynman rules can be used for tree-level calculations.

Loop calculations are more subtle. We have imposed the gauge condition $\chi(x) = 0$, which corresponds to inserting a functional delta function $\prod_x \delta(\chi(x))$ into the path integral. In order to integrate over χ, we must make a change of integration variables from $\text{Re}\,\varphi$ and $\text{Im}\,\varphi$ to ρ and χ; this is simply a transformation from cartesian to polar coordinates, analogous to $dx\,dy = r\,dr\,d\phi$. So we must include a factor analogous to r in the functional measure; this factor is

$$\begin{aligned}\prod_x \Big(v + \rho(x)\Big) &= \det(v + \rho) \\ &\propto \det(1 + v^{-1}\rho) \\ &\propto \int \mathcal{D}\bar{c}\,\mathcal{D}c\; e^{-im_{\text{gh}}^2 \int d^4x\; \bar{c}(1+v^{-1}\rho)c} \,. \end{aligned} \tag{85.18}$$

In the last line, we have written the functional determinant as an integral over ghost fields. We see that they have no kinetic term, and we have chosen the overall nomalization of their action so that their mass is m_{gh}, where m_{gh} is an arbitrary mass parameter. Thus the momentum-space propagator for the ghosts is simply $\tilde{\Delta}(k^2) = 1/m_{\text{gh}}^2$. We also see that there is a ghost-ghost-scalar vertex, with vertex factor $-im_{\text{gh}}^2 v^{-1}$, but there is no interaction between the ghosts and the vector field.

This seems like a fairly convenient gauge for loop calculations, but there is a complication. The fact that the ghost propagator is independent of the momentum means that additional internal ghost propagators do not help the convergence of loop-momentum integrals. The same is true of vector-field propagators; from eq. (85.17) we see that, in momentum space, the propagator scales like $1/M^2$ in the limit that all components of k become large. Thus, in unitary gauge, loop diagrams with arbitrarily many external lines diverge. This makes it difficult to establish renormalizability.

A gauge that does not suffer from this problem is a generalization of R_ξ gauge (and in fact this name has traditionally been applied only to this generalization). We begin by using a cartesian basis for φ,

$$\varphi = \tfrac{1}{\sqrt{2}}(v + h + ib) \,, \tag{85.19}$$

where h and b are real scalar fields. In terms of h and b, the potential is

$$V(\varphi) = \tfrac{1}{4}\lambda v^2 h^2 + \tfrac{1}{4}\lambda v h(h^2 + b^2) + \tfrac{1}{16}\lambda(h^2 + b^2)^2 \,, \tag{85.20}$$

and the covariant derivative of φ is

$$D_\mu\varphi = \tfrac{1}{\sqrt{2}}\Big[(\partial_\mu h + gbA_\mu) + i(\partial_\mu b - g(v{+}h)A_\mu)\Big] \,. \tag{85.21}$$

Thus the kinetic term for φ becomes

$$-(D^\mu\varphi)^\dagger D_\mu\varphi = -\tfrac{1}{2}(\partial_\mu h + gbA_\mu)^2 - \tfrac{1}{2}(\partial_\mu b - g(v{+}h)A_\mu)^2 \,. \tag{85.22}$$

Expanding this out, and rearranging, we get

$$\begin{aligned}-(D^\mu\varphi)^\dagger D_\mu\varphi = {}& -\tfrac{1}{2}\partial_\mu h\partial_\mu h - \tfrac{1}{2}\partial_\mu b\partial_\mu b - \tfrac{1}{2}g^2v^2A^\mu A_\mu + gvA^\mu\partial_\mu b \\ & + gA^\mu(h\partial_\mu b - b\partial_\mu h) \\ & - g^2vhA^\mu A_\mu - \tfrac{1}{2}g^2(h^2+b^2)A^\mu A_\mu \,. \end{aligned} \tag{85.23}$$

The first line on the right-hand side of eq. (85.23) contains all the terms that are quadratic in the fields. The first two are the kinetic terms for the h and b fields. The third is the mass term for the vector field. The fourth is an annoying cross term between the vector field and the derivative of b.

In abelian gauge theory, in the absence of spontaneous symmetry breaking, we fix R_ξ gauge by adding to $\mathcal{L}$ the gauge-fixing and ghost terms

$$\mathcal{L}_{\rm gf} + \mathcal{L}_{\rm gh} = -\tfrac{1}{2}\xi^{-1}G^2 - \bar{c}\,\frac{\delta G}{\delta\theta}\,c \,, \tag{85.24}$$

where $G = \partial^\mu A_\mu$, and $\theta(x)$ parameterizes an infinitesimal gauge transformation,

$$A_\mu \to A_\mu - \partial_\mu\theta \,, \tag{85.25}$$

$$\varphi \to \varphi - ig\theta\varphi \,. \tag{85.26}$$

With $G = \partial^\mu A_\mu$, we have $\delta G/\delta\theta = -\partial^2$. Thus the ghost fields have no interactions, and can be ignored.

In the presence of spontaneous symmetry breaking, we choose instead

$$G = \partial^\mu A_\mu - \xi gvb \,, \tag{85.27}$$

which reduces to $\partial^\mu A_\mu$ when $v = 0$. Multiplying out G^2, we have

$$\begin{aligned}\mathcal{L}_{\rm gf} &= -\tfrac{1}{2}\xi^{-1}\partial^\mu A_\mu\partial^\nu A_\nu + gvb\partial^\mu A_\mu - \tfrac{1}{2}\xi g^2v^2b^2 \\ &= -\tfrac{1}{2}\xi^{-1}\partial^\mu A_\nu\partial^\nu A_\mu - gvA_\mu\partial^\mu b - \tfrac{1}{2}\xi g^2v^2b^2 \,, \end{aligned} \tag{85.28}$$

where we integrated by parts in the first two terms to get the second line. Note that the second term on the second line of eq. (85.28) cancels the annoying last term on the first line of eq. (85.23). Also, the last term on the second line of eq. (85.28) gives a mass $\xi^{1/2}M$ to the b field.

We must still evaluate $\mathcal{L}_{\rm gh}$. To do so, we first translate eq. (85.26) into

$$h \to h + g\theta b\,, \tag{85.29}$$

$$b \to b - g\theta(v+h)\,. \tag{85.30}$$

Then we have

$$\frac{\delta G}{\delta\theta} = -\partial^2 + \xi g^2 v(v+h)\,. \tag{85.31}$$

From eq. (85.24) we see that the ghost lagrangian is

$$\begin{aligned}\mathcal{L}_{\rm gh} &= -\bar{c}\Big[-\partial^2 + \xi g^2 v(v+h)\Big]c \\ &= -\partial^\mu\bar{c}\partial_\mu c - \xi g^2 v^2 \bar{c}c - \xi g^2 v h\bar{c}c\,.\end{aligned} \tag{85.32}$$

We see from the second term that the ghost has acquired the same mass as the b field, $\xi^{1/2}M$.

Now let us examine the vector field. Including $\mathcal{L}_{\rm gf}$, the terms in $\mathcal{L}$ that are quadratic in the vector field can be written as

$$\mathcal{L}_0 = -\tfrac{1}{2}A_\mu\Big[g^{\mu\nu}(-\partial^2+M^2) + (1-\xi^{-1})\partial^\mu\partial^\nu\Big]A_\nu\,. \tag{85.33}$$

In momentum space, this reads

$$\tilde{\mathcal{L}}_0 = -\tfrac{1}{2}\tilde{A}_\mu(-k)\Big[(k^2+M^2)g^{\mu\nu} - (1-\xi^{-1})k^\mu k^\nu\Big]\tilde{A}_\nu(k)\,. \tag{85.34}$$

The kinematic matrix is

$$\begin{aligned}\Big[\ldots\Big] &= (k^2+M^2)g^{\mu\nu} - (1-\xi^{-1})k^\mu k^\nu \\ &= (k^2+M^2)\Big(P^{\mu\nu}(k) + k^\mu k^\nu/k^2\Big) - (1-\xi^{-1})k^\mu k^\nu \\ &= (k^2+M^2)P^{\mu\nu}(k) + \xi^{-1}(k^2+\xi M^2)k^\mu k^\nu/k^2\,,\end{aligned} \tag{85.35}$$

where $P^{\mu\nu}(k) = g^{\mu\nu} - k^\mu k^\nu/k^2$ projects onto the transverse subspace; $P^{\mu\nu}(k)$ and $k^\mu k^\nu/k^2$ are orthogonal projection matrices. Using this fact, it is easy to invert eq. (85.35) to get the propagator for the massive vector field in R_ξ gauge,

$$\tilde{\Delta}^{\mu\nu}(k) = \frac{P^{\mu\nu}(k)}{k^2+M^2-i\epsilon} + \frac{\xi\,k^\mu k^\nu/k^2}{k^2+\xi M^2-i\epsilon}\,. \tag{85.36}$$

We see that the transverse components of the vector field propagate with mass M, while the longitudinal component propagates with the same mass as the b and ghost fields, $\xi^{1/2}M$.

Eq. (85.36) simplifies greatly if we choose $\xi = 1$; then we have

$$\tilde{\Delta}^{\mu\nu}(k) = \frac{g^{\mu\nu}}{k^2 + M^2 - i\epsilon} \quad (\xi = 1) \,. \tag{85.37}$$

On the other hand, leaving ξ as a free parameter allows us to check that all ξ dependence cancels out of any physical scattering amplitude. Since their masses depend on ξ, the ghosts, the b field, and the longitudinal component of the vector field must all represent unphysical particles that do not appear in incoming or outgoing states.

To summarize, in R_ξ gauge we have the physical h field with mass-squared $m_h^2 = \frac{1}{2}\lambda v^2$ and propagator $1/(k^2 + m_h^2)$, the unphysical b field with propagator $1/(k^2 + \xi M^2)$, the ghost fields $\bar{c}$ and c with propagator $1/(k^2 + \xi M^2)$, and the vector field with the propagator of eq. (85.36). For external vectors, the polarizations are still given by eq. (85.13), and obey the sum rules of eq. (85.16). The mass parameter M is given by $M = gv$. The interactions of these fields are governed by

$$\begin{aligned}\mathcal{L}_1 = &-\tfrac{1}{4}\lambda v h(h^2 + b^2) - \tfrac{1}{16}\lambda(h^2 + b^2)^2 \\ &+ gA^\mu(h\partial_\mu b - b\partial_\mu h) - gvhA^\mu A_\mu - \tfrac{1}{2}g^2(h^2 + b^2)A^\mu A_\mu \\ &- \xi g^2 v h \bar{c} c \,.\end{aligned} \tag{85.38}$$

It is interesting to consider the limit $\xi \to \infty$. In this limit, the vector propagator in R_ξ gauge, eq. (85.36), turns into the massive vector propagator of eq. (85.17),

$$\tilde{\Delta}^{\mu\nu}(k) = \frac{g^{\mu\nu} + k^\mu k^\nu/M^2}{k^2 + M^2 - i\epsilon} \quad (\xi = \infty) \,. \tag{85.39}$$

The b field becomes infinitely heavy, and we can drop it. (Equivalently, its propagator goes to zero.) The ghost fields also become infinitely heavy, but we must be more careful with them because their interaction term, the last line of eq. (85.38), also contains a factor of ξ. The vertex factor for this interaction is $-i\xi g^2 v = -i(\xi M^2)v^{-1}$. Note that this is the same vertex factor that we found in unitary gauge for the interaction between the ρ field and the ghost fields; see eq. (85.18) and take $m_{\rm gh}^2 = \xi M^2$. Thus we cannot drop the ghost fields, but we can take their propagator to be $1/m_{\rm gh}^2$ rather than $1/(k^2 + m_{\rm gh}^2)$, since $k^2 \ll m_{\rm gh}^2 = \xi M^2$ in the limit $\xi \to \infty$. This is the ghost propagator that we found in unitary gauge. We conclude that

R_ξ gauge in the limit $\xi \to \infty$ is equivalent to unitary gauge. Of course, in this limit, we reencounter the problems with divergent diagrams that led us to consider alternative gauge choices in the first place. For practical loop calculations, R_ξ gauge with $\xi = 1$ is typically the most convenient.

In the next section, we consider R_ξ gauge for nonabelian theories.

86

Spontaneously broken nonabelian gauge theory

Prerequisite: 85

In the previous section, we worked out the lagrangian for a U(1) gauge theory with spontaneous symmetry breaking in R_ξ gauge. In this section, we extend this analysis to a general nonabelian gauge theory.

As in section 85, it will be convenient to work with real scalar fields. We therefore decompose any complex scalar fields into pairs of real ones, and organize all the real scalar fields into a big list ϕ_i, $i = 1, \ldots, N$. These real scalar fields form a (possibly reducible) representation R of the gauge group. Let T^a be the gauge-group generator matrices that act on ϕ; they are linear combinations of the generators of the SO(N) group that rotates all components of ϕ_i into each other. Because these SO(N) generators are hermitian and antisymmetric, so are the T^as. Thus $i(T^a)_{ij}$ is a real, antisymmetric matrix.

The lagrangian for our theory can now be written as

$$\mathcal{L} = -\tfrac{1}{2}D^\mu\phi D_\mu\phi - V(\phi) - \tfrac{1}{4}F^{a\mu\nu}F^a_{\mu\nu} \,, \tag{86.1}$$

where

$$(D_\mu\phi)_i = \partial_\mu\phi_i - ig_a A^a_\mu (T^a)_{ij}\phi_j \tag{86.2}$$

is the covariant derivative, and the adjoint index a runs over all generators of all gauge groups. Because ϕ_i and A^a_μ are real fields, and $i(T^a)_{ij}$ is a real matrix, $(D_\mu\phi)_i$ is real.

Now we suppose that the potential $V(\phi)$ is minimized when ϕ has a VEV

$$\langle 0|\phi_i(x)|0\rangle = v_i \,. \tag{86.3}$$

A generator T^a is unbroken if $(T^a)_{ij}v_j = 0$, and broken if $(T^a)_{ij}v_j \neq 0$.

Each broken generator results in a massless Goldstone boson. To see this, we note that the potential must be invariant under a global gauge transformation,

$$V((1-i\theta^a T^a)\phi) = V(\phi)\,. \tag{86.4}$$

Expanding to linear order in the infinitesimal parameter θ, we find

$$\frac{\partial V}{\partial \phi_j}(T^a)_{jk}\phi_k = 0\,. \tag{86.5}$$

We differentiate eq. (86.5) with respect to ϕ_i to get

$$\frac{\partial^2 V}{\partial \phi_i \partial \phi_j}(T^a)_{jk}\phi_k + \frac{\partial V}{\partial \phi_j}(T^a)_{ji} = 0\,. \tag{86.6}$$

Now set $\phi_i = v_i$; then $\partial V/\partial \phi_i$ vanishes, because $\phi_i = v_i$ minimizes $V(\phi)$. Also, we can identify

$$(m^2)_{ij} = \left.\frac{\partial^2 V}{\partial \phi_i \partial \phi_j}\right|_{\phi_i = v_i} \tag{86.7}$$

as the mass-squared matrix for the scalars (after spontaneous symmetry breaking). Thus eq. (86.6) becomes

$$(m^2)_{ij}(T^a v)_j = 0\,. \tag{86.8}$$

We see that if $T^a v \neq 0$, then $T^a v$ is an eigenvector of the mass-squared matrix with eigenvalue zero. So there is a zero eigenvalue for every linearly independent broken generator.

Let us write

$$\phi_i(x) = v_i + \chi_i(x)\,, \tag{86.9}$$

where χ_i is a real scalar field. The covariant derivative of ϕ becomes

$$(D_\mu \phi)_i = \partial_\mu \chi_i - i g_a A^a_\mu (T^a)_{ij}(v+\chi)_j\,. \tag{86.10}$$

It is now convenient to define a set of real antisymmetric matrices

$$(\tau^a)_{ij} \equiv i g_a (T^a)_{ij}\,, \tag{86.11}$$

and the real rectangular matrix

$$F^a{}_i \equiv (\tau^a)_{ij} v_j\,. \tag{86.12}$$

We can now write

$$(D_\mu \phi)_i = \partial_\mu \chi_i - A^a_\mu (F^a + \tau^a \chi)_i\,. \tag{86.13}$$

The kinetic term for ϕ becomes

$$-\tfrac{1}{2}D^\mu\phi D_\mu\phi = -\tfrac{1}{2}\partial^\mu\chi_i\partial_\mu\chi_i - \tfrac{1}{2}(F^a{}_iF^b{}_i)A^{a\mu}A^b_\mu + F^a{}_iA^a_\mu\partial^\mu\chi_i$$
$$+ A^a_\mu\chi_i(\tau^a)_{ij}\partial^\mu\chi_j - A^{a\mu}A^b_\mu F^a{}_i(\tau^b)_{ij}\chi_j$$
$$- \tfrac{1}{2}A^{a\mu}A^b_\mu\chi_i(\tau^a\tau^b)_{ij}\chi_j \,. \tag{86.14}$$

We see (from the second term on the right-hand side) that the mass-squared matrix for the vector fields is

$$(M^2)^{ab} = F^a{}_iF^b{}_i = (FF^{\mathrm{T}})^{ab} \,. \tag{86.15}$$

A theorem of linear algebra states that every real rectangular matrix can be written as

$$F^a{}_i = S^{ab}(M^b\delta^b{}_j)R_{ji} \,, \tag{86.16}$$

where S and R are orthogonal matrices, and the diagonal entries M^a are real and nonnegative. From eq. (86.15) we see that these diagonal entries are the masses of the vector fields. The vector fields of definite mass are then given by $\tilde{A}^a_\mu = S^{ba}A^b_\mu$.

Now we are ready to fix R_ξ gauge. To do so, we add to $\mathcal{L}$ the gauge-fixing and ghost terms

$$\mathcal{L}_{\mathrm{gf}} + \mathcal{L}_{\mathrm{gh}} = -\tfrac{1}{2}\xi^{-1}G^aG^a - \bar{c}^a\,\frac{\delta G^a}{\delta\theta^b}\,c^b \,, \tag{86.17}$$

where we choose

$$G^a = \partial^\mu A^a_\mu - \xi F^a{}_i\chi_i \,. \tag{86.18}$$

Then we have

$$\mathcal{L}_{\mathrm{gf}} = -\tfrac{1}{2}\xi^{-1}\partial^\mu A^a_\mu\partial^\nu A^a_\nu + F^a{}_i\chi_i\partial^\mu A^a_\mu - \tfrac{1}{2}\xi(F^a{}_iF^a{}_j)\chi_i\chi_j$$
$$= -\tfrac{1}{2}\xi^{-1}\partial^\mu A^a_\nu\partial^\nu A^a_\mu - F^a{}_iA^a_\mu\partial^\mu\chi_i - \tfrac{1}{2}\xi(F^a{}_iF^a{}_j)\chi_i\chi_j \,. \tag{86.19}$$

We integrated by parts in the first two terms to get the second line. Note that the second term on the second line of eq. (86.19) cancels the annoying last term on the first line of eq. (86.14). Also, the last term on the second line of eq. (86.19) makes a contribution to the mass-squared matrix for the χ fields,

$$\xi(M^2)_{ij} = \xi F^a{}_iF^a{}_j = \xi(F^{\mathrm{T}}F)_{ij} \,. \tag{86.20}$$

Eq. (86.16) tells us that the eigenvalues of this matrix are $\xi^{1/2}M^a$, where M^a are the vector-boson masses. The mass-squared matrix ξM^2 should be added to the mass-squared matrix m^2 that we get from the potential, eq. (86.7). Note that eqs. (86.8) and (86.12) imply that $(m^2)_{ij}F^a{}_j = 0$; eq. (86.20) then yields $(m^2)_{ij}(\xi M^2)_{jk} = 0$. Thus these two contributions to the mass-squared matrix of the scalar fields live in orthogonal subspaces. The scalar fields of definite mass are $\tilde{\chi}_i = R_{ij}\chi_j$, where the block of R in the m^2 subspace is chosen to diagonalize m^2. The m^2 subspace consists of the physical, massive scalars, and the ξM^2 subspace consists of the unphysical Goldstone bosons; these are the fields that would be set to zero in unitary gauge.

We must still evaluate $\mathcal{L}_{\text{gh}}$. To do so, we recall that $\theta^a(x)$ parameterizes an infinitesimal gauge transformation,

$$A^a_\mu \to A^a_\mu - D^{ab}_\mu \theta^b \,, \tag{86.21}$$

$$\chi_i \to -\theta^a (\tau^a)_{ij}(v+\chi)_j \,. \tag{86.22}$$

Thus we have

$$\begin{aligned}\frac{\delta G^a}{\delta \theta^b} &= -\partial^\mu D^{ab}_\mu + \xi F^a{}_j (\tau^b)_{jk}(v+\chi)_k \\ &= -\partial^\mu D^{ab}_\mu + \xi F^a{}_j F^b{}_j + \xi F^a{}_j (\tau^b)_{jk}\chi_k \\ &= -\partial^\mu D^{ab}_\mu + \xi (M^2)^{ab} + \xi F^a{}_j (\tau^b)_{jk}\chi_k \,,\end{aligned} \tag{86.23}$$

and so the ghost lagrangian is

$$\mathcal{L}_{\text{gh}} = -\partial^\mu \bar{c}^a D^{ab}_\mu c^b - \xi (M^2)^{ab}\bar{c}^a c^b - \xi F^a{}_j (\tau^b)_{jk}\chi_k \bar{c}^a c^b \,. \tag{86.24}$$

The ghost fields of definite mass are $\tilde{c}^a = S^{ba}c^b$ and $\tilde{\bar{c}}^a = S^{ba}\bar{c}^b$.

The complete gauge-fixed lagrangian is now given by eqs. (86.1), (86.14) (86.19), and (86.24). We can rewrite it in terms of the fields of definite mass. This results in the replacements

$$F^a{}_i \to M^a \delta^a{}_i \,, \tag{86.25}$$

$$(\tau^a)_{ij} \to (R\tau^b R^{\mathrm{T}})_{ij} S^{ba} \,, \tag{86.26}$$

$$f^{abc} \to f^{deg} S^{da} S^{eb} S^{gc} \tag{86.27}$$

throughout $\mathcal{L}$. The Feynman rules then follow in the usual way.

Problems

86.1) Let φ_i be a complex scalar field in a complex representation R of the gauge group. Under an infinitesimal gauge transformation, we have

$\delta\varphi_i = -i\theta^a (T_{\mathrm{R}}^a)_i{}^j \varphi_j$. Let us write $\varphi_i = \frac{1}{\sqrt{2}}(\phi_i + i\phi_{i+d(\mathrm{R})})$, where ϕ_i is a real scalar field with the index i running from 1 to $2d(\mathrm{R})$. Then, under an infinitesimal gauge transformation, we have $\delta\phi_i = -i\theta^a (\mathcal{T}^a)_{ij}\phi_j$.

a) Express $\mathcal{T}^a$ in terms of the real and imaginary parts of T_{R}^a.

b) Show that the $\mathcal{T}^a$ matrices satisfy the appropriate commutation relations.

87

The Standard Model: gauge and Higgs sector

Prerequisite: 84

We now turn to the construction of the *Standard Model* of elementary particles, also called the *Glashow–Weinberg–Salam model*. This is the complete (except for gravity) quantum field theory that appears to describe our world. It can be succinctly specified as a gauge theory with gauge group SU(3) × SU(2) × U(1), with left-handed Weyl fields in three copies of the representation $(1,2,-\frac{1}{2}) \oplus (1,1,+1) \oplus (3,2,+\frac{1}{6}) \oplus (\overline{3},1,-\frac{2}{3}) \oplus (\overline{3},1,+\frac{1}{3})$, and a complex scalar field in the representation $(1,2,-\frac{1}{2})$. Here the last entry of each triplet gives the value of the U(1) charge, known as *hypercharge*. The lagrangian includes all terms of mass dimension four or less that are allowed by the gauge symmetries and Lorentz invariance.

We will construct the Standard Model over several sections. We begin with the *electroweak* part of the gauge group, SU(2) × U(1), and the complex scalar field φ, known as the *Higgs field*, in the representation $(2,-\frac{1}{2})$. The Higgs field acquires a nonzero VEV that spontaneously breaks SU(2) × U(1) to U(1); the unbroken U(1) is identified as electromagnetism.

We begin with the covariant derivative of the Higgs field φ,

$$(D_\mu\varphi)_i = \partial_\mu\varphi_i - i[g_2A^a_\mu T^a + g_1B_\mu Y]_i{}^j\varphi_j \;, \tag{87.1}$$

where $T^a = \frac{1}{2}\sigma^a$ and $Y = -\frac{1}{2}I$; Y is the hypercharge generator. It will prove useful to write out $g_2A^a_\mu T^a + g_1B_\mu Y$ in matrix form,

$$g_2A^a_\mu T^a + g_1B_\mu Y = \frac{1}{2}\begin{pmatrix} g_2A^3_\mu - g_1B_\mu & g_2(A^1_\mu - iA^2_\mu) \\ g_2(A^1_\mu + iA^2_\mu) & -g_2A^3_\mu - g_1B_\mu \end{pmatrix}. \tag{87.2}$$

Now suppose that φ has a potential

$$V(\varphi) = \tfrac{1}{4}\lambda(\varphi^\dagger\varphi - \tfrac{1}{2}v^2)^2 \,. \tag{87.3}$$

This potential gives φ a nonzero VEV. We can make a global gauge transformation to bring this VEV entirely into the first component, and furthermore make it real, so that

$$\langle 0|\varphi(x)|0\rangle = \frac{1}{\sqrt{2}}\begin{pmatrix} v \\ 0 \end{pmatrix} . \tag{87.4}$$

The kinetic term for φ is $-(D^\mu\varphi)^\dagger D_\mu\varphi$. After replacing φ by its VEV, we find a mass term for the gauge fields,

$$\mathcal{L}_{\text{mass}} = -\tfrac{1}{8}v^2\,(1,\ 0)\begin{pmatrix} g_2A^3_\mu - g_1B_\mu & g_2(A^1_\mu - iA^2_\mu) \\ g_2(A^1_\mu + iA^2_\mu) & -g_2A^3_\mu - g_1B_\mu \end{pmatrix}^2 \begin{pmatrix} 1 \\ 0 \end{pmatrix} . \tag{87.5}$$

To diagonalize this mass-squared matrix, we first define the *weak mixing angle*

$$\theta_{\text{W}} \equiv \tan^{-1}(g_1/g_2)\,, \tag{87.6}$$

and the fields

$$W^\pm_\mu \equiv \tfrac{1}{\sqrt{2}}(A^1_\mu \mp iA^2_\mu)\,, \tag{87.7}$$

$$Z_\mu \equiv c_{\text{W}}A^3_\mu - s_{\text{W}}B_\mu\,, \tag{87.8}$$

$$A_\mu \equiv s_{\text{W}}A^3_\mu + c_{\text{W}}B_\mu\,, \tag{87.9}$$

where $s_{\text{W}} \equiv \sin\theta_{\text{W}}$, $c_{\text{W}} \equiv \cos\theta_{\text{W}}$. In terms of these fields, eq. (87.5) becomes

$$\begin{aligned} \mathcal{L}_{\text{mass}} &= -\tfrac{1}{8}g_2^2v^2\,(1,\ 0)\begin{pmatrix} \frac{1}{c_{\text{W}}}Z_\mu & \sqrt{2}\,W^+_\mu \\ \sqrt{2}\,W^-_\mu & \dots \end{pmatrix}^2 \begin{pmatrix} 1 \\ 0 \end{pmatrix} . \\ &= -(g_2v/2)^2\,W^{+\mu}W^-_\mu - \tfrac{1}{2}(g_2v/2c_{\text{W}})^2\,Z^\mu Z_\mu \\ &= -M^2_{\text{W}}W^{+\mu}W^-_\mu - \tfrac{1}{2}M^2_{\text{Z}}Z^\mu Z_\mu\,, \end{aligned} \tag{87.10}$$

where we have identified

$$M_{\text{W}} = g_2v/2\,, \tag{87.11}$$

$$M_{\text{Z}} = M_{\text{W}}/\cos\theta_{\text{W}}\,. \tag{87.12}$$

The observed masses of the $W^\pm$ and Z^0 particles are $M_{\rm W} = 80.4\,\text{GeV}$ and $M_{\rm Z} = 91.2\,\text{GeV}$. Eq. (87.12) then implies $\cos\theta_{\rm W} = 0.882$, or, as it is more usually expressed, $\sin^2\theta_{\rm W} = 0.223$.[1]

Note that the A_μ field remains massless; this signifies that there is an unbroken U(1) subgroup. We will identify this unbroken U(1) with the gauge group of electromagnetism.

Before introducing leptons and quarks (which we do in sections 88 and 89), let us work out the complete lagrangian for the gauge and Higgs fields, in unitary gauge. This is sufficient for tree-level calculations.

The two complex components of the φ field yield four real scalar fields; three of these become the longitudinal components of the $W^\pm$ and Z^0. The remaining scalar field must be able to account for shifts in the overall scale of φ. Thus we can write, in unitary gauge,

$$\varphi(x) = \frac{1}{\sqrt{2}}\begin{pmatrix} v + H(x) \\ 0 \end{pmatrix}, \tag{87.13}$$

where H is a real scalar field; the corresponding particle is the *Higgs boson.* The potential now reads

$$V(\varphi) = \tfrac{1}{4}\lambda v^2 H^2 + \tfrac{1}{4}\lambda v H^3 + \tfrac{1}{16}\lambda H^4 \,. \tag{87.14}$$

We see that the mass of the Higgs boson is given by $m_{\rm H}^2 = \frac{1}{2}\lambda v^2$. (As of this writing, the Higgs boson has not been observed; the lower limit on its mass is $m_{\rm H} > 115\,\text{GeV}$.) The kinetic term for H comes from the kinetic term for φ, and is the usual one for a real scalar field, $-\frac{1}{2}\partial^\mu H \partial_\mu H$. Finally, recall that the mass term for the gauge fields, eq. (87.10), is proportional to v^2. Hence it should be multiplied by a factor of $(1 + v^{-1}H)^2$.

Now we have to work out the kinetic terms for the gauge fields. We have

$$\mathcal{L} = -\tfrac{1}{4}F^{a\mu\nu}F^a_{\mu\nu} - \tfrac{1}{4}B^{\mu\nu}B_{\mu\nu}\,, \tag{87.15}$$

where

$$F^1_{\mu\nu} = \partial_\mu A^1_\nu - \partial_\nu A^1_\mu + g_2(A^2_\mu A^3_\nu - A^2_\nu A^3_\mu)\,, \tag{87.16}$$

$$F^2_{\mu\nu} = \partial_\mu A^2_\nu - \partial_\nu A^2_\mu + g_2(A^3_\mu A^1_\nu - A^3_\nu A^1_\mu)\,, \tag{87.17}$$

$$F^3_{\mu\nu} = \partial_\mu A^3_\nu - \partial_\nu A^3_\mu + g_2(A^1_\mu A^2_\nu - A^1_\nu A^2_\mu)\,, \tag{87.18}$$

$$B_{\mu\nu} \equiv \partial_\mu B_\nu - \partial_\nu B_\mu \,. \tag{87.19}$$

[1] Of course, this number is only meaningful once a renormalization scheme has been specified. We are implicitly using an on-shell scheme in which $\theta_{\rm W}$ is *defined* by the relation $\cos\theta_{\rm W} = M_{\rm W}/M_{\rm Z}$, where $M_{\rm W}$ and $M_{\rm Z}$ are the actual particle masses. The relation $g_1 = g_2\tan\theta_{\rm W}$ is then subject to loop corrections that depend on the precise definitions adopted for g_1 and g_2. In the $\overline{\rm MS}$ scheme, on the other hand, $\theta_{\rm W}$ is defined by eq. (87.6), and for $\mu = M_{\rm Z}$, we have $\sin^2\theta_{\rm W} = 0.231$.

Next, form the combinations $F^1_{\mu\nu} \pm iF^2_{\mu\nu}$. Using eq. (87.7), we find

$$\tfrac{1}{\sqrt{2}}(F^1_{\mu\nu} - iF^2_{\mu\nu}) = D_\mu W^+_\nu - D_\nu W^+_\mu \,, \tag{87.20}$$

$$\tfrac{1}{\sqrt{2}}(F^1_{\mu\nu} + iF^2_{\mu\nu}) = D^\dagger_\mu W^-_\nu - D^\dagger_\nu W^-_\mu \,, \tag{87.21}$$

where we have defined a covariant derivative that acts on W^+_μ,

$$\begin{aligned} D_\mu &\equiv \partial_\mu - ig_2 A^3_\mu \\ &= \partial_\mu - ig_2(s_{\rm W} A_\mu + c_{\rm W} Z_\mu) \,. \end{aligned} \tag{87.22}$$

If we identify A_μ as the electromagnetic vector potential, and assign electric charge $Q = +1$ (in units of the proton charge) to the W^+, then we see from eq. (87.22) that we must identify the electromagnetic coupling constant e as

$$e = g_2 \sin\theta_{\rm W} \,. \tag{87.23}$$

Here we are adopting the convention that e is positive. (In our treatment of quantum electrodynamics, we used the convention that e is negative, but that is less convenient in the present context.)

We also have

$$\begin{aligned} F^3_{\mu\nu} &= \partial_\mu A^3_\nu - \partial_\nu A^3_\mu - ig_2(W^+_\mu W^-_\nu - W^+_\nu W^-_\mu) \\ &= s_{\rm W} F_{\mu\nu} + c_{\rm W} Z_{\mu\nu} - ig_2(W^+_\mu W^-_\nu - W^+_\nu W^-_\mu) \,, \end{aligned} \tag{87.24}$$

$$B_{\mu\nu} = c_{\rm W} F_{\mu\nu} - s_{\rm W} Z_{\mu\nu} \,, \tag{87.25}$$

where $F_{\mu\nu} = \partial_\mu A_\nu - \partial_\nu A_\mu$ is the usual electromagnetic field strength, and

$$Z_{\mu\nu} \equiv \partial_\mu Z_\nu - \partial_\nu Z_\mu \tag{87.26}$$

is the abelian field strength associated with the Z_μ field.

Now we can assemble all of this into the complete lagrangian for the electroweak gauge fields and the Higgs boson in unitary gauge. We will express g_2 in terms of e and $\theta_{\rm W}$ via $g_2 = e/\sin\theta_{\rm W}$, and λ in terms of $m_{\rm H}$ and v via $\lambda = 2m^2_{\rm H}/v^2$. We ultimately get

$$\begin{aligned} \mathcal{L} = &-\tfrac{1}{4}F^{\mu\nu}F_{\mu\nu} - \tfrac{1}{4}Z^{\mu\nu}Z_{\mu\nu} - D^{\dagger\mu}W^{-\nu}D_\mu W^+_\nu + D^{\dagger\mu}W^{-\nu}D_\nu W^+_\mu \\ &+ ie(F^{\mu\nu} + \cot\theta_{\rm W} Z^{\mu\nu})W^+_\mu W^-_\nu \\ &- \tfrac{1}{2}(e^2/\sin^2\theta_{\rm W})(W^{+\mu}W^-_\mu W^{+\nu}W^-_\nu - W^{+\mu}W^+_\mu W^{-\nu}W^-_\nu) \\ &- (M^2_{\rm W} W^{+\mu}W^-_\mu + \tfrac{1}{2}M^2_{\rm Z} Z^\mu Z_\mu)(1 + v^{-1}H)^2 \\ &- \tfrac{1}{2}\partial^\mu H \partial_\mu H - \tfrac{1}{2}m^2_{\rm H} H^2 - \tfrac{1}{2}m^2_{\rm H} v^{-1} H^3 - \tfrac{1}{8}m^2_{\rm H} v^{-2} H^4 \,, \end{aligned} \tag{87.27}$$

where

$$D_\mu = \partial_\mu - ie(A_\mu + \cot\theta_W Z_\mu) \,. \tag{87.28}$$

With the W^+_μ field assigned electric charge $Q = +1$, this lagrangian exhibits manifest electromagnetic gauge invariance. The full underlying $\mathrm{SU}(2) \times \mathrm{U}(1)$ gauge invariance is not manifest, however, because we have fixed unitary gauge.

Reference notes

Discussions of the Standard Model in R_ξ gauge can be found in *Cheng & Li*, and *Ramond II.*

Problems

87.1) Find the generator Q of the unbroken $\mathrm{U}(1)$ subgroup as a linear combination of the T^as and Y.

87.2) a) Ignoring loop corrections, find the numerical values of v, g_1, and g_2. Take $e^2/4\pi = \alpha(M_Z) = 1/127.90$ and $\sin^2\theta_W = 0.231$.

b) The *Fermi constant* is defined (at tree level) as

$$G_F \equiv \frac{e^2}{4\sqrt{2}\sin^2\theta_W M_W^2} \,. \tag{87.29}$$

Find its numerical value in GeV^{-2}.

c) Express G_F in terms of v.

87.3) In this problem we will work out the generator matrices introduced in section 86 for the case of the Standard Model.

a) Write the Higgs field as

$$\varphi = \frac{1}{\sqrt{2}} \begin{pmatrix} \phi_1 + i\phi_3 \\ \phi_2 + i\phi_4 \end{pmatrix} . \tag{87.30}$$

where ϕ_i is a real scalar field. Express the $\mathrm{SU}(2)$ generators T^a and the hypercharge generator Y as 4×4 matrices $\mathcal{T}^a$ and $\mathcal{Y}$ that act on ϕ_i. Hint: see problem 86.1.

b) Compute the matrix $F^a{}_i$, defined in eq (86.12).

c) Compute the mass-squared matrix for the vector fields, $(M^2)^{ab} = F^a{}_i F^b{}_i$, and find its eigenvalues.

87.4) Work out the Feynman rules for the lagrangian of eq. (87.27).

87.5) Assume that $m_H > 2M_Z$, and compute (at tree level) the decay rate of the Higgs boson into W^+W^- and Z^0Z^0 pairs. Express your answer in GeV for $m_H = 200\,\mathrm{GeV}$.

88

The Standard Model: lepton sector

Prerequisites: 75, 87

Leptons are spin-one-half particles that are singlets of the color group. There are six different flavors of lepton; see Table 88.1 The six flavors are naturally grouped into three *families* or *generations*: e and ν_e, μ and ν_μ, τ and ν_τ.

Let us begin by describing a single lepton family, the electron and its neutrino. We introduce left-handed Weyl fields ℓ and $\bar{e}$ in the representations $(2, -\frac{1}{2})$ and $(1, +1)$ of SU(2) × U(1). Here the bar over the e in the field $\bar{e}$ is *part of the name of the field*, and does not denote *any sort* of conjugation. The covariant derivatives of these fields are

$$(D_\mu \ell)_i = \partial_\mu \ell_i - i g_2 A^a_\mu (T^a)_i{}^j \ell_j - i g_1 (-\tfrac{1}{2}) B_\mu \ell_i \,, \tag{88.1}$$

$$D_\mu \bar{e} = \partial_\mu \bar{e} - i g_1 (+1) B_\mu \bar{e} \,, \tag{88.2}$$

and their kinetic terms are

$$\mathcal{L}_{\text{kin}} = i \ell^{\dagger i} \bar{\sigma}^\mu (D_\mu \ell)_i + i \bar{e}^\dagger \bar{\sigma}^\mu D_\mu \bar{e} \,. \tag{88.3}$$

The representation $(2, -\frac{1}{2}) \oplus (1, +1)$ for the left-handed Weyl fields is complex; hence the gauge theory is chiral, and therefore parity violating.

We cannot write down a mass term involving ℓ and/or $\bar{e}$ because there is no gauge-group singlet contained in any of the products

$$\begin{aligned} &(2, -\tfrac{1}{2}) \otimes (2, -\tfrac{1}{2}) \,, \\ &(2, -\tfrac{1}{2}) \otimes (1, +1) \,, \\ &(1, +1) \otimes (1, +1) \,. \end{aligned} \tag{88.4}$$

However, we are able to write down a Yukawa coupling of the form

$$\mathcal{L}_{\text{Yuk}} = -y \varepsilon^{ij} \varphi_i \ell_j \bar{e} + \text{h.c.} \,, \tag{88.5}$$

Table 88.1. The six flavors of lepton. Q is the electric charge in units of the proton charge. Each charged flavor is represented by a Dirac field, each neutral flavor by a Majorana field (or, equivalently, a left-handed Weyl field). Neutrino masses are exactly zero in the Standard Model.

Name	Symbol	Mass (MeV)	Q
electron	e	0.511	-1
electron neutrino	ν_e	0	0
muon	μ	105.7	-1
muon neutrino	ν_μ	0	0
tau	τ	1777	-1
tau neutrino	ν_τ	0	0

where φ is the Higgs field in the $(2,-\frac{1}{2})$ representation that we introduced in the last section, and y is the Yukawa coupling constant. A gauge-invariant Yukawa coupling is possible because there is a singlet on the right-hand side of

$$(2,-\tfrac{1}{2}) \otimes (2,-\tfrac{1}{2}) \otimes (1,+1) = (1,0) \oplus (3,0) \,. \tag{88.6}$$

There are no other gauge-invariant terms involving ℓ or $\bar{e}$ that have mass dimension four or less. Hence there are no other terms that we could add to $\mathcal{L}$ while preserving renormalizability.

We add eqs. (88.3) and (88.5) to the lagrangian for φ and the gauge fields that we worked out in the last section. In unitary gauge, we replace φ_1 with $\frac{1}{\sqrt{2}}(v+H)$, where H is the real scalar field representing the physical Higgs boson, and φ_2 with zero. The Yukawa term becomes

$$\mathcal{L}_{\text{Yuk}} = -\tfrac{1}{\sqrt{2}}y(v+H)(\ell_2\bar{e} + \text{h.c.}) \,. \tag{88.7}$$

It is now convenient to assign new names to the SU(2) components of ℓ,

$$\ell = \begin{pmatrix} \nu \\ e \end{pmatrix} . \tag{88.8}$$

(We will rely on context to distinguish the field e from the electromagnetic coupling constant e.) Then eq. (88.7) becomes

$$\mathcal{L}_{\rm Yuk} = -\tfrac{1}{\sqrt{2}} y(v+H)(e\bar{e} + \bar{e}^\dagger e^\dagger)$$
$$= -\tfrac{1}{\sqrt{2}} y(v+H)\overline{\mathcal{E}}\mathcal{E}\,, \tag{88.9}$$

where we have defined a Dirac field for the electron,

$$\mathcal{E} \equiv \begin{pmatrix} e \\ \bar{e}^\dagger \end{pmatrix}. \tag{88.10}$$

We see that the electron has acquired a mass

$$m_e = \frac{yv}{\sqrt{2}}\,. \tag{88.11}$$

The neutrino has remained massless.

We can describe the neutrino with a Majorana field

$$\mathcal{N} \equiv \begin{pmatrix} \nu \\ \nu^\dagger \end{pmatrix}. \tag{88.12}$$

However, it is often more convenient to work with

$$\mathcal{N}_{\rm L} \equiv P_{\rm L}\mathcal{N} = \begin{pmatrix} \nu \\ 0 \end{pmatrix}, \tag{88.13}$$

where $P_{\rm L} = \frac{1}{2}(1-\gamma_5)$. We can think of $\mathcal{N}_{\rm L}$ as a Dirac field; for example, the neutrino kinetic term $i\nu^\dagger\bar{\sigma}^\mu\partial_\mu\nu$ can be written as $i\overline{\mathcal{N}}_{\rm L}\partial\!\!\!/\mathcal{N}_{\rm L}$.

Now we return to eqs. (88.1) and (88.2), and express the covariant derivatives in terms of the $W^\pm_\mu$, Z_μ, and A_μ fields. From our results in section 87, we have

$$g_2 A^1_\mu T^1 + g_2 A^2_\mu T^2 = \frac{g_2}{\sqrt{2}} \begin{pmatrix} 0 & W^+_\mu \\ W^-_\mu & 0 \end{pmatrix} \tag{88.14}$$

and

$$g_2 A^3_\mu T^3 + g_1 B_\mu Y = \tfrac{e}{s_{\rm W}}(s_{\rm W}A_\mu + c_{\rm W}Z_\mu)T^3 + \tfrac{e}{c_{\rm W}}(c_{\rm W}A_\mu - s_{\rm W}Z_\mu)Y$$
$$= e(A_\mu + \cot\theta_{\rm W} Z_\mu)T^3 + e(A_\mu - \tan\theta_{\rm W} Z_\mu)Y$$
$$= e(T^3+Y)A_\mu + e(\cot\theta_{\rm W}T^3 - \tan\theta_{\rm W}Y)Z_\mu\,. \tag{88.15}$$

Since we identify A_μ as the electromagnetic field and e as the electromagnetic coupling constant (with the convention that e is positive), we identify

$$Q = T^3 + Y \tag{88.16}$$

as the generator of electric charge. Then, since

$$T^3\nu = +\tfrac{1}{2}\nu\,, \quad T^3 e = -\tfrac{1}{2}e\,, \quad T^3\bar{e} = 0\,, \tag{88.17}$$

$$Y\nu = -\tfrac{1}{2}\nu\,, \quad Ye = -\tfrac{1}{2}e\,, \quad Y\bar{e} = +\bar{e}\,, \tag{88.18}$$

we see from eq. (88.16) that

$$Q\nu = 0\,, \qquad Qe = -e\,, \qquad Q\bar{e} = +\bar{e}\,. \tag{88.19}$$

This is just the set of electric charge assignments that we expect for the electron and the neutrino. Then (since the action of Q on the fields is more familiar than the action of Y) it is convenient to replace Y in eq. (88.15) with $Q - T^3$. We find

$$\begin{aligned} g_2 A^3_\mu T^3 + g_1 B_\mu Y &= eQA_\mu + e[(\cot\theta_{\rm W} + \tan\theta_{\rm W})T^3 - \tan\theta_{\rm W} Q]Z_\mu \\ &= eQA_\mu + \tfrac{e}{s_{\rm W}c_{\rm W}}(T^3 - s^2_{\rm W}Q)Z_\mu\,. \end{aligned} \tag{88.20}$$

In terms of the four-component fields, we have

$$(g_2 A^3_\mu T^3 + g_1 B_\mu Y)\mathcal{E} = \Big[-eA_\mu + \tfrac{e}{s_{\rm W}c_{\rm W}}(-\tfrac{1}{2}P_{\rm L} + s^2_{\rm W})Z_\mu\Big]\mathcal{E}\,, \tag{88.21}$$

$$(g_2 A^3_\mu T^3 + g_1 B_\mu Y)\mathcal{N}_{\rm L} = \tfrac{e}{s_{\rm W}c_{\rm W}}(+\tfrac{1}{2})Z_\mu \mathcal{N}_{\rm L}\,. \tag{88.22}$$

Using eqs. (88.14) and (88.21)–(88.22) in eqs. (88.1)–(88.3), we find the couplings of the gauge fields to the leptons,

$$\mathcal{L}_{\rm int} = \tfrac{1}{\sqrt{2}}g_2 W^+_\mu J^{-\mu} + \tfrac{1}{\sqrt{2}}g_2 W^-_\mu J^{+\mu} + \tfrac{e}{s_{\rm W}c_{\rm W}}Z_\mu J^\mu_{\rm Z} + eA_\mu J^\mu_{\rm EM}\,, \tag{88.23}$$

where we have defined the currents

$$J^{+\mu} \equiv \overline{\mathcal{E}}_{\rm L}\gamma^\mu \mathcal{N}_{\rm L}\,, \tag{88.24}$$

$$J^{-\mu} \equiv \overline{\mathcal{N}}_{\rm L}\gamma^\mu \mathcal{E}_{\rm L}\,, \tag{88.25}$$

$$J^\mu_{\rm Z} \equiv J^\mu_3 - s^2_{\rm W} J^\mu_{\rm EM}\,, \tag{88.26}$$

$$J^\mu_3 \equiv \tfrac{1}{2}\overline{\mathcal{N}}_{\rm L}\gamma^\mu \mathcal{N}_{\rm L} - \tfrac{1}{2}\overline{\mathcal{E}}_{\rm L}\gamma^\mu \mathcal{E}_{\rm L}\,, \tag{88.27}$$

$$J^\mu_{\rm EM} \equiv -\overline{\mathcal{E}}\gamma^\mu \mathcal{E}\,. \tag{88.28}$$

In many cases, we are interested in scattering amplitudes for leptons whose momenta are all well below the $W^\pm$ and Z^0 masses. In this case, we can integrate the $W^\pm_\mu$ and Z_μ fields out of the path integral, as discussed in section 29. We get the leading term (in a double expansion in powers of the gauge couplings and inverse powers of $M_{\rm W}$ and $M_{\rm Z}$) by ignoring the kinetic energy and other interactions of the $W^\pm_\mu$ and Z_μ fields, solving the equations of motion for them that follow from $\mathcal{L}_{\rm mass} + \mathcal{L}_{\rm int}$, where $\mathcal{L}_{\rm int}$ is given by eq. (88.23) and

$$\mathcal{L}_{\rm mass} = -M_{\rm W}^2 W^{+\mu} W_\mu^- - \tfrac{1}{2} M_{\rm Z}^2 Z^\mu Z_\mu \, , \tag{88.29}$$

and finally substituting the solutions back into $\mathcal{L}_{\rm mass} + \mathcal{L}_{\rm int}$. This is equivalent to evaluating tree-level Feynman diagrams with a single $W^\pm$ or Z^0 exchanged, with the propagator $g^{\mu\nu}/M_{\rm W,Z}^2$. The result is

$$\begin{aligned} \mathcal{L}_{\rm eff} &= \frac{g_2^2}{2M_{\rm W}^2} J^{+\mu} J_\mu^- + \frac{e^2}{2 s_{\rm W}^2 c_{\rm W}^2 M_{\rm Z}^2} J_{\rm Z}^\mu J_{{\rm Z}\mu} \\ &= \frac{e^2}{2 s_{\rm W}^2 M_{\rm W}^2} (J^{+\mu} J_\mu^- + J_{\rm Z}^\mu J_{{\rm Z}\mu}) \\ &= 2\sqrt{2}\, G_{\rm F} (J^{+\mu} J_\mu^- + J_{\rm Z}^\mu J_{{\rm Z}\mu}) \, . \end{aligned} \tag{88.30}$$

We used $e = g_2 \sin\theta_{\rm W}$ and $M_{\rm W} = M_{\rm Z} \cos\theta_{\rm W}$ to get the second line, and we defined the *Fermi constant*

$$G_{\rm F} \equiv \frac{e^2}{4\sqrt{2} \sin^2\theta_{\rm W} M_{\rm W}^2} \tag{88.31}$$

in the third line. We can use $\mathcal{L}_{\rm eff}$ to compute the tree-level scattering amplitude for processes like $\nu_e e^- \to \nu_e e^-$; we leave this to the problems.

Having worked out the interactions of a single lepton generation, we now examine what happens when there is more than one of them. Let us consider the fields ℓ_{iI} and $\bar{e}_I$, where $I = 1, 2, 3$ is a generation index. The kinetic term for all these fields is

$$\mathcal{L}_{\rm kin} = i \ell_I^{\dagger i} \bar{\sigma}^\mu (D_\mu)_i{}^j \ell_{jI} + i \bar{e}_I^\dagger \bar{\sigma}^\mu D_\mu \bar{e}_I \, , \tag{88.32}$$

where the repeated generation index is summed. The most general Yukawa term we can write down now reads

$$\mathcal{L}_{\rm Yuk} = -\varepsilon^{ij} \varphi_i \ell_{jI} y_{IJ} \bar{e}_J + \text{h.c.} \, , \tag{88.33}$$

where y_{IJ} is a complex 3×3 matrix, and the generation indices are summed. We can make unitary transformations in generation space on the fields: $\ell_I \to L_{IJ} \ell_J$ and $\bar{e}_I \to \bar{E}_{IJ} \bar{e}_J$, where L and $\bar{E}$ are independent unitary matrices. The kinetic terms are unchanged, and the Yukawa matrix y is replaced with $L^{\rm T} y \bar{E}$. We can choose L and $\bar{E}$ so that $L^{\rm T} y \bar{E}$ is diagonal with positive real entries y_I. The charged leptons $\mathcal{E}_I$ then have masses $m_{e_I} = y_I v/\sqrt{2}$, and the neutrinos remain massless. In the currents, eqs. (88.24)–(88.28), we simply add a generation index I to each field, and sum over it.

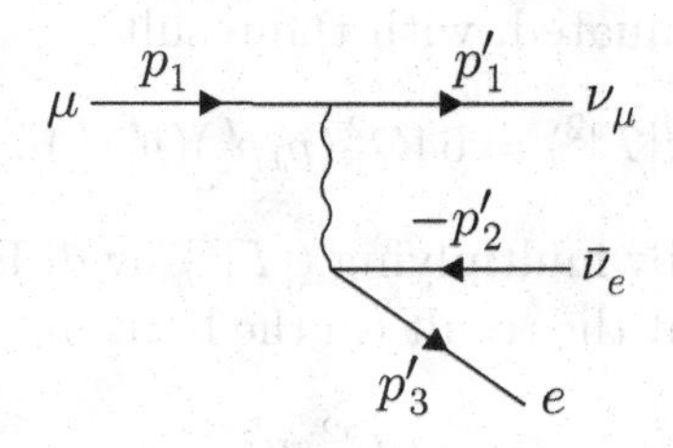

Figure 88.1. Feynman diagram for muon decay. The wavy line is a W propagator.

Let us work out the details for one process of particular importance: muon decay, $\mu^- \to e^-\overline{\nu}_e\nu_\mu$. Let the four-component fields be $\mathcal{E}$ for the electron, $\mathcal{M}$ for the muon, $\mathcal{N}_e$ for the electron neutrino, and $\mathcal{N}_m$ for the muon neutrino. Only the *charged currents* $J^\pm_\mu$ are relevant; the *neutral current* $J^\mu_{\rm Z}$ and the electromagnetic current $J^\mu_{\rm EM}$ do not contribute. Ignoring the τ terms in the charged currents, we have

$$J^{+\mu} = \overline{\mathcal{E}}_{\rm L}\gamma^\mu \mathcal{N}_{e{\rm L}} + \overline{\mathcal{M}}_{\rm L}\gamma^\mu \mathcal{N}_{m{\rm L}} \,, \tag{88.34}$$

$$J^{-\mu} = \overline{\mathcal{N}}_{e{\rm L}}\gamma^\mu \mathcal{E}_{\rm L} + \overline{\mathcal{N}}_{m{\rm L}}\gamma^\mu \mathcal{M}_{\rm L} \,. \tag{88.35}$$

The relevant term in the effective interaction is

$$\mathcal{L}_{\rm eff} = 2\sqrt{2}\,G_{\rm F}(\overline{\mathcal{E}}_{\rm L}\gamma^\mu \mathcal{N}_{e{\rm L}})(\overline{\mathcal{N}}_{m{\rm L}}\gamma_\mu \mathcal{M}_{\rm L}) \,. \tag{88.36}$$

This can be simplified by means of a Fierz identity (see problem 36.3),

$$\mathcal{L}_{\rm eff} = -4\sqrt{2}\,G_{\rm F}(\overline{\mathcal{M}^{\rm C}}P_{\rm L}\,\mathcal{N}_e)(\overline{\mathcal{E}}P_{\rm R}\mathcal{N}^{\rm C}_m) \,. \tag{88.37}$$

Assigning momenta as shown in fig. 88.1, and using the usual Feynman rules for incoming and outgoing particles and antiparticles, the scattering amplitude is

$$\begin{aligned} \mathcal{T} &= -4\sqrt{2}\,G_{\rm F}(u_1^{\rm T}\mathcal{C}P_{\rm L}v_2')(\overline{u}_3'P_{\rm R}\mathcal{C}\overline{u}_1'^{\rm T}) \\ &= -4\sqrt{2}\,G_{\rm F}(\overline{v}_1P_{\rm L}v_2')(\overline{u}_3'P_{\rm R}v_1') \,. \end{aligned} \tag{88.38}$$

Taking the complex conjugate, and using $\overline{P_{\rm L}} = P_{\rm R}$, we find

$$\mathcal{T}^* = -4\sqrt{2}\,G_{\rm F}(\overline{v}_2'P_{\rm R}v_1)(\overline{v}_1'P_{\rm L}u_3') \,. \tag{88.39}$$

Multiplying eqs. (88.38) and (88.39), summing over final spins and averaging over the initial spin, we get

$$\begin{aligned} \langle|\mathcal{T}|^2\rangle = \tfrac{1}{2}(4\sqrt{2})^2G_{\rm F}^2\,&{\rm Tr}[(-\not{p}_1-m_\mu)P_{\rm L}(-\not{p}_2')P_{\rm R}] \\ &\times {\rm Tr}[(-\not{p}_3'+m_e)P_{\rm R}(-\not{p}_1')P_{\rm L}] \,. \end{aligned} \tag{88.40}$$

The traces are easily evaluated, with the result

$$\langle|\mathcal{T}|^2\rangle = 64G_{\rm F}^2(p_1p_2')(p_1'p_3') \,. \tag{88.41}$$

We get the decay rate Γ by multiplying $\langle|\mathcal{T}|^2\rangle$ by $d\text{LIPS}_3(p_1)$ and integrating over $p'_{1,2,3}$. We worked out the result (in the limit $m_e \ll m_\mu$) in problem 11.3,

$$\Gamma = \frac{G_{\rm F}^2 m_\mu^5}{192\pi^3} \,. \tag{88.42}$$

After including one-loop corrections from electromagnetism, and accounting for the nonzero electron mass, the measured muon decay rate is used to determine the value of $G_{\rm F}$, with the result $G_{\rm F} = 1.166 \times 10^{-5}\,\text{GeV}^{-2}$.

Reference notes

Lepton phenomenology is covered in more detail in *Cheng & Li*, *Georgi*, *Peskin & Schroeder*, *Quigg*, and *Ramond II*.

Problems

88.1) Verify the claim made immediately after eq. (88.6).

88.2) Show that a neutrino always has negative helicity, and that an antineutrino always has positive helicity. Hint: see section 75.

88.3) Show that the sum of eqs. (88.32) and (88.33), when rewritten in terms of fields of definite mass, has a global symmetry U(1) × U(1) × U(1). The corresponding charges are called *electron number*, *muon number*, and *tau number*; the sum of the charges is the *lepton number*. List the value of each charge for each Dirac field $\mathcal{E}_I$ and $\mathcal{N}_{{\rm L}I}$.

88.4) Compute $\langle|\mathcal{T}|^2\rangle$ for muon decay using eq. (88.36), without making the Fierz transformation to eq. (88.37), and verify eq. (88.41).

88.5) a) Write down the term in $\mathcal{L}_{\rm eff}$ that is relevant for $\nu_\mu e^- \to \nu_\mu e^-$. Express your answer in the form

$$\mathcal{L}_{\rm eff} = \tfrac{1}{\sqrt{2}} G_{\rm F}\,\overline{\mathcal{N}}\gamma^\mu(1-\gamma_5)\mathcal{N}\,\overline{\mathcal{E}}\gamma_\mu(C_{\rm V}-C_{\rm A}\gamma_5)\mathcal{E} \,, \tag{88.43}$$

where $\mathcal{N}$ is the muon neutrino field, and determine the values of $C_{\rm V}$ and $C_{\rm A}$.

b) Repeat part (a) for $\nu_e e^- \to \nu_e e^-$.

c) Compute $\langle|\mathcal{T}|^2\rangle$ as a function of the Mandelstam variables, the electron mass, and $C_{\rm V}$ and $C_{\rm A}$.

88.6) Compute the rates for the decay processes $W^+ \to e^+\nu_e$, $Z^0 \to e^+e^-$, and $Z^0 \to \overline{\nu}_e\nu_e$. Neglect the electron mass. Express your results in GeV.

88.7) *Anomalous dimension of the Fermi constant.* The coefficient of the effective interaction for muon decay, eq. (88.36), is subject to renormalization by quantum electrodynamic processes. In particular, we can compute its anomalous dimension γ_G, defined via

$$\mu \frac{d}{d\mu} G_{\rm F}(\mu) = \gamma_G(\alpha) G_{\rm F}(\mu) \,, \qquad (88.44)$$

where $\alpha = e^2/4\pi$ is the fine-structure constant in the $\overline{\rm MS}$ scheme with renormalization scale μ.

a) Argue that it is $G_{\rm F}(M_{\rm W})$ that is given by eq. (88.31).

b) Multiply eq. (88.36) by a renormalizing factor Z_G, and define

$$\ln(Z_G/Z_2) = \sum_{n=1}^{\infty} \frac{\mathcal{G}_n(\alpha)}{\varepsilon^n} \,, \qquad (88.45)$$

where Z_2 is the renormalizing factor for a field of unit charge in spinor electrodynamics. Show that

$$\gamma_G(\alpha) = \alpha \mathcal{G}_1'(\alpha) \,. \qquad (88.46)$$

c) If $\gamma_G(\alpha) = c_1\alpha + O(\alpha^2)$ and $\beta(\alpha) = b_1\alpha^2 + O(\alpha^3)$, show that

$$G_{\rm F}(\mu) = \left[\frac{\alpha(\mu)}{\alpha(M_{\rm W})}\right]^{c_1/b_1} G_{\rm F}(M_{\rm W}) \qquad (88.47)$$

for $\mu < M_{\rm W}$. (For $\mu > M_{\rm W}$, we should not be using an effective interaction.)

d) If $\alpha(\mu)\ln(M_{\rm W}/\mu) \ll 1$, show that eq. (88.47) becomes

$$G_{\rm F}(\mu) = \Big[1 - c_1\alpha(\mu)\ln(M_{\rm W}/\mu)\Big] G_{\rm F}(M_{\rm W}) \,. \qquad (88.48)$$

e) Use a Fierz identity to rewrite eq. (88.36) in *charge retention form*,

$$\mathcal{L}_{\rm eff} = 2\sqrt{2}\, Z_G G_{\rm F} (\overline{\mathcal{E}}_{\rm L}\gamma^\mu \mathcal{M}_{\rm L})(\overline{\mathcal{N}}_{m\rm L}\gamma_\mu \mathcal{N}_{e\rm L}) \,. \qquad (88.49)$$

f) Consider the process of muon decay with an extra photon connecting the μ and e lines. Work in Lorenz gauge, and with the four-fermion vertex provided by eq. (88.49). Use your results from problem 62.2 to show that, in this gauge, there is no $O(\alpha)$ contribution to Z_G in the $\overline{\rm MS}$ scheme.

g) Use your result from part (d), and your result for Z_2 in Lorenz gauge from problem 62.2, to show that $c_1 = 0$, and hence that $G_{\rm F}(\mu) = G_{\rm F}(M_{\rm W})$ at the one-loop level.

89

The Standard Model: quark sector

Prerequisite: 88

Quarks are spin-one-half particles that are triplets of the color group. There are six different flavors of quark; see Table 83.1. The six flavors are naturally grouped into three *families* or *generations*: u and d, c and s, t and b.

Let us begin by describing a single quark family, the up and down quarks. We introduce left-handed Weyl fields q, $\bar{u}$, and $\bar{d}$ in the representations $(3, 2, +\frac{1}{6})$, $(\bar{3}, 1, -\frac{2}{3})$, and $(\bar{3}, 1, +\frac{1}{3})$ of $\mathrm{SU}(3) \times \mathrm{SU}(2) \times \mathrm{U}(1)$. Here the bar over the letter in the fields $\bar{u}$ and $\bar{d}$ is *part of the name of the field*, and does not denote *any sort* of conjugation. The covariant derivatives of these fields are

$$(D_\mu q)_{\alpha i} = \partial_\mu q_{\alpha i} - ig_3 A^a_\mu (T^a_3)_\alpha{}^\beta q_{\beta i} - ig_2 A^a_\mu (T^a_2)_i{}^j q_{\alpha j} \\ - ig_1(+\tfrac{1}{6})B_\mu q_{\alpha i}\,, \tag{89.1}$$

$$(D_\mu \bar{u})^\alpha = \partial_\mu \bar{u}^\alpha - ig_3 A^a_\mu (T^a_{\bar{3}})^\alpha{}_\beta \bar{u}^\beta - ig_1(-\tfrac{2}{3})B_\mu \bar{u}^\alpha\,, \tag{89.2}$$

$$(D_\mu \bar{d})^\alpha = \partial_\mu \bar{d}^\alpha - ig_3 A^a_\mu (T^a_{\bar{3}})^\alpha{}_\beta \bar{d}^\beta - ig_1(+\tfrac{1}{3})B_\mu \bar{d}^\alpha\,. \tag{89.3}$$

We rely on context to distinguish the SU(3) gauge fields from the SU(2) gauge fields. The kinetic terms for q, $\bar{u}$, and $\bar{d}$ are

$$\mathcal{L}_{\rm kin} = iq^{\dagger\alpha i}\bar{\sigma}^\mu (D_\mu q)_{\alpha i} + i\bar{u}^\dagger_\alpha \bar{\sigma}^\mu (D_\mu \bar{u})^\alpha + i\bar{d}^\dagger_\alpha \bar{\sigma}^\mu (D_\mu \bar{d})^\alpha\,. \tag{89.4}$$

The representation $(3, 2, +\frac{1}{6}) \oplus (\bar{3}, 1, -\frac{2}{3}) \oplus (\bar{3}, 1, +\frac{1}{3})$ for the left-handed Weyl fields is complex; hence the gauge theory is chiral, and therefore parity violating.

We cannot write down a mass term involving q, $\bar{u}$, and/or $\bar{d}$ because there is no gauge-group singlet contained in any of the products of their representations. But we are able to write down Yukawa couplings of the form

$$\mathcal{L}_{\text{Yuk}} = -y'\varepsilon^{ij}\varphi_i q_{\alpha j}\bar{d}^\alpha - y''\varphi^{\dagger i} q_{\alpha i}\bar{u}^\alpha + \text{h.c.}\,, \tag{89.5}$$

where φ is the Higgs field in the $(1, 2, -\frac{1}{2})$ representation that we introduced in section 87, and y' and y'' are the Yukawa coupling constants. These gauge-invariant Yukawa couplings are possible because there are singlets on the right-hand sides of

$$(1, 2, -\tfrac{1}{2}) \otimes (3, 2, +\tfrac{1}{6}) \otimes (\bar{3}, 1, +\tfrac{1}{3}) = (1, 1, 0) \oplus \ldots\,, \tag{89.6}$$

$$(1, 2, +\tfrac{1}{2}) \otimes (3, 2, +\tfrac{1}{6}) \otimes (\bar{3}, 1, -\tfrac{2}{3}) = (1, 1, 0) \oplus \ldots\,. \tag{89.7}$$

There are no other gauge-invariant terms involving q, $\bar{u}$, or $\bar{d}$ that have mass dimension four or less. Hence there are no other terms that we could add to $\mathcal{L}$ while preserving renormalizability.

In unitary gauge, we replace φ_1 with $\frac{1}{\sqrt{2}}(v + H)$, where H is the real scalar field representing the physical Higgs boson, and φ_2 with zero. The Yukawa term becomes

$$\mathcal{L}_{\text{Yuk}} = -\tfrac{1}{\sqrt{2}} y'(v + H) q_{\alpha 2}\bar{d}^\alpha - \tfrac{1}{\sqrt{2}} y''(v + H) q_{\alpha 1}\bar{u}^\alpha + \text{h.c.}\,. \tag{89.8}$$

It is now convenient to assign new names to the SU(2) components of q,

$$q = \begin{pmatrix} u \\ d \end{pmatrix}. \tag{89.9}$$

Then eq. (89.8) becomes

$$\begin{aligned}\mathcal{L}_{\text{Yuk}} &= -\tfrac{1}{\sqrt{2}} y'(v + H)(d_\alpha \bar{d}^\alpha + \bar{d}^\dagger_\alpha d^{\dagger\alpha}) - \tfrac{1}{\sqrt{2}} y''(v + H)(u_\alpha \bar{u}^\alpha + \bar{u}^\dagger_\alpha u^{\dagger\alpha}) \\ &= -\tfrac{1}{\sqrt{2}} y'(v + H)\overline{\mathcal{D}}^\alpha \mathcal{D}_\alpha - \tfrac{1}{\sqrt{2}} y''(v + H)\overline{\mathcal{U}}^\alpha \mathcal{U}_\alpha\,, \end{aligned} \tag{89.10}$$

where we have defined Dirac fields for the down and up quarks,

$$\mathcal{D}_\alpha \equiv \begin{pmatrix} d_\alpha \\ \bar{d}^\dagger_\alpha \end{pmatrix}, \qquad \mathcal{U}_\alpha \equiv \begin{pmatrix} u_\alpha \\ \bar{u}^\dagger_\alpha \end{pmatrix}. \tag{89.11}$$

We see from eq. (89.10) that the up and down quarks have acquired masses

$$m_d = \frac{y'v}{\sqrt{2}}\,, \qquad m_u = \frac{y''v}{\sqrt{2}}\,. \tag{89.12}$$

Now we return to eqs. (89.1)–(89.3), and express the covariant derivatives in terms of the $W^\pm_\mu$, Z_μ, and A_μ fields. From our results in section 88, we have

$$g_2 A^1_\mu T^1 + g_2 A^2_\mu T^2 = \frac{g_2}{\sqrt{2}} \begin{pmatrix} 0 & W^+_\mu \\ W^-_\mu & 0 \end{pmatrix}, \tag{89.13}$$

$$g_2 A^3_\mu T^3 + g_1 B_\mu Y = eQA_\mu + \tfrac{e}{s_\text{W} c_\text{W}}(T^3 - s^2_\text{W} Q) Z_\mu \ , \tag{89.14}$$

where

$$Q = T^3 + Y \tag{89.15}$$

is the generator of electric charge. Then, since

$$T^3 u = +\tfrac{1}{2} u \ , \quad T^3 d = -\tfrac{1}{2} d \ , \quad T^3 \bar{u} = 0 \ , \qquad T^3 \bar{d} = 0 \ , \tag{89.16}$$

$$Yu = +\tfrac{1}{6} u \ , \quad Yd = +\tfrac{1}{6} d \ , \quad Y\bar{u} = -\tfrac{2}{3} \bar{u} \ , \quad Y\bar{d} = +\tfrac{1}{3} \bar{d} \ , \tag{89.17}$$

we see from eq. (89.15) that

$$Qu = +\tfrac{2}{3} u, \qquad Qd = -\tfrac{1}{3} d \ , \quad Q\bar{u} = -\tfrac{2}{3} \bar{u} \ , \quad Q\bar{d} = +\tfrac{1}{3} \bar{d} \ . \tag{89.18}$$

This is just the set of electric charge assignments that we expect for the up and down quarks. In terms of the four-component fields, we have

$$(g_2 A^3_\mu T^3 + g_1 B_\mu Y)\mathcal{U} = \Big[+\tfrac{2}{3} e A_\mu + \tfrac{e}{s_\text{W} c_\text{W}}(+\tfrac{1}{2} P_\text{L} - \tfrac{2}{3} s^2_\text{W}) Z_\mu \Big] \mathcal{U} \ , \tag{89.19}$$

$$(g_2 A^3_\mu T^3 + g_1 B_\mu Y)\mathcal{D} = \Big[-\tfrac{1}{3} e A_\mu + \tfrac{e}{s_\text{W} c_\text{W}}(-\tfrac{1}{2} P_\text{L} + \tfrac{1}{3} s^2_\text{W}) Z_\mu \Big] \mathcal{D} \ . \tag{89.20}$$

Using eqs. (89.13) and (89.19)–(89.20) in eqs. (89.1)–(89.4), we find the coupings of the electroweak gauge fields to the quarks,

$$\mathcal{L}_\text{int} = \tfrac{1}{\sqrt{2}} g_2 W^+_\mu J^{-\mu} + \tfrac{1}{\sqrt{2}} g_2 W^-_\mu J^{+\mu} + \tfrac{e}{s_\text{W} c_\text{W}} Z_\mu J^\mu_\text{Z} + e A_\mu J^\mu_\text{EM} \ , \tag{89.21}$$

where we have defined the currents

$$J^{+\mu} \equiv \overline{\mathcal{D}}_\text{L} \gamma^\mu \mathcal{U}_\text{L} \ , \tag{89.22}$$

$$J^{-\mu} \equiv \overline{\mathcal{U}}_\text{L} \gamma^\mu \mathcal{D}_\text{L} \ , \tag{89.23}$$

$$J^\mu_\text{Z} \equiv J^\mu_3 - s^2_\text{W} J^\mu_\text{EM} \ , \tag{89.24}$$

$$J^\mu_3 \equiv \tfrac{1}{2} \overline{\mathcal{U}}_\text{L} \gamma^\mu \mathcal{U}_\text{L} - \tfrac{1}{2} \overline{\mathcal{D}}_\text{L} \gamma^\mu \mathcal{D}_\text{L} \ , \tag{89.25}$$

$$J^\mu_\text{EM} \equiv +\tfrac{2}{3} \overline{\mathcal{U}} \gamma^\mu \mathcal{U} - \tfrac{1}{3} \overline{\mathcal{D}} \gamma^\mu \mathcal{D} \ . \tag{89.26}$$

Having worked out the interactions of a single quark generation, we now examine what happens when there is more than one of them. Let us consider the fields $q_{\alpha i I}$, $\bar{u}^{\alpha}_{I}$, and $\bar{d}^{\alpha}_{I}$, where $I = 1, 2, 3$ is a generation index. The kinetic term for all these fields is

$$\mathcal{L}_{\text{kin}} = iq^{\dagger\alpha i I}\bar{\sigma}^{\mu}(D_{\mu})_{\alpha i}{}^{\beta j}q_{\beta j I} + i\bar{u}^{\dagger}_{\alpha I}\bar{\sigma}^{\mu}(D_{\mu})^{\alpha}{}_{\beta}\bar{u}^{\beta}_{I} + i\bar{d}^{\dagger}_{\alpha I}\bar{\sigma}^{\mu}(D_{\mu})^{\alpha}{}_{\beta}\bar{d}^{\beta}_{I}, \tag{89.27}$$

where the repeated generation index is summed. The most general Yukawa term we can write down now reads

$$\mathcal{L}_{\text{Yuk}} = -\varepsilon^{ij}\varphi_{i}q_{\alpha j I}y'_{IJ}\bar{d}^{\alpha}_{J} - \varphi^{\dagger i}q_{\alpha i I}y''_{IJ}\bar{u}^{\alpha}_{J} + \text{h.c.}\,, \tag{89.28}$$

where y'_{IJ} and y''_{IJ} are complex 3×3 matrices, and the generation indices are summed. In unitary gauge, this becomes

$$\mathcal{L}_{\text{Yuk}} = -\tfrac{1}{\sqrt{2}}(v + H)d_{\alpha I}y'_{IJ}\bar{d}^{\alpha}_{J} - \tfrac{1}{\sqrt{2}}(v + H)u_{\alpha I}y''_{IJ}\bar{u}^{\alpha}_{J} + \text{h.c.}\,. \tag{89.29}$$

We can make unitary transformations in generation space on the fields: $d_I \to D_{IJ}d_J$, $\bar{d}_I \to \bar{D}_{IJ}\bar{d}_J$, $u_I \to U_{IJ}u_J$, and $\bar{u}_I \to \bar{U}_{IJ}\bar{u}_J$, where U, D, $\bar{U}$ and $\bar{D}$ are independent unitary matrices. The kinetic terms are unchanged (except for the couplings to the $W^{\pm}$, as we will discuss momentarily), and the Yukawa matrices y' and y'' are replaced with $D^{\mathrm{T}}y'\bar{D}$ and $U^{\mathrm{T}}y''\bar{U}$. We can choose D, $\bar{D}$, U, and $\bar{U}$ so that $D^{\mathrm{T}}y'\bar{D}$ and $U^{\mathrm{T}}y''\bar{U}$ are diagonal with positive real entries y'_I and y''_I. The down quarks $\mathcal{D}_I$ then have masses $m_{d_I} = y'_I v/\sqrt{2}$, and the up quarks $\mathcal{U}_I$ have masses $m_{u_I} = y''_I v/\sqrt{2}$. In the neutral currents J^{μ}_{3} and J^{μ}_{EM}, we simply add a generation index I to each field, and sum over it. The charged currents are more complicated, however; they become

$$J^{+\mu} = \overline{\mathcal{D}}_{\text{L}I}(V^{\dagger})_{IJ}\gamma^{\mu}\mathcal{U}_{\text{L}J}\,, \tag{89.30}$$

$$J^{-\mu} = \overline{\mathcal{U}}_{\text{L}I}V_{IJ}\gamma^{\mu}\mathcal{D}_{\text{L}J}\,, \tag{89.31}$$

where $V \equiv U^{\dagger}D$ is the *Cabibbo–Kobayasi–Maskawa matrix* (or *CKM matrix* for short). Note that we did not have this complication in the lepton sector, because there we had only one Yukawa term.

A 3×3 unitary matrix has nine real parameters. However, we are still free to make the independent phase rotations $\mathcal{D}_I \to e^{i\alpha_I}\mathcal{D}_I$ and $\mathcal{U}_I \to e^{i\beta_I}\mathcal{U}_I$, as these leave the kinetic and mass terms invariant. These phase changes allow us to make the first row and column of V_{IJ} real, eliminating five of the nine parameters. The remaining four can be chosen as θ_1 (the *Cabibbo angle*), θ_2, θ_3, and δ, where

$$V = \begin{pmatrix} c_1 & +s_1 c_3 & +s_1 s_3 \\ -s_1 c_2 & c_1 c_2 c_3 - s_2 s_3 e^{i\delta} & c_1 c_2 s_3 + s_2 c_3 e^{i\delta} \\ -s_1 s_2 & c_1 s_2 c_3 + c_2 s_3 e^{i\delta} & c_1 s_2 s_3 - c_2 c_3 e^{i\delta} \end{pmatrix}, \tag{89.32}$$

and $c_i = \cos\theta_i$ and $s_i = \sin\theta_i$. The measured values of these angles are $s_1 = 0.224$, $s_2 = 0.041$, $s_3 = 0.016$, and $\delta = 40°$. Note that the charged currents now have some terms with a phase factor $e^{i\delta}$, and some without. Since the time-reversal operator T is antiunitary ($T^{-1}iT = -i$), the charged currents do not transform in a simple way under time reversal. This implies that the charged current terms in $\mathcal{L}_{\rm int}$ are not time-reversal invariant; hence the electroweak interactions violate time-reversal symmetry. Since CPT is always a good symmetry, time-reversal violation is equivalent to CP violation; δ is therefore sometimes called the *CP violating phase.*

At high energies, we can use our results to compute electroweak contributions to scattering amplitudes involving quarks. This is because, at high energies, the SU(3) coupling g_3 is weak; we can, for example, reliably compute the decay rates of the $W^\pm$ and Z^0 into quarks, because $\alpha_3(M_{\rm Z}) \equiv g_3^2(M_{\rm Z})/4\pi = 0.12$ is small enough to make QCD loop corrections a few-percent effect.

To understand low-energy processes such as neutron decay, we must first write the currents in terms of hadron fields. We take this up in the next section. For now, we simply note that the terms in the charged currents that involve only up and down quarks are

$$J^{+\mu} = c_1 \overline{\mathcal{D}}_{\rm L}\gamma^\mu \mathcal{U}_{\rm L} \,, \tag{89.33}$$

$$J^{-\mu} = c_1 \overline{\mathcal{U}}_{\rm L}\gamma^\mu \mathcal{D}_{\rm L} \,, \tag{89.34}$$

where c_1 is the cosine of the Cabibbo angle.

Reference notes

Electroweak interactions of quarks are discussed in more detail in *Cheng & Li*, *Georgi*, *Peskin & Schroeder*, *Quigg*, and *Ramond II*.

Problems

89.1) Verify the claims made immediately after eqs. (89.4) and (89.7).

89.2) Compute the rates for the decay processes $W^+ \to u\bar{d}$, $Z^0 \to \bar{u}u$, and $Z^0 \to \bar{d}d$. Neglect the quark masses. Express your results in GeV. Combine your answers with those of problem 88.6, and sum over generations to get the total decay rates for the $W^\pm$ and Z^0. You can neglect the masses of all quarks and leptons except the top quark, and take $\theta_2 = \theta_3 = 0$.

89.3) Show that the Standard Model is anomaly free. Hint: you must consider 3–3–3, 2–2–2, 3–3–1, 2–2–1, and 1–1–1 anomalies, where the number denotes the gauge group of one of the external gauge fields in the triangle diagram. Why do we not need to worry about the unlisted combinations?

89.4) Compute the leading term in the beta function for each of the three gauge couplings of the Standard Model.

89.5) After integrating out the $W^{\pm}$ fields, we get an effective interaction between the hadron and lepton currents that includes

$$\mathcal{L}_{\text{eff}} = 2\sqrt{2}\, Z_C C(\overline{\mathcal{E}}_{\text{L}}\gamma^{\mu}\mathcal{N}_{e\text{L}})(\overline{\mathcal{U}}_{\text{L}}\gamma^{\mu}\mathcal{D}_{\text{L}}) \,, \tag{89.35}$$

where we have defined $C \equiv c_1 G_{\text{F}}$ at a renormalization scale $\mu = M_{\text{W}}$, and Z_C is a renormalizing factor. This interaction contributes to neutron decay; see section 90. In this problem, following the analysis of problem 88.7, we will compute the anomalous dimension γ_C of C due to one-loop photon and gluon exchange.

a) Use Fierz identities to show that eq. (89.35) can be rewritten as

$$\mathcal{L}_{\text{eff}} = 2\sqrt{2}\, Z_C C(\overline{\mathcal{E}}_{\text{L}}\gamma^{\mu}\mathcal{D}_{\text{L}})(\overline{\mathcal{U}}_{\text{L}}\gamma^{\mu}\mathcal{N}_{e\text{L}}) \,, \tag{89.36}$$

and also as

$$\mathcal{L}_{\text{eff}} = -4\sqrt{2}\, Z_C C(\overline{\mathcal{D}}^{\text{C}} P_{\text{L}} \mathcal{N}_e)(\overline{\mathcal{E}} P_{\text{R}} \mathcal{U}^{\text{C}}) \,. \tag{89.37}$$

b) Working in Lorenz gauge and using the results of problem 88.7, show that gluon exchange does not make a one-loop contribution to Z_C.

c) Show that only a photon connecting the e and u lines makes a one-loop contribution to Z_C.

d) Note that $\overline{\mathcal{E}} P_{\text{R}} \mathcal{U}^{\text{C}} = e^{\dagger} u^{\dagger}$, and compare this with $\overline{\mathcal{E}}\mathcal{E} = e^{\dagger}\bar{e}^{\dagger} + \text{h.c.}$. Argue that the photon-exchange contribution to Z_C is given by the one-loop contribution to Z_m in spinor electrodynamics in Lorenz gauge, with the replacement $(-1)(+1)e^2 \to (-1)(+\frac{2}{3})e^2$.

e) Let $\gamma_C(\alpha) = c_1\alpha + \ldots$, where $\alpha = e^2/4\pi$, and find c_1. (This c_1 should not be confused with the cosine of the Cabibbo angle.)

90

Electroweak interactions of hadrons

Prerequisites: 83, 89

Now that we know how quarks couple to the electroweak gauge fields, we can use this information to obtain amplitudes for various processes involving hadrons. We will focus on three of the most important: neutron decay, $n \to pe^- \bar{\nu}$; charged pion decay, $\pi^- \to \mu^- \bar{\nu}_\mu$; and neutral pion decay, $\pi^0 \to \gamma\gamma$.

Recall from section 83 the chiral lagrangian for pions and nucleons,

$$\begin{aligned}\mathcal{L} = &-\tfrac{1}{4} f_\pi^2 \operatorname{Tr} \partial^\mu U^\dagger \partial_\mu U + v^3 \operatorname{Tr}(MU + M^\dagger U^\dagger) \\ &+ i\overline{N}\partial\!\!\!/ N - m_N \overline{N}(U^\dagger P_{\rm L} + UP_{\rm R})N \\ &- \tfrac{1}{2}(g_{\rm A}-1)i\overline{N}\gamma^\mu(U\partial_\mu U^\dagger P_{\rm L} + U^\dagger\partial_\mu U P_{\rm R})N \,, \end{aligned} \tag{90.1}$$

where $U(x) = \exp[2i\pi^a(x)T^a/f_\pi]$, π^a is the pion field, N is the nucleon field, f_π is the pion decay constant, M is the quark mass matrix, v^3 is the value of the quark condensate, m_N is the nucleon mass, and $g_{\rm A}$ is the axial vector coupling. The electroweak gauge group SU(2) $\times$ U(1) is a subgroup of the $\mathrm{SU(2)_L \times SU(2)_R \times U(1)_V}$ flavor group that we have in the limit of zero quark mass. It will prove convenient to go through the formal procedure of gauging the full flavor group, and only later identifying the electroweak subgroup. We therefore define matrix-valued gauge fields $l_\mu(x)$ and $r_\mu(x)$ that transform as

$$l_\mu \to L l_\mu L^\dagger + iL\partial_\mu L^\dagger \,, \tag{90.2}$$

$$r_\mu \to R r_\mu R^\dagger + iR\partial_\mu R^\dagger \,. \tag{90.3}$$

Here $L(x)$ and $R(x)$ are 2×2 unitary matrices that correspond to a general $\mathrm{SU(2)_L \times SU(2)_R \times U(1)_V}$ gauge transformation; we restrict the U(1) part of the transformation to the vector subgroup by requiring $\det L = \det R$ and

$\mathrm{Tr}\, l_\mu = \mathrm{Tr}\, r_\mu$. The transformation rules for the pion and nucleon fields are

$$U \to LUR^\dagger\,, \tag{90.4}$$

$$N_\mathrm{L} \to LN_\mathrm{L}\,, \tag{90.5}$$

$$N_\mathrm{R} \to RN_\mathrm{R}\,, \tag{90.6}$$

where $N_\mathrm{L} \equiv P_\mathrm{L} N$ and $N_\mathrm{R} \equiv P_\mathrm{R} N$ are the left- and right-handed parts of the nucleon field.

We can make the chiral lagrangian gauge invariant (except for terms involving the quark masses) by replacing ordinary derivatives with appropriate covariant derivatives. We determine the covariant derivative of each field by requiring it to transform in the same way as the field itself; for example, $D_\mu U \to L(D_\mu U)R^\dagger$. We thus find

$$D_\mu U = \partial_\mu U - il_\mu U + iU r_\mu\,, \tag{90.7}$$

$$D_\mu U^\dagger = \partial_\mu U^\dagger + iU^\dagger l_\mu - ir_\mu U^\dagger\,, \tag{90.8}$$

$$D_\mu N_\mathrm{L} = (\partial_\mu - il_\mu)N_\mathrm{L}\,, \tag{90.9}$$

$$D_\mu N_\mathrm{R} = (\partial_\mu - ir_\mu)N_\mathrm{R}\,. \tag{90.10}$$

Making the substitution $\partial \to D$ in $\mathcal{L}$, we learn how the pions and nucleons couple to these gauge fields.

As in section 83, it is more convenient to work with the nucleon field $\mathcal{N}$, defined via

$$N = (uP_\mathrm{L} + u^\dagger P_\mathrm{R})\mathcal{N}\,, \tag{90.11}$$

where $u^2 = U$. Making this transformation, we ultimately find

$$\begin{aligned}\mathcal{L} = -\tfrac{1}{4}f_\pi^2\,\mathrm{Tr}(&\partial^\mu U^\dagger \partial_\mu U - il^\mu U \overleftrightarrow{\partial}_\mu U^\dagger - ir^\mu U^\dagger \overleftrightarrow{\partial}_\mu U \\ &+ l^\mu l_\mu + r^\mu r_\mu - 2l^\mu U r_\mu U^\dagger) \\ &+ v^3\,\mathrm{Tr}(MU + M^\dagger U^\dagger) + i\overline{\mathcal{N}}\partial\!\!\!/\mathcal{N} - m_N\overline{\mathcal{N}}\mathcal{N} \\ &+ \overline{\mathcal{N}}(v\!\!\!/ + \tfrac{1}{2}\tilde{l\!\!/} + \tfrac{1}{2}\tilde{r\!\!/})\mathcal{N} - g_\mathrm{A}\overline{\mathcal{N}}(a\!\!\!/ + \tfrac{1}{2}\tilde{l\!\!/} - \tfrac{1}{2}\tilde{r\!\!/})\gamma_5\mathcal{N}\,,\end{aligned} \tag{90.12}$$

where

$$v_\mu \equiv \tfrac{1}{2}i[u^\dagger(\partial_\mu u) + u(\partial_\mu u^\dagger)]\,, \tag{90.13}$$

$$a_\mu \equiv \tfrac{1}{2}i[u^\dagger(\partial_\mu u) - u(\partial_\mu u^\dagger)]\,, \tag{90.14}$$

$$\tilde{l}_\mu \equiv u^\dagger l_\mu u\,, \tag{90.15}$$

$$\tilde{r}_\mu \equiv u r_\mu u^\dagger\,. \tag{90.16}$$

It is now convenient to set

$$l_\mu = l^a_\mu T^a + b_\mu \,, \tag{90.17}$$

$$r_\mu = r^a_\mu T^a + b_\mu \,. \tag{90.18}$$

We have normalized b_μ so that the corresponding charge is the baryon number. The SU(2) gauge fields of the Standard Model can now be identified as

$$g_2 A^a_\mu = l^a_\mu \,, \tag{90.19}$$

and the electromagnetic gauge field as

$$eA_\mu = l^3_\mu + r^3_\mu + \tfrac{1}{2} b_\mu \,. \tag{90.20}$$

Eq. (90.20) follows from reconciling eqs. (90.9) and (90.10) with the requirement that the electromagnetic covariant derivatives of the proton field $p = \mathcal{N}_1$ and the neutron field $n = \mathcal{N}_2$ be given by $(\partial_\mu - ieA_\mu)p$ and $\partial_\mu n$.

We can now find the hadronic parts of the currents that couple to the gauge fields by differentiating $\mathcal{L}$ with respect to them, and then setting them to zero. We find

$$\begin{aligned} J^{a\mu}_{\rm L} &= (\partial\mathcal{L}/\partial l^a_\mu)\Big|_{l=r=0} \\ &= \tfrac{1}{4} i f_\pi^2 \,{\rm Tr}\, T^a U \overleftrightarrow{\partial^\mu} U^\dagger + \tfrac{1}{2}\overline{\mathcal{N}} u^\dagger T^a \gamma^\mu (1 - g_{\rm A}\gamma_5) u \mathcal{N} \\ &= +\tfrac{1}{2} f_\pi \partial^\mu \pi^a - \tfrac{1}{2}\varepsilon^{abc}\pi^b \partial^\mu \pi^c + \tfrac{1}{2}\overline{\mathcal{N}} T^a \gamma^\mu (1 - g_{\rm A}\gamma_5)\mathcal{N} + \ldots \,, \end{aligned} \tag{90.21}$$

$$\begin{aligned} J^{a\mu}_{\rm R} &= (\partial\mathcal{L}/\partial r^a_\mu)\Big|_{l=r=0} \\ &= \tfrac{1}{4} i f_\pi^2 \,{\rm Tr}\, T^a U^\dagger \overleftrightarrow{\partial^\mu} U + \tfrac{1}{2}\overline{\mathcal{N}} u T^a \gamma^\mu (1 + g_{\rm A}\gamma_5) u^\dagger \mathcal{N} \\ &= -\tfrac{1}{2} f_\pi \partial^\mu \pi^a - \tfrac{1}{2}\varepsilon^{abc}\pi^b \partial^\mu \pi^c + \tfrac{1}{2}\overline{\mathcal{N}} T^a \gamma^\mu (1 + g_{\rm A}\gamma_5)\mathcal{N} + \ldots \,, \end{aligned} \tag{90.22}$$

$$\begin{aligned} J^{\mu}_{\rm B} &= (\partial\mathcal{L}/\partial b_\mu)\Big|_{l=r=0} \\ &= \overline{\mathcal{N}}\gamma^\mu \mathcal{N} \,. \end{aligned} \tag{90.23}$$

In the third lines of eqs. (90.21) and (90.22), we have expanded in inverse powers of f_π. We can now identify the currents that couple to the physical

$W_\mu^\pm$, Z_μ, and A_μ fields as

$$\begin{aligned} J^{+\mu} &= c_1(J_{\rm L}^{1\mu} - iJ_{\rm L}^{2\mu}) \\ &= \tfrac{1}{\sqrt{2}}c_1(f_\pi\partial^\mu\pi^+ + i\pi^0\overleftrightarrow{\partial}_\mu\pi^+) + \tfrac{1}{2}c_1\overline{n}\gamma^\mu(1-g_{\rm A}\gamma_5)p + \ldots\ , \end{aligned} \tag{90.24}$$

$$\begin{aligned} J^{-\mu} &= c_1(J_{\rm L}^{1\mu} + iJ_{\rm L}^{2\mu}) \\ &= \tfrac{1}{\sqrt{2}}c_1(f_\pi\partial^\mu\pi^- - i\pi^0\overleftrightarrow{\partial^\mu}\pi^-) + \tfrac{1}{2}c_1\overline{p}\gamma^\mu(1-g_{\rm A}\gamma_5)n + \ldots\ , \end{aligned} \tag{90.25}$$

$$J_{\rm Z}^\mu = J_3^\mu - s_{\rm W}^2 J_{\rm EM}^\mu\ , \tag{90.26}$$

$$\begin{aligned} J_3^\mu &= J_{\rm L}^{3\mu} \\ &= \tfrac{1}{2}(f_\pi\partial^\mu\pi^0 + i\pi^+\overleftrightarrow{\partial^\mu}\pi^-) \\ &\quad + \tfrac{1}{4}\overline{p}\gamma^\mu(1-g_{\rm A}\gamma_5)p - \tfrac{1}{4}\overline{n}\gamma^\mu(1-g_{\rm A}\gamma_5)n + \ldots\ , \end{aligned} \tag{90.27}$$

$$\begin{aligned} J_{\rm EM}^\mu &= J_{\rm L}^{3\mu} + J_{\rm R}^{3\mu} + \tfrac{1}{2}J_{\rm B}^\mu \\ &= i\pi^+\overleftrightarrow{\partial^\mu}\pi^- + \overline{p}\gamma^\mu p + \ldots\ , \end{aligned} \tag{90.28}$$

where c_1 is the cosine of the Cabibbo angle, and the interactions are specified by

$$\mathcal{L}_{\rm int} = \tfrac{1}{\sqrt{2}}g_2W_\mu^+J^{-\mu} + \tfrac{1}{\sqrt{2}}g_2W_\mu^-J^{+\mu} + \frac{e}{s_{\rm W}c_{\rm W}}Z_\mu J_{\rm Z}^\mu + eA_\mu J_{\rm EM}^\mu\ . \tag{90.29}$$

For low-energy processes involving $W^\pm$ or Z^0 exchange, we can use the effective current-current interaction that we derived in section 88,

$$\mathcal{L}_{\rm eff} = 2\sqrt{2}\,G_{\rm F}(J^{+\mu}J_\mu^- + J_{\rm Z}^\mu J_{{\rm Z}\mu})\ . \tag{90.30}$$

We should include both hadronic and leptonic contributions to the currents.

Consider charged pion decay, $\pi^- \to \mu^-\overline{\nu}_\mu$. The relevant terms in the charged currents (neutral currents do not contribute) are

$$J^{-\mu} = \tfrac{1}{\sqrt{2}}c_1f_\pi\partial^\mu\pi^-\ , \tag{90.31}$$

$$J^{+\mu} = \tfrac{1}{2}\overline{\mathcal{M}}\gamma^\mu(1-\gamma_5)\mathcal{N}_m\ , \tag{90.32}$$

where $\mathcal{M}$ is the muon field and $\mathcal{N}_m$ is the muon neutrino field. The relevant term in the effective interaction is then

$$\mathcal{L}_{\rm eff} = G_{\rm F}c_1f_\pi\partial^\mu\pi^-\overline{\mathcal{M}}\gamma_\mu(1-\gamma_5)\mathcal{N}_m\ . \tag{90.33}$$

The corresponding decay amplitude is

$$\mathcal{T} = G_{\rm F}c_1f_\pi k^\mu\,\overline{u}_1\gamma_\mu(1-\gamma_5)v_2\ , \tag{90.34}$$

where the four-momenta of the pion, muon, and antineutrino are k, p_1, and p_2. Eq. (90.34) can be simplified by using $\not{k} = \not{p}_1 + \not{p}_2$ along with $\overline{u}_1\not{p}_1 = -m_\mu \overline{u}_1$ and $\not{p}_2 v_2 = 0$; we get

$$\mathcal{T} = -G_F c_1 f_\pi m_\mu \overline{u}_1 (1-\gamma_5) v_2 \ . \tag{90.35}$$

We see that $\mathcal{T}$ is proportional to the muon mass; since $m_\mu \gg m_e$, decay to $\mu^- \overline{\nu}_\mu$ is preferred over decay to $e^- \overline{\nu}_e$.

Squaring $\mathcal{T}$ and summing over final spins, we find

$$\begin{aligned} \langle |\mathcal{T}|^2 \rangle &= (G_F c_1 f_\pi m_\mu)^2 (-8 p_1 \cdot p_2) \\ &= 4 (G_F c_1 f_\pi m_\mu)^2 (m_\pi^2 - m_\mu^2) \ . \end{aligned} \tag{90.36}$$

We used $-2p_1 \cdot p_2 = p_1^2 + p_2^2 - (p_1 + p_2)^2 = -m_\mu^2 + 0 + m_\pi^2$ to get the second line. We now have

$$\begin{aligned} \Gamma &= \frac{1}{2m_\pi} \int d\text{LIPS}_2(k) \langle |\mathcal{T}|^2 \rangle \\ &= \frac{|\mathbf{p}_1|}{8\pi m_\pi^2} \langle |\mathcal{T}|^2 \rangle \\ &= \frac{G_F^2 c_1^2 f_\pi^2 m_\mu^2 m_\pi}{4\pi} \left(1 - \frac{m_\mu^2}{m_\pi^2} \right)^2 , \end{aligned} \tag{90.37}$$

where we used $|\mathbf{p}_1| = (m_\pi^2 - m_\mu^2)/2m_\pi$ to get the last line. Since we determine the value of G_F from the decay rate of the muon (see section 88), the charged pion decay rate allows us to fix the value of $c_1 f_\pi$.

The value of c_1 can be determined from the rate for the decay process $\pi^- \to \pi^0 e^- \overline{\nu}_e$, which we will calculate in problem 90.6. The relevant hadronic term in the charged current is

$$J^{-\mu} = -\tfrac{1}{\sqrt{2}} i c_1 \pi^0 \overleftrightarrow{\partial^\mu} \pi^- \ , \tag{90.38}$$

which depends on c_1 but not f_π. Comparison with experiment then yields $c_1 = 0.974$. We note that the key feature of eq. (90.38) is that it involves spin-zero hadrons that are members of an isospin triplet; eq. (90.38) applies to any such hadrons, including nuclei. Thus c_1 can also be measured in *superallowed Fermi decays*, which take a nucleus from one spin-zero state to another spin-zero state with the same parity in the same isospin multiplet. With c_1 fixed, the pion decay rate yields $f_\pi = 92.4\,\text{MeV}$.

Next we consider neutron decay, $n \to pe^-\overline{\nu}_e$. The relevant terms in the charged currents (neutral currents do not contribute) are

$$J^{-\mu} = \tfrac{1}{2}c_1\overline{p}\gamma^\mu(1-g_{\rm A}\gamma_5)n\,, \tag{90.39}$$

$$J^{+\mu} = \tfrac{1}{2}\overline{\mathcal{E}}\gamma^\mu(1-\gamma_5)\mathcal{N}_e\,, \tag{90.40}$$

where $\mathcal{E}$ is the electron field and $\mathcal{N}_e$ is the electron neutrino field. The relevant term in the effective interaction is then

$$\mathcal{L}_{\rm eff} = \tfrac{1}{\sqrt{2}}G_{\rm F}c_1\overline{p}\gamma^\mu(1-g_{\rm A}\gamma_5)n\overline{\mathcal{E}}\gamma_\mu(1-\gamma_5)\mathcal{N}_e\,. \tag{90.41}$$

Consider a neutron with four-momentum $p_n = (m_n, \mathbf{0})$, and spin up along the z axis; the decay amplitude is

$$\mathcal{T} = \tfrac{1}{\sqrt{2}}G_{\rm F}c_1[\overline{u}_p\gamma^\mu(1-g_{\rm A}\gamma_5)u_n][\overline{u}_e\gamma_\mu(1-\gamma_5)v_{\overline{\nu}}]\,, \tag{90.42}$$

where $u_n\overline{u}_n = \frac{1}{2}(1-\gamma_5\not{z})(-\not{p}_n+m_n)$. We take the absolute square of $\mathcal{T}$ and sum over the final spins. Since the maximum available kinetic energy is $m_n - m_p - m_e = 0.782\,\text{MeV} \ll m_p$, the proton is nonrelativistic, and we can use the approximations $p_p{\cdot}p_e \simeq -m_pE_e$ and $p_p{\cdot}p_{\overline{\nu}} \simeq -m_pE_{\overline{\nu}}$ in addition to the exact formulae $p_n{\cdot}p_e = -m_nE_e$ and $p_n{\cdot}p_{\overline{\nu}} = -m_nE_{\overline{\nu}}$. After a tedious but straightforward calculation, we find

$$\begin{aligned}\langle|\mathcal{T}|^2\rangle &= 16\,G_{\rm F}^2c_1^2(1+3g_{\rm A}^2)\,m_nm_pE_eE_{\overline{\nu}}\\ &\quad\times\left[1+a\frac{\mathbf{p}_e{\cdot}\mathbf{p}_{\overline{\nu}}}{E_eE_{\overline{\nu}}}+A\frac{\hat{\mathbf{z}}{\cdot}\mathbf{p}_e}{E_e}+B\frac{\hat{\mathbf{z}}{\cdot}\mathbf{p}_{\overline{\nu}}}{E_{\overline{\nu}}}\right],\end{aligned} \tag{90.43}$$

where the *correlation coefficients* are given by

$$a = \frac{1-g_{\rm A}^2}{1+3g_{\rm A}^2}\,,\qquad A = \frac{2g_{\rm A}(1-g_{\rm A})}{1+3g_{\rm A}^2}\,,\qquad B = \frac{2g_{\rm A}(1+g_{\rm A})}{1+3g_{\rm A}^2}\,. \tag{90.44}$$

When we integrate over the final momenta to get the total decay rate, the correlation terms vanish, and so the rate is proportional to $G_{\rm F}^2c_1^2(1+3g_{\rm A}^2)$. Since we get the value of $G_{\rm F}^2c_1^2$ from the rate for $\pi^- \to \pi^0e^-\overline{\nu}_e$, the neutron decay rate allows us to determine $1+3g_{\rm A}^2$. To get the sign of $g_{\rm A}$, we need a measurement of either A or B. (The antineutrino three-momentum can be determined from the electron and proton three-momenta.) The result is that $g_{\rm A} = +1.27$. The measured values of the three correlation coefficients are all consistent with eq. (90.44).

Finally, we consider the decay of the neutral pion into two photons, $\pi^0 \to \gamma\gamma$. None of the terms in our chiral lagrangian, eq. (90.12), couple a single π^0 to two photons. Therefore, without adding more terms, this process does not occur at tree level. However, at the one-loop level, we have the diagrams

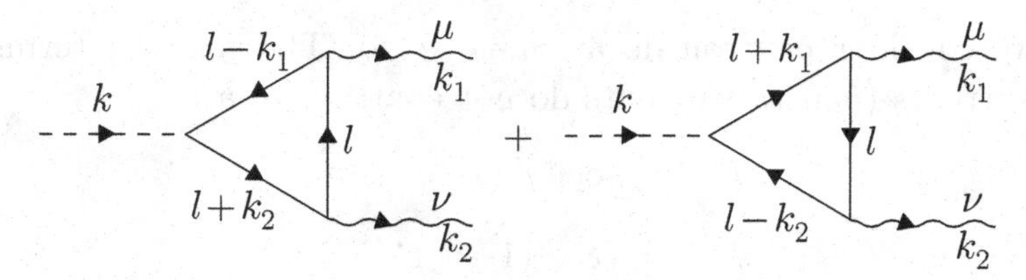

Figure 90.1. One-loop diagrams contributing to $\pi^0 \to \gamma\gamma$. The solid line is a proton.

of fig. 90.1; a proton circulates in the loop. Let us evaluate these diagrams. In section 83, we found that the coupling of the π^0 to the nucleons is given by

$$\mathcal{L}_{\pi^0\overline{N}N} = -\tfrac{1}{2}(g_{\rm A}/f_\pi)\partial_\mu\pi^0(\overline{p}\gamma^\mu\gamma_5 p - \overline{n}\gamma^\mu\gamma_5 n)\ . \tag{90.45}$$

This leads to a $\pi^0\overline{p}p$ vertex factor of $\frac{1}{2}(g_{\rm A}/f_\pi)k_\rho\gamma^\rho\gamma_5$. The diagrams in fig. 90.1 are then identical to the diagrams we evaluated in section 76, and so the one-loop decay amplitude is

$$i\mathcal{T}_{\text{1-loop}} = \tfrac{1}{2}(g_{\rm A}/f_\pi)(ie)^2\varepsilon_{1\mu}\varepsilon_{2\nu}k_\rho C^{\mu\nu\rho}(k_1,k_2,k)\ , \tag{90.46}$$

where

$$k_\rho C^{\mu\nu\rho}(k_1,k_2,k) = -\frac{i}{2\pi^2}\varepsilon^{\mu\nu\alpha\beta}k_{1\alpha}k_{2\beta}\ . \tag{90.47}$$

Here we have chosen to renormalize so as to have $k_{1\mu}C^{\mu\nu\rho}(k_1,k_2,k) = 0$ and $k_{2\nu}C^{\mu\nu\rho}(k_1,k_2,k) = 0$; this is required by electromagnetic gauge invariance. Combining eqs. (90.46) and (90.47), we get

$$\mathcal{T}_{\text{1-loop}} = -\frac{g_{\rm A}e^2}{4\pi^2 f_\pi}\,\varepsilon^{\alpha\mu\beta\nu}k_{1\alpha}\varepsilon_{1\mu}k_{2\beta}\varepsilon_{2\nu}\ . \tag{90.48}$$

This result is subject to higher-loop corrections. Note that diagrams with extra internal pion lines attached to the nucleon loop are not suppressed by any small expansion parameter. Thus we cannot trust the overall coefficient in eq. (90.48).

Note that this amplitude would arise at tree level from an interaction of the form $\mathcal{L}_{\pi^0\gamma\gamma} \propto \pi^0\varepsilon^{\alpha\mu\beta\nu}F_{\mu\alpha}F_{\nu\beta}$. If we integrate out the nucleon fields to get an effective lagrangian for the pions and photons alone, such a term should appear.

There is a problem, however. The $\mathrm{SU(2)_L \times SU(2)_R \times U(1)_V}$ symmetry of the effective lagrangian implies that a pion field that has no derivatives acting on it must be accompanied by at least one factor of a quark mass. For example, we could have $\mathcal{L}_{\pi^0\gamma\gamma} \propto i\,\mathrm{Tr}(MU - M^\dagger U^\dagger)\varepsilon^{\alpha\mu\beta\nu}F_{\mu\alpha}F_{\nu\beta}$. The problem is that there are no quark-mass factors in eq. (90.48). So we have an apparent contradiction between our explicit one-loop result, and a general argument based on symmetry.

This contradiction is resolved by noting that the electromagnetic gauge field results in an anomaly in the axial current $J_{\rm A}^{3\mu} \equiv J_{\rm L}^{3\mu} - J_{\rm R}^{3\mu}$. In terms of the quark doublet

$$\mathcal{Q} = \begin{pmatrix} \mathcal{U} \\ \mathcal{D} \end{pmatrix}, \tag{90.49}$$

this current is

$$\begin{aligned} J_{\rm A}^{3\mu} &= \overline{\mathcal{Q}} T^3 \gamma^\mu \gamma_5 \mathcal{Q} \\ &= \tfrac{1}{2} \overline{\mathcal{U}} \gamma^\mu \gamma_5 \mathcal{U} - \tfrac{1}{2} \overline{\mathcal{D}} \gamma^\mu \gamma_5 \mathcal{D} , \end{aligned} \tag{90.50}$$

where we have suppressed the color indices. Using our results in sections 76 and 77, the anomalous divergence of this current is given by

$$\partial_\mu J_{\rm A}^{3\mu} = -\frac{e^2}{16\pi^2} \mathrm{Tr}(T^3 Q^2) \varepsilon^{\mu\nu\rho\sigma} F_{\mu\nu} F_{\rho\sigma} , \tag{90.51}$$

where

$$Q = \begin{pmatrix} +\frac{2}{3} & 0 \\ 0 & -\frac{1}{3} \end{pmatrix} \tag{90.52}$$

is the electric charge matrix acting on the quark fields, and the trace includes a factor of three for color; we thus have

$$\mathrm{Tr}(T^3 Q^2) = 3\Big(\tfrac{1}{2}(+\tfrac{2}{3})^2 - \tfrac{1}{2}(-\tfrac{1}{3})^2\Big) = +\tfrac{1}{2} , \tag{90.53}$$

and so

$$\partial_\mu J_{\rm A}^{3\mu} = -\frac{e^2}{32\pi^2} \varepsilon^{\mu\nu\rho\sigma} F_{\mu\nu} F_{\rho\sigma} . \tag{90.54}$$

This formula is exact in the limit of zero quark mass.

Now using eqs. (90.21) and (90.22), we can write the axial current in terms of the pion fields as

$$\begin{aligned} J_{\rm A}^{3\mu} &\equiv J_{\rm L}^{3\mu} - J_{\rm R}^{3\mu} \\ &= f_\pi \partial^\mu \pi^0 + \ldots . \end{aligned} \tag{90.55}$$

(We do not include the nucleon contribution because we are considering the effective lagrangian for pions and photons after integrating out the nucleons.) From eq. (90.55) we have $\partial_\mu J_{\rm A}^{3\mu} = f_\pi \partial^2 \pi^0 + \ldots$; combining this with eq. (90.54), we get

$$-\partial^2\pi^0 = \frac{e^2}{32\pi^2 f_\pi}\,\varepsilon^{\mu\nu\rho\sigma}F_{\mu\nu}F_{\rho\sigma} + O(f_\pi^{-2})\,. \tag{90.56}$$

This equation of motion would follow from an effective lagrangian that included an interaction term of the form

$$\mathcal{L}_{\pi^0\gamma\gamma} = \frac{e^2}{32\pi^2 f_\pi}\,\pi^0\varepsilon^{\mu\nu\rho\sigma}F_{\mu\nu}F_{\rho\sigma}\,. \tag{90.57}$$

This interaction leads to a $\pi^0 \to \gamma\gamma$ decay amplitude of

$$\mathcal{T} = -\frac{e^2}{4\pi^2 f_\pi}\,\varepsilon^{\mu\nu\rho\sigma}k_{1\mu}\varepsilon_{1\nu}k_{2\rho}\varepsilon_{2\sigma}\,. \tag{90.58}$$

This amplitude receives no higher-order corrections in e^2, but is subject to quark-mass corrections; these are suppressed by powers of $m_\pi^2/(4\pi f_\pi)^2$. Comparing eq. (90.58) with eq. (90.48), we see the one-loop result (which receives unsuppressed corrections) is too large by a factor of $g_{\rm A} = 1.27$.

Squaring $\mathcal{T}$, summing over final spins, integrating over $d\text{LIPS}_2(k)$, and multiplying by a symmetry factor of one half (because there are two identical particles in the final state), we ultimately find that the decay rate is

$$\Gamma = \frac{\alpha^2 m_\pi^3}{64\pi^3 f_\pi^2}\,. \tag{90.59}$$

This prediction is in agreement with the experimental result, which has an uncertainty of about 7%.

Reference notes

Electroweak interactions of hadrons are treated in *Georgi* and *Ramond II*.

Problems

90.1) Verify that the covariant derivatives in eqs. (90.7)–(90.10) transform appropriately.

90.2) Verify that substituting eq. (90.11) into eq. (90.1) yields eq. (90.12).

90.3) Compute the rate for the decay process $\tau^- \to \pi^-\nu_\tau$. Look up the measured value and compare with your result.

90.4) a) Verify eq. (90.43).

b) Compute the total neutron decay rate. Given the measured neutron lifetime $\tau = 886\,\text{s}$, and using $G_{\rm F} = 1.166 \times 10^{-5}\,\text{GeV}^{-2}$ and $c_1 = 0.974$, compute $g_{\rm A}$. Your answer is about 4% too high, because we neglected loop corrections, and the Coulomb interaction between the outgoing electron and proton.

90.5) Use your results from problems 88.7 and 89.5 to show that the neutron decay rate is enhanced by a factor of $1 + \frac{2}{\pi}\alpha \ln(M_{\rm W}/m_p)$. How much of the 4% discrepancy is accounted for by this effect?

90.6) Compute the rate for the decay process $\pi^- \to \pi^0 e^- \bar{\nu}_e$. Note that, since $m_{\pi^+} - m_{\pi^0} = 4.594\,\text{MeV} \ll m_{\pi^0}$, the outgoing π^0 is nonrelativistic. Compare your calculated rate with the measured value of $0.397\,\text{s}^{-1}$ to determine c_1. Your answer is about 1% too low, due to neglect of loop corrections.

90.7) Verify eq. (90.59). Express Γ in eV.

91

Neutrino masses

Prerequisite: 89

Recall from sections 88 and 89 that a single generation of quarks and leptons consists of left-handed Weyl fields $q_{\alpha i}$, $\bar{u}^{\alpha}$, $\bar{d}^{\alpha}$, ℓ_i, and $\bar{e}$ in the representations $(3, 2, +\frac{1}{6})$, $(\bar{3}, 1, -\frac{2}{3})$, $(\bar{3}, 1, +\frac{1}{3})$, $(1, 2, -\frac{1}{2})$, and $(1, 1, +1)$ of the gauge group $\mathrm{SU}(3) \times \mathrm{SU}(2) \times \mathrm{U}(1)$. The Higgs field is a complex scalar φ_i in the representation $(1, 2, -\frac{1}{2})$. The Yukawa couplings among these fields that are allowed by the gauge symmetry are

$$\mathcal{L}_{\text{Yuk}} = -y\varepsilon^{ij}\varphi_i\ell_j\bar{e} - y'\varepsilon^{ij}\varphi_i q_{\alpha j}\bar{d}^{\alpha} - y''\varphi^{\dagger i}q_{\alpha i}\bar{u}^{\alpha} + \text{h.c.} \,. \tag{91.1}$$

After the Higgs field acquires its VEV, these three terms give masses to the electron, down quark, and up quark, respectively. The neutrino remains massless. Thus, massless neutrinos are a prediction of the Standard Model.

However, there is now good experimental evidence that the three neutrinos actually have small masses. The data implies that the mass of the heaviest neutrino is in the range from 0.04 to 0.5 eV. To account for this, we must extend the Standard Model.

Let us continue to consider a single generation. We introduce a new left-handed Weyl field $\bar{\nu}$ in the representation $(1, 1, 0)$; this field does not couple to the gauge fields at all, and its kinetic term is simply $i\bar{\nu}^{\dagger}\bar{\sigma}^{\mu}\partial_{\mu}\bar{\nu}$. (The bar over the ν in the field $\bar{\nu}$ is *part of the name of the field*, and does not denote *any sort* of conjugation.) With this new field, we can introduce a new Yukawa coupling of the form

$$\mathcal{L}_{\nu\,\text{Yuk}} = -\tilde{y}\varphi^{\dagger i}\ell_i\bar{\nu} + \text{h.c.} \,. \tag{91.2}$$

In unitary gauge, this becomes

$$\mathcal{L}_{\nu\,\text{Yuk}} = -\tfrac{1}{\sqrt{2}}\tilde{y}(v + H)(\nu\bar{\nu} + \bar{\nu}^{\dagger}\nu^{\dagger}) \,. \tag{91.3}$$

We see that the neutrino mass is $\tilde{m} = \tilde{y}v/\sqrt{2}$.

If this was the end of the story, we would have no understanding of why the neutrino mass is so much less than the other first-generation quark and lepton masses; we would simply have to take $\tilde{y}$ much less than y, y', and y''.

However, because $\bar{\nu}$ is in a real representation of the gauge group, we are allowed by the gauge symmetry to add a mass term of the form

$$\mathcal{L}_{\bar{\nu}\,\text{mass}} = -\tfrac{1}{2}M(\bar{\nu}\bar{\nu} + \bar{\nu}^\dagger\bar{\nu}^\dagger)\,. \tag{91.4}$$

Here M is an arbitrary mass parameter. In particular, it could be quite large.

Adding eqs. (91.3) and (91.4), we find a mass matrix of the form

$$\mathcal{L}_{\nu\,\bar{\nu}\,\text{mass}} = -\tfrac{1}{2}\,(\nu \quad \bar{\nu}) \begin{pmatrix} 0 & \tilde{m} \\ \tilde{m} & M \end{pmatrix} \begin{pmatrix} \nu \\ \bar{\nu} \end{pmatrix} + \text{h.c.}\,. \tag{91.5}$$

If we take $M \gg \tilde{m}$, then the eigenvalues of this mass matrix are M and $-\tilde{m}^2/M$. (The sign of the smaller eigenvalue can be absorbed into the phase of the corresponding eigenfield.) Thus, if $\tilde{m}$ is of the order of the electron mass, then $\tilde{m}^2/M$ is less than $1\,\text{eV}$ if M is greater than $10^3\,\text{GeV}$. So $\tilde{y}$ can be of the same order as the other Yukawa couplings, provided M is large. This is called the *seesaw mechanism* for getting small neutrino masses. The eigenfield corresponding to the smaller eigenvalue is mostly ν, and the eigenfield corresponding to the larger eigenvalue is mostly $\bar{\nu}$.

Another way to get this result is to integrate out the heavy $\bar{\nu}$ field at the beginning of our analysis. We get the leading term (in an expansion in inverse powers of M) by ignoring the kinetic energy of the $\bar{\nu}$ field, solving the equation of motion for it that follows from $\mathcal{L}_{\bar{\nu}\,\text{mass}} + \mathcal{L}_{\nu\,\text{Yuk}}$, and finally substituting the solution back into $\mathcal{L}_{\bar{\nu}\,\text{mass}} + \mathcal{L}_{\nu\,\text{Yuk}}$. The result is

$$\begin{aligned}\mathcal{L}_{\nu\text{Yuk+mass}} &= \frac{\tilde{y}^2}{2M}\Big[(\varphi^{\dagger i}\ell_i)(\varphi^{\dagger j}\ell_j) + \text{h.c.}\Big] \\ &= -\tfrac{1}{2}m_\nu(\nu\nu + \nu^\dagger\nu^\dagger)(1 + H/v)^2\,,\end{aligned} \tag{91.6}$$

where

$$m_\nu \equiv -\frac{\tilde{m}^2}{M} = -\frac{\tilde{y}^2 v^2}{2M}\,. \tag{91.7}$$

Again, we can absorb the minus sign in eq. (91.7) by making the field redefinition $\nu \to i\nu$.

The seesaw mechanism has a straightforward extension to three generations. Let us consider the fields ℓ_{iI}, $\bar{e}_I$, and $\bar{\nu}_I$, where $I = 1, 2, 3$ is a generation index. The most general Yukawa and mass terms we can write down now read

$$\mathcal{L}_{\text{Yuk+mass}} = -\varepsilon^{ij}\varphi_i \ell_{jI} y_{IJ} \bar{e}_J - \varphi^{\dagger i}\ell_{iI}\tilde{y}_{IJ}\bar{\nu}_J - \tfrac{1}{2}M_{IJ}\bar{\nu}_I\bar{\nu}_J + \text{h.c.}\,, \tag{91.8}$$

where y_{IJ} and $\tilde{y}_{IJ}$ are complex 3×3 matrices, M_{IJ} is a complex *symmetric* 3×3 matrix, and the generation indices are summed. In unitary gauge, this becomes

$$\mathcal{L}_{\text{Yuk+mass}} = -\tfrac{1}{\sqrt{2}}(v{+}H)e_I y_{IJ}\bar{e}_J - \tfrac{1}{\sqrt{2}}(v{+}H)\nu_I\tilde{y}_{IJ}\bar{\nu}_J - \tfrac{1}{2}M_{IJ}\bar{\nu}_I\bar{\nu}_J + \text{h.c.}\,. \tag{91.9}$$

We can now integrate out the $\bar{\nu}_I$ fields; eq. (91.9) is then replaced with

$$\mathcal{L}_{\text{Yuk+mass}} = -\tfrac{1}{\sqrt{2}}(v{+}H)e_I y_{IJ}\bar{e}_J - \tfrac{1}{2}(m_\nu)_{IJ}(\nu_I\nu_J + \nu_I^\dagger\nu_J^\dagger)(1{+}H/v)^2\,, \tag{91.10}$$

where we have defined the complex symmetric neutrino mass matrix

$$(m_\nu)_{IJ} \equiv -\tfrac{1}{2}v^2(\tilde{y}^{\mathrm{T}}M^{-1}\tilde{y})_{IJ}\,. \tag{91.11}$$

We can make unitary transformations in generation space on the fields: $e_I \to E_{IJ}e_J$, $\bar{e}_I \to \bar{E}_{IJ}\bar{e}_J$, and $\nu_I \to N_{IJ}\nu_J$, where E, $\bar{E}$, and N are independent unitary matrices. The kinetic terms are unchanged (except for the couplings to the $W^\pm$, as we will discuss momentarily), and the matrices y and m_ν are replaced with $E^{\mathrm{T}}y\bar{E}$ and $N^{\mathrm{T}}m_\nu N$. We can choose the unitary matrices E, $\bar{E}$, and N so that $E^{\mathrm{T}}y\bar{E}$ and $N^{\mathrm{T}}m_\nu N$ are diagonal with positive real entries y_I and m_{ν_I}. The neutrinos $\mathcal{N}_I$ then have masses m_{ν_I}, and the charged leptons $\mathcal{E}_I$ have masses $m_{e_I} = y_I v/\sqrt{2}$. In the neutral currents J_3^μ and J_{EM}^μ, we simply add a generation index I to each field, and sum over it. The charged currents are more complicated, however; they become

$$J^{+\mu} = \overline{\mathcal{E}}_{\text{L}I}(X^\dagger)_{IJ}\gamma^\mu\mathcal{N}_{\text{L}J}\,, \tag{91.12}$$

$$J^{-\mu} = \overline{\mathcal{N}}_{\text{L}I}X_{IJ}\gamma^\mu\mathcal{E}_{\text{L}J}\,, \tag{91.13}$$

where $X \equiv N^\dagger E$ is the analog in the lepton sector of the CKM matrix V in the quark sector.

One difference, though, between X and V is that the phases of the Majorana $\mathcal{N}_I$ fields are fixed by the requirement that the neutrino masses are real and positive. Thus we cannot change these phases to make the first column of X real, as we did with V. We *are* allowed to change the phases of the Dirac $\mathcal{E}_I$ fields, so we can make the first row of X real. Thus X has $9 - 3 = 6$ parameters, two more than the CKM matrix V.

The presence of X in the charged currents leads to the phenomenon of *neutrino oscillations.* A neutrino that is produced by scattering an electron off a target will be a linear combination $X_{IJ}\nu_J$ of the neutrinos of definite mass. The different mass eigenstates propagate at different speeds, and then (in a subsequent scattering) may become (if there is enough energy) muons or taus rather than electrons. It is the observation of neutrino oscillations that leads us to believe that neutrinos do, in fact, have mass.

Reference notes

Neutrino masses are discussed in detail in *Ramond II.*

Problems

91.1) Show that introducing neutrino masses via the seesaw mechanism results in lepton number no longer being conserved.

92

Solitons and monopoles

Prerequisite: 84

Consider a real scalar field φ with lagrangian

$$\mathcal{L} = -\tfrac{1}{2}\partial^\mu\varphi\partial_\mu\varphi - V(\varphi)\,, \tag{92.1}$$

with

$$V(\varphi) = \tfrac{1}{8}\lambda(\varphi^2 - v^2)^2\,. \tag{92.2}$$

As we discussed in section 30, this potential yields two ground states or vacua, corresponding to the classical field configurations $\varphi(x) = +v$ and $\varphi(x) = -v$. After shifting the field by its VEV (either $+v$ or $-v$), we find that the particle mass is $m = \lambda^{1/2}v$.

Let us consider this theory in two spacetime dimensions (one space dimension x and time t). In this case, φ and v are dimensionless, and λ has dimensions of mass squared. In the quantum theory, the coupling is weak if $\lambda \ll m^2$.

The case of one space dimension is interesting for the following reason. The boundary of one-dimensional space consists of two points, $x = -\infty$ and $x = +\infty$. This topology of the spatial boundary is mirrored by the topology of the space of vacuum field configurations, which also consists of two points, $\varphi(x) = -v$ and $\varphi(x) = +v$. In each vacuum, *both* spatial boundary points ($x = -\infty$ and $x = +\infty$) are mapped to the *same* field value (either $-v$ or $+v$). This is a *trivial map*. More interesting is the *identity map*, where $x = -\infty$ is mapped to $\varphi = -v$, and $x = +\infty$ is mapped to $\varphi = +v$. This map does *not* correspond to a vacuum; the field must smoothly interpolate between $\varphi = -v$ at $x = -\infty$ and $\varphi = +v$ at $x = +\infty$, and this requires energy. The interesting question is whether it can be done at the cost of a *finite* amount of energy.

To make these notions more precise, we will look for a minimum energy, time-independent solution of the classical field equations, with the boundary conditions

$$\lim_{x\to\pm\infty} \varphi(x) = \pm v\,. \tag{92.3}$$

The total energy is given by

$$E = \int_{-\infty}^{+\infty} dx \left[\tfrac{1}{2}\dot{\varphi}^2 + \tfrac{1}{2}\varphi'^2 + V(\varphi)\right]. \tag{92.4}$$

The solution of interest is time independent, so we can set $\dot{\varphi} = 0$. We can also rewrite the remaining terms in E as

$$\begin{aligned} E &= \int_{-\infty}^{+\infty} dx \left[\tfrac{1}{2}\Big(\varphi' - \sqrt{2V(\varphi)}\Big)^2 + \sqrt{2V(\varphi)}\,\varphi'\right] \\ &= \int_{-\infty}^{+\infty} dx\, \tfrac{1}{2}\Big(\varphi' - \sqrt{2V(\varphi)}\Big)^2 + \int_{-v}^{+v} \sqrt{2V(\varphi)}\, d\varphi \\ &= \int_{-\infty}^{+\infty} dx\, \tfrac{1}{2}\Big(\varphi' - \sqrt{2V(\varphi)}\Big)^2 + \tfrac{2}{3}(m^2/\lambda)m\,. \end{aligned} \tag{92.5}$$

Since the first term in eq. (92.5) is positive, the minimum possible energy is $M \equiv \frac{2}{3}(m^2/\lambda)m$; this is much larger than the particle mass m if the theory is weakly coupled ($\lambda \ll m^2$). Requiring the first term in eq. (92.5) to vanish yields $\varphi' = \sqrt{2V(\varphi)}$, which is easily integrated to get

$$\varphi(x) = v \tanh\Big(\tfrac{1}{2}m(x - x_0)\Big)\,, \tag{92.6}$$

where x_0 is a constant of integration. The energy density is localized near $x = x_0$, and goes to zero exponentially fast for $|x - x_0| > 1/m$.

This solution is a *soliton*, a solution of the classical field equations with an energy density that is localized in space, and that does not dissipate or change its shape with time. In this case (and in all cases of interest to us), its existence is related to the topology of the boundary of space and the topology of the set of vacua, and the existence of a nontrivial map from the boundary of space to the set of vacua.

Given eq. (92.6), we can get other soliton solutions by making a Lorentz boost; these solutions take the form

$$\varphi(x,t) = v \tanh\Big(\tfrac{1}{2}\gamma m(x - x_0 - \beta t)\Big)\,, \tag{92.7}$$

where $\gamma = (1-\beta^2)^{-1/2}$; their energy is $E = \gamma M = (p^2 + M^2)^{1/2}$, where $M = \frac{2}{3}(m^2/\lambda)m$ is the energy of the soliton at rest, and $p = \gamma\beta M$ is the momentum of the soliton, found by integrating the momentum density $T^{01} = \varphi'\partial^0\varphi$.

We see that the soliton behaves very much like a particle. We may expect that, in the quantum theory, the soliton will correspond to a new species of particle with mass M, in addition to the elementary field excitation with mass m.

The soliton solution is still interesting if there is more than one spatial dimension. In that case, eq. (92.6) describes a *domain wall*, a structure that is localized in one particular spatial direction, but extended in the others. The wall has a surface tension (energy per unit transverse area) given by $\sigma = \frac{2}{3}m^3/\lambda$.

Having found a theory that has a soliton that is localized in *one* spatial direction, let us try to find a theory that has a soliton that is localized in *two* spatial directions. In two space dimensions, the spatial boundary has the topology of a circle, denoted by the symbol S^1. There is no smooth nontrivial map from a circle to two points; continuity of the map requires the entire circle to be mapped into one of the two points. But there do exist smooth nontrivial maps from one circle to another circle, as we will discuss momentarily.

So, we would like to find a theory whose vacua have the topology of a circle. To this end, let us consider a complex scalar field $\varphi(x)$, with lagrangian

$$\mathcal{L} = -\partial^\mu\varphi^\dagger\partial_\mu\varphi - V(\varphi)\,, \tag{92.8}$$

where

$$V(\varphi) = \tfrac{1}{4}\lambda(\varphi^\dagger\varphi - v^2)^2\,. \tag{92.9}$$

The vacuum field configurations are

$$\varphi(x) = ve^{i\alpha}\,, \tag{92.10}$$

where α is an arbitrary angle. This angle specifies a point on a circle, and so the space of vacua does indeed have the topology of S^1.

Let us write $\mathbf{x} = r(\cos\phi, \sin\phi)$; then the angle ϕ specifies a point on the spatial circle at infinity. We can specify a map from the spatial circle to the vacuum circle by giving α as a function of ϕ. In order for $\varphi(x)$ to be single valued, this function must obey $\alpha(\phi + 2\pi) = \alpha(\phi) + 2\pi n$, where the integer n is the *winding number* of the map: we wind around the vacuum circle n times for every one time that we wind around the spatial circle. (If n is negative, the vacuum winding is opposite in direction to the spatial winding.) An example of a map with winding number n is $U(\phi) \equiv e^{i\alpha(\phi)} = e^{in\phi}$. Setting

$n = 0$ then yields the trivial map, $n = 1$ the identity map, and $n = -1$ the inverse of the identity map.

Given a smooth map $U(\phi)$, its winding number can be written as

$$n = \frac{i}{2\pi} \int_0^{2\pi} d\phi \, U \partial_\phi U^\dagger \,, \tag{92.11}$$

where $U^\dagger$ is the complex conjugate of U. To verify that eq. (92.11) agrees with our previous definition, we first check that plugging in our example map indeed yields the correct value of the winding number. We then show that the right-hand side of eq. (92.11) is invariant under smooth deformations of $U(\phi)$; see problem 92.2. Thus any $U(\phi)$ that can be smoothly deformed to $e^{in\phi}$ has winding number n.

Next, we want to look for a finite-energy solution of the classical field equations for the theory specified by eqs. (92.8) and (92.9), with the boundary condition

$$\lim_{r \to \infty} \varphi(r, \phi) = vU(\phi) \,, \tag{92.12}$$

with $U(\phi)$ corresponding to a map with nonzero winding number. We therefore make the ansatz

$$\varphi(r, \phi) = v f(r) e^{in\phi} \,, \tag{92.13}$$

where $f(r)$ is a real function that obeys $f(\infty) = 1$. We must also have $f(0) = 0$ so that $\nabla\varphi(r, \phi)$ is well defined at $r = 0$.

Alas, it is easy to see that there is no finite-energy solution of this form. The gradient of the field is

$$\nabla\varphi = v \Big[f'(r)\hat{r} + inr^{-1} f(r)\hat{\phi} \Big] e^{in\phi} \,, \tag{92.14}$$

and the gradient energy density is

$$|\nabla\varphi|^2 = v^2 \Big[f'(r)^2 + n^2 r^{-2} f(r)^2 \Big] \,. \tag{92.15}$$

At large r, $f(r)$ must approach one; then the integral over the second term in eq. (92.15) diverges logarithmically,

$$\int d^2x \, |\nabla\varphi|^2 \sim 2\pi n^2 v^2 \int^\infty \frac{dr}{r} \,. \tag{92.16}$$

So the energy is infinite. This is, in fact, a very general result, known as *Derrick's theorem*: with scalar fields only, there are no finite-energy, time-independent solitons that are localized in more than one dimension; see problem 92.1. The problem is that the gradient energy always diverges at large distances from the putative soliton's core.

To get solitons that are localized in more than one dimension, we must introduce gauge fields. Note that the lagrangian of eq. (92.8) has a global U(1) symmetry. Let us gauge this U(1) symmetry, so that the lagrangian becomes

$$\mathcal{L} = -(D^\mu\varphi)^\dagger D_\mu\varphi - V(\varphi) - \tfrac{1}{4}F^{\mu\nu}F_{\mu\nu}\ , \tag{92.17}$$

where

$$D_\mu\varphi = \partial_\mu\varphi - ieA_\mu\varphi\ , \tag{92.18}$$

and $V(\varphi)$ is still given by eq. (92.9). The gauge symmetry is therefore spontaneously broken, and the mass of the vector particle is $m_{\rm V} = ev$. The mass of the scalar particle is $m_{\rm S} = \lambda^{1/2}v$.

The gradient energy density of the scalar field is now

$$|\vec{D}\varphi|^2 = |(\nabla - ie\mathbf{A})\varphi|^2\ . \tag{92.19}$$

Thus we have the opportunity to choose $\mathbf{A}$ so as to partially cancel the badly behaved second term in eq. (92.14). To see how to do this, recall that a gauge transformation in this theory takes the form

$$\varphi \to U\varphi\ , \tag{92.20}$$

$$A_\mu \to UA_\mu U^\dagger + \tfrac{i}{e}U\partial_\mu U^\dagger\ , \tag{92.21}$$

where U is a 1×1 unitary matrix that is a function of spacetime. As $r\to\infty$, our ansatz for φ, eq. (92.13), corresponds to a gauge transformation of a vacuum, $\varphi = v$, by $U = e^{in\phi}$. The corresponding transformation of $A_\mu = 0$ is

$$\begin{aligned}\lim_{r\to\infty}\mathbf{A}(r,\phi) &= \tfrac{i}{e}U\nabla U^\dagger \\ &= \frac{n}{er}\,\hat{\phi}\ .\end{aligned} \tag{92.22}$$

Before making the gauge transformation, we have $\varphi = v$ and $A_\mu = 0$, and so $D_\mu\varphi = 0$; by gauge invariance, this must be true after the transformation as well. Indeed, it is easy to check that, with $\mathbf{A}$ given by eq. (92.22), we have $(\nabla - ie\mathbf{A})ve^{i\phi} = 0$.

For $n \neq 0$, the gauge transformation $U = e^{in\phi}$ is *large*: it cannot be smoothly deformed to $U = 1$. This implies that we cannot extend it from $r = \infty$ into the interior of space without meeting an *obstruction*, a point where $U(r,\phi)$ is ill defined. For example, the simplest attempt at such an extension, $U(r,\phi) = e^{in\phi}$, is ill defined at $r = 0$. Near the obstruction, the fields φ and $\mathbf{A}$ must deviate from a gauge transformation of a vacuum. This deviation costs energy, and results in a soliton.

Our ansatz for a soliton in the theory specified by eq. (92.17) is then

$$\varphi(r,\phi) = vf(r)U(\phi)\ , \tag{92.23}$$

$$\mathbf{A}(r,\phi) = \tfrac{i}{e}a(r)U(\phi)\nabla U^\dagger(\phi)\ , \tag{92.24}$$

where $U(\phi) = e^{in\phi}$, and we require $f(\infty) = a(\infty) = 1$ (so that the solution approaches a large gauge transformation of a vacuum as $r \to \infty$) and $f(0) = a(0) = 0$ (so that $\mathbf{A}$ and $\nabla\varphi$ are well defined at $r = 0$). For $n = 1$, this soliton is a *Nielsen–Olesen vortex.*

The nonzero vector potential results in a perpendicular magnetic field

$$\begin{aligned}\mathbf{B} &= \nabla \times \mathbf{A} \\ &= \frac{1}{r}\left(\frac{\partial}{\partial r}(rA_\phi) - \frac{\partial}{\partial \phi}A_r\right)\hat{z} \\ &= \frac{n}{e}\,\frac{a'(r)}{r}\,\hat{z}\ .\end{aligned} \tag{92.25}$$

The corresponding magnetic flux is

$$\begin{aligned}\Phi &= \int d\mathbf{S}\cdot\mathbf{B} \\ &= \lim_{r\to\infty}\int d\boldsymbol{\ell}\cdot\mathbf{A} \\ &= \frac{i}{e}\lim_{r\to\infty} a(r)\int_0^{2\pi} d\phi\, U\partial_\phi U^\dagger \\ &= \frac{2\pi n}{e}\ .\end{aligned} \tag{92.26}$$

Here the second line follows from Stokes' theorem, the third from eq. (92.24), and the fourth from eq. (92.11).

The energy of the soliton is

$$E = \int d^2x\left[|(\nabla - ie\mathbf{A})\varphi|^2 + V(\varphi) + \tfrac{1}{2}\mathbf{B}^2\right]. \tag{92.27}$$

Substituting in our ansatz, eqs. (92.23) and (92.24), we get

$$E = 2\pi v^2\int_0^\infty dr\, r\left[f'^2 + \frac{n^2}{r^2}(a-1)^2f^2 + \tfrac{1}{4}\lambda v^2(f^2-1)^2 + \frac{n^2}{e^2v^2r^2}\,a'^2\right]. \tag{92.28}$$

It is convenient to define a dimensionless radial coordinate $\rho \equiv evr = m_{\rm V} r$. Let us also define $\beta^2 \equiv \lambda/e^2 = m_{\rm S}^2/m_{\rm V}^2$. Then eq. (92.28) becomes

$$E = 2\pi v^2 \int_0^\infty d\rho\,\rho \left[f'^2 + \frac{n^2}{\rho^2}(a-1)^2 f^2 + \tfrac{1}{4}\beta^2 (f^2-1)^2 + \frac{n^2}{\rho^2} a'^2 \right] , \qquad (92.29)$$

where a prime now denotes a derivative with respect to ρ. We can find the equations obeyed by $f(\rho)$ and $a(\rho)$ either by substituting the ansatz into the equations of motion, or by applying the variational principle directly to eq. (92.29). Either way, the result is

$$f'' + \frac{f'}{\rho} - \frac{n^2 f}{\rho^2}(1-a)^2 + \tfrac{1}{2}\beta^2(1-f^2)f = 0 , \qquad (92.30)$$

$$a'' - \frac{a'}{\rho} + (1-a)f^2 = 0 , \qquad (92.31)$$

with the boundary conditions $a(0) = f(0) = 0$ and $a(\infty) = f(\infty) = 1$.

Eqs. (92.30) and (92.31) have no closed-form solution. However, for $\rho \ll 1$, we can show that $a(\rho) \sim \rho^2$ and $f(\rho) \sim \rho^n$; and for $\rho \gg 1$, that $1 - a(\rho) \sim e^{-\rho}$ and $1 - f(\rho) \sim e^{-c\rho}$, where $c = \min(\beta, 2)$; see problem 92.4. For n and β of order one, the integral in eq. (92.29) also results in a number of order one, and so we have $E \sim 2\pi v^2$. For $\beta > 1$, it is possible to prove a *Bogomolny bound*, $E > 2\pi v^2|n|$. In this case, a soliton with winding number n is unstable against breaking up into $|n|$ solitons, each with winding number one (or minus one, if n is negative).

Once we have our soliton solution, we can translate and/or boost it; thus we expect the soliton to behave like a particle in two space dimensions. In three space dimensions, the soliton becomes a *Nielsen–Olesen string* (also called a *gauge string*), a structure that is localized in two directions, but extended in the third. Such strings can bend, and even form closed loops. In certain unified theories (see section 97), gauge strings may have formed in the early universe; they are then called *cosmic strings.*

Now let us try to find a soliton that is localized in *three* spatial directions. In three space dimensions, the spatial boundary has the topology of a two-dimensional sphere S^2. There are smooth nontrivial maps from S^2 to S^2, as we will discuss momentarily, so let us look for a theory whose vacua have the topology of S^2.

Consider three real scalar fields φ^a, $a = 1, 2, 3$, with lagrangian

$$\mathcal{L} = -\tfrac{1}{2}\partial^\mu \varphi^a \partial_\mu \varphi^a - V(\varphi) , \qquad (92.32)$$

where

$$V(\varphi) = \tfrac{1}{8}\lambda(\varphi^a\varphi^a - v^2)^2 . \qquad (92.33)$$

The vacuum field configurations are

$$\varphi^a(x) = v\hat{\varphi}^a \ , \tag{92.34}$$

where $\hat{\varphi}$ is an arbitrary unit vector. This unit vector specifies a point on a two-sphere, and so the space of vacua does indeed have the topology of S^2.

Let us write $\mathbf{x} = r(\sin\theta\cos\phi, \sin\theta\sin\phi, \cos\theta)$; then the polar and azimuthal angles θ and ϕ specify a point on the spatial two-sphere at infinity. We can specify a map from the spatial two-sphere to the vacuum two-sphere by giving $\hat{\varphi}$ as an (appropriately periodic) function of θ and ϕ. We can define a winding number n that counts the number of times the vacuum two-sphere covers the spatial two-sphere, with n negative if the orientation is reversed. An example of a map with winding number n can be constructed by taking the polar angle of $\hat{\varphi}$ to be θ, and the azimuthal angle to be $n\phi$. Setting $n = 1$ then yields the identity map, and $n = -1$ the inverse of the identity map.

Given a smooth map $\hat{\varphi}_a(\theta, \phi)$, its winding number can be written as

$$n = \frac{1}{8\pi}\int d^2\theta\, \varepsilon^{abc}\varepsilon^{ij}\,\hat{\varphi}^a\partial_i\hat{\varphi}^b\partial_j\hat{\varphi}^c \ , \tag{92.35}$$

where $d^2\theta = d\theta\, d\phi$, $\partial_1 = \partial/\partial\theta$, $\partial_2 = \partial/\partial\phi$, and $\varepsilon^{12} = -\varepsilon^{21} = +1$. To verify that eq. (92.35) agrees with our previous definition, we first check that plugging in our example map indeed yields the correct value of the winding number; see problem 92.5. We then show that the right-hand side of eq. (92.35) is invariant under smooth deformations of $\hat{\varphi}_a(\theta, \phi)$; see problem 92.6. It is also worthwhile to note that the right-hand side of eq. (92.35) is invariant under a change of coordinates, because the jacobian for $d^2\theta$ is canceled by the jacobian for $\partial_1\partial_2$. This is of course closely related to the invariance under smooth deformations, since one way to make such a deformation is via a coordinate change.

Next, we want to look for a finite-energy solution of the classical field equations with nonzero winding number, but we already know that these will not exist unless we introduce gauge fields. We therefore take φ^a to be in the adjoint representation of an SU(2) gauge group. The lagrangian is now

$$\mathcal{L} = -\tfrac{1}{2}(D^\mu\varphi)^a(D_\mu\varphi)^a - V(\varphi) - \tfrac{1}{4}F^{a\mu\nu}F^a_{\mu\nu} \ , \tag{92.36}$$

where

$$(D_\mu\varphi)^a = \partial_\mu\varphi^a + e\varepsilon^{abc}A^b_\mu\varphi^c \ , \tag{92.37}$$

$$F^a_{\mu\nu} = \partial_\mu A^a_\nu - \partial_\nu A^a_\mu + e\varepsilon^{abc}A^b_\mu A^c_\nu \ , \tag{92.38}$$

and $V(\varphi)$ is given by eq. (92.33). We have called the gauge coupling e for reasons that will become clear in a moment.

The gauge symmetry is spontaneously broken to U(1). If we take the vacuum field configuration to be $\varphi^a = v\delta^{a3}$, then the A^3_μ field remains massless; we will think of it as the electromagnetic field. The complex vector fields $W^{\pm}_\mu = (A^1_\mu \mp iA^2_\mu)/\sqrt{2}$ get a mass $m_{\rm W} = ev$, and have electric charge $\pm e$. (This is the reason for calling the gauge coupling e.) This theory, known as the *Georgi–Glashow model*, was once considered as an alternative to the Standard Model of electroweak interactions (but is now ruled out, because it does not have a Z^0 boson).

When the vacuum field configuration is $\varphi^a = v\delta^{a3}$, the electromagnetic field strength is $F_{\mu\nu} = \partial_\mu A^3_\nu - \partial_\nu A^3_\mu$. We can write down a gauge-invariant expression that reduces to $F_{\mu\nu}$ when we set $\varphi^a = v\delta^{a3}$; this expression is

$$F_{\mu\nu} = \hat{\varphi}^a F^a_{\mu\nu} - e^{-1}\varepsilon^{abc}\hat{\varphi}^a (D_\mu\hat{\varphi})^b (D_\nu\hat{\varphi})^c \,. \tag{92.39}$$

Here $\hat{\varphi}^a = \varphi^a/|\varphi|$, where $|\varphi| = (\varphi^a\varphi^a)^{1/2}$. We can, in fact, use eq. (92.39) as the definition of the electromagnetic field strength at any spacetime point where $|\varphi| \neq 0$. (If $|\varphi| = 0$, the SU(2) symmetry is unbroken, and there is no gauge-invariant way to pick out a particular component of the nonabelian field strength $F^a_{\mu\nu}$.) If we substitute in eqs. (92.37) and (92.38), and make repeated use of $\hat{\varphi}^a\hat{\varphi}^a = 1$ and the identity $\varepsilon^{abc}\varepsilon^{ade} = \delta^{bd}\delta^{ce} - \delta^{be}\delta^{cd}$, it is possible to rewrite eq. (92.39) as

$$F_{\mu\nu} = \partial_\mu(\hat{\varphi}^a A^a_\nu) - \partial_\nu(\hat{\varphi}^a A^a_\mu) - e^{-1}\varepsilon^{abc}\hat{\varphi}^a \partial_\mu\hat{\varphi}^b \partial_\nu\hat{\varphi}^c \,. \tag{92.40}$$

In particular, the magnetic field is

$$\begin{aligned} B^i &= \tfrac{1}{2}\varepsilon^{ijk}F_{jk} \\ &= \varepsilon^{ijk}\partial_j(\hat{\varphi}^a A^a_k) - (2e)^{-1}\varepsilon^{ijk}\varepsilon^{abc}\hat{\varphi}^a \partial_j\hat{\varphi}^b \partial_k\hat{\varphi}^c \,. \end{aligned} \tag{92.41}$$

Let us consider the magnetic flux through a sphere at spatial infinity; this is given by $\Phi = \int d\mathbf{S}\cdot\mathbf{B}$, where $dS_k = r^2 \sin\theta \, d\theta \, d\phi \, \hat{x}_k$, and $\hat{x} = \mathbf{x}/r$ is a radially outward unit vector. The first term in eq. (92.41) for $\mathbf{B}$ is $\nabla \times (\hat{\varphi}^a \mathbf{A}^a)$; since this is a curl, it has zero divergence, and therefore zero surface integral. From eq. (92.35), we see that the second term in eq. (92.41) results in

$$\Phi = -\frac{4\pi n}{e} \,. \tag{92.42}$$

This flux implies that any soliton with nonzero winding number is a *magnetic monopole* with magnetic charge $Q_{\rm M} = \Phi$. (In Heaviside–Lorentz units, the Coulomb field of an electric point charge $Q_{\rm E}$ is $E_i = Q_{\rm E}\hat{x}_i/4\pi r^2$, and so the total electric flux is $Q_{\rm E}$. We adopt the same convention for magnetic charge.)

If we add a field in the fundamental representation of SU(2), then the component fields have electric charges $\pm\frac{1}{2}e$. This is the smallest electric charge we can get, and all possible electric charges are integer multiples of it. Eq. (92.42) tells us that all possible magnetic charges are integer multiples of $4\pi/e$. Thus the possible electric and magnetic charges obey the *Dirac charge quantization condition*, which is

$$Q_{\rm E}Q_{\rm M} = 2\pi k\ , \tag{92.43}$$

where k is an integer. This condition can be derived from general considerations of the quantum properties of monopoles.

Now let us turn to the explicit construction of a soliton solution. The simplest case to consider is provided by the identity map (which has winding number $n=1$); the soliton we will find is the *'t Hooft–Polyakov monopole.*

The boundary condition on the scalar field is

$$\lim_{r\to\infty} \varphi^a(\mathbf{x}) = vx^a/r\ . \tag{92.44}$$

We can find the appropriate boundary condition on the gauge field by requiring $(D_\mu\varphi)^a = 0$ in the limit of large r. This condition yields

$$\partial_i(x^a/r) + e\varepsilon^{abc}A_i^b x^c/r = 0\ . \tag{92.45}$$

We have $\partial_i(x^a/r) = (r^2\delta_{ai} - x_a x_i)/r^3$. Next we multiply by $rx_j\varepsilon^{jda}$, and use the identity $\varepsilon^{jda}\varepsilon^{abc} = \delta^{jb}\delta^{dc} - \delta^{jc}\delta^{db}$ to get

$$\varepsilon^{dij}x_j + e(x^d x_j A_i^j - r^2 A_i^d) = 0\ . \tag{92.46}$$

If we ignore the first term in the parentheses, we find $A_i^d = \varepsilon^{dij}x_j/er^2$. But then $x_d A_i^d = 0$, and so the first term in the parentheses vanishes. Thus we have found the needed asymptotic behavior of A_i^a. Our ansatz is therefore

$$\varphi^a(\mathbf{x}) = vf(r)x^a/r\ , \tag{92.47}$$

$$A_i^a(\mathbf{x}) = a(r)\varepsilon^{aij}x_j/er^2\ . \tag{92.48}$$

We require $f(\infty) = a(\infty) = 1$ (so that A_i^a and φ^a have the desired asymptotic limits) and $f(0) = a(0) = 0$ (so that A_i^a and φ^a are well defined at $r=0$).

The total energy of the soliton (which we will call M, because it is the mass of the monopole) is given by

$$M = \int d^3x \left[\tfrac{1}{2}B_i^a B_i^a + \tfrac{1}{2}(D_i\varphi)^a(D_i\varphi)^a + V(\varphi)\right] . \tag{92.49}$$

The nonabelian magnetic field is

$$B_i^a = \tfrac{1}{2}\varepsilon_{ijk}F_{jk}^a$$
$$= \varepsilon_{ijk}\partial_j A_k^a + \tfrac{1}{2}e\varepsilon_{ijk}\varepsilon^{abc}A_j^b A_k^c\,. \tag{92.50}$$

If we write eq. (92.48) as $A_i^a = \varepsilon^{aij}K_j$, then after some manipulation we find that eq. (92.50) becomes $B_i^a = \partial_a K_i - \delta_{ai}\partial_j K^j + eK_a K_i$. Plugging in $K_i = a(r)x_i/er^2$ then yields

$$B_i^a = -\frac{1}{e}\left[\frac{a'}{r}\Big(\delta_{ai} - \hat{x}_a\hat{x}_i\Big) + \frac{2a - a^2}{r^2}\,\hat{x}_a\hat{x}_i\right]. \tag{92.51}$$

The magnetic field energy then becomes

$$\tfrac{1}{2}B_i^a B_i^a = \frac{1}{2e^2r^4}\big[2r^2a'^2 + (2a - a^2)^2\big]\,. \tag{92.52}$$

The covariant derivative of the scalar field is

$$(D_i\varphi)^a = v\left[\frac{(1-a)f}{r}\Big(\delta_{ai} - \hat{x}_a\hat{x}_i\Big) + f'\,\hat{x}_a\hat{x}_i\right]. \tag{92.53}$$

The scalar gradient energy density then becomes

$$\tfrac{1}{2}(D_i\varphi)^a(D_i\varphi)^a = \frac{v^2}{2r^2}\big[2(1-a)^2f^2 + r^2f'^2\big]\,. \tag{92.54}$$

The scalar potential energy density is

$$V(\varphi) = \tfrac{1}{8}\lambda v^4(f^2 - 1)^2\,. \tag{92.55}$$

We can plug eqs. (92.52), (92.54), and (92.55) into eq. (92.49), and then use the variational principle to get the second-order differential equations obeyed by $f(r)$ and $a(r)$.

We can get a lower bound on M by performing a trick analogous to the one we used in eq. (92.5). We write

$$\tfrac{1}{2}B_i^a B_i^a + \tfrac{1}{2}(D_i\varphi)^a(D_i\varphi)^a = \tfrac{1}{2}[B_i^a + (D_i\varphi)^a]^2 - B_i^a(D_i\varphi)^a\,. \tag{92.56}$$

We can apply the distribution rule for covariant derivatives (see problem 70.5) to rewrite the last term as $B_i^a(D_i\varphi)^a = \partial_i(B_i^a\varphi^a) - (D_iB_i)^a\varphi^a$. Then we note that the Bianchi identity (see problem 70.6) implies $(D_iB_i)^a = 0$. Thus $B_i^a(D_i\varphi)^a = \partial_i(B_i^a\varphi^a)$, and this is a total divergence. Then Gauss's theorem yields

$$\int d^3x\,\partial_i(B_i^a\varphi^a) = \int dS_i\,B_i^a\varphi^a\,, \tag{92.57}$$

where the integral is over the surface at spatial infinity. On this surface, we have $\varphi^a = v\hat{x}^a$.

Next we use eq. (92.39). At spatial infinity, the covariant derivatives of φ vanish; thus we have $B_i^a \varphi^a = vB_i$, where B_i is the magnetic field of electromagnetism. We can now see that the right-hand side of eq. (92.57) evaluates to $v\Phi$, where $\Phi = Q_{\rm M} = -4\pi n/e$ is the magnetic charge of the monopole.

In our case, $n = 1$ and $Q_{\rm M}$ is negative; thus the last term in eq. (92.56) integrates to $v|Q_{\rm M}|$. (For the case of positive $Q_{\rm M}$, we can swap the plus and minus signs in eq. (92.56) to get the same result.) Thus the mass of the monopole, eq. (92.49), can be written as

$$M = \frac{4\pi|n|v}{e} + \int d^3x \left[\tfrac{1}{2}[B_i^a + (\operatorname{sign} n)(D_i\varphi)^a]^2 + V(\varphi)\right] . \qquad (92.58)$$

Both terms in the integrand of eq. (92.58) are positive, and so we have a *Bogomolny bound* on the mass of the monopole. For $\lambda > 0$, a monopole with winding number n is unstable against breaking up into $|n|$ monopoles, each with winding number one (or minus one, if n is negative).

Using $m_{\rm W} = ev$ and $\alpha = e^2/4\pi$, we can write the Bogomolny bound as

$$M \geq \frac{m_{\rm W}}{\alpha}|n| . \qquad (92.59)$$

Since $\alpha \ll 1$, the monopole is much heavier than the W boson.

Alas, the Georgi–Glashow model, which has monopole solutions, is not in accord with nature, while the Standard Model, which is in accord with nature, does not have monopole solutions. This is because, in the Standard Model, electric charge is a linear combination of an SU(2) generator and the U(1) hypercharge generator. Nothing prevents us from introducing an SU(2) singlet field with an arbitrarily small hypercharge. Such a field would have an arbitrarily small electric charge (in units of e), and then the Dirac charge quantization condition would preclude the existence of magnetic monopoles.

This disappointing situation is remedied in unified theories (see section 97), where the gauge group has a single nonabelian factor like SU(5). In unified theories, the monopole mass is of order $m_{\rm X}/\alpha$, where $m_{\rm X}$ is the mass of a superheavy vector boson; typically $m_{\rm X} \sim 10^{15}\,$GeV.

Returning to the Georgi–Glashow model, we can saturate the Bogomolny bound if we consider the formal limit of $\lambda \to 0$; then $V(\varphi)$ vanishes. (This limit is formal because we need a nonzero potential to fix the magnitude of φ at infinity.) Then we saturate the bound if $B_i^a = -(\operatorname{sign} n)(D_i\varphi)^a$. In the case of the 't Hooft–Polyakov monopole, we have $n = 1$, and B_i^a and $(D_i\varphi)^a$ are given by eqs. (92.51) and (92.53). Matching the coefficients of $\delta_{ai} - \hat{x}_a\hat{x}_i$ and $\hat{x}_a\hat{x}_i$ yields a pair of first-order differential equations. These look nicer if we introduce the dimensionless radial coordinate $\rho \equiv evr = m_{\rm W}r$; then we find

$$a' = (1-a)f \;, \tag{92.60}$$

$$f' = (2a - a^2)/\rho^2 \;, \tag{92.61}$$

where a prime now denotes a derivative with respect to ρ. These equations have a closed-form solution,

$$a(\rho) = 1 - \frac{\rho}{\sinh \rho} \;, \tag{92.62}$$

$$f(\rho) = \coth \rho - \frac{1}{\rho} \;. \tag{92.63}$$

This is the *Bogomolny–Prasad–Sommerfeld* (or BPS for short) solution. A soliton that saturates a Bogomolny bound is generically called a *BPS soliton.*

Reference notes

Discussions of solitons, and their relation to the theory of homotopy groups, can be found in *Coleman* and *Weinberg II.*

Problems

92.1) *Derrick's theorem* says that, in a theory with scalar fields only, there are no solitons localized in more than one dimension. To prove this, consider a theory in D space dimensions with a set of real scalar fields φ_i; any complex scalar fields are written as a pair of real ones. The lagrangian is $\mathcal{L} = -\frac{1}{2}\partial^\mu\varphi_i\partial_\mu\varphi_i - V(\varphi_i)$, with $V(\varphi_i) \ge 0$. Suppose we have a soliton solution $\varphi_i(\mathbf{x})$; its energy is $E = T + U$, where $T = \frac{1}{2}\int d^D x\,(\nabla\varphi_i)^2$ and $U = \int d^D x\, V(\varphi_i)$.

a) Now consider $\varphi_i(\mathbf{x}/\alpha)$, where α is a positive real number. Show that, for this field configuration, the energy is $E(\alpha) = \alpha^{D-2}T + \alpha^D U$.

b) Argue that we must have $E'(1) = 0$.

c) Use this to prove the theorem.

92.2) The winding number n for a map from $\mathrm{S}^1 \to \mathrm{S}^1$ is given by eq. (92.11), where $U^\dagger U = 1$. We will prove that n is invariant under an infinitesimal deformation of U. Since any smooth deformation can be made by compounding infinitesimal ones, this will prove that n is invariant under any smooth deformation.

a) Consider an infinitesimal deformation of U, $U \to U + \delta U$. Show that $\delta U^\dagger = -U^{\dagger 2}\delta U$.

b) Use this result to show that $\delta(U\partial_\phi U^\dagger) = -\partial_\phi(U^\dagger \delta U)$.

c) Use this to show that $\delta n = 0$.

92.3) Show that if $U_n(\phi)$ and $U_k(\phi)$ are maps from $S^1 \to S^1$ with winding numbers n and k, then $U_n(\phi)U_k(\phi)$ is a map with winding number $n+k$. Hint: consider smoothly deforming $U_n(\phi)$ to equal one for $0 \le \phi \le \pi$. How should $U_k(\phi)$ be deformed?

92.4) Verify the statements made about the solutions to eqs. (92.30) and (92.31) in the limit of large and small ρ.

92.5) Use eq. (92.35) to compute the winding number for the map specified by $\hat{\varphi} = (\sin\theta\cos n\phi, \sin\theta\sin n\phi, \cos\theta)$.

92.6) The winding number n for a map from $S^2 \to S^2$ is given by eq. (92.35), where $\hat{\varphi}^a\hat{\varphi}^a = 1$. We will prove that n is invariant under an infinitesimal deformation of $\hat{\varphi}$. Since any smooth deformation can be made by compounding infinitesimal ones, this will prove that n is invariant under any smooth deformation.

a) Consider an infinitesimal deformation of $\hat{\varphi}$, $\hat{\varphi} \to \hat{\varphi} + \delta\hat{\varphi}$. Show that $\hat{\varphi}\cdot\delta\hat{\varphi} = 0$ and that $\hat{\varphi}\cdot\partial_i\hat{\varphi} = 0$.

b) Use these results to show that $\varepsilon^{abc}\delta\hat{\varphi}^a\partial_i\hat{\varphi}^b\partial_j\hat{\varphi}^c = 0$.

c) Use this to show that $\delta n = 0$.

93

Instantons and theta vacua

Prerequisite: 92

Consider SU(2) gauge theory, with gauge fields only. The classical field configuration corresponding to the ground state is $F^a_{\mu\nu} = 0$. This implies that the vector potential A^a_μ is a gauge transformation of zero, $A_\mu = A^a_\mu T^a = \frac{i}{g} U\partial_\mu U^\dagger$.

Let us restrict our attention to gauge transformations that are time independent, $U = U(\mathbf{x})$. This fixes *temporal gauge*, $A_0 = 0$. We will also impose the boundary condition that $U(\mathbf{x})$ approaches a particular constant matrix as $|\mathbf{x}| \to \infty$, independent of direction. This is equivalent to adding a spatial "point at infinity" where U has a definite value; space then has the topology of a three-dimensional sphere S^3.

Can every $U(\mathbf{x})$ be smoothly deformed into every other $U(\mathbf{x})$? If the answer is *yes*, then all these field configurations are gauge equivalent, and they correspond to a single quantum vacuum state. If the answer is *no*, then there must be more than one quantum vacuum state. To see why, suppose that $U(\mathbf{x})$ and $\tilde{U}(\mathbf{x})$ cannot be smoothly deformed into each other. The associated vector potentials, $A_\mu = \frac{i}{g} U\partial_\mu U^\dagger$ and $\tilde{A}_\mu = \frac{i}{g}\tilde{U}\partial_\mu \tilde{U}^\dagger$, are both gauge transformations of zero, and so both $F_{\mu\nu}$ and $\tilde{F}_{\mu\nu}$ vanish. However, if we try to smoothly deform A_μ into $\tilde{A}_\mu$, we must pass through vector potentials that are *not* gauge transformations of zero, and whose field strengths therefore do *not* vanish. These nonzero field strengths imply nonzero energy: there is an energy barrier between the field configurations A_μ and $\tilde{A}_\mu$. Therefore, they represent two different minima of the hamiltonian in the space of classical field configurations. Different minima in the space of classical field configurations correspond to different vacuum states in the quantum theory.

It turns out that every $U(\mathbf{x})$ can *not* be smoothly deformed into every other $U(\mathbf{x})$; the field configurations specified by $U(\mathbf{x})$ are classified by a winding number. To see this, we first note that any 2×2 special unitary matrix U can be written in the form

$$U = a_4 + i\vec{a}\cdot\vec{\sigma} \,, \tag{93.1}$$

where a_4 and the three-vector $\vec{a}$ are real, and

$$\vec{a}^2 + a_4^2 = 1 \,; \tag{93.2}$$

see problem 93.1. Thus $a_\mu \equiv (\vec{a}, a_4)$ specifies a euclidean four-vector of unit length, $a_\mu a_\mu = 1$, and hence a point on a three-sphere. We will call this the vacuum three-sphere. Since our boundary conditions give space the topology of a three-sphere, $U(\mathbf{x})$ provides a map from the spatial three-sphere to the vacuum three-sphere. We can define a winding number n that counts the number of times the vacuum three-sphere covers the spatial three-sphere, with n negative if the orientation is reversed.

It is convenient to specify the spatial three-sphere by a euclidean four-vector $z_\mu \equiv (\vec{z}, z_4)$ of unit length, $z_\mu z_\mu = 1$. An explicit relation between z_μ and $\mathbf{x}$ can be constructed by (for example) stereographic projection: we take $\hat{z} = \vec{z}/|\vec{z}| = \hat{x}$, and $|\vec{z}| = 2r/(1{+}r^2)$, $z_4 = (1{-}r^2)/(1{+}r^2)$, where $r = |\mathbf{x}|$. Then we can construct an example of a map from the spatial S^3 to the vacuum S^3 with winding number n by taking the two polar angles of a_μ to be equal to the two polar angles of z_μ, and the azimuthal angle of a_μ to be equal to n times the azimuthal angle of z_μ. (The polar angles run from 0 to π, and the azimuthal angle from 0 to 2π.)

Given a smooth map $U(\mathbf{x})$, its winding number can be written as

$$n = \frac{-1}{24\pi^2} \int d^3x \, \varepsilon^{ijk} \, \mathrm{Tr}[(U\partial_i U^\dagger)(U\partial_j U^\dagger)(U\partial_k U^\dagger)] \,. \tag{93.3}$$

Here we have used the original $\mathbf{x}$ coordinates, but we could also use the angles that specify z_μ: the integral in eq. (93.3) is invariant under a change of coordinates, because the jacobian for d^3x is canceled by the jacobian for $\partial_1\partial_2\partial_3$. To verify that eq. (93.3) agrees with our previous definition, we first check that plugging in our example map indeed yields the correct value of the winding number; see problem 93.5. We then show that the right-hand side of eq. (93.3) is invariant under smooth deformations of $U(\mathbf{x})$; see problem 93.3.

So, we have concluded that SU(2) gauge theory has an infinite number of classical field configurations of zero energy, distinguished by an integer n, and separated by energy barriers. This is analogous to a scalar field theory with a potential

$$V(\varphi) = \lambda v^4[1 - \cos(2\pi\varphi/v)] \,. \tag{93.4}$$

This potential has minima at $\varphi = nv$, where n is an integer. Let $|n\rangle$ be the quantum state corresponding to the minimum at $\varphi = nv$. Generically, between two quantum states $|n\rangle$ and $|n'\rangle$ that are separated by an energy barrier, there is a tunneling amplitude of the form

$$\langle n'|H|n\rangle \sim e^{-S}, \tag{93.5}$$

where H is the hamiltonian, and S is the euclidean action for a classical solution of the euclidean field equations that mediates between the field configuration corresponding to n at $t = -\infty$, and the field configuration corresponding to n' at $t = +\infty$. In the scalar field theory, this solution is independent of $\mathbf{x}$. Thus, S scales like the volume of space V, and so $\langle n'|H|n\rangle$ vanishes in the infinite volume limit. The minima of eq. (93.4) therefore remain exactly degenerate in the quantum theory.

Things are different in the SU(2) gauge theory. In this case, there is a classical solution of the euclidean field equations that mediates between states with winding numbers n and n', and that has an action that stays fixed and finite in the infinite-volume limit. The value of this action is $S = |n'-n|S_1$, where $S_1 = 8\pi^2/g^2$, and g is the Yang–Mills coupling constant. For $n' = n + 1$, this solution is the *instanton*. The instanton is localized in all four euclidean directions. For $n' = n - 1$, the solution is the *anti-instanton*. For $|n' - n| > 1$, the solution is a *dilute gas* of $|n' - n|$ instantons (or anti-instantons, if $n' - n$ is negative) distributed throughout euclidean spacetime.

We will shortly construct the instanton and examine its properties, but first we study the consequences of its existence. For SU(2) gauge theory, eq. (93.5) reads

$$\langle n'|H|n\rangle \sim e^{-|n'-n|S_1} \,. \tag{93.6}$$

These matrix elements depend only on $n' - n$, and so H can be diagonalized by *theta vacua* of the form

$$|\theta\rangle = \sum_{n=-\infty}^{+\infty} e^{-in\theta}|n\rangle \;; \tag{93.7}$$

see problem 93.2. For weak coupling, $S_1 \gg 1$, and so we can neglect all matrix elements of H except those with $n' = n \pm 1$. Then we find that the energy of a theta vacuum is proportional to $-\cos\theta$. (We are of course free to add a constant to H so that the lowest lying state, the theta vacuum with $\theta = 0$, has energy zero.)

We have derived these results in the weak-coupling regime. However, we are discussing properties of low-energy states, and the gauge coupling becomes large at low energies. Therefore we must consider the theory to be in the strong-coupling regime. How does this affect our conclusions?

The topological properties of the gauge fields are independent of the value of the coupling constant, so we still expect vacuum states labeled by the winding number n to exist. We also expect that $\langle n'|H|n\rangle$ will depend only on $|n' - n|$. To see this, consider making a gauge transformation by $U_k(\mathbf{x})$, where $U_k(\mathbf{x})$ has winding number k. The product of two maps with winding numbers n and k is a map with winding number $n + k$; see problem 93.4. Thus, making a gauge transformation by $U_k(\mathbf{x})$ converts a field configuration with winding number n to one with winding number $n + k$. In the quantum theory, the gauge transformation is implemented by a unitary operator $\mathcal{U}_k$, and we should have

$$\mathcal{U}_k|n\rangle = |n{+}k\rangle \; . \tag{93.8}$$

On the other hand, the hamiltonian, which is built out of field strengths, must be invariant under time-independent gauge transformations:

$$\mathcal{U}_k H \mathcal{U}_k^\dagger = H \; . \tag{93.9}$$

Inserting factors of $I = \mathcal{U}_k^\dagger \mathcal{U}_k$ on either side of H in $\langle n'|H|n\rangle$, and using eqs. (93.8) and (93.9), we find

$$\langle n'|H|n\rangle = \langle n'{+}k|H|n{+}k\rangle \; . \tag{93.10}$$

We conclude that $\langle n'|H|n\rangle$ depends only on $n' - n$. We can also note that winding number is reversed by parity, $P|n\rangle = |{-n}\rangle$, and that the Yang–Mills hamiltonian is parity invariant, $PHP^{-1} = H$, to conclude similarly that $\langle n'|H|n\rangle = \langle -n'|H|-n\rangle$. Thus $\langle n'|H|n\rangle$ depends only on $|n' - n|$.

The fact that $\langle n'|H|n\rangle$ depends only on $|n' - n|$ tells us that the theta vacua are still eigenstates of H. Furthermore, their energies must be a periodic, even function of θ. Of course, the eigenvalues of H should scale with the volume of space V. Then, on dimensional grounds, we have

$$H|\theta\rangle = V\Lambda_{\text{QCD}}^4 f(\theta)|\theta\rangle \; , \tag{93.11}$$

where Λ_{QCD} is the scale where the gauge coupling becomes strong. The function $f(\theta)$ must obey $f(\theta + 2\pi) = f(\theta)$ and $f(-\theta) = f(\theta)$. We expect the minimum of $f(\theta)$ to be at $\theta = 0$.

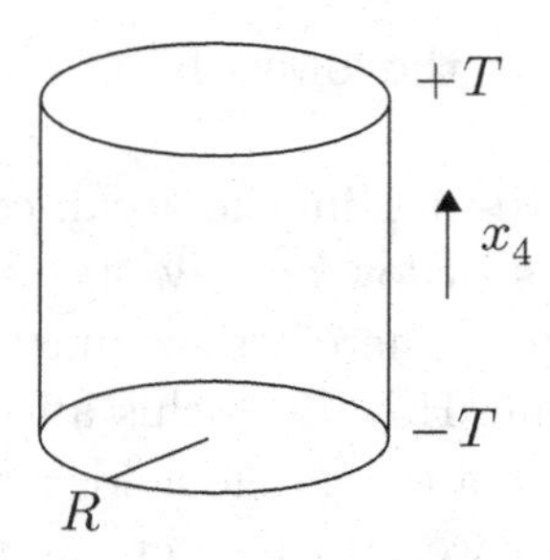

Figure 93.1. The boundary in euclidean spacetime. We have a field configuration with winding number n_- on the cap at $x_4 = -T$, and one with winding number n_+ on the cap at $x_4 = +T$. On the cylindrical surface at $|\mathbf{x}| = R$, the field vanishes.

We turn now to the solutions of the euclidean field equations. At euclidean time $x_4 = -T$, we set $A_\mu(\mathbf{x}) = \frac{i}{g} U_-(\mathbf{x}) \partial_\mu U_-^\dagger(\mathbf{x})$, where $U_-(\mathbf{x})$ has winding number n_-. Similarly, at euclidean time $x_4 = +T$, we set $A_\mu(\mathbf{x}) = \frac{i}{g} U_+(\mathbf{x}) \partial_\mu U_+^\dagger(\mathbf{x})$, where $U_+(\mathbf{x})$ has winding number n_+. At $|\mathbf{x}| = R$, for $-T \le x_4 \le T$, we set the boundary condition $A_\mu = 0$. This is equivalent to $A_\mu = \frac{i}{g} U \partial_\mu U^\dagger$ with $\partial_\mu U^\dagger = 0$; we therefore set $U(\mathbf{x})$ to a constant matrix at $|\mathbf{x}| = R$. We want to take T and R to infinity at the end of the calculation.

We have now specified $U(\mathbf{x}, x_4)$ on the cylindrical boundary of four-dimensional spacetime shown in fig. 93.1. This boundary is topologically a three-sphere. The winding number of the map on this three-sphere is $n_+ - n_-$. We see this by using eq. (93.3), and noting that the cylindrical wall makes no contribution (because $\partial_\mu U^\dagger = 0$ there), the upper cap contributes n_+, and the lower cap contributes $-n_-$; the sign is negative because the orientation of the cap as part of the boundary is reversed from its original orientation.

Since we are interested in large R and T, and since the shape of the boundary should not matter in this limit, we instead consider the boundary to be a three-sphere at $\rho \equiv (x_\mu x_\mu)^{1/2} = \infty$. On this boundary, we have a map $U(\hat{x})$, where $\hat{x}_\mu = x_\mu / \rho$; this map has winding number $n \equiv n_+ - n_-$.

Our first task will be to construct a *Bogomolny bound* on the euclidean action

$$S = \tfrac{1}{2} \int d^4x \, \mathrm{Tr}(F^{\mu\nu} F_{\mu\nu}) \tag{93.12}$$

of a field that obeys the boundary condition

$$\lim_{\rho \to \infty} A_\mu(x) = \frac{i}{g} U(\hat{x}) \partial_\mu U^\dagger(\hat{x}) \,, \tag{93.13}$$

where $U(\hat{x})$ is a map with winding number n. The field strength is given in terms of the vector potential by

$$F_{\mu\nu} = \partial_\mu A_\nu - \partial_\nu A_\mu - ig[A_\mu, A_\nu] \,. \tag{93.14}$$

We begin by defining the polar angles χ and ψ, and the azimuthal angle ϕ, via

$$\hat{x}_\mu = (\sin\chi \sin\psi \cos\phi,\ \sin\chi \sin\psi \sin\phi,\ \sin\chi \cos\psi,\ \cos\chi) \,. \tag{93.15}$$

(We use ψ rather than θ in order to avoid any possible confusion with the vacuum angle.) Next we write the winding number in terms of these angles,

$$n = \frac{-1}{24\pi^2} \int_0^\pi d\chi \int_0^\pi d\psi \int_0^{2\pi} d\phi \; \varepsilon^{\alpha\beta\gamma} \, \mathrm{Tr}[(U\partial_\alpha U^\dagger)(U\partial_\beta U^\dagger)(U\partial_\gamma U^\dagger)] \,, \tag{93.16}$$

where α, β, γ run over χ, ψ, ϕ; $\partial_\phi = \partial/\partial\phi$, etc.; and $\varepsilon^{\chi\psi\phi} = +1$. Now we write eq. (93.16) as a surface integral over a surface at infinity in four-dimensional euclidean space,

$$n = \frac{1}{24\pi^2} \int dS_\mu \, \varepsilon^{\mu\nu\sigma\tau} \, \mathrm{Tr}[(U\partial_\nu U^\dagger)(U\partial_\sigma U^\dagger)(U\partial_\tau U^\dagger)] \,, \tag{93.17}$$

where $\partial_\mu = \partial/\partial x^\mu$, and $\varepsilon^{1234} = +1$. (This implies that $\varepsilon^{\rho\chi\psi\phi} = -1$, as can be checked by computing the jacobian for this change of coordinates; that is why the overall minus sign disappeared.) Now we use eq. (93.13) to write the winding number in terms of the vector potential,

$$n = \frac{ig^3}{24\pi^2} \int dS_\mu \, \varepsilon^{\mu\nu\sigma\tau} \, \mathrm{Tr}(A_\nu A_\sigma A_\tau) \,. \tag{93.18}$$

Next, we will write this surface integral as a volume integral.

To do so, we first define the *Chern–Simons current*,

$$J^\mu_{\rm CS} \equiv 2\, \varepsilon^{\mu\nu\sigma\tau} \, \mathrm{Tr}(A_\nu F_{\sigma\tau} + \tfrac{2}{3} ig A_\nu A_\sigma A_\tau) \,. \tag{93.19}$$

This current is not gauge invariant, but the relative coefficient of its two terms has been chosen so that its divergence *is* gauge invariant,

$$\begin{aligned} \partial_\mu J^\mu_{\rm CS} &= \varepsilon^{\mu\nu\sigma\tau} \, \mathrm{Tr}(F_{\mu\nu} F_{\sigma\tau}) \\ &= 2 \, \mathrm{Tr}(\tilde{F}^{\mu\nu} F_{\mu\nu}) \,, \end{aligned} \tag{93.20}$$

where

$$\tilde{F}^{\mu\nu} \equiv \tfrac{1}{2} \varepsilon^{\mu\nu\sigma\tau} F_{\sigma\tau} \tag{93.21}$$

is the *dual field strength*.

On the surface at infinity, the vector potential is a gauge transformation of zero, and so the field strength $F_{\mu\nu}$ vanishes there. Thus we can use eq. (93.19) to write eq. (93.18) as

$$n = \frac{g^2}{32\pi^2} \int dS_\mu \, J^\mu_{\text{CS}} \; . \tag{93.22}$$

Using Gauss's theorem, this becomes

$$n = \frac{g^2}{32\pi^2} \int d^4x \; \partial_\mu J^\mu_{\text{CS}} \; . \tag{93.23}$$

Finally, we use eq. (93.20) to get

$$n = \frac{g^2}{16\pi^2} \int d^4x \; \text{Tr}(\tilde{F}^{\mu\nu} F_{\mu\nu}) \; . \tag{93.24}$$

Thus we have expressed the winding number as a (four-dimensional) volume integral of a gauge-invariant expression.

Now it is easy to construct a Bogomolny bound. We first note that $\tilde{F}^{\mu\nu}\tilde{F}_{\mu\nu} = F^{\mu\nu}F_{\mu\nu}$, and hence

$$\tfrac{1}{2}\text{Tr}(\tilde{F}_{\mu\nu} \pm F_{\mu\nu})^2 = \text{Tr}(F^{\mu\nu}F_{\mu\nu}) \pm \text{Tr}(\tilde{F}^{\mu\nu}F_{\mu\nu}) \; . \tag{93.25}$$

The left-hand side of eq. (93.25) is nonnegative, and so we have

$$\int d^4x \; \text{Tr}(F^{\mu\nu}F_{\mu\nu}) \ge \left| \int d^4x \; \text{Tr}(\tilde{F}^{\mu\nu}F_{\mu\nu}) \right| . \tag{93.26}$$

The left-hand side of eq. (93.26) is twice the euclidean action, while the right-hand side is, according to eq. (93.24), $16\pi^2|n|/g^2$. Thus we have

$$S \ge 8\pi^2|n|/g^2 \; . \tag{93.27}$$

Eq. (93.27) gives us the minimum value of the euclidean action for a solution of the euclidean field equations that mediates between a vacuum configuration with winding number n_- at $x_4 = -\infty$ and a vacuum configuration with winding number $n_+ = n_- + n$ at $x_4 = +\infty$.

From eq. (93.25) we see that we can saturate the bound in eq. (93.27) if and only if

$$\tilde{F}_{\mu\nu} = (\text{sign}\, n) F_{\mu\nu} \; . \tag{93.28}$$

We can find an explicit solution of eq. (93.28) for a map with winding number $n = 1$; this solution is the instanton.

We take the map on the boundary to be the identity map,

$$U(\hat{x}) = \frac{x_4 + i\vec{x}\cdot\vec{\sigma}}{\rho}$$

$$= \begin{pmatrix} \cos\chi + i\sin\chi\cos\psi & i\sin\chi\sin\psi\, e^{-i\phi} \\ i\sin\chi\sin\psi\, e^{+i\phi} & \cos\chi - i\sin\chi\cos\psi \end{pmatrix}, \tag{93.29}$$

which has $n = 1$. We then make the ansatz

$$A_\mu(x) = \tfrac{i}{g} f(\rho) U(\hat{x}) \partial_\mu U^\dagger(\hat{x}) \;, \tag{93.30}$$

where $f(\infty) = 1$ (so that the solution obeys the boundary condition) and $f(0) = 0$ (so that A_μ is well defined at $\rho = 0$). Using eq. (93.14), we find that the field strength is

$$F_{\mu\nu} = \tfrac{i}{g}\Big[(\partial_\mu f)U\partial_\nu U^\dagger + f\partial_\mu U\partial_\nu U^\dagger + f^2(U\partial_\mu U^\dagger)(U\partial_\nu U^\dagger) - (\mu{\leftrightarrow}\nu)\Big] \,. \tag{93.31}$$

Using $(\partial_\mu U^\dagger)U = -U^\dagger\partial_\mu U$ in the third term, eq. (93.31) can be simplified to

$$F_{\mu\nu} = \tfrac{i}{g}\Big[(\partial_\mu f)U\partial_\nu U^\dagger + f(1{-}f)\partial_\mu U\partial_\nu U^\dagger - (\mu{\leftrightarrow}\nu)\Big] \,. \tag{93.32}$$

In three-spherical coordinates, we have

$$\boldsymbol{\partial} = \hat{\rho}\,\partial_\rho + \hat{\chi}\,\rho^{-1}\partial_\chi + \hat{\psi}\,(\rho\sin\chi)^{-1}\partial_\psi + \hat{\phi}\,(\rho\sin\chi\sin\psi)^{-1}\partial_\phi \;, \tag{93.33}$$

where $\partial_\phi = \partial/\partial\phi$, etc., and $\hat{\rho}$ is the same radial unit vector as $\hat{x}$. Note that f is a function of ρ only, while U is a function of the angles only. We therefore have

$$F_{\rho\chi} = \tfrac{i}{g} f'\rho^{-1} U\partial_\chi U^\dagger \;, \tag{93.34}$$

$$F_{\psi\phi} = \tfrac{i}{g} f(1{-}f)(\rho^2\sin^2\!\chi\sin\psi)^{-1}(\partial_\psi U\partial_\phi U^\dagger - \partial_\phi U\partial_\psi U^\dagger) \,. \tag{93.35}$$

Next we note that eq. (93.21) implies $\widetilde{F}_{\rho\chi} = -F_{\psi\phi}$ (since $\varepsilon^{\rho\chi\psi\phi} = -1$), and since $\widetilde{F}_{\mu\nu} = F_{\mu\nu}$ for the instanton solution, we have $F_{\rho\chi} = -F_{\psi\phi}$. Thus the right-hand side of eq. (93.34) equals minus the right-hand side of eq. (93.35). We then use separation of variables to conclude that

$$\rho f' = cf(1{-}f) \;, \tag{93.36}$$

$$U\partial_\chi U^\dagger = -c^{-1}(\sin^2\!\chi\sin\psi)^{-1}(\partial_\psi U\partial_\phi U^\dagger - \partial_\phi U\partial_\psi U^\dagger) \;, \tag{93.37}$$

where c is the separation constant. If we plug eq. (93.29) into eq. (93.37), we find that it is satisfied if $c = 2$. The solution of eq. (93.36) is then

$$f(\rho) = \frac{\rho^2}{\rho^2 + a^2} \,, \tag{93.38}$$

where a, the *size of the instanton*, is a constant of integration. The instanton solution is also parameterized by the location of its center; we have used the spacetime origin, but translation invariance allows us to displace it.

If we consider initial and final states whose winding numbers differ by more than one, we can construct a mediating solution by patching together instantons (or anti-instantons) whose centers are widely separated on the scale set by their sizes. Each instanton (or anti-instanton) contributes $S_1 = 8\pi^2/g^2$ to the action, and so the minimum total action is $|n_+ {-} n_-|S_1$.

To better understand the role of the θ parameter, let us consider the euclidean path integral, with the boundary condition that we start with a state with winding number n_- at $x_4 = -\infty$, and end with a state with winding number n_+ at $x_4 = +\infty$. The only field configurations that contribute are those with winding number $n_+ - n_-$. We can therefore write

$$Z_{n_+ \leftarrow n_-}(J) = \int \mathcal{D}A_{n_+ - n_-}\, e^{-S + JA} \,, \tag{93.39}$$

where JA is short for $\int d^4x \,\mathrm{Tr}(J^\mu A_\mu)$, and the subscript on the field differential means that we integrate only over fields with that winding number. We see that $Z_{n_+ \leftarrow n_-}(J)$ depends only on $n_+ - n_-$, and not separately on n_+ and n_-. This is in accord with our previous conclusion that $\langle n'|H|n\rangle$ depends only on $n' - n$.

Suppose now that we are interested in starting with a particular theta vacuum $|\theta\rangle$, and ending with a (possibly different) theta vacuum $|\theta'\rangle$. Then, we see from eq. (93.7) that the corresponding path integral is

$$Z_{\theta' \leftarrow \theta}(J) = \sum_{n_-, n_+} e^{i(n_+\theta' - n_-\theta)} Z_{n_+ \leftarrow n_-}(J) \,. \tag{93.40}$$

Let $n_+ = n_- + n$, so that $n_+\theta' - n_-\theta = n_-(\theta' - \theta) + n\theta'$. Since $Z_{n_+ \leftarrow n_-}(J)$ depends only on n, n_- appears in eq. (93.40) only through a factor of $e^{in_-(\theta' - \theta)}$. Summing over n_- then generates $\delta(\theta' - \theta)$, which implies that the value of θ is time independent. (Of course, we already knew this, because θ labels energy eigenstates.) We now have

$$Z_{\theta' \leftarrow \theta}(J) = \delta(\theta' - \theta) \sum_n e^{in\theta} \int \mathcal{D}A_n\, e^{-S + JA} \,. \tag{93.41}$$

We can drop the delta function, and just define

$$Z_\theta(J) \equiv \sum_n e^{in\theta} \int \mathcal{D}A_n \, e^{-S+JA} \,. \tag{93.42}$$

Next, we combine the sum over n and the integral over A_n into an integral over all A. To account for the factor of $e^{in\theta}$, we use eq. (93.24); we get

$$Z_\theta(J) = \int \mathcal{D}A \exp \int d^4x \, \mathrm{Tr}\left[-\frac{1}{2}F^{\mu\nu}F_{\mu\nu} + \frac{ig^2\theta}{16\pi^2}\tilde{F}^{\mu\nu}F_{\mu\nu} + J^\mu A_\mu\right] . \tag{93.43}$$

The vacuum angle θ now appears as the coefficient of an extra term in the Yang–Mills lagrangian.

We can write the path integral in Minkowski space by setting $x_4 = it$. The extra term contains one derivative with respect to x_4, and thus picks up a factor of $-i$. Also, $\varepsilon^{1234} = +1$ but $\varepsilon^{1230} = -1$. Putting all this together, we get

$$Z_\theta(J) = \int \mathcal{D}A \exp i \int d^4x \, \mathrm{Tr}\left[-\frac{1}{2}F^{\mu\nu}F_{\mu\nu} - \frac{g^2\theta}{16\pi^2}\tilde{F}^{\mu\nu}F_{\mu\nu} + J^\mu A_\mu\right] \tag{93.44}$$

in Minkowski space. We see that the extra term is gauge invariant, Lorentz invariant, hermitian, and has a dimensionless coefficient. We therefore could have included it when we first considered Yang–Mills theory. We did not do so because this term is a total divergence; see eq. (93.20). We have always dropped total divergences from the lagrangian, because they do not affect the equations of motion or the Feynman rules. In this case, however, the new term does change the quantum physics, as we have seen. We will explore this in more detail in the next section.

So far, we have only discussed SU(2) gauge theory, without scalar or fermion fields. What can we say more generally?

Adding scalar fields has no effect on our analysis. Changing from SU(2) to another simple nonabelian group also has no effect; instanton solutions always reside in an SU(2) subgroup. If the gauge group is U(1), there are no instantons, and hence no vacuum angle. If the gauge group includes more than one nonabelian factor, then there is an independent vacuum angle for each of these factors.

On the other hand, adding fermions can significantly change the physics. We take this up in the next section.

Reference notes

Instantons are treated in more detail in *Coleman* and *Weinberg II*.

Problems

93.1) Show that any 2×2 special unitary matrix U can be written in the form $U = a_4 + i\vec{a}\cdot\vec{\sigma}$, where a_4 and the three-vector $\vec{a}$ are real, and $\vec{a}^2 + a_4^2 = 1$.

93.2) Verify that a state of the form of eq. (93.7) is an eigenstate of any hamiltonian with matrix elements of the form $\langle n'|H|n\rangle = f(n' - n)$.

93.3) The winding number n for a map from $\mathrm{S}^3 \to \mathrm{S}^3$ is given by eq. (93.3), where $U^\dagger U = 1$. We will prove that n is invariant under an infinitesimal deformation of U. Since any smooth deformation can be made by compounding infinitesimal ones, this will prove that n is invariant under any smooth deformation.

a) Consider an infinitesimal deformation of U, $U \to U + \delta U$. Show that $\delta U^\dagger = -U^\dagger \delta U U^\dagger$, and hence that $\delta(U\partial_k U^\dagger) = -U\partial_k(U^\dagger \delta U)U^\dagger$.

b) Show that

$$\delta n = \frac{-3}{24\pi^2} \int d^3x\, \varepsilon^{ijk}\, \mathrm{Tr}[(U\partial_i U^\dagger)(U\partial_j U^\dagger)\delta(U\partial_k U^\dagger)] \,. \tag{93.45}$$

Plug in your result from part (a), and integrate ∂_k by parts. Show that the resulting integrand vanishes. Hint: make repeated use of $U\partial_i U^\dagger = -\partial_i U U^\dagger$, and the antisymmetry of ε^{ijk}.

93.4) Show that if $U_n(\mathbf{x})$ and $U_k(\mathbf{x})$ are maps from $\mathrm{S}^3 \to \mathrm{S}^3$ with winding numbers n and k, then $U_n(\mathbf{x})U_k(\mathbf{x})$ is a map with winding number $n + k$. Hint: consider smoothly deforming $U_n(\mathbf{x})$ to equal one for $x_3 < 0$. How should $U_k(\mathbf{x})$ be deformed?

93.5) Use eq. (93.16) to compute the winding number for the map given in eq. (93.29), and for a generalization where ϕ is replaced by $n\phi$.

93.6) Use eq. (93.16) to compute the winding number for U^n, where U is the map given in eq. (93.29). Hint: first show that U can be written in the form $U = \exp[i\vec{\chi}\cdot\vec{\sigma}]$, where $\vec{\chi}$ is a three-vector that you should specify. Is your result in accord with the theorem of problem 93.4?

94

Quarks and theta vacua

Prerequisites: 77, 83, 93

Consider quantum chromodynamics with one flavor of massless quark, represented by a Dirac field Ψ in the fundamental representation of the gauge group SU(3). The path integral is

$$Z = \int \mathcal{D}A\, \mathcal{D}\Psi\, \mathcal{D}\overline{\Psi} \,\exp i\int d^4x\left[i\overline{\Psi}\not{D}\Psi - \frac{1}{4}F^{a\mu\nu}F^a_{\mu\nu} - \frac{g^2\theta}{32\pi^2}\tilde{F}^{a\mu\nu}F^a_{\mu\nu}\right]; \tag{94.1}$$

for the sake of brevity, we have not written the source terms explicitly.

In addition to the SU(3) gauge symmetry, there is a $\mathrm{U}(1)_\mathrm{V} \times \mathrm{U}(1)_\mathrm{A}$ global symmetry of the quark action. However, the $\mathrm{U}(1)_\mathrm{A}$ symmetry is anomalous. As we saw in section 77, under a $\mathrm{U}(1)_\mathrm{A}$ transformation

$$\Psi \to e^{-i\alpha\gamma_5}\Psi\,, \tag{94.2}$$

$$\overline{\Psi} \to \overline{\Psi}e^{-i\alpha\gamma_5}\,, \tag{94.3}$$

the integration measure picks up a phase factor,

$$\mathcal{D}\Psi\,\mathcal{D}\overline{\Psi} \to \exp\left[-i\int d^4x\,\frac{g^2\alpha}{16\pi^2}\tilde{F}^{a\mu\nu}F^a_{\mu\nu}\right]\mathcal{D}\Psi\,\mathcal{D}\overline{\Psi}\,. \tag{94.4}$$

Using this in eq. (94.1), we see that the effect of a $\mathrm{U}(1)_\mathrm{A}$ transformation is to *change the value of the theta angle* from θ to $\theta + 2\alpha$. Since the value of θ can be changed by making a $\mathrm{U}(1)_\mathrm{A}$ transformation (which is simply a change of the dummy integration variable in the path integral), we must conclude that Z does not depend on θ. Apparently (and surprisingly), adding a massless quark has turned the theta angle into a physically irrelevant, unobservable parameter.

How do we reconcile this with our analysis in the previous section, where we concluded that instanton-mediated tunneling amplitudes make the vacuum energy density depend on θ? The answer is that when we perform the integral over the quark field in eq. (94.1), we get a functional determinant; the path integral becomes

$$Z = \int \mathcal{D}A \, \det(i\rlap{\,/}D) e^{iS} e^{in\theta} \,, \tag{94.5}$$

where S is the Yang–Mills action and n is the winding number. We see from eq. (94.5) that Z would be independent of θ if gauge fields with nonzero winding number did not contribute. This will be the case if $\det(i\rlap{\,/}D)$ vanishes for gauge fields with $n \neq 0$. We conclude that $i\rlap{\,/}D$ must have a zero eigenvalue or *zero mode* whenever the gauge field has nonzero winding number. This renders Z independent of θ. Now consider adding a mass term for the quark. If we write the Dirac field Ψ in terms of two left-handed Weyl fields χ and ξ,

$$\Psi = \begin{pmatrix} \chi \\ \xi^\dagger \end{pmatrix} , \tag{94.6}$$

the mass term reads

$$\mathcal{L}_{\text{mass}} = -m\chi\xi - m^*\xi^\dagger\chi^\dagger \,. \tag{94.7}$$

We have allowed m to be complex: $m = |m|e^{i\phi}$. In terms of Ψ, eq. (94.7) can be written as

$$\mathcal{L}_{\text{mass}} = -|m|\,\overline{\Psi} e^{-i\phi\gamma_5} \Psi \,. \tag{94.8}$$

A $\text{U}(1)_\text{A}$ transformation changes ϕ to $\phi + 2\alpha$. Since θ simultaneously changes to $\theta + 2\alpha$, we see that $\phi - \theta$, or equivalently $me^{-i\theta}$, is unchanged. Thus, the path integral can (and does) depend on $me^{-i\theta}$, but not on m and θ separately.

With more quark fields, the mass term is $\mathcal{L} = -M_{ij}\chi_i\xi_j + \text{h.c.}$; a $\text{U}(1)_\text{A}$ transformation changes the phase of every χ_i and ξ_i by $e^{i\alpha}$, and so every matrix element of M picks up a factor of $e^{2i\alpha}$. Simultaneously, θ changes to $\theta + 2N\alpha$, where N is the number of quark fields. Thus $(\det M)e^{-i\theta}$ is invariant under a $\text{U}(1)_\text{A}$ transformation.

To understand the effects of the theta angle on hadronic physics, we turn to the effective lagrangian (for the case of two light flavors) that we developed in section 83,

$$\begin{aligned}\mathcal{L} = &-\tfrac{1}{4}f_\pi^2 \operatorname{Tr} \partial^\mu U^\dagger \partial_\mu U + v^3 \operatorname{Tr}(MU + M^\dagger U^\dagger) \\ &+ i\overline{N}\partial\!\!\!/ N - m_N \overline{N}(U^\dagger P_{\rm L} + U P_{\rm R})N \\ &- \tfrac{1}{2}(g_{\rm A}-1)i\overline{N}\gamma^\mu(U\partial_\mu U^\dagger P_{\rm L} + U^\dagger \partial_\mu U P_{\rm R})N \\ &- c_1\overline{N}(MP_{\rm L} + M^\dagger P_{\rm R})N - c_2\overline{N}(U^\dagger M^\dagger U^\dagger P_{\rm L} + UMUP_{\rm R})N \\ &- c_3 \operatorname{Tr}(MU + M^\dagger U^\dagger)\,\overline{N}(U^\dagger P_{\rm L} + UP_{\rm R})N \\ &- c_4 \operatorname{Tr}(MU - M^\dagger U^\dagger)\,\overline{N}(U^\dagger P_{\rm L} - UP_{\rm R})N\ , \end{aligned} \tag{94.9}$$

where U is a 2×2 special unitary matrix field representing the pions, N is the field for the nucleon doublet,

$$M = \begin{pmatrix} m_u & 0 \\ 0 & m_d \end{pmatrix} e^{-i\theta/2} \tag{94.10}$$

is the quark mass matrix, v^3 is the value of the quark condensate, and $g_{\rm A}$ and c_i are numerical constants. (The terms in the last three lines were introduced in problem 83.8.)

For the case of $\theta = 0$, the potential

$$V(U) = -v^3 \operatorname{Tr} MU + \text{h.c.} \tag{94.11}$$

is minimized by $U = I$, and we can expand about this point in powers of the pion fields, as we did in section 83. However, for nonzero θ, the minimum of $V(U)$ occurs at $U = U_0$, where U_0 is diagonal (because M is) and has unit determinant (because U is required to have unit determinant). We can therefore write

$$U_0 = \begin{pmatrix} e^{+i\phi} & 0 \\ 0 & e^{-i\phi} \end{pmatrix} . \tag{94.12}$$

We determine ϕ by minimizing

$$V(U_0) = -2v^3 \Big[m_u \cos(\phi - \tfrac{1}{2}\theta) + m_d \cos(\phi + \tfrac{1}{2}\theta) \Big]\ ; \tag{94.13}$$

the result is

$$\tan\phi = \frac{m_u - m_d}{m_u + m_d}\, \tan(\tfrac{1}{2}\theta)\ . \tag{94.14}$$

As we will see shortly, experimental results require the value of $|\theta|$ to be less than 10^{-9}; therefore, we can work to first order in an expansion in powers of θ. For $\theta \ll 1$, eqs. (94.12) and (94.14) can be written in the elegant form

$$MU_0 = M_0 - i\theta\tilde{m}I + O(\theta^2)\ , \tag{94.15}$$

where

$$M_0 = \begin{pmatrix} m_u & 0 \\ 0 & m_d \end{pmatrix} \tag{94.16}$$

is the quark mass matrix with θ set to zero, I is the identity matrix, and

$$\tilde{m} = \frac{m_u m_d}{m_u + m_d} \tag{94.17}$$

is the reduced mass of the up and down quarks.

We can now expand in powers of the pion fields. Though it is not at all obvious, it turns out that the most convenient way to define the pion fields is by writing

$$U(x) = u_0 u^2(x) u_0 \ , \tag{94.18}$$

where $u_0^2 = U_0$ and $u(x) = \exp[i\pi^a(x)T^a/f_\pi]$. We also define $\tilde{U}(x) = u^2(x)$, and a new nucleon field $\mathcal{N}$ via

$$N = (u_0 u P_{\rm L} + u_0^\dagger u^\dagger P_{\rm R})\mathcal{N} \ . \tag{94.19}$$

Substituting eqs. (94.18) and (94.19) into eq. (94.9), and using $u_0 M u_0 = MU_0$ (which follows because u_0 and M are both diagonal, and hence commute), we ultimately get

$$\begin{aligned}\mathcal{L} = &-\tfrac{1}{4}f_\pi^2 \operatorname{Tr} \partial^\mu \tilde{U}^\dagger \partial_\mu \tilde{U} + v^3 \operatorname{Tr}[(MU_0)\tilde{U} + (MU_0)^\dagger \tilde{U}^\dagger] \\ &+ i\overline{\mathcal{N}}\partial\!\!\!/\mathcal{N} - m_N\overline{\mathcal{N}}\mathcal{N} + \overline{\mathcal{N}}v\!\!\!/\mathcal{N} - g_{\rm A}\overline{\mathcal{N}}a\!\!\!/\gamma_5\mathcal{N} \\ &- \tfrac{1}{2}c_+\overline{\mathcal{N}}[u(MU_0)u + u^\dagger(MU_0)^\dagger u^\dagger]\mathcal{N} \\ &+ \tfrac{1}{2}c_-\overline{\mathcal{N}}[u(MU_0)u - u^\dagger(MU_0)^\dagger u^\dagger]\gamma_5\mathcal{N} \\ &- c_3 \operatorname{Tr}[(MU_0)\tilde{U} + (MU_0)^\dagger\tilde{U}^\dagger]\overline{\mathcal{N}}\mathcal{N} \\ &+ c_4 \operatorname{Tr}[(MU_0)\tilde{U} - (MU_0)^\dagger\tilde{U}^\dagger]\overline{\mathcal{N}}\gamma_5\mathcal{N} \ ,\end{aligned} \tag{94.20}$$

where $v_\mu = \frac{1}{2}i[u^\dagger(\partial_\mu u) + u(\partial_\mu u^\dagger)]$, $a_\mu = \frac{1}{2}i[u^\dagger(\partial_\mu u) - u(\partial_\mu u^\dagger)]$, and $c_\pm = c_1 \pm c_2$. Eq. (94.20) is exactly what we found in section 83, except that the quark mass matrix M has been replaced everywhere by MU_0.

We can now use eq. (94.15) to get the $O(\theta)$ contributions to $\mathcal{L}$. Using the fact that $\operatorname{Tr}(\tilde{U} - \tilde{U}^\dagger)$ vanishes in the case of two light flavors, we find

$$\begin{aligned}\mathcal{L}_\theta = -i\theta\tilde{m}\Big[&-\tfrac{1}{2}c_+\overline{\mathcal{N}}(\tilde{U} - \tilde{U}^\dagger)\mathcal{N} + \tfrac{1}{2}c_-\overline{\mathcal{N}}(\tilde{U} + \tilde{U}^\dagger)\gamma_5\mathcal{N} \\ &+ c_4 \operatorname{Tr}(\tilde{U} + \tilde{U}^\dagger)\,\overline{\mathcal{N}}\gamma_5\mathcal{N}\Big] \ .\end{aligned} \tag{94.21}$$

Expanding in powers of the pion fields yields

$$\mathcal{L}_\theta = -i\theta\tilde{m}(c_- + 4c_4)\overline{\mathcal{N}}\gamma_5\mathcal{N} - (\theta c_+\tilde{m}/f_\pi)\pi^a\overline{\mathcal{N}}\sigma^a\mathcal{N} + \ldots \ , \tag{94.22}$$

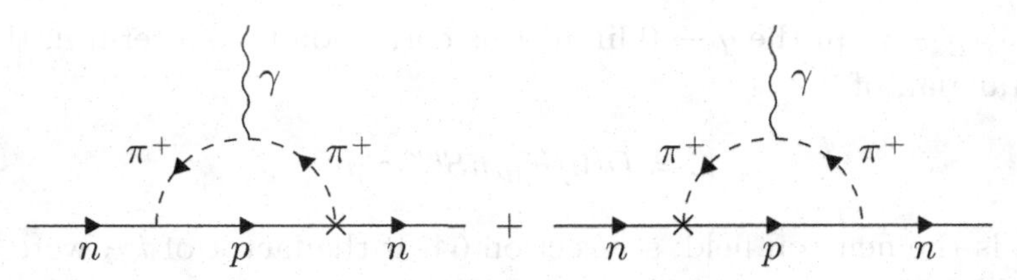

Figure 94.1. Diagrams contributing to the electric dipole moment of the neutron that are enhanced by a chiral log. The CP violating vertex is denoted with a cross.

where we used $T^a = \frac{1}{2}\sigma^a$. We can eliminate the first term with a field redefinition of the form $\mathcal{N} \to e^{-i\alpha\gamma_5}\mathcal{N}$. This generates some new terms in eq. (94.20), but all have at least two factors of quark masses, and hence can be neglected. The second term in eq. (94.22) provides a pion-nucleon coupling that violates both parity P and time-reversal T (equivalently, CP).

The value of c_+ can be fixed by baryon mass differences. The c_+ term in eq. (94.20) makes a contribution of $c_+(m_u - m_d)$ to the proton-neutron mass difference, $m_p - m_n = -1.3\,\text{MeV}$. Using $m_u = 1.7\,\text{MeV}$ and $m_d = 3.9\,\text{MeV}$ yields $c_+ = 0.6$. However, there is a comparable electromagnetic contribution to the proton-neutron mass difference. We get a better estimate from the masses of baryons with strange quarks, $c_+(m_s - \frac{1}{2}m_u - \frac{1}{2}m_d) = m_{\Xi^0} - m_{\Sigma^0} = 122\,\text{MeV}$; using $m_s = 76\,\text{MeV}$ yields $c_+ = 1.7$. (All these values assume the $\overline{\text{MS}}$ renormalization scheme with $\mu = 2\,\text{GeV}$.)

For comparison with the interaction of eq. (94.22), the dominant (P and CP conserving) pion-nucleon interaction comes from the last term in the second line of eq. (94.20), and is

$$\mathcal{L}_{\pi\overline{N}N} = (g_{\rm A}/f_\pi)\partial_\mu\pi^a\overline{\mathcal{N}}T^a\gamma^\mu\gamma_5\mathcal{N}\ , \tag{94.23}$$

where $g_{\rm A} = 1.27$. As we did in section 83, we can integrate by parts to put the derivative on the nucleon fields, and then (for on-shell nucleons) use the Dirac equation to get

$$\mathcal{L}_{\pi\overline{N}N} = -i(g_{\rm A}m_N/f_\pi)\pi^a\overline{\mathcal{N}}\sigma^a\gamma_5\mathcal{N}\ . \tag{94.24}$$

The strongest limit on a CP violating pion-nucleon coupling comes from measurements of the electric dipole moment of the neutron. The Feynman diagrams of fig. 94.1 contribute to an amplitude of the form

$$\mathcal{T} = -2iD(q^2)\varepsilon^*_\mu(q)\overline{u}_{s'}(\mathbf{p}')S^{\mu\nu}q_\nu i\gamma_5 u_s(\mathbf{p})\ , \tag{94.25}$$

where $q = p' - p$. In the $q \to 0$ limit, this corresponds to a term in the effective lagrangian of

$$\mathcal{L} = D(0) F_{\mu\nu} \overline{n} S^{\mu\nu} i\gamma_5 n \; , \tag{94.26}$$

where n is the neutron field; see section 64. If the factor of $i\gamma_5$ were absent, this would represent a contribution of $D(0)$ to the magnetic dipole moment of the neutron. To account for the factor of $i\gamma_5$, we use

$$S^{\mu\nu} i\gamma_5 = -\tfrac{1}{2}\varepsilon^{\mu\nu\rho\sigma} S_{\rho\sigma} \tag{94.27}$$

to see that eq. (94.26) is equivalent to

$$\mathcal{L} = -D(0) \tilde{F}_{\mu\nu} \overline{n} S^{\mu\nu} n \; , \tag{94.28}$$

where $\tilde{F}_{\mu\nu} = \frac{1}{2}\varepsilon^{\mu\nu\rho\sigma} F_{\rho\sigma}$ is the dual field strength. Since $\tilde{\mathbf{B}} = -\mathbf{E}$, eq. (94.26) represents a contribution of $D(0)$ to the *electric* dipole moment of the neutron d_n.

The salient feature of the diagrams shown in fig. 94.1 is that the photon line attaches to the pion line. These diagrams are enhanced by a *chiral log* $\ln(\Lambda^2/m_\pi^2) \sim 4.2$, where $\Lambda \sim 4\pi f_\pi$ is the ultraviolet cutoff in the effective theory. No other contributing diagrams have this enhancement; it is an infrared effect, due to the light pion. Of course, 4.2 is not an impressively large number, and so we cannot be certain that the remaining contributions are not significant. These contributions depend on coefficients in the effective lagrangian that are not well determined by other experimental results.

Using

$$\pi^a \sigma^a = \begin{pmatrix} \pi^0 & \sqrt{2}\pi^+ \\ \sqrt{2}\pi^- & -\pi^0 \end{pmatrix} , \tag{94.29}$$

we can write the charged-pion terms in eqs. (94.24) and (94.21) as

$$\mathcal{L}_{\pi \overline{N} N} = -i\sqrt{2}(g_{\rm A} m_N / f_\pi)(\pi^+ \overline{p}\gamma_5 n + \pi^- \overline{n}\gamma_5 p) \, , \tag{94.30}$$

$$\mathcal{L}_{\theta\pi \overline{N} N} = -\sqrt{2}(\theta c_+ \tilde{m}/f_\pi)(\pi^+ \overline{p} n + \pi^- \overline{n} p) \; . \tag{94.31}$$

From these we read off the pion-nucleon vertex factors. We label the internal momenta as shown in fig. 94.2; for small q, ℓ is a pion momentum that should be cut off at $\Lambda \sim 4\pi f_\pi$. Because the terms of interest have a chiral log produced by an infrared divergence at small ℓ, we can treat ℓ as much less than p and p'. Thus the internal proton is nearly on-shell, which justifies the use of eq. (94.24) for the CP conserving interaction.

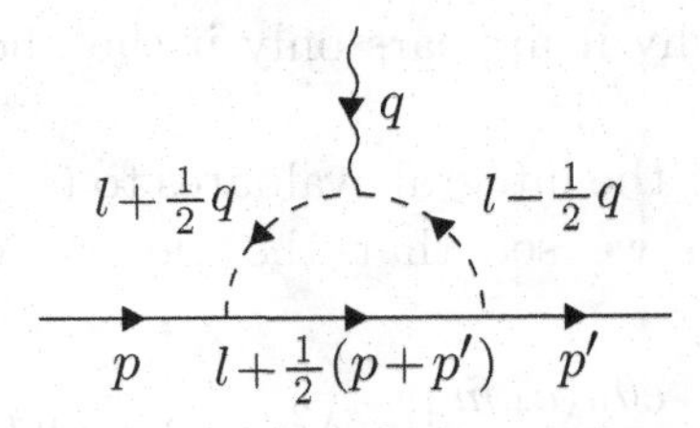

Figure 94.2. Momentum flow in the diagrams of fig. 94.1.

The diagrams of fig. 94.1 yield an amplitude of

$$iT = \left(\tfrac{1}{i}\right)^3 (ie)(\sqrt{2}g_A m_N/f_\pi)(-i\sqrt{2}\theta c_+ \tilde{m}/f_\pi)\varepsilon^*_\mu \int_0^\Lambda \frac{d^4\ell}{(2\pi)^4}$$
$$\times \frac{(2\ell^\mu)\,\overline{u}'[(-\not{\ell}-\not{\bar{p}}+m_N)\gamma_5+\gamma_5(-\not{\ell}-\not{\bar{p}}+m_N)]u}{((\ell+\bar{p})^2+m_N^2)((\ell+\frac{1}{2}q)^2+m_\pi^2)((\ell-\frac{1}{2}q)^2+m_\pi^2)}\,, \tag{94.32}$$

where $\bar{p} = \frac{1}{2}(p'+p)$. Using $\{\gamma^\mu, \gamma_5\} = 0$, the spinor factors in the numerator simplify to

$$\overline{u}'[\ldots]u = 2m_N\,\overline{u}'\gamma_5 u\,. \tag{94.33}$$

From the spinor properties established in section 38, it is easy to check that $\overline{u}'\gamma_5 u$ vanishes when $p' = p$; thus $\overline{u}'\gamma_5 u$ must be $O(q)$, and so we can set $q = 0$ everywhere else. Also, taking $\ell \ll p$, we can set $(\ell+\bar{p})^2 + m_N^2 = 2p{\cdot}\ell$ in the denominator of eq. (94.32). We now have

$$\mathcal{T} = 4(e\theta g_A c_+ \tilde{m} m_N^2/f_\pi^2)\varepsilon^*_\mu \int_0^\Lambda \frac{d^4\ell}{(2\pi)^4}\frac{(2\ell^\mu)\,\overline{u}'\gamma_5 u}{(2p{\cdot}\ell)(\ell^2+m_\pi^2)^2}\,. \tag{94.34}$$

Integrating over the direction of ℓ results in

$$\frac{\ell^\mu}{p{\cdot}\ell} \to \frac{p^\mu}{p^2} = -\frac{p^\mu}{m_N^2}\,. \tag{94.35}$$

Next we use the Gordon identity (see problem 38.4)

$$p^\mu\,\overline{u}'\gamma_5 u = \overline{u}'S^{\mu\nu}q_\nu i\gamma_5 u + O(q^2)\,, \tag{94.36}$$

which verifies that $\overline{u}'\gamma_5 u$ is linear in q. We now have

$$\mathcal{T} = -4(e\theta g_A c_+ \tilde{m}/f_\pi^2)\varepsilon^*_\mu\,\overline{u}'S^{\mu\nu}q_\nu i\gamma_5 u \int_0^\Lambda \frac{d^4\ell}{(2\pi)^4}\frac{1}{(\ell^2+m_\pi^2)^2}\,. \tag{94.37}$$

In the limit $m_\pi \to 0$, the integral diverges at small ℓ, generating a chiral log. This infrared divergence can only arise from diagrams with two pion propagators, which is why it appears only if the photon is attached to the pion.

After a Wick rotation, the integral evaluates to $(i/16\pi^2)\ln(\Lambda^2/m_\pi^2)$. Comparing with eq. (94.25), we see that the electric dipole moment of the neutron is

$$d_n = \frac{e\theta g_{\rm A} c_+ \tilde{m}}{8\pi^2 f_\pi^2}\Big[\ln(\Lambda^2/m_\pi^2) + O(1)\Big] \,. \qquad (94.38)$$

Putting in numbers ($g_{\rm A} = 1.27$, $c_+ = 1.7$, $\tilde{m} = 1.2\,{\rm MeV}$), we find

$$d_n = 3.2 \times 10^{-16}\,\theta\, e\,{\rm cm} \,. \qquad (94.39)$$

The experimental upper limit is $|d_n| < 6.3 \times 10^{-26}\,e\,{\rm cm}$, and so we must have $|\theta| < 2 \times 10^{-10}$.

Such a small value for a fundamental parameter cries out for an explanation; this is the *strong CP problem*. Several solutions have been proposed. (1) The up quark mass may actually be zero, since a massless quark renders θ unobservable (and effectively zero). This requires higher-order corrections in the quark masses to account for the masses of the pseudogoldstone bosons. (2) The fundamental lagrangian may be CP invariant, and the observed CP violation in weak interactions due to spontaneous breaking of CP symmetry. (3) The theta parameter may be promoted to a field, the *axion*, which would minimize its energy by rolling to $\theta = 0$; see problem 94.2. All of these solutions have interesting physical consequences.

Reference notes

Quarks and theta vacua are discussed in *Coleman*, *Ramond II*, and *Weinberg II*.

Problems

94.1) Carry out the field redefinition discussed after eq. (94.22), and verify that all new terms generated in the lagrangian are suppressed by at least two powers of quark masses.

94.2) Consider adding to the Standard Model a massless quark, represented by a pair of Weyl fermions χ and ξ in the 3 and $\overline{3}$ representations of SU(3). Also add a complex scalar Φ in the singlet representation. Assume that these fields have a Yukawa interaction of the form $\mathcal{L}_{\rm Yuk} = y\Phi\chi\xi + {\rm h.c.}$, where y is the Yukawa coupling constant. Assume that the scalar potential $V(\Phi)$ depends only on $\Phi^\dagger\Phi$.

a) Show that the lagrangian is invariant under a *Peccei–Quinn transformation* $\chi \to e^{i\alpha}\chi$, $\xi \to e^{i\alpha}\xi$, $\Phi \to e^{-2i\alpha}\Phi$, all other fields unchanged.

b) Show that this global $U(1)_{PQ}$ symmetry is anomalous, and that $\theta \to \theta + 2\alpha$ under a $U(1)_{PQ}$ transformation.

c) Suppose that $V(\Phi)$ has its minimum at $|\Phi| = f/\sqrt{2}$, with $f \neq 0$. Show that this gives a mass to the quark we introduced.

d) Write $\Phi = 2^{-1/2}(f + \rho)e^{ia/f}$, where ρ and a are fields. Argue that, in eq. (94.10), we should replace θ with $\theta + a/f$, and add to eq. (94.9) a kinetic term $-\frac{1}{2}\partial^\mu a \partial_\mu a$ for the a field.

e) Show that the minimum of $V(U)$, defined in eq. (94.11), is at $U = I$ and $a = -f\theta$. Show that P and CP are conserved at this minimum.

f) The particle corresponding to the a field is the *axion*; compute its mass, assuming $f \gg f_\pi$.

g) Note that if f is large, the extra quark becomes very heavy, and the axion becomes very light. Show that couplings of the axion to the hadrons are all suppressed by a factor of $1/f$.

95

Supersymmetry

Prerequisite: 69

Supersymmetry is a continuous symmetry that mixes up bosonic and fermionic degrees of freedom. A supersymmetric theory (in four spacetime dimensions) has a set of *supercharges* Q_{aA}, where a is a left-handed spinor index, and A is an internal index that runs from 1 to $\mathcal{N}$, where the allowed values of $\mathcal{N}$ are 1, 2, and 4. The supercharges can be obtained as integrals over d^3x of the time component of a *supercurrent*. The supercurrent is found via the Noether procedure, once we have identified the set of supersymmetry transformations that leaves the action invariant.

The supercharges Q_{aA} and their hermitian conjugates $Q^\dagger_{\dot{a}A}$, together with the generators of the Poincaré group P^μ and $M^{\mu\nu}$, obey a *supersymmetry algebra*

$$[Q_{aA}, P^\mu] = 0\,, \tag{95.1}$$

$$[Q^\dagger_{\dot{a}A}, P^\mu] = 0\,, \tag{95.2}$$

$$[Q_{aA}, M^{\mu\nu}] = (S^{\mu\nu}_{\rm L})_a{}^c Q_{cA}\,, \tag{95.3}$$

$$[Q^\dagger_{\dot{a}A}, M^{\mu\nu}] = (S^{\mu\nu}_{\rm R})_{\dot{a}}{}^{\dot{c}} Q^\dagger_{\dot{c}A}\,, \tag{95.4}$$

$$\{Q_{aA}, Q_{bB}\} = Z_{AB}\varepsilon_{ab}\,, \tag{95.5}$$

$$\{Q_{aA}, Q^\dagger_{\dot{a}B}\} = -2\delta_{AB}\sigma^\mu_{a\dot{a}}P_\mu\,. \tag{95.6}$$

Eqs. (95.1) and (95.2) simply say that the supercharges are conserved, and eqs. (95.3) and (95.4) simply say that their spinor indices are indeed spinor indices. In eq. (95.5), $Z_{AB} = -Z_{BA}$ must commute with Q_{aA}, P^μ, and $M^{\mu\nu}$, and so represents a *central charge* in the supersymmetry algebra. We will be concerned only with the case of $\mathcal{N} = 1$ supersymmetry: the index A then takes on only one value (and so can be dropped), and $Z_{AB} = 0$.

$\mathcal{N}=1$ supersymmetric theories are most easily formulated in *superspace*, where we augment the usual spacetime coordinate x^μ with an anticommuting left-handed spinor coordinate θ_a and its right-handed complex conjugate $\theta^*_{\dot{a}}$. We define *superfields* $\Phi(x,\theta,\theta^*)$ that are functions of all these coordinates.

The energy-momentum vector generates translations of the usual spacetime coordinate x^μ in the usual way,

$$[\Phi(x,\theta,\theta^*),P^\mu] = -i\partial^\mu\Phi(x,\theta,\theta^*) \, . \tag{95.7}$$

By analogy, we would expect

$$[\Phi(x,\theta,\theta^*),Q_a] = -i\mathcal{Q}_a\Phi(x,\theta,\theta^*) \, , \tag{95.8}$$

$$[\Phi(x,\theta,\theta^*),Q^\dagger_{\dot{a}}] = -i\mathcal{Q}^*_{\dot{a}}\Phi(x,\theta,\theta^*) \, , \tag{95.9}$$

where $\mathcal{Q}_a$ and $\mathcal{Q}^*_{\dot{a}}$ are appropriate differential operators. To figure out what they should be, we first introduce the anticommuting derivatives $\partial_a \equiv \partial/\partial\theta^a$ and $\partial^*_{\dot{a}} \equiv \partial/\partial\theta^{*\dot{a}}$, which obey $\partial_a\theta^c = \delta_a{}^c$ and $\partial^*_{\dot{a}}\theta^{*\dot{c}} = \delta_{\dot{a}}{}^{\dot{c}}$. Note, however, that complex conjugation should reverse the order of a product of Grassmann variables, in order to maintain consistency with hermitian conjugation. Then we have $\delta_a{}^c = (\partial_a\theta^c)^* = \theta^{*\dot{c}}(\partial_a)^* = -(\partial_a)^*\theta^{*\dot{c}}$, which implies

$$(\partial_a)^* = -\partial^*_{\dot{a}} \, . \tag{95.10}$$

Thus, our first guess for the differential operators in eqs. (95.8) and (95.9) is $\mathcal{Q}_a = \partial_a$ and $\mathcal{Q}^*_{\dot{a}} = -\partial^*_{\dot{a}}$. However, this choice is inconsistent with $\{Q_a, Q^\dagger_{\dot{a}}\} = -2\sigma^\mu_{a\dot{a}}P_\mu$.

An alternative that avoids this pitfall is

$$\mathcal{Q}_a = +\partial_a + i\sigma^\mu_{a\dot{c}}\theta^{*\dot{c}}\partial_\mu \, , \tag{95.11}$$

$$\mathcal{Q}^*_{\dot{a}} = -\partial^*_{\dot{a}} - \; i\theta^c\sigma^\mu_{c\dot{a}}\partial_\mu \, . \tag{95.12}$$

These obey the anticommutation relations

$$\{\mathcal{Q}_a,\mathcal{Q}_b\} = \{\mathcal{Q}^*_{\dot{a}},\mathcal{Q}^*_{\dot{b}}\} = 0 \, , \tag{95.13}$$

$$\{\mathcal{Q}_a,\mathcal{Q}^*_{\dot{a}}\} = -2i\sigma^\mu_{a\dot{a}}\partial_\mu \, . \tag{95.14}$$

It is straightforward to check that eqs. (95.5)–(95.12) are now mutually compatible. In particular, the Jacobi identity

$$\{[\Phi,Q],Q^\dagger\} + \{[\Phi,Q^\dagger],Q\} - [\Phi,\{Q,Q^\dagger\}] = 0 \tag{95.15}$$

is satisfied.

Next we introduce the *supercovariant derivatives*

$$\mathcal{D}_a = +\partial_a - i\sigma^\mu_{a\dot{c}}\theta^{*\dot{c}}\partial_\mu \,, \tag{95.16}$$

$$\mathcal{D}^*_{\dot{a}} = -\partial^*_{\dot{a}} + \ i\theta^c\sigma^\mu_{c\dot{a}}\partial_\mu \,. \tag{95.17}$$

These obey

$$\{\mathcal{D}_a, \mathcal{D}_b\} = \{\mathcal{D}^*_{\dot{a}}, \mathcal{D}^*_{\dot{b}}\} = 0 \,, \tag{95.18}$$

$$\{\mathcal{D}_a, \mathcal{D}^*_{\dot{a}}\} = 2i\sigma^\mu_{a\dot{a}}\partial_\mu \,, \tag{95.19}$$

$$\{\mathcal{D}_a, \mathcal{Q}_b\} = \{\mathcal{D}_a, \mathcal{Q}^*_{\dot{b}}\} = \{\mathcal{D}^*_{\dot{a}}, \mathcal{Q}_b\} = \{\mathcal{D}^*_{\dot{a}}, \mathcal{Q}^*_{\dot{b}}\} = 0 \,. \tag{95.20}$$

Because of eq. (95.20), we could impose the condition $\mathcal{D}_a\Phi = 0$ or $\mathcal{D}^*_{\dot{a}}\Phi = 0$ on a superfield, and this condition would be preserved by the supersymmetry transformations of eqs. (95.8) and (95.9). A superfield that obeys

$$\mathcal{D}^*_{\dot{a}}\Phi(x, \theta, \theta^*) = 0 \tag{95.21}$$

is a *left-handed chiral superfield.* Its hermitian conjugate $\Phi^\dagger(x, \theta, \theta^*)$ obeys

$$\mathcal{D}_a\Phi^\dagger(x, \theta, \theta^*) = 0 \,, \tag{95.22}$$

and is a *right-handed chiral superfield.*

We can solve eq. (95.21) by introducing

$$y^\mu = x^\mu - i\theta^c\sigma^\mu_{c\dot{c}}\theta^{*\dot{c}} \,, \tag{95.23}$$

and noting that

$$\mathcal{D}^*_{\dot{a}}\theta_a = 0 \quad \text{and} \quad \mathcal{D}^*_{\dot{a}}y^\mu = 0 \,. \tag{95.24}$$

(When verifying $\mathcal{D}^*_{\dot{a}}y^\mu = 0$, remember that there is a minus sign from pulling the $\partial^*_{\dot{a}}$ through θ^c.) Thus, any superfield $\Phi(y, \theta)$ that is a function of y and θ only is a left-handed chiral superfield.

We can expand $\Phi(y, \theta)$ in powers of θ; because θ is an anticommuting variable with a two-valued index, we have $\theta_a\theta_b\theta_c = 0$, and so the expansion terminates after the quadratic term. We thus have

$$\Phi(y, \theta) = A(y) + \sqrt{2}\theta\psi(y) + \theta\theta F(y) \,, \tag{95.25}$$

where $A(y)$ and $F(y)$ are complex scalar fields, $\psi_a(y)$ is a left-handed Weyl field, and we have used our standard index-suppression conventions: $\theta\psi = \theta^a\psi_a$ and $\theta\theta = \theta^a\theta_a$. The factor of root-two is conventional.

We can now substitute in eq. (95.23), and continue to expand in powers of θ and θ^*. Making use of the spinor identities

$$\theta_a\theta_b = +\tfrac{1}{2}\theta\theta\varepsilon_{ab}\,, \qquad \theta^a\theta^b = -\tfrac{1}{2}\theta\theta\varepsilon^{ab}\,, \tag{95.26}$$

$$\theta^*_{\dot a}\theta^*_{\dot b} = -\tfrac{1}{2}\theta^*\theta^*\varepsilon_{\dot a\dot b}\,, \qquad \theta^{*\dot a}\theta^{*\dot b} = +\tfrac{1}{2}\theta^*\theta^*\varepsilon^{\dot a\dot b}\,, \tag{95.27}$$

where $\theta\theta = \theta^a\theta_a$ and $\theta^*\theta^* = (\theta\theta)^* = \theta^*_{\dot a}\theta^{*\dot a}$, along with the Fierz identity

$$(\theta\sigma^\mu\theta^*)(\theta\sigma^\nu\theta^*) = -\tfrac{1}{2}\theta\theta\theta^*\theta^* g^{\mu\nu}\,, \tag{95.28}$$

we find

$$\begin{aligned}\Phi(x,\theta,\theta^*) = A(x) &+ \sqrt{2}\theta\psi(x) + \theta\theta F(x) - i(\theta\sigma^\mu\theta^*)\partial_\mu A(x)\\ &- \tfrac{1}{\sqrt{2}}i\theta\theta\theta^*\bar\sigma^\mu\partial_\mu\psi(x) + \tfrac{1}{4}\theta\theta\theta^*\theta^*\partial^2 A(x)\,.\end{aligned} \tag{95.29}$$

Let us investigate the properties of a left-handed chiral superfield under a supersymmetry transformation, given by eqs. (95.8) and (95.9). It is easiest to use the y and θ coordinates, since

$$\begin{aligned}\mathcal{Q}_a\theta^b &= \delta_a{}^b\,, \qquad \mathcal{Q}_a y^\mu = 0\,,\\ \mathcal{Q}^*_{\dot a}\theta^b &= 0\,, \qquad \mathcal{Q}^*_{\dot a}y^\mu = -2i\theta^c\sigma^\mu_{c\dot a}\,.\end{aligned} \tag{95.30}$$

We thus have

$$\begin{aligned}\mathcal{Q}_a\Phi(y,\theta) &= \partial_a\Phi(y,\theta)\\ &= \sqrt{2}\psi_a(y) + 2\theta_a F(y)\,,\end{aligned} \tag{95.31}$$

$$\begin{aligned}\mathcal{Q}^*_{\dot a}\Phi(y,\theta) &= -2i\theta^c\sigma^\mu_{c\dot a}\partial_\mu\Phi(y,\theta)\\ &= -2i\theta^c\sigma^\mu_{c\dot a}\partial_\mu A(y) + i\sqrt{2}\theta\theta\partial_\mu\psi^c(y)\sigma^\mu_{c\dot a}\,,\end{aligned} \tag{95.32}$$

where ∂_μ is with respect to y; we used eq. (95.26) to get the last line. We can now find the supersymmetry transformations of the component fields A, ψ, and F by matching powers of θ on each side of eqs. (95.8) and (95.9). Remembering that the Q and $Q^\dagger$ operators anticommute with θ and θ^*, we get

$$[A,Q_a] = -i\sqrt{2}\psi_a\,, \qquad [A,Q^\dagger_{\dot a}] = 0\,, \tag{95.33}$$

$$\{\psi_c,Q_a\} = -i\sqrt{2}\varepsilon_{ac}F\,, \qquad \{\psi_c,Q^\dagger_{\dot a}\} = -\sqrt{2}\sigma^\mu_{c\dot a}\partial_\mu A\,, \tag{95.34}$$

$$[F,Q_a] = 0\,, \qquad [F,Q^\dagger_{\dot a}] = \sqrt{2}\partial_\mu\psi^c\sigma^\mu_{c\dot a}\,, \tag{95.35}$$

where all component fields have spacetime argument y. However, y is arbitrary, and so we are free to replace it with x.

Eq. (95.35) is the most important: it tells us that the supersymmetry transformation of the F field is a total derivative. Therefore, $\int d^4x\, F(x)$ is invariant under a supersymmetry transformation, and hence could be a term in the action of a supersymmetric theory.

The product of two left-handed chiral superfields is another left-handed chiral superfield; this is obvious from eq. (95.25), and the fact that the θ expansion always terminates with the quadratic term. For two chiral superfields $\Phi_1(y,\theta)$ and $\Phi_2(y,\theta)$, we have

$$\Phi_1\Phi_2 = A_1A_2 + \sqrt{2}\theta(A_1\psi_2 + A_2\psi_1) + \theta\theta(A_1F_2 + A_2F_1 - \psi_1\psi_2)\ . \quad (95.36)$$

More generally, given a set of left-handed chiral superfields Φ_i, we can consider a function of them $W(\Phi)$; this function is itself a left-handed chiral superfield. Its F term (the coefficient of $\theta\theta$) is

$$W(\Phi)\Big|_F = \frac{\partial W(A)}{\partial A_i}F_i - \frac{1}{2}\frac{\partial^2 W(A)}{\partial A_i \partial A_j}\psi_i\psi_j\ , \quad (95.37)$$

where repeated indices are summed. The spacetime integral of this term (like the spacetime integral of any F term) is invariant under supersymmetry, and hence could be a term in the action of a supersymmetric theory. In this case, the function $W(\Phi)$ is called the *superpotential.*

We still need kinetic terms. To get them, we first investigate the properties of a *vector superfield.* A vector superfield $V(x,\theta,\theta^*)$ is hermitian,

$$[V(x,\theta,\theta^*)]^\dagger = V(x,\theta,\theta^*)\ , \quad (95.38)$$

but is not subject to any other constraint. Its component expansion is

$$\begin{aligned} V(x,\theta,\theta^*) = {} & C(x) + \theta\chi(x) + \theta^*\chi^\dagger(x) + \theta\theta M(x) + \theta^*\theta^* M^\dagger(x) \\ & + \theta\sigma^\mu\theta^* v_\mu(x) + \theta\theta\theta^*\lambda^\dagger(x) + \theta^*\theta^*\theta\lambda(x) \\ & + \tfrac{1}{2}\theta\theta\theta^*\theta^* D(x)\ , \end{aligned} \quad (95.39)$$

where C and D are real scalar fields, M is a complex scalar field, χ and λ are left-handed Weyl fields, and v_μ is a real vector field.

Following the analysis that led to eq. (95.35), we find

$$[D, Q_a] = -\sigma^\mu_{a\dot c}\partial_\mu\lambda^{\dagger\dot c}\ , \qquad [D, Q^\dagger_{\dot a}] = +\partial_\mu\lambda^c\sigma^\mu_{c\dot a}\ . \quad (95.40)$$

We see that the supersymmetry transformation of the D component of a vector superfield is a total derivative. Therefore, $\int d^4x\, D(x)$ is invariant under a supersymmetry transformation, and hence could be a term in the action of a supersymmetric theory.

Consider the product of a left-handed chiral superfield $\Phi(x,\theta,\theta^*)$, as given by eq. (95.29), and its hermitian conjugate

$$\begin{aligned}\Phi^\dagger(x,\theta,\theta^*) &= A^\dagger(x) + \sqrt{2}\theta^*\psi^\dagger(x) + \theta^*\theta^* F^\dagger(x) + i(\theta\sigma^\mu\theta^*)\partial_\mu A^\dagger(x) \\ &\quad + \tfrac{1}{\sqrt{2}} i\theta^*\theta^*\partial_\mu\psi^\dagger(x)\bar\sigma^\mu\theta + \tfrac{1}{4}\theta\theta\theta^*\theta^*\partial^2 A^\dagger(x)\ . \end{aligned} \tag{95.41}$$

The product $\Phi^\dagger\Phi$ is obviously hermitian, and so is a vector superfield. After considerable use of eqs. (95.26)–(95.28), we find that the D term (the coefficient of $\theta\theta\theta^*\theta^*$) of this vector superfield is

$$\begin{aligned}\Phi^\dagger\Phi\Big|_D = &-\tfrac{1}{2}\partial^\mu A^\dagger\partial_\mu A + \tfrac{1}{4}A\partial^2 A^\dagger + \tfrac{1}{4}A^\dagger\partial^2 A \\ &+ \tfrac{1}{2}i\psi^\dagger\bar\sigma^\mu\partial_\mu\psi - \tfrac{1}{2}i\partial_\mu\psi^\dagger\bar\sigma^\mu\psi \\ &+ F^\dagger F\ . \end{aligned} \tag{95.42}$$

The spacetime integral of this term (like the spacetime integral of any D term) is invariant under supersymmetry, and hence could be a term in the action of a supersymmetric theory. After some integrations by parts, and dropping total divergences, we find

$$\Phi^\dagger\Phi\Big|_D = -\partial^\mu A^\dagger\partial_\mu A + i\psi^\dagger\bar\sigma^\mu\partial_\mu\psi + F^\dagger F\ . \tag{95.43}$$

We see that we have standard kinetic terms for the complex scalar field A and the left-handed Weyl field ψ. We also have a term with no derivatives for the complex scalar field F. The F field is therefore called an *auxiliary field*.

If we consider a set of left-handed chiral superfields Φ_i, we get a hermitian, supersymmetric action if we take as the lagrangian

$$\mathcal{L} = \Phi_i^\dagger\Phi_i\Big|_D + \Big(W(\Phi)\Big|_F + \text{h.c.}\Big)\ , \tag{95.44}$$

where the index in the first term is summed. Since F_i appears only quadratically and without derivatives, we can easily perform the path integral over it. The result is equivalent to solving the classical equation of motion for F_i,

$$\frac{\partial\mathcal{L}}{\partial F_i} = F_i^\dagger + \frac{\partial W(A)}{\partial A_i} = 0\ , \tag{95.45}$$

and substituting the solution back into the lagrangian; the result is

$$\begin{aligned}\mathcal{L} = &-\partial^\mu A_i^\dagger\partial_\mu A_i + i\psi_i^\dagger\bar\sigma^\mu\partial_\mu\psi_i \\ &- \left|\frac{\partial W(A)}{\partial A_i}\right|^2 - \frac{1}{2}\left[\frac{\partial^2 W(A)}{\partial A_i\partial A_j}\psi_i\psi_j + \text{h.c.}\right]\ , \end{aligned} \tag{95.46}$$

where the indices are summed in each term.

As an example, let us consider a single left-handed chiral superfield, with superpotential

$$W(A) = \tfrac{1}{2}mA^2 + \tfrac{1}{6}gA^3 \,. \tag{95.47}$$

This is the *Wess–Zumino model.* The scalar potential is

$$\begin{aligned} V(A) &= |\partial W/\partial A|^2 \\ &= m^2 A^\dagger A + \tfrac{1}{2}gm(A^\dagger A^2 + A^{\dagger 2}A) + \tfrac{1}{4}g^2(A^\dagger A)^2 \,. \end{aligned} \tag{95.48}$$

We see that the scalar has mass m. The last term in eq. (95.46) becomes

$$\mathcal{L}_{\text{mass+Yuk}} = -\tfrac{1}{2}m\psi\psi - \tfrac{1}{2}gA\psi\psi + \text{h.c.} \,. \tag{95.49}$$

We see that the fermion also has mass m, and a Yukawa interaction with the scalar. The Yukawa couping is related (by supersymmetry) to the cubic and quartic self-interactions of the scalar.

Next we would like to introduce gauge fields. Recall that the vector superfield $V(x,\theta,\theta^*)$ has among its components a real vector field $v_\mu(x)$ that could be identified as an abelian gauge field. (Later we will add an adjoint index to the superfield in order to get a nonabelian gauge field.)

We need to generalize the notion of a gauge transformation to superfields. We begin by noting that if Ξ is a left-handed chiral superfield, then $i(\Xi^\dagger - \Xi)$ is a vector superfield. We then define a *supergauge transformation*

$$V \to V + i(\Xi^\dagger - \Xi) \,. \tag{95.50}$$

We will attempt to construct actions that are invariant under eq. (95.50).

Following the pattern of eqs. (95.29) and (95.41), we write

$$\begin{aligned} \Xi(x,\theta,\theta^*) &= B(x) + \theta\xi(x) + \theta\theta G(x) \\ &\quad - i(\theta\sigma^\mu\theta^*)\partial_\mu B(x) + \ldots \,, \end{aligned} \tag{95.51}$$

$$\begin{aligned} \Xi^\dagger(x,\theta,\theta^*) &= B^\dagger(x) + \theta^*\xi^\dagger(x) + \theta^*\theta^* G^\dagger(x) \\ &\quad + i(\theta\sigma^\mu\theta^*)\partial_\mu B^\dagger(x) + \ldots \,. \end{aligned} \tag{95.52}$$

If we set $B = \frac{1}{2}(b + ia)$, where a and b are real scalar fields, we find

$$i(\Xi^\dagger - \Xi) = a - i\theta\xi + i\theta^*\xi^\dagger - i\theta\theta G + i\theta^*\theta^* G^\dagger - (\theta\sigma^\mu\theta^*)\partial_\mu b + \ldots \,. \tag{95.53}$$

From eq. (95.39), we see that the supergauge transformation of eq. (95.50) results in

$$\begin{aligned} C &\to C + a \,, \\ \chi &\to \chi - i\xi \,, \\ M &\to M - iG \,, \\ v_\mu &\to v_\mu - \partial_\mu b \,. \end{aligned} \tag{95.54}$$

The last of these is the usual abelian gauge transformation. The first three allow us to *gauge away* the C, χ, and M components of a vector superfield. That is, we can make a supergauge transformation with $a = -C$, $\xi = -i\chi$ and $G = -iM$; in this gauge, known as *Wess–Zumino gauge*, the vector superfield becomes

$$V = (\theta\sigma^\mu\theta^*)v_\mu + \theta\theta\theta^*\lambda^\dagger + \theta^*\theta^*\theta\lambda + \tfrac{1}{2}\theta\theta\theta^*\theta^* D\,. \tag{95.55}$$

Note that we still have the freedom to make the supergauge transformation of eq. (95.50) with $B(x) = \frac{1}{2}b(x)$, and that this still implements the ordinary abelian gauge transformation of eq. (95.54).

Now consider a left-handed chiral superfield Φ that has charge $+1$ under a U(1) gauge group. We take the kinetic term for Φ to be

$$\mathcal{L}_{\text{kin}} = \Phi^\dagger e^{-2gV}\Phi\Big|_D\,, \tag{95.56}$$

where g is the gauge coupling. The vector superfield $\Phi^\dagger e^{-2gV}\Phi$ is clearly invariant under the supergauge transformation

$$\Phi \to e^{-2ig\Xi}\,\Phi\,, \tag{95.57}$$

$$\Phi^\dagger \to \Phi^\dagger e^{+2ig\Xi^\dagger}\,, \tag{95.58}$$

$$V \to V + i(\Xi^\dagger - \Xi)\,. \tag{95.59}$$

Let us evaluate eq. (95.56) in Wess–Zumino gauge, where we have

$$V^2 = -\tfrac{1}{2}\theta\theta\theta^*\theta^* v^\mu v_\mu\,, \tag{95.60}$$

$$V^3 = 0\,. \tag{95.61}$$

The exponential factor in eq. (95.56) becomes

$$\begin{aligned} e^{-2gV} &= 1 - 2g(\theta\sigma^\mu\theta^*)v_\mu - 2g\theta\theta\theta^*\lambda^\dagger - 2g\theta^*\theta^*\theta\lambda \\ &\quad - \theta\theta\theta^*\theta^*(gD + g^2 v^\mu v_\mu)\,. \end{aligned} \tag{95.62}$$

The relevant terms in $\Phi^\dagger\Phi$ are

$$\begin{aligned} \Phi^\dagger\Phi = A^\dagger A &+ \sqrt{2}\theta^*\psi^\dagger A + \sqrt{2}\theta\psi A^\dagger \\ &+ (\theta\sigma^\mu\theta^*)(\psi^\dagger\bar\sigma_\mu\psi - iA^\dagger\partial_\mu A + iA\partial_\mu A^\dagger) \\ &+ \ldots + \theta\theta\theta^*\theta^*(\Phi^\dagger\Phi)_D\,, \end{aligned} \tag{95.63}$$

where we used $2(\theta^*\psi^\dagger)(\theta\psi) = (\theta\sigma^\mu\theta^*)(\psi^\dagger\bar\sigma_\mu\psi)$ to get the first term in the second line, and $(\Phi^\dagger\Phi)_D$ is given by eq. (95.42).

Combining eqs. (95.62) and (95.63), taking the D term, and performing the same integrations by parts that led to eq. (95.43), we find

$$\Phi^\dagger e^{-2gV}\Phi\Big|_D = -(D^\mu A)^\dagger D_\mu A + i\psi^\dagger\bar\sigma^\mu D_\mu\psi + F^\dagger F \\ + \sqrt{2}g\psi^\dagger\lambda^\dagger A + \sqrt{2}gA^\dagger\lambda\psi - gA^\dagger DA\,, \qquad (95.64)$$

where $D_\mu = \partial_\mu - igv_\mu$ is the usual gauge covariant derivative acting on a field of charge $+1$.

We still need a kinetic term for the vector superfield. To get it, we first introduce a superfield that carries a left-handed spinor index,

$$W_a \equiv \tfrac{1}{4}\mathcal{D}^*_{\dot a}\mathcal{D}^{*\dot a}\mathcal{D}_a V\,. \qquad (95.65)$$

Since the two components of $\mathcal{D}^*$ anticommute, we have $\mathcal{D}^*_{\dot a}\mathcal{D}^*_{\dot b}\mathcal{D}^*_{\dot c} = 0$. Thus W_a obeys $\mathcal{D}^*_{\dot a}W_a = 0$, and is therefore a left-handed chiral superfield. Furthermore, W_a is invariant under the supergauge transformation of eq. (95.50). To see this, we first note that $\mathcal{D}^*_{\dot a}\mathcal{D}^{*\dot a}\mathcal{D}_a$ annihilates $\Xi^\dagger$ (because $\mathcal{D}_a$ does). Then, we use eq. (95.19) to write

$$\mathcal{D}^*_{\dot a}\mathcal{D}^{*\dot a}\mathcal{D}_a = -(\mathcal{D}^*_{\dot a}\mathcal{D}_a + 2i\sigma^\mu_{a\dot a}\partial_\mu)\mathcal{D}^{*\dot a}\,. \qquad (95.66)$$

Thus, $\mathcal{D}^*_{\dot a}\mathcal{D}^{*\dot a}\mathcal{D}_a$ also annihilates Ξ (because $\mathcal{D}^{*\dot a}$ does). Therefore, W_a is invariant under eq. (95.50).

Since W_a is a left-handed chiral superfield, it has an expansion in the form of eq. (95.25). To find the component fields of W_a, we set $x = y + i\theta\sigma^\mu\theta^*$ in eq. (95.55), and expand in θ and θ^*. The result is

$$V = (\theta\sigma^\mu\theta^*)v_\mu + \theta\theta\theta^*\lambda^\dagger + \theta^*\theta^*\theta\lambda + \tfrac{1}{2}\theta\theta\theta^*\theta^*(D - i\partial^\mu v_\mu)\,, \qquad (95.67)$$

where all component fields have spacetime argument y. From eq. (95.24), we see that $\mathcal{D}^*_{\dot a} = -\partial^*_{\dot a}$ when it acts on a function of y, θ, and θ^*. We also have

$$\mathcal{D}_a = (\mathcal{D}_a\theta^c)\partial_c + (\mathcal{D}_a\theta^{*\dot c})\partial^*_{\dot c} + (\mathcal{D}_a y^\mu)\partial_\mu = \partial_a + 0 - 2i\sigma^\mu_{a\dot a}\theta^{*\dot a}\partial_\mu\,, \qquad (95.68)$$

where ∂_μ is with respect to y. Using eq. (95.68), we find

$$\mathcal{D}_a V = \theta^*\theta^*\Big[\lambda_a + \theta_a(D - i\partial\cdot v) - i(\sigma^\mu\bar\sigma^\nu\theta)_a\partial_\mu v_\nu + i\theta\theta(\sigma^\mu\partial_\mu\lambda^\dagger)_a\Big] \\ + \ldots\,. \qquad (95.69)$$

When we act on $\mathcal{D}_aV$ with $\mathcal{D}^*_{\dot a}\mathcal{D}^{*\dot a} = \partial^*_{\dot a}\partial^{*\dot a}$ to get W_a, only the coefficient of $\theta^*\theta^*$ survives. Since $\partial^*_{\dot a}\partial^{*\dot a}(\theta^*\theta^*) = 4$, we find

$$W_a = \lambda_a + \theta_a(D - i\partial\cdot v) - i(\sigma^\mu\bar\sigma^\nu\theta)_a\partial_\mu v_\nu + i\theta\theta(\sigma^\mu\partial_\mu\lambda^\dagger)_a\,. \qquad (95.70)$$

We can simplify eq. (95.70) by using the identity

$$(\sigma^\mu\bar\sigma^\nu)_a{}^b = -g^{\mu\nu}\delta_a{}^b - 2i(S_{\rm L}^{\mu\nu})_a{}^b \,. \tag{95.71}$$

Remembering that $S_{\rm L}^{\mu\nu}$ is antisymmetric on $\mu \leftrightarrow \nu$, and defining the field strength

$$F_{\mu\nu} \equiv \partial_\mu v_\nu - \partial_\nu v_\mu \,, \tag{95.72}$$

we get

$$W_a = \lambda_a + \theta_a D - (S_{\rm L}^{\mu\nu})_a{}^c\theta_c F_{\mu\nu} + i\theta\theta\sigma^\mu_{a\dot a}\partial_\mu\lambda^{\dagger\dot a} \,. \tag{95.73}$$

We see that W_a involves the vector field v_μ only through its gauge-invariant field strength $F_{\mu\nu}$. Since W_a is supergauge invariant, this is to be expected.

Next, consider the F term of W^aW_a. This term is Lorentz invariant, and its spacetime integral (like the spacetime integral of any F term) is invariant under supersymmetry, and hence could be a term in the action of a supersymmetric theory. Working out the components, we find

$$W^aW_a\Big|_F = 2i\lambda^a\sigma^\mu_{a\dot a}\partial_\mu\lambda^{\dagger\dot a} - \tfrac{1}{2}{\rm Tr}(S_{\rm L}^{\mu\nu}S_{\rm L}^{\rho\sigma})F_{\mu\nu}F_{\rho\sigma} + D^2 \,, \tag{95.74}$$

where ${\rm Tr}(S_{\rm L}^{\mu\nu}S_{\rm L}^{\rho\sigma}) = (S_{\rm L}^{\mu\nu})_a{}^c(S_{\rm L}^{\rho\sigma})_c{}^a$. To get the spin matrices into this form, we used the fact that $(S_{\rm L}^{\mu\nu})_{ac}$ is symmetric on $a \leftrightarrow c$. Now we use the identity

$${\rm Tr}(S_{\rm L}^{\mu\nu}S_{\rm L}^{\rho\sigma}) = \tfrac{1}{2}(g^{\mu\rho}g^{\nu\sigma} - g^{\mu\sigma}g^{\nu\rho}) - \tfrac{1}{2}i\varepsilon^{\mu\nu\rho\sigma} \tag{95.75}$$

to get

$$W^aW_a\Big|_F = 2i\lambda^a\sigma^\mu_{a\dot a}\partial_\mu\lambda^{\dagger\dot a} - \tfrac{1}{2}F^{\mu\nu}F_{\mu\nu} - \tfrac{1}{2}i\tilde F^{\mu\nu}F_{\mu\nu} + D^2 \,, \tag{95.76}$$

where $\tilde F^{\mu\nu} = \frac{1}{2}\varepsilon^{\mu\nu\rho\sigma}F_{\rho\sigma}$. We can now identify the kinetic term for the vector superfield as

$$\begin{aligned}\mathcal{L}_{\rm kin} &= \tfrac{1}{4}W^aW_a\Big|_F + \text{h.c.} \\ &= i\lambda^\dagger\bar\sigma^\mu\partial_\mu\lambda - \tfrac{1}{4}F^{\mu\nu}F_{\mu\nu} + \tfrac{1}{2}D^2 \,. \end{aligned}\tag{95.77}$$

We integrated by parts and dropped total divergences to get the second line. We see that we have the standard kinetic terms for the gauge field v_μ and the *gaugino* field λ, while D is an auxiliary field.

All of this generalizes in a straightforward way to the nonabelian case. We define a matrix-valued vector superfield $V = V^aT_{\rm R}^a$, and a matrix-valued chiral superfield $\Xi = \Xi^aT_{\rm R}^a$, where a is an adjoint group index. A chiral superfield Φ in the representation R still transforms according to eqs. (95.57) and (95.58), but for the vector field we have

$$e^{-2gV} \to e^{-2ig\Xi^\dagger} e^{-2gV} e^{+2ig\Xi} \; ; \tag{95.78}$$

this reduces to eq. (95.59) in the abelian case.

The field-strength superfield is now

$$W_a = -\tfrac{1}{8g} \mathcal{D}^*_{\dot{a}} \mathcal{D}^{*\dot{a}} e^{+2gV} \mathcal{D}_a e^{-2gV} \; . \tag{95.79}$$

Under a supergauge transformation,

$$W_a \to e^{-2ig\Xi}\, W_a\, e^{+2ig\Xi} \; . \tag{95.80}$$

Eq. (95.73) still holds, but the derivative that acts on $\lambda^\dagger$ is now the gauge covariant derivative for the adjoint representation, and $F^{\mu\nu}$ now includes the usual nonabelian commutator term. These changes also apply to eq. (95.77), where we must also trace over the group indices and (for proper normalization) divide by the index $T(\mathrm{R})$.

Reference notes

Introductions to supersymmetry can be found in *Martin*, *Siegel*, *Weinberg III*, and *Wess & Bagger*.

Problems

95.1) Use eq. (95.6) to show that the hamiltonian is positive semidefinite, and that a state with zero energy must be annihilated by all the supercharges.

95.2) Supersymmetry is spontaneously broken if the ground state $|0\rangle$ is not annihilated by all the supercharges.

a) Use the first of eqs. (95.34) to show that supersymmetry is spontaneously broken if $\langle 0|F|0\rangle \neq 0$.

b) Compute $\{\lambda_a, Q_b\}$. Use the result to show that supersymmetry is spontaneously broken if $\langle 0|D|0\rangle \neq 0$.

95.3) Consider a supersymmetric theory with three chiral superfields A, B, and C, and a superpotential $W = mBC + \kappa A(C^2 - v^2)$, where m and v are parameters with dimensions of mass, and κ is a dimensionless coupling constant. This is the *O'Raifeartaigh model.*

a) Show that one or more F components is nonzero at the minimum of the potential, and hence that supersymmetry is spontaneously broken.

b) Show that the potential is minimized along a line in field space, and find the masses of the particles at an arbitrary point on this line. You should find that there is a massless *Goldstone fermion* or *goldstino* that is related by supersymmetry to the linear combination of F fields that gets a nonzero vacuum expectation value.

95.4) *Supersymmetric quantum electrodynamics.* Consider a supersymmetric U(1) gauge theory with chiral superfields Φ and $\overline{\Phi}$ with charges $+1$ and -1, respectively. Here the bar over the Φ in the field $\overline{\Phi}$ is *part of the name of the field,* and does not denote *any sort* of conjugation. We include a gauge-invariant superpotential $W = m\overline{\Phi}\Phi$.

a) Work out the lagrangian in terms of the component fields.

b) Eliminate the auxiliary fields F, $\overline{F}$, and D.

95.5) In a supersymmetric gauge theory with a U(1) factor, we can add a *Fayet–Iliopoulos term* $\mathcal{L}_{\text{FI}} = e\xi D$ to the lagrangian, where D is the auxiliary field for the U(1) gauge field, and ξ is a parameter with dimensions of mass-squared.

a) Explain why adding this term preserves supersymmetry. Explain why the corresponding gauge field cannot be nonabelian.

b) Add this term to the SQED lagrangian that you found in problem 94.4, and eliminate the auxiliary fields.

c) Minimize the resulting potential. Show that supersymmetry is spontaneously broken if ξ is in a certain range.

95.6) *R symmetry.* Given a supersymmetric gauge theory (abelian or nonabelian), consider a global U(1) transformation that changes the phase of the gaugino fields, $\lambda_a \to e^{-i\alpha}\lambda_a$; we say that the gauginos have *R charge* $+1$.

a) If R symmetry is to be a good symmetry of the lagrangian, what relation must hold between the R charges of the scalar and fermion components of a chiral superfield that couples to the gauge fields?

b) In addition, what conditions must be placed on the superpotential?

c) Identify the R symmetry, if any, of supersymmetric quantum electrodynamics.

96

The Minimal Supersymmetric Standard Model

Prerequisites: 89, 95

Having seen how to construct a general supersymmetric gauge theory in section 95, we can now write down a supersymmetric version of the Standard Model.

To do so, we introduce vector superfields for the gauge group SU(3) × SU(2) × U(1), which include the usual gauge bosons along with their fermionic partners, the *gauginos*; chiral superfields L_{iI}, $\bar{E}_I$, $Q_{\alpha i I}$, $\bar{U}^{\alpha}_I$, and $\bar{D}^{\alpha}_I$ that include three generations of quarks and leptons along with their scalar partners, the *squarks* and *sleptons*; and chiral superfields H_i and $\bar{H}_i$ in the representations $(1, 2, -\frac{1}{2})$ and $(1, 2, +\frac{1}{2})$, which include two copies of the usual Higgs field along with their fermionic partners, the *higgsinos*. We give all these fields supersymmetric, gauge invariant kinetic terms, and a superpotential

$$W = -y_{IJ}\varepsilon^{ij}H_i L_{jI}\bar{E}_J - y'_{IJ}\varepsilon^{ij}H_i Q_{\alpha jI}\bar{D}^{\alpha}_J - y''_{IJ}\varepsilon^{ij}\bar{H}_j Q_{\alpha iI}\bar{U}^{\alpha}_J - \mu\varepsilon^{ij}\bar{H}_i H_j \;, \tag{96.1}$$

where μ is a mass parameter. This superpotential generates the usual Yukawa couplings, among others, and gives a positive mass-squared to both Higgs fields. We forbid terms of the form $\varepsilon^{ij}\bar{H}_j L_{iI}$ (which are allowed by the gauge symmetry) by invoking a discrete symmetry, *R parity*. Under R parity, all Standard Model fields (including both Higgs scalars) are taken to be even, and all superpartner fields (gauginos, higgsinos, squarks, sleptons) are taken to be odd.

Supersymmetry is clearly not an exact symmetry of the real world, and so if present must be spontaneously broken. Correct phenomenology requires supersymmetry breaking to be triggered by fields other than those listed above. There are many possibilities for the dynamics of the fields in this *hidden sector*, and an exploration of them is beyond our scope. However, we can parameterize the effects of the hidden sector via a *spurion field*. This is a constant chiral superfield of the form

$$S = m_S^2\,\theta\theta\;, \tag{96.2}$$

where m_S is the *supersymmetry breaking scale.* We couple S to the quark, lepton, and Higgs superfields via D terms of the form

$$\begin{aligned}\mathcal{L}_{\text{spur},\,D} = m_M^{-2}\,S^\dagger S\Big\{&C^{(H)}H^\dagger H + C^{(\bar{H})}\bar{H}^\dagger\bar{H}\\ &+\sum_{IJ}\Big[C^{(L)}_{IJ}L^\dagger_I L_J + C^{(\bar{E})}_{IJ}\bar{E}^\dagger_I\bar{E}_J + C^{(Q)}_{IJ}Q^\dagger_I Q_J\\ &\qquad + C^{(\bar{U})}_{IJ}\bar{U}^\dagger_I\bar{U}_J + C^{(\bar{D})}_{IJ}\bar{D}^\dagger_I\bar{D}_J\Big]\Big\}\Big|_D\;. \end{aligned}\tag{96.3}$$

Here we have suppressed all but generation indices; each $C^{(\Phi)}_{IJ}$ is a dimensionless hermitian matrix in generation space. The parameter m_M is the *messenger scale.* Eq. (96.3) gives masses of order m_S^2/m_M to the scalars.

We also couple S to the chiral gauge superfields $W^{(i)}_a$ via F terms of the form

$$\mathcal{L}_{\text{spur, gauge}} = m_M^{-1}S\sum_{i=1,2,3}C^{(i)}W^{(i)a}W^{(i)}_a\Big|_F + \text{h.c.}\;, \tag{96.4}$$

where the sum is over the three gauge-group factors; a is a spinor index. Eq. (96.4) gives masses of order m_S^2/m_M to the gauginos.

Finally, we couple S to the chiral superfields via F terms of the form

$$\mathcal{L}_{\text{spur},\,F} = m_M^{-1}S\sum_{A=1}^{28}C_A W_A\Big|_F + \text{h.c.}\;, \tag{96.5}$$

where W_A is one of the 28 gauge-invariant terms in the superpotential. In particular, eq. (96.5) includes a mass term of the form $\bar{H}H + \text{h.c.}$ for the Higgs scalars.

Eqs. (96.1) and (96.3)–(96.5) specify the *Minimal Supersymmetric Standard Model*, or MSSM for short. Obviously, it has a complicated phenomenology that is beyond our scope to explore in detail. One point worth noting is that R parity implies that the lightest superpartner, or LSP, is absolutely stable.

Reference notes

Supersymmetric versions of the Standard Model are discussed in *Martin*, *Ramond II*, and *Weinberg III*.

Problems

96.1) Explain why we need two Higgs doublets.

96.2) a) Write the mass terms for the Higgs scalars as

$$\mathcal{L}_{\text{Higgs mass}} = -\,m_1^2 H^\dagger H - m_2^2 \bar{H}^\dagger \bar{H} - m_3^2 \varepsilon^{ij}(\bar{H}_i H_j + \text{h.c.})\ , \tag{96.6}$$

and compute the quartic terms in H and $\bar{H}$ that arise from eliminating the auxiliary SU(2) and U(1) D fields.

b) Find the conditions on the mass parameters in eq. (96.6) in order for the potential to be bounded below.

c) Find the conditions on the mass parameters in eq. (96.6) in order to have spontaneous breaking of the SU(2) $\times$ U(1) symmetry.

d) Show that there are five physical Higgs particles, two charged and three neutral.

e) Let $\tan\beta \equiv \bar{v}/v$ be the ratio of the two Higgs VEVs. Show that

$$\tan\beta + \cot\beta = (m_1^2 + m_2^2)/m_3^2\ . \tag{96.7}$$

97

Grand unification

Prerequisite: 89

The Standard Model is based on the gauge group $\mathrm{SU}(3) \times \mathrm{SU}(2) \times \mathrm{U}(1)$, with left-handed Weyl fields in three copies of the representation $(1, 2, -\frac{1}{2}) \oplus (1, 1, +1) \oplus (3, 2, +\frac{1}{6}) \oplus (\bar{3}, 1, -\frac{2}{3}) \oplus (\bar{3}, 1, +\frac{1}{3})$, and a complex scalar field in the representation $(1, 2, -\frac{1}{2})$. The lagrangian includes all terms of mass dimension four or less that are allowed by the gauge symmetries and Lorentz invariance.

To complete the specification of the Standard Model, we need twenty real numbers: the three gauge couplings; the three diagonal entries of each of the three diagonalized Yukawa coupling matrices (for the up quarks, the down quarks, and the charged leptons); the four angles in the CKM mixing matrix for the quarks; the vacuum angles for the SU(3) and SU(2) gauge groups; the scalar quartic coupling; and the scalar mass-squared.

The goal of *grand unification* is to construct a more compact model with fewer parameters by supposing that the Standard Model is the result of the spontaneous breaking of a larger gauge symmetry. The simplest model along these lines is the *Georgi–Glashow SU(5)* model. Its starting point is the *grand unified* gauge group SU(5). We include a scalar field $\Phi = \Phi^a T^a$ in the adjoint or 24 representation, and assume that the scalar potential for this field results in a vacuum expectation value (VEV) of the form $\langle 0|\Phi|0\rangle = \mathrm{diag}(-\frac{1}{3}, -\frac{1}{3}, -\frac{1}{3}, +\frac{1}{2}, +\frac{1}{2})V$. As we saw in section 84, this VEV spontaneously breaks the gauge symmetry down to $\mathrm{SU}(3) \times \mathrm{SU}(2) \times \mathrm{U}(1)$. The generator of the unbroken U(1) subgroup is $T^{24} = c\,\mathrm{diag}(-\frac{1}{3}, -\frac{1}{3}, -\frac{1}{3}, +\frac{1}{2}, +\frac{1}{2})$, where $c^2 = 3/5$. It will prove convenient to write

$$T^{24} = \sqrt{\tfrac{3}{5}}\,Y\,, \tag{97.1}$$

and express the U(1) charge as the value of Y rather than the value of T^{24}. We also note that the SU(5) breaking scale V must be considerably larger than the SU(2) × U(1) breaking scale $v \sim 250\,\text{GeV}$ in order to suppress the observable effects of the extra gauge fields.

Under the SU(3) × SU(2) × U(1) subgroup, the fundamental and antifundamental representations of SU(5) transform as

$$5 \to (3,1,-\tfrac{1}{3}) \oplus (1,2,+\tfrac{1}{2})\,, \tag{97.2}$$

$$\overline{5} \to (\overline{3},1,+\tfrac{1}{3}) \oplus (1,2,-\tfrac{1}{2})\,. \tag{97.3}$$

Next we use eq. (97.2) to find that the product $5 \otimes 5$ transforms as

$$\begin{aligned} 5 \otimes 5 \to\; & (6,1,-\tfrac{2}{3})_{\text{S}} \oplus (3,2,+\tfrac{1}{6})_{\text{S}} \oplus (1,3,+1)_{\text{S}} \\ & \oplus (\overline{3},1,-\tfrac{2}{3})_{\text{A}} \oplus (3,2,+\tfrac{1}{6})_{\text{A}} \oplus (1,1,+1)_{\text{A}}\,, \end{aligned} \tag{97.4}$$

where the subscripts indicate the symmetric and antisymmetric parts of the product. In terms of SU(5), we have

$$5 \otimes 5 = 15_{\text{S}} \oplus 10_{\text{A}}\,. \tag{97.5}$$

Comparing eqs. (97.4) and (97.5), we find

$$10 \to (\overline{3},1,-\tfrac{2}{3}) \oplus (3,2,+\tfrac{1}{6}) \oplus (1,1,+1)\,. \tag{97.6}$$

From eqs. (97.3) and (97.6), we see that *one generation of quark and lepton fields fits exactly into the representation* $\overline{5} \oplus 10$ *of* SU(5).

We therefore define a left-handed Weyl field ψ^i in the $\overline{5}$ representation of SU(5), and a left-handed Weyl field $\chi_{ij} = -\chi_{ji}$ in the 10 representation. The gauge covariant derivatives of these fields are

$$\begin{aligned} (D_\mu\psi)^i &= \partial_\mu\psi^i - ig_5 A^a_\mu (T^a_{\overline{5}})^i{}_j\psi^j \\ &= \partial_\mu\psi^i + ig_5 A^a_\mu (T^a)_j{}^i\psi^j\,, \end{aligned} \tag{97.7}$$

$$\begin{aligned} (D_\mu\chi)_{ij} &= \partial_\mu\chi_{ij} - ig_5 A^a_\mu (T^a_{10})_{ij}{}^{kl}\chi_{kl} \\ &= \partial_\mu\chi_{ij} - ig_5 A^a_\mu [(T^a)_i{}^k\chi_{kj} + (T^a)_j{}^l\chi_{il}]\,, \end{aligned} \tag{97.8}$$

where g_5 is the SU(5) gauge coupling and T^a is the generator matrix in the fundamental representation. The kinetic terms for these fields are

$$\mathcal{L}_{\text{kin}} = i\psi^\dagger_i\bar{\sigma}^\mu(D_\mu\psi)^i + \tfrac{1}{2}i\chi^{\dagger ij}\bar{\sigma}^\mu(D_\mu\chi)_{ij}\,, \tag{97.9}$$

where the implicit sum over i and j is unrestricted; this necessitates the prefactor of one-half in the second term to avoid double counting. The interaction terms with the gauge fields then work out to be

$$\mathcal{L}_{\rm int} = -g_5 \Big[\psi_i^\dagger (A_\mu^{\rm T})^i{}_j \bar\sigma^\mu \psi^j + \chi^{\dagger ji} (A_\mu)_i{}^k \bar\sigma^\mu \chi_{kj} \Big] \;, \tag{97.10}$$

where $A_\mu = A_\mu^a T^a$ is the matrix-valued gauge field, and $A_\mu^{\rm T}$ is its transpose. Note that we have written the factors in matrix-multiplication order (with a trace for the second term).

We can identify the components of ψ^i and χ_{ij} as

$$\psi^i = (\;\; \bar d^r \quad \bar d^b \quad \bar d^g \quad e \quad -\nu \;\;) \;, \tag{97.11}$$

$$\chi_{ij} = \begin{pmatrix} 0 & \bar u^g & -\bar u^b & u_r & d_r \\ -\bar u^g & 0 & \bar u^r & u_b & d_b \\ \bar u^b & -\bar u^r & 0 & u_g & d_g \\ -u_r & -u_b & -u_g & 0 & \bar e \\ -d_r & -d_b & -d_g & -\bar e & 0 \end{pmatrix} , \tag{97.12}$$

where r, b, and g stand for the three colors (red, blue, and green). We can also write the gauge fields as

$$A^a T^a = \begin{pmatrix} G_r{}^r - \frac13 cB & G_r{}^b & G_r{}^g & \frac{1}{\sqrt2} X_r^1 & \frac{1}{\sqrt2} X_r^2 \\ G_b{}^r & G_b{}^b - \frac13 cB & G_b{}^g & \frac{1}{\sqrt2} X_b^1 & \frac{1}{\sqrt2} X_b^2 \\ G_g{}^r & G_g{}^b & G_g{}^g - \frac13 cB & \frac{1}{\sqrt2} X_g^1 & \frac{1}{\sqrt2} X_g^2 \\ \frac{1}{\sqrt2} X_1^{\dagger r} & \frac{1}{\sqrt2} X_1^{\dagger b} & \frac{1}{\sqrt2} X_1^{\dagger g} & \frac12 W^3 + \frac12 cB & \frac{1}{\sqrt2} W^+ \\ \frac{1}{\sqrt2} X_2^{\dagger r} & \frac{1}{\sqrt2} X_2^{\dagger b} & \frac{1}{\sqrt2} X_2^{\dagger g} & \frac{1}{\sqrt2} W^- & -\frac12 W^3 + \frac12 cB \end{pmatrix} , \tag{97.13}$$

where the Lorentz index has been omitted. Here B is the hypercharge gauge field, and W^3 and $\sqrt2\, W^\pm = W^1 \pm iW^2$ are the SU(2) gauge fields. The gluon fields $G_i{}^j$ are subject to the constraint $G_r{}^r + G_b{}^b + G_g{}^g = 0$. The X_α^i fields correspond to the broken generators of SU(5), and hence become massive; here i is an SU(2) index and α is an SU(3) index. As we saw in section 84, the X_α^i fields transform as $(3, 2, -\frac56)$ under SU(3) × SU(2) × U(1), and their mass is $M_{\rm X} = \frac{5}{6\sqrt2} g_5 V$.

If we substitute eqs. (97.11)–(97.13) into eq. (97.10), we find the usual interactions of the SU(3) × SU(2) × U(1) gauge fields with the quarks and leptons, but with

$$\sqrt{\tfrac53}\, g_1 = g_2 = g_3 = g_5 \;. \tag{97.14}$$

These relations among the gauge couplings hold in the $\overline{\rm MS}$ renormalization scheme; later we will discuss a modified scheme that is more appropriate at energies well below $M_{\rm X}$.

We also find the couplings of the X field to quarks and leptons; these work out to be

$$\begin{aligned}\mathcal{L}_{X,\rm int} &= -\tfrac{1}{\sqrt{2}}g_5\Big[X^{\dagger\alpha}_{1\mu}(\bar{d}^{\dagger}_{\alpha}\bar{\sigma}^{\mu}e - \bar{e}^{\dagger}\bar{\sigma}^{\mu}d_{\alpha} + u^{\dagger\beta}\bar{\sigma}^{\mu}\bar{u}^{\gamma}\varepsilon_{\alpha\beta\gamma}) \\ &\qquad + X^{\dagger\alpha}_{2\mu}(-\bar{d}^{\dagger}_{\alpha}\bar{\sigma}^{\mu}\nu + \bar{e}^{\dagger}\bar{\sigma}^{\mu}u_{\alpha} + d^{\dagger\beta}\bar{\sigma}^{\mu}\bar{u}^{\gamma}\varepsilon_{\alpha\beta\gamma})\Big] + \text{h.c.} \\ &= -\tfrac{1}{\sqrt{2}}g_5 X^{\dagger\alpha}_{i\mu}(\varepsilon^{ij}\bar{d}^{\dagger}_{\alpha}\bar{\sigma}^{\mu}\ell_j - \varepsilon^{ij}\bar{e}^{\dagger}\bar{\sigma}^{\mu}q_{j\alpha} + q^{\dagger\beta i}\bar{\sigma}^{\mu}\bar{u}^{\gamma}\varepsilon_{\alpha\beta\gamma}) + \text{h.c.} \\ &\equiv -\tfrac{1}{\sqrt{2}}g_5 X^{\dagger\alpha}_{i\mu}J^{i\mu}_{\alpha} + \text{h.c.}\,, \end{aligned} \tag{97.15}$$

where the last line defines the current J^{μ} that couples to the X_{μ} field. (To include more than one generation, we add a generation index I to each quark and lepton field, and sum over it.) The most interesting feature of eq. (97.15) is that the first two terms in the current have baryon number $B = +\frac{1}{3}$ and lepton number $L = +1$, while the third term has $B = -\frac{2}{3}$ and $L = 0$. Thus exchange of an X boson can violate baryon and lepton number conservation, leading to phenomena such as proton decay. Proton decay has not been observed; the limit on the rate $1/\tau$ for $p \to e^{+}\pi^{0}$ is $\tau > 10^{33}\,$yr. A rough estimate of $1/\tau$ from eq. (97.15) is $g_5^4 m_p^5/8\pi M_{\rm X}^4$. Taking $g_5 \sim g_2 \sim 0.6$, we find that we must have $M_{\rm X} > 3\times 10^{15}\,$GeV.

We still need a scalar field in the representation $(1,2,-\frac{1}{2})$. The smallest complete representation of SU(5) that includes this piece is the $\overline{5}$. Call the corresponding field H^{i}; we can identify its components as

$$H^{i} = (\,\phi^{r},\;\; \phi^{b},\;\; \phi^{g},\;\; \varphi^{-},\;\; -\varphi^{0}\,)\,. \tag{97.16}$$

The possible gauge-invariant Yukawa couplings with the ψ^{i} and χ_{ij} fields are

$$\mathcal{L}_{\rm Yuk} = -\,yH^{i}\psi^{j}\chi_{ij} - \tfrac{1}{8}y''\varepsilon^{ijklm}H^{\dagger}_{i}\chi_{jk}\chi_{lm} + \text{h.c.}\,; \tag{97.17}$$

with three generations, y and y'' become matrices in generation space. We can write out $\mathcal{L}_{\rm Yuk}$ using eqs. (97.11), (97.12), and (97.16); the result is

$$\begin{aligned}\mathcal{L}_{\rm Yuk} &= -\,y\varepsilon^{ij}\varphi_i\ell_j\bar{e} - y\varepsilon^{ij}\varphi_i q_{\alpha j}\bar{d}^{\alpha} - y''\varphi^{\dagger i}q_{\alpha i}\bar{u}^{\alpha} \\ &\quad - y\varepsilon_{\alpha\beta\gamma}\phi^{\alpha}\bar{d}^{\beta}\bar{u}^{\gamma} - y\varepsilon^{ij}\phi^{\alpha}q_{\alpha i}\ell_j - y''\phi^{\dagger}_{\alpha}\bar{u}^{\alpha}\bar{e} + \text{h.c.}\,. \end{aligned} \tag{97.18}$$

The terms on the first line are those of the Standard Model, except that the down-quark Yukawa coupling matrix (called y' in section 89) is the same as the charged-lepton Yukawa coupling matrix (called y in section 88). Since the quark and lepton masses are directly proportional to the Yukawa couplings, eq. (97.18) predicts

$$m_b = m_\tau\ , \qquad m_s = m_\mu\ , \qquad m_d = m_e\ . \tag{97.19}$$

These relations hold in the $\overline{\text{MS}}$ renormalization scheme; later we will discuss a modified scheme that is more appropriate at energies well below M_{X}. The terms on the second line of eq. (97.18) are the couplings of the colored scalar field ϕ to the quarks and leptons; we see that these couplings, like those of the X_μ field, violate baryon and lepton number conservation. Since first-generation Yukawa couplings are smaller than gauge couplings by a factor of 10^5, the limit on M_ϕ from proton decay is roughly $M_\phi > 10^{10}$ GeV.

To compute M_ϕ, we need the complete scalar potential. For simplicity, we assume a Z_2 symmetry under $\Phi \leftrightarrow -\Phi$. Then the most general renormalizable potential is

$$\begin{aligned} V(\Phi, H) = &-\tfrac{1}{2}m_\Phi^2 \operatorname{Tr}\Phi^2 + \tfrac{1}{4}\lambda_1 \operatorname{Tr}\Phi^4 + \tfrac{1}{4}\lambda_2(\operatorname{Tr}\Phi^2)^2 \\ &+ m_H^2 H^\dagger H + \tfrac{1}{4}\kappa_1 (H^\dagger H)^2 - \tfrac{1}{2}\kappa_2 H^\dagger \Phi^2 H\ . \end{aligned} \tag{97.20}$$

We take all the parameters (m_Φ^2, m_H^2, λ_1, λ_2, κ_1, κ_2) to be positive. We found in problem 84.1 that in this case the first line of eq. (97.20) is minimized by $\Phi = \text{diag}(-\frac{1}{3}, -\frac{1}{3}, -\frac{1}{3}, +\frac{1}{2}, +\frac{1}{2})V$, with $V^2 = 36m_\Phi^2/(7\lambda_1 + 30\lambda_2)$. From the first and third terms on the second line of eq. (97.20), we find that the masses-squared of the $\varphi \sim (1, 2, -\frac{1}{2})$ and $\phi \sim (\overline{3}, 1, +\frac{1}{3})$ scalar fields are

$$m_\varphi^2 = m_H^2 - \tfrac{1}{8}\kappa_2 V^2\ , \tag{97.21}$$

$$M_\phi^2 = m_H^2 - \tfrac{1}{18}\kappa_2 V^2\ . \tag{97.22}$$

We want $m_\varphi^2 \sim -(100\,\text{GeV})^2$ and $M_\phi^2 > +(10^{10}\,\text{GeV})^2$. This requires m_H^2 to be equal to $\frac{1}{8}\kappa_2 V^2$ to at least sixteen significant digits (but not exactly). There is no obvious reason for the parameters in eq. (97.20) to satisfy this odd relation; we would more naturally expect m_φ^2 and M_ϕ^2 to have the same order of magnitude (and furthermore to have $\kappa_2 \sim g_5^2$ and hence $M_\phi \sim M_{\text{X}} > 10^{15}$ GeV). This is the *fine tuning problem* of grand unified theories. More generally, we can ask why the breaking scale of the grand unified group is so much larger than the breaking scale of the electroweak subgroup; this is the *gauge hierarchy problem.*

Since the X_μ and ϕ fields are so heavy, we can integrate them out, generating effective interactions among the quarks and leptons.[1] For the X_μ field, we get the leading term (in a double expansion in powers of g_5 and inverse powers of $M_{\rm X}$) by ignoring the kinetic energy and other interactions of the X_μ field, solving the equations of motion for X_μ that follow from $\mathcal{L}_{X,\rm mass} + \mathcal{L}_{X,\rm int}$, where $\mathcal{L}_{X,\rm int}$ is given by eq. (97.15) and

$$\mathcal{L}_{X,\rm mass} = -M_{\rm X}^2 X_{i\mu}^{\dagger\alpha} X_\alpha^{i\mu} \, , \tag{97.23}$$

and finally substituting the solutions back into $\mathcal{L}_{X,\rm mass} + \mathcal{L}_{X,\rm int}$. This is equivalent to evaluating tree-level Feynman diagrams with a single X exchanged. The result is

$$\mathcal{L}_{X,\rm eff} = \frac{1}{2M_{\rm X}^2} J_{i\mu}^{\dagger\alpha} J_\alpha^{i\mu} \, . \tag{97.24}$$

Keeping only fields from the first generation, we find that the baryon- and lepton-number violating terms in $\mathcal{L}_{X,\rm eff}$ are

$$\begin{aligned}\mathcal{L}_{X,\rm eff}^{|\Delta B|=1} &= \frac{g_5^2}{2M_{\rm X}^2} \varepsilon^{ij} \varepsilon^{\alpha\beta\gamma} (\bar{d}_\alpha^\dagger \bar{\sigma}^\mu \ell_i - \bar{e}^\dagger \bar{\sigma}^\mu q_{i\alpha}) \bar{u}_\beta^\dagger \bar{\sigma}_\mu q_{j\gamma} + \text{h.c.} \\ &= -\frac{g_5^2}{M_{\rm X}^2} \varepsilon^{ij} \varepsilon^{\alpha\beta\gamma} \Big[(\ell_i q_{j\gamma})(\bar{d}_\alpha^\dagger \bar{u}_\beta^\dagger) + (\bar{e}^\dagger \bar{u}_\gamma^\dagger)(q_{i\alpha} q_{j\beta}) \Big] + \text{h.c.} \, ,\end{aligned} \tag{97.25}$$

where the second line follows from a Fierz identity and a relabeling of the color indices in the second term. We can treat the ϕ field similarly; see problem 97.2. We will compute the decay rate for $p \to e^+\pi^0$ from eq. (97.25) in problem 97.4.

Once the heavy fields have been integrated out, we can apply the $\overline{\rm MS}$ renormalization scheme to the theory with the remaining light fields. This is not the same as $\overline{\rm MS}$ for the original theory, because now only light fields circulate in loops. (Loops of heavy fields contribute to corrections to the effective interactions.) With the usual fields of the Standard Model, we find from our results in sections 66 and 73 that the one-loop beta functions for the three gauge couplings are given by

$$\mu \frac{d}{d\mu} g_i = \frac{b_i}{16\pi^2} g_i^3 + O(g_i^5) \, , \tag{97.26}$$

[1] We should also integrate out the heavy components of Φ, which transform as $(8,1,0) \oplus (1,3,0) \oplus (1,1,0)$ under SU(3) × SU(2) × U(1), but these do not couple directly to quarks and leptons.

with

$$b_3 = -11 + \tfrac{4}{3}n \,, \tag{97.27}$$

$$b_2 = -\tfrac{22}{3} + \tfrac{4}{3}n + \tfrac{1}{6} \,, \tag{97.28}$$

$$b_1 = +\tfrac{20}{9}n + \tfrac{1}{6} \,, \tag{97.29}$$

where $n = 3$ is the number of generations; the $+\frac{1}{6}$ contributions to b_2 and b_1 are from the φ field. These formulae apply for $\mu < M_{\rm X}$. (We are assuming that the heavy scalars do not have masses much less than $M_{\rm X}$.) For $\mu \geq M_{\rm X}$, we must restore the heavy fields, and then eq. (97.14) applies. If we now neglect the higher-loop corrections, integrate eq. (97.26) for each coupling, impose eq. (97.14) at $\mu = M_{\rm X}$, and set $g_2 = e/\sin\theta_{\rm W}$ and $g_1 = e/\cos\theta_{\rm W}$, we find for $\mu < M_{\rm X}$ that

$$\frac{1}{\alpha_3(\mu)} = \frac{1}{\alpha_5(M_{\rm X})} + \frac{b_3}{2\pi}\ln(M_{\rm X}/\mu) \,, \tag{97.30}$$

$$\frac{\sin^2\theta_{\rm W}(\mu)}{\alpha(\mu)} = \frac{1}{\alpha_5(M_{\rm X})} + \frac{b_2}{2\pi}\ln(M_{\rm X}/\mu) \,, \tag{97.31}$$

$$\frac{\cos^2\theta_{\rm W}(\mu)}{\alpha(\mu)} = \frac{5/3}{\alpha_5(M_{\rm X})} + \frac{b_1}{2\pi}\ln(M_{\rm X}/\mu) \,. \tag{97.32}$$

The quantities on the left-hand sides are measured at $\mu = M_{\rm Z}$ to be

$$\alpha_3(M_{\rm Z}) = 0.1187 \pm 0.0020 \,, \tag{97.33}$$

$$1/\alpha(M_{\rm Z}) = 127.91 \pm 0.02 \,, \tag{97.34}$$

$$\sin^2\theta_{\rm W}(M_{\rm Z}) = 0.23120 \pm 0.00015 \,. \tag{97.35}$$

We can use eqs. (97.30)–(97.32) to solve for $1/\alpha_5(M_{\rm X})$ and $\ln(M_{\rm X}/M_{\rm Z})$ in terms of the known parameters $\alpha(M_{\rm Z})$ and $\alpha_3(M_{\rm Z})$; the result is

$$\frac{1}{\alpha_5(M_{\rm X})} = \frac{1}{b_1{+}b_2{-}\frac{8}{3}b_3}\left(\frac{-b_3}{\alpha(M_{\rm Z})} + \frac{b_1{+}b_2}{\alpha_3(M_{\rm Z})}\right) , \tag{97.36}$$

$$\ln(M_{\rm X}/M_{\rm Z}) = \frac{2\pi}{b_1{+}b_2{-}\frac{8}{3}b_3}\left(\frac{1}{\alpha(M_{\rm Z})} - \frac{8/3}{\alpha_3(M_{\rm Z})}\right) . \tag{97.37}$$

Plugging in eqs. (97.27)–(97.29) and eqs. (97.33)–(97.34), we find $1/\alpha_5(M_{\rm X}) = 41.5$ and $M_{\rm X} = 7 \times 10^{14}$ GeV; two-loop corrections lower $M_{\rm X}$ to 4×10^{14} GeV. This value of $M_{\rm X}$ is about an order of magnitude below the lower limit imposed by proton decay.

We can also use eqs. (97.30)–(97.32) to express $\sin^2\theta_{\rm W}(M_{\rm Z})$ in terms of $\alpha(M_{\rm Z})$ and $\alpha_3(M_{\rm Z})$; the result is

$$\sin^2\theta_{\rm W}(M_{\rm Z}) = \frac{1}{b_1{+}b_2{-}\frac{8}{3}b_3}\left(b_2{-}b_3 + (b1{-}\tfrac{5}{3}b_2)\frac{\alpha(M_{\rm Z})}{\alpha_3(M_{\rm Z})}\right). \tag{97.38}$$

This is a prediction of the SU(5) model that we can test. Plugging in eqs. (97.27)–(97.29) and eqs. (97.33)–(97.34), we find $\sin^2\theta_{\rm W}(M_{\rm Z}) = 0.207$. Two-loop corrections raise this to 0.210 ± 0.001. Comparing with eq. (97.35), we see that the SU(5) prediction is too low by about 10%.

The situation improves considerably if we consider the Minimal Supersymmetric Standard Model, discussed in section 96. In this case the beta-function coefficients become

$$b_3 = -9 + 2n\,, \tag{97.39}$$

$$b_2 = -6 + 2n + 1\,, \tag{97.40}$$

$$b_1 = +\tfrac{10}{3}n + 1\,. \tag{97.41}$$

Now we find $\sin^2\theta_{\rm W}(M_{\rm Z}) = 0.231$, with two-loop corrections raising it to 0.234; there are, however, numerous sources of uncertainty related to the masses of the supersymmetric particles. We also find $M_{\rm X} = 2\times 10^{16}\,{\rm GeV}$; this result (which is not changed significantly by two-loop corrections) is high enough to avoid too-rapid proton decay.

Next, let us consider the predicted equality of the down quark and charged lepton masses, eq. (97.19). These relations are subject to renormalization; see problem 97.5. However, the one-loop corrections from gauge-boson exchange cancel in the predicted ratios

$$\frac{m_e}{m_\mu} = \frac{m_d}{m_s}\,, \qquad \frac{m_\mu}{m_\tau} = \frac{m_s}{m_b}\,. \tag{97.42}$$

Alas, these predictions are not satisfied, in the first case by an order of magnitude. Resolving this problem requires either a more complicated set of Higgs fields, and/or including higher-dimension, nonrenormalizable terms in the lagrangian that are suppressed by inverse powers of some mass scale larger than $M_{\rm X}$, such as the Planck mass $M_P = 1.2\times 10^{19}\,{\rm GeV}$. To get neutrino masses in the SU(5) model (see section 91), we must add left-handed Weyl fields $\bar\nu_I$ that are singlets of SU(5), and couple them to the neutrinos via $\mathcal{L}_{\rm Yuk} = -\tilde y_{IJ} H_i^\dagger \psi_I^i \bar\nu_J$.

A more elegant scheme starts with SO(10) as the grand unified gauge group. SO(10) has a complex sixteen-dimensional *spinor representation* that transforms as $1 \oplus \bar 5 \oplus 10$ under the SU(5) subgroup; thus, each generation of fermions (including $\bar{\nu}$) fits into a single Weyl field in this 16 representation. A scalar field in the 10 representation is needed for the Standard Model Higgs field, and additional scalars in higher dimensional representations (such as the 45-dimensional adjoint representation and the 16 representation) are needed to break SO(10) down to SU(3) $\times$ SU(2) $\times$ U(1).

A great variety of grand unified models can be constructed, with and without supersymmetry. Which, if any, are relevant to the natural world is a question yet to be answered.

Problems

97.1) Is the gauge symmetry of the SU(5) model anomalous? If it is, modify the model to turn it into a consistent quantum field theory. Prerequisite: section 75. Hint: see problem 70.4.

97.2) Compute the $\Delta B = \pm 1$ terms in the effective lagrangian that arise from ϕ exchange.

97.3) Let us write eq. (97.25) as

$$\mathcal{L}_{\text{eff}} = -\, Z_{C_1} C_1 \mathcal{O}_1 - Z_{C_2} C_2 \mathcal{O}_2 + \text{h.c.}\,, \tag{97.43}$$

where $\mathcal{O}_1 \equiv \varepsilon^{ij}\varepsilon^{\alpha\beta\gamma}(\ell_i q_{j\gamma})(\bar{d}^\dagger_\alpha \bar{u}^\dagger_\beta)$ and $\mathcal{O}_2 \equiv \varepsilon^{ij}\varepsilon^{\alpha\beta\gamma}(\bar{e}^\dagger \bar{u}^\dagger_\gamma)(q_{i\alpha} q_{j\beta})$, Z_{C_1} and Z_{C_2} are renormalizing factors, and C_1 and C_2 are coefficients that depend on the $\overline{\text{MS}}$ renormalization scale μ. At $\mu = M_{\text{X}}$, we have $C_1(M_{\text{X}}) = C_2(M_{\text{X}}) = 4\pi\alpha_5(M_{\text{X}})/M_{\text{X}}^2$.

a) Working in Lorenz gauge, and using the results of problems 88.7 and 89.5, show that the one-loop contribution to Z_{C_1} from gauge-boson exchange is given by Z_m in spinor electrodynamics in Lorenz gauge, with

$$\begin{aligned}(-1)(+1)e^2 \to \Big[0 + \varepsilon^{\alpha'\beta'\gamma}(T_3^a)_{\alpha'}{}^{\alpha}(T_3^a)_{\beta'}{}^{\beta}/\varepsilon^{\alpha\beta\gamma}\Big]\,g_3^2 \\ + \Big[\varepsilon^{i'j'}(T_2^a)_{i'}{}^{i}(T_2^a)_{j'}{}^{j}/\varepsilon^{ij} + 0\Big]\,g_2^2 \\ + \Big[(-\tfrac{1}{2})(+\tfrac{1}{6}) + (+\tfrac{1}{3})(-\tfrac{2}{3})\Big]\,g_1^2\,.\end{aligned} \tag{97.44}$$

Evaluate these coefficients.

b) Similarly, compute the one-loop contribution to Z_{C_2} from gauge-boson exchange.

c) Compute the corresponding anomalous dimensions γ_1 and γ_2 of C_1 and C_2.

d) Compute the numerical values of $C_1(\mu)$ and $C_2(\mu)$ at $\mu = 2\,\text{GeV}$. For simplicity, take the top quark mass equal to M_{Z}, and all other quark masses less than $2\,\text{GeV}$. Ignore electromagnetic renormalization below M_{Z}.

97.4) Consider the proton decay mode with the most stringent experimental bound, $p \to e^+\pi^0$. The terms in eq. (97.43) relevant for this mode are

$$\mathcal{L}_{\text{eff}} = C_1\varepsilon^{\alpha\beta\gamma}(eu_\gamma)(\bar{d}^\dagger_\alpha\bar{u}^\dagger_\beta) + 2C_2\varepsilon^{\alpha\beta\gamma}(\bar{e}^\dagger\bar{u}^\dagger_\gamma)(d_\alpha u_\beta) + \text{h.c.}\,. \tag{97.45}$$

Note that, under the $\text{SU}(2)_\text{L} \times \text{SU}(2)_\text{R}$ global symmetry of QCD that we discussed in section 83, the operator $u(\bar{d}^\dagger\bar{u}^\dagger)$ transforms as the first component of a $(2,1)$ representation, while the operator $\bar{u}^\dagger(du)$ is related by parity, and transforms as the first component of a $(1,2)$ representation. At low energies, we can replace these operators (up to an overall constant factor) with hadron fields with the same properties under Lorentz and $\text{SU}(2)_\text{L} \times \text{SU}(2)_\text{R} \times \text{U}(1)_\text{V}$ transformations.

a) Show that $P_\text{L}(u\mathcal{N})_1$ and $P_\text{R}(u^\dagger\mathcal{N})_1$ transform appropriately. Here $u = \exp[i\pi^a T^a/f_\pi]$, where π^a is the triplet of pion fields, and $\mathcal{N}_i$ is the Dirac field for the proton-neutron doublet.

b) Show that the low-energy version of eq. (97.45) is then

$$\mathcal{L}_{\text{eff}} = C_1 A\,\overline{\mathcal{E}^\text{C}}P_\text{L}(u\mathcal{N})_1 + 2C_2 A\,\overline{\mathcal{E}^\text{C}}P_\text{R}(u^\dagger\mathcal{N})_1 + \text{h.c.}\,, \tag{97.46}$$

where $\mathcal{E}^\text{C}$ is the charge conjugate of the Dirac field for the electron (in other words, $\mathcal{E}^\text{C}$ is the Dirac field for the positron), and A is a constant with dimensions of mass-cubed. Lattice calculations have yielded a value of $A = 0.0090\,\text{GeV}^3$ for $\mu = 2\,\text{GeV}$.

c) Write out the terms in eq. (97.46) that contain the proton field and either zero or one π^0 fields.

d) Compute the amplitude for $p \to e^+\pi^0$. Note that there are two contributing Feynman diagrams: one where eq. (97.46) supplies the proton-positron-pion vertex, and one where the proton emits a pion via the interaction in eq. (88.30), and then converts to a positron via the no-pion terms in eq. (97.46). Neglect the positron mass. Hint: your result should be proportional to $1 + g_\text{A}$.

e) Compute the spin-averaged decay rate for $p \to e^+\pi^0$. Use the values of C_1 and C_2 for $\mu = 2\,\text{GeV}$ that you computed in problem 97.3. How does your answer compare with the naive estimate we made earlier? Hint: your result should be proportional to $C_1^2 + 4C_2^2$.

97.5) Consider the Yukawa couplings for the down quark and charged lepton of one generation,

$$\mathcal{L}_{\text{Yuk}} = -Z_y y\varepsilon^{ij}\varphi_i\ell_j\bar{e} - Z_{y'}y'\varepsilon^{ij}\varphi_i q_{\alpha j}\bar{d}^\alpha\,, \tag{97.47}$$

where we have included renormalizing factors.

a) Consider one-loop contributions from gauge-boson exchange to $Z_{y'}/Z_y$. Show that the only contributions of this type that do not cancel in the ratio are those where the gauge boson connects the two fermion lines.

b) Show that, in Lorenz gauge, these contributions to $Z_{y'}$ and Z_y are given by Z_m in spinor electrodynamics in Lorenz gauge, with a replacement analogous to eq. (97.44) that you should specify.
c) Let $r \equiv y'/y$, and compute the anomalous dimension of r.
d) Take $r(M_{\rm X}) = 1$, and evaluate $r(M_{\rm Z})$. For simplicity, take the top quark mass equal to $M_{\rm Z}$.
e) Below $M_{\rm Z}$, treat the top quark as heavy, and neglect the small electromagnetic contribution to the anomalous dimension of r. Compute $r(m_b)$, where $m_b = m_b(m_b) = 4.3\,\text{GeV}$ is the bottom quark mass parameter. Use your results to predict the tau lepton mass. How does your prediction compare with its observed value, $m_\tau = 1.8\,\text{GeV}$?

Bibliography

M. V. Berry and K. E. Mount, *Semiclassical approximations in wave mechanics*, Rep. Prog. Phys. **35**, 315 (1972), available online at http://www.phy.bris.ac.uk/people/berry_mv/publications.html.
Lowell S. Brown, *Quantum Field Theory* (Cambridge 1994).
Ta-Pei Cheng and Ling-Fong Li, *Gauge Theory of Elementary Particles* (Oxford 1988).
Sidney Coleman, *Aspects of Symmetry* (Cambridge 1985).
John Collins, *Renormalization* (Cambridge 1984).
Howard Georgi, *Weak Interactions and Modern Particle Theory* (Benjamin/Cummings 1984).
Claude Itzykson and Jean-Bernard Zuber, *Quantum Field Theory* (McGraw-Hill 1980).
Stephen P. Martin, *A Supersymmetry Primer*, available online at http://arxiv.org/abs/hep-ph/9709356.
MILC Collaboration, Phys. Rev. D **70**, 114501 (2004), available online at http://arxiv.org/abs/hep-lat/0407028.
T. Muta, *Foundations of Quantum Chromodynamics* (World Scientific 1998).
Abraham Pais, *Inward Bound* (Oxford 1986).
Michael E. Peskin and Daniel V. Schroeder, *An Introduction to Quantum Field Theory* (Westview 1995).
Chris Quigg, *Gauge Theories of the Strong, Weak and Electromagnetic Interactions* (Westview 1983).
Pierre Ramond (I), *Field Theory: A Modern Primer* (Addison-Wesley 1990).
Pierre Ramond (II), *Journeys Beyond the Standard Model* (Perseus 1999).
Warren Siegel, *Fields*, available online at http://arxiv.org/abs/hep-th/9912205.
Jan Smit, *Introduction to Quantum Fields on a Lattice* (Cambridge 2002).
George Sterman, *An Introduction to Quantum Field Theory* (Cambridge 1993).
Steven Weinberg, *The Quantum Theory of Fields*, Volumes I, II, and III (Cambridge 1995).
Julius Wess and Jonathan Bagger, *Supersymmetry and Supergravity* (Princeton 1992).
Anthony Zee, *Quantum Field Theory in a Nutshell* (Princeton 2003).

Errata for this book are maintained at http://www.physics.ucsb.edu/~mark/qft.html.
Current experimental results on elementary particles are available from the Particle Data Group at http://pdg.lbl.gov.
Essentially all papers on elementary particle theory written after 1992 can be found in the hep-th (high energy physics, theory), hep-ph (high energy physics, phenomenology), and hep-lat (high energy physics, lattice) archives at http://arXiv.org.

Index

Printed in the United States
by Baker & Taylor Publisher Services